Shing F. Fung
Astronomy Program
University of Maryland
College Park, Md.
20742
U.S.A.

ANNUAL REVIEW OF ASTRONOMY AND ASTROPHYSICS

ANNUAL REVIEW OF ASTRONOMY AND ASTROPHYSICS

GEOFFREY BURBIDGE, *Editor*
University of California, San Diego

DAVID LAYZER, *Associate Editor*
Harvard College Observatory

JOHN G. PHILLIPS, *Associate Editor*
University of California, Berkeley

VOLUME 15

1977

ANNUAL REVIEWS INC. 4139 EL CAMINO WAY PALO ALTO, CALIFORNIA 94306

ANNUAL REVIEWS INC.
Palo Alto, California, USA

International Standard Book Number: 0–8243–0915–4
Library of Congress Catalog Number: 63–8846

REPRINTS

The conspicuous number aligned in the margin with the title of each article in this volume is a key for use in ordering reprints. Available reprints are priced at the uniform rate of $1 each postpaid. The minimum acceptable reprint order is 10 reprints and/or $10.00, prepaid. A quantity discount is available.

FILMSET BY TYPESETTING SERVICES LTD, GLASGOW, SCOTLAND
PRINTED AND BOUND IN THE UNITED STATES OF AMERICA

PREFACE

This volume of the *Annual Review of Astronomy and Astrophysics* was planned at a meeting of the Editorial Committee which was held on May 3, 1975, in Palo Alto, California. Committee members present were A. G. W. Cameron, Leo Goldberg, Arno Penzias, David Layzer, John Phillips and George Wallerstein; George Blumenthal attended as a guest and gave us extensive help.

Once again I would like to thank the Associate Editors for carrying out the scientific editorial work very effectively. I also wish to thank the Assistant Editor Beverly Stein, who worked with me until her departure in the summer of 1976, and Rosalie West, who has succeeded her. I would also like to acknowledge the very considerable assistance given by my secretary in La Jolla, Dawn Pedersen.

THE EDITOR

SOME RELATED ARTICLES APPEARING IN OTHER ANNUAL REVIEWS

From the *Annual Review of Earth and Planetary Sciences,* Volume 5 (1977)

Composition of the Mantle and Core, Don L. Anderson

Theoretical Foundations of Equations of State for the Terrestrial Planets, Leon Thomsen

Cratering and Obliteration History of Mars, Clark R. Chapman and Kenneth L. Jones

From the *Annual Review of Fluid Mechanics,* Volume 9 (1977)

Hydrodynamics of the Universe, Ya. B. Zel'dovich

From the *Annual Review of Nuclear Science,* Volume 27 (1977)

The Weak Neutral Current and its Effects in Stellar Collapse, Daniel Z. Freedman, David N. Schramm, and David L. Tubbs

Element Production in the Early Universe, David N. Schramm and R. V. Wagoner

CONTENTS

About Dogma in Science, and Other Recollections of an Astronomer, *E. J. Öpik* 1

Recent Observations of Pulsars, *J. H. Taylor and R. N. Manchester* 19

The Origin of Solar Activity, *E. N. Parker* 45

Seyfert Galaxies, *Daniel W. Weedman* 69

Mercury, *D. E. Gault, J. A. Burns, P. Cassen, and R. G. Strom* 97

Consequences of Mass Transfer in Close Binary Systems, *Hans-Christoph Thomas* 127

Large-Scale Solar Magnetic Fields, *Robert Howard* 153

The Interaction of Supernovae with the Interstellar Medium, *Roger A. Chevalier* 175

Transition Probability Data for Molecules of Astrophysical Interest, *R. W. Nicholls* 197

Recent Theories of Galaxy Formation, *J. Richard Gott III* 235

Formation and Destruction of Dust Grains, *E. E. Salpeter* 267

The Galactic Center, *J. H. Oort* 295

Mass and Energy Flow in the Solar Chromosphere and Corona, *George L. Withbroe and Robert W. Noyes* 363

Jupiter's Magnetosphere, *C. F. Kennel and F. V. Coroniti* 389

Theories of Spiral Structure, *Alar Toomre* 437

Interstellar Scattering and Scintillation of Radio Waves, *Barney J. Rickett* 479

Clusters of Galaxies, *Neta A. Bahcall* 505

Extragalactic X-Ray Sources, *Herbert Gursky and Daniel A. Schwartz* 541

INDEXES

Author Index 569

Subject Index

Cumulative Index of Contributing Authors, Volumes 11–15

Cumulative Index of Chapter Titles, Volumes 11–15

ERRATUM

In Volume 14 (1976)—*The Interstellar Magnetic Field* by Carl Heiles

My discussion of the Wilkinson & Smith (1974) paper was incomplete and incorrect, and its sense should be modified as follows. The decrease in percentage polarization with increasing wavelength shows that significant Faraday rotation occurs in the region within which the synchrotron emission itself is generated; in fact, the data are well represented by a model in which little or no rotation occurs outside the emitting region. However, it does not necessarily follow that an increase in magnetic field alone is responsible for the increased synchrotron emissivity, because the large Faraday rotation within the region could equally well result from an increased ionization of the ambient interstellar gas by an excess of relativistic electrons or a possible associated excess of low-energy cosmic rays.

Wilkinson, A., Smith, F. G. 1974. MNRAS 167:593.

ANNUAL REVIEWS INC. is a nonprofit corporation established to promote the advancement of the sciences. Beginning in 1932 with the *Annual Review of Biochemistry,* the Company has pursued as its principal function the publication of high quality, reasonably priced Annual Review volumes. The volumes are organized by Editors and Editorial Committees who invite qualified authors to contribute critical articles reviewing significant developments within each major discipline.

Annual Reviews Inc. is administered by a Board of Directors whose members serve without compensation.

Annual Reviews are published in the following sciences: Anthropology, Astronomy and Astrophysics, Biochemistry, Biophysics and Bioengineering, Earth and Planetary Sciences, Ecology and Systematics, Energy, Entomology, Fluid Mechanics, Genetics, Materials Science, Medicine, Microbiology, Nuclear Science, Pharmacology and Toxicology, Physical Chemistry, Physiology, Phytopathology, Plant Physiology, Psychology, and Sociology. The *Annual Review of Neuroscience* will begin publication in 1978. In addition, two special volumes have been published by Annual Reviews Inc.: *History of Entomology* (1973) and *The Excitement and Fascination of Science* (1965).

Ann. Rev. Astron. Astrophys. 1977. 15 : 1–17

ABOUT DOGMA IN SCIENCE, AND OTHER RECOLLECTIONS OF AN ASTRONOMER

E. J. Öpik
Armagh Observatory, Armagh BT61 9DG, Northern Ireland and University of Maryland, College Park, Maryland 20742

Ernst J. Öpik.

Ann. Rev. Astron. Astrophys. 1977. 15: 1–17

ABOUT DOGMA IN SCIENCE, AND OTHER RECOLLECTIONS OF AN ASTRONOMER

*2105

E. J. Öpik
Armagh Observatory, Armagh BT61 9DG, Northern Ireland and University of Maryland, College Park, Maryland 20742

DEFINITIONS

The notion of dogma, or the unquestionable acceptance of certain propositions—the doctrines—is usually associated in our minds with religious (or antireligious) beliefs. Yet it has a much wider application. However alien to science, and not widespread there, it *de facto* sometimes infiltrates the realm of research. Usually based on some recognized authority and accepted in a group or "mini-establishment" of true believers, scientific dogma lacks the punitive aspects of its religious counterpart and therefore is open either to ultimate destruction when it is proven wrong, or to logically justified acceptance when finally it is vindicated by facts. Dogma differs from a hypothesis by the refusal of its adherents even to consider the aspects of its validity. Legitimate disagreement or controversy creates dogma when arguments are no longer listened to. Although usually belonging to the realm of theoretical models where direct experiment (or observation) is not possible, dogmatism may sometimes induce its followers to misquotation or misrepresentation of the most undisputable facts, even of the statements made in print by their opponents: if the statements themselves may be subject to doubt or erroneous, the fact of the printed word is indisputable. As shown in the following, my astronomical experience has met several examples of such a "prejudiced blindness." In any case, these misquotations are not intentional but seem to be caused by a specific "dogmatic" superficiality, something like knowing in advance what the other fellow would say and therefore not listening to him, or not reading properly his work.

A UNIVERSITY TREK TO CENTRAL ASIA

Nevertheless, sometimes the dogmatic counterpart may listen, and even may see the light. In the following, a dramatic episode of this kind is described.

In the beginning of 1919, when Bolshevik rule was firmly established in most of Russia, with some fighting still continuing on the fringes, the new rulers decided to found a university in Tashkent, the capital of the just reconquered Russian possessions of Central Asia (so called, although they are rather in the West of the continent). Over 100 professors and other teaching staff with their families volunteered to leave starving Moscow and to start a new life in the food-rich but otherwise risky Asiatic surroundings. As the only astronomer in the group, I was to be Chairman of astronomy and put new life into Tashkent Observatory. A former military geodetic observatory, it had been reorganized by V. V. Stratonoff, but after three decades of respectable research activity it had somehow fallen into disarray during the revolution, and it was now to be made an integral part of the new Turkestan University.

Rail communications in Russia were at that time in a state of disorganization, and so it was no wonder that our legendary trek of 3000 km from Moscow to Tashkent took 70 days, from end of January to beginning of April, 1919. The obstacles were chiefly on the first half of the journey; after entering Asia behind the river Ural the sponsorship of the Tashkent authorities helped us to make the second half of the journey in only three days. There was no coal or fuel storage, so we had to saw and split the raw timber for the locomotive ourselves. Beyond the Volga, there are oak forests, and, to our surprise, freshly felled raw oak turned out to be excellent fuel despite its wet appearance! In March, at a small town on the border of Europe and Asia, we were prevented by the local Soviet from proceeding further for a whole fortnight: it was an act of sheer ingratitude. When we arrived in the town, the Soviet, aware of the presence of so many learned persons, asked us to give a lecture and perhaps a variety show to the populace of the isolated township, and we readily agreed. There had just been a splendid display of aurora, with streamers reaching to the zenith, and secret rumours were spreading that these were the artillery flashes of the cavalry of Dutov, advancing from the north to wipe out the Communists (who were not too much liked by the population). So I gave a lecture on aurorae, geomagnetism and solar activity, and then a concert followed. There was quite a good tenor among the professors; I accompanied him at the piano, and other items of entertainment followed.

As a result of our success, the next day we got a letter from the Soviet requesting a repeat of our performance, with a threat that they would not let us proceed unless we complied. We did not bow to the threat and, in protest, refused to deal with them; instead we sat out in our carriages until a strong order from Tashkent forced the local Soviet to lift its embargo. At the next station we again lost three days: our engine with its fuel—the fruit of many hours of our hard work—was stolen by a trainload of Red Marines while we were asleep. After having obtained another engine and prepared the fuel again, we put out sentries overnight (I was one of them), and not in vain: another migrant group approached us at night, intending to take over the fruits of our labour, but they retreated without a fight. The next morning we went on without further adventures—changing engines and preparing fuel, no longer of regular logs but from ragged, thorny, yet very dry saksaoul shrub of the Kirghiz semidesert steppes; "goblins, not fuel" as some of us resentfully remarked. But it burned well and carried us in three days through all Central Asia to our final destination. Before departure, on the meter-thick snow cover, frozen hard on the surface under the intermittent action of the springtime sun and the frost at night, we celebrated in jubilation and I performed a wild improvised dance in long leaps—I called it the dance of the polar bear: my only solo dance in my life.

A CONFRONTATION RESOLVED

After arrival in April, 1919, I served for two years on the faculty of the newly organized Turkestan University, with my main concern being the revitalization of the Tashkent Observatory. Placed on the outskirts of Tashkent at an elevation of only 440 m above sea level, but near the Central Asian plateaus and the highest mountain ranges of the world, the Observatory enjoyed a most favorable astronomical climate. Besides astronomy and astrophysics, it comprised meteorology and seismology as well—the latter being of especial local importance in view of the frequent earthquakes (one night I was almost thrown out of my shaking bed, but the tremor stopped, and I—quite lightmindedly—did not run out into the safety of open space, as the other inhabitants did). The observatory grounds occupied a vast area, and the single detached structures—laboratories (I had a living room alone in the Astrophysics Laboratory), telescope domes, and seismic and meteorological instruments as well as living quarters—were widely spread over the area without a proper view of each other. Most of Russia was already under the Bolshevik (Communist) rule, after the second or Red revolution of October 1917, and this included also Central Asia. In the transition time, a local physics teacher, V. N. Milovanoff, consented to take over the Directorship of the Observatory, and after my arrival he stayed on and became mainly concerned with administration, while I became Vice-Director and looked after the scientific side. The changing circumstances, especially the takeover by the new rulers while law and order were far from being guaranteed over the vast, sparsely populated expanses of Central Asia, created many administrative problems.

Seismology was represented by one person, G. V. Popoff, an athletically-built bearded religious fanatic who lived with his middle-aged housekeeper Maria Abramovna. She looked also after me, in so far as laundry and cooking of lamb chop was concerned; I bought the provisions myself (Tashkent was at that time a cornucopia for food, the envy of starving Russia). Popoff volunteered to meet the visitors, much to our relief, and hundreds flocked to his popular lectures, which were held on open ground. What he was lecturing about, we did not know—he was competent enough to deal with all the aspects of astronomy, meteorology, seismology. Then suddenly we got a shock: a letter came from the local Commissariat of Education informing us that Popoff was preaching religion under the cover of scientific popularizations, and that the people were flocking to his lectures because of their religious superstitions. The letter insisted that this was contrary to the proclaimed antireligious principles of the ruling party and was also nonscientific. Therefore, the letter concluded, Popoff was not worthy of keeping a scientific post and must be immediately dismissed.

We were deeply worried by this. Not only because we believed in humanitarian considerations and the freedom of speech and thought, but also because Popoff was a very much wanted member of the staff whom we did not wish to lose, Milovanoff and I decided to plead for him at the Commissariat. This was indeed a dangerous enterprise, in view of the political situation and the sensitivity of the new rulers to violations of the Marxist dogma, but we did not mind the risk, and it went off quite happily. The Commissar, a young man of about my age (26) named Dvolaitsky, received us in the entrance hall of his office. We pointed out that Popoff was a good scientist and the only seismologist available, and that, while supplies of photographic registering paper were cut off by the revolution, he managed to resuscitate the almost forgotten mechanical method of using smoked paper scratched by the seismograph needle (with levers in between) and thus continued a makeshift uninterrupted registration and recording of Earth tremors. "Well," said the Commissar, addressing me, "you are an astronomer and you should know that astronomy has proven that there is no God." The naïveté of this sentence was obviously genuine—not just an officially imposed party attitude—and seemed to invite discussion. I burst into laughter and explained to the perplexed Commissar that, while belief or disbelief in God is a matter of inner feeling and personal freedom, science is powerless to prove or disprove His existence. We had quite a long talk, and after listening to me attentively Dvolaitsky then decreed: all right, Popoff may remain in office, but he must limit his activity to purely scientific professional functions and he shall not be allowed to lecture to the public any further.

THE AFTERMATH

After a couple of years—I had already left Tashkent at that time for my homeland Estonia—I learned that Popoff was not at all grateful and almost strangled Milovanoff, the Director. Popoff apparently did not appreciate what was done for him and considered Milovanoff the culprit who terminated his lecturing. One day, when Milovanoff descended underground to have a glance at the seismic laboratory, he was accused of "spying" and attacked by Popoff. A passerby heard the screams and rescued the Director.

Many years later, during Stalin's purges of his colleagues, among the lists of comrades—Bolsheviks put to death by their overlord—I recognized the name of Dvolaitsky. The name is uncommon—I never have seen it before or afterwards. It was almost certainly the name of my Tashkent Commissar of Education, a top Communist who listened to reason against the party dogma. Could this have been the cause of his "liquidation" by Stalin?

A WARNING

In 1940/41, at Tartu Observatory, during the first Communist occupation of Estonia, I had another experience along similar lines, although not so dramatic. I gave a course of popular lectures to the public at Tartu Observatory, with philosophical digression into the mysteries of existence and the meaning or purposefulness of the Universe. People flocked to my lectures and demonstrations in about the same manner as to Popoff's in Tashkent, and when I finished my course, I was specially thanked by representatives of the audience for idealistically lifting them up from Earth, nearer to Heaven. Although religion was never explicitly mentioned in my lectures, somebody apparently reported on my idealistic approach, and one evening, while I was outdoors explaining the constellations to my audience, an important-looking young man approached and warned me that I was on a slippery path and that I must avoid the themes of God and religion; otherwise the lectures would have to be stopped. "You know," he said, "many years ago there was an old professor at Tartu who said that when searching the heavens his telescope had never shown him either angels or God." This time I did not burst into laughter—a more sinister threat was behind this naïve sentence. Neither was the course of my lectures interrupted.

DOGMATISM IN SCIENCE

While religious—or antireligious—abstract dogma stands beyond the reach of science, in science itself the dogmatic approach has played, and is still playing, a conspicuous role, but with a difference: the preconceived ideas in science are subject to verification or rejection through further research. They remain "dogma" only as their adherents refuse to consider alternatives to their doctrines, rejecting criticism beforehand. A group of scientists proclaiming such a dogma would thus form an "establishment" or "mini-establishment" (depending on the extent of the group), an extreme case being a single person persistently building on a certain unproven assumption.

Scientific dogma as here defined may prove correct, and even when ultimately disproved, it may serve a useful purpose, as a stimulant to research and the accumulation of facts. Often it may also be harmful, as an obstacle to freedom of research, especially when influencing editors and reviewers of scientific magazines or directors of institutions.

The discovery of America by Christopher Columbus, who underestimated the size of our globe and until his death firmly believed that he had reached the outskirts of Asia, is a case to this effect. It has been often said that he would not have embarked on the voyage had he known the actual distance to be covered. Yet—and the analogy may also hold in research—Columbus had other indications, such as old stories, washed-up twigs of unfamiliar trees, etc., which implied that a continent must exist not too far away in the West. Even knowing the true dimensions of the Earth, he could have concluded that land was not too far away across the Atlantic, and he could still have gone out in search of it, Asia or not.

Scientific dogma or "mini-dogma" is still usually based on some recognized authority whose pronouncements are unconditionally accepted by his followers or, at least, by himself. Newton's laws of mechanics and gravitation are a splendid example of dogma justified by centuries of research and still basically valid, despite the *corrigenda* introduced by the Theory of Relativity. On the contrary, the dogma that lunar craters are volcanoes, which was maintained for so long by the mini-establishment of (chiefly) amateur lunar observers, has been proven wrong, although this does not detract from the value of their observations.

In the following are briefly described some cases of astronomical dogma that are part of my personal experience. Of course, by a kind of "natural selection," only examples of misjudgment or misinterpretation are pointed out, while those instances that proved correct are not mentioned, simply because it is difficult to distinguish between the roles of dogma and of critical research in such cases.

STELLAR STRUCTURE

While Emden produced purely hydrostatic models of gaseous spheres, the genius of Eddington put life into them by introducing the concepts of energy transfer, chiefly by radiation, and of energy generation. Eddington's merits in deciphering the internal structure of stars are incomparable and his work, especially in laying out the physical principles, still serves as a basis for continuing research, despite his failure in explaining the structure of giants. And the failure itself was not in the lack of physical or mathematical methods of approach; everything was already contained in Eddington's own papers and equations. Yet Eddington's models were conveniently assumed to be of uniform chemical composition, which became a doctrine for himself and for the mini-establishment of his followers and which, as we know now, cannot produce "inflated" or giant stellar structures. Against the well-known fact that the recognized nuclear energy source—the conversion of hydrogen into helium—creates a concentration of matter with heavier molecular weight around the stellar core, Eddington produced the von Zeipel concept of forced rotational convection in operation in the hydrodynamically stable radiative-equilibrium layers (1). However, he failed to put numbers into his equations, and when I did it (2), it turned out that the time scale of would-be mixing for the Sun is of the order of 10^{14}–10^{15} years, 10,000 to 100,000 times longer than the age of the solar system or the time scale of nuclear reactions. This refers to most slowly or moderately rotating stars, especially to giants, and the assumption that the helium produced from hydrogen in the hot central regions would somehow get mixed into the entire volume of the star is untenable, and by a large margin. It should be emphasized that in this respect I did not add a single bit to Eddington's admirable creation; I only performed the calculation according to his prescribed formulae. Eddington's failure to pursue the consequences of his own theory can only be explained in terms of a "blind spot," a dogmatic refusal to abandon his model of convenience—that of uniform chemical composition. My next step was the numerical integration of "composite" nonuniform models, which was much more complicated than the application of Eddington-Emden's homology formulae. However, it was realistic and brought the reward of explaining the structure of giant stars, with high central temperatures and densities that are adequate for advanced nuclear reactions, but with large radii and low mean densities (3).

MIXING LENGTH

A model of convection, initiated by Schmidt and Prandtl, pictures the vertical transport of excess heat in a gaseous or liquid medium through the symbol of a "mixing length," L_m, such that a hotter element ("a bubble"), while rising over this length, does not exchange heat with the colder surroundings and delivers all its excess at the top. It is matched by a similar cold bubble descending from the top and absorbing its prescribed share of excess heat only when arriving at the bottom. Such a lateral isolation of the moving "bubbles" or streams could be achieved only by a miracle, and I devised a realistic model of convective transport (4, 5), taking into account lateral exchange, which agrees with laboratory experiments within $\pm 20\%$, while the mixing-length model predicts a transport by $+3000\%$ in excess of the true value. In a model of cellular convection, the rising current is, of course, always warmer than the descending current *at the same potential level*, but, because of lateral exchange, this is only one tenth of the total adiabatic temperature excess between the extreme levels; the rising current is gradually precooled through lateral contact with its gradually preheated descending counterpart. The stream velocity, proportional to the square root of the equipotential temperature difference, is reduced to one third, and the real convective transport is thus reduced in a ratio of $\frac{1}{10} \times \frac{1}{3} = \frac{1}{30}$th of the mixing-length prediction, while dimensionally the transport equation remains unaltered.

The mixing-length symbolism, probably meant only as a simile, was grasped by a school of astrophysicists in its literal sense and used for numerical applications. In the deep stellar interiors, with their high temperatures and densities, the super-efficient mixing-length model would require deviations from the adiabatic temperature gradient of about one part in three million, while the realistic model would require about one part in one hundred thousand—both small enough not to be reckoned with in the calculations of stellar models. However, in the outer layers near the stellar surface the difference is enormous. In this context, the dogma of the mixing-length has become an obstacle to progress, as can be seen from the following incident.

In 1970, D. J. Mullan—a pupil of mine who has now risen to prominence with numerous researches, especially in the physics of stellar atmospheres, offered for publication in *Astronomy and Astrophysics* (the European Journal) a paper on "Cellular Convection in Stellar Envelopes." By applying my theory, he explained a score of various spectral traits, created by the bottleneck of inefficient convection in stellar atmospheres, which could not be accounted for by the mixing-length doctrine, with its over-efficient transport. On the report of a referee who disagreed with my theory, the paper was rejected. The fact of rejection on such grounds, even if disputable, is in itself very ominous. Although the story had a happy end—the paper was then published without much delay by the Royal Astronomical Society (6) and represents undoubtedly a gem of a contribution to the knowledge of stellar atmospheres—the attitude of the referee (a staunch believer in the mixing-length, yet officially anonymous) was characteristically dogmatic. By misrepresenting—apparently from unwittingly misreading the texts but perhaps true to his creed—the unwanted alternative to the mixing length, it is a remarkable example of wishful thinking. Here are a few citations from the referee's report, which was communicated to the author. He (or she) writes (exact excerpts are in quotation marks, with my comments following):

1. In Öpik's investigation, "The largest part of the convective upward heat transport is assumed to be transported down again since the matter presumably

cannot get rid of its surplus energy." Quite contrarily, I show that the contact heat transport is so powerful that the excess heat escapes laterally into the downward current before reaching its ultimate destination and that matter gets rid of its surplus energy much too readily and sooner than in the mixing-length analogue. Further, convective transport is not to be equated to the total heat content of matter moving upwards, but is only the net difference delivered at the top, and this, once delivered, cannot "be transported down again," a physically meaningless suggestion and something I nowhere had intimated.

2. The referee then continues, "This then leads to the conclusion that a hot gas stream ($\Delta T > 0$, $\Delta \rho < 0$) will sink in the atmosphere, which seems impossible to the referee." The naive term "hot" betrays the root of his misunderstanding: there is nothing like absolutely "hot" or "cold," nor is there any absolute definition of the temperature excess, ΔT. What matters is the difference between the warmer rising current and the colder descending current at the same potential level, and for the "sinking" stream ΔT is always less than 0 or colder, although the difference is decreased ten times compared to the mixing-length figment, while, compared to the bottom level, both rising and descending currents are colder because of adiabatic expansion.

This would suffice, though there is more to it. The quotations as cited above are typically similar to the dogmatic criticism of misrepresented tenets of another faith, often as wholehearted as it is prejudiced; unwillingness to get acquainted with the actual pronouncements of the other side is common to both. Such an attitude, though alien to science, is nevertheless there as the consequence of human weakness, and we have to reckon with it as a fact.

As an outstanding example of dogmatism in science, the mixing-length syndrome still persists and comprises an influential mini-establishment, though it is harmless as long as it is confined to stellar interiors and keeps clear of the surface layers, or of planetary atmospheres.

CRATERING

Impact cratering is important in the process of shaping the surfaces of planets, especially of the Moon and Mars, as well as in cosmogonic processes of building the planets from aggregates of smaller stray bodies. Until lately experiments at cosmic velocities were not available, and I developed a theory based on first principles which now, when compared with the experimental data, turns out to allow the calculation of crater volume, diameter, depth, and ellipticity of craters with an unexpected accuracy of better than 20% in linear measure and at all velocities (7). No empirical parameters are used. The crater volume is essentially proportional to the *translational momentum* of the projectile, with a *corrigendum* for vapor formation and a secondary shock caused by it, while the cohesive strength of the target enters as an independent variable determining the coefficients of proportionality. The success of this a priori theory, now empirically verified over a range of velocities from 2 to 20,000 m sec^{-1} without using ad hoc coefficients of proportionality, is its main justification. Consideration of the *radial momentum*, created in the target by the intruding projectile, is the main feature of the theory.

Regrettably, experimenters in hypervelocity impact have used and are still using kinetic energy as the argument for interpolation, even without proper regard for the cohesive strength of the target. While this procedure may be practically satisfactory for representing experiments with the same materials within a limited range of velocity and a more or less constant mass of the projectile, extrapolation by kinetic energy beyond the experimental range may lead to errors of many orders of magnitude: mv^2 can never be made to correspond uniquely to mv. Actually, because of vapor formation and second shock, the velocity exponent for crater volume may be about $\frac{4}{3}$, while at constant velocity the crater volume must be proportional to the mass of the projectile. An intermediate formula for cosmic velocities, something like $mv^{4/3}$, could be suggested, instead of mv. My theory actually allows for this, although without the mathematical oversimplification. The empiricists, however, having discovered the lesser power of velocity but insisting on kinetic energy as the argument, would put the crater volume proportional to $(mv^2)^{2/3} = m^{2/3} v^{4/3}$, with the unnatural $\frac{2}{3}$ power for mass (instead of 1). There are more details to it, to be looked for in the relevant publications.

At present, however, a mini-establishment exists around the doctrine of kinetic energy as the impact argument. This is an impediment to progress and, until the successful correct theory (especially with regard to cohesive strength) is applied, extrapolations and speculations on the cosmogonic role of cratering are subject to major pitfalls.

Estimates of the mass of projectiles that produced meteor craters on the Earth and Moon offer a relevant example. For the Arizona crater, estimates based on the doctrine of kinetic energy were about 40 times lower than the mass corresponding to the criterion of translational momentum, and independently confirmed by the depth of penetration. If such were the efficiency of meteorite impacts (which actually waste most of their energy on the inelastic radial shock and heating of the target material), the number of craters in lunar maria would be by almost two orders of magnitude higher than observed, amounting to saturation cratering and equal to that on the continentes. It has been shown (8, 7) that, with the observed population of stray bodies in the solar system and my theory of cratering, the frequency of small and medium-sized craters in the maria is closely accounted for by impacts during the past 4,500 million years, while larger craters show an excess, accountable by survival of premare craters through the event of mare lava flooding (itself a result of a major impact). This in itself is a most impressive confirmation of the cratering theory, obtained well before experiments with hypervelocity (and non-hypervelocity) impacts were made on Earth.

By arguing ad absurdum, we could say that, if energy were directly relevant to the size of a crater, a bonfire lit on a rock surface should lead to "progressive cratering" because heat is also kinetic energy. Cratering is the result of action of forces, and action is in direct relationship to momentum. In a kind of transfiguration of momentum, the translational momentum of the projectile creates radial momentum of the displaced target (rock) material in a constant ratio of from two to five (depending on velocity as determining the secondary shock from vaporization); the total momentum of the symmetric radial shock is, of course, zero, while the translational momentum of the projectile is absorbed by the main body (planet).

In the target, the inelastic shock conserves its momentum separately in each radial direction as long as destruction of the solid material and hydrodynamic flow takes place. The velocity, U, of the radial motion thus decreases as the volume and mass involved increase, until the hydrodynamic resistance ρU^2 (ρ = density) becomes equal to the crushing strength of the material, s. This determines the limit of destruction and the volume of the crater.

Agreement with experiment and with observation (frequency of lunar craters) completely supports this theory (which, of course, is more complicated than could be sketched here). It can only be wished that the dogmatic eclipse of the realistic cratering theory by the kinetic energy scaling would be lifted and that the overlooked and neglected *perfect* theory be raised to its proper place, for realistic dealings with cosmic or cosmogonic events. (Years ago it was found that the theory correctly predicts the armour-blasting properties of artillery shells, a "practical" confirmation of, unfortunately, too sombre associations).

LUNAR AND MARTIAN VOLCANOES

Despite all the eloquent statistical arguments of interplanetary astronomy, the thesis that lunar craters are presumably of volcanic origin was quite widespread, chiefly among amateur astronomers and professional geologists. Decades ago some of them even tried to deny a meteoritic origin of the Arizona crater. The dogma was very strong and was proclaimed by a considerable group or establishment of true believers. At present it has been completely destroyed, at least as concerns the Moon, by direct space exploration and landings, and need not interfere with scientific progress any further. Of course, Mars has now become the refuge of planetary volcanoes, although no longer in a dogmatic sense. It is conceded that most of the Martian craters also originated from impact, because their frequency (surface density) corresponds to statistical expectations for the fringe of the asteroidal belt. But a few large structures are still called "calderas," perhaps wishfully implying their volcanic origin. From the total evidence available, I still prefer to consider them impact craters, surviving an immense lava flooding in the Martian northern hemisphere from an impact of a large asteroid "in the beginning" (9). Since they are similar to the larger surviving lunar craters, and since they are placed on elevated ground well above the average level, such survival in the midst of a lava sea is quite plausible. The matter, however, cannot yet be considered as finally settled. Besides, from the slowness of erosion on Mars—which on Earth is a necessary link in mountain building—Mars cannot yet have entered the phase of volcanism that may be billions of years ahead.

ANCIENT MARTIAN "RIVERS"

The identification of some gigantic meandering cracks as the beds of ancient rivers (of water or lava) on Mars is in danger of becoming a mini-dogma, misleading and perhaps impeding progress. The only reason for such an identification is their meandering shape and formation of systems of succursals closely reminiscent of terrestrial river systems. I have pointed out that exactly similar meandering and branching systems of cracks are omnipresent on asphalt or concrete sidewalks, e.g. on all university campuses I have visited (10, 9). These are caused by the pressure of encroaching vehicles without any relation to fluid flow, and the cracks or clefts on the Martian surface are most probably of similar origin, caused by the pressure of readjustments of the deeper crust on which the top layer rests. Water rivers are definitely excluded—with large amounts of water, cloud and snow formation (at present solar luminosity, though it was *lower* in the past) would have depressed the mean global temperature on Mars from the present low temperature of $-42°C$ to $-62°C$, equal to the coldest Siberian midwinter. Water would everywhere be completely frozen under such circumstances, unable to flow and create rivers. Although lava flow, dubious as it is, cannot be excluded by such an argument, cracks on the surface of a thermally evolving planet must inevitably arise, and before looking for the "rivers," let us look for the few real cracks: they are there, relegating the "rivers" to the realm of fantasy until better confirmation is available. The lunar "rilles," which also were regarded as traces of liquid flow, are now more definitely identified as cracks or rifts (10), and this may serve as an analogy for Mars.

ORIGIN OF METEORITES

The physical and mineralogical structure of meteorites implies that they are collision fragments of asteroidal or sublunar sized bodies. Consideration of encounter probabilities ensures that collisions between members of the asteroid belt do happen; hence there is a possibility that meteorites actually arrive from the asteroid belt, their orbits being changed by the velocity imparted at collision and by subsequent planetary perturbations. The newly determined density of the largest asteroid, Ceres, indeed confirms the hypothesis that asteroids are compact stony bodies and not fluffy objects like cometary nuclei. This falls now in line with the hypothesis of the asteroidal origin of meteorites, which is now seemingly becoming a mini-dogma, accepted without further doubts. Yet it has been shown (11) that contemporary collisions and perturbations cannot account either for the yield or for the orbits of meteorites, which resemble those of short-period comets brought inside Jupiter's orbit by nongravitational forces. The orbits of the so-called Apollo class of "asteroids" belong to the same type, which suggests that they are extinct remnants of disintegrated gigantic cometary nuclei. The apparent conclusion, to be substituted for the fruitless dogma, would relegate their origin to collisions among quite another class of primeval asteroids and to the dawn of the solar system. The original fragments would then have become incorporated into the ices of accreting comet nuclei, and been ejected by planetary (Jupiter) perturbations to the outskirts of the solar system (Oort's sphere of comets), where they would be stabilized by stellar perturbations and sent back to the inner regions of the solar system by similar perturbations. After being captured by Jupiter into short-period orbits, the "rocket-effect" of evaporating gases would cause some of the orbits to shrink (namely those with retrograde rotation of the nuclei) and thus to escape Jupiter's dangerous vicinity. With evaporation of the volatiles, the meteoritic fragments or the Apollo "asteroids" are released into our interplanetary surroundings, to be ultimately removed by planetary encounters within a lifetime of the order of 100 million years. This model is also in harmony with the cosmic-ray ages of meteorites, which represent the time since they were released from shielding inside the cometary nuclei, and not the time of their collisional break-up. If this were so, their relevance to the origins of the solar system would be greatly enhanced.

TIDAL ORIGIN OF THE SOLAR SYSTEM

The hypothesis by Chamberlain and Moulton, so diligently pursued by Jeans, that a close tidal encounter of the Sun with another star led to the formation of the solar system, has enjoyed widespread (though not universal) acceptance and is still on the books, despite its improbability bordering on impossibility. It offers an example of dogmatic attitude with formation of its peculiar establishment group, which is especially strong in popular writings. Such an encounter, of course, is quite possible but extremely improbable. Further, as pointed out by Russell, the hot gases ejected from the Sun could not condense into planets, especially not during the short time of the stellar passage, but would instead disperse into space. Their angular momentum (which raises the cardinal challenge to all cosmogonic theories) at ejection would be short of the requested value by a factor of the order of 20. This difficulty, and then putting the planets into circular, regularly spaced orbits, created formidable problems that Jeans attempted to answer through appropriate perturbations by the passing star, a gigantic mathematical task never convincingly concluded. The problem is in itself interesting and it was worthwhile to treat it, but physically the outcome of the encounter would have been the ejection of uncondensed gaseous matter, which would at first have formed a nebula. Any wisps of gas, ejected into various intersecting or interpenetrating orbits, would then through collision settle into a circularly rotating aggregate with conservation of the sum of angular momentum, i.e. a primeval nebula from which later the planets could have condensed. Yet this returns us to the nebular hypothesis, and there is no way whereby we can distinguish between a nebula directly condensed from interstellar space, and one formed in the tidal encounter—except for the criterion of probability. And in this respect the solar system gives an answer. The systems of the satellites of the outer planets show the same kind of regularity exemplified by the mother system itself: a coplanar succession of near-circular orbits with a more or less regular spacing (Titius-Bode Law). Instead of the extremely low probability of stellar encounters, the formation of solar or planetary satellite systems appears to be the rule rather than an exception. The nebular hypothesis is thus able to account for everything. While the improbability of a tidal encounter can be partly brushed away by assuming that it happened when the stars were much closer together (this, however, would require an improbably high age for the solar system), the ensuing regular spacing of the planets (or the satellites) requires the intervention of another improbable configuration, so that the idea of a tidal encounter can hardly be maintained in this context. Of course, there could have been tidal encounters during the early phases of evolution of our stellar universe, and the theoretical work done in this respect is not quite in vain. Yet the outcome of such encounters could be very different from the formation of something resembling the solar system.

PUBLICATION BIAS AND EXCHANGE

When, in 1938, my papers (3) on "Stellar Structure and Stellar Evolution" (with calculations of unmixed stellar models and those of giant stars) were printed in the *Publications of the Tartu (Dorpat) Observatory* in Estonia, I soon afterwards received a letter from George Gamow, underlining the importance of my work but reproaching me for publishing in such an "obscure" place, wherefore—in his opinion—progress in the study of stellar structure must have been unnecessarily delayed.

The view that, by all means, publication must be achieved in the internationally recognized "important" journals pervaded and still pervades the astronomical establishment and especially the young generation; the latter of course for obvious practical reasons. Yet the fact that Gamow—within a year—got hold of my papers, and that others soon continued on these lines (sometimes referring, sometimes not, to my work), is the best answer. Tartu Observatory, in the centuries-old astronomical tradition of exchange of publications (possibly explained by the fact that we are all dealing with the same cosmic laboratory called the Universe) exchanged its publications with all astronomical institutes of the world, so that the work did not remain unknown to those who cared (physics and most other branches of science do not adhere to such a tradition). And, as to the economical side, the publication costs in Estonia were very low (as they are also at Armagh where the same tradition of exchange is continued). Although, for instance, the editorial setup of the *Astrophysical Journal* was friendly toward my aspirations, the page charges and reprint costs for some 200 pages of mathematical and tabular material would have been absolutely prohibitive. Also, it was certain that full-length publication of such extended papers would not be possible. I had already had previous experience with editors and reviewers requiring great reductions in size, to the detriment of detail which is so essential in pioneering work.

I wish here to emphasize further the importance of the traditional exchange of publications. Not all observatories (especially the smaller ones) are in a position to subscribe to all the "important" journals. Also, a search in libraries for the relevant articles a scientist may need would involve unnecessary psychological effort and waste of time. Thus, because of human weakness, communication between scientists would considerably suffer unless, as in the astronomical tradition, reprints and independent publications of an observatory are systematically numbered and kept in one place. In such a case it is easy and even rather tempting to look among the systematized publications of another astronomical institution for the collected printed papers of a colleague who is known to work on a definite subject. Theoretically, the convenience may appear to be irrelevant, but practically it is of utmost value.

The institutions that do not follow the tradition of exchange usually send out Lists of Reprints, available on demand. This procedure disregards the fact that any research of value is not meant to satisfy the interests of individuals of today, but should address itself also to the future. It cannot be known in advance which work will be of relevance within decades (or even centuries) to come. A selection made to satisfy the temporary interests of today may and certainly will miss the works of relevance for tomorrow. A library always contains more works than would be ever needed or read: but it is impossible to foretell the interests of the future, and the collection must necessarily be very much more complete than the actual needs that may arise.

In the matter of exchange, the giving hand should take the initiative. As his moral and vocational obligation, a creative spirit must make his results known, at least where similar work is, or could be done. It is like seeding. Few of the seeds may fall on fertile soil, yet nobody can predict for certain which of them will grow. And, where there is no seeding, there will be no growth.

THE CRITICS: EDITORIAL REFEREEING AND CENSORSHIP

Another reason for not always publishing in the accepted "important" journals is a danger of being rejected, either because of a sincere failure of the editorial apparatus to appreciate new developments (often because of the fear of appearing ridiculous), or because of dogmatic and personal prejudices. These last two should not enter into the editorial judgment, which should be as impartial as possible, yet actually they sometimes do.

As an example of editorial changes that do not infringe on impartiality, I would cite a case with *The Irish Astronomical Journal,* where I am Editor, and where, in an article on the "Lunar Surface" offered by Patrick Moore (12), he voiced his support for the volcanic theory of lunar craters. Although completely disagreeing, it was not my business to interfere except for one minor change: the two groups with opposing views were called "authorities" by Patrick Moore—one, favoring the volcanic origin, consisting chiefly of amateur astronomers, the other, siding with the impact origin, consisting of professional scientists (one of them a Nobel laureate); I had the word "authorities" changed to "authors," with the author's consent.

In my experience, however, editors have not always been so impartial, although usually they are. An article entitled "The Optical Oblateness of Mars" and based on my microphotometric measurements of Mount Wilson photographs during the Opposition of 1958 was accepted by *Icarus* for publication, on the condition that I omit two concluding pages and a figure, actually containing my chief results. The photometrically measured diameters showed a consistent variation with areographic latitude, closely similar in the two colors—the blue and the yellow—and were interpreted as revealing climatic zones of atmospheric circulation similar to those on Earth. I maintained namely that what we measure as the limb is the top of a dust layer that is mainly responsible for the reflectivity of the atmosphere, and not its gas (this is now confirmed by the Viking 1976 landings on Mars). An upward current would lift the dust up, a downward current would carry it down. The measurements showed an equatorial uplift, a subtropical depression as for the anticyclone trade-winds, again an increased diameter in the zone of middle latitudes corresponding to the terrestrial temperate zone of westerlies, and again a subpolar—apparently anticyclonic—depression; only the polar diameters (all relative to an equipotential surface) were increased, contrary to the expectation of an anticyclonic depression, but this clearly seemed to be caused by snow or ice crystals of high albedo replacing the yellow dusty haze of the other latitudes. Now, on this *observational* evidence (whatever its interpretation, right or wrong), the Editor of *Icarus* and my professed friend had put his veto! Another journal was then ready to accept publication, but on the evidence of previous rejection the Editor changed his mind. The article was ultimately published in The Irish Astronomical Journal (13).

From my long experience, both with my own papers, and with those sent to me for refereeing, I have a feeling that the "recognized" journals usually accept without difficulty papers with a middle-of-the-road content, useful contributions to research which already has established itself or accumulations of additional new material. Papers that make little sense are mostly rejected, but some of them are slipping through. As to pioneering work, papers of this kind often are running the risk of rejection or of excessive curtailment.

The practice of anonymous referees is much to blame for editorial malfunctioning. A referee for a scientific journal is a scientist, morally committed to seek and openly proclaim the truth as he feels it, and he should never hide behind the screen of anonymity. As referee, I always send the author an exact carbon copy of my letter to the Editor with all my comments. Among about 150 of such reviews, I have received many letters of thanks (for my suggestions) and only one of what practically amounted to abuse; in most cases, however, there was no reaction. There were a few cases when I was asked to be arbiter in an unfavorable referee report, and I succeeded in rehabilitating some authors from unfair criticism by anonymous colleagues who appeared to think that they alone were entitled to write about a certain subject. Anonymity in refereeing is like kicking somebody in the dark, without a chance of response; it "protects" the reviewer but not the author. The sooner this scourge of anonymity is abandoned, the better for the honest pursuit of research. If fewer referees can be found when there is no anonymity, it will be only to an advantage: those who consent will be a more qualified selection for the job of critics.

We may ask here how many geniuses have been crushed by the unsympathetic and prejudiced attitude of editors and critics in the sciences as well as in the arts, and remained unknown forever? The late discovery of forgotten geniuses testifies to the existence of a graveyard of misunderstood or mishandled originalities which did not fit into the "establishment" of the critics. The sad record of George Bizet, who died in desperation witnessing the failure of his "Carmen" in the eyes of the critics and the Paris public influenced by them, serves as a reminder: "Carmen" has become unquestionably the most popular opera of all times. By independently printing in the "obscure" Tartu or Armagh publications, my work has ultimately made itself known. Would I have been forced to limit myself to the "recognized" big journals, much of it—possibly some most original contributions—could have remained buried forever.

THE BIRTH OF A MYTH

A remarkable article by Leighton (14) supplied with artistically rendered humorous illustrations almost true to life, emphasizes the fact that scientists seldom listen to others and, if they are not dozing during lectures, they may either be preoccupied with their own thoughts or enter into private discussions. Although partly explained by acoustical difficulties, this attitude may not be limited to conference lectures.

The following example describes a case to the point. The printed word is an indubitable fact, irrespective of whether or not its statements are subject to debate. Yet here we have a critic who first completely misrepresents a published work, and then—rightly—sets out to destroy this figment of his own imagination which, incidentally, is just the opposite of what the author of the criticized work was saying.

In the universally recognized international journal *Science,* R. K. Ulrich (15) refers unfavorably to my work (16) on stellar structure and variations of solar luminosity. He purports to describe my model and calls it "physically untenable," but the described model is not mine—only the critic's invention and, so to speak, the very opposite of what I had proposed. In my reply (17) I point out that, while I consider inward diffusion of *hydrogen* into the core depleted by hydrogen burning, he objects to inward diffusion of the *heavy elements* which in my model are not diffusing at all. In my model, turbulent mixing suddenly transports more hydrogen to the core, triggering thus an increase in the nuclear energy output, this being the most important point in my theory of variability. Yet Ulrich never mentions "hydrogen" by name or "nuclear energy generation" in this context. While I trace the heavy-element content in the Sun to interstellar *diffuse* matter (dust) during the process of star formation by accretion, thus predating all the development stages of the future Sun, the critic insists that I am putting them into the core by internal diffusion inside the Sun! etc. etc. Possibly, the words "diffuse" matter and negligent reading, with a mind concerned with gas diffusion, may have led him to this gross misrepresentation.

Now, as often happens, other authors could rely on the second-hand information of such a source, and a ready-made legend or myth, perhaps a new dogma could emerge, something of the sort "Öpik wants diffusion of the heavy elements in the Sun to be responsible for its variability, which of course is too slow on the time scale of stellar evolution." Note that a similar, perhaps not so extreme misstatement about "hot bubbles going down" has been mentioned above in connection with the ill-conceived notion of the "mixing length."

It is not a question of whether I am right or wrong in my theory, but only of what I had actually said in print, thus of the complete distortion of an undisputable fact. In this case—as probably in many others—the editorial reviewer system has goofed, while the critical author, instead of a straightforward apology and admission of fault, in a "reply" (18) just vaguely expresses some of his own views on solar models and restricts himself to considering the (irrelevant and practically nonexistent) diffusion of the heavy elements "relevant to hydrogen," instead of considering the diffusion of hydrogen itself into the depleted core.

FOR A POSTSCRIPT

These nonsystematic recollections, based on personal experience, are concerned chiefly but not exclusively with preconceived notions and dogmatism. The research topics mentioned above as examples are of necessity those close to the author's scientific activity; he considers the points raised in their connection of great importance in the study of the Universe, but by no means implies that he is always right. He sincerely wishes that his words may not completely remain a lonely cry in the wilderness, but may perhaps at some time help someone in the impartial search for truth.

Literature Cited

1. Eddington, A. S. 1929. *MNRAS* 90:54
2. Öpik, E. J. 1951. *MNRAS* 111:278
3. Öpik, E. J. 1938. *Tartu Obs. Publ.* 30, No. 3 (118 pp.), No. 4 (48 pp.)
4. Öpik, E. J. 1950. *MNRAS* 110:559
5. Öpik, E. J. 1953. *Geophys. Bull. (Dublin)*, No. 8, 14 pp.
6. Mullan, D. J. 1971. *MNRAS* 154:467
7. Öpik, E. J. 1969. *Ann. Rev. Astron. Astrophys.* 7:473
8. Öpik, E. J. 1960. *MNRAS* 120:404
9. Öpik, E. J. 1973. *Irish Astron. J.* 11:85
10. Öpik, E. J. 1969. *Irish Astron J.* 9:79
11. Öpik, E. J. 1968. *Irish Astron. J.* 8:185
12. Moore, P. 1965. *Irish Astron. J.* 7:106
13. Öpik, E. J. 1973. *Irish Astron J.* 11:1
14. Leighton, R. B. 1971. *Phys. Today* 24: No. 4, 30
15. Ulrich, R. K. 1975. *Science* 190:619
16. Öpik, E. J. 1965. *Icarus* 4:289
17. Öpik, E. J. 1976. *Science* 191:1292
18. Ulrich, R. K. 1976. *Science* 191:1293

Ann. Rev. Astron. Astrophys. 1977. 15: 19–44

RECENT OBSERVATIONS OF PULSARS

×2106

J. H. Taylor
Department of Physics and Astronomy, University of Massachusetts, Amherst, Massachusetts, 01003

R. N. Manchester
CSIRO, Division of Radiophysics, Sydney, Australia and Department of Physics and Astronomy, University of Massachusetts, Amherst, Massachusetts, 01003

INTRODUCTION

It is now almost ten years since pulsars were discovered (Hewish et al. 1968), and the fascination that these objects have provided for the observer and theorist alike continues to ensure them a fair share of astronomical attention. Since 1968 more than 400 papers have been written on pulsar observations or their interpretation, and this work has contributed significantly to our knowledge of galactic structure, stellar evolution, and the physics of very dense matter. Although remarkable progress has been made in understanding the nature of pulsars, and a basic model—a rapidly rotating, highly magnetized neutron star—has gained general acceptance, the mechanism responsible for the observed radio emission is still not understood.

Our intention in this review is not to attempt to discuss all the known properties of pulsars; rather, we have concentrated on two main areas in which there has been substantial progress during the last few years and which look promising for future work. The second section is devoted to a summary of the existing observational material relevant to the pulse emission mechanism. In general, we have not attempted to describe the possible interpretations of the data or the various models that have been proposed for generation of the pulsed emission. The third section deals with the important problem of the origin and evolution of pulsars. The observed variations, both regular and irregular, in pulsar periods are first described. Recent extensive surveys have detected a large number of new pulsars, many at large distances from the Sun. We review the analysis of these results, which give the galactic distribution of pulsars and their luminosity function. Finally, we describe recent observations suggesting that, for many pulsars, the characteristic age is an overestimate of the true age. These results, together with the derived galactic distribution, imply a relatively high birthrate for pulsars.

OBSERVED PULSE CHARACTERISTICS

Integrated Pulse Profiles

PULSE SHAPE One of the most important characteristics of a pulsar is the shape of its integrated profile, i.e. the mean pulse shape obtained by synchronously summing a large number of pulses. Each pulsar has a unique and identifiable pulse profile which, in general, remains stable in shape. Figure 1 shows integrated profiles for

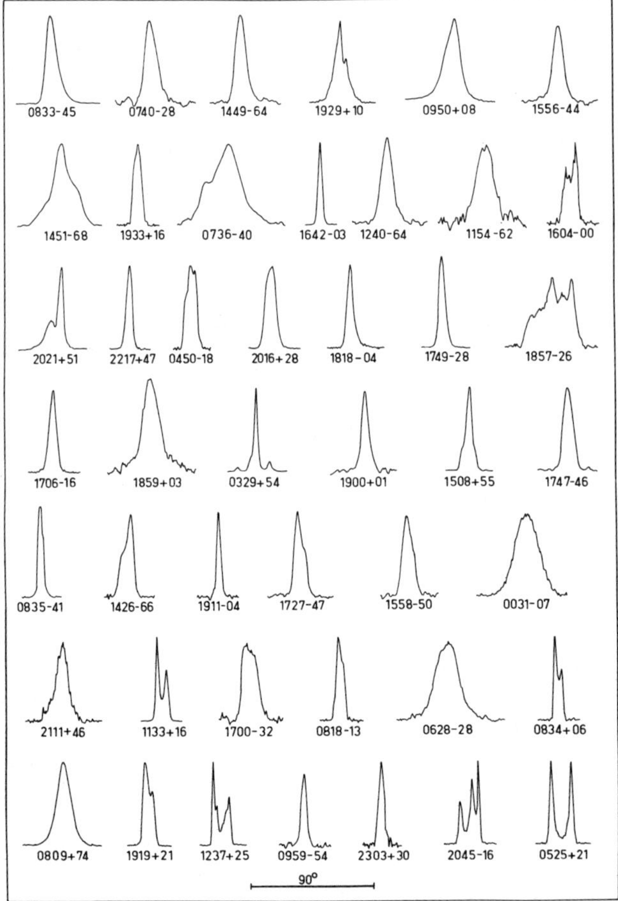

Figure 1 Integrated pulse profiles for 45 pulsars, all plotted on the same longitude scale indicated by the 90° bar at the bottom of the figure. All profiles were recorded at frequencies between 400 and 650 MHz and they are plotted in order of increasing pulse period (Manchester & Taylor 1977).

45 pulsars, in most cases recorded at a frequency near 400 MHz. In many pulsars the profile is dominated by a single peak or *component,* which typically has a width of 5–10° of longitude (where 360° of longitude equals the pulsar period). Other pulsars have two or more components that may be either partially overlapping or well separated. In pulsars with separated or resolved components the profile shape often has a symmetrical, basically "double" form with two outer components separated by a region of weaker emission.

As shown in Figure 1, the pulse emission is confined to a rather narrow longitude range in most pulsars. The equivalent width (profile area divided by peak intensity) is typically about 10° of longitude, although for double profiles the component separation is often closer to 20°. There are, however, two classes of profile for which the emission is not so narrowly confined. In the first of these, a secondary pulse or *interpulse* is located approximately 180° or half the period from the main pulse. Six pulsars are known to have interpulses, the best-known being the Crab pulsar, PSR 0531+21, (Rankin et al. 1970). The southern pulsar, PSR 1055–52, is remarkable in that it has a two-component main pulse and an interpulse with the relative component separations very similar to those of the Crab pulsar (McCulloch et al. 1976). With one exception (PSR 0904+77), all pulsars possessing interpulses have relatively short periods. In the other class, the profile is confined to one region of longitude but is much wider than usual. For example, PSR 1541+09 has a half-power width of 33° and significant emission over more than 180° (Backer, Boriakoff & Manchester 1973); PSR 1911+03 has the widest known profile with a half-power width of 75° (Mohanty & Balasubramanian 1975).

Several authors have recognized that pulsars may be divided into classes on the basis of their pulse profile. Taylor & Huguenin (1971) proposed two such classes: Type S (for Simple), in which the profile is dominated by a single component, and Type C (for Complex) in which there are two or more components of comparable intensity. Because of the frequency dependence of relative component amplitudes (described below), profiles at frequencies around 400 MHz are generally adopted for classification purposes. Similar schemes have been proposed by Backer (1976) and Roberts (1976a). Backer suggests division into five classes: S (single component), DU (unresolved double), DR (resolved double), T (triple or three-component) and M (multiple or more than three components). The significance of these classes derives largely from the fact that other pulse properties such as period, period derivative, polarization, and fluctuation spectra correlate with the classifications (Huguenin, Manchester & Taylor 1971, Roberts 1976a). These correlations are further described below.

SPECTRAL PROPERTIES Pulsars have been observed at radio frequencies as low as 10 MHz (Bruck & Ustimenko 1973, 1976) and as high as 15 GHz (Downs, Reichley & Morris 1973). At frequencies greater than a few hundred megahertz the spectral index is always negative with slopes generally in the range -1 to -3 (Sieber 1973, Backer & Fisher 1974). Examples of the several different spectral forms observed are given in Figure 2. Of the 27 sources studied by Sieber (1973), 13 had essentially straight (i.e. power-law) spectra, 7 had spectra that could be described by two power-law segments, the higher frequency one always having the more negative slope, and 8 showed evidence for an intrinsic low-frequency turnover (i.e. one not resulting from interstellar effects). These spectra all refer to the profile as a whole; higher resolution observations (Manchester 1971, Lyne, Smith & Graham 1971, Backer 1972) show that different components of multiple-component profiles normally have different spectral indices. The component with the steepest spectrum occurs at different positions within the profile for different pulsars. Spectral index differences are typically in the range 0.1 to 0.4. The precursor component of the Crab pulsar profile is exceptional in this regard, having a spectral index $\alpha \lesssim -5$ at frequencies above about 500 MHz compared to $\alpha \approx -3$ for the rest of the profile (Boriakoff & Payne 1974).

Where a profile has two or more components, the separation of these components in longitude is also frequency dependent. As illustrated in Figure 2, the variation can normally be divided into two power-law segments, where the one at lower frequencies has a negative index (typically about -0.25) and the one at higher frequencies either is almost flat or has a positive index ($+0.11$ for PSR 0834+06). The break

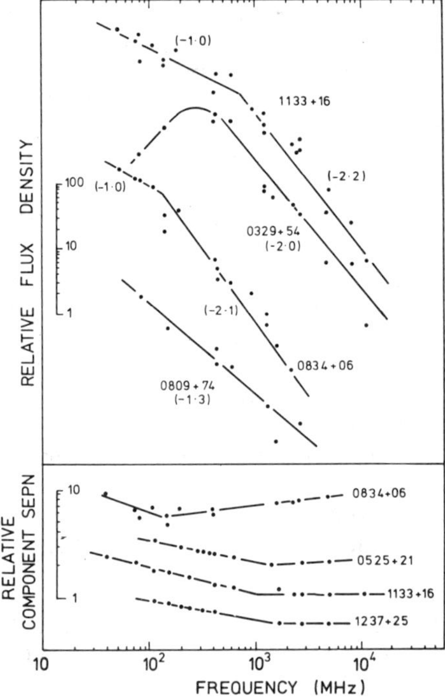

Figure 2 Upper part: relative mean flux densities plotted against frequency for four pulsars (Sieber 1973). Spectral indices for the power-law segments are given in parentheses. Lower part: relative component separations as a function of frequency for four Type C pulsars (Sieber, Reinecke & Wielebinski 1975).

between the two segments is often at about the same frequency as the similar break in the spectrum; for example, the break frequencies for PSR 0834+06 and PSR 1133+16 are about 150 MHz and 1 GHz respectively (Figure 2). In multiple-component profiles the relative spacing of the components is generally independent of frequency. Pulsars where this is not true (e.g. PSR 0329+54) appear to have overlapping emission from two independent regions.

There is some evidence that component widths are also less at higher frequencies. For example, Sieber, Reinecke & Wielebinski (1975) find that component widths at 4.9 GHz are typically 15% narrower than those at 2.7 GHz.

POLARIZATION The emission from pulsars shows a wide range of fractional polarization with linear dominating over circular in most cases (Manchester 1971, Lyne, Smith & Graham 1971, Hamilton et al. 1977a, b). Polarization properties are illustrated for two integrated profiles in Figure 3. For PSR 0833-45 the average degree of linear polarization is 82% and circular polarization is 6%. In about 75% of the pulsars with significant linear polarization, the position angle changes continuously through the profile, often with the approximately linear variation seen for PSR 0833-45 (Figure 3). For most Type C pulsars, the position-angle variation has a characteristic shape with an approximately constant position angle in the wings of the profile and a rapid variation near the center. Observations at different frequencies show that, at a given longitude, the position-angle gradient is independent of frequency; at low frequencies the total range of variation is generally larger because of the wider profile, but it is always less than or about 180°.

The second profile in Figure 3, for PSR 1857-26, is an example showing a

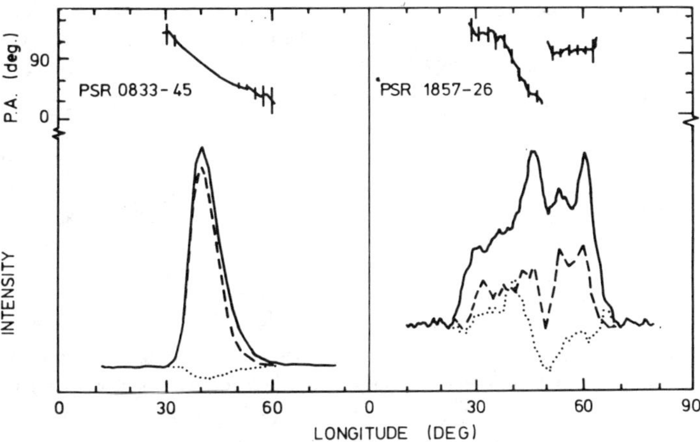

Figure 3 Integrated profiles and polarization characteristics for two pulsars, recorded at a frequency of 631 MHz (Hamilton et al. 1977b). In each case the full line represents the total intensity profile, the dashed line the linearly polarized part, and the dotted line the circularly polarized part. The position angle of the linear part is shown with error bars ($\pm 2\ \sigma$) on every second point.

discontinuous position-angle variation. For this source and others like it, the position angle has one or more discontinuous steps with the magnitude of step generally being close to 90°. The fractional polarization always has a minimum at the longitude of the step, which suggests an overlap of two components with orthogonal position angles. This effect is clearly related to the orthogonally polarized sub-pulses seen in most pulsars as described below. If the trailing position-angle segment for PSR 1857-26 is moved by 90°, then the position-angle variation is continuous and, in fact, has the shape characteristic of Type C pulsars. The implication is that, for the trailing part of the profile, the emission is in a mode whose position angle relative to some vector on the star (presumably the magnetic field) is orthogonal to that of the rest of the profile.

Observations at various frequencies show that for pulsars, in contrast to most radio sources, the degree of linear polarization is greatest at low frequencies (Manchester, Taylor & Huguenin 1973). The fractional polarization is generally constant below some break frequency (generally a few hundred megahertz) and approximately inversely proportional to frequency at higher frequencies. For some pulsars (e.g. PSR 0833-45), the degree of circular polarization increases with frequency, but in others it is less at higher frequencies (Hamilton et al. 1977b).

PROFILE STABILITY One of the principal reasons for the importance of the integrated profile as a pulsar characteristic is its great stability with time. With the exception of mode changing (see below), there is no evidence for a secular change in the intrinsic shape of the integrated profile or its polarization characteristics for any pulsar (Rankin, Payne & Campbell 1974, Helfand, Manchester & Taylor 1975). However, the mean pulse energy does change with time; McLean (1973) found that (after removing the effects of interstellar scattering) mean pulse energies often varied by about a factor of two from day to day, and much larger variations, by a factor of ten or more, are observed over longer intervals (Huguenin, Taylor & Helfand 1973, Rankin, Payne & Campbell 1974, Lyne & Thorne 1975).

Mode changing is a phenomenon, first observed in PSR 1237+25 by Backer (1970a), in which the profile abruptly changes to a second stable form, remains in that form for some interval, and then abruptly reverts to its normal form. For PSR 1237+25, these mode changes typically occur every few hours and last for several minutes. A more complicated form of mode changing with several stable modes has been observed for PSR 0329+54 by Lyne (1971), Hesse (1973), and Hesse, Sieber & Wielebinski (1973).

HIGH-FREQUENCY OBSERVATIONS Up to this point this review has been concerned solely with the radio properties of pulsars. It is of course well known that the Crab pulsar emits infrared, optical, X-ray, and γ-ray pulses synchronously with the radio emission. Integrated profiles for four of these frequency regimes are shown in Figure 4. Although the pulse clearly has the same basic shape (with a main pulse and interpulse) at all frequencies, the profiles differ considerably in detail. The radio main pulse has a precursor component that is absent at higher frequencies; the components become wider at higher frequencies and the ratio of interpulse energy to main pulse energy is greater at higher frequencies. Spectral measurements indicate that different emission mechanisms operate in the radio and higher frequency regimes. As mentioned above, the radio spectrum is steep; its extrapolation falls well below the measured intensity at infrared and optical frequencies where the spectrum is approximately flat. At X-ray energies between 1 and 20 keV, the spectral index steepens once again to about -0.6 (Kestenbaum et al. 1976) and to approximately -1.1 at higher energies (Laros, Matteson & Pelling 1973, Walraven et al. 1975).

Despite many searches, the only other pulsar to have been detected unquestionably at frequencies outside the radio band is the Vela pulsar, PSR 0833-45. Satellite observations by Thompson et al. (1975) at energies greater than 34 MeV show that this pulsar emits a broad two-component profile remarkably similar to the Crab γ-ray profile (Figure 4). However, unlike the Crab pulsar, the Vela pulsar does not emit either of the γ-ray pulses in phase with the radio pulse. At X-ray frequencies there have been several reports of pulsed emission (e.g. Harnden & Gorenstein 1973), but more recent and more sensitive searches (Moore, Agrawal & Garmire 1974, Rappaport et al. 1974, Pravdo et al. 1976) have failed to find any evidence for pulsations.

The optical emission from the Crab pulsar has rather weak linear polarization compared with the radio pulse (Kristian et al. 1970). Both the main pulse and the interpulse are about 15% polarized near the wings and have a minimum of polariza-

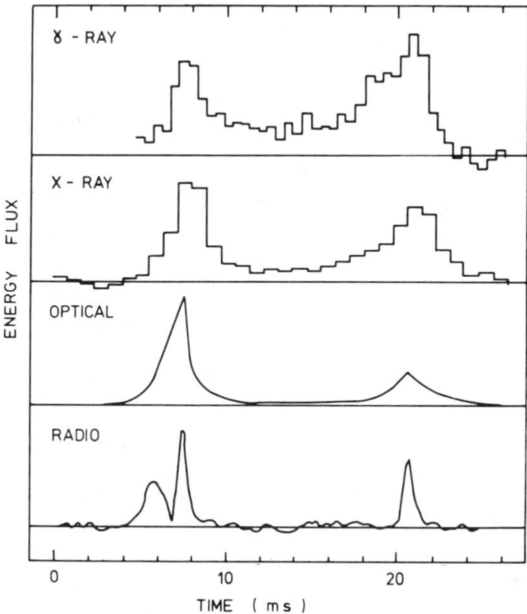

Figure 4 Integrated profiles for the Crab pulsar, PSR 0531+21, from radio to γ-ray frequencies. Data are from Manchester (1971), Warner, Nather & MacFarlane (1969), Rappaport et al. (1971) and Kurfess (1971).

tion near the pulse peak. The position-angle variation appears to be similar to that characteristic of Type C pulsars, but may be more complicated near the minimum owing to the uncertainty in the interstellar polarization (Ferguson, Cocke & Gehrels 1974). A limit of 0.07% has been set on the circular polarization at the peak of the main pulse by Cocke, Muncaster & Gehrels (1971).

Individual Pulses

Unlike the integrated profile, which is very stable, individual pulses vary greatly in shape, intensity, and polarization. The short time scale of these variations shows that they are almost certainly associated with the emission process rather than with the interstellar medium, so their study is an important aspect of pulsar astronomy.

SUBPULSES Individual pulses typically consist of one or more *subpulses* that have a characteristic width of 3–10° of longitude. For a given pulsar this width is either independent of frequency or at most weakly dependent on frequency. Subpulses usually have a rather simple, almost Gaussian, shape and may fall anywhere within the integrated profile; the form of the integrated profile results from variation with longitude of both the strength and the frequency of occurrence of subpulses. Observations of series of individual pulses show that the pulse modulation indices are generally between 0.5 and 3, are different for different components of the integrated profile, and are larger at lower frequencies (Backer 1973, Taylor, Manchester & Huguenin 1975). These pulse-to-pulse fluctuations are highly correlated at frequencies separated by several hundred megahertz, which shows that subpulse modulation is a broadband process. Sporadic narrow-band emission at frequencies around 100 MHz has been reported by Udal'tsov & Zlobin (1974) and Zlobin & Udal'tsov (1975). The narrow features appear in about 10% of pulses from PSRs 0329+54, 0834+06, 0950+08, and 1919+21 and typically have a bandwidth of a few megahertz.

Histograms of pulse intensities (Smith 1973, Hesse & Wielebinski 1974, Ritchings 1976) are often bimodal, with one of the maxima at pulse intensity zero. The missing or "null" pulses, first discussed by Backer (1970b), are weaker than normal pulses by at least a factor of 100 and tend to come in groups of from just a few missing pulses up to 100 or more. Ritchings (1976) has shown that the fraction of null pulses observed is greater in long-period pulsars, especially those with small-period derivatives. This suggests that a large fraction of null pulses may indicate an advanced state of senility for a pulsar.

Spectral analyses of the pulse-to-pulse energy fluctuations show that some pulsars exhibit secondary periodicities (e.g. Taylor & Huguenin 1971). Observed modulation frequencies range from about 0.1 cycles/period up to very close to the Nyquist value, 0.5 cycles/period. Strong features in the fluctuation spectrum occur preferentially in longer-period (usually Type C) pulsars and are sometimes narrow with a fractional width ($\Delta f/f$) of only a few percent. Pulsars with periods less than 0.75 sec (usually Type S) generally have featureless fluctuation spectra that either are flat or rise gradually toward low frequencies. Computation of independent fluctuation spectra for different pulse longitude regions (Backer 1973, Schönhardt & Sieber 1973,

Backer, Rankin & Campbell 1975) show that the periodic modulations are frequently confined to just part of the profile. For example, in PSR 1237+25, a Type C pulsar with five components, the periodic modulation is present only in the outer components (1 and 5). A similar symmetry of fluctuation characteristics about the profile center is observed in other Type C pulsars. A relationship between the periodic fluctuations and mode changing in PSR 1237+25 has been found by Taylor, Manchester & Huguenin (1975). When this pulsar is in its second mode (in which components 4 and 5 are weak), the periodic modulation of component 1 disappears.

DRIFTING SUBPULSES One of the more curious aspects of pulsar morphology is the phenomenon known as drifting subpulses, first observed in PSR 2016+28 by Drake & Craft (1968). For most pulsars the precise longitudes of successive subpulses do not appear to be causally related; however, there exists a class of pulsars, called Type D by Taylor & Huguenin (1971), in which subpulses in successive periods drift systematically across the integrated profile. Both directions of drift are observed, although in pulsars where the effect is most prominent (e.g. PSRs 0031-07, 0809+74, and 2016+28), the drift is always from the trailing edge of the profile toward the leading edge, i.e. toward earlier longitudes. The separation in longitude between successive bands of subpulses, denoted by P_2, is usually less than the width of the integrated profile, so that two subpulses from adjacent drift bands are often observed in one pulse. A third period, P_3, is defined as the interval between successive appearances of subpulses at a given longitude. The definition of P_2 and P_3 is schematically illustrated in Figure 5. For some pulsars P_3 is variable (e.g. for

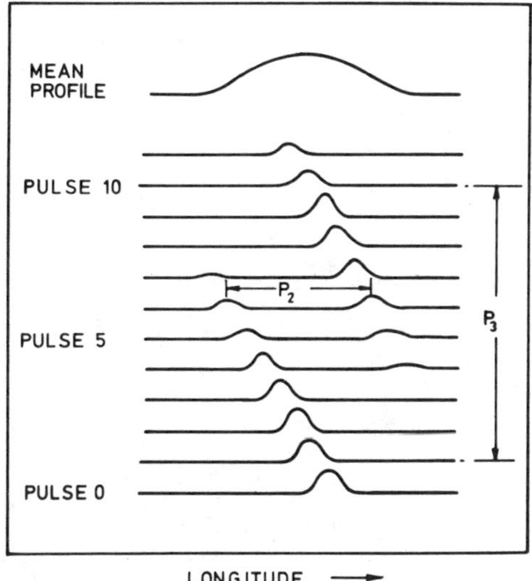

Figure 5 Schematic diagram illustrating the definition of the drifting subpulse parameters P_2 and P_3.

PSR 2016+28, from 3 to 15 periods) but the subpulse spacing, P_2, is more stable. Hence subpulses drift across the profile more slowly when P_3 is large. For PSR 0031-07, P_3 appears to be restricted to one of three approximately harmonically related values, 4, 7, and 13 periods respectively (Huguenin, Taylor & Troland 1970).

For pulsars such as PSR 0809+74 (e.g. Page 1973) where P_3 is stable, there is a corresponding feature at frequency P_3^{-1} in the fluctuation spectrum. Variation of the phase of this periodic component as a function of longitude can reveal drifting behavior that is otherwise difficult to detect (Backer 1973, Backer, Rankin & Campbell 1975). Especially for pulsars where P_3 is close to two periods (the Nyquist value), there is an ambiguity in the drift direction resulting from possible aliasing of the periodic component (Sieber & Oster 1975). For example, in PSR 0943+10, a P_3 of 2.11 periods corresponds to drift toward later longitudes at a rate of 4.0° per period, whereas for the aliased P_3 of 1.90 periods the drift is towards earlier longitudes at a rate of 4.4° per period. The ambiguity in drift direction can only be resolved by correlation of some feature down the drift band. So far

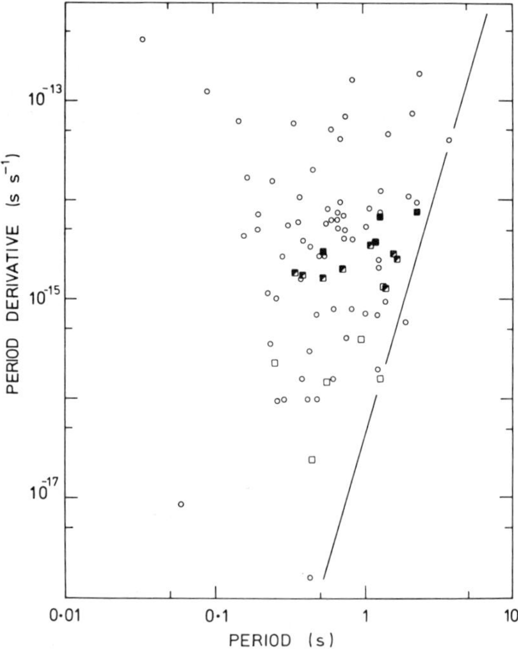

Figure 6 Period derivative plotted against period for 87 pulsars. Pulsars that have drifting subpulses (Type D) are plotted as squares. The direction of subpulse drift is toward earlier longitudes if the square is open, toward later longitudes if it is filled, and if the square is half-filled, both directions of drift are seen. The sloping line apparently represents a period beyond which no pulsed emission is possible.

such a correlation has been clearly observed only for PSR 2016+28 (Backer 1973, Cordes 1975a). By means of spectral analysis, Sieber & Oster (1975) have detected two different drifting systems in subpulses from PSR 2303+30; system I has a P_3 of 2.10 or 1.91 periods and system II has P_3 equal to 2.3 or 1.8 periods. This pulsar and PSR 1944+17 are also rather unusual in that they tend to have a series of pulses with drifting subpulses separated by a null region from another series with no apparent drifting (Rankin, Campbell & Backer 1974, Backer, Rankin & Campbell 1975).

Like pulse nulling, drifting subpulses seem to be a characteristic of old age in pulsars. As shown in Figure 6, Type D pulsars generally have periods relatively close to the cutoff line, which has a slope of 5 in this diagram (Lyne, Ritchings & Smith 1975). This figure also shows that the direction of drifting is closely related to the period derivative (Ritchings & Lyne 1975): drifting in Type D pulsars with small derivatives is toward earlier longitudes; for intermediate derivatives it can be in either direction; and for large derivatives it is generally toward later longitudes.

Observations at various frequencies have shown that, at least for the pulsars with strong drifting subpulses, the subpulse separation P_2 varies with frequency approximately as $\nu^{-0.25}$ (Taylor, Manchester & Huguenin 1975), that is, in a similar way to the component separation in Type C profiles. In a study of drifting behavior in PSR 1919+21, Cordes (1975b) has shown that there are apparently two modes of emission, one with drifting subpulses and one without. The relative strengths of these two modes is a function of frequency, where the drifting subpulses are dominant at frequencies below about 150 MHz and negligible at frequencies above about 450 MHz.

MICROSTRUCTURE Observations with high time resolution, usually involving some form of coherent dispersion removal, have shown that, in some pulsars, rapid time variations known as micropulses exist within subpulses (Hankins 1971, 1972). This microstructure typically has a time scale of a few hundred microseconds or 0.2 to 0.3° of longitude, and has many of the same characteristics as subpulses. For example, micropulses tend to occur at random longitudes within a subpulse just as subpulses tend to occur randomly within the integrated profile. Also, as for subpulses, the characteristic width of micropulses appears to be independent of frequency (Hankins 1972, Rickett, Hankins & Cordes 1975). As shown in Table 1 (Cordes 1975c) microstructure has been detected in five out of eight pulsars for which observations have been made. It is not clear at this stage why only some pulsars have microstructure.

Observations of PSR 1919+21 (Cordes 1975b) show that, as for the drifting subpulses, microstructure is most intense at low frequencies. It therefore appears that the microstructure is imposed on the drifting mode of emission in this pulsar, but not on the nondrifting mode that dominates at higher frequencies.

A quasiperiodic modulation of microstructure intensity has been observed in several pulsars. Hankins (1971) gave an example of a pulse from PSR 0950+08 in which the microstructure is modulated with a period of about 600 μsec. A similar quasiperiodic modulation with a period of about 900 μsec has been detected in

Table 1 Pulsar microstructure

PSR	Period (sec)	Microstructure timescale, τ_μ (μsec)	(deg)
1929+10	0.227	none	—
0950+08	0.253	175	0.25
2020+28	0.343	none	—
2016+28	0.558	290	0.19
1133+16	1.188	575	0.17
0834+06	1.274	1050	0.30
1919+21	1.337	1220	0.33
1237+25	1.382	none	—

microstructure from PSR 2016+28 (Boriakoff 1976). Cordes (1975c) has shown that the phase of this 900 μsec modulation is correlated in successive subpulses down a drift band whereas the actual micropulse modulation is not. Microstructure with a similar quasiperiodic modulation has been observed for PSR 1133+16 at a frequency of 1420 MHz by Ferguson et al. (1976) and at 430 MHz by Cordes (1975c).

Simultaneous observations of PSR 0950+08 at 318 and 111 MHz by Rickett, Hankins & Cordes (1975) have shown that the microstructure at these two frequencies is correlated and hence that the micropulse emission process is broadband. These authors also investigated the dynamic spectra of micropulses over a band of 125 kHz total width. The micropulses were found to be deeply modulated with a characteristic bandwidth of a few tens of kHz, with the modulation quite different for different micropulses, ruling out an interstellar origin.

POLARIZATION Observations show that, in general, subpulses and micropulses are more highly polarized than integrated profiles, but that their polarization varies from pulse to pulse leading to depolarization of the integrated profile. Linear polarization generally dominates over circular as in integrated profiles; continuous variation of position angle through subpulses is common but the total swing is usually less than 30° (Manchester, Taylor & Huguenin 1975). Polarization parameters illustrating these properties are shown in Figure 7 for several consecutive pulses from PSR 0329+54. In some pulsars, e.g. PSR 1237+25, all subpulses at a given longitude have essentially the same position angle, but the degree of polarization is variable. For this pulsar, subpulses occurring at the longitudes of components 1 and 5 are more highly polarized than those occurring at other longitudes. Although the average polarization of subpulses is not strongly frequency dependent, their position-angle stability is generally less at higher frequencies. This is the primary reason for the depolarization of integrated profiles with increasing frequency.

The third principal mechanism by which integrated polarizations are reduced is the occurrence of orthogonally polarized subpulses (Lyne, Smith & Graham

1971, Manchester, Taylor & Huguenin 1975). Histograms of subpulse position angle show that, at certain longitudes in most pulsars, there are two preferred values separated by approximately 90°. As may be seen in Figure 7, such orthogonal subpulses occur in the fourth (trailing) component of PSR 0329+54. Subpulses with the orthogonal position angle are similar to normal subpulses in all other respects and appear to occur randomly, often overlapping normal subpulses. In some pulsars the orthogonal component that predominates changes at some point across the profile, leading to 90° discontinuities in the integrated profile polarization (Figure 3).

Similar orthogonal polarization is observed in micropulses (Cordes 1975c). An example of a pulse from PSR 1133+16 recorded at a frequency of 430 MHz showing 90° position-angle discontinuities between micropulses is shown in Figure 8. Cordes finds that such 90° transitions are usually accompanied by a change in the sense of circular polarization, so the two polarization states are truly orthogonal, separated by 180° on the Poincaré sphere. In general, micropulse polarization characteristics are very similar to those of subpulses. Micropulses are commonly more highly polarized than the subpulses that contain them, implying some fluctuation in position angle and sense of circular polarization. However, where subpulse polarization is low, the polarization of micropulses is generally relatively low also.

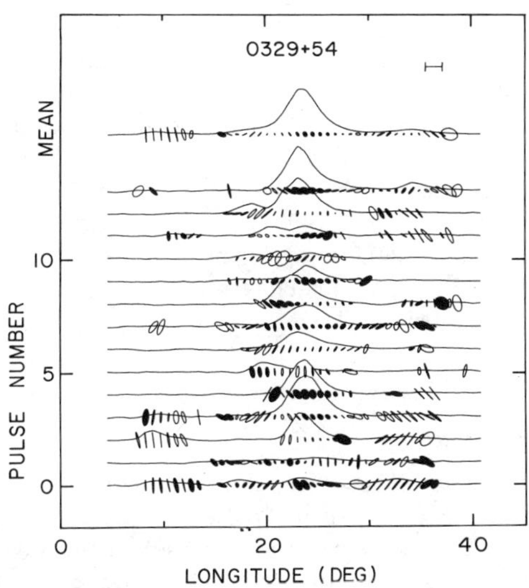

Figure 7 Polarization characteristics of 14 consecutive pulses from PSR 0329+54, recorded at a frequency of 410 MHz, and of the integrated profile obtained by summing these pulses. Ellipses representing the state of polarization of the signal at each longitude are plotted under a line representing the pulse total intensity. The shape and orientation of the ellipse is that of the polarization ellipse, with the length of the major axis proportional to the degree of polarization: a filled ellipse indicates left-circular polarization and an open one right-circular polarization. The bar in the upper right corner shows the major axis length corresponding to complete polarization (Manchester, Taylor & Huguenin 1975).

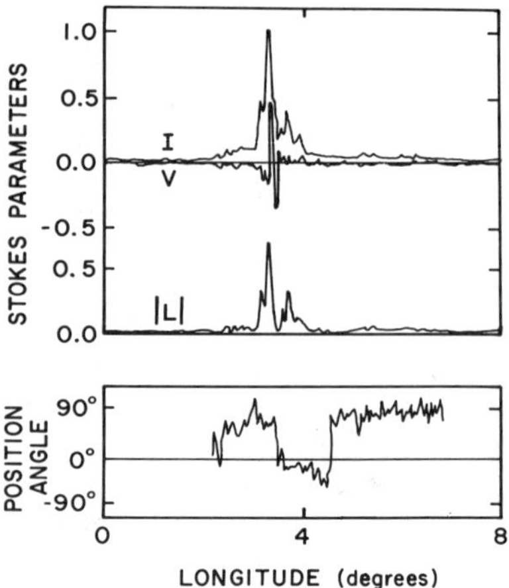

Figure 8 An individual pulse from PSR 1133+16 showing the total intensity, I, the linearly polarized intensity $|L| = (U^2+Q^2)^{1/2}$, the circularly polarized intensity V and the position angle of the linear polarization. The observing frequency was 430 MHz and the data are plotted with a time resolution of 128 μsec (Cordes 1975c).

As a class, Type D pulsars have weakly polarized integrated profiles. Observations of Type D pulsars show that in general there is a swing of polarization position angle across the subpulse and that the position-angle variation remains stable with respect to the subpulse as it drifts across the integrated profile (Rankin, Campbell & Backer 1974, Manchester, Taylor & Huguenin 1975, Cordes 1975c). In many cases the range of variation is close to 90° and in some cases there appears to be a position-angle discontinuity of close to 90° near the center of the subpulse. Drift of the subpulse across the profile therefore results in cancellation of most of the polarization of the integrated profile.

EVOLUTION OF PULSARS

Pulsar Timing Observations

ANALYSIS Accurate measurements of a pulsar period and its variations are accomplished by recording the times of arrival of pulses at a telescope, applying corrections related to the motion of the telescope around the solar system barycenter, and then fitting a mathematical model to the data. The parameters of the model may include the pulsar period P (or angular frequency $\Omega = 2\pi/P$) and one or more of the time derivatives ($\dot{P}$, $\ddot{P}$,...). In addition, with sufficiently well-sampled data, it is possible to solve for the pulsar's celestial coordinates (α, δ) and proper motion (μ_α, μ_δ). By using total intensity data (sum of two orthogonal polarizations) or one sense of circular polarization and averaging at least a few hundred pulses to obtain a stable integrated profile, pulse arrival times at the telescope can usually be obtained to an accuracy of between 10 and 500 μsec by fitting a standard profile to the data (e.g. Manchester & Peters 1972, Groth 1975). Effects of the Earth's motions are removed by referring arrival times to the solar system barycenter using an ephemeris that is in most cases derived from planetary radar observations (e.g. Ash et al. 1967). Corrections must also be made for the annual variation in the rate of terrestrial clocks caused by changes in distance between Earth and Sun.

The reduction of observed pulse arrival times to yield values of P, $\dot{P}$, α, δ, ... is usually done by means of a linearized least-squares fitting procedure that solves for small corrections to the parameters in question. In the one known instance of a pulsar in a binary system (Hulse & Taylor 1975a), the orbital parameters are also included in the least-squares fit; the solution then yields an accurate description of the orbit and its secular changes as well as the behavior of the pulsar (Taylor et al. 1976). A thorough discussion of both the timing corrections and the fitting procedure has recently been published by Blandford & Teukolsky (1976).

In the absence of inherent timing irregularities much larger than the typical measurement uncertainty of ~ 200 μsec, it is possible to obtain, from an accumulation of several years of data on a given pulsar, measurement of the period to within $\pm 10^{-12}$ sec and of the period derivative to $\pm 10^{-18}$ sec sec^{-1}. The accuracy of the derived positions is in most cases limited by uncertainty in the orientation of the planetary coordinate system, currently believed to be about $\pm 0''.1$ (Groth 1975).

The observed distribution of pulsar periods, given in Figure 9, extends from 0.03 to 4.0 sec with the most common value being close to 0.6 sec. Examination of possible selection effects shows that they are small (Taylor & Manchester 1977), so the distribution in Figure 9 closely approximates the true distribution of pulsar

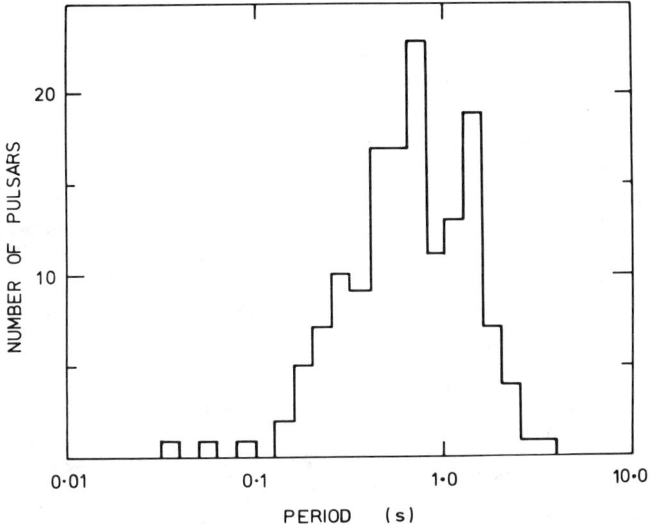

Figure 9 Distribution in period of the 149 known pulsars.

periods. The distribution is apparently bimodal with a gap at periods close to 1.0 sec. This gap shows up in independent surveys (Hulse & Taylor 1975b, Davies, Lyne & Seiradakis 1977) and may be related to division of pulsars into Type S and Type C classes.

SECULAR VARIATIONS At the time of writing nearly 90 pulsars have measured period derivatives. Most of these derivatives have been obtained from full timing solutions (e.g. Manchester & Peters 1972, Richards, Rankin & Zeissig 1974, Gullahorn et al. 1976), but 36 were obtained by measuring periods in two sessions spaced by precisely one year, thereby avoiding errors resulting from uncertainty in the assumed pulsar position (Lyne, Ritchings & Smith 1975). All observed derivatives are positive, that is, the pulsars are all slowing down. The rates vary between about 4.2×10^{-13} sec sec^{-1} or 36 nsec day^{-1} for the Crab pulsar and 1.6×10^{-18} sec sec^{-1} or 0.14 psec day^{-1} for PSR 1952+29. Except for the Crab and Vela pulsars (at the top left of Figure 6), there is little correlation between period and period derivative. It is notable that the binary pulsar has a very small derivative (8.8×10^{-18} sec sec^{-1}) despite the fact that its period of 59 msec is the second shortest known (Taylor et al. 1976).

In most theoretical models, the braking torque is proportional to some power n of the rotation frequency, so that

$$\dot{\Omega} = -K\Omega^n. \tag{1}$$

The parameter n, known as the braking index, can in principle be determined from a measurement of the period or frequency second derivative, because differentiation of (1) yields the relation

$$n = \ddot{\Omega}\Omega/\dot{\Omega}^2. \tag{2}$$

The validity of the braking law (1) can be checked if the third derivative is measurable, because differentiating once again gives

$$n(2n-1) = \dddot{\Omega}\Omega^2/\dot{\Omega}^3. \tag{3}$$

If the pulsar magnetic field is dipolar in form, the braking index n is equal to 3 (Ostriker & Gunn 1969, Goldreich & Julian 1969). The value will be greater than 3 if multipole electromagnetic or gravitational radiation is an important energy loss process, if the magnetic field is decaying, or if the magnetic axis is moving toward alignment with the rotation axis. On the other hand, it will be less than three if the field lines are deformed radially outward (by, for example, inertial effects) or if the magnetic axis is moving toward counteralignment (i.e. perpendicular to the rotation axis).

Unfortunately, the value of $\ddot{\Omega}$, and hence of n, has been measured for only one pulsar, the one in the Crab Nebula. Groth (1975) obtains the value $n = 2.515 \pm 0.005$, somewhat less than the "canonical" value of 3. This suggests that the radial variation of magnetic field strength between the pulsar and the light cylinder is somewhat slower than that characterizing a dipole (r^{-3}), and/or that the magnetic axis is moving toward counter-alignment. The presence of an interpulse in the Crab pulse profile

(Figure 4) suggests that the magnetic and rotational axes are already almost perpendicular. For other pulsars the frequency second derivative has not been measurable because insufficient time is spanned by the measurements and/or because inherent timing irregularities exist in the pulsar. The same comment applies to the term $\dddot{\Omega}$ for the Crab pulsar.

If all pulsars followed similar evolutionary tracks, the common braking index could be estimated from a plot of P vs $\dot{P}$ such as that shown in Figure 6. However, the wide scatter of points in Figure 6 proves that all pulsars do not evolve along the same path. For $n = 3$, a given pulsar will evolve along a line of constant $P\dot{P}$, i.e. a line of slope -1 in Figure 6. The data and Equation (1) then imply a range of about five orders of magnitude in the parameter $B_0^2 R^6/I$, where B_0 is the surface magnetic field strength, R is the neutron star radius, and I is the moment of inertia (Greenstein 1972). Several explanations for the wide scatter of points exist: (*a*) pulsars may be born with different masses, so that R^6/I covers a wide range; (*b*) pulsars may be born with different surface field strengths; (*c*) the magnetic fields may decay significantly on time scales of the order of 10^6 yr; or any combination of these possibilities may be true. Lyne, Ritchings & Smith (1975) have interpreted Figure 6 as indicative of magnetic decay, chiefly because the observed number of short-period pulsars with small-period derivatives is smaller than would be expected if pulsars followed straight evolutionary paths with $n = 3$. Although the statistical basis for this conclusion is not very strong, and although theoretical work (Ewart, Guyer & Greenstein 1975) suggests that magnetic decay on such short time scales is unlikely, independent arguments presented below suggest that pulsars must cease to radiate on a time scale of a few million years. Decay of the magnetic field would have this effect.

IRREGULAR VARIATIONS In addition to systematic period variations that can be fitted by Equation (1), many pulsars have unpredictable period changes of significant magnitude. These changes appear to be of two main types: (*a*) sudden, apparently discontinuous decreases in period, and (*b*) small irregular fluctuations or period "noise." By far the largest of these irregularities are the jumps (glitches) observed in the period of the Vela pulsar. Three of these events have occurred over a seven-year interval (Radhakrishnan & Manchester 1969, Reichley & Downs 1969, 1971, Manchester, Goss & Hamilton 1976). In each case the pulsation frequency increased by a fractional amount $\Delta\Omega/\Omega \approx 2 \times 10^{-6}$, and at the same time the slowdown rate increased by an amount $\Delta\dot{\Omega}/\dot{\Omega} \approx 10^{-2}$ before slowly relaxing toward its original value. The post-jump decay of the period appears to be consistent with the "two-component" model for neutron stars (Baym et al. 1969), in which much of the mass of the neutron star is in a superfluid state. This model predicts an exponential decay of $\dot{\Omega}$ toward its pre-jump value; data for the third Vela discontinuity (Manchester et al. 1976) indicate that the decay time-constant is about 450 days and that the period will remain offset from the extrapolated pre-jump value by about 80% of the initial period decrease.

Similar but much smaller discontinuities have been observed in the Crab pulsar period in 1969 (Boynton et al. 1972) and 1975 (Lohsen 1975). Groth (1975) has argued that a third event in 1971 reported by Lohsen (1972) is most likely a statistical fluctuation in the period noise. For the 1969 and 1975 events the fractional increases in pulsation frequency $\Delta\Omega/\Omega$ were respectively 9×10^{-9} and 3.7×10^{-8}, the fractional increases in derivative $\Delta\dot{\Omega}/\dot{\Omega}$ were about 2×10^{-3} in both cases, and the decay times were respectively 5 days and 16 days. Although the post-jump period variations for these two jumps are separately consistent with the two-component neutron star model, the fact that the ratio of $\Delta\dot{\Omega}/\dot{\Omega}$ to $\Delta\Omega/\Omega$ differed for the two events is inconsistent with this model (Greenstein 1976).

Small-period discontinuities ($\Delta\Omega/\Omega \sim 10^{-10}$) have been reported for several longer-period pulsars by Manchester & Taylor (1974) and Gullahorn et al. (1976). In these cases also, the post-jump decay does not seem to be consistent with the two-component model. It is possible that, as for the 1971 Crab event, these changes represent fluctuations in the period noise rather than discrete events of the Vela class.

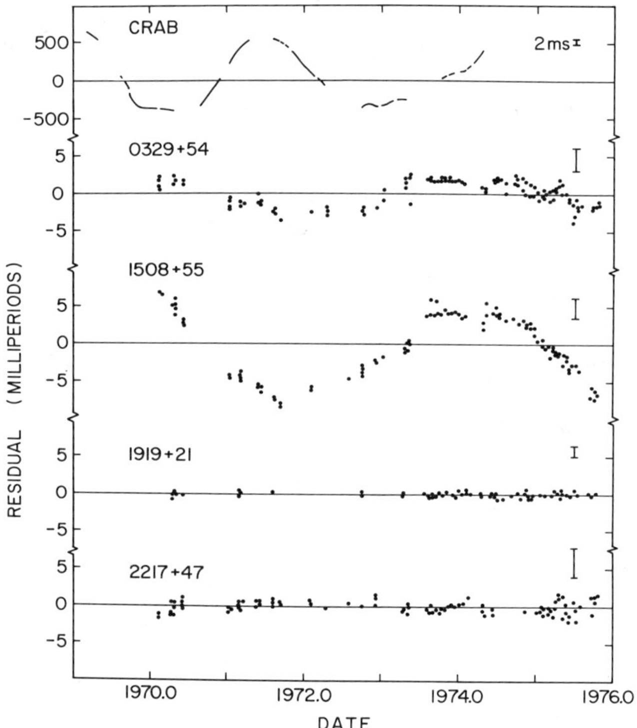

Figure 10 Timing residuals for five pulsars. For the Crab pulsar the data are residuals after fitting for the pulsation frequency, Ω, and its first and second derivatives, with equal weights assigned to all observations (data from Groth 1975). For the other four pulsars the residuals were obtained by fitting for the pulsation frequency, its first derivative, and the pulsar position (Manchester & Taylor 1977).

In addition to occasional glitches, pulsar periods are in some cases continuously subject to small fluctuations that far exceed the uncertainties of measurement. These fluctuations are manifested as systematic timing residuals, i.e. differences between observed arrival times and those predicted by a Taylor series in Ω and its derivative(s). Plots of residuals over an interval of about six years are shown for five different pulsars in Figure 10. In the case of the Crab pulsar, shown at the top, a cubic polynomial has been removed from the data, and for the other pulsars a parabolic curve has been removed. In all but two cases, systematic timing residuals are seen. Obviously the residuals can be reduced to low values by increasing the number of parameters fitted. However, subsequent observations are not correctly predicted by such fits, and the number of parameters required increases with the length of the data span. Furthermore, the implied braking indices are implausibly large and not constant. Boynton et al. (1972) and more recently Groth (1975) have interpreted the timing behavior of the Crab pulsar in terms of a random series of small frequency jumps with stationary statistical properties. From five years of timing data, Groth (1975) finds that the parameter $R\langle\Delta\Omega^2\rangle$, where R is the average rate at which the frequency jumps occur, and $\langle\Delta\Omega^2\rangle$ is their variance, to be about 2×10^{-21} sec^{-3}. Since the individual jumps are not resolved with daily observations, $R \gtrsim 10^{-4}$ sec^{-1}, so the rms size of the frequency jumps must be $\langle\Delta\Omega^2\rangle^{1/2} \lesssim 5\times10^{-9}$ sec^{-1}. Groth finds no statistically significant evidence for time variations in the parameters describing either the secular slowdown or the random irregularities.

The systematic residuals found for other pulsars (Figure 10) seem to have similar characteristics to those of the Crab pulsar. For PSR 0329+54 the parameter $R\langle\Delta\Omega^2\rangle$ has a value of about 2×10^{-27} sec^{-3} (Manchester & Taylor 1974). Generally, period noise seems to be greater in younger pulsars; the two pulsars in Figure 10 with no detectable irregularities both have relatively small values of the parameter $\dot{P}P^{-5}$, i.e. they are close to the "turn-off line" in Figure 6.

Galactic Distribution

The number of known pulsars (149) is now sufficiently large and the galactic distribution of these pulsars sufficiently widespread to permit meaningful analyses of the spatial distribution of pulsars within the Galaxy and of the pulsar luminosity function. Evaluation of these distributions requires an understanding of the effects of selection in pulsar surveys. About three-fourths of the known pulsars were detected in one or more of three major surveys, the Molonglo survey (Large & Vaughan 1971), the Jodrell Bank survey (Davies, Lyne & Seiradakis 1972, 1973), and the University of Massachusetts-Arecibo survey (Hulse & Taylor 1975b), for which selection effects are reasonably well understood. Data from the Arecibo survey have been analyzed by Roberts (1976b), that from the Jodrell Bank survey by Davies, Lyne & Sieradakis (1977), and that from both surveys by Taylor & Manchester (1977) to determine the pulsar luminosity function and the distribution and total number of pulsars in the Galaxy.

DISTANCES AND Z-DISTRIBUTION In order to scale these distributions, the distances to the pulsars must be known. It is fortunate that each pulsar comes equipped with a built-in distance indicator, its dispersion measure. Observations of neutral-hydrogen absorption (e.g. Gómez-Gonzáles & Guélin 1974, Ables & Manchester 1976) show that the mean interstellar electron density is close to 0.03 cm^{-3} over a substantial portion of the Galaxy and hence that, at least on a statistical basis, the dispersion measure is a reasonably accurate distance indicator.

There is good evidence (e.g. Bridle & Venugopal 1969, Falgarone & Lequeux 1973, Readhead & Duffett-Smith 1975) that the equivalent half-thickness of the interstellar electron layer is of the order of 1000 pc, much larger than that of most other galactic constituents including pulsars. Therefore the distribution of DM sin $|b|$, the "z-component" of dispersion measure, essentially represents the z-distribution of pulsars. For an assumed mean electron density $\langle n_e \rangle = 0.03$ cm^{-3} at $z = 0$, the pulsar distribution is such that $\langle z^2 \rangle^{1/2} \approx 270$ pc and $\langle |z| \rangle \approx 230$ pc. At present the data are not sufficient to determine whether the distribution follows an exponential law or, say, a Gaussian.

R-DISTRIBUTION Evidence that the density of pulsars is a strong function of galactocentric radius, R, came first from the Jodrell Bank survey. Lyne (1974) showed that there was a strong longitude dependence in the number of pulsars detected, with most being found in directions towards the galactic center. Despite its high sensitivity, the Arecibo survey failed to detect many pulsars with dispersion measures greater than 300 cm^{-3} pc, implying a cutoff in the pulsar distribution at about $R = 10$ kpc (Hulse & Taylor 1975b). The distribution of pulsars with respect to R obtained from the combined results of these two surveys after evaluation of selection effects (Taylor & Manchester 1977) is shown in Figure 11.

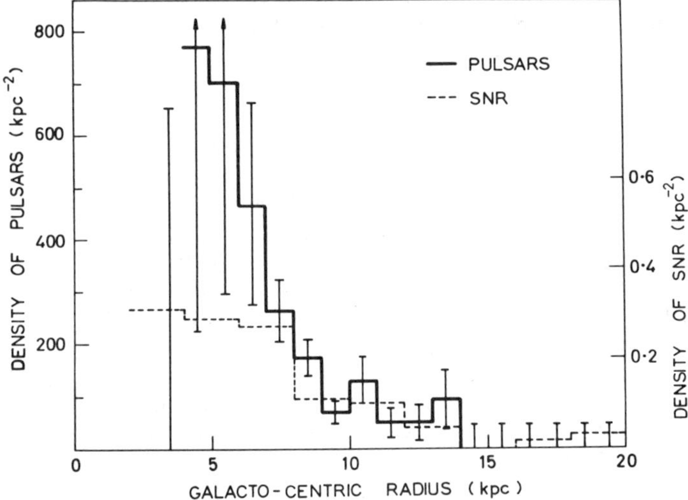

Figure 11 Distribution of pulsars as a function of galactocentric radius (Taylor & Manchester 1977). Also plotted is the distribution of supernova remnants as derived by Clark & Caswell (1976).

The plotted distribution is in units of pulsars per square kiloparsec, projected on to the galactic plane; that is, it has already been integrated over z. The pulsar density is seen to increase rapidly inside the solar circle, reaching a maximum possibly as large as 5 times the local density at $R \approx 6$ kpc. At smaller values of R the distribution is not well-determined, but may decrease somewhat. Outside the solar circle the density falls fairly rapidly, with no pulsars being found at $R > 14$ kpc. Davies, Lyne & Seiradakis (1977) find a similar distribution from their analysis of the Jodrell Bank data.

LUMINOSITY FUNCTION AND GALACTIC POPULATION The pulsar luminosity function $\Phi(L)$ obtained from these analyses, where $\Phi(L)\, dL$ is the number of pulsars in a given volume with luminosity between L and $L+dL$, is well represented by a power law of the form $\Phi(L) = kL^{-\gamma}$ over the observed range of luminosities. As the true radio luminosity of pulsars cannot be measured, a pseudo luminosity, L, is defined as the observed mean 400 MHz flux density multiplied by the square of the pulsar distance (or dispersion measure). Taylor & Manchester (1977) obtain a slope $\gamma = 1.12 \pm 0.03$ for logarithmic luminosity intervals, in good agreement with the value of 0.96 obtained by Davies, Lyne & Seiradakis (1977).

Assuming a cutoff in the luminosity function at about the lowest luminosities observed (~ 3 mJy kpc^2), and a mean electron density $\langle n_e \rangle = 0.03$ cm^{-3}, Taylor & Manchester (1977) find that the density of observable pulsars in the solar neighborhood is 90 ± 15 pulsars per square kiloparsec. This result is rather dependent on $\langle n_e \rangle$; assuming $\langle n_e \rangle = 0.025$ cm^{-3}, Davies, Lyne & Seiradakis (1977) obtain a density of about 50 kpc^{-2}. Integration over the radial distribution shown in Figure 11 gives a total number of observable pulsars in the Galaxy of $(1.3 \pm 0.4) \times 10^5$ for $\langle n_e \rangle = 0.03$ cm^{-3} or 2×10^4 for $\langle n_e \rangle = 0.02$ cm^{-3}. Davies, Lyne & Seiradakis obtain a similar result, $(1\text{–}3) \times 10^5$ pulsars. The much lower figure of 4000 derived by Roberts (1976b) appears to be a considerable underestimate. If only a fraction of all active pulsars are seen because of beaming effects, then the total number of active pulsars in the Galaxy may be greater than 5×10^5.

Ages and Origins

CHARACTERISTIC AGES Before one can adequately assess the importance of pulsars as a galactic population, it is necessary to have some estimate of their active lifetime. If pulsars evolve with a braking index $n = 3$, the number in equal logarithmic period intervals would be expected to increase as P^2. The distribution shown in Figure 9 therefore implies that many pulsars turn off at periods as short as 0.7 or 0.8 sec. If a pulsar was born spinning rapidly ($P_0^2 \ll P^2$) and has since slowed down according to Equation (1), then its age is given by

$$\tau = -\frac{\Omega}{(n-1)\dot{\Omega}} = \frac{P}{(n-1)\dot{P}} \qquad (n \neq 1). \tag{4}$$

For $n = 3$, this age, $P/(2\dot{P})$, is known as the characteristic age. For the Crab pulsar, the characteristic age $\tau = 1240$ yr is close to the 922 yr elapsed since the supernova outburst; similarly, the characteristic age of PSR 0833-45, 1.1×10^4 yr, is consistent with estimates of (1–3) × 10^4 yr for the age of the Vela supernova remnant obtained from semi-empirical relations involving linear diameter and surface brightness (e.g. Clark & Caswell 1976). These facts suggest that, at least for the shorter-period pulsars, characteristic ages are good estimates of the true ages. A histogram of the known characteristic ages is given in Figure 12. If one assumes that characteristic ages are reliable measures of the true age, this figure shows immediately that many pulsars have active lifetimes of just one or two million years. However, the distribution has a long tail, some of which is not plotted in Figure 12; for example, PSR 1952+29 has $\tau = 4.3 \times 10^9$ yr. There is considerable evidence that these larger characteristic ages are gross overestimates of the true ages.

KINEMATIC AGES It has been shown conclusively that pulsars are high-velocity objects. Direct measurements of proper motion have now been made for nine pulsars (Trimble 1971, Manchester, Taylor & Van 1974, Anderson, Lyne & Peckham 1975, Backer & Sramek 1976), and for these the mean transverse velocity is greater than 200 km sec^{-1}. If pulsars are formed from a parent population of small-scale height, then, as first pointed out by Gunn & Ostriker (1970), the value of $\langle |z| \rangle$ for any sample of pulsars is a monotonically increasing function of the mean age of the sample. The expected relation is graphed in Figure 13 for three combinations of assumed initial scale height and average z-velocity (Taylor & Manchester 1977). The data points in Figure 13 represent values of $\langle \tau \rangle$ and $\langle |z| \rangle$ for five groups of 13 to 19 pulsars each, which together comprise all pulsars with known characteristic ages. It is clear that unless the pulsars with $\tau \gtrsim 2 \times 10^6$ yr have z-velocities much smaller than 100 km sec^{-1}, they cannot be as old as their characteristic ages. In other words, the kinematically derived mean age of pulsars,

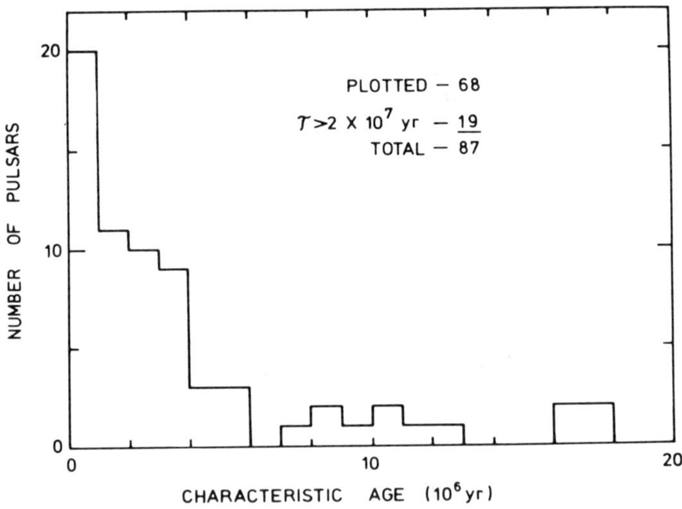

Figure 12 Distribution of pulsar characteristic ages.

$\langle|z|\rangle/\langle|v_z|\rangle \approx 2\times10^6$ yr, is considerably less than their mean characteristic age of about 5×10^7 yr. Magnetic decay with a timescale of 10^6 yr (Lyne, Ritchings & Smith 1975) would result in a large increase in τ after a few million years and would be consistent with the distribution in Figure 12. As an alternative, pulsars could be born with periods comparable to their present value and with different magnetic field strengths and/or masses. In this case, some other mechanism would be responsible for the turn-off in pulsed emission after a few million years.

PULSAR BIRTHRATE A parameter vital to the question of the origin of pulsars is the birthrate. From the histogram in Figure 12, a birthrate of about 20 pulsars per million years is necessary to generate a population of 100 objects (cf. Davies, Lyne & Seiradakis 1977). Therefore to generate a galactic population of say 10^5 pulsars, one pulsar birth every 50 yr is required. If the true mean age of pulsars is only 2×10^6 yr, as suggested above, then to maintain a density of nearly 100 observable pulsars per square kiloparsec in the solar neighborhood, the required birthrate is about 2.5×10^{-5} kpc^{-2} yr^{-1}, corresponding to one every 40 yr in the Galaxy (Taylor & Manchester 1977). This birthrate would be an overestimate if the true pulsar luminosity function does not continue to rise down to $L = 3$ mJy kpc^{-2}, or if the pulsars are somewhat more distant or older than we have assumed. However, we have not yet allowed for the fact that nearly all pulsar theories involve highly beamed radio emission, such that a given pulsar has a probability of only $\sim20\%$ of being observable at the Earth (see, for example, Smith 1970, Roberts 1976a).

If we allow for this effect, the computed rate greatly exceeds the birthrate for galactic supernovae deduced from observations of their radio remnants (Ilovaisky & Lequeux 1972, Clark & Caswell 1976), typically 50–150 years. It is comparable to the highest possible rates derived for our galaxy from observations of supernovae

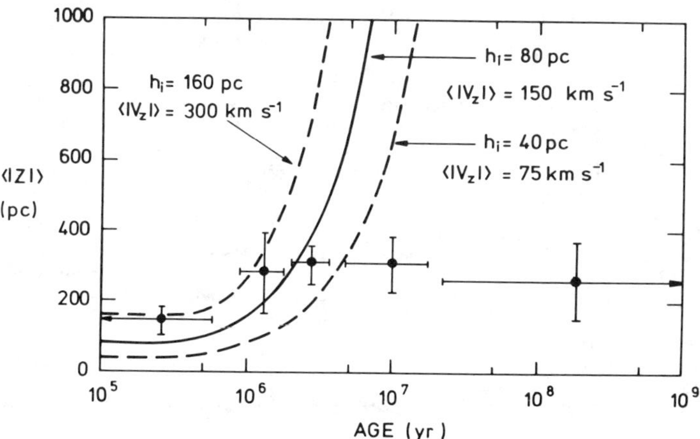

Figure 13 Expected $\langle|z|\rangle$ for pulsars of initial scale height h_i and mean z-component of velocity $\langle|v_z|\rangle$, and the observed $\langle|z|\rangle$ for five groups of pulsars in different characteristic age intervals (Taylor & Manchester 1977).

in external galaxies (Tammann 1974). It would seem that one must consider seriously the possibility that some pulsars—perhaps the majority—are born in circumstances not accompanied by a supernova explosion.

Ostriker, Richstone & Thuan (1974) and Biermann & Tinsley (1974) have shown that the death rate of stars with mass $\gtrsim 2.5\ M_{\odot}$ is about $10^{-4}\ yr^{-1}\ kpc^{-2}$ in the solar neighborhood, about equal to the pulsar birthrate if 80% of pulsars are unobservable because of beaming effects. We note that the scale height of young pulsars (Figure 13) is more consistent with progenitors of mass 2.5–4 $M_{\odot}$ than, say, $>6\ M_{\odot}$. Stars of mass less than 4 $M_{\odot}$ are generally thought to evolve into white dwarfs, but the present information seems to require that many of them must (at least eventually) become pulsars. Better determination of the low end of the pulsar luminosity function, as well as theoretical work on the final evolutionary tracks of massive stars, should help to resolve these problems.

Literature Cited

Ables, J. G., Manchester, R. N. 1976. *Astron. Astrophys.* 50:177–84

Anderson, B., Lyne, A. G., Peckham, R. J. 1975. *Nature* 258:215–17

Ash, M. E., Shapiro, I. I., Smith, W. B. 1967. *Astron. J.* 72:338–50

Backer, D. C. 1970a. *Nature* 228:1297–98

Backer, D. C. 1970b. *Nature* 228:42–43

Backer, D. C. 1972. *Ap. J.* 174:L157–61

Backer, D. C. 1973. *Ap. J.* 182:245–76

Backer, D. C. 1976. *Ap. J.* 209:895–907

Backer, D. C., Boriakoff, V., Manchester, R. N. 1973. *Nature Phys. Sci.* 243:77–78

Backer, D. C., Fisher, J. R. 1974. *Ap. J.* 189: 137–45

Backer, D. C., Rankin, J. M., Campbell, D. B. 1975. *Ap. J.* 197:481–87

Backer, D. C., Sramek, R. A. 1976. *Astron. J.* 81:430–32

Baym, G., Pethick, C., Pines, D., Ruderman, M. 1969. *Nature* 224:872–74

Biermann, P., Tinsley, B. M. 1974. *Astron. Astrophys.* 30:1–12

Blandford, R. D., Teukolsky, S. A. 1976. *Ap. J.* 205:580–91

Boriakoff, V. 1976. *Ap. J. Lett.* 208:L43–46

Boriakoff, V., Payne, R. R. 1974. *Astrophys. Lett.* 15:175–76

Boynton, P. E., Groth, E. J., Hutchinson, D. P., Nanos, G. P., Partridge, R. B., Wilkinson, D. T. 1972. *Ap. J.* 175:217–41

Bridle, A. H., Venugopal, V. R. 1969. *Nature* 224:545–47

Bruck, Yu. M., Ustimenko, B. Yu. 1973. *Nature Phys. Sci.* 242:58–59

Bruck, Yu. M., Ustimenko, B. Yu. 1976. *Nature Phys. Sci.* 260:766–67

Clark, D. H., Caswell, J. L. 1976. *MNRAS* 174:267–305

Cocke, W. J., Muncaster, G. W., Gehrels, T. 1971. *Ap. J. Lett.* 169:L119–21

Cordes, J. M. 1975a. *Ap. J.* 208:944–54

Cordes, J. M. 1975b. *Ap. J.* 195:193–202

Cordes, J. M. 1975c. PhD thesis. Univ. Calif., San Diego, pp. 1–173

Davies, J. G., Lyne, A. G., Seiradakis, J. H. 1972. *Nature* 240:229–30

Davies, J. G., Lyne, A. G., Seiradakis, J. H. 1973. *Nature Phys. Sci.* 244:84–85

Davies, J. G., Lyne, A. G., Seiradakis, J. H. 1977. *MNRAS*. In press

Downs, G. S., Reichley, P. E., Morris, G. A. 1973. *Ap. J. Lett.* 181:L143–46

Drake, F. D., Craft, H. D. 1968. *Nature* 220:231–35

Ewart, G. M., Guyer, R. A., Greenstein, G. 1975. *Ap. J.* 202:238–47

Falgarone, E., Lequeux, J. 1973. *Astron. Astrophys.* 25:253–60

Ferguson, D. C., Cocke, W. J., Gehrels, T. 1974. *Ap. J.* 190:375–80

Ferguson, D. C., Graham, D. A., Jones, B. B., Seiradakis, J. H., Wielebinski, R. 1976. *Nature* 260:25–27

Goldreich, P., Julian, W. H. 1969. *Ap. J.* 157:869–80

Gómez-Gonzáles, J., Guélin, M. 1974. *Astron. Astrophys.* 32:441–46

Greenstein, G. 1972. *Ap. J.* 177:251–53

Greenstein, G. 1976. *Ap. J.* 208:836–48

Groth, E. J. 1975. *Ap. J. Suppl. Ser.,* No. 293, 29:431–65

Gullahorn, G. E., Payne, R. R., Rankin, J. M., Richards, D. W. 1976. *Ap. J. Lett.* 205: L151–53

Gunn, J. E., Ostriker, J. P. 1970. *Ap. J.* 160:979–1002

Hamilton, P. A., McCulloch, P. M., Ables,

J. G., Komesaroff, M. M. 1977a. *MNRAS*. In press
Hamilton, P. A., McCulloch, P. M., Manchester, R. N., Ables, J. G. 1977b. *MNRAS*. Submitted for publication
Hankins, T. H. 1971. *Ap. J.* 169:487–94
Hankins, T. H. 1972. *Ap. J. Lett.* 177:L11–15
Harnden, F. R., Gorenstein, P. 1973. *Nature* 241:107–8
Helfand, D. J., Manchester, R. N., Taylor, J. H. 1975. *Ap. J.* 198:661–70
Hesse, K. H. 1973. *Astron. Astrophys.* 27: 373–77
Hesse, K. H., Sieber, W., Wielebinski, R. 1973. *Nature Phys. Sci.* 245:57–58
Hesse, K. H., Wielebinski, R. 1974. *Astron. Astrophys.* 31:409–13
Hewish, A., Bell, S. J., Pilkington, J. D. H., Scott, P. F., Collins, R. A. 1968. *Nature* 217:709–13
Huguenin, G. R., Manchester, R. N., Taylor, J. H. 1971. *Ap. J.* 169:97–104
Huguenin, G. R., Taylor, J. H., Helfand, D. J. 1973. *Ap. J. Lett.* 181:L139–42
Huguenin, G. R., Taylor, J. H., Troland, T. H. 1970. *Ap. J.* 162:727–35
Hulse, R. A., Taylor, J. H. 1975a. *Ap. J. Lett.* 195:L51–53
Hulse, R. A., Taylor, J. H. 1975b. *Ap. J. Lett.* 201:L55–59
Ilovaisky, S. A., Lequeux, J. 1972. *Astron. Astrophys.* 20:347–56
Kestenbaum, H. L., Ku, W., Novick, R., Wolff, R. S. 1976. *Ap. J. Lett.* 203:L57–61
Kristian, J., Visvanathan, N., Westphal, J. A., Snellen, G. H. 1970. *Ap. J.* 162:475–83
Kurfess, J. D. 1971. *Ap. J. Lett.* 168:L39–42
Large, M. I., Vaughan, A. E. 1971. *MNRAS* 151:277–87
Laros, J. G., Matteson, J. L., Pelling, R. M. 1973. *Nature Phys. Sci.* 246:109–11
Lohsen, E. 1972. *Nature Phys. Sci.* 236:70–71
Lohsen, E. 1975. *Nature* 258:688–89
Lyne, A. G. 1971. *MNRAS* 153:27P–32P
Lyne, A. G. 1974. *Galactic Radio Astronomy, IAU Symp. 60,* ed. F. J. Kerr, S. C. Simonson, pp. 87–95. Dordrecht: Reidel
Lyne, A. G., Ritchings, R. T., Smith, F. G. 1975. *MNRAS* 171:579–97
Lyne, A. G., Smith, F. G., Graham, D. A. 1971. *MNRAS* 153:337–82
Lyne, A. G., Thorne, D. J. 1975. *MNRAS* 172:97–108
Manchester, R. N. 1971. *Ap. J. Suppl.,* No. 199, 23:283–322
Manchester, R. N., Goss, W. M., Hamilton, P. A. 1976. *Nature* 259:291–92
Manchester, R. N., Hamilton, P. A., Goss, W. M., Newton, L. M. 1976. *Proc. Astron. Soc. Aust.* 3:81–83
Manchester, R. N., Peters, W. L. 1972. *Ap. J.* 173:221–26
Manchester, R. N., Taylor, J. H. 1974. *Ap. J. Lett.* 191:L63–65
Manchester, R. N., Taylor, J. H. 1977. *Pulsars,* pp. 1–275. San Francisco: Freeman
Manchester, R. N., Taylor, J. H., Huguenin, G. R. 1973. *Ap. J. Lett.* 179:L7–10
Manchester, R. N., Taylor, J. H., Huguenin, G. R. 1975. *Ap. J.* 196:83–102
Manchester, R. N., Taylor, J. H., Van, Y. Y. 1974. *Ap. J. Lett.* 189:L119–22
McCulloch, P. M., Hamilton, P. A., Ables, J. G., Komesaroff, M. M. 1976. *MNRAS* 175:71P–75P
McLean, A. I. O. 1973. *MNRAS* 165:133–48
Mohanty, D. K., Balasubramanian, V. 1975. *MNRAS* 171:17P–18P
Moore, W. E., Agrawal, P. C., Garmire, G. 1974. *Ap. J. Lett.* 189:L117–18
Ostriker, J. P., Gunn, J. E. 1969. *Ap. J.* 157:1395–1417
Ostriker, J. P., Richstone, D. O., Thuan, T. X. 1974. *Ap. J. Lett.* 188:L87–89
Page, C. G. 1973. *MNRAS* 163:29–40
Pravdo, S. H., Becker, R. H., Boldt, E. A., Holt, S. S., Rothschild, R. E., Serlemitsos, P. J., Swank, J. H. 1976. *Ap. J. Lett.* 208: L67–69
Radhakrishnan, V., Manchester, R. N. 1969. *Nature* 222:228–29
Rankin, J. M., Campbell, D. B., Backer, D. C. 1974. *Ap. J.* 188:609–13
Rankin, J. M., Comella, J. M., Craft, H. D., Richards, D. W., Campbell, D. B., Counselman, C. C. 1970. *Ap. J.* 162:707–25
Rankin, J. M., Payne, R. R., Campbell, D. B. 1974. *Ap. J. Lett.* 193:L71–74
Rappaport, S., Bradt, H., Doxsey, R., Levine, A., Spada, G. 1974. *Nature* 251: 471–72
Rappaport, S., Bradt, H., Mayer, W. 1971. *Nature Phys. Sci.* 229:40–42
Readhead, A. C. S., Duffett-Smith, P. J. 1975. *Astron. Astrophys.* 42:151–53
Reichley, P. E., Downs, G. S. 1969. *Nature* 222:229–30
Reichley, P. E., Downs, G. S. 1971. *Nature Phys. Sci.* 234:48
Richards, D. W., Rankin, J. M., Zeissig, G. A. 1974. *Nature* 251:37–39
Rickett, B. J., Hankins, T. H., Cordes, J. M. 1975. *Ap. J.* 201:425–30
Ritchings, R. T. 1976. *MNRAS* 176:249–263
Ritchings, R. T., Lyne, A. G. 1975. *Nature* 257:293–94
Roberts, D. H. 1976a. *Ap. J.* 207:949–61
Roberts, D. H. 1976b. *Ap. J.* 205:L29–33
Schönhardt, R. E., Sieber, W. 1973. *Astrophys. Lett.* 14:61–64
Sieber, W. 1973. *Astron. Astrophys.* 28:237–52

Sieber, W., Oster, L. 1975. *Astron. Astrophys.* 38:325–27
Sieber, W., Reinecke, R., Wielebinski, R. 1975. *Astron. Astrophys.* 38:169–82
Smith, F. G. 1970. *MNRAS* 149:1–15
Smith, F. G. 1973. *MNRAS* 161:9P–10P
Tammann, G. A. 1974. *Supernovae and Supernova Remnants*, ed. C. B. Cosmovici. pp. 155–85. Dordrecht: Reidel
Taylor, J. H., Huguenin, G. R. 1971. *Ap. J.* 167:273–91
Taylor, J. H., Hulse, R. A., Fowler, L. A., Gullahorn, G. E., Rankin, J. M. 1976. *Ap. J. Lett.* 206:L53–58
Taylor, J. H., Manchester, R. N. 1977. *Ap. J.* 215, No. 3
Taylor, J. H., Manchester, R. N., Huguenin, G. R. 1975. *Ap. J.* 195:513–28
Thompson, D. J., Fichtel, C. E., Kniffen, D. A., Ögelman, H. B. 1975. *Ap. J. Lett.* 200: L79–82
Trimble, V. 1971. *The Crab Nebula, IAU Symp. 46*, ed. R. D. Davies, F. G. Smith. pp. 12–21. Dordrecht: Reidel
Udal'tsov, V. A., Zlobin, V. N. 1974. *Astron. Astrophys.* 37:21–25
Walraven, G. D., Hall, R. D., Meegan, C. A., Coleman, P. L., Shelton, D. H., Haymes, R. C. 1975. *Ap. J.* 202:502–10
Warner, B., Nather, R. E., MacFarlane, M. 1969. *Nature* 222:233–38
Zlobin, V. N., Udal'tsov, V. A. 1975. *Astron. Zh.* 52:1139–42

Ann. Rev. Astron. Astrophys. 1977. 15: 45–68

THE ORIGIN OF SOLAR ACTIVITY

E. N. Parker

Department of Physics, Department of Astronomy and Astrophysics,
University of Chicago, Chicago, Illinois 60637

1 THE IMPLICATIONS OF SOLAR ACTIVITY

One would expect from elementary physical principles that a massive, isolated, slowly rotating, self-gravitating, gently heated globe of gas like the Sun would be entirely placid, the epitome of celestial tranquillity. That the Sun is active, with continual emergence of magnetic field through the surface, continual eruption of gases, and wide-spread superheating, is a source of wonder—and a challenge—to the theoretical physicist.

Solar activity ranges from isolated, concentrated flux tubes and spicules in the supergranule boundaries, to the ephemeral bipolar regions and X-ray bright spots, to the classical bipolar magnetic regions with sunspots, flares, and the newly discovered coronal transients erupting into space. This activity is magnetic in origin. The magnetic fields are thought to be produced by convection and circulation in the outer envelope (2×10^5 km) of the Sun. The convection and circulation, as well as the nonuniform rotation of the Sun, are consequences of the high opacity and high effective specific heat in the outer envelope. This article reviews the basic physical problems and principles of solar activity, insofar as they are presently understood, concentrating on the general question of why magnetic fields produce activity. It is evident from the magnetic antics displayed by the Sun that there is much yet to be learned of the dynamical properties of large-scale magnetic fields embedded in highly conducting gases. The problem posed by the observed magnetic activity of the Sun is of fundamental importance to all of astrophysics, because although activity is observed in many distant stars and galaxies, the Sun is the only star whose activity can be seen in enough detail to guide us toward an understanding.

Perhaps the best point at which to begin the inquiry is the fact that magnetic flux tubes continually emerge through the surface of the Sun, at a rate of the order of 10^{15} Mx sec^{-1}. The resulting bipolar magnetic regions have a net mean east-west orientation in the separate northern and southern hemispheres. Hence their emergence represents a net loss of flux from the respective hemisphere (Parker 1975e). We conclude that there must be a very effective source of magnetic flux somewhere below the surface. It is believed that the combined effects of the nonuniform rotation and the cyclonic convection provide the magnetic field in the cyclical pattern indicated by the eleven-year magnetic cycle of the Sun (see review by Parker 1970b). Many theoretical models have been worked out to demonstrate the various possibilities for the generation of field in the Sun, so that the basic principles seem to be well in hand. But the development of a specific model for the Sun must await knowledge of the qualitative nature of its circulation and nonuniform rotation. A crucial question, for instance, is whether the angular velocity increases or decreases from the surface downward through the convective zone. Indeed, the convection and circulation within the Sun present the outstanding problem on which so many other questions depend for their answers. We find that the study of several effects, of which the magnetic field is only one, requires a general knowledge of the internal circulation. Unfortunately that knowledge is not yet in hand.

The most startling revelation on convection and circulation, and the generation of magnetic fields, is the reinvestigation (Eddy 1976a) of the complete absence of solar activity in the last half of the 17th century (Maunder 1894). All traces of activity, including sunspots and an enhanced corona, were absent for most of that period. The C^{14} production rate for that period bounced up to a relatively high level because of freer entry of cosmic rays into the solar system. The C^{14} record now extends back seven millennia, to about 5000 BC, and Eddy (1976b) points out from this record that solar activity was also largely absent during the 15th century AD and during the 4th, 7th, and 14th centuries BC. Evidently at those times the convection and circulation within the Sun were of a sufficiently different form from the present that little or no magnetic field was produced, so that the activity fell to negligible levels. Tranquillity is evidently a possibility after all. It is interesting to note that Eddy, Gilman & Trotter (1976) have found from a study of old records of sunspot motions that the high angular velocity of the equatorial surface of the Sun was concentrated much closer to the equator just prior to the cessation of activity in 1645 AD.

There is, then, a disquieting conclusion that seems difficult to avoid. The heat delivered to the surface of the Sun, to be radiated into space, is put there by the convection and circulation beneath the surface. Hence if there is some change in the general circulation, it is difficult to see how the brightness of the solar surface can remain completely unaffected. Very modest changes in the circulation can have large effects. A vertical flow of 10 m sec^{-1} in the photosphere has a one-percent effect on the local surface brightness, while 50 m sec^{-1} is regularly observed (Howard & Harvey 1970, Howard 1971, 1972). At a depth of 2×10^4 km, 10^{-2} cm sec^{-1} produces a one-percent effect, visible at the surface with a time constant of about 50 years. Fortunately the historical record holds information on solar luminosity. Earth is, in reality, nothing more than a complex probe for monitoring the brightness of the Sun at a distance of 1 AU. Eddy (1976a, b) points out how closely the epochs of reduced mean annual temperature at the surface of Earth match the intermittent centuries of inactivity at the Sun. Fortunately the cold epochs amount to no more than a degree or so over periods of 50 to 100 years. The long-term variations of terrestrial temperature, over 10^3–10^4 years, suggest that there may be larger variations of luminosity (and hence circulation) over longer periods of time. But at present the C^{14} record extends back no farther than 5000 BC, so there is no way to explore the question.

We emphasize again, then, the fundamental nature of circulation and convection in the outer envelope of the Sun. It is the origin of the magnetic variability, and presumably also the brightness variability, of the Sun. A few brief remarks are in order, then, on the present state of knowledge of convection and circulation. A great body of theoretical work has been carried out, largely with the Boussinesq approximation, in which the density of the ambient fluid is taken to be independent of height. In fact the density varies from the bottom to the top of the convective zone by a factor of 10^6. Conventional analytical and numerical methods are unable to cope effectively with the problem of large density variations. As an illustration of the uncertainties with which we are presently confronted, the existing theoretical hydrodynamical models of global meridional circulation and nonuniform rotation lead to a pole-equator temperature difference $\Delta T/T$ of the same order as the difference $\Delta\Omega/\Omega$ in angular velocity (Kippenhahn 1963, Yoshimura 1972, Gilman 1972, 1974, Gierasch 1974). On the other hand, it is observed (Dicke 1974, Hill & Stebbins 1975) that the surface brightness of the Sun is uniform to at least one part in 10^3 between the pole and the equator. Durney (1975) suggests alternative hydrodynamic models involving circulation on rigidly rotating cylindrical surfaces. The question is by no means settled, but it shows how sensitive is the distribution of surface brightness to the circulation. Earth is warmed principally by the equatorial regions (half of the visible disk lies between latitudes $\pm 24°$). Thus, even with no overall change in luminosity, a pole–equator brightness difference of two parts in one hundred would produce about a one-percent change in net brightness as seen from Earth. A small net change in luminosity, or a slight redistribution of brightness, seems hard to avoid when there is a major change in circulation.

2 ORIGIN OF THE MAGNETIC FIELDS

The copious generation of magnetic fields beneath the surface of the Sun is implied by the rapid rate of emergence of magnetic flux through the visible surface and the subsequent escape into space (Sheeley et al. 1975, Parker 1975e) of 10^{23}–10^{24} Mx in a typical eleven-year period of activity. The mean rate of generation is, then, of the order of 10^{15} Mx sec^{-1}. The field is presumably generated by the combined effects of the nonuniform rotation ($\partial\Omega/\partial r$, $\partial\Omega/\partial\theta$) and the cyclonic component of the convective motions in the rotating Sun (Parker 1955b, 1957a). The operation of the solar dynamo was described at length in an earlier review (Parker 1970b) and is not elaborated further here. A number of authors (Parker 1955b, 1957a, 1970a, c, 1971, 1972b, 1975b, Steenbeck, Krause & Radler 1966, Steenbeck & Krause 1969, Leighton 1969, Gilman 1969, Deinzer & Stix 1974, Deinzer, Kusserow & Stix 1974, Stix 1974, 1976, Yoshimura 1972, 1973, Gubbins 1974) have explored a wide variety of kinematic dynamo models so that we have a pretty good idea of the fields produced by the various circulations and nonuniform rotation.

The loss of magnetic flux from the Sun is largely a result of turbulent diffusion and magnetic buoyancy (Parker 1973c, 1975e). The buoyancy of a magnetic tube of flux submerged in the Sun leads to a rapid rate of rise (Parker 1955a, 1974d, 1975b) at nearly the Alfvén speed. In order of magnitude, the escape time of a field B from a depth h, where the density is $\rho(h)$, is

$$t = O\{h[4\pi\rho(h)]^{1/2}/B\}. \tag{1}$$

Thus, using the number 10^2 G again, the buoyant escape from a depth of 2×10^4 km (where $\rho = 4 \times 10^{-3}$ gm cm^{-3}) is only 4×10^6 sec, or about two months. This is to be compared with the characteristic generation time of five to ten years, indicated by the eleven-year magnetic cycle of the Sun. It is evident that the magnetic fields of the Sun can be generated only in the lower levels of the convective zone (where $h = 2 \times 10^5$ km and $\rho = 0.2$ gm cm^{-3}), from which the escape time is of the order of ten years. The magnetic fields originate in the deep (Parker 1975b), and, once escaping from the basement levels, pass quickly to the surface, where they appear as bipolar regions, both large and small. What is more, we conclude that the fields in the lower convective zone cannot be much stronger than the general order of 10^2 G. For if they were, then they would not remain even for five to ten years near the bottom of the convective zone (see also Parker 1973d). Nor can the fields be much weaker than 10^2 G, because the flux emerging in one cycle is of the order of 10^{23}–10^{24} Mx, presumably a measure of the total flux beneath the surface at any given time. The convective zone, with a depth of $h = 2 \times 10^5$ km, has a meridional cross section of $\frac{1}{2}\pi R_{\odot}h \cong 2 \times 10^{21}$ cm^2 from the equator to the pole, so that an azimuthal field of 10^2 G implies a total flux of the order of 2×10^{23} Mx. This is barely enough.

As mentioned above, a fundamental question is the gradient of the angular velocity beneath the surface of the Sun. The direction of migration of the azimuthal fields in the Sun is determined by the product of the radial gradient of the angular velocity and the helicity $\langle \mathbf{v} \cdot \mathrm{curl}\ \mathbf{v} \rangle$ of the convective cells. The magnetic regions that we see at the surface are, quite literally, a trail of magnetic bubbles that rise from the depths and allow us to follow the generation of field. They show that the fields have opposite signs in the northern and southern hemispheres and migrate toward the equator. Now from elementary considerations on Coriolis forces one expects that a rising and expanding convective cell rotates less rapidly than the Sun, i.e. rotates backward relative to its surroundings (Steenbeck, Krause & Radler 1966). This sense of cyclonic rotation, together with an inward (downward) increase of angular velocity, generates azimuthal fields that migrate toward the equator, in agreement with observation. However, we must not overlook the fact that most hydrodynamic models of the circulation and nonuniform rotation of the Sun produce an inward *decrease* of the angular velocity. The models give an internal rotation that is the simplest configuration consistent with the observed rotation of the surface: the fluid near the axis rotates at the slower speed of the polar regions, and a surface shell at the equator rotates faster. Then the angular velocity decreases inward and, with the expected sense of rotation of the convective cells, produces an azimuthal field that migrates away from the equator toward high latitudes, opposite to the observed migration. The resolution of the difficulty is not clear.

Perhaps the hydrodynamic models, suggesting $\partial\Omega/\partial r > 0$, are incorrect in their simple view of the circulation in terms of uniform density, when in fact the density varies so much from top to bottom. Or perhaps the angular velocity does not vary much with depth but does vary with latitude. In that case the direction of migration of the waves of field is vertical rather than horizontal, so that the horizontal migration depends upon which way the waves deflect upon arriving at the surface (Parker 1972b). Durney (1975) has pointed out that recent work of Yoshimura (1975) indicates that the direction of rotation of the cyclonic motions at the bottom of the convective zone may be reversed from that suggested by the elementary considerations of Coriolis forces on the rising expanding cell. Hence, the combination of an inward decrease of the angular velocity with a forward rotating rising cell predicts migration from high to low latitudes, as is observed. We have already pointed out that the dynamo can function only in the lowest levels of the convective zone, so that picture would hang together in a consistent manner. A number of possibilities are explored recently by Stix (1976), including the $\omega \times \mathbf{j}$ dynamo of Radler (1969), which functions independently of the helicity. It is clear, however, that further understanding of the solar dynamo must wait for dynamical models of the internal circulation to give a definite sign for both $\partial\Omega/\partial r$ and the helicity of the cyclonic convection in the lower part of the convective zone.

3 THE BEHAVIOR OF SOLAR MAGNETIC FIELDS

Why do magnetic fields continually rise to the surface of the Sun, and, arriving at the surface, why are they so concentrated, so dissipative, and so active? How is it that a large-scale field of complex form almost immediately evolves toward the configuration of the simple potential field once it emerges through the surface? The general magnetic configuration of an active region may change drastically in 10^5 sec, with transient eruptions originating in 10^2–10^3 sec. Neutral point annihilation (see Section 7) is believed to be responsible for the explosive (10^2 sec) solar flare, but it has become clear in the last few years that there is more to the activity. No less puzzling than the explosive dissipation is the tendency for all magnetic fields extending through the surface of the Sun to concentrate into intense, isolated flux tubes, in direct opposition to their own considerable pressure $B^2/8\pi$. The expectation is that a magnetic field expands to fill all the available volume as a consequence of $B^2/8\pi$. Yet the general field of the Sun, whether in quiet regions where the mean field is only a few gauss, or in active regions where it is 20–100 G, is not a continuous field. Recent work shows it to be composed of isolated, widely separated flux tubes, each of perhaps 3×10^{18} Mx compressed to mean field densities of 500–2000 G, to form a flux tube with a radius of only 200–400 km (Sheeley 1967, Beckers & Schröter 1969, Livingston & Harvey 1969, 1971, Sawyer 1971, Simon & Noyes 1971, Howard & Stenflo 1972, Frazier & Stenflo 1972, Chapman 1973, Stenflo 1973). The concentrated tubes appear in the boundaries of supergranules, where they are presumably swept by the converging surface flows and downwelling of fluid. It remains yet to be determined whether the mean field across the tube is 1×10^3 or 2×10^3 G. The question is important because the pressure of the field increases as the square of the field, and any explanation of the concentration must account for the forces that oppose the magnetic pressure. If, for instance, the field is "only" 500 G ($B^2/8\pi = 10^4$ dynes cm^{-2}), then the dynamical forces $\frac{1}{2}\rho v^2$ of the granules (of the order of 10^4 dynes cm^{-2} for $\rho = 3 \times 10^{-7}$ gm cm^{-3} and $v = 3$ km sec^{-1}) might account for the compression (Parker 1974a). But if the fields are near 2000 G (Chapman 1973, Stenflo 1973) the pressure is sixteen times larger and can be contained only by the full pressure of the ambient gas. The spicules seem to be an eruptive phenomenon associated with the individual concentrated tube (Parker 1974a; see descriptions in Zirin & Lazareff 1975) and may perhaps be associated with the dynamical forces causing the tube to concentrate (Parker 1976b).

The hallmark of the active region is the sunspot, appearing as a cool patch on the surface of the Sun, into which the magnetic flux is gathered and compressed to 3–4 × 10^3 G. The pressure of such a field is 5×10^5 dynes cm^{-2}, equal to the ambient gas pressure one scale height (150 km) below the visible surface of the Sun. What forces gather together the individual tubes of force from a broad area of the Sun and then compress them to 3000 G in opposition of their mutual repulsion $B^2/8\pi$? Beckers & Schröter (1969) describe in detail the observed "absurd" behavior of individual flux tubes that seem to attract each other, or that march in irregular file along a well-trodden path into the sunspot. Meyer et al. (1974) suggest that the flux tubes are responding to powerful subsurface convective currents, much as an iceberg in the ocean may move in opposition to surface winds and currents because it extends downward into strong counter currents far below the surface.

In both the concentrated flux tubes in the supergranule boundaries and in the sunspot there is the very difficult question of the stability of the wasp-waisted concentration (illustrated in Figure 1) that we see at the surface (Parker 1955a). The wasp waist is confined by a reduced internal gas pressure p_i so that the

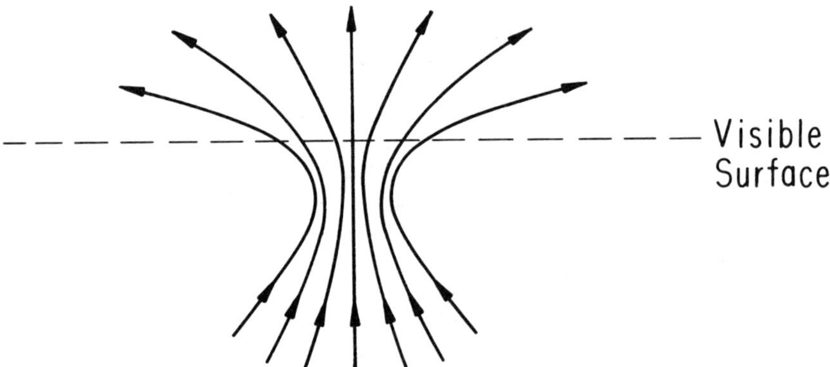

Figure 1 A sketch of the magnetic field configuration of a flux tube passing through the photosphere, showing the strong concentration of field that occurs near or below the visible surface.

external gas pressure p_e can balance the magnetic pressure. Assuming that the tube has a sharp, well-defined surface, then

$$p_e = p_i + B^2/8\pi \tag{2}$$

across that surface. The tension along the lines of force also tends to expand the waist, so that the gas pressure must further decrease from the surface to the central axis of the tube. The tension in the lines of force produces the fluting instability, in which individual flux tubes break away from the main body of the field and escape, shortening their length in the process. Somehow the environment below the surface of the Sun is able to assemble the magnetic flux tubes in opposition to this instability, and then to maintain the equilibrium for a long time thereafter (Parker 1974b, 1975a). The life span of the small individual concentrated flux tubes is not known. The individual tube may be as transient as the associated spicule. But sunspots can be very stable, lasting for a month or more. Clearly there is some force, as yet undefined, which counteracts the intrinsic instability of the field and brings about the assembly of a sunspot.

4 MAGNETIC BUOYANCY

To begin, let us imagine that the fluid motions in the convective zone have produced a strong azimuthal magnetic field B of, say, 10^2 G at a depth of $1-1.5 \times 10^5$ km beneath the surface of the Sun. Then how does that large-scale field break up to produce the magnetic debris that arrives continually at the surface of the Sun? The answer would appear to be that the gas pressure within a flux tube is lower by $B^2/8\pi$ than the pressure immediately outside. Hence, if the temperature inside is the same as it is outside, the density of the gas within the tube is reduced by

$$\Delta\rho = mB^2/8\pi kT, \tag{3}$$

$$= \tfrac{1}{2}\rho V_A^2/u^2, \tag{4}$$

where V_A is the Alfvén speed $B/(4\pi\rho_e)^{1/2}$ and u is the characteristic thermal velocity $(kT/m)^{1/2}$. It is this reduced gas density within a magnetic field that produces the phenomenon of magnetic buoyancy (Parker 1955a).

At the upper surface of a horizontal magnetic field the gas density increases by $\Delta\rho$ into the denser gas outside (above). This produces the familiar Rayleigh-Taylor instability, when a denser fluid overlies a lighter. There is an overturning of light and heavy fluid, with tongues of heavy fluid intruding downward into the lighter fluid, and vice versa. When the density difference is caused by a magnetic field, which has tension along its lines of force, the most unstable modes involve no distortion of the taut lines of force, but are confined to the two dimensions perpendicular to the lines of force. It is readily shown (see, for instance, Chandrasekhar 1961) that the growth in the linear regime is of the form $\exp(t/\tau + ikx)$ with $\tau^2 = 2\rho/gk\,\Delta\rho$. Then since $\Delta\rho$ is related to the field strength by (4) it follows that the growth rate is

$$1/\tau = (gk)^{1/2}\, V_A/2u, \tag{5}$$

where k is the wave number across the upper surface of the field (in the direction perpendicular to the field) and g is the acceleration of gravity. It is evident that the smallest wavelengths are the most unstable. Hence we can ignore the vertical scale height kT/mg of the gas. As a matter of fact, if the upper boundary of the field is not infinitely sharp, but has a characteristic thickness h, then the most unstable modes occur over scales of the general order of h. Thus, the instability causes the upper boundary of the field to break up into horizontal flux tubes with diameters of the order of h.

The individual flux tubes are buoyant, with a force

$$F = g\,\Delta\rho = B^2/8\pi\Lambda \tag{6}$$

per unit volume, where Λ is the scale height kT/mg. Hence the buoyant force per unit length for a tube of radius R is $\pi R^2 F$. The tube rises at a rate V such that the buoyant force is balanced by the aerodynamic drag. The aerodynamic drag on a circular cylinder is $CR\rho V^2$ where C is the drag coefficient and is of the general order of unity over a wide range of Reynolds numbers (Goldstein 1938). Of course, we do not expect the rising tube to maintain a circular cross section, but in order of magnitude we have $\pi R^2 F = O(CR\rho V^2)$, from which it follows (Parker 1975b) that the rate of rise is

$$V = V_A(\pi C/2)^{1/2}(R/\Lambda)^{1/2}. \tag{7}$$

The consequences of this rapid rise, at a rate of the general order of the Alfvén speed, were discussed in Section 2.

Altogether, then, the continual emergence of magnetic flux tubes through the surface of the Sun is a direct and understandable consequence of the production of magnetic flux at depth. The fields that we see are the buoyant debris escaping from the hydromagnetic dynamo deep in the convective envelope.

5 TWISTED FLUX TUBES

We might expect to find diffuse tubes of flux at the surface, whereas in fact the flux concentrates itself, in some way, into intense quasi-stable tubes of hundreds and thousands of gauss (see discussion in Section 3). The magnetic field pattern of some sunspots shows a definite twist or spiral pattern. This twisting adds an azimuthal field to the longitudinal field along the tube. The tension in the lines of force wrapped around the flux tube tends to compress the central portion of the tube, increasing the longitudinal field on the axis of the tube. But a point is soon reached in the twisting when the pressure of the azimuthal field reduces the net tension in the tube to zero and to negative values. The tube is then under compression and buckles, coiling itself into complicated patterns like an overwound rubber band (Parker 1974c). What then is the potential for twisting to produce the observed concentration of flux tubes?

To illustrate these effects with a quantitative example, note that the equation for magnetohydrostatic equilibrium of a twisted flux tube concentric with the z-axis is

$$0 = (\partial/\partial\varpi)[p+(B_z^2+B_\phi^2)/8\pi]+B_\phi^2/4\pi\varpi, \tag{8}$$

where $\varpi \equiv (x^2+y^2)^{1/2}$ denotes distance from the z-axis. Following Lüst & Schlüter (1954), note that the general solution to this equation is

$$8\pi p + B_z^2 = f(\varpi) + \tfrac{1}{2}\varpi\, df/d\varpi, \qquad B_\phi^2 = -\tfrac{1}{2}\varpi\, df/d\varpi, \tag{9}$$

where $f(\varpi)$ is an arbitrary function of ϖ subject only to the condition that $O > d\ln f/d\varpi > 2/\varpi$, so that p, B_z^2, and B_ϕ^2 are all positive. For the simple case that the confining pressure increment Δp is exerted at the surface of the tube, the uniform internal gas pressure plays no role in the formal solution and can be dropped.

For simplicity consider the initial flux tube to be of radius R_0, free of twisting, with a uniform field strength B_0 that is confined by the excess external pressure increment $\Delta p = B_0^2/8\pi$ at the surface $\varpi = R_0$. Then keeping $\Delta\rho$ constant, twist the tube uniformly so that each line of force makes one revolution about the axis of the tube in a distance λ. The lines of force lie on cylindrical shells of radius ϖ. The individual line crossing the plane $z = 0$ at the azimuthal angle ϕ_0 is given by

$$z = \varpi(\phi - \phi_0)B_z(\varpi)/B_\phi(\varpi) \tag{10}$$

and completes a revolution in a distance

$$\lambda = 2\pi\varpi B_z(\varpi)/B_\phi(\varpi). \tag{11}$$

For a uniformly twisted tube λ is independent of ϖ. Hence $B_\phi(\varpi) \propto \varpi B_z(\varpi)$, and it is readily shown that the appropriate generating function is

$$f(\varpi) = B_0^2(1+R^2/a^2)/(1+\varpi^2/a^2) \tag{12}$$

for the tube radius R. Then

$$B_z = B_0(1+R^2/a^2)^{1/2}/(1+\varpi^2/a^2), \tag{13}$$

$$B_\phi = B_z\varpi/a. \tag{14}$$

The field at the surface $\varpi = R$ maintains the fixed value B_0, as a consequence of the fixed confining pressure $\Delta\rho$. It is readily shown that $\lambda = 2\pi a$, which serves to relate a to the wavelength of the helices into which the lines are twisted. The length a decreases as the tube is increasingly twisted. The total longitudinal flux Φ is conserved so that $\Phi = \pi R_0^2 B_0$ for the untwisted tube and

$$\Phi = 2\pi \int_0^R d\varpi\, \varpi B_z(\varpi), \tag{15}$$

$$= \pi a^2 B_0(1+R^2/a^2)^{1/2} \ln(1+R^2/a^2) \tag{16}$$

subsequently. Since Φ and B_0 are constant, this relation gives the increasing tube radius R as a function of a, plotted in Figure 2.

The peak field, on the axis of the tube, is

$$B_{\text{peak}} = B_z(0) = B_0(1+R^2/a^2)^{1/2}, \tag{17}$$

and increases as shown in Figure 2. The mean square longitudinal field is unaffected by the twisting: $\langle B_z^2 \rangle = B_0^2$. The mean longitudinal field *declines*, of course, because the constant longitudinal flux Φ is spread out over the increasing radius R,

$$\langle B_z \rangle = (2/R^2) \int_0^R d\varpi \varpi B_z(\varpi) \tag{18}$$

$$= B_0 R_0^2/R^2. \tag{19}$$

The declining mean field is plotted in Figure 2.

Now the stress density in the ambient gas outside the tube is just the uniform pressure p_e while inside it is the net tension $(B_z^2 - B_\phi^2)/8\pi - (p_e - \Delta p)$. Hence the net stress difference in the tube is the tension $(B_z^2 - B_\phi^2)/8\pi + \Delta p$. The total tension carried by the tube is then

$$\mathcal{T} = 2\pi \int_0^R d\varpi \varpi [(B_z^2 - B_\phi^2)/8\pi + \Delta p], \tag{20}$$

$$= \tfrac{1}{8} B_0^2 R^2 [3 - (a^2/R^2 + 1) \ln(1 + R^2/a^2)]. \tag{21}$$

The tension declines monotonically with increased twisting, from $\frac{1}{4} B_0^2 R_0^2$ in the initial untwisted tube, to zero for $a = 0.295 R_0 (R = 3.97a)$, and to negative values as the twisting increases further, shown in Figure 2. The tension reaches zero when the

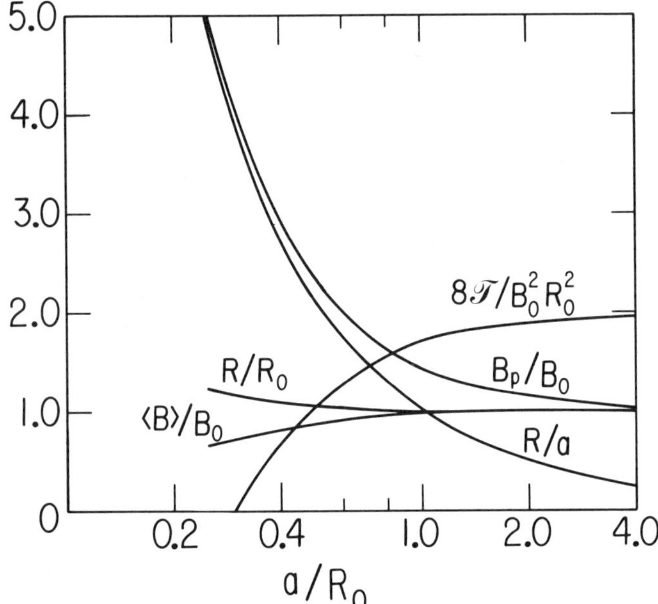

Figure 2 A plot of the radius R [from (16)], the mean longitudinal field $\langle B_z \rangle$ [from (19)], the peak field [from (17)], the total tension $\mathcal{T}$ [from (21)], and the ratio of radius R to wavelength of $2\pi a$ of the helical lines of force for a twisted flux tube confined by a fixed pressure. The horizontal axis is the wavelength of the helical lines in terms of the radius R_0 when the tube is without twisting and the field is B_0.

ratio of azimuthal to longitudinal field at the surface ($B_\phi/B_z = \varpi/a$) increases to 3.97 (Parker 1974a, 1976a). Unless artificial constraints are imposed, the tube buckles and spirals. Adjacent coils of the buckling tube soon come in contact, leading to neutral point annihilation and the rapid reconnection of the lines of force. The tube cannot be twisted significantly beyond $a = 0.295R_0$, if indeed it can be twisted as much as that.

Altogether, twisting a flux tube increases the peak field on the axis, perhaps by as much as a factor of 4.1 before the tube buckles. But the mean field declines to $0.73B_0$. Thus in the isolated flux tubes in supergranule boundaries, twisting adversely affects the concentration of flux to the high mean values that are observed. The concentration of the mean field can be effected, then, only by the confining gas pressure increment Δp. (Parker 1955a, 1974a, b, 1976a). If the tube is twisted, a larger confining pressure is needed. As a matter of fact, it will be shown in Section 6 that the field in the concentrated portion of a flux tube is not twisted significantly. The observed twisting in the field above a sunspot does not extend into the intense fields of the spot itself.

The question is often raised as to whether twisting a flux tube may stabilize the tube against the fluting instability because the lines of force wrap around the tube so as to prevent the tube from splitting lengthwise into two or more separate tubes. Clearly the tension in the azimuthal field must have this effect, but again the possibility of stabilization is excluded by the fact that the twisting does not extend into the compressed part of a flux tube where the instability originates. It is unfortunate, but twisting a tube seems to contribute adversely to concentration of the tube, and adds but little to the stability (Parker 1975a, 1976a). However, it does play an important role in the activity of the flux tube, i.e. it has a runaway destabilizing effect. That question is pursued in the next section.

6 VARIATION OF TWISTING ALONG A FLUX TUBE

Consider how the torsion in a flux tube varies with the radius of the tube from one location along the tube to another. It is sufficient, to establish the basic principles, to consider a slender tube of radius $R(z)$ for which the radius varies only slowly along the tube $\partial R/\partial z \ll 1$. Speaking in terms of average field strengths across the tube radius, conservation of flux requires that the mean longitudinal field B_z varies with the radius as $B_z \propto 1/R^2$. The total torque transmitted along the tube by the tension in the lines of force is also conserved. To demonstrate the implications, note that the azimuthal stress exerted by the field across any perpendicular section $z = constant$ is $B_z B_\phi/4\pi$. The lever arm is R so that the torque L is proportional to $B_z B_\phi R^3$. Since $B_z R^2$ is constant, it follows that $L \propto B_\phi R$ and hence

$$B_\phi \propto 1/R. \tag{22}$$

Hence, the angle θ between the spiral lines of force and the axis of the tube varies as

$$\tan\theta = B_\phi/B_z \propto R \tag{23}$$

and increases with increasing tube radius. The twisting is weak in the concentrated regions of the tube where R is small. Thus, while twisting has desirable properties by which one might hope to explain the remarkable stability of the sunspot, it also has properties that prevent it from appearing where we need it, in the most concentrated sections of the flux tube.

Indeed, the equilibrium of a twisted tube of varying radius has some very remarkable properties, which play a role in the observed activity of flux tubes extending up through the surface of the Sun. To go beyond simple considerations of mean values of B_z and B_ϕ, consider the idealized circumstance in which a twisted flux tube extends along the z-axis from $-\infty$ to $+\infty$. For $z < -h$ the radius of the tube has the uniform value c, while for $z > +h$ it has the uniform value $C(>c)$. In the finite section $-h < z < +h$ the radius varies in any arbitrary way $R(z)$ subject only to the requirement that $R(-h) = c$, $R(+h) = C$. The situation is sketched in Figure 3.

How is the field distribution at $z = +\infty$ related to the distribution at $z = -\infty$? There are three exact integrals of the equilibrium equations

$$\nabla \cdot \mathbf{B} = 0, \qquad 4\pi \nabla p = (\nabla \times \mathbf{B}) \times \mathbf{B}, \tag{24}$$

that provide an answer in a very simple way. Consider the field $b_\phi(\varpi)$, $b_z(\varpi)$ in the annulus $(\varpi, \varpi + d\varpi)$ at $z = -\infty$. These lines of force map into an expanded annulus, which is conveniently denoted by $(\Pi, \Pi + d\Pi)$ at $z = +\infty$ as sketched in Figure 3, where the field is $B_\phi(\Pi)$ and $B_z(\Pi)$. To treat the simplest example, suppose again that the gas pressure is uniform across the radius of the tube at $z = -\infty$. Then since $(\nabla \times \mathbf{B}) \times \mathbf{B}$ has no component parallel to $\mathbf{B}$, the pressure is uniform along each line of force and hence exactly uniform across the radius at $z = +\infty$. This is the first exact integral of (24) along the lines of force from $z = -\infty$ to $z = +\infty$. At $z = \pm\infty$, where the field has only a ϕ and a z-component, the equilibrium fields satisfy (8), and hence (9). So write them

$$b_z^2 = f + \tfrac{1}{2}\varpi df/d\varpi, \qquad b_\phi^2 = -\tfrac{1}{2}\varpi df/d\varpi, \tag{25}$$

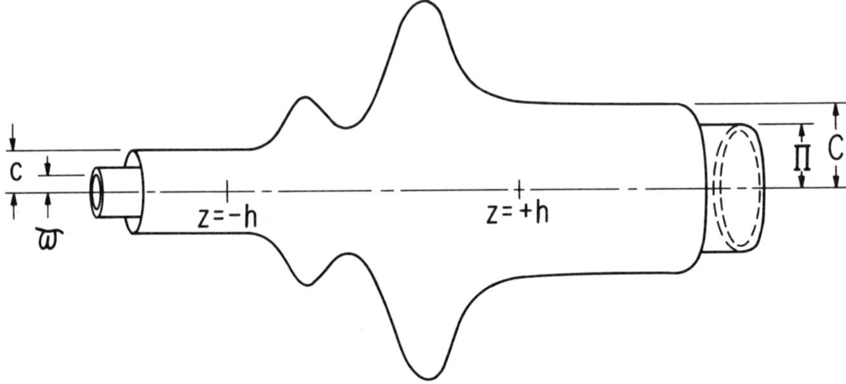

Figure 3 A sketch of the boundary confining the twisted flux tube and the annulus $(\varpi, \varpi + d\varpi)$ at $z = -\infty$, which extends to $(\Pi, \Pi + d\Pi)$ at $z = +\infty$.

and

$$B_z^2 = F + \tfrac{1}{2}\Pi dF/d\Pi, \qquad B_\phi^2 = -\tfrac{1}{2}\Pi dF/d\Pi, \tag{26}$$

with $f(\varpi)$ given and $F(\Pi)$ to be computed (or vice versa).

The second exact integral is conservation of flux in the annulus, so that we have the exact relation

$$B_z(\Pi)\Pi\, d\Pi = b_z(\varpi)\varpi d\varpi. \tag{27}$$

The third exact integral is conservation of torque along each annulus, so that

$$B_z(\Pi)B_\phi(\Pi)\Pi^2 d\Pi = b_z(\varpi)b_\phi(\varpi)\varpi^2 d\varpi. \tag{28}$$

Altogether, then, the three integrals provide the complete set of equations (25) through (28). To solve these equations divide (27) into (28) obtaining

$$B_\phi(\Pi)d\Pi = b_\phi(\varpi)d\varpi. \tag{29}$$

Square (27) and (29) and write the fields in terms of their generating functions, with the radial variables $u = \varpi^2/b^2$ and $U = \Pi^2/b^2$ where b is a characteristic length to be determined later to suit our convenience. The result is

$$b_z^2 = duf/du, \qquad b_\phi^2 = -udf/du,$$

$$B_z^2 = dUF/dU, \qquad B_\phi^2 = -UdF/dU,$$

and

$$F + UdF/dU = (du/dU)^2(f + udf/du), \tag{30}$$

$$U^2 dF/dU = u^2 df/du. \tag{31}$$

Eliminating $F(U)$ between these two equations leads to

$$\frac{du}{dU}\left[2\frac{duf}{du}\frac{d^2u}{dU^2} + \frac{d^2uf}{du^2}\left(\frac{du}{dU}\right)^2 - \frac{1}{U}\frac{d}{du}u^2\frac{df}{du}\right] = 0. \tag{32}$$

Thus, choosing a form $f(u)$ for the field distribution at $z = -\infty$ where the radius of the tube is c, the integration of (32) over the radius of the tube provides the relation between U and u, mapping the annulus ϖ at $z = -\infty$ into the annulus Π at $z = +\infty$. Then (30) or (31) yields the field distribution $F(U)$ in the tube at $z = +\infty$ where the radius is C. A variety of examples have been worked out and are available in the literature (Parker 1974c, 1976c). It is sufficient for purposes of illustration to consider a tube that is twisted only very slightly where the tube radius has the small value c at $z = -\infty$. Then, for a uniform twist, $f(u)$ is given by (11). Since (32) is linear in f, constant factors can be dropped. With the tube twisted only slightly, $c/a \ll 1$, put $b = a$ so that (12) is $f(u) = 1/(1+u) \cong 1-u$. Thus the simple form

$$f(u) = 1 - u$$

represents a tube of small radius ($c \ll b$) with a small degree of uniform twisting.

In this case (31) reduces to

$$\frac{du}{dU}\left[(1-2u)\frac{d^2u}{dU^2}-\left(\frac{du}{dU}\right)^2+\frac{u}{U}\right]=0.$$

To illustrate the interesting effects, suppose that at $z=+\infty$ the tube is greatly expanded so that $C \gg c$. Then the expanded radius Π, corresponding to the radius ϖ at $z=-\infty$, is very large compared to ϖ, i.e. $u \ll U$, $du/dU \ll 1$. In the limit of small u and small u/U the equation takes the asymptotic form

$$\frac{du}{dU}\left(\frac{d^2u}{dU^2}+\frac{u}{U}\right)=0.$$

Equating the two factors separately to zero yields the two solutions[1]

$$u = constant \tag{33}$$

$$u = u_1 U^{1/2} J_1(2U^{1/2})/U_1^{1/2} J_1(2U_1^{1/2}), \tag{34}$$

where J_1 is the Bessel function of the first kind and $u_1 = c^2/b^2$, $U_1 = C^2/b^2$. It is readily shown, then, that

$$F(U) = u_1^2[J_0^2(2U^{1/2})+J_1^2(2U^{1/2})]/U_1 J_1^2(2U_1^{1/2}) \tag{35}$$

and

$$B_z = u_1 J_0(2U^{1/2})/U_1^{1/2} J_1(2U_1^{1/2}), \tag{36}$$

$$B_\phi = u_1 J_1(2U^{1/2})/U_1^{1/2} J_1(2U_1^{1/2}). \tag{37}$$

Now u is a single-valued, monotonically increasing function of U. What is more, B_z is a positive quantity, because b_z is assumed to be positive at $z=-\infty$. But du/dU and B_z fall to zero at the first zero of J_0 at $U = 1.4458$, $\Pi/b = 1.2024$. At larger U, B_z becomes negative, which is impossible because all the flux in the tube is positive.

But there is nothing to prevent us from expanding the tube beyond $\Pi = 1.2024b$. We need only reduce the confining pressure increment Δp to a suitably low level. What happens, then, if $C > 1.2024b$? The answer is that the appropriate solution beyond $1.2024b$ is then (33), for which

$$F(U) = u_1^2/U, \tag{38}$$

and the field is exactly azimuthal. This is merely a quantitative illustration of the general qualitative conclusion (23) that the mean pitch angle θ increases with tube radius R. In fact the pitch angle of the outermost lines of force increases to $\pi/2$ when the expansion of the tube reaches $\Pi = 1.2024b$. Further expansion produces a purely azimuthal sheath, the lines of force in the sheath coming entirely from the surface $\varpi = c$ [see discussion and description in Parker (1974c, 1976c)]. The mean

[1] Another representative form for $f(u)$ is $J_0^2(2u^{1/2})+J_1^2(2u^{1/2})$. It too represents a uniformly twisted tube for small u and has the additional advantage that (34) is the general solution of (31) and not merely the asymptotic solution for large U/u, with $u^{1/2}J_1(2u^{1/2}) \propto U^{1/2}J_1(2U^{1/2})$.

magnetic tension $(B_z^2 - B_\phi^2)/4\pi$ falls to zero as C increases to $1.2024b$ and is negative for large C. The tube becomes unstable, etc.

To apply this result to a flux tube in the real world, note that the three exact integrals—free extension of gas pressure along the lines of force, conservation of flux, and conservation of torque—apply to any straight flux tube, whether long and slender, or short and thick. The representation of flux and torque may of course be very complicated where the radius of the tube is varying rapidly along the tube. But if the radius varies only slowly ($\partial R/\partial z \ll 1$), then (25) and (26) give a good approximation to the local fields [neglecting terms $O(\partial R/\partial z)$ compared to one]. Then (32) again relates u to U, and integration over the tube radius R gives an approximation to the variation of U with u, and $F(U)$. The details may be in error by $\partial R/\partial z$ compared to one, but the qualitative conclusion that the field becomes azimuthal as the radius increases is inescapable.[2]

The important point for the flux tubes observed in the Sun is that the helical coils of field migrate (at the Alfvén speed) from the compressed regions along the tube to the expanded regions, so that the twisting is not available in the concentrated part of the tube, either to stabilize the tube against the fluting instability, or to increase the peak field on the axis of the tube. The twisting passes to the expanded apex of a re-entrant tube where the tube bulges above the surface of the Sun. The twisting that is sometimes observed in the field over a sunspot, up to pitch angles of the order of $\pi/4$, indicates that the compressed field down inside the spot is twisted relatively little, as indicated by (23), (36), and (37). The degree of twisting $B_\phi/B_z = J_1(2U^{1/2})/J_0(2U^{1/2})$ declines approximately linearly as the radius at the tube declines.

It is hard to imagine how a flux tube in the turbulent gas beneath the surface of the Sun can fail to be twisted in some degree. So the tubes emerging through the surface of the Sun must be twisted. What we find is that the twisting is rapidly transmitted to the expanded portions of the tube. Since that is above the surface of the Sun, it is the portion of the tube that we observe. We have shown, however, that with increasing twist the tube becomes unstable to buckling, as the total tension declines and changes to compression. The buckling (see Section 8) soon leads to field annihilation, where separate coils come into contact. The rapid reconnection of lines between coils reduces the twisting, of course, and the result is an active dynamic balance, with twisting continually coming up from below into the expanded apex of the tube, where it is continually dissipated by neutral point annihilation. Thus we can see why the fields above the surface of the Sun are perpetually active. Such phenomena as the X-ray bright spots (Golub et al. 1974, Golub, Krieger & Vaiana 1975), ephemeral bipolar regions (Harvey, Martin & Martin 1975), and the plages around the concentrated tubes in the supergranule boundaries may be caused, in whole or in part, by the dynamical dissipation (Parker 1975c, d) of accumulating azimuthal field in the expanded portion of the flux tube.

This perhaps is the appropriate place to return once again to the unanswered question of the remarkable stability of the wasp-waisted sunspot, in the face of the

[2] Wilson (1977) urges that the mathematics is inapplicable, and the conclusion incorrect.

interchange fluting instability (see Figure 1). There is little or no twisting of the field in the concentrated portion of the flux tubes, where the azimuthal field might have bound the tube together to inhibit the splitting into many separate tubes. So we must think of other things. We note again the observation of Beckers & Schröter (1969, Rust 1976a) that the many individual separate flux tubes (magnetic knots) observed in and around an evolving active region seem to be attracted to the nearest sunspot of the same magnetic polarity. Such selective behavior can be understood only if the regions of like polarity are already connected beneath the surface. The picture would be that the field of an active region is organized (by unspecified forces) at some depth beneath the surface into an upward tube and a downward tube, and presumably strongly concentrated. Above the level of strong concentration the relatively untwisted tubes break up in many separate filaments that become mixed in some degree by the convection on their way to the surface. Hence at the surface the newly emerged tubes present a chaotic pattern. But when the separate tubes become sufficiently concentrated, their increased magnetic buoyancy and decreased aerodynamic drag tend to straighten them up, restoring them to their own kind at the visible surface. All of this depends, of course, on the organization being imposed in the depths of the Sun, if not at the surface, but forces that we do not understand at the moment.

7 THE CONCENTRATION OF FIELD INTO ISOLATED FLUX TUBES

Let us return to the problem, pointed out in Section 3, of the remarkable concentration of the photospheric fields into isolated, intense flux tubes. A reduced gas pressure within the magnetic flux tube is the only effect producing strong compression of the field (by the external pressure). As noted in Section 3, we do not know what stabilizes the compressed portion of the tube (Figure 1), but we can ask what it is that reduces the internal gas pressure. It was pointed out some years ago (Parker 1955a) that a small reduction of the internal temperature extending vertically over several scale heights may have a large effect. Thus, for instance, if the ambient temperature has the uniform value T, the ambient pressure varies with height z as

$$p(z) = p(0)\exp(-z/\Lambda),$$

where Λ is the scale height kT/mg. If the temperature within the flux tube is $T-\delta T$ with $\delta T \ll T$, the scale height is $\Lambda-\delta\Lambda$, with $\delta\Lambda = k\delta T/mg$. The gas pressure declines upward along the lines of force within the tube as

$$p_i(z) = [p(z)p_i(0)/p(0)]\exp[-(z/\Lambda)\delta\Lambda/\Lambda].$$

Hence the magnetic field pressure within the tube varies as

$$\begin{aligned} B^2(z)/8\pi &= p(z) - p_i(z) \\ &= p(z)\{1-[1-B^2(0)/8\pi p(0)]\exp[-(z/\Lambda)\delta T/T]\} \\ &\cong p(z)\{1-\exp-(z/\Lambda)\delta T/T\} \end{aligned} \tag{39}$$

if $B^2(0) \ll 8\pi p(0)$. The gas within the tube does not stand as high in the gravitational field as the gas outside, leaving the upper portion of the tube partially evacuated, and so concentrating the field. Note that the magnetic pressure increases monotonically in height relative to the ambient gas pressure and asymptotically approaches the gas pressure at large z. But note too that the magnetic pressure in absolute units reaches a maximum in one scale height $z = \Lambda$, where it has a value $[p(0)/e]\, \delta T/T$. It is suggested, then, that the intense field of the sunspot is a direct consequence of the reduced temperature within the spot (Parker 1955a, Schlüter & Temesvary 1958). A temperature reduction of about one third produces a strong concentration within two or three scale heights. Presumably the strong concentration is effected in the first $1-2 \times 10^3$ km beneath the photosphere. The field pressure is a large fraction of the ambient gas pressure just a few scale heights beneath the surface, easily accounting for the observed $2-4 \times 10^3$ G.

This brings us to the cause of the cooling. The inhibition of convection surely decreases the heat transport (Biermann 1941), with the unfortunate property of blocking the heat flow immediately beneath the sunspot, thereby raising the temperature there. That is the opposite of the cooling required to concentrate the field (Parker 1974b, 1975a, 1976a). Something more is needed, and the only evident possibility is the conversion of heat into Alfvén waves via the strong convective instability in the 2×10^3 km beneath the visible surface. The conversion to mechanical waves would have to be very efficient, approaching the ideal thermodynamic efficiency of the convective heat engine, because about 80% of the heat must be converted to account for the reduced radiative flux from the cool sunspot umbra. Observations of line profiles in sunspots (see, for instance, Phillis 1975 and the very recent results of Beckers 1976) admit the possibility of Alfvén waves of small scale ($1-5 \times 10^2$ km) upward from the sunspot (with a presumably equal downward flux, for a total of about 5×10^{10} ergs sec^{-1}). But they do not provide direct confirmation. So at the time of writing the question is open. If it could be shown that the generation of Alfvén waves is not the principal cause of cooling, then the concentration of magnetic field that makes up the sunspot is without explanation at the present time. No viable theoretical alternative has been proposed.[3] So it will be very interesting, indeed, to see if Alfvén waves are the principal cause of cooling of a sunspot.

It has been suggested (Parker 1976a, Roberts 1976) that perhaps the concentrated flux tubes of $1-2 \times 10^3$ G that appear in the supergranule boundaries are to be understood as miniature sunspots, with reduced temperatures of, say, 10%, over a diameter of a few hundred kilometers. They would scarcely be visible, particularly as the active cooling does not extend all the way to the surface.

Altogether, then, it is conjectured that whenever a magnetic field extends through the 10^3 km layer of very strong convective overstability (see, for instance, the convective zone model of Spruit 1974) immediately beneath the photosphere, the vigorous production of Alfvén waves automatically cools the gas within the field,

[3] The Biermann theory of sunspot cooling has recently been defended without elucidation by Cowling (1976).

so that gravity evacuates the flux tube and we see only strong, concentrated fields at the visible surface. The conjecture remains to be proved by direct observation.[4]

In a search for viable alternatives we have examined a number of dynamical effects, involving turbulent pumping and rapid, intermittent, or steady, flow along the flux tube, that may contribute to compression of the field up to the equipartition value $B^2 = 4\pi\rho v^2$ in the supergranule boundaries (Parker 1974a, 1976a, b). Evaluation of these theoretical possibilities awaits more definitive estimates of B and of the associated gas velocity v in the flux tubes. For instance, 10^3 G at the visible surface of the Sun, where $\rho \cong 3 \times 10^{-7}$ gm cm^{-3}, would be associated with gas motions of 5 km sec^{-1}. (See, for instance, the remarks of Deubner 1975, 1976.) It must be remembered that there is a spicule associated with each of the individual flux tubes in the supergranule boundaries, strongly suggesting that the tubes are in some way dynamically active (see, for instance, Parker 1964, 1974a). The dynamical effects may be a byproduct, or an integral part, of the concentration of field. If, for instance, we look at other eruptive phenomena, the white light coronal transients are evidently dominated by magnetic forces (Dulk et al. 1976). The vigorous and newly discovered macrospicules, occurring in coronal holes (Bohlin et al. 1975, Bohlin 1976) suggest magnetic effects by the systematic variation of their inclination with latitude. Yet they are broad ($3-15 \times 10^3$ km) and arise in coronal holes where conventional magnetographs show only weak fields. It is far from clear what conceptual similarities and differences exist between these many forms of solar eruption.

8 DYNAMICAL DISSIPATION OF MAGNETIC FIELDS

It was pointed out some years ago (Sweet 1958, Parker 1957b, 1963) that when two oppositely directed magnetic fields are pressed together, they rapidly annihilate each other unless we can, somehow, contrive to maintain enough fluid pressure between to keep them apart. That is to say, if the gas pressure in either field is p_i, so that the total pressure is $p_i + B^2/8\pi$, it requires the enhanced gas pressure $p_e = p_i + B^2/8\pi$ to keep the fields apart. There is usually no reason for the pressure p_e to be much greater than p_i, so the gas squeezes out from between the fields and the gradient between the two opposite fields increases without limit, in the manner familiar to the reader. If the electrical conductivity of the fluid is σ (resistive diffusion coefficient $\eta = c^2/4\pi\sigma$), and the opposite fields, each of magnitude B, are pressed together over a width L, then from the characteristic diffusion velocity $u = \eta/l$ over a scale l, and conservation of the fluid that is squirted out from between the fields with a velocity comparable to the Alfvén speed $V_A = B/(4\pi\rho)^{1/2}$, it follows that under steady conditions the fields merge and annihilate at a velocity u of the order of

$$u = V_A/R_m^{1/2}, \tag{40}$$

where R_m is the magnetic Reynolds number $V_A L/\eta$. The ordinary resistive diffusion velocity over a scale L is $\eta/L = V_A/R_m$, so the merging is accelerated from simple

[4] The question of the stability remains to be resolved whether the concentration is produced by cooling or not.

ohmic dissipation by the large factor $R_n^{1/2}$. But the rate is still very small compared to the Alfvén speed when, as in most astrophysical fields, the overall magnetic Reynolds number is very large. It remained for Petschek (1964) to point out that the fluid motions and the magnetic field may arrange themselves in such a form that the front L over which the fields are pressed together may be much smaller than the scale of the fields themselves. He showed that a plausible case can be made for the hydrodynamics to arrange itself in such a way that L may shrink and the merging speed may have any value up to something of the order of $V_A/\ln R_m$. Subsequent theoretical investigations (Sonnerup 1970, 1971) have shown that in very special circumstances the merging speed may be almost as large as V_A no matter how large σ is, and more generally that merging speeds of $10^{-2}-10^{-1}\,V_A$ can be expected under a wide variety of circumstances (Priest 1971a, b, 1972, Fukao & Tsuda 1973, Vasyliunas 1975, Roberts & Priest 1975). The excitation of microturbulence (Linhart 1960, Alfvén & Carlquist 1967, Hamberger & Friedman 1969, Hamberger & Jancarik 1970, Coppi & Mazzucato 1971, Coppi & Friedland 1971, Smith & Priest 1972, Priest & Heyvaerts 1974, Coppi 1975) in the plasma (enormously increasing the effective resistive diffusion coefficient), the onset of the resistive tearing mode instability (Furth, Killeen & Rosenbluth 1963, Jaggi 1963, Cross & Van Hoven 1971, Biskamp & Schindler 1971), and large-scale hydromagnetic exchange instabilities (Parker 1973a, b) may further enhance the merging rate. On this basis, the mutual annihilation of two opposite fields through rapid reconnection of the lines of force becomes an important astrophysical phenomenon. We have already mentioned its role in the active dissipation of buckling flux tubes, etc. The "neutral point annihilation," as it is often called, has been developed with particular attention to the solar flare, where it is believed to accomplish the explosive conversion of magnetic energy into fast particles, as much as 10^{32} ergs in 10^3 sec in a large flare. More recently the subject has come up in the theory of the gigantic flares of some of the faint M-dwarfs (Luyten 1949, Joy & Humason 1949, Lovell 1971, Moffett 1975) and the geomagnetic tail (Dungey 1961). A variety of highly stressed magnetic configurations have been investigated theoretically (Tanaka & Nakagawa 1973, Sturrock 1974, Low 1974, Low & Nakagawa 1975), and several authors have pointed out magnetic configurations that may be susceptible to the sudden onset of instability suggested by the explosive onset of the solar flare. Flares seem to occur most frequently where new fields emerge through the surface in the midst of old fields (Rust 1976a, b, Heyvaerts & Priest 1976). Recently Low (1977) has shown circumstances under which increasing shear brings the field to a state of strain beyond which there is no further equilibrium. The evolution of the field in the increasing shear (Low & Nakagawa 1975) resembles in many respects the observed accumulation of strain and magnetic energy observed in photospheric fields prior to the onset of a flare (Prata 1972, Rust 1973, Rust & Bar 1973, Zirin & Tanaka 1973, Zirin 1974). Hence the formal example treated by Low may be an illustration of the evolutionary path leading to the solar flare. The effect merits further investigation, particularly into the nature of the nonequilibrium, to see whether the topology of the field is such as to produce neutral point annihilation and rapid reconnection.

The rapid advance of X-ray and γ-ray astronomy in the last few years (Vorpahl 1972, Chupp et al. 1973, Glencross, Dorling & Herring 1974, Kane 1974, Rust &

Hegwer 1975) has provided insight into the detailed superheating and acceleration of electrons within the flare. The direct observation of isotopic abundances among the ions that are accelerated to high energies, and escape to be observed at Earth, provides further information (Serlemitsos & Balasubrahmanyan 1975, Garcia-Munoz, Mason & Simpson 1975) on conditions in the flare. The remarkably high abundance of the spallation product He^3, without a significant amount of the expected companion product H^2, is particularly puzzling. The subject of the complex and highly individualistic solar flare has now developed to such a degree that it merits an entire review of its own (see, for instance, Sweet 1969). Recent papers by Martres, Soru-Escaut & Rayrole (1973), Rust, Nakagawa & Neupert (1975), Syrovatski (1976), and Heyvaerts & Priest (1976) [see also the review volumes arranged by De Jager, Obayashi & Svestka (1976) and Massey, Sweet & Gabriel (1976)] provide an instructive synthesis of existing observational knowledge and general theory of field annihilation, tracing the development of the "typical" flare through a preflare heating phase, wherein reconnection of opposite fields first gets underway. Then, when the electric current density becomes sufficiently large to excite micro-instabilities, the explosive phase begins. Heyvaerts & Priest (1976) suggest that the field configuration then evolves into a large neutral sheet (current sheet) that produces the main phase of the flare, and eventually decays away when the reconnection of lines has suitably reduced the nonequilibrium components.

Rather than try to go into the many fascinating details of the solar flare—not to mention the equally fascinating details of the X-ray bright spots, the spicules, and many other effects passed over in so cursory a fashion—we present a theorem that provides some insight into the very general occurrence of dynamical nonequilibrium of magnetic fields and their continual activity in the Sun. Up to the present time theoretical studies of magnetic fields have dealt largely with magnetohydrostatic equilibrium in conducting fluids. The equations are nonlinear, but with the introduction of one form of symmetry or another, many elegant solutions have been worked out, exhibiting the rich variety of symmetric equilibria that magnetic fields might assume. What has not been realized, however, is the fact that the symmetries and invariance introduced into the problem to make the mathematics tractable exclude the major class of field topologies, most of which have no static equilibrium. Magnetohydrostatic equilibrium exists only for fields with suitable invariance or symmetry. All other topologies, i.e. most field topologies, can exist only in a dynamical state and are subject to rapid reconnection somewhere within their volume. The reconnection soon destroys the maverick components of the field that break the symmetry, so that the field approaches equilibrium if left to itself. But in the Sun where the lines of force all extend into the convective zone beneath the surface, the field is never left alone. The lines of force are continually manipulated by the convection, the twisted components continually rise up through the surface into the apex of each re-entrant flux tube, and the general symmetry necessary to avoid rapid reconnection is never achieved. The fields are always active in some degree.

To establish the theorem, let us first be clear as to the circumstances under which equilibrium occurs in the magnetic fields in the Sun, or other stars, or in the

Galaxy. There are no rigid boundaries to grab the field and steady it. We are interested, then, in the internal equilibrium of a magnetic field and fluid far from any boundaries. Hence, from the mathematical point of view, we are interested in equilibrium throughout an infinite space. We look for nonsymmetric equilibria, in the neighborhood of known symmetric or invariant configurations, assuming that any nonsymmetric equilibria are analytic functions of the parameter ε that measures their deviation from symmetry. Thus in the simplest case, we consider a magnetic field $\mathbf{B}$ in the neighborhood of the uniform field $\mathbf{e}_z B_0$. Write $\mathbf{B} = \mathbf{e}_z B_0 + \varepsilon \mathbf{b}(x, y, z)$. This field is produced from the uniform field by winding, braiding, and twisting the lines of force about each other, as sketched in Figure 4, so that their direction deviates slightly from the z-direction by angles $O(\varepsilon)$. Thus, no true knots are possible in a single line, but clearly, any variation of winding pattern along the lines of force can be achieved by first wrapping one bundle of lines about another for a way, and then dividing the bundles and wrapping in a new pattern. The general hydromagnetic equilibrium Equation (24) reduces to

$$\nabla(p + B_0 b_z/4\pi) = B_0\, \partial \mathbf{b}/\partial z, \qquad \nabla \cdot \mathbf{b} = 0 \tag{41}$$

upon neglecting terms second order in ε. The divergence of this equation yields

$$\nabla^2 (p + B_0 b_z/4\pi) = 0. \tag{42}$$

We seek a solution for p and $\mathbf{b}$ that is finite and bounded in the neighborhood of the origin, and which remains bounded at all distances from the origin. The only solution of Laplace's equation with this property is the constant,

$$p + B_0 b_z/4\pi = \textit{constant}. \tag{43}$$

All other solutions diverge linearly, or exponentially, in one direction or another. With (43), (41) reduces to

$$\partial \mathbf{b}/\partial z = 0. \tag{44}$$

As a matter of fact the equations can be solved to all orders in ε (Parker 1972a) with the same result (see also Yu 1973). A search in the neighborhood of the general two-dimensional equilibrium field $\partial \mathbf{B}/\partial z = 0$ yields (44) again (Parker 1976c). Thus, apart from the possibility of isolated topologies, a necessary condition for equilibrium is that the winding pattern of the lines of force must be invariant

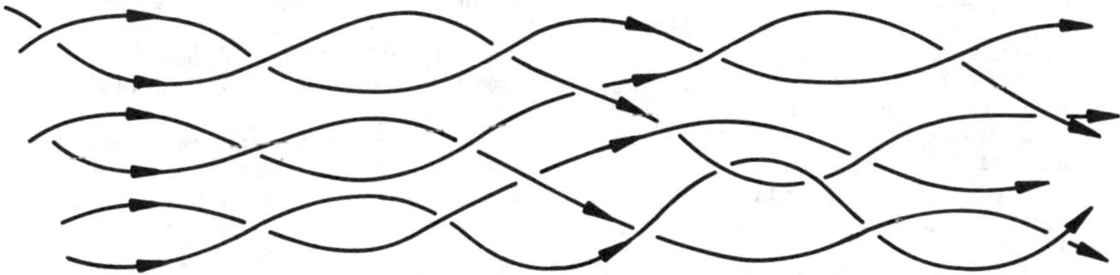

Figure 4 A sketch of the lines of force of the field $\mathbf{e}_z B_0 + \varepsilon \mathbf{b}$ showing a pattern of winding that varies along the general direction of the field.

along the principal direction of the field. The lines of force may be twisted about each other in any way that we wish, but that pattern of twisting must extend the full length of the field. If one bundle of flux is wound about a second, and farther along about a third, etc, then there is no hydrostatic equilibrium. There is no hydrostatic equilibrium when three or more tubes are braided, etc.

The nonequilibrium may be understood from the fact that the fluid pressure extends uniformly along the magnetic lines of force ($\mathbf{B}\cdot\nabla p = 0$). Now for any given winding pattern, there is a suitable pressure distribution for equilibrium, viz $\nabla p = -\nabla B^2/8\pi + (\mathbf{B}\cdot\nabla)\mathbf{B}/4\pi$. But if farther along the winding pattern changes, the pressure, extending uniformly along the lines of force into the new pattern, will not have the necessary form for equilibrium of the new pattern. Somewhere the pressure will be inadequate to keep apart some opposite field components, and rapid reconnection will occur. The rapid reconnection destroys the variant patterns and the field configuration relaxes toward the invariant form (44).

We expect that the magnetic fields produced in the convecting Sun and other objects in the astrophysical universe are generally prevented, by the fluid motions that create them, from achieving the perfect topological symmetry necessary for the existence of an equilibrium. Hence we expect them to be active, perpetually in a state of reconnection. This seems to be the cause of the continual activity of the magnetic fields that come up to us through the surface of the Sun. We may presume that it is the cause of the activity observed in more distant stars and galaxies.

Literature Cited

Alfvén, H., Carlquist, P. 1967. *Solar Phys.* 1:220

Beckers, J. M. 1976. *Ap. J.* 203:739

Beckers, J. M., Schröter, E. H. 1969. *Solar Phys.* 4:192, 303

Biermann, L. 1941. *Vierteljahresschr. Astron. Ges.* 76:194

Biskamp, D., Schindler, K. 1971. *Plasma Phys.* 13:1013

Bohlin, J. D. 1976. In *Proc. Int. Symp. Sol.-Terr. Phys., June 1976, Boulder, Colo.*, ed. D. J. Williams

Bohlin, J. D., Vogel, S. N., Purcell, J. D., Sheeley, N. R., Tousey, R., Van Hoosier, M. E. 1975. *Ap. J. Lett.* 197:L133

Chandrasekhar, S. 1961. *Hydrodynamic and Hydromagnetic Stability*, Chap. 10. Oxford: Clarendon

Chapman, G. A. 1973. *Ap. J.* 191:255

Chupp, E. L., Forrest, D. J., Higbie, P. R., Suri, A. N., Tsai, C., Dunphy, P. P. 1973. *Nature* 241:333

Coppi, B. 1975. *Ap. J.* 195:545

Coppi, B., Friedland, A. B. 1971. *Ap. J.* 169: 379

Coppi, B., Mazzucato, E. 1971. *Phys. Fluids* 44:134

Cowling, T. G. 1976. *MNRAS* 177:409

Cross, M. A., Van Hoven, G. 1971. *Phys. Rev. A* 4:2347

Deinzer, W., Kusserow, H. U. V., Stix, M. 1974. *Astron. Astrophys.* 36:69

Deinzer, W., Stix, M. 1974. *Astron. Astrophys.* 12:111

De Jager, C., Obayashi, T., Svestka, Z. 1976. *Solar Phys.* 41, No. 1

Deubner, F. 1975. *IAU Symp. No. 71, Prague, 25–29 August*

Deubner, F. 1976. *Astron. Astrophys.* 47:475

Dicke, R. H. 1974. *Ap. J.* 190:187

Dulk, G. A., Smerd, S. F., MacQueen, R. M., Gosling, J. T., Magun, A., Stewart, R. T., Sheridan, K. V., Robinson, R. D., Jaques, S. 1976. *Solar Phys.* 49:369

Dungey, J. W. 1961. *Phys. Rev. Lett.* 6:47

Durney, B. 1975. *Ap. J.* 204:589

Eddy, J. A. 1976a. *Science* 192:1189 (No. 4245, 18 June)

Eddy, J. A. 1976b. *The sun since the bronze age.* Invited paper, Int. Symp. Sol.-Terr. Phys. (Am. Geophys. Union) 17 June, Boulder, Colo.

Eddy, J. A., Gilman, P. A., Trotter, D. E. 1976. *Sol. Phys.* 46:3

Frazier, E. N., Stenflo, J. O. 1972. *Sol. Phys.* 27:330

Fukao, S., Tsuda, T. 1973. *Planet. Space Sci.* 21:1151; *J. Plasma Phys.* 9:409

Furth, H. P., Killeen, J., Rosenbluth, M. N. 1963. *Phys. Fluids* 6:459

Garcia-Munoz, M., Mason, G. M., Simpson, J. A. 1975. *Ap. J. Lett.* 201:L141

Gierasch, P. J. 1974. *Ap. J.* 190:199

Gilman, P. A. 1969. *Sol. Phys.* 8:316; 9:3

Gilman, P. A. 1972. *Sol. Phys.* 27:3

Gilman, P. A. 1974. *Ann. Rev. Astron. Astrophys.* 12:47

Glencross, W. M., Dorling, E. B., Herring, J. R. 1974. *Sol. Phys.* 38:183

Goldstein, S. 1938 *Modern Developments in Fluid Dynamics,* Vol. 1, pp. 15, 16. Oxford: Clarendon

Golub, L., Krieger, A. S., Silk, J. K., Timothy, A. F., Vaiana, G. S. 1974. *Ap. J. Lett.* 189:L93

Golub, L., Krieger, A. S., Vaiana, G. S. 1975. *Sol. Phys.* 42:131

Gubbins, D. 1974. *Rev. Geophys. Space Phys.* 12:137

Hamberger, S. M., Friedman, M. 1969. *Phys. Rev. Lett.* 21:674

Hamberger, S. M., Jancarik, J. 1970. *Phys. Rev. Lett.* 25:999

Harvey, K. L., Martin, J. W., Martin, S. F. 1975. *Sol. Phys.* 40:87

Heyvaerts, J., Priest, E. R. 1976. *Solar Phys.* 47:223

Hill, H. A., Stebbins, R. T. 1975. *Ap. J.* 200: 471

Howard, R. 1971. *Sol. Phys.* 16:21

Howard, R. 1972. *Sol. Phys.* 24:123

Howard, R., Harvey, J. W. 1970. *Sol. Phys.* 20:3

Howard, R., Stenflo, J. O. 1972. *Sol. Phys.* 22:402

Jaggi, R. K. 1963. *J. Geophys. Res.* 68:4429

Joy, A. H., Humason, M. L. 1949. *Publ. Astron. Soc. Pac.* 61:133

Kane, S. R. 1974. *Coronal Disturbances,* IAU Symposium No. 57, p. 105. Dordrecht: Reidel

Kippenhahn, R. 1963. *Ap. J.* 137:664

Leighton, R. B. 1969. *Ap. J.* 156:1

Linhart, J. G. 1960. *Plasma Physics,* p. 173. Amsterdam: North Holland

Livingston, W., Harvey, J. W. 1969. *Sol. Phys.* 10:294

Livingston, W., Harvey, J. W. 1971. In *Solar Magnetic Fields,* IAU Symp. No. 43, ed. R. Howard, p. 51. Dordrecht: Reidel

Lovell, B. 1971. *Q. J. Roy. Astron. Soc.* 12:98

Low, B. C. 1974. *Ap. J.* 193:243

Low, B. C. 1977. *Ap. J.* In press

Low, B. C., Nakagawa, Y. 1975. *Ap. J.* 199: 237

Lüst, R., Schlüter, A. 1954. *Z. Astrophys.* 34: 363

Luyten, W. J. 1949. *Ap. J.* 109:532

Martres, M. J., Soru-Escaut, I., Rayrole, J. 1973. *Sol. Phys.* 32:365

Massey, H., Sweet, P. A., Gabriel, A. H. 1976. *Philos. Trans. R. Soc. London* 281:293–513

Maunder, E. W. 1894. *J. Br. Astron. Assoc.* 5:47

Meyer, F., Schmidt, H. U., Weiss, N. O., Wilson, P. R. 1974. *MNRAS* 169:35

Moffett, T. J. 1975. *Ap. J. Suppl.* 29:1

Parker, E. N. 1955a. *Ap. J.* 121:491

Parker, E. N. 1955b. *Ap. J.* 122:293

Parker, E. N. 1957a. *Proc. Nat. Acad. Sci. USA* 43:8

Parker, E. N. 1957b. *J. Geophys. Res.* 62: 509

Parker, E. N. 1963. *Ap. J. Suppl.* No. 77, Vol. 8:177

Parker, E. N. 1964. *Ap. J.* 140:1170

Parker, E. N. 1970a. *Ap. J.* 160:383

Parker, E. N. 1970b. *Ann. Rev. Astron. Astrophys.* 8:1

Parker, E. N. 1970c. *Ap. J.* 162:665

Parker, E. N. 1971. *Ap. J.* 163:279; 164:491

Parker, E. N. 1972a. *Ap. J.* 174:499

Parker, E. N. 1972b. *Ap. J.* 176:213

Parker, E. N. 1973a. *Ap. J.* 180:247

Parker, E. N. 1973b. *J. Plasma Phys.* 9:49

Parker, E. N. 1973c. *Astrophys. Space Sci.* 22:279

Parker, E. N. 1973d. *Ap. J.* 186:665

Parker, E. N. 1974a. *Ap. J.* 189:563; 190: 429

Parker, E. N. 1974b. *Sol. Phys.* 36:249; 37:127

Parker, E. N. 1974c. *Ap. J.* 191:245

Parker, E. N. 1974d. *Astrophys. Space Sci.* 31:261

Parker, E. N. 1975a. *Sol. Phys.* 40:275; 291

Parker, E. N. 1975b. *Ap. J.* 198:205

Parker, E. N. 1975c. *Ap. J.* 201:494

Parker, E. N. 1975d. *Ap. J.* 201:502

Parker, E. N. 1975e. *Ap. J.* 202:523

Parker, E. N. 1976a. *Ap. J.* 204:259

Parker, E. N. 1976b. *Ap. J.* 210:810, 816

Parker, E. N. 1976c. *Astrophys. Space Sci.* 44:107

Petschek, H. E. 1964. *The Physics of Solar Flares,* AAS-NASA Symp. (NASA SP-50), ed. W. N. Hess, p. 425

Phillis, G. L. 1975. *Sol. Phys.* 41:71

Prata, S. W. 1972. *Sol. Phys.* 25:36

Priest, E. R. 1971a. *Q. J. Mech. Appl. Math.* 25:319

Priest, E. R. 1971b. *MNRAS* 159:389

Priest, E. R. 1972. *Ap. J.* 181:227

Priest, E. R., Heyvaerts, J. 1974. *Sol. Phys.* 36:433

Radler, K. H. 1969. *Monatsber. Dtsch. Akad. Wiss. Berlin.* 11:194. English transl. in P.

H. Roberts, M. Stix, 1971. *The Turbulent Dynamo*. NCAR Tech. Note TN/IA-60
Roberts, B. 1976. *Ap. J.* 204:263
Roberts, B. Priest, E. R. 1975. *J. Plasma Phys.* 14:417
Rust, D. M. 1973. *Sol. Phys.* 33:205
Rust, D. M. 1976a. *Philos. Trans. R. Soc. London Ser. A.* 281:353
Rust, D. M. 1976b. *Philos. Trans. R. Soc. London Ser. A.* 281:427
Rust, D. M., Bar, V. 1973. *Sol. Phys.* 33:445
Rust, D. M., Hewner, F. 1975. *Sol. Phys.* 40:141
Rust, D. M., Nakagawa, Y., Neupert, W. M. 1975. *Sol. Phys.* 41:397
Sawyer, C. 1971. In *Solar Magnetic Fields,* IAU Symp. No. 43, ed. R. Howard, p. 316. Dordrecht: Reidel
Schlüter, A., Temesvary, S. 1958. In *Electromagnetic Phenomena in Cosmic Physics,* IAU Symp. No. 6, ed. B. Lehnert, p. 263
Serlemitsos, A. T., Balasubrahmanyan, V. K. 1975. *Ap. J.* 198:195
Sheeley, N. R. 1967. *Sol. Phys.* 1:171
Sheeley, N. R., Bohlin, J. D., Brueckner, G. E., Purcell, J. D., Scherrer, V., Tousey, R. 1975. *Sol. Phys.* 40:103
Simon, G. W., Noyes, R. W. 1971. In *Solar Magnetic Fields,* IAU Symp. No. 43, ed. R. Howard, p. 663. Dordrecht: Reidel
Smith, D. F., Priest, E. R. 1972. *Ap. J.* 176: 487
Sonnerup, B. U. O. 1970. *J. Plasma Phys.* 4:161
Sonnerup, B. U. O. 1971. *J. Geophys. Res.* 76:8211
Spruit, H. C. 1974. *Sol. Phys.* 34:277
Steenbeck, M., Krause, F. 1969. *Astron. Nach.* 291:49, 271
Steenbeck, M., Krause, F., Radler, K. H. 1966. *Z. Naturforsch.* 21a:369
Stenflo, J. O. 1973. *Sol. Phys.* 32:41
Stix, M. 1974. *Astron. Astrophys.* 37:121
Stix, M. 1976. *Astron. Astrophys.* 47:243
Sturrock, P. 1974. In *Flare-Related Magnetic Field Dynamics,* ed. Y. Nakagawa, D. M. Rust. Boulder, Colorado, NCAR
Sweet, P. A. 1958. *Nuovo Cimento, Suppl.* (10) 8:188
Sweet, P. A. 1969. *Ann. Rev. Astron. Astrophys.* 7:149
Syrovatski, S. I. 1976. *Lett. Sov. Astron.* 2:1
Tanaka, K., Nakagawa, Y. 1973. *Sol. Phys.* 33:187
Vasyliunas, V. M. 1975. *Rev. Geophys. Space Sci.* 13:303
Vorpahl, J. A. 1972. *Sol. Phys.* 26:397
Wilson, P. R. 1977. *Ap. J.* In press
Yoshimura, H. 1972. *Sol. Phys.* 22:20; *Ap. J.* 178:863
Yoshimura, H. 1973. *Sol. Phys.* 33:131
Yoshimura, H. 1975. *Ap. J. Suppl.* 29:467 (No. 294)
Yu, G. 1973. *Ap. J.* 181:1003
Zirin, H. 1974. *Vistas Astron.* 16:1
Zirin, H., Lazareff, B. 1975. *Sol. Phys.* 41: 425
Zirin, H., Tanaka, K. 1973. *Sol. Phys.* 32: 173

Ann. Rev. Astron. Astrophys. 1977. 15: 69–95

SEYFERT GALAXIES ×2108

Daniel W. Weedman
Dyer Observatory, Vanderbilt University, Nashville, Tennessee 37235

1 INTRODUCTION

This review appears nearly a decade after the Tucson conference "Seyfert Galaxies and Related Objects" held in February, 1968. At that time, realization of the similarities between the QSOs and the nuclei of Seyfert galaxies led 56 astrophysicists to contribute their ideas regarding Seyfert galaxies [*Astronomical Journal,* Vol. 73 (1968), pp. 836–943]. A number of questions were posed regarding the energetic events in Seyfert galaxies. Most of these remain unanswered. Progress has been made in increasing the sample of Seyfert galaxies, in describing the physical conditions in their nuclei, and in delineating the relation between Seyferts and QSOs. Two basic aspects of the Seyfert galaxy problem have been emphasized. One is the nature of redshifts and whether a study of Seyfert galaxies leads to proof that QSO redshifts, by analogy, are cosmological. The other is the search for an explanation of the extraordinary energy source that powers the nuclei of Seyfert galaxies. There are many reasons to believe that the same mechanism of energy generation operates within most Seyfert nuclei as in the QSOs. To a certain extent, this has been a frustrating research problem because of the lack of progress in deciphering the fundamental mechanisms involved. On the other hand, far more observational research is being directed toward Seyfert galaxies than ever before, so this deserves to be chronicled.

This review makes no attempt to incorporate the numerous observations and theories concerning the general problem of galactic nuclei, even though this is certainly relevant to Seyfert galaxies. Some judgment had to be used as to the point when the review was sufficiently comprehensive to encourage new researchers to consider the problems but not so encyclopaedic as to be intimidating. An earlier review in this series provided a comprehensive discussion of galactic nuclei including data on Seyferts known at the time (Burbidge 1970). Somewhat more recent summaries of observational work on Seyfert galaxies are given by Osterbrock (1971), Sargent (1971), and Ulrich (1974). A comprehensive theoretical review is that of Saslaw (1974). Listings of Seyfert galaxies and notes on the morphology and spectra of individual objects are given by Khachikian & Weedman (1974) and Adams (1977). The relation between Seyferts and other emission-line galaxies has recently been discussed by van den Bergh (1975a) and Weedman (1976a). Osterbrock (1977a) compares the properties of Seyfert and radio galaxies. The role of Seyfert galaxies in the redshift controversy is discussed by Weedman (1976b). The present review emphasizes the description of the Seyfert galaxy phenomenon as synthesized from recent observational studies; most of this summary is based on work done since 1970. Unfortunately, we have so far been unable to explain much of what we see.

2 SEYFERT GALAXIES CURRENTLY KNOWN

After looking at only a few galaxies through a telescope, the distinctive properties of Seyfert galaxies are readily apparent. An ordinary galaxy, spiral or elliptical, has a concentration of light in the center which is usually called a "nucleus." In such cases, there is a gradual brightening of the galaxy toward the nucleus, and the surface brightness at the center is much fainter than that of a star whose magnitude is comparable to the total magnitude of the galaxy. In any photograph, no matter how short the exposure, such a galaxy looks fuzzier than a star. A Seyfert galaxy, however, can readily be recognized at the telescope (which greatly simplifies the use of finding charts) because it looks like a bright star surrounded by a faint envelope. In short-exposure photographs, it cannot usually be distinguished from a star. Consequently, the morphological definition of a Seyfert galaxy is the presence of an unresolved, very bright nucleus. The extent to which the nucleus dominates the galaxy depends both on the nucleus/galaxy luminosity ratio and the distance to the system. Beyond redshifts of about 0.1, objects bright enough to have been identified as Seyferts probably have a nucleus so dominant in photographs that the galactic envelope is unclassifiable. While the lowest redshift Seyferts have been classified morphologically, there have been only limited attempts to define Seyfert galaxies on the basis of their photographic appearance (de Vaucouleurs & de Vaucouleurs 1968, Morgan et al. 1971). There are few direct measures of the actual sizes of Seyfert nuclei. The largest (NGC 1068) has emission lines arising over a region 8″ in diameter (Walker 1968) although the continuum source is unresolved. Photographs of NGC 4151 from a balloon-borne telescope give an upper limit of 0″.08 to the diameter of the continuum source (Schwarzschild 1973). VLBI results for 3C 120 measure individual radio sources, which presumably are no smaller than the optical source, to be as small as 0″.001 (Kellermann et al. 1973). Many Seyfert nuclei are variable in brightness on time scales of years, which sets limits to their sizes if light-travel-time arguments are used.

There is also a spectroscopic definition of a Seyfert galaxy, which requires that they have strong and broad emission lines in their spectra (Seyfert 1943). Such lines are never seen in galaxies without bright nuclei, but there are some galaxies with bright nuclei that do not have such emission lines. In my experience, there are no galaxies without Seyfert spectroscopic characteristics whose nuclei are as bright relative to the galactic disk as they are in Seyferts. However, this opinion has not been quantitatively verified. Certainly, the presence of a compact luminosity source is a necessary prerequisite for having broad emission lines. This spectroscopic definition seems uncomfortably qualitative until it is realized that there are few ambiguous cases. Many galaxies have emission lines in their spectra, but these are generally narrow lines with widths less than a few hundred km sec^{-1}. Seyfert galaxies have emission lines with full widths of 10^3 km sec^{-1} or more, which means that only in a few cases is it necessary to argue whether an emission line is broad enough to call a galaxy a Seyfert. Astrophysically, the broad emission lines are just one indication of the activity that distinguishes Seyfert galaxies. There are other properties that are found to correlate with the Seyfert definition, such as the nature of the continuous spectra.

It is inappropriate to include too many parameters within a definition, so for this review a Seyfert galaxy is considered to be any object appearing nonstellar (nebulous) in photographs and having broad emission lines in its spectrum. This can include N galaxies as well as objects called QSOs, if they are accompanied by surrounding nebulosity. The implication will be that the nebulosity is in fact the disk of a galaxy, but this requires proof and so is irrelevant to the observational definition. Because the most homogeneous photographic data generally available for galaxies is that of the *National Geographic Society—Palomar Observatory Sky Survey*, the working definition of a Seyfert galaxy in this review requires that it appear nonstellar on the *Sky Survey*. This means in practice that it must have a diameter greater than about 7″ for the nebulosity around the nucleus to be visible. We therefore are not including objects occasionally called galaxies because improved photographs show nebulosity associated with them (e.g. Kristian 1973). In Table 1, a list is given of 88 Seyfert galaxies that satisfy our working definition. This list is an update of that given by Khachikian & Weedman (1974) with 23 galaxies having been added and 6 deleted because there was uncertainty in their initial classification as Seyferts. Other than such updating, the differences between Table 1 and the Seyfert listing by Vorontsov-Velyaminov & Ivanisevich (1974) arise because the latter authors include a number of narrow-line galaxies. *The listing in Table 1 is conservative, generally omitting galaxies referred to by spectroscopists as probably or possibly being Seyferts* unless *UBV* photometry also showed the characteristic Seyfert colors. Galaxies are included only if a published reference contains their Seyfert designation, although a few unpublished Seyferts exist in various private collections. Redshifts are given in km sec^{-1}, as *cz*. The *UBV* colors are given for the smallest apertures used so as to represent primarily the nuclei. Magnitudes are not listed because they are dependent both on aperture and epoch. Most are between 14 mag and 16 mag although a few NGC objects are as bright as 11 mag. Angular diameters θ'' are the maximum diameters that I measured on the *Sky Survey* prints, except for a few southern objects for which other photographs were used. References to photometry, spectroscopy, and other studies are usually included only for work published since 1970. No morphological properties are listed in Table 1 because these are summarized by Adams (1977). This list is intended to include those Seyfert galaxies reported in the literature before Sept. 1, 1976, but will certainly be incomplete when published. For example, two new Markarian lists (through the 800 series) were in press as of this date and may contain perhaps 20 more Seyfert galaxies.

3 SURVEYS AND MORPHOLOGY

In 1943, Seyfert studied six galaxies that had been noted to have unusual emission lines in the collection of Mt. Wilson spectra accumulated for redshift studies. The sample did not increase significantly for 25 years, until the Markarian and Zwicky

Table 1 Listing of Seyfert galaxies

Object	RA (1950)	δ	Class	cz	θ''	Chart	$(B-V)$	$(U-B)$	UBV	Spectra	IR or Radio
Mkn 335	$00^h03^m45.1^s$	$+19°55'27''$	1[a]	7500	16	1	0.41	−0.70	6	7, 5, 9, 23, 160	8
III Zw 2	00 08.0	+10 42	1	26930	12	2	0.52	−0.70	5, 135	2, 5, 73, 23, 160	161
Zwicky	00 39.5	+40 03	1	30780	8	3	—	—	136	3, 73	—
Mkn 348	00 46 04.4	+31 41 00	2	4200	77	1	0.95	+0.17	6	7, 5, 9, 162	8, 33, 47, 60, 155, 161
I Zw 1	00 51.0	+12 25	1	18280	22	4	0.36	−0.80	4, 5, 93, 135	10, 5, 11, 23, 73, 132, 160	8, 45, 51
Mkn 352	00 57 08.6	+31 33 27	1[a]	4500	38	1	0.44	−0.66	6	7, 5, 9, 23, 160	8
Tololo	01 09	−38.3	2	3300	135	12	—	—	—	12	—
Mkn 1	01 13 19.5	+32 49 33	2[a]	4800	34	13	0.90	+0.09	6	5, 14, 15, 30, 162	8, 30, 33, 45, 47, 155, 161
II Zw 1	01 19.5	−01 18	1	16240	15	4	0.57	−0.42	4, 5	4, 5, 23, 160	—
Akn 42	01 21.9	+31 55	1	10800	38	16	—	—	—	17	—
Mkn 358	01 23 45.1	+31 21 13	1	13750	42	1	0.79	−0.32	5	7, 18, 23, 160	—
4C 29.6	02 04 08.9	+29 16 40	1	32700		101	—	—	—	102	—
Mkn 590	02 12.0	−01 00	1	8100	55	20	—	—	—	19, 160	—
Akn 79	02 14.3	+38 11	1	6000	60	16	—	—	—	17	—
Akn 81	02 20.4	+31 58	1	10500	32	16	—	—	—	17	—
NGC 985	02 32.1	−08 59	1	12950	45	72	0.36	−0.95	21	21	—
NGC 1068	02 40.1	−00 14	2[a]	1090	380	72	1.00	−0.01	22, 77, 26, 124	73, 52, 57, 96, 71, 128, 138, 118, 35, 94, 42, 49, 162	59, 48, 45, 34, 77, 107, 108, 110, 111, 127, 134, 125, 161
Mkn 372	02 46 30.9	+19 05 54	2	9300	23	1	1.05	−0.03	6	7, 5, 9, 162	8
NGC 1275	03 16.5	+41 20	?	5290	68	72	0.64	−0.23	22, 77, 124	89, 73, 71, 9, 5, 94, 79, 81	77, 45, 44, 8, 77, 112, 91, 137, 161
Mkn 609	03 22.9	−06 19	1	9600	17	24	—	—	—	19, 23	—

III Zw 55	03 38.7	−01 27	2	7380	13	4	0.84	+0.17	4	4, 5, 9, 162	8
NGC 1566	04 18.9	−55 04	1[a]	1170	420	25	0.76	−0.04	26, 27, 133	25, 28, 73, 131	117, 133, 161
3C 120	04 30.0	+05 15	1	9900	42	2	0.58	−0.75	6, 77, 99, 124, 130	58, 56, 9, 73, 23, 106, 160	8, 45, 77, 41, 43, 140, 141, 161
Mkn 618	04 34.0	−10 28	1	10200	68	24	—	—	—	23, 160	—
Akn 120	05 13.6	−00 12	1	9900	60	16	—	—	—	17	—
Mkn 3	06 09 48.1	+71 03 00	2[a]	4110	44	13	1.15	+0.15	6	5, 15, 14, 9, 30, 162	30, 33, 47, 155, 161
Mkn 6	06 45 43.4	+74 29 07	1[a]	5290	52	13	0.97	+0.02	6	55, 9, 15, 30, 87, 88, 122, 162	30, 33, 47, 105, 161
Mkn 374	06 55 33.9	+54 15 53	1	13200	41	1	0.70	−0.38	6	7, 5, 9, 23, 160	8, 161
Mkn 376	07 10 35.8	+45 47 07	1	16800	18	1	0.55	−0.58	5	31, 9, 32, 23, 160	8
Mkn 9	07 32 42.0	+58 53 00	1	12040	22	13	0.54	−0.64	6	30, 14, 9, 23	30, 8, 45
Mkn 78	07 37 55.9	+65 17 43	2[a]	11260	19	38	0.99	−0.31	6	36, 37, 9, 162	161
Mkn 79	07 38 46.9	+49 55 47	1	6580	86	38	0.47	−0.74	39, 6, 66	37, 30, 9, 23, 5, 115, 160	30, 8, 45, 51, 161
Mkn 10	07 43 07.4	+61 03 23	1	8790	95	13	0.67	−0.55	39, 6, 124	14, 30, 23, 160	30, 45
Mkn 382	07 52 03.2	+39 19 07	1	10200	42	1	0.51	−0.63	5	32, 23, 160	8
Mkn 110	09 21 44.4	+52 30 14	1[a]	10800	47	38	0.76	−0.67	6	40, 5, 9, 160	8, 161
Zwicky	09 34.5	+01 20	1	15150	—	—	—	—	—	139	—
3C 227	09 45 06.3	+07 39 17	1[a]	25600	8	29	0.98	−0.36	65	54, 23, 160	161
Mkn 124	09 45 24.3	+50 43 26	1	16970	18	38	0.61	−0.47	66	37, 30, 9, 160	30, 8, 161
Akn 223	09 54.7	+07 26	1	6600	21	16	—	—	—	67	—
Mkn 141	10 15 38.7	+64 13 14	1	11700	19	38	0.71	−0.32	6, 66	40, 9, 160	—
NGC 3227	10 20.7	+20 07	2[a]	1000	330	72	0.82	−0.11	22, 77	69, 9, 71, 73, 5, 118, 160	45, 8, 77, 34, 125, 127, 126, 161
Mkn 142	10 22 23.1	+51 55 40	1	13500	24	38	0.44	−0.58	66	40, 160	8
Ton 524a	10 28.8	+29 06	1	18000	10	121	—	—	—	121	—
Mkn 34	10 30 52.2	+60 17 20	2[a]	15300	39	13	1.06	+0.07	6	15, 9, 30, 162	30, 45, 161
Akn 253	10 41.4	−01 01	1	7800	34	16	—	—	—	67	—

Table 1—*continued*

Object	RA (1950)	δ	Class	cz	θ''	Chart	$(B-V)$	$(U-B)$	UBV	Spectra	IR or Radio
NGC 3516	11 03.4	+72 50	1[a]	2780	80	72	0.77	−0.18	22, 77, 123, 125	84, 73, 68, 71, 9, 119, 128, 138, 160	45, 77, 126, 161
VV 144	11 22 48.0	+54 39 26	1	6150	16	13	0.82	−0.23	6	4, 14, 73, 160	—
Mkn 176	11 29 54.0	+53 13 27	2	8080	25	38	0.89	+0.30	6	37, 5, 9, 55, 162	8, 51, 161
NGC 3783	11 36.5	−37 28	1[a]	2740	72	72	0.56	−0.70	25	25, 28	61, 117
Mkn 42	11 51 05.3	+46 29 20	1	7200	44	13	0.79	−0.19	14	75, 9, 162	8
NGC 4051	12 00.6	+44 48	1	700	280	72	0.73	−0.38	22, 77	71, 73, 5, 118, 128, 138	45, 8, 77, 125, 127, 126, 161
NGC 4151	12 08.0	+39 41	1	990	450	72	0.53	−0.72	22, 77, 124	82, 71, 55, 86, 35, 128, 109, 114, 138, 83, 53, 100, 95, 73	77, 45, 44, 34, 8, 108, 120, 125, 127, 126, 161
Mkn 50	12 20 50.9	+02 57 20	1	6910	16	13	—	—	—	74	—
NGC 4507	12 32.9	−39 38	2	3310	68	72	—	—	—	62	117
Mkn 231	12 54 05.0	+57 08 37	1	12300	27	142	0.84	+0.15	6	78, 142, 143, 5, 149	45, 44, 33, 47, 51, 90, 155, 161
X Comae	12 57 58	+28 40.2	1	27600	10	103	—	—	104, 116	104	—
Mkn 236	12 58 18.0	+61 55 27	1	15600	30	144	0.66	−0.45	5	145, 160	—
Mkn 64	13 04 47.7	+34 40 23	1	55200	8	13	0.41	−0.74	146	146	—
3C 287.1	13 30 21.1	+02 16 11	1	64700	8	29	0.92	−0.15	148	147	161
Mkn 268	13 38 54.2	+30 37 47	2	12300	36	144	0.94	+0.28	6	149, 5, 9, 162	8, 161
Mkn 270	13 39 40.7	+67 55 33	2	2700	45	144	0.92	+0.31	6	149, 5, 162	161
Mkn 69	13 43 51.3	+29 53 03	1	22870	11	13	—	—	—	74, 160	—
IC 4329A	13 46.2	−30 03	1[a]	4140	70	80	0.96	+0.51	80	80, 28	61, 161
Mkn 279	13 51 51.9	+69 33 13	1	9220	43	144	0.69	−0.45	6	75, 5, 9, 149, 160	161
Mkn 463	13 53 39.8	+18 36 40	2	15140	94	151	—	—	—	150	33, 47, 161
Mkn 464	13 53 45.1	+38 48 54	1	15300	13	151	—	—	—	150	—

NGC 5548	14 15.7	+25 22	1[a]	4990	80	72	0.46	−0.82	77, 123, 124, 22	118, 71, 68, 70, 9, 73, 128, 160	45, 77, 126, 8, 161
Mkn 474	14 33 06.0	+48 52 47	1	12300	22	151	0.97	−0.18	5	152	8
Mkn 478	14 40 04.6	+35 38 53	1	23700	16	151	0.33	−0.84	5	19, 160	8
4C 35.37	15 31 44.9	+35 54 18	1	46950	8	101	—	—	—	102	—
Mkn 290	15 34 45.4	+58 04 00	1	9240	14	144	0.60	−0.62	6	75, 149, 160	8, 161
Mkn 486	15 35 21.5	+54 43 04	1	11700	25	151	0.43	−0.68	5	19, 9, 160	8
Mkn 291	15 52 54.1	+19 20 20	1	10500	35	144	—	—	—	149, 160	—
Mkn 298	16 03 21.7	+17 56 03	2	10350	42	144	1.01	−0.09	6	75, 149, 162	161
Mkn 504	16 59.2	+29 29	1	10800	19	151	0.85	−0.47	5	153, 160	8
Mkn 506	17 20.7	+30 56	1	12900	34	151	0.74	−0.55	5	19, 5, 160	8
3C 382	18 33 11.9	+32 39 18	1[a]	17580	15	29	—	—	—	54, 64, 160	—
3C 390.3	18 45 38.7	+79 43 10	1[a]	17100	12	29	0.68	−0.69	65, 97, 135	54, 64, 85, 23, 92, 160	161
NGC 6764	19 06.9	+50 51	2	2405	128	72	—	—	—	154, 162	161
NGC 6814	19 39.9	−10 25	1	1590	136	72	1.12	+0.37	77, 113, 5	46, 28	45, 77, 8
Mkn 509	20 41.5	−10 54	1	10650	17	20	0.23	−0.93	8	19, 55, 23, 160	8, 51
II Zw 136	21 30.0	+09 56	1	18510	30	156	0.29	−0.90	4, 135, 124, 6	4, 5, 132, 23, 160	51, 8
Mkn 304	22 14 45.2	+13 59 27	1	19950	11	1	0.36	−0.86	5	7, 5, 23, 160	8
NGC 7469	23 00.7	+08 36	1[a]	5020	89	72	0.54	−0.72	77, 124, 22	73, 94, 28, 159, 50, 5, 118, 23, 98, 71, 148, 160	45, 77, 8, 34, 161
Mkn 315	23 01 35.6	+22 21 10	1	11830	19	1	0.82	−0.09	5	18, 7, 5, 162	161
NGC 7603	23 16.4	−00 01	1[a]	8800	55	72	0.72	−0.21	6	129, 76, 5, 23, 9	—
Pks	23 49 22.5	−01 26.0	1	52200	12	158	—	—	—	157	157, 161
Mkn 541	23 53.5	+07 15	1	12300	38	20	—	—	—	19, 23, 160	

[a] Asymmetries have been noted in line profiles.

Table 1—*continued*

References to Table

1. Markarian & Lipovetsky 1971
2. Arp 1968
3. Zwicky et al. 1970
4. Sargent 1970a
5. Khachikian & Weedman 1974
6. Weedman 1973
7. Arakelian et al. 1972b
8. Stein & Weedman 1976
9. Adams & Weedman 1975
10. Sargent 1968
11. Phillips 1976
12. Smith 1975
13. Markarian 1967
14. Weedman & Khachikian 1969
15. Weedman 1970
16. Arakelian 1975
17. Arakelian et al. 1975
18. Huchra & Sargent 1973
19. Kopilov et al. 1974
20. Markarian & Lipovetsky 1973
21. de Vaucouleurs & de Vaucouleurs 1975
22. Zasov & Lyutyi 1973
23. Weedman 1976c
24. Markarian & Lipovetsky 1974
25. Osmer et al. 1974a
26. Smith et al. 1972
27. de Vaucouleurs 1973
28. Martin 1974
29. Wyndham 1966
30. Neugebauer et al. 1976
31. Osterbrock 1976
32. Arakelian et al. 1972c
33. Kojoian et al. 1976
34. de Bruyn & Willis 1974
35. Glaspey et al. 1976
36. Adams 1973
37. Sargent 1972
38. Markarian 1969a
39. Dibay & Lyutyi 1971
40. Arakelian et al. 1970
41. Schilizzi et al. 1975
42. Richstone & Morton 1975
43. Seielstad 1974
44. Rieke & Low 1975a
45. Rieke & Low 1972
46. Ulrich 1971
47. Sramek & Tovmassian 1975
48. Rieke & Low 1975b
49. Eilek et al. 1973
50. Anderson 1973
51. Allen 1976
52. Shields & Oke 1975a
53. Netzer & Penston 1976
54. Osterbrock et al. 1976
55. Osterbrock & Koski 1976
56. Shields et al. 1972
57. Angel et al. 1976
58. Phillips & Osterbrock 1975
59. Telesco et al. 1976
60. Sramek & Tovmassian 1974a
61. Kleinmann & Wright 1974
62. Martin 1976
63. Griffin 1963
64. Osterbrock et al. 1975
65. Sandage 1967
66. Dibay 1970
67. Doroshenko & Terebizh 1975
68. Ulrich 1972a
69. Rubin & Ford 1968
70. Anderson 1971
71. Anderson 1970
72. Sulentic & Tifft 1973
73. Burbidge 1970
74. Sargent 1970b
75. Khachikian & Weedman 1971
76. Tohline & Osterbrock 1976
77. Penston et al. 1974
78. Boksenberg et al. 1977
79. Pronik 1974a
80. Disney 1973
81. Pronik 1974b
82. Boksenberg et al. 1975a
83. Weedman 1971
84. Boksenberg et al. 1975b
85. Penston & Penston 1973
86. Anderson 1974a
87. Adams 1972a
88. Pronik & Chuvaev 1972
89. Shields & Oke 1975b
90. Joyce et al. 1975

91. DeYoung et al. 1973
92. Burbidge & Burbidge 1971
93. Usher et al 1971
94. Wampler 1971
95. Nussbaumer & Osterbrock 1970
96. Schild 1972
97. Shen et al 1972
98. Ulrich 1972b
99. Usher 1972
100. Ulrich 1973
101. Olsen 1970
102. Sargent 1973
103. Bond 1973
104. Bond & Sargent 1973
105. Sramek & Tovmassian 1974b
106. Shields 1974
107. Jameson et al. 1974
108. Stein et al. 1974
109. Anderson 1974b
110. Knacke & Capps 1974
111. Jones & Stein 1975
112. Niell et al. 1975
113. MacPherson 1972
114. Netzer 1974
115. Yankulova et al. 1974
116. Barbieri 1973
117. Glass 1973
118. Yankulova 1975
119. Collin-Souffrin et al. 1973
120. Davies 1973
121. Robinson & Wampler 1973
122. Khachikian 1973
123. de Vaucouleurs & de Vaucouleurs 1972
124. Lyutyi 1973
125. Lewis 1972
126. Van der Kruit 1971
127. Allen et al. 1971
128. Andrillat & Souffrin 1971
129. Arp 1971
130. Jurkevich et al. 1971
131. Pastoriza & Gerola 1970
132. Oke & Shields 1976
133. Quintana et al. 1975
134. Simon & Dyck 1975
135. Selmes et al. 1975
136. Barbieri et al. 1976
137. Miley & Perola 1975
138. Andrillat & Collin-Souffrin 1975
139. Ulrich 1975
140. Shapiro et al. 1973
141. Kellermann et al. 1973
142. Adams & Weedman 1972
143. Adams 1972b
144. Markarian 1969b
145. Arakelian et al. 1972a
146. Bracessi et al. 1968
147. Sandage 1966
148. Pronik 1975
149. Arakelian et al. 1971
150. Denisyuk & Lipovetsky 1974
151. Markarian & Lipovetsky 1972
152. Arp & Khachikian 1973
153. Arakelian et al. 1973
154. Rubin et al. 1975
155. Sramek & Tovmassian 1976
156. Fairall 1968
157. Searle & Bolton 1968
158. Bolton & Ekers 1966
159. Shectman & MacAlpine 1975
160. Osterbrock 1977b
161. de Bruyn & Wilson 1976
162. Koski 1976

surveys. Beginning in 1967, Markarian has published seven lists containing about 700 galaxies (with two more lists in press as of this writing) distinguished by their unusually strong ultraviolet continua (Markarian 1967, 1969a, 1969b, Markarian & Lipovetsky 1971, 1972, 1973, 1974). (Improved positions for some of these are given by Peterson 1973.) These objects have a space density of about one per ten square degrees and were found on low-dispersion, objective prism survey plates obtained with the 1-m Schmidt at Byurakan Observatory. Subsequent studies showed that the majority of the Markarian galaxies have emission-line spectra (see references in Weedman 1976a). The Markarian survey reaches to 17 mag but seems to be complete only to about 16 mag (Huchra & Sargent 1973). Approximately 10% of the Markarian galaxies are Seyferts, and the subsequent spectroscopic discovery observations of the Seyferts thus found are referenced in Table 1. In contrast to Markarian's spectroscopic survey techniques, Zwicky compiled lists of compact galaxies based on their morphological appearance on the *Sky Survey* plates. He selected objects that were almost stellar, having soft edges or faint associated nebulosity. These lists were never published, but spectroscopic surveys primarily by Sargent (1970a) of a number of Zwicky's compact galaxies revealed several Seyferts.

Other surveys analogous to these are now underway and have already provided some Seyfert galaxies. A southern hemisphere objective prism survey is being conducted with the 61-cm Curtis Schmidt telescope at Cerro Tololo (Smith 1975, 1976, Smith et al. 1976). The Tololo survey has higher resolution than the Markarian survey and selects objects because they have emission lines in their spectra. An important variety of emission-line galaxies have been found this way; observations so far available indicate that similar sorts of galaxies are found in the Tololo lists as in the Markarian lists. The Tololo survey as instituted and carried out by M. G. Smith covered approximately 3000 square degrees and resulted in the discovery of about 500 emission-line galaxies, whose publication is beginning (Smith et al. 1976). Another portion of the southern sky ($-20° \leqq \delta \leqq 8°$) is being surveyed in the same way with the Curtis Schmidt by astronomers from the University of Michigan, which owns this telescope. Several hundred emission-line galaxies have been found in this region, and they are accessible from the north (G. MacAlpine, personal communication). Large-scale follow-up studies of all these Tololo galaxies, expected to yield numerous new Seyferts, are underway.

Morphologically oriented surveys have also been conducted recently. A list of 591 galaxies of high surface brightness has been prepared by Arakelian (1975). He estimated the surface brightnesses by comparing the magnitude measures in the Zwicky catalogs of galaxies (*not* the Zwicky compact galaxies) to the diameter measures in the Vorontsov-Velyaminov morphological catalogs of galaxies. Six new Seyferts have already been found in Arakelian's lists (Table 1). A spectroscopic survey of southern galaxies classified as having bright nuclei was carried out by Martin (1976), who found one new Seyfert in this way.

It is important, of course, to study the galaxies with bright nuclei that are not called Seyferts in hopes of finding just when and how the Seyfert activity turns on. We do not know, for example, whether there are galaxies with narrow emission lines that have nonthermal continuous spectra like the Seyferts. Certainly there are bright nucleus galaxies like NGC 4385 (Mkn 52) and NGC 7714 (Mkn 538) which have very strong but narrow emission lines with intensity ratios like those in conventional H II regions. The *UBV* colors correspond to collections of hot stars (Weedman 1973, Huchra 1976), but some of these galaxies have infrared luminosities comparable to the Seyferts (Rieke & Low 1972). Another intriguing result is that interacting galaxies seem to have strong nuclear emission lines unexpectedly often. Representative examples are NGC 2992 (Osmer et al. 1974b) and the aforementioned NGC 7714 (Burbidge 1968). Adams (1977) also noted a probable surplus of interacting systems among the Seyferts themselves. The large samples of galaxies from the Markarian and Tololo surveys may allow us to determine how such galaxies relate to the Seyfert problem.

From the number of Seyfert galaxies among bright Shapley-Ames galaxies, it was concluded that 1% of spiral galaxies are Seyferts (de Vaucouleurs & de Vaucouleurs 1968). This estimate remains valid for the larger sample now available, because Huchra & Sargent (1973) concluded that 5% of all galaxies were Markarian galaxies, of which 10% are Seyferts. Consequently, about 1 galaxy in 200 (of all types) is a Seyfert. Why? Do 1% of spiral galaxies remain Seyferts forever, or does each galaxy spend 1% of its life in the Seyfert phase? The current research relevant to this problem is into the morphological appearance of Seyfert galaxies to determine whether the presence of a Seyfert nucleus is correlated with any other property of a galaxy.

The most striking empirical result is a deficiency of ellipticals among the Seyferts. There are ten galaxies now known to be Seyferts that are bright enough to be Shapley-Ames galaxies. Van den Bergh (1975b) has emphasized the fact that all ten are spirals. A comprehensive morphological study of 80 Seyfert galaxies was recently completed by Adams (1977). He finds that Seyferts occur in a broad distribution of ordinary and barred spirals but that there are very few ellipticals. Counting those galaxies with amorphous but unclassifiable envelopes, Adams finds a maximum allowable limit of 10% for the fraction of Seyferts that could be ellipticals, but the more reasonable limit is 5%. By contrast, 33% of the galaxies in Humason et al. (1956) are ellipticals. Because many of the Seyfert galaxies have such small angular sizes as to be unclassifiable, it is possible that none are ellipticals. The only Seyfert that can be called a giant elliptical because of its dominant appearance in a cluster is NGC 1275. It has many other properties that are unique among Seyferts, so much so that it is clearly an anomalous object. (Had it not been included in Seyfert's original study, it probably would not have been grouped with the other galaxies called Seyferts today, but more likely would be considered a peculiar radio galaxy.) Adams and van den Bergh discuss the implications of this elliptical deficiency for our understanding of QSOs, as the morphology of Seyferts is in dramatic contrast to that of the radio galaxies.

Previous studies had pointed out the peculiar appearance of certain classical Seyferts, especially the presence of faint annular structures (Burbidge et al. 1963, Hodge 1968). Adams adds some examples of these as well. Such effects could be important for demonstrating interactions between the disk and the nucleus, although they would not show whether the nuclear activity triggered changes in the disk or vice-versa. A possible correlation between the spectroscopic properties of the nuclei and the appearance of the galactic disk was suggested by Khachikian & Weedman (1971). To some degree, Adams confirms this in the sense that no Seyferts of spectroscopic class 2 (see below) have spiral structure as conspicuous as that in some of class 1. The only suggested explanation so far for the various peculiarities referenced is that mass outflow from the nuclei of the Seyferts is somehow affecting the galactic disks.

The above-mentioned morphological studies, which demonstrate the spiral nature of many Seyfert galaxies, are of special importance regarding the redshift controversy. Continuity arguments demonstrating a luminosity overlap between Seyferts and QSOs provide empirical evidence for cosmological redshifts of QSOs, if the Seyfert redshifts are cosmological. It is therefore imperative to show that some "Seyfert galaxies" whose cosmological luminosities would compare to QSOs really are distant galaxies. The presence of spiral arms is an acceptable morphological indication that a nebulous object really is a galaxy. For more amorphous envelopes, it has been suggested that absorption lines must be detected in the envelopes before they can be considered as galaxies made of stars (Burbidge 1973). Such absorption lines are difficult to detect. No such lines are found even in the outer disk of the spiral Seyfert NGC 4151 (Simkin 1975). However, extensive photometric studies of the disks of the larger Seyferts show that their colors correspond to those expected for galaxies of stars (Penston et al. 1974, Dibay & Lyutyi 1971, Walker et al. 1974, Zasov & Lyutyi 1973). For higher redshift Seyferts, a morphological demonstration that real galaxies are present is difficult. Adams (1977) finds that only 21% of all Seyferts are unresolved or have amorphous main bodies, but that this fraction is 80% for those with redshifts above 20,000 km sec^{-1}. This is as expected, of course, if the Seyferts are really galaxies whose angular size decreases in proportion to their distance.

4 EMISSION-LINE SPECTRA

Within the past five years, extraordinary progress has been made in the accumulation of spectrophotometric data for Seyfert galaxies. This is because of the multichannel spectrum scanners now available for most large telescopes. Technical details of these instruments can be found in some of the references included in Table 1, primarily the work involving Osterbrock, Oke, or Boksenberg. Earlier studies of bright Seyferts have been repeated with higher accuracy, and precision studies of faint spectral features as well as extensive observations of previously unstudied objects have begun. Perhaps a measure of the progress in our knowledge of Seyfert galaxies is the fact that it is no longer feasible to discuss each one individually in a review. Consequently, recent spectrophotometric studies of the various Seyferts are referenced in Table 1. The review that follows is a synthesis of the generalities that can be deduced from this data.

As the sample of Seyfert galaxies grew, it became obvious that all such objects by no means had identical spectra. The Seyferts were therefore subclassified strictly on the basis of their spectroscopic properties in an attempt to identify those subgroups with similar physical properties. This, of course, is the motivation for any astronomical classification scheme. Because Seyfert galaxies are characterized by strong and broad emission lines, the nature of these lines was used as the classification criterion. A simple division into classes 1 and 2 (Sy 1 and Sy 2), which depended only on *relative* emission-line widths, was used by Weedman (1970, 1973) and Khachikian & Weedman (1971, 1974). Illustrations of spectra and line profiles are given in these references. The Sy 1 have broad hydrogen lines but narrower forbidden lines, such as the classical Seyfert NGC 5548. Other permitted lines are also broad when the Balmer lines are broad though they do not necessarily have the same profiles (Boksenberg et al. 1975a, Osterbrock 1976, Osterbrock et al. 1976). The only such lines observed are He I λ5876, He II λ4686, occasional blends of Fe II lines, and OI λ8446. These features are so weak relative to the Balmer lines that they have not been measured in all Sy 1. It is probably appropriate to infer that the other permitted lines arise in much the same volume as do the Balmer lines. Comprehensive spectrophotometric data on 40 Sy 1 galaxies are given by Osterbrock (1977b). This includes relative line intensities as well as line profile widths.

The Sy 2 are like NGC 1068, having hydrogen and forbidden lines of the same width, up to about 10^3 km sec^{-1}. The classification into Sy 1 or Sy 2 is most easily made just by comparing Hβ with an adjacent [OIII] line. The simplest explanation for the spectroscopic difference between Sy 1 and Sy 2 is that the broad Balmer lines characterising Sy 1 arise in a different part of the nucleus than do the narrower forbidden lines, whereas all lines arise together in Sy 2. There is little ambiguity as to which galaxies are Sy 1, where the Balmer lines often have full widths of 10^4 km sec^{-1}. The problem in deciding whether to call a galaxy a Seyfert is in defining the line width at which an Sy 2 is distinguished from other emission-line galaxies. Those few objects classified Sy 2 whose line profiles have been measured show full widths at half maximum intensity greater than 500 km sec^{-1} (Weedman 1970), whereas such widths for representative narrow-line galaxies (from the same study) did not exceed 200 km sec^{-1}. The ambiguity between Sy 2 and narrow-emission-line galaxies is most serious in low-resolution spectroscopic surveys. For this reason, there are probably a number of Sy 2 in the Markarian and Tololo surveys that have not yet been observed with sufficient resolution to be classified and so are not included in Table 1. To date, the most detailed study of Sy 2 galaxies is by Koski (1976).

After the profile classification was adopted, other properties of the nuclei were found to correlate well with the Sy 1 and Sy 2 classes. A useful secondary classification is the Balmer-line to forbidden-line intensity ratio, especially Hβ:[OIII]. The forbidden lines are unusually strong in Sy 2, with an average Hβ:λ5007 ratio of 0.1, whereas Hβ and λ5007 typically have the same total flux for Sy 1 (Adams & Weedman 1975). For the narrow-emission-line galaxies, Hβ is also comparable to λ5007 in most cases. Correlations with *UBV* colors also exist because the continuous spectra for Sy 1, Sy 2, and narrow-line galaxies are different. It is then possible to use *UBV* colors as a secondary classification (Markarian 1973). Color-color diagrams showing results for Seyfert galaxies along with comparative colors of narrow-line galaxies are summarized in Weedman (1973) and Adams & Weedman (1975). Extensive *UBV* observations of narrow-line galaxies have been made by Huchra (1976), and these can be compared to the Seyfert colors in Table 1. The *UBV* colors provide reasonable distinctions in most cases among Sy 1, Sy 2, and narrow-line galaxies even if high-resolution line profiles are not available.

It is the Sy 1 that have power-law continuous spectra and the resulting ultraviolet excesses like QSOs. This correlation between the presence of very broad Balmer emission lines and a nonthermal continuum is one of the more interesting results established. Searle & Sargent (1968) pointed out the similar equivalent widths of the broad Balmer lines for Seyferts then known and showed that these were too small for the continuum to be emission from the same gas producing the Balmer lines. Adams & Weedman (1975) showed quantitatively that Balmer-line luminosities and continuum luminosities scale together for Sy 1 over six magnitudes of luminosity. They concluded that the Balmer lines therefore arise primarily from ionization by a radiative source that also produces the visible continuum, in accordance with models such as those of Williams & Weymann (1968). Extensions of the observed power-law continua into the ionizing ultraviolet generally provide enough photons to account for the Balmer-line strengths (Osterbrock 1971, 1976, Stein & Weedman 1976). The nature of the continuous spectra in Sy 2 is not as clear because they are affected significantly by starlight. The probable presence of heavy dust reddening and a possible nonthermal contribution means that the nature of Sy 2 continua is not as well understood as Sy 1.

If cosmological redshifts are assumed, the observed forbidden-line luminosities of Sy 1 and Sy 2 range over comparable values, whereas the Balmer lines can be up to ten times more luminous in Sy 1 (Adams & Weedman 1975). Osterbrock & Koski (1976) have emphasized that the broad Balmer-line wings provide the distinction between Sy 1 and Sy 2. The Balmer-line cores in Sy 1 have similar profiles and intensity ratios relative to forbidden lines as in the Sy 2. So, observationally at least, an Sy 2 could be transformed to an Sy 1 by adding a strong power-law continuum accompanied by the broad Balmer-line wings. Conversely, removing these would change Sy 1 to Sy 2. As Osterbrock & Koski put it, "Galaxies in which the broad Balmer line component is relatively strong are classified as Sy 1, while galaxies in which it is invisible are classified as Sy 2. . . . Physically, different Seyfert galaxies must have different relative amounts of the material and physical conditions that emit these two components." They suggest that the Sy 1 have a higher proportion of very dense gas in which the broad wings arise. A similar conclusion was stated by Neugebauer et al. (1976). However, the ease with which an Sy 2 could be converted to an Sy 1 is not necessarily evidence that there is an evolutionary relation between them. They may be totally unrelated phenomena. One particularly troubling aspect is the implication that Sy 2 may be heavily obscured by dust while Sy 1 are not. In this case, the intrinsic Balmer-line luminosities of Sy 2 would be comparable to Sy 1, and the forbidden lines typically ten times more luminous (Adams & Weedman 1975). Under such circumstances, addition of more Balmer emission would not make the intrinsic properties of Sy 2 match Sy 1.

5 BALMER DECREMENT AND LINE PROFILES

The problem of reddening is directly related to the interpretation of the Balmer decrement in Seyfert galaxies. The Balmer emission in most astrophysical plasmas, such as H II regions and planetary nebulae, arises from radiative ionization followed by recombination. Downward cascades within the newly recombined atoms produce emission lines whose intensity ratios can be calculated. For a nebula optically thick only in the Lyman lines, the ratio of Balmer-line intensities (the Balmer decrement) is $H\alpha : H\beta : H\gamma = 2.8 : 1.0 : 0.47$ for temperatures near 10^4 °K (Miller 1974). In Seyfert galaxies, the decrement is invariably steeper than this, meaning that the redder Balmer lines are relatively more intense than predicted. There are two categories of explanation. One is that some mechanism selectively populates hydrogen energy levels from the ground up, which increases the relative intensity of the redder lines in a series. Such a mechanism could be collisional excitation, radiative excitation by a continuum, or Balmer-line self absorption. That such effects are important under the high-density conditions found in Seyfert nuclei has been shown by Netzer (1975) and Shields (1974). The alternative circumstance is that the Balmer lines are reddened by dust so that the observed decrement is steepened. If sufficient reddening is present to produce the observed decrement, there must be a lot of absorption by the same dust. In some cases, several magnitudes of absorption would be needed in the visible spectrum. This would mean that Seyfert nuclei are even brighter than they appear. Of course, the absorbed energy would have to come out somewhere in the spectrum, and the dust could be expected to reradiate in the infrared. The fact that Seyfert nuclei are strong infrared sources might mean that there is dust in some.

To resolve the decrement problem, it is therefore necessary to consider the effects that reddening would have both on the lines and the continuum. The most accurate spectrophotometric studies to date of the decrements in Sy 1, through the $H\gamma$ line, have been unable to account for the decrements with any assumed amount of normal reddening (Osterbrock et al. 1975, Osterbrock 1976). If dust reddening were responsible for the steep decrements, the continua on average should become redder as the decrement increased. However, there are no empirical correlations between the $H\alpha : H\beta$ ratio and the continuum colors, either in the optical or in the optical vs. the infrared (Adams & Weedman 1975, Stein & Weedman 1976). All of these authors therefore concluded that in the Sy 1 nuclei radiative recombination was not the only contributor to excitation of the Balmer lines. Some other of the effects mentioned above are probably important in the region where the broad Balmer lines arise and account for the unexpectedly steep decrement.

This explanation does not necessarily apply to Sy 2, however. Virtually all of the conclusions about Sy 2 depend on NGC 1068, which has emission-line profiles, relative intensities, and luminosities that are representative of other Sy 2. NGC 1068 also has a steep Balmer decrement; the Sy 2 generally have steeper decrements than the Sy 1. In NGC 1068, relative [SII] line intensities show clearly that some dust reddening is present (Wampler 1968). Such lines are measurable in Sy 2 because of the greater relative intensity of the forbidden lines in these Seyferts.

Substantial recent results are also available relative to another problem of spectroscopic interest, which is the cause of the broadening of the line profiles. For the Sy 2, the similar asymmetric profiles in both Balmer and forbidden lines are considered as proof that real mass motions cause Doppler broadening. The classical study of NGC 1068 showing discrete clouds moving at different velocities set the tone of this interpretation (Walker 1968). The gas sometimes escapes the nucleus and can spread through large volumes, having a kinetic energy of 10^{56} ergs. The source of this energy is unknown. The situation was not so well resolved for Sy 1, however, as a difficulty arises if the very broad Balmer-line wings are attributed to Doppler broadening. This is the containment problem, because the emitting gas has to remain in a nucleus on the order of a parsec in diameter while moving at 10^4 km sec^{-1}. The travel time across the nucleus would be only 10^2 years, so the gas must be replaced as it flows out, or it must somehow be bound. To avoid worrying about this, the broad wings were attributed to electron scattering, i.e. scattering between the Balmer-line photons and the thermal electrons (e.g. Mathis 1970, Weymann 1970). This explanation was acceptable for line wings that are smooth and symmetric. However, there are now many unquestionable examples of Sy 1 with asymmetric or structured Balmer-line wings (Table 1) so it seems necessary to attribute these to mass motions of some sort.

The mechanism that accelerates and contains the gas is unknown. The empirical association between the continuum luminosity source and the broad lines implies that the gas may be accelerated by radiation pressure, such as described by Mathews (1974) and Blumenthal & Mathews (1975). An early suggestion by Woltjer (1959) was that the broad profiles are caused by rotation, which would overcome the containment problem. However, Woltjer assumed that the gas rotated about the nucleus with a radius about 100 pc and inferred large nuclear masses both from assigning the line width to rotational velocity and by assuming the nuclear luminosity came from stars. Rotation curves of Seyfert galaxies deduced later by the Burbidges showed no sign of the large masses needed. The rotation models were therefore not pursued, especially after it was decided that the nuclear luminosities were not from stars and that the nuclei of Sy 1 could be much smaller than 100 pc. Rotation models may nevertheless prove useful if applied to accretion disks around massive, condensed objects (Hills 1975). On a much smaller luminosity scale, such models explain the broad Balmer lines in cataclysmic variable stars (e.g. Warner 1973). (I am indebted to J. S. Gallagher for pointing this out.) Oke & Shields (1976) argue, however, that the Fe II emission in some Sy 1 implies that the gas cannot be in ordered rotation around the continuum source. Regardless of whether the high gas velocities are in random filamentary motions or ordered rotation, the association between gas and luminosity source is further emphasized by the correlation between increasing profile width and increasing nuclear luminosity found in some Sy 1 by Osterbrock et al. (1976).

An extensive set of models for explaining line profiles has been produced by Ptak, Stoner and collaborators (e.g. Ptak & Stoner 1973, Stoner et al. 1974, Hubbard 1975,

Ptak & Stoner 1975, MacAlpine 1974). These are the most detailed attempts so far to account for the line profiles, especially in trying to fit the asymmetries. They describe their model as follows (Ptak & Stoner 1973): A point source injects high-energy protons into a partially ionized gas cloud. The protons are decelerated by interactions, and "after they have slowed to speeds of a few times the orbital velocities of bound electrons in the atoms of the cloud, the cross-section for charge transfer during collisions with the atoms becomes large enough that the streamers spend a significant amount of time as atoms themselves, and can emit photons characteristic of hydrogen." These models still face the same dilemma as other mass motion models, in that there must be an unknown source for supplying and accelerating the protons. Osterbrock et al. (1976), Osterbrock & Koski (1977), and Katz (1975) contend that the proton-streaming models cannot explain the great width of some profiles, or the relative widths of He I, He II, and hydrogen profiles.

It has been accepted for some time that Sy 1 nuclei have a very inhomogeneous distribution of gas, with the broad Balmer-line wings arising in a denser gas than where the Balmer cores and forbidden lines arise (e.g. Oke & Sargent 1968). Because wings have never been seen on any forbidden lines, it is concluded that the Balmer wings arise in gas with $N_e \gtrsim 10^7$ cm^{-3} so that forbidden line radiation is suppressed by collisional de-excitations. Such two-phase models for Sy 1 nuclei have received substantial confirmation from the observation that permitted Fe II emission lines are often found in Sy 1 spectra, but are not accompanied by forbidden Fe II (Sargent 1968, Phillips & Osterbrock 1975, Osterbrock 1976, Phillips 1976, Oke 1972, Boksenberg et al. 1975a, Adams 1975, Oke & Shields 1976). This requires $N_e \gtrsim 10^8$ cm^{-3}. Because of the high densities, only a small mass of gas is needed to produce the broad Balmer-line emission. The luminosity of recombination emission depends on the product (mass) × (density), so the value derived for the mass is determined by that assumed for the density. Adopting $N_e \geqq 10^7$ cm^{-3}, Adams & Weedman (1975) found that no more than 4×10^3 $M_\odot$ of ionized hydrogen was ever required to account for the broad lines in Sy 1. Osterbrock et al. (1976) increased N_e to 10^9 and therefore lowered their maximum mass to 26 $M_\odot$. Clearly, only small amounts of gas are needed to produce the Balmer lines if a sufficient ionization or excitation mechanism is present. Proportionally larger gas masses are needed for the emission in the forbidden-line volume, however, because the density is $\lesssim 10^4$ cm^{-3}. The summary scenario of an Sy 1 nucleus is then a highly localized continuum source, accompanied by high-density clouds or filaments that produce the broad Balmer lines. All of this is immersed in a volume, perhaps 10^6 times greater, that contains the lower-density gas producing the Balmer-line cores and the forbidden lines.

6. CONTINUUM RADIATION AND LUMINOSITIES

It is the continuum source that provides the extraordinary luminosity of Seyfert galaxies. The continuous spectra do not look like the composite stellar spectra that characterize other galactic nuclei. At least three different radiation mechanisms —stars, nonthermal radiation such as synchrotron radiation, and reradiation from heated dust—have to be combined in various ways to explain Seyfert continua (e.g. Neugebauer et al. 1976). Absorption lines that presumably arise in stars are seen in the spectra of some Seyferts, even bright Sy 1 (Osterbrock et al. 1976). However, the energy distribution in the continua cannot be explained with starlight. Most Seyfert nuclei have power-law spectra, $f_\nu \propto \nu^{-\alpha}$, so most of the radiation is attributed to a nonthermal source. This seems to be so for all Sy 1. For the Sy 2, there are indications that the continua may be due primarily to stars heavily reddened by dust, although this is not conclusive. Certainly, a larger fraction of stellar radiation is present in Sy 2 spectra than in Sy 1 (Osterbrock 1976, Neugebauer et al. 1976).

The first explicit demonstration of a Seyfert power-law spectrum was by Oke & Sargent (1968). In the visible spectrum, the many Seyferts observed since usually have indices $1.0 < \alpha < 2.0$, although α can be as large as 2.7 (Oke 1972, Shields et al. 1972, Osterbrock 1976, Osterbrock et al. 1976, Neugebauer et al. 1976). As infrared observations have accumulated, it has become clear that most Seyferts are strong infrared sources that can be accounted for by extensions of the power-law spectra (Penston et al. 1974, Neugebauer et al. 1976, Stein & Weedman 1976). The infrared continua are important to observe in detail because they contain the bulk of the bolometric luminosity (Rieke & Low 1972). However, there is at present a highly inhomogeneous collection of infrared data for Seyfert galaxies. In some cases, the inference has been made that the existence of strong infrared radiation means that objects are astrophysically similar. This is not the case, and it is important to learn how to distinguish between nonthermal infrared radiation and dust reradiation. This is significant even when comparing Sy 1 and Sy 2.

The infrared continuum of NGC 1068 indicates the difficulties in interpretation. It is the only Seyfert galaxy whose infrared spectrum has been followed until turnover, at about 100 μ (Telesco et al. 1976). This turnover mimics synchrotron self-absorption so the entire spectrum might be attributed to a nonthermal source on the basis of this observation. In fact, however, the infrared spectrum is attributed to dust reradiation (Jones & Stein 1975), primarily because the infrared nucleus is resolved (Becklin et al. 1973). The detection of polarization in the optical continuum was initially taken as evidence that the source was nonthermal (Visvanathan & Oke 1968), but Jones and Stein pointed out that dust would also polarize the continuum. Very convincing evidence that dust really does account for the polarization is given by Angel et al. (1976), who found that the continuum is circularly polarized. This would be very unlikely for a nonthermal source but common for dust scattering. Furthermore, they also found the emission lines to be polarized, which is strong evidence that both lines and continuum are strongly affected by dust. We therefore have several independent evidences that the nucleus of NGC 1068 is embedded in dust. A similar conclusion applies to other Sy 2, based on less comprehensive data (Neugebauer et al. 1976). The difference between this and the circumstances deduced for Sy 1 is an important demonstration of the necessity to distinguish between at least two classes of Seyferts.

It is known that continuum radiation exists in Seyfert galaxies out to 10 μ and, in a few cases, to 35 μ (Rieke & Low 1975a). The only Seyfert for which the long wavelength cutoff has been observed is NGC 1068, whose spectrum turns over at about

100 μ. This is deduced from broad-band measures centered at 93 μ and 140 μ (Telesco et al. 1976). Seyfert galaxies are not generally strong radio sources (Kellermann 1972), as indicated by the fact that less than 10% of them are 3C objects. Sramek & Tovmassian (1975) surveyed 506 Markarian galaxies at 6 cm, of which 51 were Seyferts. Only 14% of the Seyferts were detected to a flux limit of 30 mJy. They did find that virtually all detections were of Sy 2, which was the only correlation found between optical and radio properties. A more sensitive survey by de Bruyn & Wilson (1976) with the Westerbork telescope detected 50% of the 43 Seyferts they observed, to a limit of 4 mJy. They confirm that Sy 2 are generally the stronger radio emitters but note that no extended double radio sources are associated with Seyfert galaxies. Another correlation including various Seyferts has been reported between the infrared and 21 cm continuum radiation (Van der Kruit 1971, Rieke & Low 1972). It is not obvious that there is astrophysical meaning to this because such diverse objects as Sy 1, Sy 2, narrow-line galaxies like NGC 4385 and 7714, and wierd M82 are all included. Since there is substantial evidence that the infrared radiation from some of these objects is nonthermal, but from others is dust reradiation, it does not now seem appropriate to conclude that there is necessarily a physical relation between the infrared and radio spectra.

Even for those few Seyferts like 3C 120 that are strong radio sources, the radio continuum lies substantially below the extrapolation of the infrared spectrum at 10 μ (Shields et al. 1972). Somewhere—at millimeter, submillimeter, or far-infrared wavelengths—the spectra of all Seyferts must turn over. The point at which they do so is critical for determining their bolometric luminosities. For example, estimated bolometric luminosities for Sy 1 would differ by factors of two to four depending on whether the turnover is put at 3.5 μ or 350 μ (Stein & Weedman 1976). (For the power-law index $\alpha = 1$, the same bolometric luminosity is contained within each decade of frequency, so a spectrum from 350 Å to 350 μ would be twice as luminous as one from 350 Å to 3.5 μ).

As discussed above, the infrared continuum in Sy 1 is attributed to nonthermal radiation while that in Sy 2 probably contains substantial dust reradiation. There must therefore be an extremely luminous ultraviolet source in Sy 2 such as NGC 1068 to explain the production of the emission lines and the infrared reradiation (e.g. Neugebauer et al. 1976, Jones & Stein 1975, Shields & Oke 1975a). Whether this source is thermal or nonthermal is not known, but Sy 2 can be interpreted consistently by invoking large numbers of hot stars as the radiation source (Adams & Weedman 1975, Angel et al. 1976, Harwit & Pacini 1975). Other than NGC 1068, the Seyfert that shows the strongest evidence for dust reradiation is the pathological Sy 1, Markarian 231. In addition to having a very steep infrared spectrum, Mkn 231 has the silicate signature at 10 μ that characterizes dust radiation for infrared sources (Allen 1976 and private communications therein). The initial distinction of Mkn 231 was because of the extremely strong absorption lines of Na I and Ca II in its spectrum (Arakelian et al. 1971, Adams & Weedman 1972, Adams 1972a). These lines have components and are blueshifted relative to the emission line spectrum, so they are attributed to interstellar absorption within the nucleus. There are also absorption lines attributed to early type stars at the emission-line redshift. The presence of extreme interstellar absorption, a steep Balmer decrement, and unusual continuum colors are all evidence that this galaxy is heavily dust reddened. A thorough study of Mkn 231 has been prepared by Boksenberg et al. (1977). The empirical demonstration that structured absorption lines, blueshifted relative to the galaxy, can arise in an Sy 1 is an important analogy for the absorption line QSOs. However, Boksenberg et al. (1977) emphasize that the absorption lines are broader and shallower than in most QSOs. Mkn 231 has also been reported to be the most luminous galaxy in the universe (Rieke & Low 1972). While it is of great interest to understand how Mkn 231 achieved such a distinction, it must be realized that it also has unique spectroscopic properties so that an explanation of it may not be generally applicable to other Seyferts.

Empirical evidence therefore exists that Seyfert galaxies have a luminous continuum from the far infrared down to the ionizing ultraviolet. How much further does it go? The only Seyferts so far detected as X-ray sources are NGC 1275 and 4151 (Gursky et al. 1971) and possibly NGC 3783 (Cooke et al. 1976). NGC 1275 is intrinsically much more luminous but, as remarked previously, is atypical of Seyferts and more nearly resembles the giant elliptical radio galaxies. Other Seyferts have been observed but not detected (Ulmer & Murray 1976). Baity et al. (1975) detected hard X-rays from NGC 4151 and concluded that the X-ray luminosity was comparable to that in all other parts of the spectrum. These X-rays are too strong to arise from an extrapolation of the power-law spectrum in the optical, but they could arise from a similar nonthermal mechanism such as synchrotron radiation. Other suggested alternatives are that the X-rays are Compton scattered optical photons or are thermal emission from a very hot gas (Baity et al. 1975), but the soft X-ray data imply that the X-rays are not thermal and probably arise in the same volume as the optical nonthermal spectrum (Ives et al. 1976). Baity et al. emphasize that NGC 4151 could not be typical of Seyfert galaxies or else the contribution from Seyferts alone would give a diffuse X-ray background 100 times greater than observed. On the basis of this, it seems that Seyferts are not in general strong X-ray sources.

7 RELATION TO QSOs

The similarities between the nuclei of Seyfert galaxies and QSOs have been pointed out many times, and numerous efforts have been made to demonstrate a continuity between these objects (e.g. Sandage 1971, Lynden-Bell 1971, Notni & Richter 1972, Arakelian 1971, Weedman 1976b). It is the Sy 1 nuclei whose properties most closely resemble the QSOs. It is argued that as long as compact sources of high luminosity are known to exist in the nuclei of Seyfert galaxies, there should be no objection to interpreting QSOs as distant, more luminous examples of such sources. The Seyfert nuclei are therefore used to demonstrate empirically that QSOs really have cosmological redshifts. The most recent presentation of this continuity argument is by Khachikian & Weedman (1977). They consider the magnitude-redshift and angular diameter-redshift diagrams for Seyfert galaxies. There is substantial scatter in the former (the Hubble diagram), which is attributed to the differences in absolute luminosity among the various Seyfert nuclei. Assuming a linear redshift-distance relation gives a scatter $\sigma(\Delta m) = 1.04$ mag for Seyferts compared to Sandage's 0.28 mag for first ranked ellipticals. It is argued that angular diameters should be a more reasonable measure of relative distance for Seyferts since such diameters are determined by the galactic disks rather than the nuclei. The angular diameters measured from the *Sky Survey* decrease with redshift and have a scatter about the mean line of $\sigma(\Delta \log \theta) = 0.21$ for Sy 1. This compares to 0.11 for first-ranked ellipticals. It is therefore argued that, while Seyferts are not as homogeneous as first-ranked ellipticals, they still show good evidence for a linear redshift-distance relation.

There is not yet agreement, however, that the extended nebulosities associated with "Seyfert galaxy" nuclei are really the disks of galaxies (Burbidge 1973). The morphological studies discussed previously (Section 3) have demonstrated that a lot of Seyferts are spiral galaxies, but such demonstrations are difficult for the high redshift Seyferts that are critical in providing a transition to QSOs. There are a few Sy 1 that look like spiral galaxies whose luminosities would approach those of cosmologically redshifted QSOs. These include Mkn 79 and 618, NGC 7469, IC 4329A, I Zwl, and II Zw 136.

If cosmological redshifts are accepted for Seyferts and assumed for QSOs, there is an overlap in luminosity between the nuclei of Sy 1 and the QSOs. This is summarized in Khachikian & Weedman (1977). The luminosity indicators in their redshift-luminosity diagram are the broad hydrogen lines from Sy 1 and QSOs. Such lines arise only in the nuclei and are roughly proportional to the luminosity of the nonthermal continuum that carries most of the energy. The Hβ lines of lower redshift objects are related to the Lα lines in high redshift QSOs by assuming an intrinsic Lα:Hβ ratio of 40 (Davidson 1972). Luminosities are given for 37 Sy 1 and 62 QSOs using fluxes collected in Weedman (1976c) and Osmer & Smith (1976). The optically brightest QSOs are presumably included because nine of the high redshift QSOs were discovered from their strong Lα emission on objective prism survey plates (Smith 1976). The Sy 1 nuclei cover a luminosity range of 10^3, and most QSOs are less than 10 times brighter than the brightest Seyfert. The entire phenomenon, from faintest Sy 1 to brightest QSO, covers a factor of 10^5 in luminosity. As the bolometric luminosities are approximately 10^3 times the Hβ luminosity (Weedman 1976c), this corresponds to a luminosity range of 10^{43} ergs $\sec^{-1}$ to 10^{48} ergs $\sec^{-1}$ (for $q_0 = 0$ and $H_0 = 50$ km $\sec^{-1}$ Mpc^{-1}). Improvements in these numbers can be expected as more far-infrared data are accumulated (e.g. Rieke & Low 1975a) and as better composite spectra of Sy 1 and QSOs are assembled (e.g. Chan & Burbidge 1975, Baldwin 1975).

A key link between the nuclei of Seyfert galaxies and the QSOs is the fact that both can be variable in luminosity. As a consequence of the small radiating volumes implied by this variability, there are severe constraints on theoretical models of the radiation sources. Extensive considerations of such compact, nonthermal sources have been presented by Jones et al. (1974a,b) and Burbidge et al. (1974). Results relevant to this problem are also reviewed by Stein et al. (1976) in the context of the BL Lacertae objects. The energy-density paradox that arises in a luminous source that varies rapidly was first pointed out by Hoyle et al. (1966). Simply put, the problem is that electrons producing photons via synchrotron radiation will lose their energy in subsequent Compton scatterings with the very photons they created. The radiation source quickly quenches itself unless there is continuous injection of high-energy electrons and/or rapid source expansion (see references in Stein et al. 1976). Recent VLBI measures of the nuclei of the Sy 1 3C 120 and other radio sources do indicate relativistic expansions, or at least separations of individual radio-source components at relativistic speeds (Schilizzi et al. 1975, Wittels et al. 1976). The references cited include numerous references to earlier work in which such expansion was suspected (e.g. Shapiro et al. 1973, Kellermann et al. 1973 for 3C 120). The important progress in the recent work has been the demonstration that source contraction is never observed. This means that "Christmas-tree" models, in which the changes in source structure are attributed to flaring at random places in the source, are ruled out.

The long-term monitoring of Seyfert galaxies with different techniques has shown that variability is a common property. At one time or another, observers of all parts of the spectrum have reported a variable Seyfert galaxy. As such variability is the only measure of source size for most Seyfert nuclei, it is certainly important to know about. Data on the optical variability of Seyferts has been assembled by Cannon et al. (1971), Lyutyi (1973), Penston et al. (1974), Lyutyi & Pronik (1975), and Scott et al. (1976). Other studies of individual objects are included in the *UBV* references in Table 1. (Some useful comparison star sequences are in Penston et al. 1971.) One Seyfert (X Comae) was even discovered because of its variability (Bond 1973). The problem in interpreting the results is the tedious accumulation of data required before correlations can be searched for. Are there, for example, any correlations between variability and source luminosity (e.g. Elliot & Shapiro 1974)? It appears at the moment that the optical continua of all Sy 1, like the QSOs, are variable if observed closely enough. It is of great importance to demonstrate whether the continua of Sy 2 are variable as it is not yet known whether or not these sources are nonthermal.

Reports have existed for some time that the emission lines can also vary in Seyfert galaxies. Such variations have been reported for NGC 1068, 1275, 1566, 3227, 3516, 4151, 5548, 7469, 7603, 3C 390.3 and Markarian 6 (references to Table 1 and Lyutyi & Pronik 1975). Such variations are important for understanding the volume of the ionized gas and the location of the gas relative to the variable continuum source. Unavoidably, most of the observations from which line variability has been deduced are very inhomogeneous. Results from different observers using different spectrographs have been compared and changes claimed. Sometimes, the emission lines have been monitored with narrow-band interference filters without obtaining spectra. NGC 1068, in which the emission arises from a resolved nucleus, has been observed with entrance apertures smaller than the nucleus. Consequently, much of the data regarding emission-line variability is hard to evaluate with certainty. It is particularly perplexing when such results are reported for NGC 1068 in which low-density gas is thought to encompass a volume several hundred parsecs in diameter. It is important to continue attempts to confirm such results, however, especially because there are convincing cases in which line variability has been carefully monitored under the same observing conditions. The best examples seem to be the emission-line variations in NGC 7603 (Tohline & Osterbrock 1976) and the variable interstellar absorption lines in NGC 4151 (Anderson 1974a). As systematic monitoring of any sort requires diligence and patience, it is likely to be quite some time before sufficient data is accumulated to understand the nature of variability in Seyfert nuclei.

8 CONCLUSIONS

Mark Twain said that the nice thing about science is that one gets such wholesale returns of conjecture from such a trifling investment of fact. Amazingly, he said it without even being exposed to modern astrophysics. The purpose of this review has been to summarize the facts now known about Seyfert galaxies, but the reason for their existence remains one of the most pressing astrophysical mysteries. The primary difficulty in explaining Seyfert nuclei is in accounting for their energies. Many have bolometric luminosities as great as 10^{45} ergs sec^{-1}. Even this is 10^{12} solar luminosities, which has to arise in a volume about a parsec in diameter. Such a nucleus can be visualized by imagining the Crab Nebula radiating 10^7 as much energy as it does, since the absolute size, nonthermal radiation, and filamentary structure of the Crab likely resemble that of Seyfert nuclei. Obviously, some very efficient mechanism of energy production that can arise in a small volume is called for. The various possibilities for a theoretical explanation of the energy source(s) are reviewed by Saslaw (1974) and Burbidge (1970). To the extent that there is now a developing consensus, it seems to be in the direction of using gravity to explain things. This is not so much because of improvements in the theory of galactic nuclei, but is because observations of stellar X-ray sources have demonstrated empirically that gravitational accretion onto compact objects is a potent energy source (e.g. Blumenthal & Tucker 1974). Admittedly, such accretion is on a much smaller scale than needed in Seyferts. However, if a 10^7 $M_\odot$ black hole or even 10^7 neutron stars were inserted into a galactic nucleus, most of the activity in Seyfert nuclei could be made to happen as a consequence of accretion onto these objects. But an explanation of how the required compact objects get into the nuclei requires comprehensive knowledge of galactic evolution, and that's another story.

I thank Vanderbilt University and the National Science Foundation for continued support of research on the problems reviewed herein. As it is customary to acknowledge one's working environment, I wish to report that this review was prepared at the Dyer Observatory, in the quiet and pleasant Tennessee forest.

Literature Cited

Adams, T. F. 1972a. *Ap. J.* (*Lett.*) 176: L1
Adams, T. F. 1972b. *Ap. J.* (*Lett.*) 172: L101
Adams, T. F. 1973. *Ap. J.* 179: 417
Adams, T. F. 1975. *Ap. J.* 196: 675
Adams, T. F. 1977. *Ap. J. Suppl.* 33: 19
Adams, T. F., Weedman, D. W. 1972. *Ap. J.* (*Lett.*) 173: L109
Adams, T. F., Weedman, D. W. 1975. *Ap. J.* 199: 19
Allen, D. A. 1976. *Ap. J.* 207: 367
Allen, R. J., Darchy, B. F., Lauqué, R. 1971.

Astron. Astrophys. 10:198
Anderson, K. S. 1970. *Ap. J.* 162:743
Anderson, K. S. 1971. *Ap. J.* 169:449
Anderson, K. S. 1973. *Ap. J.* 182:369
Anderson, K. S. 1974a. *Ap. J.* 189:195
Anderson, K. S. 1974b. *Ap. J.* 187:445
Andrillat, Y., Collin-Souffrin, S. 1975. *Astron. Astrophys.* 43:419
Andrillat, Y., Souffrin, S. 1971. *Astron. Astrophys.* 11:286
Angel, J. R. P., Stockman, H. S., Woolf, N. J., Beaver, E. A., Martin, P. G. 1976. *Ap. J.* (*Lett.*) 206:L5
Arakelian, M. A. 1971. *Astrofizika* 7:457
Arakelian, M. A. 1975. *Soobshch. Byurak. Obs.* 47:3
Arakelian, M. A., Dibay, E. A., Yesipov, V. F. 1972a. *Astrofizika* 8:33
Arakelian, M. A., Dibay, E. A., Yesipov, V. F. 1972b. *Astrofizika* 8:177
Arakelian, M. A., Dibay, E. A., Yesipov, V. F. 1972c. *Astrofizika* 8:329
Arakelian, M. A., Dibay, E. A., Yesipov, V. F. 1973. *Astrofizika* 9:325
Arakelian, M. A., Dibay, E. A., Yesipov, V. F. 1975. *Astrofizika* 11:377
Arakelian, M. A., Dibay, E. A., Yesipov, V. F., Markarian, B. E. 1970. *Astrofizika* 6:357
Arakelian, M. A., Dibay, E. A., Yesipov, V. F., Markarian, B. E. 1971. *Astrofizika* 7:177
Arp, H. C. 1968. *Ap. J.* 152:1101
Arp, H. C. 1971. *Ap. Lett.* 7:221
Arp, H. C., Khachikian, E. Ye. 1973. *Astrofizika* 9:509
Baity, W. A., Jones, T. W., Wheaton, W. A., Peterson, L. E. 1975. *Ap. J.* (*Lett.*) 199:L5
Baldwin, J. A. 1975. *Ap. J.* 201:26
Barbieri, C. 1973. *Ap. Lett.* 14:231
Barbieri, C., Romano, G., Rosino, L. 1976. *Astron. Astrophys.* 47:153
Becklin, E. E., Matthews, K., Neugebauer, G., Wynn-Williams, C. G. 1973. *Ap. J.* (*Lett.*) 186:L69
Blumenthal, G. R., Mathews, W. G. 1975. *Ap. J.* 198:517
Blumenthal, G. R., Tucker, W. H. 1974. *Ann. Rev. Astron. Astrophys.* 12:23
Boksenberg, A., Carswell, R. F., Allen, D. A., Fosbury, R. A. E., Penston, M. V., Sargent, W. L. W. 1977. *MNRAS* 178:451
Boksenberg, A., Shortridge, K., Allen, D. A., Fosbury, R. A. E., Penston, M. V., Savage, A. 1975a. *MNRAS* 173:381
Boksenberg, A., Shortridge, K., Fosbury, R. A. E., Penston, M. V., Savage, A. 1975b. *MNRAS* 172:289
Bolton, J. G., Ekers, J. 1966. *Aust. J. Phys.* 19:713
Bond, H. E. 1973. *Ap. J.* (*Lett.*) 181:L23
Bond, H. E., Sargent, W. L. W. 1973. *Ap. J.* (*Lett.*) 185:L109
Bracessi, A., Lynds, C. R., Sandage, A. R. 1968. *Ap. J.* (*Lett.*) 152:L105
Burbidge, E. M. 1968. *Astron. J.* 73:890
Burbidge, E. M., Burbidge, G. R. 1971. *Ap. J.* (*Lett.*) 163:L21
Burbidge, G. R. 1970. *Ann. Rev. Astron. Astrophys.* 8:369
Burbidge, G. R. 1973. *Nature Phys. Sci.* 246:17
Burbidge, G. R., Burbidge, E. M., Sandage, A. R. 1963. *Rev. Mod. Phys.* 35:947
Burbidge, G. R., Jones, T. W., O'dell, S. L. 1974. *Ap. J.* 193:43
Cannon, R. D., Penston, M. V., Brett, R. A. 1971. *MNRAS* 152:79
Chan, Y.-W. T., Burbidge, E. M. 1975. *Ap. J.* 198:45
Collin-Souffrin, S., Alloin, D., Andrillat, Y. 1973. *Astron. Astrophys.* 22:343
Cooke, B. A., Elvis, M., Maccacaro, T., Ward, M. J., Fosbury, R. A. E., Penston, M. V. 1976. *MNRAS* 177:121P
Davidson, K. 1972. *Ap. J.* 171:213
Davies, R. D. 1973. *MNRAS* 161:25P
de Bruyn, A. G., Willis, A. G. 1974. *Astron. Astrophys.* 33:351
de Bruyn, A. G., Wilson, A. S. 1976. *Astron. Astrophys.* 53:93
Denisyuk, E. K., Lipovetsky, V. A. 1974. *Astrofizika* 10:315
de Vaucouleurs, G. 1973. *Ap. J.* 181:31
de Vaucouleurs, G., de Vaucouleurs, A. 1968. *Astron. J.* 73:858
de Vaucouleurs, G., de Vaucouleurs, A. 1972. *Ap. Lett.* 12:1
de Vaucouleurs, G., de Vaucouleurs, A. 1975. *Ap. J.* (*Lett.*) 197:L1
DeYoung, D. S., Roberts, M. S., Saslaw, W. C. 1973. *Ap. J.* 185:809
Dibay, E. A. 1970. *Astrofizika* 6:350
Dibay, E. A., Lyutyi, V. M. 1971. *Astrofizika* 7:169
Disney, M. J. 1973. *Ap. J.* (*Lett.*) 181:L55
Doroshenko, V. T., Terebizh, V. U. 1975. *Astrofizika* 11:631
Eilek, J. A., Auman, J. R., Ulrych, T. J., Walker, G. A. H., Kuhi, L. V. 1973. *Ap. J.* 182:363
Elliot, J. L., Shapiro, S. L. 1974. *Ap. J.* (*Lett.*) 192:L3
Fairall, A. P. 1968. *Publ. Astron. Soc. Pac.* 80:235
Glaspey, J. W., Eilek, J. A., Fahlman, G. G., Auman, J. R. 1976. *Ap. J.* 203:335
Glass, I. S. 1973. *MNRAS* 164:155
Griffin, R. F. 1963. *Astron. J.* 68:421
Gursky, H., Kellogg, E. M., Leong, C., Tananbaum, H., Giacconi, R. 1971. *Ap. J.*

(*Lett.*) 165:L43
Harwit, M., Pacini, F. 1975. *Ap. J.* (*Lett.*) 200:L127
Hills, J. G. 1975. *Nature* 254:295
Hodge, P. W. 1968. *Astron. J.* 73:846
Hoyle, F., Burbidge, G. R., Sargent, W. L. W. 1966. *Nature* 209:751
Hubbard, R. 1975. *Ap. J.* 198:57
Huchra, J. P. 1976. PhD thesis. Calif. Inst. of Tech.
Huchra, J., Sargent, W. L. W. 1973. *Ap. J.* 186:433
Humason, M. L., Mayall, N. U., Sandage, A. R. 1956. *Astron. J.* 61:97
Ives, J. C., Sanford, P. W., Penston, M. V. 1976. *Ap. J.* (*Lett.*) 207:L159
Jameson, R. F., Longmore, A. J., McLinn, J. A., Woolf, N. J. 1974. *Ap. J.* 190:353
Jones, T. W., O'dell, S. L., Stein, W. A. 1974a. *Ap. J.* 188:533
Jones, T. W., O'dell, S. L., Stein, W. A. 1974b. *Ap. J.* 192:261
Jones, T. W., Stein, W. A. 1975. *Ap. J.* 197: 297
Joyce, R. R., Knacke, R. F., Simon, M., Young, E. 1975. *Publ. Astron. Soc. Pac.* 87: 683
Jurkevich, I., Usher, P. D., Shen, B. S. P. 1971. *Astrophys. Space Sci.* 10:402
Katz, A. 1975. *Ap. J.* 198:255
Kellermann, K. I. 1972. In *External Galaxies and Quasi-Stellar Objects, IAU Symp. 44,* ed. D. S. Evans, p. 190. Dordrecht: Reidel
Kellermann, K. I., Clark, B. G., Jauncey, D. L., Broderick, J. J., Shaffer, D. B., Cohen, M. H., Niell, A. E. 1973. *Ap. J.* (*Lett.*) 183:L51
Khachikian, E. Ye. 1973. *Astrofizika* 9:139
Khachikian, E. Ye., Weedman, D. W. 1971. *Astrofizika* 7:389
Khachikian, E. Ye., Weedman, D. W. 1974. *Ap. J.* 192:581
Khachikian, E. Ye., Weedman, D. W. 1977. In *Redshifts and Expansion of the Universe, IAU Colloq. 37,* ed. C. Balkowski, B. Westerlund. In press
Kleinmann, D. E., Wright, E. L. 1974. *Ap. J.* (*Lett.*) 191:L19
Knacke, R. F., Capps, R. W. 1974. *Ap. J.* (*Lett.*) 192:L19
Kojoian, G., Sramek, R. A., Dickinson, D. F., Tovmassian, H., Purton, C. R. 1976. *Ap. J.* 203:323
Kopilov, I. M., Lipovetsky, V. A., Pronik, V. I., Chuvaev, K. K. 1974. *Astrofizika* 10:483
Koski, A. T. 1976. PhD thesis. Univ. Calif., Santa Cruz
Kristian, J. 1973. *Ap. J.* (*Lett.*) 179:L61
Lewis, B. M. 1972. In *External Galaxies and Quasi-Stellar Objects, IAU Symp. 44,* ed. D. S. Evans, p. 267. Dordrecht: Reidel
Lynden-Bell, D. 1971. *MNRAS* 155:119
Lyutyi, V. M. 1973. *Sov. Astron. AJ* 16:763
Lyutyi, V. M., Pronik, V. I. 1975. In *Variable Stars and Stellar Evolution,* ed. L. Plaut, p. 591. Dordrecht: Reidel
MacAlpine, G. M. 1974. *Ap. J.* 193:37
MacPherson, G. J. 1972. *Publ. Astron. Soc. Pac.* 84:392
Markarian, B. E. 1967. *Astrofizika* 3:55
Markarian, B. E. 1969a. *Astrofizika* 5:443
Markarian, B. E. 1969b. *Astrofizika* 5:581
Markarian, B. E. 1973. *Astrofizika* 9:5
Markarian, B. E., Lipovetsky, V. A. 1971. *Astrofizika* 7:511
Markarian, B. E., Lipovetsky, V. A. 1972. *Astrofizika* 8:155
Markarian, B. E., Lipovetsky, V. A. 1973. *Astrofizika* 9:487
Markarian, B. E., Lipovetsky, V. A. 1974. *Astrofizika* 10:307
Martin, W. L. 1974. *MNRAS* 168:109
Martin, W. L. 1976. *MNRAS* 175:633
Mathews, W. G. 1974. *Ap. J.* 189:23
Mathis, J. S. 1970. *Ap. J.* 162:761
Miley, G. K., Perola, G. C. 1975. *Astron. Astrophys.* 45:223
Miller, J. S. 1974. *Ann. Rev. Astron. Astrophys.* 12:331
Morgan, W. W., Walborn, N. R., Tapscott, J. W. 1971. In *Nuclei of Galaxies,* ed. D. J. K. O'Connell, p. 27. New York: Am. Elsevier
Netzer, H. 1974. *MNRAS* 169:579
Netzer, H. 1975. *MNRAS* 171:395
Netzer, H., Penston, M. V. 1976. *MNRAS* 174:319
Neugebauer, G., Becklin, E. E., Oke, J. B., Searle, L. 1976. *Ap. J.* 205:29
Niell, A. E., Kellermann, K. I., Clark, B. G., Shaffer, D. B. 1975. *Ap. J.* (*Lett.*) 197: L109
Notni, P., Richter, G. M. 1972. *Astron. Nachr.* 294:95
Nussbaumer, H., Osterbrock, D. E. 1970. *Ap. J.* 161:811
Oke, J. B. 1972. In *External Galaxies and Quasi-Stellar Objects, IAU Symp. 44,* ed. D. S. Evans, p. 139. Dordrecht: Reidel
Oke, J. B., Sargent, W. L. W. 1968. *Ap. J.* 151:807
Oke, J. B., Shields, G. A. 1976. *Ap. J.* 207: 713
Olsen, E. T. 1970. *Astron. J.* 75:764
Osmer, P. S., Smith, M. G. 1976. *Ap. J.* 210:267
Osmer, P. S., Smith, M. G., Weedman, D. W. 1974a. *Ap. J.* 189:187
Osmer, P. S., Smith, M. G., Weedman, D. W. 1974b. *Ap. J.* 192:279
Osterbrock, D. E. 1971. In *Nuclei of Galaxies,*

ed. D. J. K. O'Connell, p. 151. New York: Am. Elsevier
Osterbrock, D. E. 1976. *Ap. J.* 203:329
Osterbrock, D. E. 1977a. *IAU Symp. 74*. In press
Osterbrock, D. E. 1977b. *Ap. J.* In press
Osterbrock, D. E., Koski, A. T. 1976. *MNRAS* 176:61P
Osterbrock, D. E., Koski, A. T., Phillips, M. M. 1975. *Ap. J.* (*Lett.*) 197:L41
Osterbrock, D. E., Koski, A. T., Phillips, M. M. 1976. *Ap. J.* 206:898
Pastoriza, M., Gerola, H. 1970. *Ap. Lett.* 6:155
Penston, M. J., Penston, M. V., Sandage, A. 1971. *Publ. Astron. Soc. Pac.* 83:783
Penston, M. V., Penston, M. J. 1973. *MNRAS* 162:109
Penston, M. V., Penston, M. J., Selmes, R. A., Becklin, E. E., Neugebauer, G. 1974. *MNRAS* 169:357
Peterson, S. D. 1973. *Astron. J.* 78:811
Phillips, M. M. 1976. *Ap. J.* 208:37
Phillips, M. M., Osterbrock, D. E. 1975. *Publ. Astron. Soc. Pac.* 86:949
Pronik, I. I. 1974a. *Sov. Astron.—AJ* 18:717
Pronik, I. I. 1974b. *Sov. Astron.—AJ* 18:271
Pronik, I. I. 1975. *Sov. Astron.—AJ* 19:293
Pronik, V. I., Chuvaev, K. K. 1972. *Astrofizika* 8:187
Ptak, R., Stoner, R. E. 1973. *Ap. J.* 185:121
Ptak, R., Stoner, R. E. 1975. *Ap. J.* 200:558
Quintana, H., Kaufmann, P., Sersic, J. L. 1975. *MNRAS* 173:57P
Richstone, D. O., Morton, D. C. 1975. *Ap. J.* 201:289
Rieke, G. H., Low, F. J. 1972. *Ap. J.* (*Lett.*) 176:L95
Rieke, G. H., Low, F. J. 1975a. *Ap. J.* (*Lett.*) 200:L67
Rieke, G. H., Low, F. J. 1975b. *Ap. J.* 199: L13
Robinson, L. B., Wampler, E. J. 1973. *Ap. J.* (*Lett.*) 179:L79
Rubin, V. C., Ford, W. K. Jr. 1968. *Ap. J.* 154:431
Rubin, V. C., Thonnard, N., Ford, W. K. Jr. 1975. *Ap. J.* 199:31
Sandage, A. 1966. *Ap. J.* 145:1
Sandage, A. 1967. *Ap. J.* (*Lett.*) 150:L9
Sandage, A. 1971. In *Nuclei of Galaxies*, ed. D. J. K. O'Connell, p. 271. New York: Am. Elsevier
Sargent, W. L. W. 1968. *Ap. J.* (*Lett.*) 151: L31
Sargent, W. L. W. 1970a. *Ap. J.* 160:405
Sargent, W. L. W. 1970b. *Ap. J.* 159:765
Sargent, W. L. W. 1971. In *Nuclei of Galaxies*, ed. D. J. K. O'Connell, p. 81. New York: Am. Elsevier
Sargent, W. L. W. 1972. *Ap. J.* 173:7
Sargent, W. L. W. 1973. *Ap. J.* (*Lett.*) 182: L13
Saslaw, W. C. 1974. In *The Formation and Dynamics of Galaxies, IAU Symp. 58*, ed. J. R. Shakeshaft, p. 305. Dordrecht: Reidel
Schild, R. E. 1972. *Ap. J.* 178:617
Schilizzi, R. T., Cohen, M. H., Romney, J. D., Shaffer, D. B., Kellermann, K. I., Swenson, G. W. Jr., Yen, J. L., Rinehart, R. 1975. *Ap. J.* 201:263
Schwarzschild, M. 1973. *Ap. J.* 182:357
Scott, R. L., Leacock, R. J., McGimsey, B. Q., Smith, A. G., Edwards, P. L., Hackney, K. R., Hackney, R. L. 1976. *Astron. J.* 81:7
Searle, L., Bolton, J. G. 1968. *Ap. J.* (*Lett.*) 154:L101
Searle, L., Sargent, W. L. W. 1968. *Ap. J.* 153:1003
Seielstad, G. A. 1974. *Ap. J.* 193:55
Selmes, R. A., Tritton, K. P., Wordsworth, R. W. 1975. *MNRAS* 170:15
Seyfert, C. K. 1943. *Ap. J.* 97:28
Shapiro, I. I., Hinteregger, H. F., Knight, C. A., Punsky, J. J., Robertson, D. S., Rogers, A. E. E., Whitney, A. R., Clark, T. A., Marandino, G. E., Goldstein, R. M., Spitzmesser, D. J. 1973. *Ap. J.* (*Lett.*) 183:L47
Shectman, S. A., MacAlpine, G. M. 1975. *Ap. J.* (*Lett.*) 199:L85
Shen, B. S. P., Usher, P. D., Barrett, J. W. 1972. *Ap. J.* 171:457
Shields, G. A. 1974. *Ap. J.* 191:309
Shields, G. A., Oke, J. B. 1975a. *Ap. J.* 197:5
Shields, G. A., Oke, J. B. 1975b. *Publ. Astron. Soc. Pac.* 87:879
Shields, G. A., Oke, J. B., Sargent, W. L. W. 1972. *Ap. J.* 176:75
Simkin, S. M. 1975. *Ap. J.* 200:567
Simon, T., Dyck, H. M. 1975. *MNRAS* 172: 19P
Smith, M. G. 1975. *Ap. J.* 202:591
Smith, M. G. 1976. *Ap. J.* (*Lett.*) 206:L125
Smith, M. G., Aguirre, C., Zemelman, M. 1976. *Ap. J. Suppl.* 32:217
Smith, M. G., Weedman, D. W., Spinrad, H. 1972. *Ap. Lett.* 11:21
Sramek, R. A., Tovmassian, H. M. 1974a. *Ap. J.* 191:L13
Sramek, R. A., Tovmassian, H. M. 1974b. *Ap. J.* 191:633
Sramek, R. A., Tovmassian, H. M. 1975. *Ap. J.* 196:339
Sramek, R. A., Tovmassian, H. M. 1976. *Ap. J.* 207:725
Stein, W. A., Gillett, F. C., Merrill, K. M. 1974. *Ap. J.* 187:213
Stein, W. A., O'dell, S. L., Strittmatter, P. A. 1976. *Ann. Rev. Astron. Astrophys.* 14: 173

Stein, W. A., Weedman, D. W. 1976. *Ap. J.* 205:44
Stoner, R. E., Ptak, R., Ellis, D. 1974. *Ap. J.* 191:291
Sulentic, J. W., Tifft, W. G. 1973. *The Revised New General Catalogue of Nonstellar Astronomical Objects,* Tucson: Univ. Ariz. Press 384 pp.
Telesco, C. M., Harper, D. A., Loewenstein, R. F. 1976. *Ap. J.* (*Lett.*) 203:L53
Tohline, J. E., Osterbrock, D. E. 1976. *Ap. J.* (*Lett.*) 210:L117
Ulmer, M. P., Murray, S. S. 1976. *Ap. J.* 207:364
Ulrich, M.-H. 1971. *Ap. J.* (*Lett.*) 165:L61
Ulrich, M.-H. 1972a. *Ap. J.* 174:483
Ulrich, M.-H. 1972b. *Ap. J.* (*Lett.*) 171:L37
Ulrich, M.-H. 1973. *Ap. J.* 181:51
Ulrich, M.-H. 1974. In *The Formation and Dynamics of Galaxies, IAU Symp. 58,* ed. J. R. Shakeshaft, p. 279. Dordrecht: Reidel
Ulrich, M.-H. 1975. *Astron. Astrophys.* 40:337
Usher, P. D. 1972. *Ap. J.* 172:L25
Usher, P. D., Shen, B. S. P., Barrett, J. W. 1971. *Ap. J.* 165:647
van den Bergh, S. 1975a. *J. R. Astron. Soc. Can.* 69:105
van den Bergh, S. 1975b. *Ap. J.* (*Lett.*) 198:L1
Van der Kruit, P. C. 1971. *Astron. Astrophys.* 15:110
Visvanathan, N., Oke, J. B. 1968. *Ap. J.* (*Lett.*) 152:L165
Vorontsov–Velyaminov, B. A., Ivanisevich, G. 1974. *Sov. Astron.—AJ* 18:174
Walker, M. F. 1968. *Ap. J.* 151:71
Walker, M. F., Pike, C. D., McGee, J. D. 1974. *Publ. Astron. Soc. Pac.* 86:870
Wampler, E. J. 1968. *Ap. J.* (*Lett.*) 154:L53
Wampler, E. J. 1971. *Ap. J.* 164:1
Warner, B. 1973. *MNRAS* 162:189
Weedman, D. W. 1970. *Ap. J.* 159:405
Weedman, D. W. 1971. *Ap. J.* (*Lett.*) 167:L23
Weedman, D. W. 1973. *Ap. J.* 183:29
Weedman, D. W. 1976a. In *Vistas in Astron.* 20: In press
Weedman, D. W. 1976b. *Q. J. R. Astron. Soc.* 17:227
Weedman, D. W. 1976c. *Ap. J.* 208:30
Weedman, D. W., Khachikian, E. Ye. 1969. *Astrofizika* 5:113
Weymann, R. J. 1970. *Ap. J.* 160:31
Williams, R. E., Weymann, R. J. 1968. *Astron. J.* 73:895
Wittels, J. J., Cotton, W. D., Counselman, C. C. III., Shapiro, I. I., Hinteregger, H. F., Knight, C. A., Rogers, A. E. E., Whitney, A. R., Clark, T. A., Hutton, L. K., Ronnang, B. O., Rydbeck, O. E. H., Niell, A. E. 1976. *Ap. J.* (*Lett.*) 206:L75
Wyndham, J. D. 1966. *Ap. J.* 144:459
Woltjer, L. 1959. *Ap. J.* 130:38
Yankulova, I. M. 1975. *Sov. Astron.—AJ* 18:720
Yankulova, I. M., Dibay, E. A., Yesipov, V. F. 1974. *Sov. Astron.—AJ* 18:275
Zasov, A. V., Lyutyi, V. M. 1973. *Sov. Astron.—AJ* 17:169
Zwicky, F., Oke, J. B., Neugebauer, G., Sargent, W. L. W., Fairall, A. P. 1970. *Publ. Astron. Soc. Pac.* 82:93

Ann. Rev. Astron. Astrophys. 1977. 15: 97–126

MERCURY

D. E. Gault, J. A. Burns[1]*, and P. Cassen*
National Aeronautics and Space Administration, Ames Research Center, Space Science Division, Moffett Field, California 94035

and

R. G. Strom
Department of Planetary Sciences, Lunar and Planetary Laboratory, University of Arizona, Tucson, Arizona 85721

INTRODUCTION

Prior to the flight of the Mariner 10 spacecraft, Mercury was the least investigated and most poorly known terrestrial planet (Kuiper 1970, Devine 1972). Observational difficulties caused by its proximity to the Sun as viewed from Earth caused the planet to remain a small, vague disk exhibiting little surface contrast or details, an object for which only three major facts were known: 1. its bulk density is similar to that of Venus and Earth, much greater than that of Mars and the Moon; 2. its surface reflects electromagnetic radiation at all wavelengths in the same manner as the Moon (taking into account differences in their solar distances); and 3. its rotation period is in 2/3 resonance with its orbital period. Images obtained during the flyby by Mariner 10 on 29 March 1974 (and the two subsequent flybys on 21 September 1974 and 16 March 1975) revealed Mercury's surface in detail equivalent to that available for the Moon during the early 1960's from Earth-based telescopic views. Additionally, however, information was obtained on the planet's mass and size, atmospheric composition and density, charged-particle environment, and infrared thermal radiation from the surface, and most significantly of all, the existence of a planetary magnetic field that is probably intrinsic to Mercury was established.

In the following, this new information is summarized together with results from theoretical studies and ground-based observations. In the quantum jumps of knowledge that have been characteristic of "space-age" exploration, the previously obscure body of Mercury has suddenly come into sharp focus. It is very likely a differentiated body, probably contains a large Earth-like iron-rich core, and displays a surface remarkably similar to that of the Moon, which suggests a similar evolutionary history.

[1] Permanent Address: Center for Radiophysics and Space Research, Cornell University, Ithaca, New York 14853

SIZE AND MASS

Mariner 10's close flybys of Mercury have permitted the planet's mass to be ascertained accurately. Esposito et al. (1976) have deduced that the ratio of the solar mass to Mercury's is 6,023,600 $\pm$ 600 [3.3020 ($\pm$0.0037) $\times 10^{26}$ g] from the first flyby and 6,023,700 $\pm$ 300 from the third (cf Howard et al. 1974); the second encounter was at too great a distance from Mercury to provide a useful estimate of its mass. These values lie near the much larger bounds (5,972,000 $\pm$ 45,000) of previous results (cf Duncombe, Klepczynski & Seidelmann 1973, Duncombe, Seidelmann & Klepczynski 1973), the most accurate of which came from a simultaneous solution of optical and radar positions for the inner planets (Ash, Shapiro & Smith 1971).

The spacecraft orbit also defined J_2, the oblateness parameter, for the first time. Based on the third encounter, J_2 equals $(8 \pm 6) \times 10^{-5}$, whereas the orbit from the first flyby, which was less suited for analysis, gives a value of the same order (Esposito et al. 1976). For most other planets, J_2 is closely associated with rotational flattening and gives information on the central condensation of the planet. Because Mercury's spin is so slow, however, the nonhydrostatic part of J_2 is larger than the centrifugal part ($\sim 10^{-6}$) by almost two orders of magnitude, and, unhappily, little can be learned of the interior from the value of J_2 except that its strength is comparable to that of other terrestrial planets. The current data show structure in the local gravity field and P. B. Esposito (personal communication 1976) believes that the data set could perhaps generate a very rough value of J_4 with considerably more effort.

Because of its smallness and its angular proximity to the Sun (its maximum elongation being 27°), Mercury is a difficult object to study telescopically, with its disk merely 6.9″ to 10.9″ across. Thus classical estimates of its radius are necessarily imprecise. Nevertheless, they have been surprisingly correct, giving estimates of 2440 $\pm$ 7.5 km (de Vaucouleurs 1964). Refined radar ranging experiments have given an accurate mean radius of 2439 $\pm$ 1 km (Ash, Shapiro & Smith 1967, 1971), which has been confirmed by the values determined from the occultation by Mercury's disk of Mariner 10 dual frequency radio signals during first encounter: 2439.5 $\pm$ 1 km at 1.1° N, 64.7° E (nightside) and 2439 $\pm$ 1 at 67.6° N, 258.4° E (dayside) (Howard et al. 1974, Fjeldbo et al. 1976).

Accepting the Mariner 10 value for the mass and the radar measurement of the radius produces a planet with a mean density of 5.433 $\pm$ 0.012 g cm^{-3}, a value essentially equal to that of the Earth but corresponding to the highest known uncompressed density for a solar system body.

ORBIT

At present Mercury's orbit is distinguished from those of the other planets in the smallness of its semimajor axis ($a = 0.387$ au) and by its comparatively large eccentricity ($e = 0.206$) and orbital inclination relative to the ecliptic ($i = 7.0°$). With the exception of the distant Pluto, Mercury's orbit is the most inclined and elliptic of any planet. This is not a temporary condition; the secular theory of Brouwer & van Woerkom (1950), which accounts for the mutual perturbations of the planets (excluding Pluto), has been numerically integrated by Cohen, Hubbard & Oesterwinter (1973) and gives a mean $e = 0.175$ and $i = 7.2°$ over 10^7 years, values that are about three times larger than for other planets. The half amplitude of the computed periodic oscillations in e is about 0.07 and has a prominent period of about 10^6 years, while i varies $\pm 2°$ with the largest period at nearly 10^6 years and much smaller amplitude variations ($\pm 0.2°$) occurring in a period of 10^5 years.

The conventional wisdom is that the unusually large e and i are mere chance occurrences or, at the most, that the mercurian inclination shows a primordial tie to the solar equatorial plane, which is tilted at 6° to the invariable plane, and, perhaps, that the high e is caused by the high-collision velocities between proto-Mercury and planetesimals perturbed into the inner solar system by Jupiter (Kaula 1976).

Ward, Colombo & Franklin (1976), however, give another explanation for the odd e and i. They demonstrate that a secular resonance between the precession rates of the lines of apsides for Mercury and Venus would have existed in the past if and when the Sun had an oblateness $J_2 \sim 10^{-3}$, corresponding to a spin period of about five hours; nearly the same J_2 would produce another resonance between the precession rates of the lines of nodes of the two planets. During such secular resonances, Mercury's e and i are pumped up by Venus. If—but only if—the characteristic time scales for decay of the solar rotation is about 10^6 years is it possible for this mechanism to produce the presently observed e and i. From a comparison with T-Tauri and other young stars, such a rapid solar rotation and decay rate are not implausible for the early Sun. Van Flandern & Harrington (1976) boldly suggest that Mercury may be an escaped former satellite of Venus, but they are unable to find a mechanism to account for the large difference between their semimajor axes.

The prograde precession of Mercury's perihelion by 43″ more per century than can be explained by the Newtonian action of the other planets was one of the major failings of classical celestial mechanics. Einstein's general theory of relativity accounted for this discrepancy and its precise description of the motion of the inner planets still remains one of the most stringent tests to be passed by any challenger to Einstein's theory (Will 1974).

ROTATION

For nearly a century following Schiaparelli's claim in 1889 that Mercury rotated slowly, telescope observers agreed that the rotation period was synchronous with the orbital period of 88 days (see Colombo & Shapiro 1966, Smith & Reese 1968, and historical references therein). This result was overturned in 1965 by the radar measurement of a rotation period of 59 ± 5 days by Pettengill & Dyce (1965), subsequently refined to 58.65 ± 0.25 days (Goldstein 1971). After theoretical justification that solid-body tides due to the Sun's attraction might well slow the spin to a value somewhat higher than the synchronous rate because of the high orbital eccentricity (Peale & Gold 1965), optical astronomers—in one of the better recoveries ever recorded—reanalyzed historical observations to find a spin period of 58.644 ± 0.009 days (Murray, Dollfus & Smith 1972). These error bounds enclose the 58.6457-day period corresponding to a 2:3 commensurability between the axial and orbital period. Colombo (1965) had suggested that such a commensurability would result from the combined effect of tides and solar torques on the nonspherical shape of the planet, and Liu & O'Keefe (1965), Colombo & Shapiro (1966), and Goldreich & Peale (1966) showed it to be a stable motion. Klaasen (1975, 1976) obtained 58.6461 ± 0.005 days from shadows measured on consecutive Mariner 10 encounters to confirm the resonance.

Ground-based telescopes and radar and spacecraft observations have also been used to find the planet's obliquity, the angle between the rotation axis and pole of the orbit. Telescopic work generally suggests that the rotation axis is perpendicular to the orbit plane to within 3° (Murray, Dollfus & Smith 1972), in agreement with much cruder radar results (Dyce, Pettengill & Shapiro 1967). Analysis of Mariner 10 pictures (Klaasen 1976) also gives a nearly perpendicular rotation axis (obliquity = 2° with a $2.6° \times 6.5°$ error ellipse). Although a very precise determination ($< 1'$) of the obliquity would permit an evaluation of $(C-A)/C$, where $A < B < C$ are the planet's principal moments of inertia, this precision cannot be achieved without the use of a lander (Peale 1972). Besides accurately determining the obliquity, a lander could measure the amplitude of the libration in longitude which, along with accurate values for the gravitational coefficients J_2 and C_{22}, may be able to indicate the extent of a molten core (Peale 1976b).

Presumably, Mercury did not originate in this spin state. Most likely it had been rotating much faster, perhaps with a period near eight hours, like so many solar system bodies today (cf Burns 1975), but it has been subsequently slowed by solar tides (Burns 1976). Such tides despin Mercury with a characteristic decay time of a billion years for a tidal dissipation factor Q of 30. This slowing would have heated the interior by about 100° K and would have caused substantial surface strains (corresponding to a change in radius of tens of kilometers) that, if not relaxed, would result in surface stresses well above fracture limits. Surface features characteristic of this failure mode (normal faulting) are not seen today although a global structural pattern appears to be present as linear features (D. Dzurisin, personal communication). Large extensional surface strains, equivalent to an increase in radius of 13 km, are generated during core infall while smaller compressional strains ($\Delta R \approx 2$ km) subsequent to core formation accompany the planet's cooling; the latter may account for the common lobate scarps (Solomon 1976). Other important volume changes, largely unexamined, could occur because of possible changes in phase of the core.

The fact that Mercury spins slowly may explain the absence of natural satellites about it. Both Burns (1973) and Ward & Reid (1973) have pointed out that tides on a slowly spinning planet like Mercury impose drag forces on satellites that decay their orbits, causing the satellites to impact eventually onto the planet.

The most complete treatment of the 2:3 resonant rotation (Goldreich & Peale

1968 and earlier papers) demonstrates that such a rotation is stable as long as the solar torque exerted on the planet's asymmetric equatorial shape is always larger than the time averaged tidal torque, or $(B-A)/C > 10^{-7}\ Q^{-1}$. Since $J_2 \equiv [C-\frac{1}{2}(B+A)]/(MR^2)$ is likely to be comparable to $(B-A)/C$, the Mariner 10 result of $J_2 \sim 10^{-4}$ assures that the stability criterion is almost certainly satisfied. The r^{-3} dependence of the torque on the permanent deformation, together with the eccentric orbit, aligns the long equatorial axis with the Sun-Mercury line at perihelion. Since many other commensurate spin rates are also stable for reasonable choices of Q, the question arises as to why Mercury avoided capture into these as it was tidally slowed. Goldreich & Peale (1966) show that capture occurs when the rotational energy lost (by dissipation) over one libration—whether due to tidal or internal origin—exceeds the kinetic energy stored in the angular motion measured relative to that commensurate spin state. Assuming the rotational phase to be randomly distributed, the capture probability into a particular state will be the ratio of these two energies and will therefore depend on the energy loss mechanism. A comforting confirmation of the theory is the fact that the capture probability of Mercury into a 2:3 resonance is much greater than into any higher-order resonance for the most common tidal models. The putative molten core would dissipate further energy through shear losses at the core boundary. Peale & Boss (1977) point out that the escape of Mercury from capture into the 1:2 resonance requires that, if a molten core does exist, its kinematic viscosity must be comparable to that of water (0.01 cm^{-2} sec^{-1}) and $Q \lesssim 100$.

Colombo (1966) and Peale (1969) also have shown that a generalization of Cassini's laws applies to Mercury: There are three possible obliquities for which the spin axis lies stably in the plane formed by the orbit normal and the axis about which the orbit precesses (i.e. the invariable plane of the solar system). Peale (1974, 1976a) deduces from a study of the history of the obliquity that all initial conditions (except for a pathological case of slow primordial spin, large initial obliquity, and, effective core-mantle interaction) produce a final state where the spin axis lies near the normal to the orbit as observed.

ATMOSPHERE

The atmosphere is very tenuous; the surface pressure at the subsolar point is less than a few times 10^{10} mb. Such an atmosphere is essentially exospheric—the gas is expected to be collisionless right down to the planetary surface.

Helium and atomic hydrogen have been identified as constituents of Mercury's atmosphere by detection of their emission lines by the UV spectrometer aboard Mariner 10 (Broadfoot et al. 1974, 1976, Broadfoot 1976). The absence of other UV emission lines and the lack of measurable absorption during solar occultation have established upper limits for several other possible constituents: H_2, O_2, Ne, Ar, O, CO_2, H_2O, and N_2 (see Kumar 1976 for a useful comparison with the lunar atmosphere). In addition, an upper limit to the electron density of 10^3 cm^{-3} has been determined by the Mariner 10 radio occultation experiment (Fjeldbo et al. 1976).

A model helium atmosphere has been calculated by Hartle et al. (1975) (see also Hartle et al. 1973, Hodges 1974), in which neutral particles follow ballistic trajectories and possess a Maxwellian distribution of energies characterized by the local surface temperature. It is assumed that the surface is saturated in helium and therefore cannot act as a sink. Because the nightside surface is about 100°C, 450°C colder than the dayside, particles spend more time on the nightside and cause the density to be higher there. Hartle et al. find a night-to-day surface density ratio of 200. However, this ratio is apparently too large, and the predicted nightside scale height is too small to match the observations of Mariner 10, which suggests that the surface boundary condition may be more complex than assumed (Broadfoot et al. 1976).

The solar wind is a likely source of helium, for, although most of it is excluded from the magnetosphere, less than a tenth of a percent of its flux need be neutralized at the surface in order to fulfill the source requirement, if the dominant loss mechanism is thermal escape. It seems likely that this small amount could enter the magnetosphere via the tail, through the cusp regions, or by diffusion across the magnetopause. On the other hand, the radioactive decay of uranium and thorium in the crust may also be sufficient to maintain the low helium density.

The hydrogen component is interesting in that two populations appear to be present: "thermal" gas, which has an exospheric structure near that expected from equilibrium with the surface, and "nonthermal" gas, which has a much smaller scale height than the former, and predominates at altitudes below 300 km. The reactive nature of hydrogen makes the problem of identifying its sources and sinks difficult: Thomas (1974) has suggested that photolysis of water may play an important role in the hydrogen budget. Model atmospheres that include such a process have yet to be constructed.

It is possible that photo ionization followed by migration to the magnetopause, where the ions are swept away in the solar wind, is an important loss mechanism. This is particularly true for heavy gases such as neon, argon, carbon dioxide, and water, whose thermal escape times are long. Kumar (1976) has estimated upper limits to their supply rates imposed by the observed density limits, and the assumption that photo ionization followed by solar wind loss is the primary sink. The problem is complicated by the possibilities that these gases can condense on the nightside or collect in permanently shaded areas near the poles, and that surface interactions may not be in steady state. Nevertheless, in the cases of CO_2 and H_2O at least, the supply rates appear to be much lower than would be expected if Mercury were outgassing at a rate comparable to the Earth. Either Mercury is intrinsically volatile deficient (as would be consistent with Lewis' 1972 compositional model), or the planet is merely inactive, at least in its outer layers.

MAGNETIC FIELD AND MAGNETOSPHERE

The magnetometer aboard Mariner 10 revealed the existence of a weak but apparently permanent magnetic field associated with Mercury (Ness et al. 1974, 1975, 1976). Although the strength of the field is much less than that of the geomagnetic field, it is strong enough to form a magnetosphere similar to Earth's in several respects. Bow-shock, magnetosheath, and magnetopause are identifiable in the magnetic and plasma data (Ness et al. 1974, 1976, Ogilvie et al. 1974); the near-planet field is well represented by a dipole (with moment 5×10^{22} G cm^3) aligned approximately parallel to the rotation axis, and with the same polarity as the Earth's field. There are indications that the magnetosphere is stretched in the antisolar direction, possibly forming a long magnetotail with an imbedded neutral sheet (although Mariner 10 did not explore this region). Furthermore, there is evidence that Mercury's magnetosphere, like Earth's, is subject to substorms—complex transient distortions caused by interaction with the solar wind (Siscoe et al. 1975).

There is, however, an important difference between Mercury's and Earth's magnetospheres. Mercury occupies much more of its magnetosphere than does the Earth. The average distance from the center of Mercury to the nose of the magnetosphere is only 1.6 planetary radii, whereas for Earth this distance is more than 10 planetary radii. Therefore, charged particles within the magnetosphere encounter Mercury much more readily than their counterparts intercept the Earth, with the result that there is no trapped radiation region similar to the Van Allen belts. There may in fact be infrequent periods when the solar wind is strong enough to compress the field to the surface of the planet (Siscoe & Christopher 1975).

The source of Mercury's magnetic field has not been established. An active magnetohydrodynamic dynamo within the planet is a favored hypothesis, even if only by analogy with other large-scale sources of magnetism in the solar system—the Sun, Earth, and Jupiter. However, neither theory nor observations have yet provided a means of predicting the magnetic field produced by a fluid dynamo using a specified energy source; only the most rudimentary requirements can be defined at this time. It is certain that the magnetic Reynolds number $R \equiv 4\pi\sigma LV/c^2$ must exceed unity in order that fluid motions can build up magnetic field faster than it diffuses. (The electrical conductivity is denoted by σ, c is the speed of light, and L and V are typical length and velocity scales.) Applied to planetary interiors, this requirement immediately implies molten material of high electrical conductivity. For Mercury, iron is a sufficiently good conductor; molten silicates are probably not (Stevenson 1975). Rotation has long been cited—but never proved to be—a necessary condition for astrophysical dynamos. The amount necessary may be minimal; only a slight amount of angular velocity of a planetary-sized dynamo (of low viscosity) is necessary for the Coriolis force to dominate the momentum balance, the condition that properly characterizes "rapid rotation" (e.g., see Gubbins 1977). Interestingly, this theoretical result has emerged only after the discovery of the mercurian field.

The power necessary to maintain a dynamo must be greater than that dissipated within it. For Mercury, this is probably about a few times 10^{10} ergs sec^{-1} (Gubbins 1977), which is not severe. However, the energy necessary to maintain a molten core is substantial, and is discussed later.

There are other possible sources for the mercurian field besides a dynamo (Stevenson 1974). For instance, the field may be due to the permanent magnetization of iron-bearing rocks in the outer layers of the planet. Such magnetization is found in rocks from the Earth, Moon, and meteorites, and is most commonly acquired by the cooling of the rock through the Curie temperature in the presence of an ambient field. (The Curie temperature of iron is 770°C.) If there was once a dipolar field due to a dynamo that has since died out (perhaps by depletion of its energy source of solidification of the core), a dipolar remanent field would remain, its source being the residual moment retained by those layers of the planet that cooled through the Curie point while the dynamo was active. Stephenson (1976) has determined the degree of magnetization that would be necessary for such a field to be comparable to that measured by Mariner 10 and concludes that the mercurian field might indeed be a remanent one.

The interplanetary electromagnetic field can generate measurable planetary fields by induction, as occurs in the Moon (Sonett & Colburn 1968, Colburn, Sonett & Schwartz 1972). However, such an induction mechanism is incapable of providing as strong and steady a field as that observed by Mariner 10 (Herbert et al. 1976, Ness et al. 1976).

SURFACE FEATURES

Mariner 10 acquired over 2700 useful pictures of Mercury during its three encounters, which covered approximately 50% of the surface at resolutions varying from about 100 meters to 5 km. Although this portion of Mercury's surface is remarkably similar to the Moon's, there are significant differences that indicate important departures between the surface evolution of Mercury and the other terrestrial planets.

In general, the surface of Mercury is pockmarked with craters ranging in diameter from at least 100 m (highest resolution) up to basins 1000 km across (Figure 1). They present a spectrum of degradational types similar to those on the Moon, and many exhibit extensive ray systems. Large areas of lightly cratered smooth plains similar to the lunar maria fill and surround the major basins. However, the albedo of this material is similar to the more heavily cratered terrain, and therefore the contrast is considerably less than that between the lunar maria and highlands. Furthermore, mercurian crater morphology is somewhat different from that of lunar craters, probably as the result of Mercury's greater surface gravity (Gault et al. 1975). Large lobate scarps occur in most areas of Mercury and are probably of tectonic origin. These scarps are unique to Mercury and indicate that the tectonic style and history has been quite different from that of the other terrestrial planets (Strom, Trask & Guest 1975).

Major Physiographic Provinces

Mercury's surface can be divided into three major physiographic provinces: intercrater plains, heavily cratered terrain, and smooth plains. In addition to these major terrain types, several localized units can be identified. By far the most interesting is the hilly and lineated terrain which is antipodal to the large, relatively fresh Caloris basin. Figure 2 is a generalized terrain map of Mercury modified from Trask & Guest (1975) which shows the distribution of these terrain types.

The intercrater plains are probably the most widespread terrain unit on Mercury. They occur between and around clusters of large craters comprising the heavily cratered terrain, and are characterized by a level to gently rolling surface covered by a high density of superposed small craters in the size range 5–10 km (Figure 3a). Many of these small craters form chains or clusters, and individual craters are often elongated and open at one end. These characteristics are common to secondary impact craters and therefore it is probable that most of these superposed craters are of secondary origin. The age of the intercrater plains relative to the other units is of paramount importance for deciphering the surface history of Mercury. Trask & Guest (1975) consider the intercrater plains to be older than the heavily cratered terrain and therefore the oldest exposed unit on Mercury because 1. they do not appear to embay or cut individual craters of the heavily cratered terrain, and 2. the only apparent source of the many secondary impact craters on the intercrater plains appears to be the large craters and basins of the heavily cratered terrain. However, at several locations within the intercrater plains are isolated, partially buried craters similar to the "ghost rings" found in the lunar maria. This suggests that at least in part the intercrater plains may be younger than the heavily cratered terrain. More detailed studies of this terrain type are required before firm conclusions can be reached regarding its relative age.

The origin of the intercrater plains is even more uncertain than their age. A similar terrain of very restricted distribution occurs on the Moon southwest of the Nectaris basin and is called the pre-Imbrium pitted plains (Wilhelms & McCauley 1971). This terrain has been interpreted as basin ejecta by Howard, Wilhelms & Scott (1974), and a similar interpretation has been applied to the mercurian intercrater plains by Wilhelms (1976). However, a major difference between the Moon and Mercury is that the mercurian intercrater plains are the most widespread terrain on Mercury (see Figure 2) and cover a much greater area than their lunar equivalent. Furthermore, Mercury appears to have fewer multi-ring basins in the size range 200–1000 km (Murray et al. 1974) than the Moon, and the ejecta from these basins has a more restricted ballistic range than on the Moon (Gault et al. 1975) because of Mercury's greater surface gravity. This great areal distribution of intercrater plains, restricted ballistic range, and apparent paucity of source basins argue against a basin ejecta origin for the mercurian intercrater plains. If the majority of intercrater plains are the oldest unit and predate the heavily cratered terrain, then they may represent an ancient, primordial surface that does not have a lunar counterpart because of the greater density of lunar basins and greater ballistic range of ejected lunar material (Trask & Guest 1975). An alternative explanation is that if much of the intercrater plains are younger than the heavily cratered terrain, they may represent a volcanic episode intermediate in age between the formation of the heavily cratered terrain and smooth plains. A complicating factor in choosing between interpretations is the unknown degradational effects of the global-wide seismic disturbances produced as a result of the monstrous impact event that formed the 1300-km diameter Caloris basin (Guest & Gault 1976). Clearly more detailed investigations are required to decide between competing hypotheses.

The heavily cratered terrain primarily consists of clusters of closely packed, overlapping craters ranging from about thirty to several hundred kilometers in diameter (Figure 3a). Ejecta blankets and discrete secondary crater fields are generally lacking around craters comprising this terrain. The interiors of many of the craters are filled with smooth plains that are much less cratered and younger than the adjacent intercrater plains. In appearance the terrain is very similar to the lunar highlands and undoubtedly resulted from a heavy bombardment of small planetesimals early in Mercury's history.

The mercurian smooth plains form sparsely cratered, relatively level surfaces

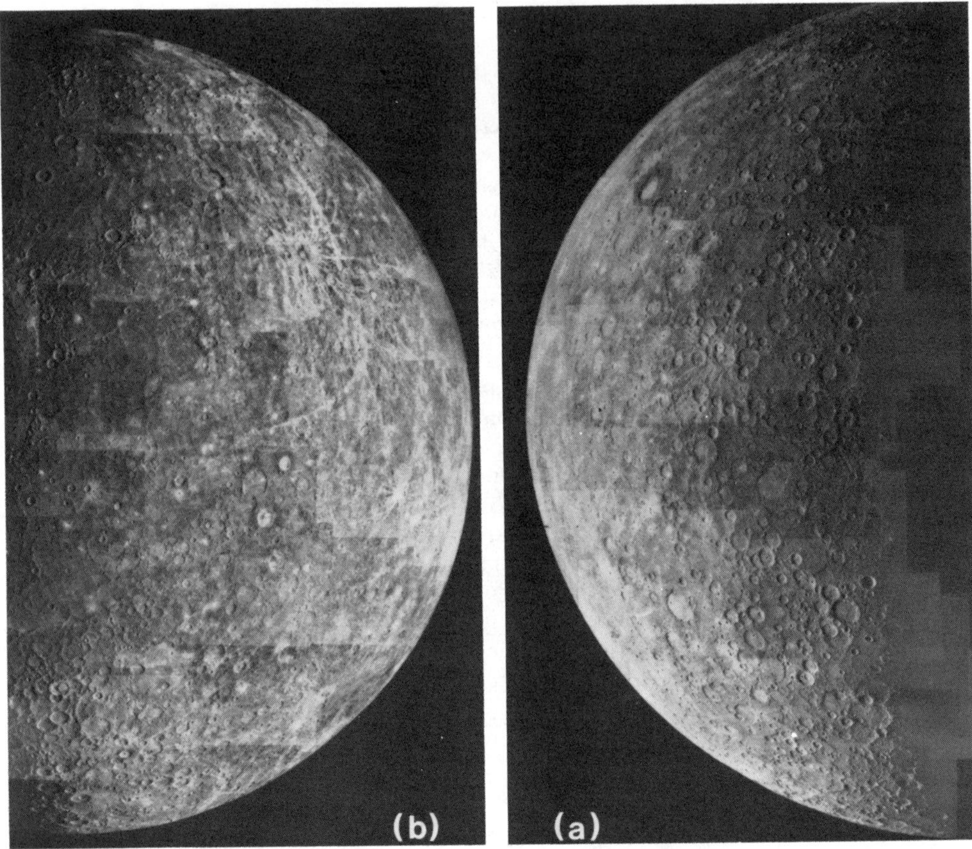

Figure 1 Photo mosaics of high-resolution pictures of Mercury obtained during Mariner 10's first encounter with the planet on 29 March 1974.
(a) "Incoming" hemisphere visible as spacecraft approached the planet. The equator lies about 20° above the center of the photo, and the evening terminator lies near 10° west longitude. The surface exhibits a heavily cratered terrain totally unlike the "outgoing" hemisphere. NASA No. 74-H-239 JPL Photo.
(b) "Outgoing" hemisphere visible after spacecraft passed planet. The equator lies about 20° below center of the photo with the morning terminator near 190° west longitude. The 1300-km diameter Caloris basin and large areas of smooth plains contrast with the older heavily cratered terrain in the incoming hemisphere. NASA No. 74-H-253 JPL Photo.

(Figure 3b), which in some respects are similar to the maria (Strom, Trask & Guest 1975) and in other respects are similar to the lunar light plains known as the Cayley formation (Wilhelms 1976). Overlap relations and the density of superimposed craters indicate that they are younger than the more densely cratered areas and among the youngest surfaces on Mercury (Trask & Guest 1975, Guest & Gault 1976). Craters in the size range 5–10 km are much less abundant than on the intercrater plains. Numerous ridges, which grossly resemble both lunar and martian wrinkle ridges, occur in the large expanses of smooth plains. However, the mercurian ridges generally lack the crenulated crests so characteristic of lunar and martian ridges. The most extensive tracks of smooth plains occur in and around the Caloris and north polar basins (Figures 1b and 2b). Other patches mostly occupy the floors of basins and large craters. Unlike the lunar maria, the albedo of the smooth plains does not contrast sharply with the surrounding heavily cratered terrain or intercrater plains (Hapke et al. 1975).

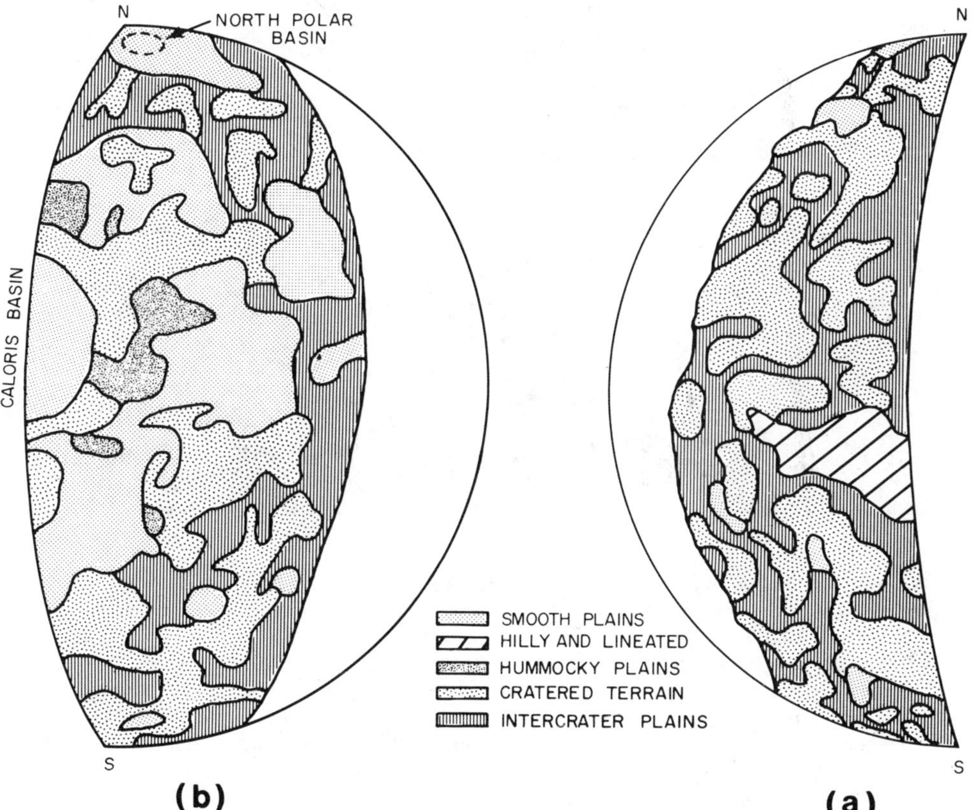

Figure 2 Generalized geological terrain map of Mercury modified from Plate 1 of Trask & Guest (1975). (a) Incoming hemisphere; (b) Outgoing hemisphere.

Two possible origins have been proposed for the smooth plains: volcanic and basin ejecta. Evidence cited in support of a volcanic origin includes: 1. the large volume of smooth plains; 2. the great differences in volume of the plains material around basins of comparable size; 3. the similarity in morphology and distribution between the smooth plains and the lunar maria; 4. stratigraphic relationships which indicate that the smooth plains are younger than many of the basins in which they lie; 5. the large volume of plains peripheral to the north polar basin, making derivation by impact from that basin unlikely; 6. contrasting albedo and

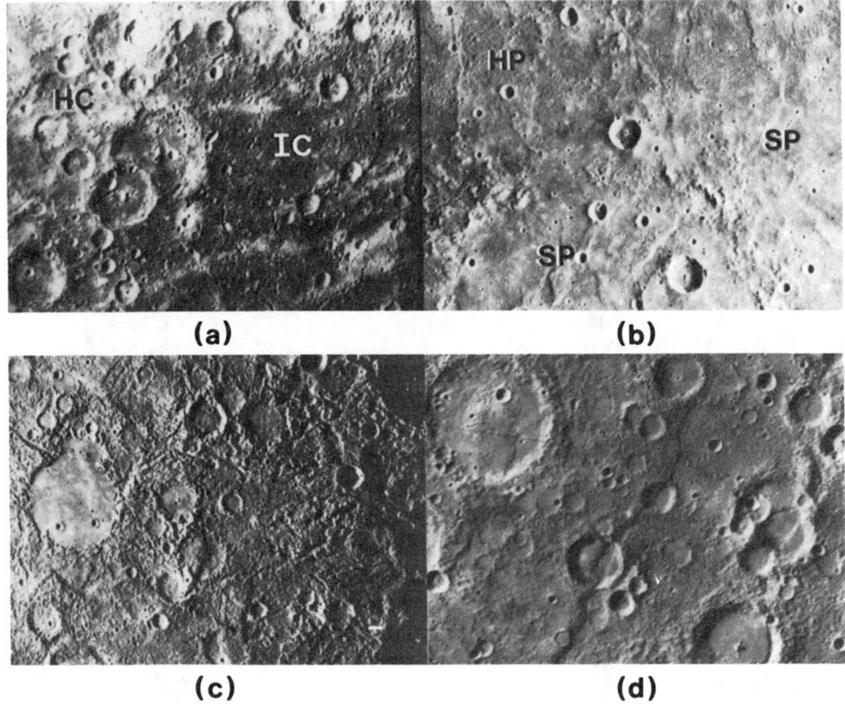

Figure 3 Examples of the major physiographic provinces. (a) Intercrater plains (IC) and heavily cratered terrain (HC) centered at 56°S, 128°W. View is about 400 km across. After Trask & Guest (1975). (b) High-resolution view about 320 km across showing smooth plains (SP) and hummocky plains (HP) about 500 km east of the Caloris basin. Part of a pre-Caloris basin 240 km in diameter in the lower part of the scene is filled with smooth plains. At the western margin of this basin is a west-facing scarp that forms the boundary between hummocky plains (west) and smooth plains (east). After Strom, Trask & Guest (1975). (c) High-resolution view of the hilly and lineated terrain that is antipodal to the Caloris basin. Scene is about 500 km across centered at 31°S, 19°W. (d) Photomosaic of one of the most prominent lobate scarps (Discovery Scarp). This feature is about 550 km long and transects two craters 55 and 35 km in diameter. Maximum height of scarp is about 3 km.

color of some of the plains material compared to the immediate surroundings; and 7. a lack of source basins for plains filling older basins (Strom, Trask & Guest 1975, Trask & Strom 1976). Wilhelms (1976), on the other hand, considers that the smooth plains more closely resemble the lunar light plains, which are probably largely basin ejecta. He cites similarities in stratigraphic relations, surface morphology, and albedo contrasts between the smooth plains and lunar light plains (Cayley formation) as evidence of a basin ejecta origin for the mercurian deposits, and suggests that the ejecta behaved more like a fluid than did lunar basin ejecta because of the greater surface gravity of Mercury. The question of mercurian volcanism should be kept open pending more conclusive observational evidence or more convincing theoretical arguments. However, the evidence for volcanism outlined above, together with thermal history models required to account for at least a partially molten core apparently necessary to explain the magnetic field, makes past mercurian volcanism a good working hypothesis.

A peculiar topography of localized distribution comprises one of the most unusual terrains on Mercury. This peculiar set of landforms occurs only in one area viewed by Mariner 10, which is antipodal to the Caloris basin. It consists of hills 5–10 km wide and 0.1–1.8 km high and several large linear valleys arranged in an orthogonal pattern (Figure 3c). The terrain includes crater rims that have been broken up into hills and depressions. However, the floors of many of these craters are occupied by smooth plains that have not been affected by the process that caused the hilly and lineated terrain. Because somewhat similar terrain occurs at the antipodal regions of the Imbrium and Orientale basins on the Moon, Schultz & Gault (1975) have postulated that the terrain may have formed at the same time as the Caloris basin by the focusing of seismic energy at the antipodal point. Moore et al (1974) have attributed the lunar terrain to clustering of basin-related secondary impacts at the antipode, but this is an unlikely process on Mercury due to the requisite ejection velocities (3.5–4 km sec^{-1}) necessary for ballistic transport to the antipode. Alternatively, Wilhelms (1976) has suggested that the intersecting linear structures are secondary crater chains formed by ejecta from mercurian basins hidden in the terminator.

Structure of the Caloris Basin

The Caloris basin is the largest structure viewed by Mariner 10 (Figure 4). It is 1300 km in diameter and resembles the lunar Imbrium basin in both size and morphology (Strom, Trask & Guest 1975). The interior of the basin is occupied by smooth plains that are highly fractured and ridged. The basin perimeter is defined by a ring of irregular mountains averaging about 2 km in height above the floor. A weak outer scarp occurs at a distance of 100–160 km from the main scarp in the northeastern part of the basin, and between these two scarps lie relatively smooth hills and domes. An extensive system of valleys and ridges radiates from the basin for a distance of about one basin diameter and strongly resembles the lunar Imbrium radial system. Both radial valleys and old craters are embayed by plains material that surrounds the basin out to three basin radii. This plains material can be divided into two units: smooth plains and hummocky plains (Trask &

Guest 1975, Strom, Trask & Guest 1975). The texture of the hummocky plains and their proximity to the basin rim (Figure 3b) suggest that they are basin ejecta excavated from the basin by the impact that formed it. As discussed previously, the smooth plains may be in part or whole volcanic flows emplaced after the Caloris event (Strom, Trask & Guest 1975, Trask & Strom 1976), or a smooth ejecta facies similar to the lunar light plains (Wilhelms 1976).

The floor structure of the Caloris basin appears to be unique; no basin on the Moon or Mars shows this type of structure. It is characterized by a high density of fractures and ridges that both show two primary orientations: one concentric with respect to the border of the basin, and the other radial. Furthermore, the intensity of the fracturing progressively increases basinward. Both fractures and ridges are probably due to readjustment of the basin floor subsequent to emplacement of the floor material.

Figure 4 Photomosaic of the Caloris basin showing the highly ridged and fractured floor of the interior, the well-developed radial system of scarps in the northeastern portion of the basin, and the extensive areas of smooth plains surrounding the basin.

Tectonic Framework

The tectonic framework of Mercury is characterized by large and widely distributed lobate scarps (Strom, Trask & Guest 1975). These features are unique to Mercury and form relatively steep escarpments that show a broadly lobate outline on a scale of a few to tens of kilometers (Figure 3d). They vary in length from about twenty to over five hundred kilometers and have heights of a few hundred meters to about three kilometers. The crests of the scarps are rounded in contrast to the sharp crests formed by normal faulting and graben on the Moon and Mars. They often transect a variety of terrain types, and in at least one case a crater rim has been offset by about 10 km and appears to have had its radius shortened. These transectional and morphological relationships indicate that the majority of the scarps are probably reverse or thrust faults due to compressive stresses. However, about 17% of the scarps are confined to smooth plains on crater floors or occur at the interior margins of old basins. The origin of these scarps is not clear; they could be flow fronts rather than fault scarps.

There are only two local areas on Mercury that show evidence of tensional stresses. One of these areas is the highly fractured floor of the Caloris basin, while the other is the hilly and lineated terrain antipodal to it. Therefore, the surface of Mercury is dominated by structures indicative of compressive stresses.

With the exception of the small percentage of scarps confined to crater floors, lobate scarps of probable tectonic origin appear to have a rather uniform distribution over that part of Mercury viewed by Mariner 10. If this region of Mercury is representative of the planet as a whole, then lobate scarps probably have a global distribution. Therefore, the entire planet appears to have been subjected to compressive stresses resulting in a general crustal shortening. Estimates of the amount of horizontal displacement represented by lobate scarps suggest that there has been a decrease in surface area of about 6.3×10^4 to 1.3×10^5 km^2, which represents a decrease in radius of about 1–2 km (Strom, Trask & Guest 1975). This value is consistent with a 2-km decrease in radius based on thermal history models, which predict a contraction of the lithosphere due to cooling after core formation (Solomon 1976).

Craters

The morphology of the mercurian craters, down to the limits of the photographic resolution, is grossly similar to that of their lunar counterparts. The craters unquestionably represent, as on the Moon, the end products of impacts by the complete size spectrum of meteoritic material, and differences between the two families of craters are attributed by Gault et al. (1975) to the factor of 2.2 difference in the respective gravitational acceleration environments. The primary morphologic elements (rim ejecta deposits, fields of satellitic secondary craters, terraces on inner walls, central peak(s) and/or inner rings of mountains) that characterize lunar craters are all associated with craters on Mercury. Moreover, the morphology changes from sharp to soft with increasing age of the features, as on the Moon, and the crater elements ultimately disappear or become barely recognizable as the crater is degraded to a low rim that is frequently partially obliterated by superposed craters. Thus, the primary degradational process on Mercury is probably the same as on the Moon; i.e. erosion and ballistic sedimentation caused by meteoritic bombardment.

The smallest craters are predominantly bowl-shaped, and with increasing size the craters exhibit a systematic progression of morphologic types beginning with the appearance of central peaks and terracing on the inner walls (Figure 5a and 5b). Further increases in size are accompanied by the appearance of complex structures of central peaks which undergo a transition into an inner ring of mountains that is

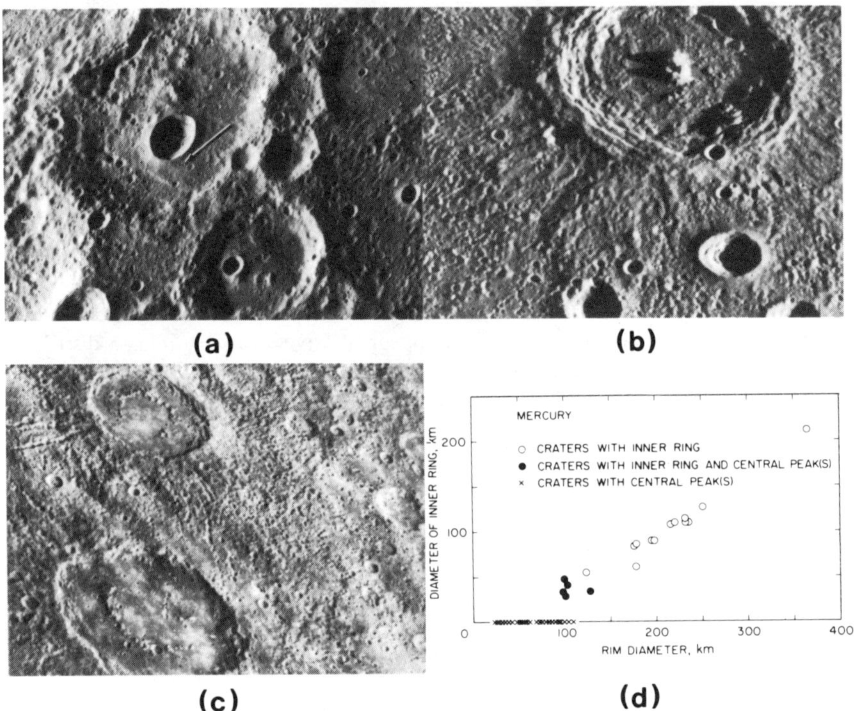

Figure 5 Examples of the morphologic progression of mercurian craters with increasing size. (a) Small bowl-shaped craters to larger features with incipient terraces and central peak(s). The fresh, sharp crater (centered) is a 20 km diameter structure with Mercury's surface longitude reference crater Hun Kal on its southern flank indicated by arrow. (b) A 98-km diameter crater illustrates the typical narrow, hummocky rim deposits, radial ridges, and surrounding extensive field of secondary craters. Interior terraces and central peaks are well-developed and typical of mercurian craters of this size. Smaller craters in foreground, about 25 km diameter, are also terraced. After Gault et al. (1975). (c) Two craters, 128 and 195 km in diameter, that have interior rings of mountains and ejecta deposits deeply scarred by chains of secondary craters. (d) Changes in interior structure of mercurian craters and inner ring diameter as a function of the rim diameter. From Gault et al. (1975).

concentric with—and approximately one half the diameter of—the main rim of the crater (Figure 5c and 5d). This progressive change in crater geometry also occurs on the Moon, but the changes from one morphologic type to another occur at smaller diameters on Mercury. Transition at smaller diameters on Mercury is consistent with differences in gravitational acceleration if the changes in morphologic type are caused by gravity-induced modifications (collapse) of the original craters of excavation (Gault et al. 1975).

The youngest craters (sharpest morphology), especially those near the limb under conditions approaching zero-phase lighting, display well-developed, extensive bright-rayed ejecta systems extending hundreds of kilometers and a few cases extending more than a thousand kilometers from their parent crater (Figure 1b). Some of these rayed craters are also centered within dark halos comprised of low albedo material. Although such prominent rays of ejecta and dark halos are common to lunar craters, it is the morphology of the main ejecta deposits on the flanks of the crater rims that distinguishes mercurian craters from their lunar cousins. For both families of craters the rim deposits consist of a hummocky facies that grades out with increasing distance from the rim into a radially ridged facies which, in turn, changes into a zone of satellitic secondary craters and discontinuous ejecta deposits (Figure 5b). On Mercury, the hummocky and radially ridged facies—the ejecta deposits termed the continuous deposits—grade into the satellitic crater field over a much shorter distance than on the Moon. Additionally, the areal density of the craters in the satellitic zone is very high compared to lunar secondary crater fields. Chains of overlapping craters resulting in long, linear grooves (subradial to the primary crater) extend across the continuous deposits almost up to the crater rim (Figure 5c), a characteristic that is rare on the Moon. The narrower continuous deposits, linear grooves, and increased areal density of the mercurian secondary craters relative to lunar craters is totally consistent with the reduced ballistic range of ejecta caused by the differences in the gravity environments between the two planetary bodies.

Size-frequency distributions of mercurian craters have been obtained as a basis for determining relative ages of the major physiographic provinces and correlating interplanetary geologic time (Murray et al. 1974, Guest & Gault 1976). The cratered terrain, comprised of both the intercrater plains and heavily cratered terrain provinces, has a crater size-frequency distribution effectively the same as for the southern highlands on the front side of the Moon. The two provinces were combined for purposes of crater counting because it is difficult to separate the two provinces into discrete areas (Guest & Gault 1976), although it is recognized that they may represent two different epochs in Mercury's history. Frequency distributions for the mercurian smooth plains are greatly reduced relative to the cratered terrains and are almost identical to those for the Apollo 14 landing site in the Fra Mauro formation. Crystallization ages of rocks from this Apollo site have been dated to be about 3.9×10^9 years (Papanastassiou & Wasserburg 1971, Turner et al. 1971), but the similarity in the mercurian smooth plains and Apollo 14 crater populations does not equate directly to similar absolute ages. Differences between the Moon and Mercury in their gravity environments, average impact velocities, and the sources (i.e. time-integral fluxes) of the bodies forming the craters have a direct effect on the resultant frequency distributions that precludes dating in absolute terms. The sources or origins of the crater-forming bodies is a subject of particular significance in the geologic history and are discussed briefly later; it is sufficient here to note that the mercurian heavily cratered terrains constitute the oldest surfaces on Mercury and if analogous to the lunar highlands must be older than 4 billion years (Murray et al. 1975).

Guest & Gault (1976) have examined the morphology of the mercurian crater populations and find that all young, morphologically sharp mercurian structures smaller than about 30 km must have been formed subsequent to the formation of the Caloris basin. Their results suggest that either some process or event degraded such craters prior to (or contemporaneously with) the formation of Caloris or, possibly, such smaller, sharp craters were never formed in the time period before the Caloris event; the observations cannot distinguish between the two alternatives and the question remains open to further investigations.

Optical and Thermal Properties

Mercury's optical and thermal emission properties (normal albedo, photometric function, polarization, spectral characteristics, thermal inertia, etc.) have been measured from Earth and are virtually identical to those of the Moon (Lyot 1929, Harris 1961, de Vaucouleurs 1964, Irvine et al. 1968, Morrison 1970, McCord & Adams 1972a, b, Dollfus & Auriere 1974, Vilas & McCord 1976, Chase et al. 1974, 1976). Observations by Mariner 10 have supplemented these Earth-based data of the integral disk in addition to providing results in the visible for surface resolutions of 20 km (Murray et al. 1974, Hapke et al. 1975) and 40–90 km in the far infrared (Chase et al. 1974, 1976).

Normal albedos at 0.554-μm wavelength were obtained by Mariner 10 for large areas of the incoming and outgoing hemispheres (Figure 1). Values of 0.16 and 0.12 were found (Hapke et al. 1975) for, respectively, areas consisting primarily of intercrater plains and smooth plains with an average value of about 0.14. For comparison, the recent polarimetric determinations by Dollfus & Auriere (1974) for the integral disk give 0.13. Although phase-angle observations by Mariner 10 were restricted to a narrow range because of trajectory and camera pointing limitations, relative brightness curves across the illuminated disk are in excellent agreement with typical lunar curves for the same phase angles. Photometrically, therefore, as major physiographic provinces, the brighter intercrater plains resemble the lunar highlands and the darker smooth plains resemble the lunar maria. Albedos vary within these provinces from 0.09 to 0.21 with the suggestion that there are two types of smooth plains; one with an albedo of about 0.13 within and around the Caloris basin and a second with an albedo of around 0.20 found in both the incoming and outgoing hemispheres of Figure 1. Mercurian rayed craters are generally brighter than lunar rayed craters with some albedos as high as 0.45 (Hapke et al. 1975). With the exception of these bright features there are no large albedo gradients on the planet comparable to the lunar highland-maria contrasts.

Polarization measurements of Mercury, first performed by Lyot (1929), are recently summarized by Dollfus & Auriere (1974). Measurements obtained for different areas of the planet and for the integral disk at different wavelengths are virtually identical to corresponding Earth-based measurements for the Moon. In particular, the negative branch of the polarization curves is indicative of the presence of a dark, fine-grained powder similar to the lunar regolith on all of Mercury's surface. Mariner 10 polarimetric observations (Hapke et al. 1975) with a surface resolution of 20 km failed to detect any anomalously high polarization that would be indicative of landforms not covered by a regolith-type soil. Consistent with the absence of large albedo gradients, variations in polarization across the surface are minimal, less than 0.02 within the value of 0.12.

Evidence for mineralogy and composition of surface materials is contained in spectral reflectance measurements in the wavelength region from about 0.3 to 2.0 μm (e.g. Adams 1968, 1974, 1975). McCord & Adams (1972a, b), using a 22-channel system operating from 0.32 to 1.05 μm, found a constant slope for the reflection spectrum of the integral disk with the exception of a weak absorption band near 0.95 μm common to pyroxenes. Because the spectral curve matches closely that for the lunar highlands and maria, they conclude that Mercury's surface is mantled with a lunar-like regolith of iron and titanium rich glasses. More recent observations with refined instrumentation (Vilas & McCord 1976) essentially re-affirm the earlier results; some differences are noted in the spectra that could be attributed to differences in composition for the two different regions of Mercury observed.

Although the Mariner 10 polarization pictures are bland, the planet's surface reveals considerable color variations (Hapke et al. 1975). The color images were formed as composites of the same scenes taken through filters having effective wavelengths of 0.355 and 0.575 μm. Areas with the same reflectance spectrum appear uniformly grey; areas of redder or bluer materials appear, respectively, as brighter or darker features. These "color" pictures reveal that all fresh bright-rayed craters are consistently bluer than their surroundings. No red-rayed craters were found, although most other bright areas generally were redder in the same manner as on the Moon. The unusual combination of high albedo and bluish color of the rayed craters suggests to Hapke et al. (1975) that, in marked contrast to the conclusions of McCord & Adams (1972a, b), the mercurian crustal material is low in titanium and iron (Fe^{+3}). The difference between the two analyses remains unresolved, but the difference may be real because the two data sets do not correspond to the same area of the planet; the Earth-based data correspond to surfaces that were behind the terminators during the Mariner 10 flybys.

Observations at longer wavelengths in the far infrared and microwave regions reinforce the lunar-like characteristics of the mercurian surface materials (e.g. Morrison 1970). Most recently, Chase et al. (1974, 1976) employed a two-channel infrared radiometer (11 and 45 μm) on Mariner 10 to scan the dark side of the planet along an approximately equatorial band with a 40- to 90-km resolution. Except for a region of strong local enhancement, values of the thermal inertia $(k\rho c)^{1/2}$ ranged from about 0.0016 to 0.0026 cal cm^{-2} sec$^{-1/2}$ °K^{-1} (k is thermal conductivity, ρc is heat capacity/volume). Such values together with derived thermal skin depths for diurnal variations, electrical skin depth, and loss tangent all fall within the range of values for the Moon.

It is interesting that the highest values of the thermal inertia occurred in the quadrant behind the morning terminator. This seems compatible with a young surface that presumably includes the hidden half of the Caloris basin structure (and external smooth plains) if it has circular symmetry. The largest values of thermal inertia (0.0031) occur in a region of local enhancement 400 km wide, part of which is within an area probed, but not evident, by radar (Zohar & Goldstein 1974). However, a second region of enhancement (40 km) appears to correlate with a similar-sized region of strong radar backscatter, suggesting a young crater surrounded by a rocky ejecta deposit. Two additional anomalies lay outside radar coverage, one of which could be another young crater.

Microwave radiometry samples of the integral mercurian surface to meter depths are in substantial agreement with observations at the shorter wavelengths; data for Mercury closely approximate that which would be expected for the Moon if it were placed in a mercurian orbit. The reader is referred to an excellent review and analysis of the literature by Morrison (1970) (see also Devine 1972). More recent microwave measurements and refined analyses are reported by Morrison & Klein (1970), Ulich, Cogdell & Davis (1973), Briggs & Drake (1973), and Cuzzi (1974).

Cartography

Acquisition of the Mariner 10 imagery has initiated a program of high-resolution mapping of Mercury (Davies & Batson 1975). The coordinate system used in this effort assumes that the equator lies in the plane of the orbit and the center of a small crater (named Hun Kal, see Figure 5a and below) defines the 20° meridian; longitude is measured from 0° to 360°, increasing to the west. For comparison, the IAU (see Davies & Batson 1975) define 0° longitude to be the subsolar meridian at the first perihelion after January 1, 1950 (Julian date 2,433,292.63) with the spin axis normal to the orbital planet. The Mariner 10 system places the IAU zero meridian at approximately 359.5° with an uncertainty of about 0.5°. The use of a body-fixed (crater) criterion for longitude is advantageous in preserving the validity of the relative positions of features on the maps until future exploration permits relating the two systems with greater accuracy than is currently possible (20 km).

The cartographic products consist of 1:5,000,000 shaded relief maps prepared by the U.S. Geological Survey, Flagstaff, Arizona. The surface of Mercury in this series is divided into 15 areas: five Mercator projections encircling the planet between 25° north and south latitudes; eight Lambert projections between 20° and 70° latitudes; and two polar stereographic projections from 65° to the poles.

Nomenclature for the topographic features on the maps, as established by the IAU (Morrison 1976), comprises six major topographic features: craters, valleys, scarps, ridges, plains, and mountains. Craters are named for authors, artists, and musicians (e.g. Homer, Renoir, Bach). Two exceptions are Kuiper and Hun Kal; the former commemorates Dr. Gerard P. Kuiper, who was a member of the Mariner 10 Imaging Team and who died prior to the spacecraft's encounters with Venus and Mercury; the latter (the number twenty in ancient Mayan) defines the

20° meridian. Valleys (vallis) are named after radio observatories, and scarps (rupes) are given names of ships associated with exploration and scientific research. Ridges (dorsum) are not named after any specific group, but plains (planitia) are assigned names for the word meaning Mercury (planet or god) in various languages.

INTERIOR

Composition and Structure

The mass and radius determinations imply that Mercury must contain a large fraction of iron—the only heavy element sufficiently abundant to account for the high density. As inferred from solar abundances, and by analogy with terrestrial, meteoritic, and lunar materials, it is presumed that silicates of iron and magnesium are also prevalent. The similarity of surface morphology to lunar counterparts suggests (see below) a mercurian mantle of differentiated silicates at least grossly like the Moon in composition. This view is borne out by ground-based and Mariner 10 observations which indicate that the radio and optical (including IR) properties of Mercury's surface are essentially the same as those of the lunar surface (Harris 1961, de Vaucouleurs 1964, Hameen-Antilla, Pikkarainen & Camichel 1970, Dollfus & Auriere 1974 and references therein, McCord & Adams 1972a, b, Murray et al. 1974, Hapke et al. 1975, Vilas & McCord 1976).

Bounds on the amount of iron present can be determined by solving the spherically symmetric hydrostatic equation, $dP/dr = -\rho Gm(r)/r^2$, with an appropriate equation of state, $P = P(\rho, T)$, and requiring the solutions to conform to the observed mass and radius of the planet. (P is the pressure, ρ the density, T the temperature, G the gravitational constant, and m the mass within radius r.) The solutions, of course, are not unique. The equation of state depends on the temperature, but for solids and liquids at planetary pressures, thermal expansion is generally less than the volume change due to compressions, so that even constant temperature models are good approximations. More important are the extent to which the planet is differentiated (specifically, how much iron has separated into a core) and possible phase changes, both of which affect the equation of state to be used at a given radius. Detailed models of Mercury have been calculated by Reynolds & Summers (1969) which include two extreme hypotheses: 1. all iron is concentrated in a core; and 2. all iron is in oxide form distributed homogeneously throughout the planet. They showed that the overall abundance of iron is rather insensitive to assumptions regarding its distribution, a result confirmed by the calculations of Siegfried & Solomon (1974) (see also Kozlavskaya 1969). The mass fractions of total iron for all models fall within the range 0.60–0.71, as compared to about 0.34–0.38 for the Earth. The density and pressure are shown in Figure 6 for two core models. The radius of a fully differentiated core is expected to be approximately three fourths the planetary radius; this result is insensitive to assumptions regarding the detailed composition of the mantle. The central pressure does not exceed 500 kbar. In a homogeneous model it may be as low as 245 kbar.

Lewis (1972) has proposed a model of the chemical compositions of the planets based on the assumptions that: 1. equilibrium chemistry prevailed as the planetary material condensed from a hot solar nebula; 2. the structure of the nebula was similar to that proposed by Cameron (1969), particularly with regard to the variation of temperature with heliocentric distance; and 3. the material which did not condense was somehow removed from the vicinity of the planet before the nebula completely cooled. This model is popular because it provides a simple explanation for the mean planetary densities. For Mercury, Lewis predicts (besides an enrichment of iron relative to magnesium and silicon with respect to solar abundances) strong depletion of sulfur, alkali metals, and volatiles. At present, our knowledge of Mercury's composition is insufficient to confirm or refute such a model, except with regard to the enhanced iron abundance.

Thermal History

The principal constraints on Mercury's thermal history drawn from the Mariner 10 observations (see section on geologic history below) are the following: 1. the surface of the planet is the product of extensive differentiation, which implies that at least the outer portion of the planet melted; 2. the magnetic field is intrinsic to the planet, and is most likely associated with an internal dynamo (either extant or extinct), which implies global differentiation involving the formation of an iron core; 3. differentiation occurred early in the planet's history (probably within the first 1/2 billion

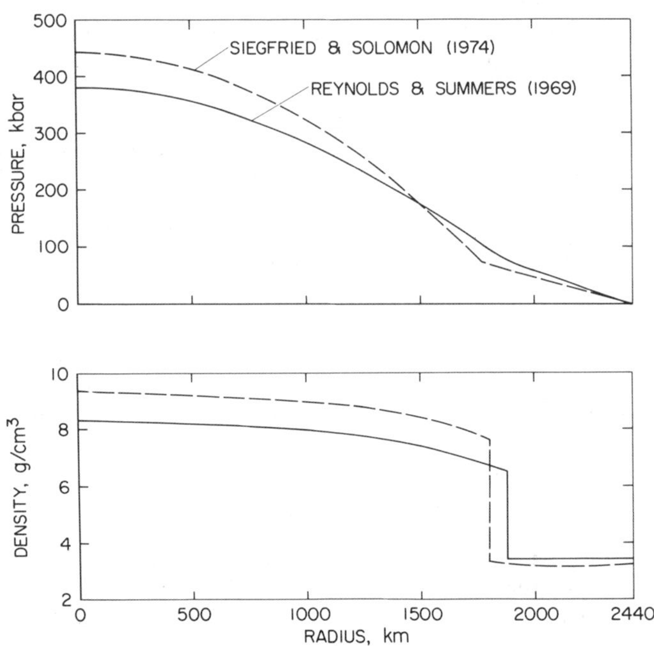

Figure 6 Pressure and density as functions of radius for two models of a differentiated Mercury. Differences are due primarily to different assumptions regarding the equations of state.

years); and 4. the surface of the planet has been largely undisturbed by tectonic processes for most of the time since differentiation. In addition, Solomon (1976) has argued that if a molten core existed at one time, it could not have completely solidified since the cessation of surface activity without resulting in contraction of the planet much greater than that which has been inferred from the compressional features described by Strom, Trask & Guest (1975).

What could be the source of energy for melting much of the planet earlier than 4 billion years ago? Mercury may have formed from material already at a relatively high temperature due to its proximity to the Sun; Lewis (1972) proposed that the planet is composed only of substances that would be condensed at 1100°C. In addition to this internal energy, gravitational energy equivalent to an average temperature increase of 10^3 °C, if all of it was retained, is released upon formation of the planet. However, much of this energy would be radiated to space. Models of the accretion process (Benfield 1950, Mizutani et al. 1972, Weidenschilling 1974) show that the amount of energy retained by the planet depends on the rate of energy deposition during formation; rapidly forming bodies retain more heat. Also, the accretional energy is deposited nonuniformly, with most of it going into the outer half of the planet. Siegfried & Solomon (1974) report that models based on the ad hoc mass accretion law of Hanks & Anderson (1969), and with an initial temperature of 1127°C, radiate away almost all of the gravitational energy if the accretion time is longer than 10^5 years. More detailed (but not necessarily more realistic) models based on the theory of the gravitational capture of small objects would retain more heat for the same accretion time. But recent work by Weidenschilling (1974, 1976) favors Safronov's (1969) estimate of accretion times for the terrestrial planets of $\sim 10^8$ years. Such slow accretion would yield negligible heating.

Accretional energy is deposited at the surface of a growing planet, and is therefore lost easily; energy deposited deep within a terrestrial planet is not lost quickly because rocks are good insulators. (In the absence of convection, the characteristic thermal transport time is greater than the age of the solar system.) It is well known that the Earth and Moon are heated from within by small amounts of long-lived radioactive elements—primarily ^{238}U, ^{235}U, ^{232}Th, and ^{40}K, which have half-lives on the order of 10^9–10^{10} years. The minimum heat source density H required to melt the center of Mercury in 1/2 billion years can be estimated by assuming that no heat escapes the region during that time. For a temperature increase of 2,000°C, $H = 8 \times 10^{-6}$ ergs cm^{-3}. This rate of heat generation is three times that which would have existed in chondritic meteorites 4.6×10^9 years ago, and four times that for the uranium-enriched, potassium-depleted Mercury implied by Lewis' model. Thus, if these radioactive elements provided the energy for early differentiation, their abundances must have substantially exceeded those found in other solar system objects, as well as current theoretical estimates.

Relatively short-lived isotopes—notably ^{26}Al (half-life $= 0.7 \times 10^6$ years) have also been suggested as important heat sources (Urey 1955a). Evidence for the existence of ^{26}Al in the early solar system has recently been discovered by Lee et al. (1976), who find a magnesium isotopic anomaly in a sample of the Allende meteorite which is most plausibly explained by the in situ decay of ^{26}Al. They point out that even a body of asteroidal size would be heated to melting if it contained the abundance of ^{26}Al inferred for their sample. Although the abundance and distribution of this isotope in early solar system bodies remains a matter for further study, it must be considered as a possible energy source for the differentiation of Mercury. (Schramm 1971 discusses the theoretical problems associated with ^{26}Al in the early solar system.)

Other sources of heating are the dissipation of tides and the extraction of energy from an early solar wind by electromagnetic induction (Sonett et al. 1968). The former has been evaluated by Burns (1976) and found to be insufficient to produce global differentiation. The latter mechanism (besides requiring the Sun to go through a T-Tauri-like phase of rapid rotation with a high magnetic field) requires that a good electrical contact be maintained between the planet and the solar wind during the heating. It is not known if these conditions existed. Under the most favorable conditions, the induction mechanism appears to be capable of raising the internal temperature of Mercury by more than 10^3 °C. (Sonett et al. 1968).

If a molten core did form early in Mercury's history, how long could it remain molten? Thermal history models of Siegfried & Solomon (1974) and Solomon (1976) indicate that an iron (or iron-nickel) core would solidify in less than 2 billion years. But the rate at which heat leaves the core is sensitive to conditions at the core-mantle boundary. In the above mentioned models (and those of Sharpe & Strangway 1976), the melting temperature is taken to be continuous and superadiabatic at the boundary. This assumption has the consequence that, as long as the core is molten at the boundary, the heat flux there must be at least $k(dT/dr)_{\text{melting}}$ which, when integrated over the boundary surface, removes energy from the core at the rate of 1.6×10^{19} ergs sec^{-1}. (The thermal conductivity of iron is denoted by k; see Liu & Bassett 1975 for the melting curve of iron at the boundary pressure 70 kbar.) On the basis of geochemical arguments (Urey 1955b), which are borne out by the low radioactive abundances found in iron meteorites (Mason 1971), it is assumed that radioactives do not exist in significant amounts in the core, and therefore cannot contribute to this power. If it is to be supplied solely by heat of fusion, solidification must occur in 1.7×10^9 years, as indeed occurs in the models.

However, Fricker et al. (1976) have pointed out that the melting temperatures of silicates are likely to be greater than that of iron at the core-mantle boundary. They have calculated an evolutionary model in which there is a discontinuous drop in melting temperature (of about 200°C) from the silicate mantle to the iron core. The discontinuity has an important effect: the temperature in the outer core increases above the melting point, while the gradient decreases, thereby reducing the heat flow from the core. In their model, a molten shell occupying the outer 500 km of core is retained to the present. The thermal gradient in the shell is subadiabatic, so the model could not explain a convectively driven dynamo. If it is required that the core be convective, the heat flux from the convecting region must be at least $k(dT/dr)_{\text{adiabatic}}$. Using recent data on the properties of iron at high pressure (G. C. Kennedy, personal communication, 1977, Liu & Bassett 1975), and models of Mercury's structure (Reynolds & Summers 1969), we find that the adiabatic gradient at the core-mantle boundary is about 1°C km^{-1}—near the melting curve gradient!—so that, again, the heat of fusion would be used up in less than 2 billion years if it were the only available source.

Another consequence of a convecting core is that the lower mantle must either be molten or solid convecting. In the absence of mantle convection, the greatest heat flow that can be expected to occur in the silicate material at the core-mantle boundary is that which would correspond to the conductive, steady-state temperature distribution with no heat sources in the mantle. If the silicate melting temperature is not too different from that of iron, this heat flow is only 6.5×10^{18} ergs sec^{-1} (Cassen et al 1976) or less than half the minimum convective flux from the core. (This disparity is a consequence of the fact that the thermal conductivity of the silicates is typically an order of magnitude less than that of iron.) The mantle must either heat up to melting, where liquid convection can carry off the heat, or convect in the solid state. Cassen et al. (1976) show that the latter is likely. In fact, mantle convection is probable even if the core is not convecting, as long as the core-mantle boundary is near the iron melting point, and the solid viscosity is similar to that of the Earth's upper mantle. This conclusion, however, provides a further disadvantage for the convecting dynamo hypothesis; mantle convection could extract as much as 3×10^{19} ergs sec^{-1} from the core, causing it to solidify in less than a billion years. Retention of heat sources throughout the mantle in an amount exceeding that found in chondritic meteorites (or that inferred for the Earth) would be necessary for the maintenance of a molten (but nonconvecting) core.

In view of the difficulty of explaining the source of energy lost by a convective dynamo, it seems worthwhile to pursue Stephenson's (1976) residual dipole model. Although an iron core might solidify rapidly, it is very unlikely that it could cool below the Curie point; therefore, the core would be paramagnetic. In this case, Stephenson finds that about 3% (by volume) or more metallic iron would probably be needed in that part of the mantle which cooled below the Curie point while the dynamo was operating. This assumes that remanence is induced in a shell somewhat greater than 200 km thick, the most that could be hoped for according to thermal history calculations. Thinner regions would require more free iron; any overturning, local reheating, or other disruption of the region would diminish the resultant dipole moment. Lunar basalts, which carry their remanence in metallic iron grains, generally contain two orders of magnitude less free iron than that suggested for Mercury. It is an open question whether or not Mercury—an iron-rich planet—might have more iron in its mantle than is found in the lunar rocks.

GEOLOGIC HISTORY

Despite some important differences, the striking duplication between the Moon and Mercury in their surface features and relative relationships suggests that the sequence of geologic events leading to their present state must have been very similar. Whether the absolute time scale for development of the surface features of Mercury (and other terrestrial planets) is the same as for the Moon remains an open question. Within the limitations of the unknown time scale, Mercury's evolution can be divided into five stages or epochs: 1. accretion and differentiation; 2. terminal heavy bombardment; 3. Caloris basin formation; 4. basin flooding; and 5. post-filling light bombardment (Murray et al. 1975).

The first epoch includes the earliest stages of the solar system, beginning with the condensation of the solar nebula into solids and the accumulation of the solids into the main mass of Mercury. The details of this accumulation are not known, but the evidence for large-scale volcanism of silicate composition grossly similar to the Moon, and the existence of an iron core inferred from the magnetic field observations, argue persuasively that Mercury is a chemically differentiated planet. This segregation into an iron core-silicate crust must have occurred very early during this epoch in order to form a lithosphere of sufficient rigidity and thickness to permit the heavily cratered terrain (formed in the next epoch) to retain its lunar appearance. Because of the absence of any evidence for eolian erosion, Murray et al. (1975) argue that any atmosphere formed during this period must have been dissipated very quickly or before the onset of the terminal heavy bombardment. Chapman (1976) disputes this interpretation from considerations of the relative rates of erosion and cratering; by inference he implies any primordial mercurian atmosphere may have been retained into the second epoch of heavy cratering.

The heavily cratered terrain, represented by large craters several tens of kilometers in diameter grading up to basins the size of Caloris, records a period of intense or prolonged bombardment by large bodies. This terrain appears to be emplaced into an older host material, the intercrater plains whose origin, as previously discussed, is very uncertain. Because of the uncertainty, a point of discussion has arisen whether the bombardment producing the heavily cratered terrain was the terminal phase of the accumulation of Mercury, or whether it was a second episode of cratering not related to accretion. Murray et al. (1975) interpret the extensive areas of intercrater plains to be degraded remnants of all topographic evidence of the saturation flux that necessarily constituted final stages of accretion. This interpretation has an important implication: if the intercrater plains are the degraded accretionary stage, then the heavily cratered terrain must record a later episode of bombardment distinct and separate from accretion.

A similar scenario has been proposed for the Moon to explain a rather sharp cutoff in crystallization ages of lunar samples at about 4 billion years (Tera, Papanastassiou & Wasserburg 1974). The proposal involves a "cataclysmic event" about 3.9 billion years ago, an event of about 10^8 years' duration (including the Imbrium and Orientale basin-forming events), and is conceived as a spike on a curve of the cratering flux that was decreasing from the high rates that occurred during accretion. Baldwin (1974), Hartmann (1975), and Chapman (1976) have pointed out the poor definition of the proposed cataclysm, but the latter also points out that the evidence does not exclude a possible cataclysm. Indeed, recent studies of lunar crater morphologies (Whitaker & Strom 1976) have been interpreted to indicate that two different populations of bodies have cratered the Moon on the appropriate time scales. Two sources for the "cataclysmic" flux have been given, both of which could have arisen as the natural consequences of the dynamical evolution of the early solar system. Wetherill (1976) has proposed that a large planetesimal formed in the vicinity of Uranus and Nepture was perturbed on a time scale of several

10^8 years into an Earth-crossing orbit, and during a close approach to either Earth or Venus was tidally disrupted. The subsequent sweepup of the fragments by the inner planets produced the "cataclysm." Chapman & Davis (1975) have suggested an alternate source from the asteroid belt; the parent body, in a highly eccentric orbit, was disrupted by collision to provide the population of fragments for the terminal heavy bombardment. In either case, as Chapman (1976) points out, if an episodic-type of cataclysmic bombardment did occur on the Moon, it must also have occurred on Mercury and the other terrestrial planets at the same time of about 4 billion years ago. Preliminary studies of mercurian crater morphology (Guest & Gault 1976) and total crater populations (Whitaker & Strom 1976) are suggestive of two families of impacting bodies in support of the Murray et al. (1975) interpretation of an episodic late heavy bombardment. However, other interpretations of the available mercurian crater data are equally plausible, and the heavily cratered terrain does not necessarily need to be interpreted as part of an inner solar system "cataclysmic" bombardment; bodies in initially almost circular Mercury-crossing orbits or in eccentric orbits with perihelia well inside Mercury's orbit would have had cratering rates on Mercury an order of magnitude greater than on the other terrestrial planets (Wetherill 1974). Further studies of Mariner 10 imagery, especially the mercurian craters, may be able to resolve the uncertainties of an episodic bombardment versus the tail end of accretion as the source of the late heavy bombardment and the heavily cratered terrain.

An important aspect of the intercrater plains and heavily cratered terrain that differs from the lunar surface is the presence of the lobate scarps of probable tectonic origin. These features appear predominantly on the intercrater plains, a pattern of occurrence that suggests that formation of the scarps was primarily during the latter stages of the first epoch, and perhaps continued into the early stages of the late heavy bombardment that produced the heavily cratered terrain. If these scarps are thrust or reverse faults arising from compressive stresses and consequent crustal shortening caused by cooling and shrinkage of the core (Strom, Trask & Guest 1975), the scarps represent an important boundary condition for thermal history calculations once an absolute time scale for intercrater plains and heavily cratered terrain can be established.

A convenient and well-delineated third point in Mercury's history is the time of the impact that formed the Caloris basin. This monstrous event modified a major fraction of the "outgoing" hemisphere in a manner similar to that forming the ringed basins on the Moon (Strom, Trask & Guest 1975). In addition to direct modifications by excavation and deposition processes, volcanic activity may have been triggered that is responsible for some of the elements of the smooth plains. Stratigraphic relationships, areal density of crater populations, and crater morphologies indicate that the short epoch of the formation of Caloris either occurred very near to, or marks the end of, the late heavy bombardment (Trask & Guest 1975, Guest & Gault 1976).

The fourth period of mercurian history started an indeterminate but probably very short interval of time after the Caloris impact. Broad expanses of smooth plains were formed and smaller, older basins were filled with similar appearing materials, most plausibly by volcanic origins (Trask & Guest 1975, Strom, Trask & Guest 1975, Trask & Strom 1976). The low areal density of craters on the smooth plains clearly indicates that the plains post date the late heavy bombardment. Small variations in the areal densities between different units of the smooth plains emphasize that the filling was not simultaneous on a global scale; although the time required for plains formation is presently uncertain, the interval was probably relatively short (Murray et al. 1975).

The fifth and final epoch that can be defined for Mercury extends from the termination of basin flooding to the present, probably a major fraction of mercurian history. It was a quiescent period during which the surface acquired a light "dusting" of meteoritic debris which produced most, if not all, of the craters now expressed on the smooth plains and the prominent rayed craters, the youngest features displayed on the surface. Degradation of craters and presumably other landforms has been minimal during this period (Guest & Gault 1976), and no tectonic deformations of the mercurian crust are evident (Murray et al. 1975). Like the Moon, Mercury currently appears to be resting quietly with no active processes to reshape its surface except an occasional random encounter with yet another fragment of interplanetary debris.

ACKNOWLEDGMENTS

The authors express their appreciation to A. L. Broadfoot, D. E. Shemansky, R. T. Reynolds, S. J. Peale, R. Greeley, and P. Dyal for helpful reviews and discussion. A special thanks to Faye L. Gray for preparing the typescript to meet an impossible deadline.

Literature Cited

Adams, J. B. 1968. *Science* 159:1453–55
Adams, J. B. 1974. *J. Geophys. Res.* 79:4229–336
Adams, J. B. 1975. *Infrared and Raman Spectroscopy of Lunar and Terrestrial Materials,* ed. C. Karr, Jr. New York: Academic. 375 pp.
Ash, M. E., Shapiro, I. I., Smith, W. B. 1967. *Astron J.* 72:338–50
Ash, M. E., Shapiro, I. I., Smith, W. B. 1971. *Science* 174:551–56
Baldwin, R. B. 1974. *Icarus* 23:157–66
Benfield, A. E. 1950. *Trans. Am. Geophys. Union* 31:53–57
Briggs, F. H., Drake, F. D. 1973. *Astrophys. J.* 182:601–07
Broadfoot, A. L. 1976. *Rev. Geophys. Space Phys.* 14:625–28
Broadfoot, A. L., Kumar, S., Belton, M. J. S., McElroy, M. B. 1974. *Science* 185:166–69
Broadfoot, A. L., Shemansky, D. E., Kumar, S. 1976. *Geophys. Res. Lett.* 3:577–80
Brouwer, D., van Woerkom, A. J. J. 1950. *Astron. Pap. Amer. Ephemeris Naut. Alm.* 13: Part 2
Burns, J. A. 1973. *Nature Phys. Sci.* 242: 23–25
Burns, J. A. 1975. *Icarus* 25:545–54
Burns, J. A. 1976. *Icarus* 28:453–58
Cameron, A. G. W. 1969. In *Meteorite Research,* ed. P. M. Millman, pp. 7–15. Dordrecht: Reidel
Cassen, P., Young, R. E., Schubert, G., Reynolds, R. T. 1976. *Icarus* 28:501–08
Chapman, C. R. 1976. *Icarus* 28:523–36
Chapman, C. R., Davis, D. R. 1975. *Science* 190:553–56
Chase, S. C., Miner, E. D., Morrison, D., Munch, G., Neugebauer, G., Schroeder, M. 1974. *Science* 185:142–45
Chase, S. C., Miner, E. D., Morrison, D., Munch, G., Neugebauer, G. 1976. *Icarus* 28:565–78
Cohen, C. J., Hubbard, E. C., Oesterwinter, C. 1973. *Celestial Mech.* 7:438–48
Colburn, D. S., Sonett, C. P., Schwartz, K. 1972. *Earth Planet Sci. Lett.* 14:325–37
Colombo, G. 1965. *Nature* 208:575
Colombo, G. 1966. *Astron. J.* 71:891–96
Colombo, G., Shapiro, I. I. 1966. *Astrophys.*

J. 145:296–307
Cuzzi, J. N. 1974. *Astrophys. J.* 189:577–86
Davies, M. E., Batson, R. M. 1975. *J. Geophys. Res.* 80:2417–30
de Vaucouleurs, G. 1964. *Icarus* 3:187–235
Devine, N. 1972. *NASA SP-8025*
Dollfus, A., Auriere, M. 1974. *Icarus* 23: 465–82
Duncombe, R. L., Klepczynski, W. J., Seidelmann, P. K. 1973. *Fundam. Cosmic Phys.* 1:119–65
Duncombe, R. L., Seidelmann, P. K., Klepczynski, W. J. 1973. *Ann. Rev. Astron. Astrophys.* 11:135–54
Dyce, R. B., Pettengill, G. H., Shapiro, I. I. 1967. *Astron. J.* 72:351–59
Esposito, P. B., Anderson, J. D., Ng, A. T. Y. 1976. Paper presented to 19th Plenary Meeting of COSPAR, Philadelphia, PA.
Fjeldbo, G., Kliore, A., Sweetnam, D., Esposito, P., Seidel, B., Howard, H. T. 1976. *Icarus* 29:439–44
Fricker, P. E., Reynolds, R. T., Summers, A. L., Cassen, P. M. 1976. *Nature* 259: 293–94
Gault, D. E., Guest, J. E., Murray, J. B., Dzurisin, D., Malin, M. C. 1975. *J. Geophys. Res.* 80:2444–60
Goldreich, P., Peale, S. J. 1966. *Astron. J.* 71:425–38
Goldreich, P., Peale, S. J. 1968. *Ann. Rev. Astron. Astrophys.* 6:287–230
Goldstein, R. M. 1971. *Astron. J.* 76:1152–54
Gubbins, D. 1977. *Icarus* 30:186–91
Guest, J. E., Gault, D. E. 1976. *Geophys. Res. Lett.* 3:121–23
Hameen-Antilla, K., Pikkarainen, T., Camichel, H. 1970. *The Moon* 1:440–48
Hanks, T. C., Anderson, D. L. 1969. *Phys. Earth Planet. Inter.* 2:19–29
Hapke, B., Danielson, G. E., Klaasen, K., Wilson, L. 1975. *J. Geophys. Res.* 80: 2431–43
Harris, D. 1961. *Planets and Satellites,* ed. G. P. Kuiper, B. Middlehurst, pp. 272–342. Chicago: Univ. Chicago Press
Hartle, R. E., Curtis, S. A., Thomas, G. E. 1975. *J. Geophys. Res.* 80:3689–92
Hartle, R. E., Ogilvie, K. W., Wu, C. S. 1973. *Planet. Space Sci.* 21:2181–92
Hartmann, W. K. 1975. *Icarus* 24: 181–87
Herbert, F., Wiskerchen, M., Sonett, C. P., Chao, J. K. 1976. *Icarus* 28:489–500
Hodges, R. E. Jr., 1974. *J. Geophys. Res.* 79:2881–85
Howard, H. T., Tyler, G. L., Esposito, P. B., Anderson, J. D., Reasenberg, R. D., Shapiro, I. I., Fjeldbo, G., Kliore, A. J., Levy, G. S., Brunn, D. L., Dickinson, R., Edelson, R. E., Martin, W. L., Postal, R. B., Seidel, B., Sesplaukis, T. T., Shirley, D. L., Stelzried, C. T., Sweetnam, D. N., Wood, G. E., Zygielbaum, A. I. 1974. *Science* 185:179–80
Howard, K. A., Wilhelms, D. E.. Scott, D. H. 1974. *Rev. Geophys. Space Phys.* 12:309–27
Irvine, W., Simon, T., Menzel, D., Pikoos, C., Young, A. 1968. *Astron. J.* 73:807–28
Kaula, W. M. 1976. *Icarus* 28:429–33
Klaasen, K. P. 1975. *J. Geophys. Res.* 80: 2145–46
Klaasen, K. P. 1976. *Icarus* 28:469–78
Kozlovskaya, S. V. 1969. *Astrophys. Lett.* 4:1–3
Kuiper, G. P. 1970. *Commun. Lunar Plan. Lab.* 8:165–74
Kumar, S. 1976. *Icarus* 28:579–92
Lee, T., Papanastassiou, D. A., Wasserburg, G. J. 1976. *Geophys. Res. Lett.* 3:109–12
Lewis, J. S. 1972. *Earth Planet Sci. Lett.* 15:286–90
Liu, H.-S., O'Keefe, J. A. 1965. *Science* 150:1717
Liu, L., Bassett, W. A. 1975. *J. Geophys. Res.* 26:3777–82
Lyot, B. 1929. *Ann. Obs. Paris* 8:169–82
Mason, B. (ed.) 1971. *Handbook of Elemental Abundances in Meteorites.* New York: Gordon & Breach. 555 pp.
McCord, T. B., Adams, J. B. 1972a. *Science* 178:745–74
McCord, T. B., Adams, J. B. 1972b. *Icarus* 17:585–88
Mizutani, H., Matsui, T., Takeushi, H. 1972. *The Moon* 4:476–89
Moore, H. J., Hodges, C. A., Scott, D. H. 1974. *Proc. Lunar Sci. Conf., 5th,* pp. 71–100
Morrison, D. 1970. *Space Sci. Rev.* 11:271–307
Morrison, D. 1976. *Icarus* 28:605–6
Morrison, D., Klein, K. M. 1970. *Ap. J.* 160:325–32
Murray, B. C., Belton, M. J. S., Danielson, G. E., Davies, M. E., Gault, D. E., Hapke, B., O'Leary, B., Strom, R. G., Suomi, V., Trask, N. J. 1974. *Science* 185:169–79
Murray, B. C., Strom, R. G., Trask, N. J., Gault, D. E. 1975. *J. Geophys. Res.* 80: 2508–14
Murray, J. B., Dollfus, A., Smith, B. 1972. *Icarus* 17:576–84
Ness, N. F., Behannon, K. W., Lepping, R. P., Whang, Y. C., Schatten, K. H. 1974. *Science* 185:151–59
Ness, N. F., Behannon, K. W., Lepping, R. P., Whang, Y. C. 1975. *J. Geophys. Res.* 80:2708–16
Ness, N. F., Behannon, K. W., Lepping, R. P., Whang, Y. C. 1976. *Icarus* 28: 479–88
Ogilvie, K. W., Scudder, J. D., Hartle, R. E., Siscoe, G. L., Bridge, H. S., Lazarus, A. J.,

Asbridge, J. R., Bame, S. J., Yeates, C. M. 1974. *Science* 185:145–51
Papanastassiou, D. A., Wasserburg, G. J. 1971. *Earth Planet. Sci. Lett.* 11:37–62
Peale, S. J. 1969. *Astron J.* 74:483–89
Peale, S. J. 1972. *Icarus* 17:168–73
Peale, S. J. 1974. *Astron. J.* 79:722–44
Peale, S. J. 1976a. *Icarus* 28:459–67
Peale, S. J. 1976b. *Nature* 262:765–66
Peale, S. J., Boss, A. P. 1977. *J. Geophys. Res.* 82:743–79
Peale, S. J., Gold, T. 1965. *Nature* 206: 1240–41
Pettengill, G. H., Dyce, R. B. 1965. *Nature* 206:1240
Reynolds, T. R., Summers, A. L. 1969. *J. Geophys. Res.* 74:2494–511
Safronov, V. S. 1969. *Evolution of the Protoplanetary Cloud and Formation of the Earth and the Planets.* Engl. transl. 1972. Israel Prog. Sci. Transl., Ltd., Jerusalem (NASA-TT-F-677). 211 pp.
Schramm, D. N. 1971. *Astrophys. Space Sci.* 13:249–66
Schultz, P. H., Gault, D. E. 1975. *The Moon* 12:159–77
Sharpe, H. N., Strangway, D. W. 1976. *Geophys. Res. Lett.* 3:285–88
Siegfried, R. W. II, Solomon, S. C. 1974. *Icarus* 23:192–205
Siscoe, G. L., Ness, N. F., Yeates, C. M. 1975. *J. Geophys. Res.* 80:4359–63
Siscoe, G., Christopher, L. 1975. *Geophys. Res. Lett.* 2:158–60
Smith, B. A., Reese, E. J. 1968. *Science* 162:1275–77
Solomon, S. C. 1976. *Icarus* 28:509–22
Sonett, C. P., Colburn, D. S. 1968. *Phys. Earth Planet. Inter.* 1:326–46
Sonett, C. P., Colburn, D. S., Schwartz, K. 1968. *Nature* 219:924–26
Stephenson, A. 1976. *Earth Planet Sci. Lett.* 28:454–58
Stevenson, D. J. 1974. *Icarus* 22:403–15
Stevenson, D. J. 1975. *Nature* 256:634
Strom, R. G., Trask, N. J., Guest, J. E. 1975. *J. Geophys. Res.* 80:2478–507
Tera, F., Papanastassiou, D., Wasserburg, G. 1974. *Earth Planet. Sci. Lett.* 22:1–21
Thomas, G. E. 1974. *Science* 183:1197–98
Trask, N. J., Guest, J. E. 1975. *J. Geophys. Res.* 80:2461–77
Trask, N. J., Strom, R. G. 1976. *Icarus* 28: 559–63
Turner, G., Huneke, J. C., Podosek, F. A., Wasserburg, G. J. 1971. *Earth Planet. Sci. Lett.* 12:19–35
Ulich, B. L., Cogdell, J. R., Davis, J. H. 1973. *Icarus* 19:59–82
Urey, H. C. 1955a. *Proc. Natl. Acad. Sci. USA* 41:127–44
Urey, H. C. 1955b. *Nature* 195:321–23
van Flandern, T. C., Harrington, R. S. 1976. *Icarus* 28:435–40
Vilas, F., McCord, T. B. 1976. *Icarus* 28: 593–99
Ward, W. R., Colombo, G., Franklin, F. A. 1976. *Icarus* 28:441–52
Ward, W. R., Reid, M. J. 1973. *MNRAS* 164:21–32
Weidenschilling, S. J. 1974. *Icarus* 22:426–35
Weidenschilling, S. J. 1976. *Icarus* 27:161–70
Wetherill, G. W. 1974. *Proc. Sov. Am. Conf. Cosmochem. Moon and Planets.* Houston: Lunar Sci. Inst.
Wetherill, G. W. 1976. *Icarus* 28:537–42
Whitaker, E. A., Strom, R. G. 1976. *Lunar Sci. VII* 933–34
Wilhelms, D. E. 1976. *Icarus* 28:551–58
Wilhelms, D. E., McCauley, J. F. 1971. *Misc. Geol. Inv. Map I–703* Reston, VA: U.S. Geol. Surv.
Will, C. M. 1974. In *Experimental Gravitation,* Varenna Course No. 56, ed. B. Bertotti, pp. 1–110. New York: Academic
Zohar, S., Goldstein, R. M. 1974. *Astron. J.* 79:85–91

Ann. Rev. Astron. Astrophys. 1977. 15: 127–51

CONSEQUENCES OF MASS TRANSFER IN CLOSE BINARY SYSTEMS

Hans-Christoph Thomas
Max-Planck-Institut für Physik und Astrophysik, 8 München 40, West Germany

INTRODUCTION

Of all aspects of stellar evolution the most important for close binary systems is that once a star has condensed out of a protostellar cloud it may repeatedly increase its dimensions by large factors. During such an expansion matter can reach a point where the attraction of a binary companion is strong enough to pull it toward the other star. Since the stellar radii that are reached in this expansion can be of the order of an astronomical unit, this kind of interaction may take place even in rather wide binaries with periods of the order of years. At least during such a stage binary evolution must follow a path different from that of a single star. There are other types of interaction between the two stars that come into play much earlier, such as energy transfer by radiation or by matter, which leaves the star as a stellar wind, and tidal friction or interaction via magnetic fields, but because of a built-in instability mass transfer is by far the most important factor. In its simplest form this instability is caused by the fact that each gram of transferred matter adds to the gravitational pull of the mass-gaining companion while weakening the attractive properties of the expanding star. Since the components of a binary are not fixed in space, their separation may change as well, depending on the amount of angular momentum transferred.

From observations of close binaries it is well known that many rotate in synchronism with their revolution. This is expected because of the tidal interaction in these systems. Hence for ratios of masses and radii that are not too extreme the rotational angular momentum of both components can be neglected compared to the orbital angular momentum. Under these conditions the individual contributions to the total angular momentum are fixed by the masses of the two stars. Nevertheless, the one-dimensional picture of single-star evolution cannot be retained, since the initial conditions must include not only mass and chemical composition, but also the distance and mass of the companion star. Further complications are met during the stages of mass transfer. However, we should not forget that binary systems in many cases provide rather accurate information about stellar masses and radii for comparison with evolutionary models.

That several reviews on the theory of close binaries have appeared in the last five years (Kopal 1971, Paczyński 1971a, Martynov 1973, Plavec 1973, Rees 1974, Kraft 1975, van den Heuvel 1976) is an indication of the great interest in this field. Many contributions were presented at the *IAU Symposium No. 73* on "Structure and Evolution of Close Binary Systems" held in 1975 in Cambridge, England. The article by Paczyński (1971a) has served as a starting point for this review, which is certainly biased by the author's interest in the theory of stellar structure applied to binary evolution.

STANDARD ASSUMPTIONS AND THEIR LIMITS

The Roche model has been quite successful in solving the problem of the internal structure of the two stars in a binary. Its assumptions are that the gravitational forces of the two stars can be replaced by those of two point masses at the center of each star and that the centrifugal force can be computed for the case of synchronous rotation around the axis through the center of mass of the system. Since stars are centrally condensed, the first assumption is likely to be fulfilled if stars have no fast rotating cores, and the second seems to be confirmed by observations for sufficiently close systems. But one should not forget that synchronous rotation is a result of tidal interaction, so the Roche potential in its original form can only be applied to processes that take place on time scales that are long compared to the time necessary for tidal synchronization. For evolution with nuclear time scales this is certainly the case. Thus, for the total potential, one obtains the well-known picture of two potential wells linked by a saddle point (see for instance Kuiper 1941, Kopal 1959, or Plavec 1964), denoted L_1. Along the line connecting the two mass centers there are two more saddle points, L_2 and L_3. But these are of a different quality, because even if we drop the assumption of synchronous rotation, there will be a saddle point similar to and near L_1, whereas L_2 and L_3 might not exist at all.

For modeling the phase of mass transfer two assumptions have been of great value in creating order out of chaos, namely the constancy of total mass and total orbital angular momentum. These assumptions have come to be known as conservative assumptions. This cannot, however, be justified by observations of binaries, since streams, discs, and rings are common phenomena in these stars, even in phases where no large-scale mass transfer is taking place. It certainly, however, was the first step to be taken. The next step, the solution of the hydrodynamical problem, has not up to now provided a definitive answer to the question of mass and angular momentum loss. So attempts are sometimes made to build models, treating these as free parameters and using plausible, but not conclusive, arguments in order to avoid useless diversifications.

Some more technical assumptions are usually made, namely: the internal structure of the stars is spherical; mass transfer starts when the volume of the star is equal to the volume inside the critical Roche surface containing L_1; the orbit is circular; the rate of mass transfer can be computed by keeping the volume of the star equal to that inside the critical Roche surface; the mass-losing star is in hydrostatic equilibrium; the mass-receiving star is in thermal equilibrium. Several of these assumptions have been criticized, and especially the last has been abandoned in several papers.

CONSERVATIVE MASS TRANSFER

The most important factors determining the outcome of mass transfer in a binary system are the mass and evolutionary state of the more massive component at the beginning of mass transfer. A logarithmic plot of the radius in solar units for this component, which is usually called the primary, against its mass in solar units is shown in Figure 1 (for references see Table 1). The lines indicate boundaries for different stages of single-star evolution. Because of the finite number of computed models and the dependence of the exact position in this diagram on chemical composition, opacity, and convection theory, these lines have a somewhat qualitative character. But they do help one to understand the outcome of mass transfer. Using the notation introduced by Kippenhahn & Weigert (1967) and Lauterborn (1970), we distinguish between cases A, B, and C, depending on whether mass transfer starts within the main sequence band, before the onset of helium burning, or after helium exhaustion.[1] The main sequence band (shaded area) is bounded by the line for chemically homogeneous hydrogen-burning models and by that for models that have practically exhausted hydrogen in the center of the star. The dependence on mass is smooth for both lines. For the case of helium burning it is important to determine whether or not the star develops a degenerate helium core. This is equivalent to asking if helium burning starts quietly or with a thermal instability called the helium flash (Schwarzschild & Härm 1962). According to Iben (1967), the helium flash occurs for masses below 2.25 solar masses. Models having degenerate cores before the helium flash are practically in thermal equilibrium and therefore have a rather simple structure, in which the whole luminosity is provided by a hydrogen-burning shell source. Central temperature, luminosity, and radius depend mainly on the mass of the degenerate core, and thus the radius for helium ignition (zigzag line) decreases only slightly with mass (after Demarque & Mengel 1971, Iben 1967). For medium masses helium burning starts when the contracting core has caused the star to move into the red giant region; the zigzag line is deduced from Iben (1965, 1966a, b) and Hofmeister (1967). Massive stars ignite helium in the blue part of the H-R diagram; here the radius has been deduced from the results of Barbaro, Chiosi & Nobili (1972) and Chiosi & Summa (1970). In these papers the problem of semiconvection is solved both for the Ledoux and the Schwarzschild criteria of convective instability. The Ledoux criterion results in radii for the thermal equilibrium models which extend the zigzag line for medium mass stars, while the Schwarzschild criterion leads to the much smaller radii drawn in Figure 1. Broken lines give the orbital period at which the mean radius of the critical Roche surface

[1] This leaves the case of mass transfer during helium burning unnamed, an opportunity for those willing to produce models of massive binaries in that stage.

R_{crit} is equal to the radius of the primary. R_{crit} has been calculated from

$$\frac{R_{crit}}{a} = 0.38 + 0.2 \log \frac{M_1}{M_2} \tag{1}$$

(Plavec 1970) together with Kepler's law, assuming a mass ratio 2:1. This diagram is similar to the one given by Plavec, Ulrich & Polidan (1973). The dots indicate

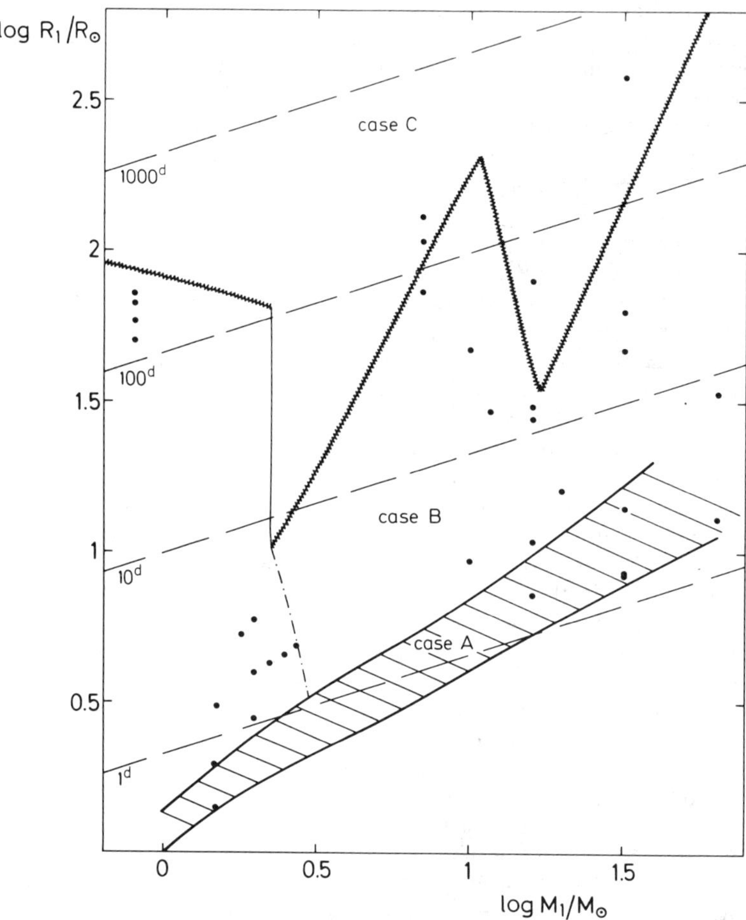

Figure 1 Distribution of radius and mass for the primary components of the systems listed in Table 1. Results of single-star evolution are used to draw boundaries for the different types of mass transfer: the main sequence band with evolution from the chemically homogeneous model to hydrogen exhaustion in the center is shaded; ignition of helium takes place at the zigzag line (the vertical line at $M_1 = 2.25\ M_\odot$ is an idealization); to the left of the dash-dotted line mass transfer in case B will produce a helium white dwarf. Dashed lines indicate periods at the beginning of mass transfer for a mass ratio of 2:1.

Table 1 Model parameters for the evolution of binary systems with mass transfer referenced in this article[a]

Case[b]	$M_1/M_\odot$	$M_2/M_\odot$	$R_1/R_\odot$	P/days	Ref.
Bc	0.8	0.78	50.1	138	1
Bc	0.8	0.78	58.3	174	1
Bc	0.8	0.78	66.5	212	1
Bc	0.8	0.78	72.3	240	1
A	1.5	0.75	1.43	0.45	2
A	1.5	0.75	1.98	0.74	2
A	1.5	0.50	1.43	0.43	3
Br	1.5	1.3	3.08	2.32	4
Bc	1.8	—	5.31	—	5
Br	2	1.5	2.83	1.15	6
Br	2	1.5	4.05	1.96	6
Bc	2	1.5	6.07	3.60	6
Br	2.25	1.25	4.31	1.96	7
Br	2.5	1	4.60	1.96	7
Br	2.75	0.75	4.93	1.96	7
Bc	7	4.5	74	80	8
Bc	7	4.5	108	142	8
Bc	7	4.5	131	190	8
Bc	7	6	108	147	8
Br	10	8	9.44	3.13	9
Br	10	9.4	47.1	33.2	10
Br	11.7	2.0	29.6	12.9	11
A	16	15	7.25	1.7	10
AB	16	15	10.8	3.04	10
Br	16	15	27.9	12.8	10
Br	16	15	30.6	14.8	10
C	16	15	79.7	61.5	10
Br	20	14	16.1	4.87	12
Br	20	10	16.1	4.71	12
Br	20	8	16.1	4.54	12
Br	20	6	16.1	4.39	12
Br	20	4	16.1	4.04	12
Br	32	2	8.57	1.3	10
Br	32	2	63.3	20	10
A	32	30	8.41	1.04	10
AB	32	30	14.1	3.21	10
Br	32	30	47.2	19.95	10
Br	32	30	63.6	30.8	10
C	32	30	380	456.1	10
A	64	60	12.9	1.84	10
Br	64	60	33.9	8.5	10

[a] The columns give type of mass transfer, mass of primary, mass of secondary, radius of primary at the beginning of mass transfer, initial period of the system, and reference numbers according to the following list:

1. Mengel, Norris & Gross (1976)
2. Webbink (1976b)
3. Webbink (1977a)
4. Yungelson (1973c)
5. Lauterborn & Weigert (1972)
6. Giannone & Giannuzzi (1970)
7. Giannone & Giannuzzi (1972)
8. Plavec, Ulrich & Polidan (1973)
9. De Grève & De Loore (1976)
10. Massevitch, Tutukov & Yungelson (1976)
11. Ziółkowski (1976a)
12. De Loore & De Grève (1975)

[b] Radiative and convective envelopes of the primary component are indicated by "r" or "c".

those mass transfer computations that are reviewed in this article; the reader is referred to Plavec et al. 1973 and Paczyński (1971a) for earlier work. The first question to be discussed concerns the production of white dwarfs. Kippenhahn & Weigert (1967) and later Ziółkowski (1970) have shown that mass transfer in case B leads to a remnant of less than 0.35 solar masses consisting of helium with a very thin hydrogen-rich layer on top, if the original primary had less than about 3 solar masses. This remnant has too small a mass to ignite helium during later evolution and so will become a white dwarf. These values are, however, only valid if mass transfer starts close to the main sequence band. We already know that close to helium ignition a star slightly above 2.25 solar masses may ignite helium quietly. To indicate this boundary, a dash-dotted line has been drawn in Figure 1. To the left of this line and below the line for the helium flash, case B mass transfer will produce an almost pure helium core that must evolve into a white dwarf (see also Giannone & Giannuzzi 1970, 1972). The details of this transfer have been analyzed by Refsdal & Weigert (1971). They present a relation between stellar mass and maximum radius attainable for a hydrogen shell-burning red giant, which is reached when the hydrogen-rich envelope of the giant contains only a few percent of the star's total mass. Up to this point the red giant can follow the increase of the critical Roche volume caused by its mass loss by increasing its core mass through hydrogen shell burning. But once the maximum radius is reached the star will, with further evolution, detach itself from the Roche lobe and therefore mass transfer will have to stop. A relation which fits the numerical results rather well is

$$\log R_{\max}/R_\odot = 4.1 + 5.3 \log M/M_\odot. \tag{2}$$

Using this relationship together with Equation 1, the final mass of the white dwarf can be computed. Changes in chemical composition and uncertainties in the mixing length parameter may reduce the constant in Equation 2 from 4.1 down to 3.6. The fast phase of mass transfer (which reverses the mass ratio) produces, in the case of extreme mass ratios, a star that has less hydrogen-rich material in its envelope than that necessary for achieving the radius $R_{\max}$. So the star, immediately after the fast phase of mass exchange, detaches from the Roche lobe, and no slow phase exists in which the star could be detected observationally. For very extreme mass ratios the two stars will come into contact even under the conservative assumptions, which makes the outcome somewhat unpredictable.

Mengel, Norris & Gross (1976) have studied this type of evolution in more detail for a binary system of 0.8 and 0.78 solar masses with five different values for the initial period, from 138 to 252 days. As can be seen from Figure 1, these models fall close to the zigzag line for helium ignition. It turns out that in the system with the highest period the helium flash starts before the onset of mass transfer, which causes the star to contract and change over to the horizontal branch. In the system with the lowest period, mass transfer leads to the production of a white dwarf, as discussed before, with a mass of 0.47 solar masses. But for the other three systems helium is ignited during the slow phase of mass transfer in which the star is trying to attain its maximum radius. The result is a horizontal branch star of reduced mass, between 0.49 and 0.55 solar masses, which will populate the blue end and the region where the subdwarf B stars are found. Mengel et al. find that, as compared to the halo population, the old disc population is better able to produce subdwarf B stars in this manner since a wider range of initial mass ratios will lead to the required type of evolution. In globular clusters, systems with such long periods will not survive collisional disruption during their main sequence life.

Two more papers study systems in the domain where mass transfer leads to a helium white dwarf, namely Yungelson (1973a), which will be discussed later, and Lauterborn & Weigert (1972). Lauterborn and Weigert investigate the process of mass transfer for a primary of 1.8 solar masses, which has a deep convective envelope. In this case the reaction of the star to mass loss is a slight increase in radius. Since, owing to the conservative assumptions made, the critical Roche radius decreases, there is no way to limit the star to this critical radius, so the rate of mass transfer cannot be obtained from this condition. They point out that for short time scales of mass transfer time-dependent convection might change the temperature gradient and large superadiabatic gradients would lead to a decrease in the radius. Clearly also the usual boundary conditions can no longer be used any more and no answer is available to the problem of angular momentum transport. If convection cannot follow the changes implied by the high rates of mass transfer, then there is no hope of maintaining synchronism and the Roche model breaks down. Paczyński & Sienkiewicz (1972) have tried to estimate the time scale of mass transfer for convective envelopes with the help of a model developed by Jendrzejec. They obtain rates up to 0.1 solar masses per year, but correcting the boundary condition for the luminosity can reduce the time scale of mass transfer to nearly the thermal time scale. Webbink (1977a) computed the evolution of a system with 1.5 and 0.5 solar masses for case A and obtained mass transfer rates that are not limited by thermal adjustment, because the mass-losing primary develops a deep convective envelope as a consequence of mass transfer. His treatment of mass transfer on dynamical time scales is similar to that of Jendrzejec. He also presented results (Webbink 1977b) for mass transfer or, better, mass loss from a system of 0.8 and 0.4 solar masses, starting with a primary practically filling its Roche lobe (which arouses one's curiosity about the possible progenitor of such a system).

Mass transfer in case B from a deep convective envelope is investigated in Plavec, Ulrich & Polidan (1973). They also use Jendrzejec's mode of mass flow, since the radii of these stellar models cannot be kept equal to the critical Roche radius. The rates of mass transfer thus obtained reach a maximum value of 0.3 solar masses per year. Thus the lifetime of the semidetached phase, which is discussed in comparison to the other phases of evolution, comes out to be rather short for two reasons: firstly, the rapid reversal of the mass ratio is short because of the nearly adiabatic structure of the convective envelope; and secondly, the slow phase of transfer, which is terminated with the ignition of helium, is short because the cores of models had already contracted to temperatures high enough for helium burning before the mass transfer. In contrast to stars with degenerate helium cores case B mass transfer does not cause the whole hydrogen-rich envelope to be lost, but terminates on a generalized helium-burning main sequence. This concept was introduced by Kippenhahn & Weigert (1967) and numerically investigated by Giannone, Kohl & Weigert (1968). In this particular case between 0.5 and 1.8 solar masses of hydrogen-rich material are left on top of the helium core, which is mainly due to the large angular momentum of the original system.

The discovery of X-ray binaries has stimulated work on the evolution of massive close binaries, starting with the suggestion by van den Heuvel & Heise (1972) that a binary with an original period of about 3 days and components of 16 and 3 solar masses might evolve into a system similar to Cen X-3 via case B mass transfer and a supernova explosion. Since it had been suggested before (Paczyński 1967b, Kippenhahn 1969) that Wolf-Rayet binaries might be produced by case B mass transfer in massive systems, van den Heuvel (1973) concluded that these may be progenitors of X-ray binaries. Detailed computations for this type of evolution were first published by Tutukov, Yungelson, Klayman, and Kraitcheva in a series of papers, which are referenced in Massevitch, Tutukov & Yungelson (1976). Mass transfer has been calculated for cases A, AB, and B, during core helium burning, with the secondary a compact object, applying both the Ledoux and the Schwarzschild criterion for convective instability. Primary masses were 10, 16, 32, and 64 solar masses. While the outcome of mass transfer in case A does not depend on the assumed criterion for convective instability, evolution is quite different for case B. Under the assumption of the Ledoux criterion almost all of the hydrogen-rich envelope is lost and the remnant moves close to the helium main sequence, coming back into the red giant region only after the exhaustion of helium in the center. With the Schwarzschild criterion the star remains in the region of the blue supergiants. But the hydrogen-rich material not lost during mass transfer, which seems to be responsible for the larger radius of the remnant, could be removed by a steady mass loss rate of 10^{-6} solar masses per year. By postulating that a star which reaches the domain of the Wolf-Rayet stars in the H-R diagram has to behave like a Wolf-Rayet star this rate of mass loss will be unavoidable once effective temperatures of about 25,000 K are reached, since this is the temperature where the coolest Wolf-Rayet stars are found (Morton 1973). Starting from the idea that this mass loss uncovers layers of different chemical composition, Tutukov & Yungelson (1973) show that systems with smaller angular momentum (case AB mass transfer) produce mainly WN stars, while for those with larger angular momentum (case B mass transfer), the WC stage lasts much longer than the WN stage. This is in agreement with the observational results of Smith (1973) that the WC's have larger separations than the WN's.

Mass transfer in massive close binaries has also been computed under the conservative assumptions by De Loore & De Grève (1975) for a primary of 20 solar masses starting at a radius of 16.1 solar radii (case B). For the secondary component masses between 14 and 4 solar masses have been assumed, since for still lower masses (in thermal equilibrium) the system would pass through a contact stage (Meyer-Hofmeister 1974). The remnant produced evolves toward the helium main sequence in the H-R diagram; its evolution has been computed as far as the ignition of carbon.

LOSS OF MASS OR ANGULAR MOMENTUM FROM THE BINARY SYSTEM

The evolution of the remnants in case B mass transfer has received a great deal of attention, not only because case B is expected to occur much more frequently than case A, but also because, in conservative case A, the semidetached phase might come to an end with the expansion of the original secondary, thereby forming a contact system with quite unpredictable evolutionary consequences. Also, as already mentioned, during case B mass transfer the two stars can get into contact if the original mass ratio was far from unity. It was also realized quite early (Benson 1970, Yungelson 1973b) that the mass-receiving star would have to radiate some additional energy. Even if matter were to arrive at the photosphere with zero velocity, gravitational settling of the accreted matter releases some energy, the amount of which has recently been calculated by Ulrich & Burger (1976), Kippenhahn & Meyer-Hofmeister (1977), Neo, Miyaji, Nomoto & Sugimoto (1977), and Webbink (1976b, 1977a, b), who find that the high mass transfer rates in the rapid phase can bring the two stars in contact. Kippenhahn & Meyer-Hofmeister (1977) define a critical period for the initial system, which is the smallest period possible in order to avoid contact during conservative mass transfer. This period is given as a function of secondary mass for different primary masses in Figure 2.

Ulrich & Burger (1976) also discussed the influence of the kinetic energy and momentum of the accreting matter on the boundary conditions of the star in the case of direct accretion and accretion from a disc. They found that only in the case of accretion from a disc does the kinetic energy released in the disc and radiated into space have a marginal effect, because part of this radiation is intercepted by the accreting star. However, a mass element at the inner edge of the disc has only converted half of its gravitational energy into heat; the other half is kinetic energy of the Keplerian motion and whether part of this can still be converted into heat or whether all of it is just added to the rotational energy of the star has not yet been investigated. Yungelson (1973b) and Webbink (1976b, 1977a, b) assumed that all of the gravitational energy released is radiated by the mass-receiving star (Webbink as one of two limiting cases) and thus produces a contact system shortly after the onset of mass transfer. Paczyński (1976) discussed a possible outcome of such systems in connection with the origin of cataclysmic variables. His crucial assumption is that once a contact binary is imbedded in a large enough common envelope, one has no reason to expect this envelope to corotate with the system; that is, the Roche potential does not apply. The viscous drag will reduce the distance of the two components until at some stage the common envelope might get lost. Flannery & Ulrich (1977) investigated the formation of a common envelope for a system of two main sequence stars with 15 and 5 solar masses and a period of 8.53 days. Mass transfer starts at $R_1 = 22.5$ solar radii and follows the pattern of case B until 0.34 solar masses are transferred to the secondary. At this point the two stars come into contact; nevertheless the authors treat the two components as physically distinct, computing the mass transfer rate from the difference in potential.

This can be justified only if the time scale of expansion is shorter than that for achieving hydrostatic equilibrium. From the effective temperatures of the two components one must conclude that the sound speed in the envelope is about 10 km sec^{-1}, and with a separation of the mass centers of about 10^7 km, hydrostatic

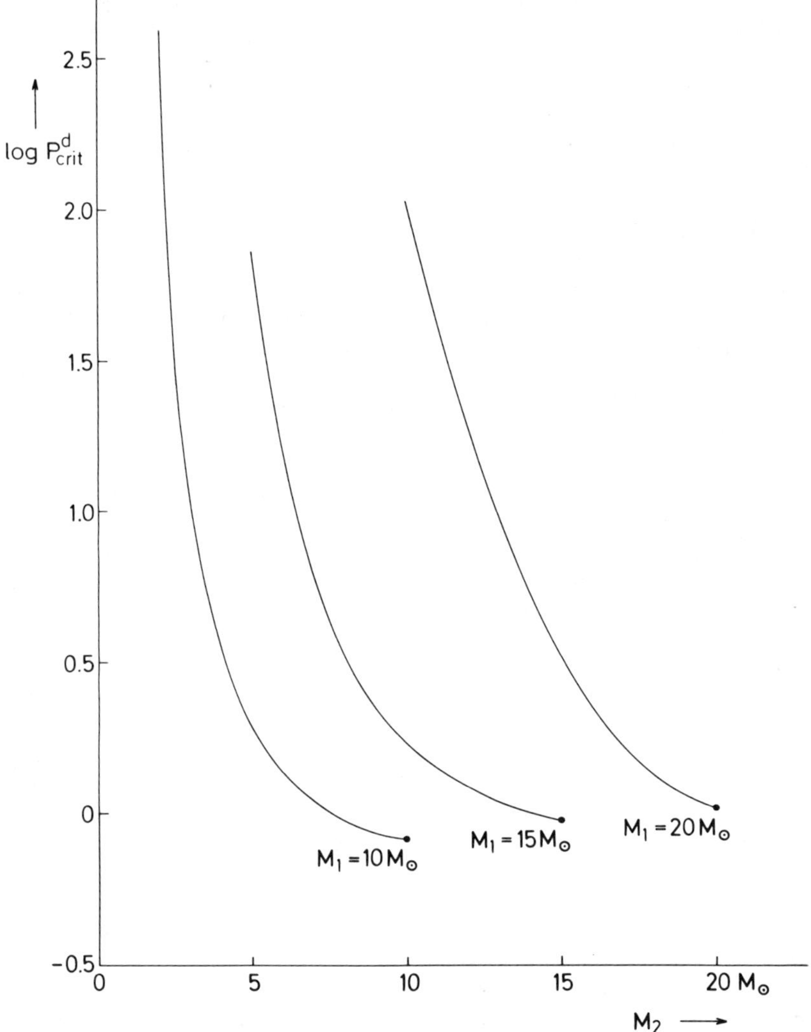

Figure 2 Critical period for contact configurations. The lowest initial period that does not result in a contact configuration during mass transfer is given as a function of secondary mass for primary masses of 10, 15, and 20 solar masses (after Kippenhahn & Meyer-Hofmeister 1977).

equilibrium across the envelope will be achieved within a few weeks. Within this time pressure and temperature will be constant on equipotential surfaces in the common envelope and meridional currents will redistribute the energy transported across the critical Roche surface from the two stars. So the structure has some similarity to that of W UMa stars (Lucy 1968), but, as discussed above, because of differential rotation the description of the centrifugal force by a potential might be invalid within the time scale dictated by the expansion of the primary star.

Arguments in favor of loss of mass or angular momentum from binary systems can also be found from comparison of theoretical models with observed stars. Kopal (1971) presents examples of semidetached systems, where the total mass is so small that both stars should be found on the main sequence. Plavec (1973) mentions U Sge and other systems, which cannot be explained by conservative case A mass transfer. Meyer-Hofmeister (1974) finds that Cen X-3 must have lost some of its orbital angular momentum, and Ritter (1976) argues that cataclysmic variables are the result of case C mass transfer, which implies loss of angular momentum. However, there have only been a few attempts to compute the hydrodynamic flow in binary systems, from Prendergast & Burbidge (1968) to the recent work of Lubow & Shu (1975) or Flannery (1975), who mainly concentrate on the formation of a disc. Lin & Pringle (1976) compute the flow of matter from particle trajectories with small-scale interactions that simulate viscosity. They find that most of the mass that is transferred reaches the other component through the disc and only between 3% and 6% leave the system. Since the formation of a disc is a result of the transfer of too much angular momentum towards the mass-receiving component, there is a continuous flow of angular momentum towards the edge of the disc. Only part of it is carried away by the mass lost from the system; the other part is fed into the orbit via tidal interaction with the outer edges of the disc. If the less massive star is losing mass by Roche lobe overflow, then mass lost from the disc may hit that star and desynchronize its rotation.

Mass loss during mass transfer has been included in only a few computations. Yungelson (1973c) followed mass transfer according to case B in a system of 1.5 and 1.3 solar masses, assuming that 25% of the transferred mass is lost from the system with $g \approx 4$ ($g = d \log J / d \log M$). This results in smaller final periods than the conservative case and seems to fit observed systems somewhat better. In a similar paper on the production of white dwarfs (Yungelson 1973b), he obtained systems with periods of only a few hours, but they can hardly be compared with cataclysmic variables, since the corresponding white dwarf masses are below 0.14 solar masses. Plavec et al. (1973), in computing the mass transfer from a seven solar mass star with a convective envelope, also considered a fraction f of the mass transferred to be lost from the system carrying a fraction g of the specific angular momentum of the system with it (a definition introduced by Paczyński & Ziolkowski 1967). For large g (>6) the orbital radius decreases faster than in the conservative case; for $g = M_2/M_1$ and $f > 0.5$ the maximum rate of mass transfer is reduced relative to the high rate obtained for the conservative case.

Drobyshevski & Reznikov (1974) followed case B mass transfer for low mass systems under a different set of assumptions, namely: the primary star is in synchronous rotation; the secondary and the disc transfer no angular momentum by tidal interaction; the specific angular momentum transferred is that at L_1; and the specific angular momentum lost is that at L_2. As a result of these assumptions 40% to 45% of the mass transferred is lost from the system together with 40% to 75% of the total angular momentum of the system.

A much simpler picture was introduced by Biermann & Hall (1973) in order to explain alternate period changes in Algol-type systems. Mass and angular momentum are transferred in two steps: in the first step the mass and angular momentum lost from the cool star of an Algol-type binary is stored as rotational angular momentum, either in a disc or in the surface layers of the hotter star (the cool star stays synchronized); in the second step the rotational angular momentum is transferred back to the orbit by tidal interaction. So the first step violates the conservative assumptions, because orbital angular momentum is changed into rotational angular momentum, but in the second step conservatism is recovered. What happens to mass transfer when the condition of synchronous rotation is dropped for the mass-losing star has been studied by Pratt & Strittmatter (1976) for the case of HZ Her. They calculated mass transfer rates under the assumption that no tidal coupling synchronizes the two stars, which results in the increase of the time scale for mass transfer from the Kelvin-Helmholtz time scale, which is relevant for the conservative case, to something close to the nuclear time scale. It seems, however, that there is a crucial difference between the case where a disc is generated by mass transfer and that where the stream of matter directly hits the surface of the companion star. To understand this, one has to remember that synchronization in binary stars takes place on a relatively long time scale if the star has a radiative envelope. This has been shown by Zahn (1975) in his work on the dynamical tide and has been supported by the observations of Levato (1976) on rotation in binaries as a function of spectral type and age on the main sequence. So angular momentum stored temporarily in nonsynchronous rotation of a stellar component with a radiative envelope will be returned to the orbit in times that are probably longer than the mass transfer phase itself. But angular momentum stored in a viscous disc will be transported to the outer edge of the disc in a much shorter time scale (depending on the viscosity) and from there returned to the orbit. Rotation was also studied for massive close binaries by Stothers (1973), who concluded that the underluminosity observed for some mass-receiving components of semidetached systems must be due to differential rotation, that is, a fast rotating core below synchronized outer layers.

It is also possible to lose angular momentum from the system with negligible mass loss. One such case has been discussed by Refsdal, Roth & Weigert (1974) and by Eggleton (1976), namely that of a stellar wind from a late-type star with a rather large Alfvén radius. This is the mechanism that explains the low rotational velocities of late-type main sequence stars (Kraft 1967a). Another example is radiation damping by gravitational waves in very close systems, as introduced by Paczyński (1967a) and recently reviewed by Faulkner (1976), but here quantitative results may still contain large errors (see Ehlers et al. 1976). Naturally the tidal influence of a third body can change the orbital angular momentum; a well-known example is the Earth-Moon system under the influence of the Sun (Goldreich & Soter 1966).

ALGOL-TYPE BINARY SYSTEMS

This prototype of close binary evolution with mass transfer, because its origin is now understood in principle, has confronted theoreticians with an increasing number of problems, as soon as they started to investigate individual systems. There is some discussion about whether the mass-losing component is still burning hydrogen in the center, in which case mass transfer had to have been nonconservative for some systems like U Sge (Plavec 1973). Arguments in favor of case A can be found there or in Stothers (1973), while Hall (1975) tries to invalidate objections against case B. But before definite conclusions can be drawn, the role of nonconservative mass transfer has to be clarified. Here U Cep can serve as an illustrative example. Period changes of both signs seem to be present in the O-C diagram analyzed by Hall (1975), assuming the light curve of U Cep is not strongly distorted (however, see Batten 1976). These are explained using the model of Biermann & Hall (1973), which leads to the conclusion that angular momentum is fed into the orbit with an average time scale of $P/\dot{P} = 1.4 \cdot 10^5$ years. The actual flow of matter must be somewhat complicated in that case, because material stored in a ring at $r_r = 0.4$ as in Hall (1975) carries about 50% of the specific angular momentum of the system, while at the L_1 point the specific angular momentum of corotating material is only 8% of that of the system. Hall also argues that the average mass transfer rate is rather high ($8 \cdot 10^{-6}$ solar masses per year), but this is a conservative deduction from the average period change observed ($1.1 \cdot 10^{-6}$ per year). Since the hotter component of U Cep rotates two to three times faster than synchronous (van den Heuvel 1970), rotational momentum might find its way back to the orbit, increasing the period. This would require a time scale for synchronization

$$\tau_{syn} = 3J_{rot}/J \cdot P/\dot{P}, \tag{3}$$

where J is the orbital angular momentum of the system, and J_{rot} is the rotational angular momentum of the hotter component. Since it will probably not be in a state of the uniform rotation, one cannot compute J_{rot}, but most likely the time scale will have to be smaller than a million years to produce the observed period change. Since the cool component evolves on the nuclear time scale of hydrogen shell burning an undersize subgiant (Kopal 1956) would be produced for a short period after the rapid phase of mass transfer (Paczyński 1971a), but according to Hall (1974) observations give no unequivocal proof for the existence of such systems. Assuming this to be the cause of the period change, no conclusions about the rate of mass transfer are possible. In any case the critical Roche lobe around the cooler component should have stopped shrinking not long ago. According to conservative assumptions, if one calculates the size of the critical Roche lobe from Equation 1, this occurs at a mass ratio of 0.78, which has to be compared with today's value of 0.67. Again one may ask how nonconservative mass transfer influences the sign of that change of the critical Roche radius of the cooler component. Defining f and g

as before (Section 4) we only need to insert Keplers law into the expression for the orbital angular momentum and the radius of the critical Roche lobe (Equation 1). Introducing α as

$$\frac{\dot{P}}{P} = -3\alpha \frac{\dot{M}_c}{M}, \tag{4}$$

where $q = M_c/M_h$ (c and h for cool and hot components, respectively) and $M = M_c + M_h$, and k as

$$\frac{\dot{R}_c}{R_c} = -k \frac{\dot{M}_c}{M}, \tag{5}$$

where R_c is the critical Roche radius of the cooler star, one obtains

$$\alpha = \frac{1}{q} - q + f\left(q + \frac{2}{3} - g\right) \tag{6}$$

and

$$k = 2\alpha - \frac{1}{3}f - \frac{0.087}{0.38 + 0.2 \log q} \frac{(1+q)(1+q-fq)}{q}. \tag{7}$$

From the fact that the cool star fills its Roche lobe, one gets $\dot{M}_c < 0$, and from the observed period increase it follows that $\alpha > 0$. So R_c decreases only if $k < 0$. In Figure 3 the lines $\alpha = 0$ and $k = 0$ are plotted in the $f-g$ plane, using $q = 0.67$ for U Cep. The area where the critical Roche radius decreases is hatched. Conservative assumptions ($f = 0$) give an increasing critical Roche radius as noted

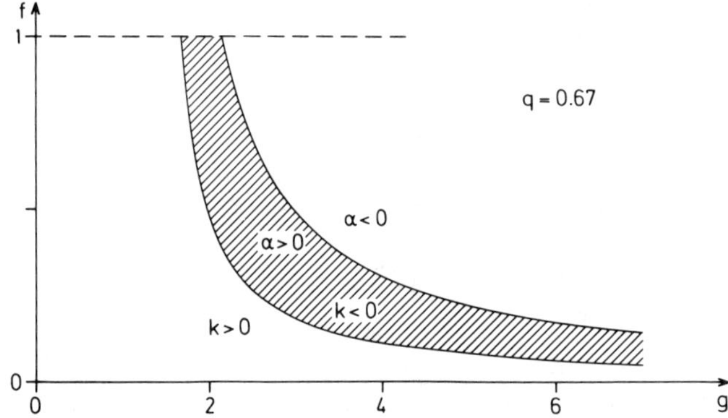

Figure 3 Sign of change of the period (parameter α) and the critical Roche radius (parameter k) during mass transfer for the nonconservative case. The mass ratio of the primary (mass-losing) component to the secondary component for U Cep ($q = 0.67$) is used. The ordinate gives the fraction of mass lost from the system; the abscissa the specific angular momentum carried away in units of the average specific angular momentum of orbital motion.

before. Even if matter carries the average specific angular momentum of the cooler star ($g \approx 1.5$), f would have to be larger than 1 to obtain $k < 0$. If, on the other hand, g is given by that of the outer Lagrangian point on the far side of the hotter star ($g \approx 5.6$), then already 7% of mass loss will cause a decrease in the critical Roche radius lobe. So while it is difficult to reconcile a decrease in the size of the critical Roche lobe of the cooler star with the average period increase, the critical Roche lobe clearly must shrink during period decreases, since independent of the parameters f and g $\alpha < 0$ always implies $k < 0$.

The determination of very accurate system parameters, such as those determined by Popper (1973) for AS Eri, have resulted in the construction of detailed theoretical models. For AS Eri, Refsdal, Roth & Weigert (1974) have computed red giant models, which take into account the influence of envelope mass and hydrogen profile beyond the dependence of luminosity on core mass. The core mass thus obtained (0.17 to 0.18 solar masses) determines the future evolution of the binary: after 100 to 200 million years and a transfer of 0.01 solar masses more, the subgiant star will have reached its maximum radius and detach from the Roche lobe, evolving towards the white dwarf stage as a white dwarf of relatively small mass (0.2 solar masses). However, this is not the endpoint of evolution, because the system has such a small period. Before the white dwarf has left its embryonic state, it will be embedded in matter from the other component, which has a main sequence lifetime of about 800 million years. So this is an example of a system that passes through a common envelope stage whose outcome is rather uncertain. Refsdal et al. also discuss the history of AS Eri under the assumption that there is a stellar wind from the subgiant, which because of a large Alfvén radius carries a high specific angular momentum. This results in the system losing angular momentum at an arbitrary rate while suffering mass loss. The originally more massive component must have started with a mass of about 1.9 solar masses, if no angular momentum was lost, but with angular momentum loss masses down to nearly half the total mass of the system (1.07 solar masses) are possible.

There are a number of other well-observed binary systems, and it is possible to draw conclusions about their evolutionary stage. Some examples are β Lyr (Ziólkowski 1976a, Wilson 1974, 1976), RY Gem (Hall & Stuhlinger 1976), X Tri (Mallama 1975), AX Mon (Plavec et al. 1973), UW UMA (McCluskey, Kondo & Morton 1975), BD +40°4220 (Bohannan & Conti 1976), HZ 22 (Greenstein 1973), BD +16°516 (Nelson & Young 1976, Vauclair 1972, Hills & Dale 1974), AG Peg (Hutchings, Cowley & Redman 1975), or in a more general context Hutchings & Hill (1971), Hutchings (1976), Hall & Neff (1976), Ziólkowski (1976b), a review about the RS CVn systems (Hall 1976), and controversial discussions of their origin (Biermann & Hall 1976, Popper & Ulrich 1977).

Much less is known about the pre-main-sequence evolution of close binary systems: Whelan (1970) investigated the stability of two fully convective stars in contact, and Yamasaki (1971) and Drobyshevski (1974) based their work on the models of Hayashi, which have been improved by Larson (1969). Larson (1972) obtained models with a ring of high density instead of a purely centrally condensed protostar, but Tscharnuter (1975) did not find any indication for such a ring, which could condense into a second star. Koch (1972) traced back the evolution of some well-observed close binaries to the point of contact, using the evolutionary tracks of Ezer & Cameron (1967), and compared the result with the predictions of fission made by Stoeckly (1965) and Roxburgh (1966) and by Bodenheimer & Ostriker (1970). Fission has also been investigated by Lebovitz (1972) and discussed within the context of the observed frequency of binary systems by Abt & Levy (1976).

W UMa SYSTEMS

For these systems the common convective envelope model invented by Lucy (1968) has been very successful in explaining the observed light curves, but the conditions it set on the internal structure of the two components were rather restrictive. Biermann & Thomas (1972) relaxed the condition of equal entropy for the convective envelopes of the two components at the expense of agreement with observed light curves. However, they obtained better agreement with the period-color relation of Eggen (1961, 1967) and the period–mass-ratio diagram of the observations discussed by Mauder (1972). Most of the models published are based on the equal entropy condition, such as the work of Moss & Whelan (1970) or of Whelan (1972a, b), Mochnacki & Doughty (1972a, b), and Ruciński (1973, 1974), in contrast to Vilhu (1973), who adopted the method of Biermann & Thomas (1972). One consequence of the original Lucy model is evolution on a modified thermal time scale, as computed by Hazlehurst & Meyer-Hofmeister (1973), whereas Biermann & Thomas (1973) obtain nuclear lifetimes. On statistical ground Kraft (1967b) argued that they should exist for nuclear time scales. Van't Veer (1975) presents both very young and very old clusters with W UMa stars. From the large number of W UMa systems in the young cluster Coll 359 ($2.5 \cdot 10^7$ years), he concludes that they must be short-lived and therefore interprets the W UMa star in NGC 188 (10^{10} years old) as one that only recently came into contact. A similar argument must hold for ε CrA, if the invisible secondary is a main sequence star (Tapia & Whelan 1975), since then the age of the system is $2 \cdot 10^9$ years. Thus far the zero age models for W UMa systems have nothing to do with mass transfer, but then Lucy (1976) and Flannery (1976) realized that the common envelope models with equal entropy are thermally unstable to mass transfer and may undergo cyclic mass transfer, alternating between a contact and a detached stage. In this way one can populate the short-period region in the period-color diagram for the so-called W-type systems, as well as obtain nuclear lifetimes for these systems (see also Webbink 1976a, Robertson & Eggleton 1977). But the final word has not yet been said, because recently Shu, Lubow & Anderson (1976) discovered the obvious, namely how stellar models with different entropies in their convective envelopes nevertheless follow Lucy's predicted light curves. It is simply the condition of hydrostatic equilibrium that demands constant temperature on equipotentials in the common envelope, the necessary energy transfer accomplished by meridional circulations, while at the Roche surface through the L_1 point a temperature jump allows the less massive component to have the higher entropy required by the models of Biermann & Thomas (1972) or Vilhu (1973).

Although no definite solution is available on the fate of the W UMa systems, the conjecture of Kraft (1965) that these are the progenitors of the U Geminorum stars, which was based on a statistical argument and the low angular momentum of both types, may not hold. Ritter (1976) argues that white dwarfs with masses above 0.45 solar masses can only be produced in case C mass transfer, because the red giant needs so much space to develop a massive enough degenerate core. This requires angular momentum loss later on, to produce the short periods of cataclysmic variables, probably through a magnetically coupled stellar wind as suggested by Eggleton (1976). However, the question of cataclysmic variables will not be discussed in any detail in this paper because the subject has been reviewed recently by Robinson (1976). Webbink (1976a) discusses the final evolution of contact systems and concludes that the secondary will be swallowed by the primary, independent of whether the system evolves in marginal contact or developes a deep common envelope and then an "excretion" disc.

EVOLUTION TOWARDS THE X-RAY STAGE

The detection of X-ray binaries revived interest in the later phases of the evolution of massive binaries, which had not been investigated because of the increased manifold of starting parameters and the rather shaky basis of mass and angular momentum conservation. But now one could aim at these systems with the least complicated evolutionary sequence one could think of. The gross properties of the X-ray sources can be explained by mass transfer from a massive primary onto a compact secondary, which may be a white dwarf, a neutron star, or a black hole, if not something worse. For a white dwarf of one solar mass and a radius of 10^9 cm an accretion rate of 10^{-5} solar masses per year is required to produce the observed luminosity of 10^{38} erg sec^{-1} out of gravitational energy alone, for a neutron star of the same mass and a radius of 10^6 cm a rate of 10^{-8} solar masses per year is sufficient. Neither of these rates fits the picture of a massive primary overflowing its Roche lobe ($\dot{M}$ between 10^{-4} and 10^{-3} solar masses per year). So it seems more likely that the primary is a blue supergiant and as such would lose mass through a stellar wind. A rate of 10^{-6} solar masses per year is expected from observations of this type of star; if 1% of that outflow is collected by a neutron star orbiting close enough, the picture is consistent. Naturally this does not fit all binary X-ray sources and more complicated setups may also be possible with a white dwarf companion, but it is worth the attempt to look for progenitors of such systems. Since the transition from a white dwarf or a less dense star to a neutron star implies a large release of gravitational energy, a supernova phenomenon is usually associated with such a transition, although this is a deduction from the association of supernova remnants with single pulsars only. Coming from the other side, Paczyński (1971a) argued that in massive binaries the original primary reaches the final stage of a supernova explosion before the companion has exhausted its hydrogen in the center. The first stage of mass transfer leading in that direction has already been discussed in Section 3. De Loore & De Grève (1975) interpolated between the results of Paczyński (1971b) and Arnett

(1972) to get models for the carbon-burning stage of the original primary and refer to Arnett (1973) for times up until the supernova explosion occurs. Twenty-thousand years after the ignition of carbon they expect the explosion to blow into space most of the 5.4 solar mass star, leaving a remnant of 1 or 2 solar masses in the form of a neutron star. The possible effects of the explosion on the companion (stripping and ablation) have been studied by Khabazin (1975), and by Wheeler, Lecar & McKee (1975), who also calculate the eccentricity and period after the explosion. Flannery & van den Heuvel (1975) obtained orbital parameters for an instantaneous kick during the explosion, therewith introducing two more unknowns. Two million years later (for an original secondary mass of 14 solar masses) the massive companion (28.6 solar masses) leaves the main sequence and the mass loss rate through a stellar wind is expected to increase from 10^{-9} solar masses per year (Rogerson & Lamers 1975) to 10^{-6} solar masses per year (Lamers, van den Heuvel & Petterson 1976), converting the binary into a strong X-ray source. The smaller secondary masses also studied by De Loore and De Grève have to be excluded from the picture, because already in the first stage of mass transfer a contact system is formed (see Figure 2). The final period of the system is then 30 days for symmetric explosion and can be reduced to 16 days with a proper kick. This again demonstrates that loss of angular momentum must have occurred at some intermediate stage.

A lower limit for the primary mass for this type of evolution was obtained by Massevitch, Tutukov & Yungelson (1976) and by De Grève & De Loore (1976), who found that a 10 solar mass star leaves a 1.66 solar mass remnant after case B mass transfer, which is reduced to 1.12 solar masses in a second stage of mass transfer after central helium burning. This remnant, because of a high density, is cooled sufficiently by plasma neutrinos, such that carbon ignition is prevented and a white dwarf of relatively high mass is formed. This is an important difference from single-star evolution where a lower limit of 4 to 6 solar masses is predicted (van den Heuvel 1975, Reimers 1976). X-ray binaries of relatively low mass like Her X-1 must therefore either have lost a large amount of mass or started mass transfer at a later stage, possibly as case *C*. This in turn leads to the conclusion that large amounts of angular momentum must have been lost from the system. But the obvious presumption that such systems are the result of contact during mass transfer is not likely, because Her X-1 has a nearly unevolved primary, while a compact core is essential to avoid coalescence of the two components (van den Heuvel 1976). But if the system started as a massive binary, which congenital defect is responsible for its deviation from the sketched line of evolution? Are we forced to accept the picture (Ergma & Tutukov 1976, Gursky 1976) of a white dwarf accreting matter fast enough so that it can collapse within a reasonable timespan, an explanation of type I supernovae proposed by Whelan & Iben (1973)? Is it more likely that the white dwarf was born with a mass above the Chandraskhar limit, supported by centrifugal forces in a state of rapid differential rotation, and has collapsed because of viscous braking (Ostriker & Bodenheimer 1968)? It seems there is only one way to make progress: trying to eliminate as many scenarios as possible. One step in this direction is the discussion of Paczyński & Ziólkowski (1975) about a model for X Per.

MASS TRANSFER IN THE X-RAY STAGE AND BEYOND

The X-ray pulses observed in some of the X-ray binaries allow very accurate determination of the orbital period because of their Doppler shift due to orbital motion. One may expect a steady decrease in the period due to mass transfer and mass loss. As always, the truth is more complicated than the theoretical expectations, as can be seen in the detailed study of pulsation period and orbital elements of Cen X-3 by Fabbiano & Schreier (1977). In principle Equations 4 and 6 can be used to obtain period changes from mass transfer and loss, but it is well known that in these massive systems the rotational angular momentum of the primary cannot be neglected even for synchronous rotation. The parameter α can be derived in the same way as before by assuming synchronous rotation of the primary and by denoting the orbital moment of inertia by I_0 and the moment of inertia of the primary by I_p. The result is

$$\alpha = \frac{\frac{1}{q} - q + f(q + \frac{2}{3} - g) + \frac{I_p}{I_0}\left(1 + \frac{1}{q} - gf\right)}{1 - 3\frac{I_p}{I_0}}. \tag{8}$$

The denominator has to be positive, otherwise synchronous orbits would be unstable (Counselman 1973). The specific angular momentum carried by the wind must be about equal to that at the surface of the corotating primary, so one has

$$g = \frac{(1+q)^2}{q}\frac{R_p^2}{a^2} \approx q\,\frac{R_p^2}{a^2}, \tag{9}$$

where R_p is the radius of the primary. For the parameters of Cen X-3 ($q = 21.8$, $R_p/a = 0.59$) and with $f \approx 1$

$$\alpha = \frac{-6.9 - 6.5\frac{I_p}{I_0}}{1 - 3\frac{I_p}{I_0}} \tag{10}$$

will be obtained, so even assuming $I_p \ll I_0$ a period change of $\dot{P}/P \approx -4 \cdot 10^{-6}$ per year results from $\dot{M} = -4 \cdot 10^{-6}$ solar masses per year (Lamers, van den Heuvel & Petterson 1976). The observed average according to Fabbiano & Schreier (1977) is -7.10^{-6} per year. For non-negligible moment of inertia of the primary the absolute value of the rate of change will be larger but an upper limit is given by the rate corresponding to the time scale of synchronization as discussed by Lecar, Wheeler & McKee (1976) and De Grève, De Loore & Sutantyo (1975). Because of the uncertainties in the theory of tidal interaction one cannot compute this limiting rate with sufficient accuracy for Cen X-3, but according to De Grève et al. I_0 is still larger than $3I_p$, so that the simple argument above may give the right order of magnitude for the period change. However, the evolution of the primary will add an additional decrease of 1 to $3 \cdot 10^{-6}$ per year, as shown by De Grève et al., whereas the influence of an accretion drag can be neglected (Alexander, Chau & Henriksen 1976). Alme & Wilson (1976) argue that for high wind velocities the accretion rate in some X-ray sources may be too small to produce the observed X-ray intensity and that radiation-driven density waves are responsible for the mass transfer. In any case at some later stage the primary fills its critical Roche lobe and mass transfer or mass loss will quickly bring the compact companion down to the photosphere of the primary. Again a common envelope system is formed and most likely there will be enough viscous energy liberated to blow the envelope out of the system (J. P. Ostriker, personal communication). A rough estimate, obtained by requiring the binding energy of envelope matter to be smaller than the energy gained from reducing the orbit of the compact star, is

$$\frac{GM_a}{a} < \frac{1}{2}\frac{GM_2}{a}\left(\frac{M_a}{4\pi a^3 \rho} - 1\right). \tag{11}$$

Here M_a is the mass inside the orbital radius a and ρ the density of the massive star at distance a from the center; M_2 is assumed to remain constant. This can also be written as

$$\frac{\bar{\rho}}{\rho} > \frac{6M_a}{M_2} + 2 \tag{12}$$

with $\bar{\rho}$ the mean density inside a. At the surface the local density will be smaller than the mean density by orders of magnitude, so there is enough energy available to bring matter outside the orbit of the compact star to escape velocity. Here it is implicitly assumed that most of the available energy is converted into mechanical energy and not radiated into space without further interaction. From the numbers given by Chiosi & Summa (1970) for a 20 solar mass star at the end of hydrogen core burning one finds that this condition is no longer fulfilled at the outer edge of the semiconvective zone, which is located at $M_r/M = 0.48$. Thus the process of blowing matter out of the system, at the rate given by the decay of the orbit, is stopped already before half of the mass is lost. Because of the high moment of inertia of the more massive star compared to the orbital moment of inertia of the compact companion, rotation will be far from synchronous; therefore, tidal interaction will decrease the orbital radius further, but then the common envelope cannot be dispersed so easily. It seems possible to convert all of the orbital angular momentum into rotational momentum, so that the neutron star sinks to the center of the core of the blue supergiant. The alternative is a very close system consisting of a helium star (or a helium white dwarf?) with a compact companion, especially if the core of the supergiant was rotating rapidly. De Loore, De Grève & De Cuyper (1975) speculated on the possibility of a second supernova explosion that would lead to the disruption of the system, resulting in two neutron stars with high space velocity, one of which is an active radio pulsar. If the supernova explosion is asymmetric, exerting a kick of sufficient strength in the proper direction, a system similar to the binary pulsar would be formed.

One way to test the suggested evolution into the X-ray binary stage is to compare lifetimes in different stages with observed frequencies and supernova rates.

This has been done by Amnuel & Guseinov (1976) and by De Cuyper et al. (1976). The results are quite contradictory, because Amnuel and Guseinov assume a rather long lifetime for binaries with a compact component and take the same lower mass limit for single and binary stars to evolve towards a supernova explosion. The first assumption is unlikely to be true, quite aside from theoretical considerations of binary evolution, because of the relatively short lifetime of stars as massive as those found in most X-ray binaries. The second is in doubt in view of the results of De Grève & De Loore (1976). De Cuyper et al. (1976) start with a total number of 5.10^8 supernova explosions in our galaxy, of which 10^7 occurred in binary systems. From a lifetime of 5.10^6 years for the stage with a compact component they obtain 3000 such systems and 30 X-ray binaries because the stellar wind phase lasts for 5.10^4 years only (the model of Alme & Wilson 1976 reduces this even further). The systems before the X-ray stage should contain a radio pulsar but it may be that a stellar wind (as observed in τ Sco by Rogerson & Lamers 1975) extinguishes the pulses during the main sequence stage. This is not a proof but rather a consistency test that may fit other types of evolution as well. It would be more interesting to learn to distinguish between supernovae in single stars and those in binaries, possibly based on the difference in chemical composition.

The discovery of the binary pulsar, besides its importance for the theory of general relativity, has added some puzzling questions to the problem of binary evolution. The main problem is that the companion of the pulsar has a rather low mass, so that the supernova explosion producing the pulsar should have disrupted the system. Flannery & van den Heuvel (1975) compute the kick parameters necessary to avoid disruption. Smarr & Blandford (1976) present several possibilities, which involve either two supernova explosions for massive systems or a white dwarf accreting matter until collapse becomes unavoidable, as suggested earlier by van Horn et al. (1975). Webbink (1975) tries to encircle the truth by elimination. He excludes hydrogen-rich progenitors because they result in much higher periods than the observed one, two white dwarfs because of problems with mass transfer required to implode one of them, and a helium star shoveling matter onto a white dwarf until it collapses because the accreted helium will quickly reach ignition temperature and prevent further accretion. This leaves one with the medium mass helium star converted into a neutron star by a supernova explosion. Coupled with a helium star or white dwarf the system has to pass a contact phase which would prevent the collapse, so a neutron star companion is to be preferred. Constraints on the components masses at different evolutionary stages are presented by Wheeler (1976).

CONCLUDING REMARKS

There are many consequences of mass transfer in binary systems. Some of them are not covered in this review, especially the structure of cataclysmic variables (reviewed by Robinson 1976), the flow of circumstellar matter, and the formation of a disc. Some of them are only touched, such as period changes or the properties of X-ray binaries. From the theoretical point of view two unsolved problems leave room for speculation: the amount of mass and angular momentum loss from binary systems and the outcome of mass transfer in contact configurations. From comparison of models with observed systems we have won a general understanding of semi-detached binaries, although much can still be learned from the detailed analysis of individual systems. The conflict between W UMa models with and without equal entropy seems to be resolved, which will certainly stimulate theoretical work on zero age contact binaries and their evolution. Concerning the X-ray binaries it is worthwhile to remember the fundamental property of a hypothesis, that is, prediction of observable facts from certain initial conditions. These predictions may allow one to reject the hypothesis, but only if it is not of the chameleon type (about equal number of free parameters and predicted facts). To show that the hypothesis is correct all other hypotheses must be proved wrong. Because there is such a wealth of observed facts about X-ray binaries, the important details of their origin and evolution may finally be brought into a consistent picture.

ACKNOWLEDGMENTS

It is a pleasure to thank all those who supplied me with preprints of their contributions, and especially P. Eggleton for lending me a copy of the proceedings of *IAU symposium No. 73* prior to publication. I am grateful to my colleagues in Munich for many clarifying discussions and to R. Kippenhahn and A. Tutukov for critically reading the manuscript. My thanks for improving the English are due to J. J. Perry. The first steps towards this review were climbed in the stimulating atmosphere of the Aspen Center for Physics.

Literature Cited

Abt, H. A., Levy, S. G. 1976. *Ap. J. Suppl.* 30:273–306

Alexander, M. E., Chau, W. Y., Henriksen, R. N. 1976. *Ap. J.* 204:879–88

Alme, M. L., Wilson, J. R. 1976. *Ap. J.* 210: 233–38

Amnuel, P. R., Guseinov, O. H. 1976. *Astron. Astrophys.* 46:163–69

Arnett, W. D. 1972. *Ap. J.* 176:699–710

Arnett, W. D. 1973. In *Explosive Nucleosynthesis,* ed. D. N. Schramm, W. D. Arnett, pp. 236–47. Austin & London: Univ. Texas Press. 301 pp.

Barbaro, G., Chiosi, C., Nobili, L. 1972. *Astron. Astrophys.* 18:186–97

Batten, A. H. 1976. In *Structure and Evolution of Close Binary Systems,* ed. P. Eggleton, S. Mitton, J. Whelan, pp. 303–10. Dordrecht-Holland: Reidel. 414 pp.

Benson, R. S. 1970. PhD thesis, Univ. California, Berkeley

Biermann, P., Hall, D. S. 1973. *Astron. Astrophys.* 27:249–53

Biermann, P., Hall, D. S. 1976. See Batten 1976, pp. 381–88

Biermann, P., Thomas, H.-C. 1972. *Astron. Astrophys.* 16:60–65

Biermann, P., Thomas, H.-C. 1973. *Astron. Astrophys.* 23:55–61

Bodenheimer, P., Ostriker, J. P. 1970. *Ap. J.* 161:1101–13

Bohannan, B., Conti, P. S. 1976. *Ap. J.* 204: 797–803

Chiosi, C., Summa, C. 1970. *Astrophys. Space Sci.* 8:478–96

Counselman, C. C., III. 1973. *Ap. J.* 180: 307–14

De Cuyper, J. P., De Grève, J. P., De Loore, C., van den Heuvel, E. P. J. 1976. *Astron. Astrophys.* 52:315–18

De Grève, J. P., De Loore, C. 1976. *Astrophys. Space Sci.* 43:35–46

De Grève, J. P., De Loore, C., Sutantyo, W. 1975. *Astrophys. Space Sci.* 38:301–12

De Loore, C., De Grève, J. P. 1975. *Astrophys. Space Sci.* 35:241–47

De Loore, C., De Grève, J. P., De Cuyper, J. P. 1975. *Astrophys. Space Sci.* 36: 219–25

Demarque, P., Mengel, J. G. 1971. *Ap. J.* 164:317–30

Drobyshevski, E. M. 1974. *Astron. Astrophys.* 36:409–13

Drobyshevski, E. M., Reznikov, B. I. 1974.

Acta Astron. 24:29–43
Eggen, O. J. 1961. *R. Obs. Bull.* 31:101–17
Eggen, O. J. 1967. *Mem. R. Astron. Soc.* 70: 111–64
Eggleton, P. P. 1976. See Batten 1976, pp. 209–12
Ehlers, J., Rosenblum, A., Goldberg, J. N., Havas, P. 1976. *Ap. J.* 208: L77–L81
Ergma, E. V., Tutukov, A. V. 1976. *Acta Astron.* 26:69–76
Ezer, D., Cameron, A. G. W. 1967. *Can. J. Phys.* 45:3429–77
Fabbiano, G., Schreier, E. J. 1977. *Ap. J.* 214: In press
Faulkner, J. 1976. See Batten 1976, pp. 193–204
Flannery, B. P. 1975. *Ap. J.* 201:661–94
Flannery, B. P. 1976. *Ap. J.* 205:217–25
Flannery, B. P., Ulrich, R. K. 1977. *Ap. J.* 212:533–40
Flannery, B. P., van den Heuvel, E. P. J. 1975. *Astron. Astrophys.* 39:61–67
Giannone, P., Giannuzzi, M. A. 1970. *Astron. Astrophys.* 6:309–17
Giannone, P., Giannuzzi, M. A. 1972. *Astron. Astrophys.* 19:298–302
Giannone, P., Kohl, K., Weigert, A. 1968. *Z. Astrophys.* 68:107–29
Goldreich, P., Soter, S. 1966. *Icarus.* 5:375–89
Greenstein, J. L. 1973. *Astron. Astrophys.* 23:1–7
Gursky, H. 1976. See Batten 1976, pp. 19–26
Hall, D. S. 1974. *Acta Astron.* 24:215–34
Hall, D. S. 1975. *Acta Astron.* 25:1–20
Hall, D. S. 1976. In *Multiple Periodic Variable Stars*, ed. W. S. Fitch, pp. 287–348. Budapest: Mitt. d. Sternwarte
Hall, D. S., Neff, S. G. 1976. See Batten 1976, pp. 283–88
Hall, D. S., Stuhlinger, T. 1976. *Acta Astron.* 26:109–16
Hazlehurst, J., Meyer-Hofmeister, E. 1973. *Astron. Astrophys.* 24:379–92
Hills, J. G., Dale, T. M. 1974. *Astron. Astrophys.* 30:135–39
Hofmeister, E. 1967. *Z. Astrophys.* 65:164–84
Hutchings, J. B. 1976. See Batten 1976, pp. 9–18
Hutchings, J. B., Cowley, A. P., Redman, R. O. 1975. *Ap. J.* 201:404–12
Hutchings, J. B., Hill, G. 1971. *Ap. J.* 167: 137–48
Iben, Jr., I. 1965. *Ap. J.* 142:1447–67
Iben, Jr., I. 1966a. *Ap. J.* 143:483–504
Iben, Jr., I. 1966b. *Ap. J.* 143:505–15
Iben, Jr., I. 1967. *Ap. J.* 147:650–63
Khabazin, Yu. G. 1975. *Sov. Astron. AJ* 19:34–37
Kippenhahn, R. 1969. *Astron. Astrophys.* 3:83–87
Kippenhahn, R., Meyer-Hofmeister, E. 1977. *Astron. Astrophys.* 54:539–42
Kippenhahn, R., Weigert, A. 1967. *Z. Astrophys.* 65:251–73
Koch, R. H. 1972. *Publ. Astron. Soc. Pac.* 84:5–24
Kopal, Z. 1956. *Ann. Astrophys.* 19:298–335
Kopal, Z. 1959. *Close Binary Systems*, pp. 125–46. New York: Wiley. 558 pp.
Kopal, Z. 1971. *Publ. Astron. Soc. Pac.* 83: 521–38
Kraft, R. P. 1965. *Ap. J.* 142:1588–93
Kraft, R. P. 1967a. *Ap. J.* 150:551–70
Kraft, R. P. 1967b. *Publ. Astron. Soc. Pac.* 79:395–413
Kraft, R. P. 1975. In *Neutron Stars, Black Holes and Binary X-Ray Sources*, ed. H. Gursky, R. Ruffini, pp. 235–55. Dordrecht-Holland: Reidel. 441 pp.
Kuiper, G. P. 1941. *Ap. J.* 93:133–77
Lamers, H. J. G. L. M., van den Heuvel, E. P. J., Petterson, J. A. 1976. *Astron. Astrophys.* 49:327–35
Larson, R. B. 1969. *MNRAS.* 145:271–95
Larson, R. B. 1972. *MNRAS.* 156:437–58
Lauterborn, D. 1970. *Astron. Astrophys.* 7: 150–59
Lauterborn, D., Weigert, A. 1972. *Astron. Astrophys.* 18:294–300
Lebovitz, N. R. 1972. *Ap. J.* 175:171–83
Lecar, M., Wheeler, J. C., McKee, C. F. 1976. *Ap. J.* 205:556–62
Levato, H. 1976. *Ap. J.* 203:680–88
Lin, D. N. C., Pringle, J. E. 1976. See Batten 1976, pp. 237–52
Lubow, S. H., Shu, F. H. 1975. *Ap. J.* 198: 383–405
Lucy, L. B. 1968. *Ap. J.* 151:1123–35
Lucy, L. B. 1976. *Ap. J.* 205:208–16
Mallama, A. D. 1975. *Acta Astron.* 25:205–14
Martynov, D. Ya. 1973. *Sov. Phys. Usp.* 15: 786–803
Massevitch, A. G., Tutukov, A. V., Yungelson, L. R. 1976. *Astrophys. Space Sci.* 40:115–33
Mauder, H. 1972. *Astron. Astrophys.* 17:1–16
McCluskey Jr., G. E., Kondo, Y., Morton, D. C. 1975. *Ap. J.* 201:607–12
Mengel, J. G., Norris, J., Gross, P. G. 1976. *Ap. J.* 204:488–92
Meyer-Hofmeister, E. 1974. *Astron. Astrophys.* 36:261–65
Mochnacki, S. W., Doughty, N. A. 1972a. *MNRAS.* 156:51–65
Mochnacki, S. W., Doughty, N. A. 1972b. *MNRAS.* 156:243–52
Morton, D. C. 1973. In *Wolf-Rayet and High-Temperature Stars*, ed. M. K. V. Bappu, J. Sahade, pp. 54–56. Dordrecht-

Holland: Reidel. 263 pp.
Moss, D. L., Whelan, J. A. J. 1970. *MNRAS*. 149: 147–65
Nelson, B., Young, A. 1976. See Batten 1976, pp. 141–46
Neo, S., Miyaji, S., Nomoto, K., Sugimoto, D. 1977. *Publ. Astron. Soc. Jpn*. 29: In press
Ostriker, J. P., Bodenheimer, P. 1968. *Ap. J*. 151: 1089–98
Paczyński, B. 1967a. *Acta Astron*. 17: 287–96
Paczyński, B. 1967b. *Acta Astron*. 17: 355–80
Paczyński, B. 1971a. *Ann. Rev. Astron. Astrophys*. 9: 183–208
Paczyński, B. 1971b. *Acta Astron*. 21: 1–14
Paczyński, B. 1976. See Batten 1976, pp. 75–80
Paczyński, B., Sienkiewicz, R. 1972. *Acta Astron*. 22: 73–91
Paczyński, B., Ziólkowski, J. 1967. *Acta Astron*. 17: 7–14
Paczyński, B., Ziólkowski, J. 1975. *Astron. Astrophys*. 40: 351–54
Plavec, M. 1964. *BAC*. 15: 165–70
Plavec, M. 1970. *Publ. Astron. Soc. Pac*. 82: 957–95
Plavec, M. 1973. In *Extended Atmospheres and Circumstellar Matter in Spectroscopic Binary Systems,* ed. A. H. Batten, pp. 216–59. Dordrecht-Holland: Reidel. 291 pp.
Plavec, M., Ulrich, R. K., Polidan, R. S. 1973. *Publ. Astron. Soc. Pac*. 85: 769–805
Popper, D. M. 1973. *Ap. J*. 185: 265–75
Popper, D. M., Ulrich, R. K. 1977. *Ap. J*. 212: L131–34
Pratt, J. P., Strittmatter, P. A. 1976. *Ap. J*. 204: L29–L33
Prendergast, K. H., Burbidge, G. R. 1968. *Ap. J*. 151: L83–L88
Rees, M. J. 1974. In *Highlights of Astronomy*. 3: 89–107
Refsdal, S., Roth, M. L., Weigert, A. 1974. *Astron. Astrophys*. 36: 113–22
Refsdal, S., Weigert, A. 1971. *Astron. Astrophys*. 13: 367–73
Reimers, D. 1976. *Astron. Astrophys*. 52: 457–59
Ritter, H. 1976. *MNRAS*. 175: 279–95
Robertson, J. A., Eggleton, P. P. 1977. *MNRAS* 179: 359–75
Robinson, E. L. 1976. *Ann. Rev. Astron. Astrophys*. 14: 119–42
Rogerson, J. B. Jr., Lamers, H. J. G. L. M. 1975. *Nature*. 256: 190
Roxburgh, I. W. 1966. *Ap. J*. 143: 111–20
Rucinski, S. M. 1973. *Acta Astron*. 23: 79–120
Rucinski, S. M. 1974. *Acta Astron*. 24: 119–51
Schwarzschild, M. Härm, R. 1962. *Ap. J*. 136: 158–65
Shu, F. H., Lubow, S. H., Anderson, L. 1976. *Ap. J*. 209: 536–46
Smarr, L. L., Blandford, R. 1976. *Ap. J*. 207: 574–88
Smith, L. F. 1973. See Morton 1973, pp. 228–34
Stoeckly, R. 1965. *Ap. J*. 142: 208–28
Stothers, R. 1973. *Publ. Astron. Soc. Pac*. 85: 363–78
Tapia, S., Whelan, J. 1975. *Ap. J*. 200: 98–105
Tscharnuter, W. 1975. *Astron. Astrophys*. 39: 207–12
Tutukov, A. V., Yungelson, L. R. 1973. *Nauchn. Inf*. 27: 58–69
Ulrich, R. K., Burger, H. L. 1976. *Ap. J*. 206: 509–14
van den Heuvel, E. P. J. 1970. In *Stellar Rotation,* ed. A. Slettebak, pp. 178–86. Dordrecht-Holland: Reidel. 355 pp.
van den Heuvel, E. P. J. 1973. *Nature Phys. Sci*. 242: 71–72
van den Heuvel, E. P. J. 1975. *Ap. J*. 196: L121–23
van den Heuvel, E. P. J. 1976. See Batten 1976, pp. 35–62
van den Heuvel, E. P. J., Heise, J. 1972. *Nature Phys. Sci*. 239: 67–69
Van Horn, H. M., Sofia, S., Savedoff, M. P., Duthie, J. G., Berg, R. A. 1975. *Science*. 188: 930–33
Van't Veer, F. 1975. *Astron. Astrophys*. 44: 437–43
Vauclair, G. 1972. *Astron. Astrophys*. 17: 437–40
Vilhu, O. 1973. *Astron. Astrophys*. 26: 267–74
Webbink, R. F. 1975. *Astron. Astrophys*. 41: 1–8
Webbink, R. F. 1976a. *Ap. J*. 209: 829–45
Webbink, R. F. 1976b. *Ap. J. Suppl*. 32: 583–601
Webbink, R. F. 1977a. *Ap. J*. 211: 486–98
Webbink, R. F. 1977b. *Ap. J*. 211: 881–89
Wheeler, J. C. 1976. *Ap. J*. 205: 578–79
Wheeler, J. C., Lecar, M., McKee, C. F. 1975. *Ap. J*. 200: 145–57
Whelan, J. A. J. 1970. *MNRAS*. 149: 167–77
Whelan, J. A. J. 1972a. *MNRAS*. 156: 115–28
Whelan, J. A. J. 1972b. *MNRAS*. 160: 63–77
Whelan, J. A. J., Iben Jr., I. 1973. *Ap. J*. 186: 1007–14
Wilson, R. E. 1974. *Ap. J*. 189: 319–29
Wilson, R. E. 1976. See Batten 1976, pp. 65–74
Yamasaki, A. 1971. *Publ. Astron. Soc. Jpn*

23:33–55
Yungelson, L. R. 1973a. *Nauchn. Inf.* 26:71–79
Yungelson, L. R. 1973b. *Nauchn. Inf.* 27:93–98
Yungelson, L. R. 1973c. *Sov. Astron. AJ.* 16:864–66
Zahn, J.-P. 1975. *Astron. Astrophys.* 41:329–44
Ziółkowski, J. 1970. *Acta Astron.* 20:213–47
Ziółkowski, J. 1976a. *Ap. J.* 204:512–15
Ziółkowski, J. 1976b. See Batten 1976, pp. 321–22

Ann. Rev. Astron. Astrophys. 1977. 15 : 153–73

LARGE-SCALE SOLAR MAGNETIC FIELDS

Robert Howard
Hale Observatories, Carnegie Institution of Washington, California Institute of Technology, Pasadena, California 91101

INTRODUCTION

George Ellery Hale detected strong magnetic fields in sunspots in 1908. For some years following that, Hale and his colleagues at Mount Wilson tried to measure what they thought should be a general dipole magnetic field of the Sun. Various numbers were quoted over the years (Hale 1913, Hale et al. 1918), but no firm results ever came from those attempts. Later analysis of the original plates indicates that no dipole field was measurable from them (Stenflo 1970).

In 1952, H. W. Babcock (1953) devised the principle of the modern solar magnetograph. This instrument, by means of a subtraction technique, avoided the major instrumental polarization problems that had plagued earlier instruments and for the first time made possible the accurate measurement of weak longitudinal magnetic fields on the solar surface. Almost all of the magnetic measurements of the photosphere in the last 25 years have been made with instruments using the principle devised by Babcock.

In an early paper, H. W. and H. D. Babcock (1955) demonstrated the existence of more than just a polar field. They found bipolar magnetic configurations associated with active regions and large, weak unipolar magnetic regions in some quiet areas of the Sun. Later work in the 1950s (Leighton 1959, Howard 1959) established that magnetic fields occupied precisely the locations of emission regions in Ca II spectroheliograms, even down to the smallest details (cf Figure 1), and that the dark Hα filaments (prominences seen against the disk) invariably lay between magnetic areas of opposite polarity. These associations confirmed the great importance of magnetic fields to solar activity and laid the basis for later studies of the appearance of active-region magnetic fields and their subsequent decay into large-scale coherent magnetic-field patterns. These expanding field patterns, in conjunction with the supergranulation (Leighton et al. 1962), proved to be important in the establishment and maintenance of the calcium (magnetic) network. This weak network pattern is now generally considered to be maintained by the action of the supergranulation motions. The nature of the large-scale distribution of magnetic

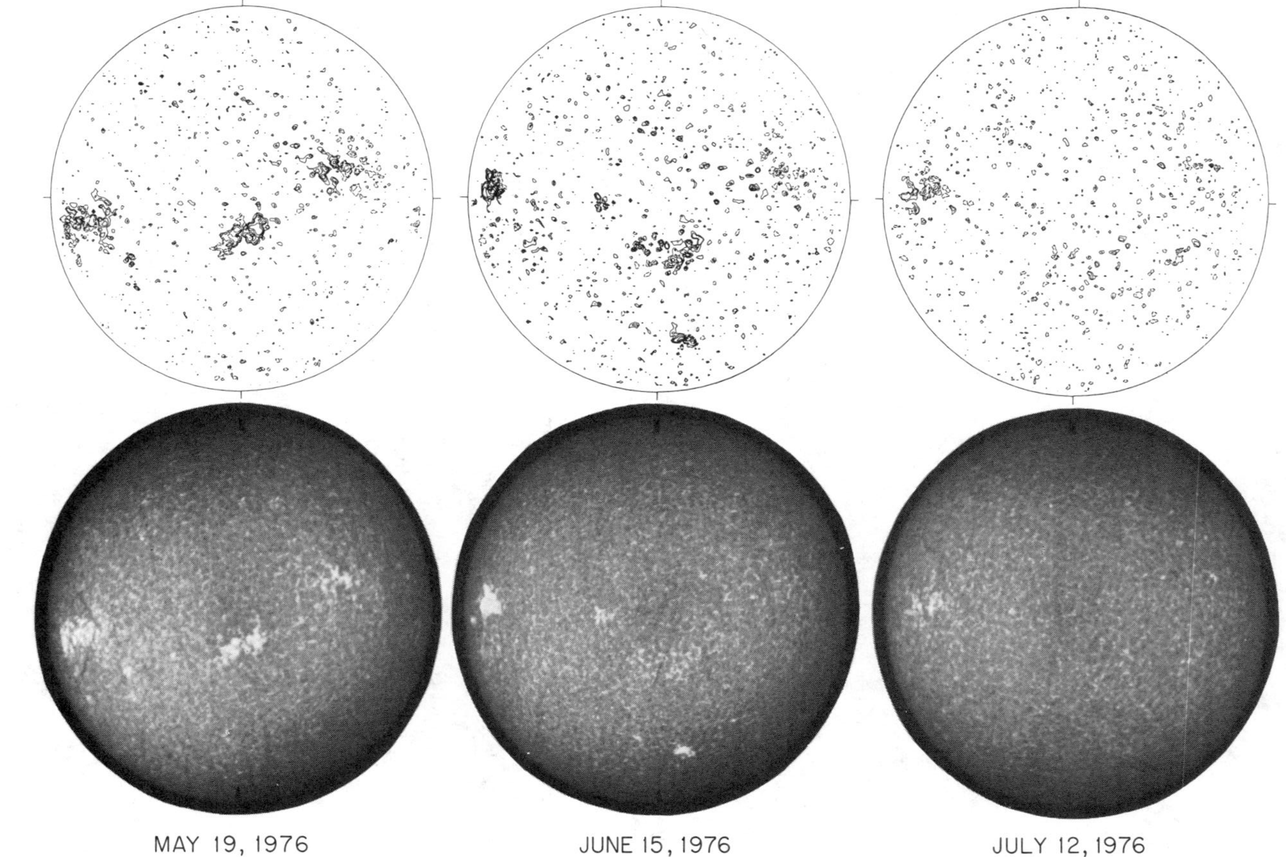

Figure 1 Magnetograms (top) and Ca II K-line spectroheliograms (bottom) for the central area of the solar disk for the dates indicated. All observations are from Mount Wilson. Note the background network pattern and the breakup of the plage and magnetic fields as the region ages.

fields on the solar surface is a result, at least to some extent, of the actions of the supergranular motions and the differential rotation (Schatten et al. 1972).

LARGE-SCALE MAGNETIC SURFACE FEATURES

The Decay of Active Regions

It has long been known that solar active regions, as seen in chromospheric lines, do not form and decay in the same manner. Their formation is generally rapid, and at the end of the growth phase the region is relatively bright, with sharp and well-defined boundaries. The decay of an active region is a slow process of spreading out, weakening, and fragmentation (Butler 1924). The supergranulation motions appear to play an important role in the process of decay.

Figure 1 shows an example of the decay and final dissolution of an active region seen in magnetograms and the Ca II K line. The eroding effects of the supergranular motions are evident. The final death of the region may be defined as occurring when the last emission patches blend with and become indistinguishable from the background chromospheric network (Bumba et al. 1968).

It should be noted that sunspots normally exist only in the earliest stages in the lifetime of an active region. The active region may live on for many weeks or even months after the last sunspot is gone. In the interval following the death of the sunspots, the active region can still be a spectacular phenomenon in the chromosphere. At times solar flares result from such spotless regions (Dodson & Hedeman 1970).

I have noted that active regions do not all decay at the same rate. Some regions decay quite rapidly, and some regions appear to last for much more than the normal lifetime. One factor in determining the lifetime is certainly the size of the region, but the nature of the surrounding fields and the nature of the magnetic connection to subsurface layers could be additional factors. This is an area where more research is needed.

Ephemeral regions (Harvey & Martin 1973, Harvey et al. 1975) are tiny, short-lived active regions without sunspots. They show as bright dots on X-ray pictures (Vaiana et al. 1973). Their latitude distribution is somewhat broader than that of ordinary active regions. They appear to follow, at least roughly, the Hale polarity law seen in active regions, but they have a broader distribution of orientation than do the active regions. Although they may be responsible for bringing to the solar surface a non-negligible fraction of the solar flux that arises from below, it seems unlikely, because of their small sizes and short lifetimes, that they contribute significantly to the large-scale magnetic patterns (Howard 1974c).

The Background Field Pattern

The total effect of the contribution of the decaying magnetic fields from all the active regions on the Sun is the formation of a large-scale pattern of alternating polarities that is remarkably long-lived (Bumba & Howard 1965, 1969). The individual areas of one polarity—but not the same field elements—may last for a year or more. As new regions are born and decay, they often provide additional magnetic flux to reinforce the existing weak fields. Smaller weak-field areas (a few tens of degrees in longitude) generally have the same polarity on either side of the equator, and may persist for nearly a year. Somewhat larger features, containing predominantly fields of one polarity, may live for up to three years. Figure 2 shows a Mount Wilson magnetic synoptic chart for the period near the most recent activity maximum (1970) and one during 1967, taken at the time of the rise to maximum.

At times scattered fields of the following polarity may combine over a longitude range of 100° or more at a latitude generally greater than that of the normal background field pattern. Such a feature is referred to as a Unipolar Magnetic Region (UMR). None are visible in Figure 2 because they occur preferentially during the decline from activity maximum. A UMR may be seen in Figure 1 of the paper by Bumba & Howard (1965). More recent UMRs, and the largest ones in recent years, may be seen in the southern hemisphere in Carrington rotations 1599 to 1606 (negative polarity), and in the northern hemisphere in rotations 1615 to 1623 (positive polarity). (The Mount Wilson synoptic charts of magnetic fields are published regularly in the *I.A.U. Quarterly Bulletin on Solar Activity*.)

The Polar Fields

The earliest photoelectric magnetograph measurements (Babcock 1953) were made to measure the "general field" of the Sun, which in those days was assumed to be a N-S dipole field. These earliest measurements indeed showed the existence of magnetic fields in the polar regions that were generally of opposite polarity at the two poles. The resemblance to a true dipole ended there because of the pattern of large-scale fields found at intermediate and low latitudes.

At the next solar maximum ($\approx$1957–58) the two polar fields changed polarity, first the south field and then a year or so later the north field (Babcock 1959).

A few years later the polar fields, which had been fairly strong during the 1950s, weakened considerably (Severny 1971, Howard 1965, 1972), so that the polarity reversal at the next solar maximum was difficult to detect. There is, in fact, some difference in the exact date of the reversal as observed at Kitt Peak and Mount Wilson, although both sets of observations agree that 1. the reversals followed the maximum of solar activity by a year or more, and 2. the reversals at the two poles took place at different times (Gillespie et al. 1973, Howard 1974a).

Sheeley (1976) has traced the history of the polar fields in this century by counting polar faculae on white-light photographs of the Sun taken at Mount Wilson. The polar faculae are known to occur at the concentrations of polar fields; thus the more polar faculae, the stronger is the polar field. Sheeley finds that the number of polar faculae go through a minimum near the maximum of each solar cycle. Figure 3 shows the facular counts for this century. One can see in Figure 3, in agreement with the magnetic results mentioned above, that the faculae decreased in number quite suddenly in the early 1960s. It is also clear from this figure that in general the polar fields reverse a year or two following solar maximum.

It should be emphasized that at least a large percentage and perhaps all of the polar field flux is contained in isolated small strong-field elements. This conclusion comes from direct observations (Howard 1959), from analogy with the low-latitude

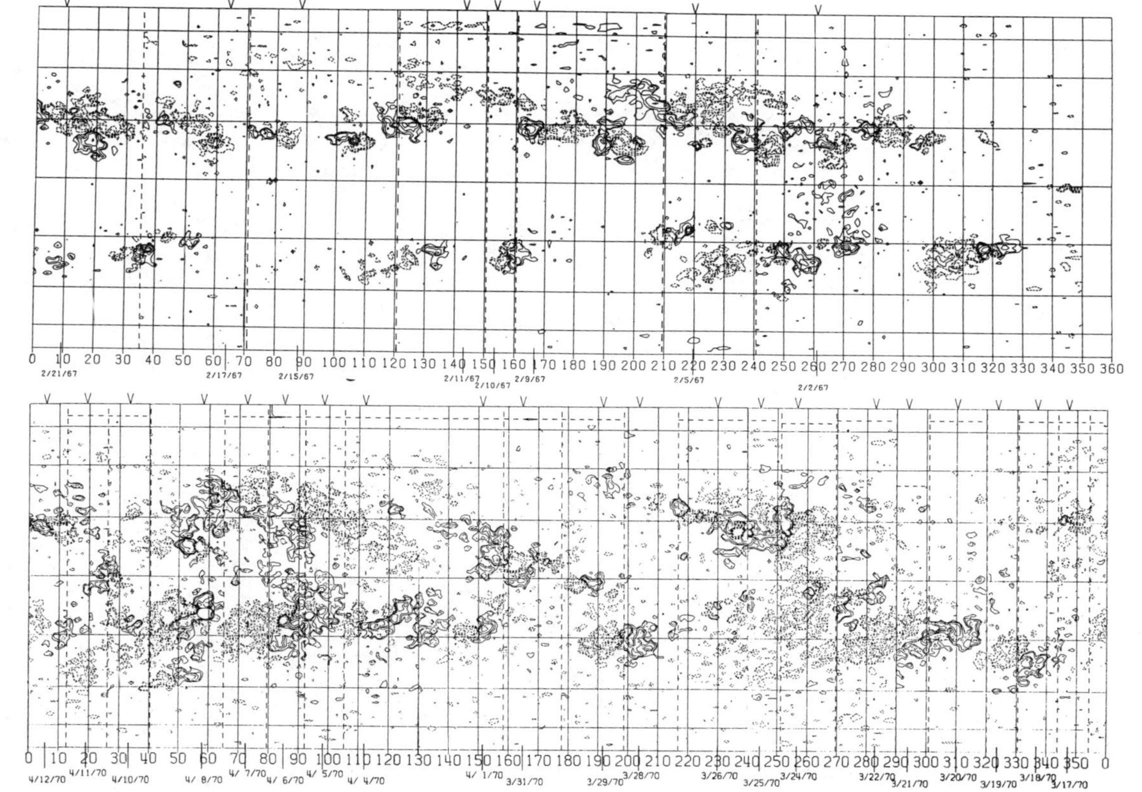

Figure 2 Magnetic contour synoptic charts from Mount Wilson data for Carrington rotations 1517 (February 1967) and 1559 (March–April 1970). These are cylindrical equal-area projections. East is at the left and the horizontal lines represent the equator, $\pm 20°$, $\pm 40°$, and $\pm 60°$. The north pole is the line at the top, and the south pole is at the bottom. The date of each observation is given at the bottom of each chart, along with the location of the central meridian at the time of the observation. The vertical dashed lines represent the dividing line between separate days' observations. No data from more than one day are averaged to make these plots. Solid lines represent positive magnetic fields (magnetic vector toward the observer) and dashed lines represent negative fields. The contours are ± 5, 10, 20, 40, 80 G.

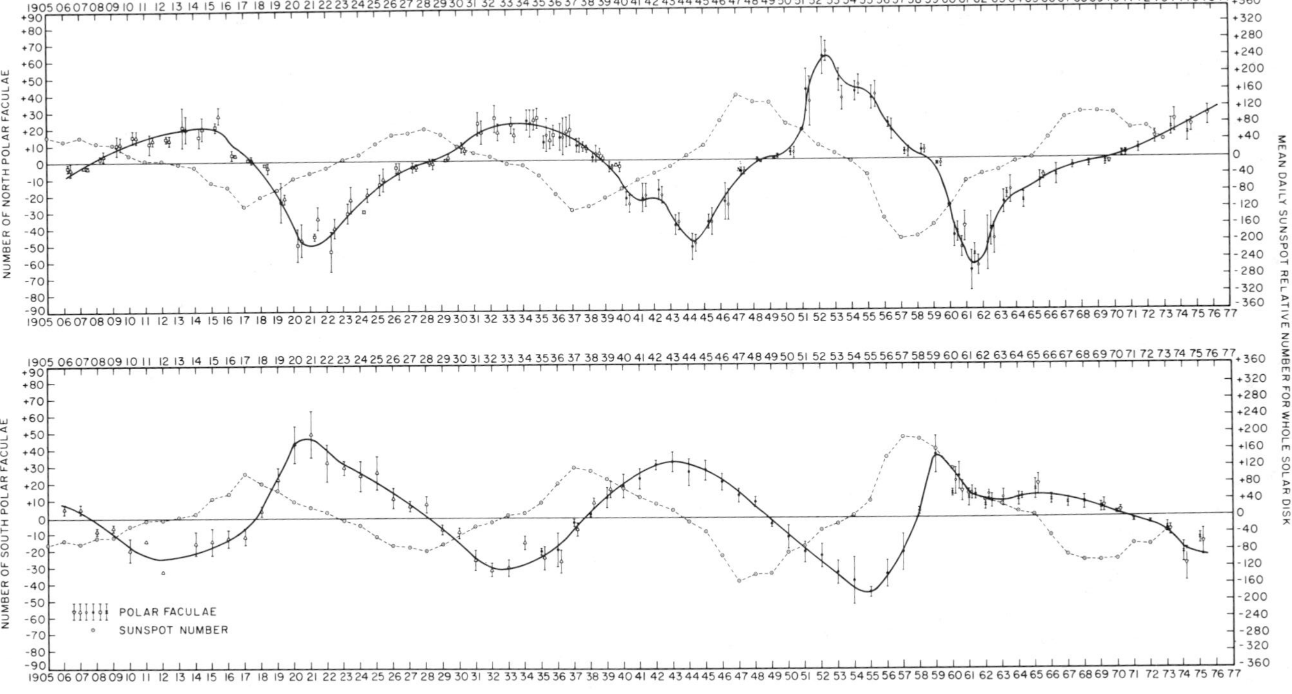

Figure 3 The numbers of north and south polar faculae, respectively, during the interval 1906–75. The numbers have been assigned polarities corresponding to the polarities of the associated polar magnetic fields. Also the numbers represent the magnetic flux normal to the entire surface of each polar cap in units of 0.3×10^{21} Mx ($\pm 50\%$). For comparison, the sunspot number for the whole solar disk has been plotted with a magnetic polarity assigned that corresponds to that of the following spots in each hemisphere. (From N. R. Sheeley Jr. 1976. *Journal of Geophysical Research* 81 : 3462, copyrighted by American Geophysical Union.)

situation (Howard & Stenflo 1972), and from the generally good agreement between Sheeley's facular counts and the measured polar fields.

The polar fields are not well measured with the solar magnetograph because of the poor angle at which the fields must be viewed. A recent analysis of the Mount Wilson data (Howard 1977a) indicates, however, that the measured magnetic fields near the poles represent fairly well the true average field strengths there. An absolute limit on the underestimation of the polar fields is about a factor of 5. Thus the true average magnetic field strength in the polar regions cannot exceed 5 or 10 G, and is most likely about 1 G. A field of 1 G spread over the polar region contains approximately the same magnetic flux as a large active region ($\approx 4 \times 10^{21}$ Mx).

Giant Regular Structures

A very large-scale ordering of the field distribution has been discussed by Bumba (1970). A giant cell-like pattern, with dimensions of the order of 400,000 km, changes its configuration slowly, but as seen in the fields of the two polarities separately, it maintains a definite cellular shape. Lifetimes of individual cells are several months (Ambrož et al. 1971).

This may represent a giant-scale analogy to the granulation and supergranulation patterns. Simon & Weiss (1968a, b) have suggested that this pattern represents a system of giant convection cells, with dimensions of the order of the depth of the convection zone in the Sun.

Bumba (1976a) has found a good correlation between the polarity of the main body of the giant structures and the interplanetary magnetic field. The ordering of the solar pattern is on about the same scale as the interplanetary sector pattern, and this also leads to a strong indication of a physical connection between the two.

Wagner & Gilliam (1976) have suggested that the large-scale convection cell pattern may be seen in the pattern of filament distribution. For a period in the summer of 1972 during which a simple pattern appeared, the longitudinal wave number they found was 5. The lifetimes of the cells they studied were 2–4 months. On the whole, although a number of indications point to a large-scale cellular pattern of magnetic fields, there is still no firm evidence of the convective nature of these patterns. No large-scale velocity fields have been conclusively associated with the magnetic patterns. However, the velocity amplitude of such motions would be quite small, and their detection will be a difficult task. A large-scale velocity pattern has been detected by Howard & Yoshimura (1976), but it has not yet been possible to demonstrate an association of this pattern with magnetic fields.

Expansion of the Field in Surface Harmonics

Altschuler et al. (1974) have carried out a surface harmonic expansion of the magnetic fields of the Sun, as measured at Mount Wilson, in terms of Legendre polynomials. The data covered the interval 1959–72. A microfilm tabulation of the harmonic coefficients through 1974 has been published recently (Altschuler et al. 1975). The most frequently occurring harmonic is that corresponding to a dipole lying in the plane of the equator. This was particularly true of the most active years of solar cycles no. 19 and 20. The north-south dipole harmonic was prominent only during

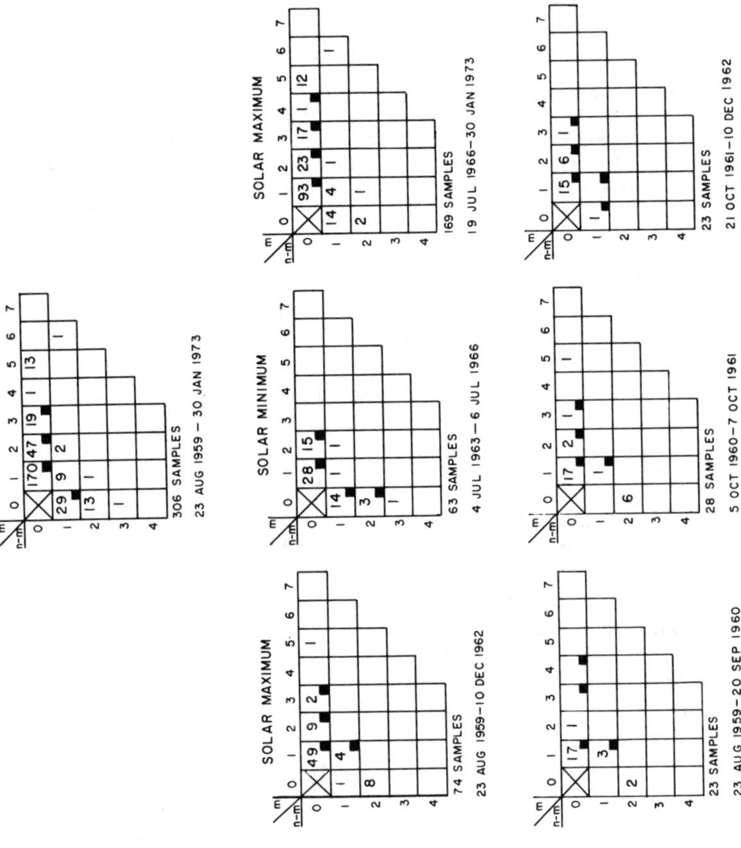
DOMINANT SURFACE HARMONIC FOR SOLAR MAGNETIC FIELD
SOLAR CYCLE
306 SAMPLES
23 AUG 1959 – 30 JAN 1973
SOLAR MAXIMUM
74 SAMPLES
23 AUG 1959–10 DEC 1962
SOLAR MINIMUM
63 SAMPLES
4 JUL 1963 – 6 JUL 1966
SOLAR MAXIMUM
169 SAMPLES
19 JUL 1966–30 JAN 1973
23 SAMPLES
23 AUG 1959–20 SEP 1960
28 SAMPLES
5 OCT 1960–7 OCT 1961
23 SAMPLES
21 OCT 1961–10 DEC 1962

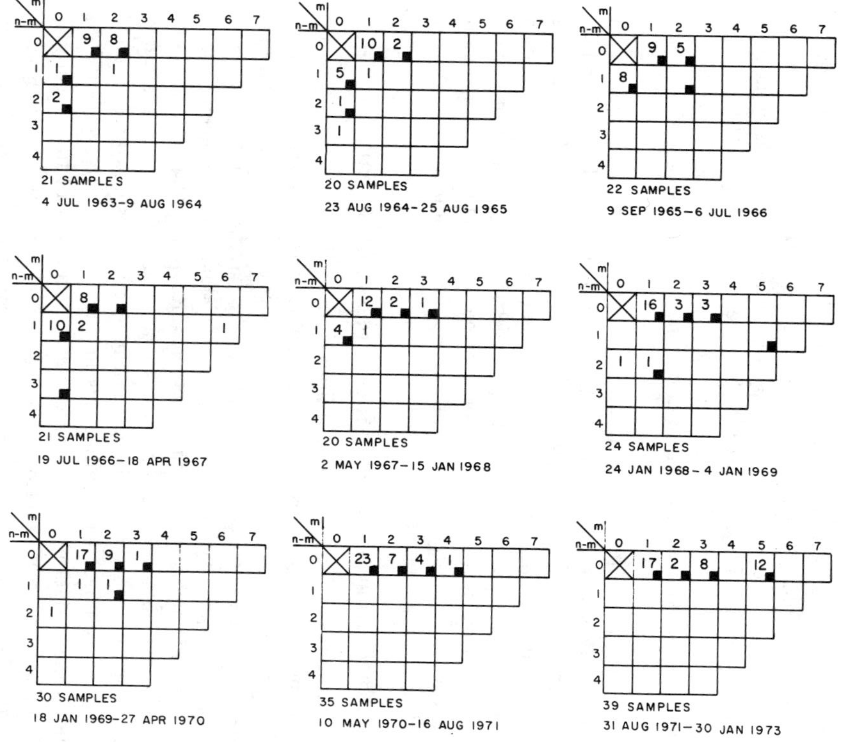

Figure 4 The dominant surface harmonics for the photospheric magnetic field are indicated for the period 1959–72. A black box in the lower right-hand corner of each space indicates harmonics that appear most often among the top 5 in importance. The numbers in the spaces give the number of times the harmonics were dominant. (From Altschuler et al. 1974.)

quiet years. A 4-sector structure was evident at various times, and occasionally a 6-sector structure could be seen. In general the contribution to the field from harmonics between 5 and 9 (the maximum calculated) was very small.

From time to time rather rapid changes (within one rotation) in the contributions of the various harmonics were seen. This indicates that either the polar field is not very deep, or that strong fluid flows connect the photosphere with deeper layers.

Figure 4 shows schematically the dominant surface harmonics for photospheric magnetic fields for the interval 1959–72. The geometrical interpretation of these harmonics is as follows. The harmonic $P_n^m(\theta) \cos m\phi$ [or $P_n^m(\theta) \sin m\phi$] is zero on 2 m different meridians equally spaced in longitude. If $m \neq 0$ there are 2 m sectors in longitude, m of positive polarity and m of negative polarity. The poles of the Sun, in this simplified representation, can have a uniform field only if $m = 0$. The quantity $n - m$ is the number of times the surface harmonic goes through zero between (but not at) the poles. The quantity n gives the index of the multipole: $n = 0$ is a monopole, $n = 1$ is a dipole, $n = 2$ is a quadrupole, etc.

Naturally the true magnetic-field distribution is much more complicated than can be represented by surface harmonics unless one includes very high-order harmonics. A glance at any synoptic chart will demonstrate this. However, the expansion in Legendre polynomials is a convenient and simple means of describing some of the large-scale characteristics of the field distribution. Whether or not the characterization by this means of, for example, "dipole" or "octupole" components of the field distribution has any bearing on the large-scale subphotospheric structure of the fields is an open question.

The Average Inclination of Magnetic-Field Lines in the Photosphere

Solar magnetographs have so far produced little if any usable transverse Zeeman-effect observations of magnetic fields outside sunspots. In general transverse measurements with solar magnetographs are roughly two orders of magnitude less sensitive than longitudinal measurements, which explains the paucity of transverse field observations. A consequence of this is that from individual observations we get no information about the inclination of magnetic-field lines to the solar surface.

One can, of course, gain information about the average east-west orientation of field lines by comparing observations of the same magnetic fields measured east and west of the central meridian; I have done this in a recent paper (Howard 1974b). Figure 5 shows the variation of the quantity β as a function of time separately for fields of the two polarities in various latitude zones in the northern hemisphere. The quantity β is defined as $\beta = (|F_E| - |F_W|)/(|F_E| + |F_W|)$, where F_E is the average magnetic flux measured east of the central meridian over one rotation, and F_W is the same quantity west of the central meridian. One degree of inclination in the east-west direction corresponds to $\beta \approx 0.01$. Positive β represents an inclination leading the rotation, i.e. from east to west with increasing height.

Figure 5 shows that in the northern hemisphere during these years the two polarities started out inclined toward each other (negative was the following polarity in the north at that time), and the inclinations gradually decreased until they were zero, or even slightly reversed by the end of the interval. Svalgaard & Wilcox (1974) found that the deviations from the average spiral angle of the interplanetary magnetic fields measured near the earth showed a roughly similar behavior in the same interval. The magnetic fields in the southern hemisphere of the Sun were weaker in this interval, and did not show quite such smooth behavior as is seen in Figure 5.

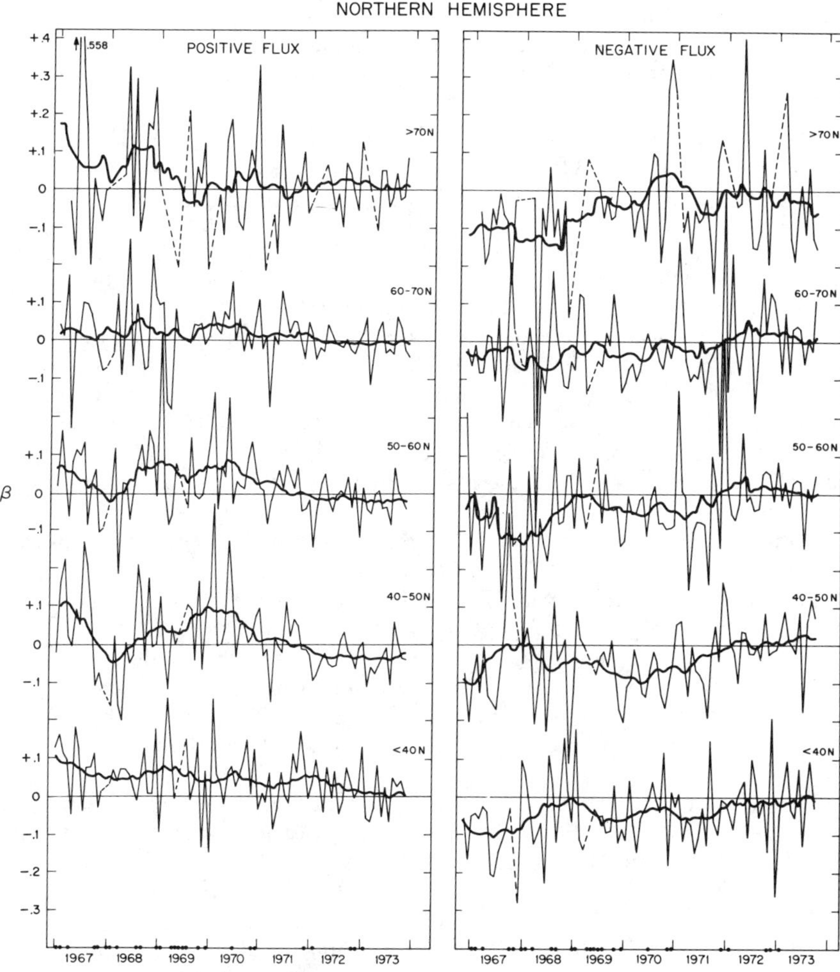

Figure 5 Values of β for the various latitude zones for each polarity separately in the northern hemisphere. Each point connected by the thin lines represents a value derived for one Carrington rotation. The thick lines are 13-rotation running means of the data. The dots along the time axis represent rotations within which two observations were separated by more than 100° in longitude, and thus could be of poor quality. The dashed lines represent missing rotations. (From Howard 1974b.)

The total magnetic flux ($F^T = |F^+| + |F^-|$) for each of the latitude zones studied showed small negative values of β, corresponding to an inclination of about 1°, trailing the rotation. A spot check of some areas in 1976 (solar minimum) shows inclinations, at low latitudes, leading the rotation (Howard 1977a).

Cross-correlation coefficients calculated for the magnetic flux of opposite polarities indicate that poleward of 40° latitude in either hemisphere the two polarities behave oppositely, i.e. an increase in the east-west inclination of positive fields is most likely to be accompanied by a decrease in inclination of negative fields. Below 40°, variations in inclination of fields of the two polarities are most likely to be parallel.

THE SOLAR ACTIVITY CYCLE AND THE LARGE-SCALE PATTERNS

The Appearance of the Background Fields Through the Cycle

In Figure 2 one may see magnetic synoptic charts from two phases of the solar activity cycle: the approach to maximum, and maximum. In Figure 6 synoptic charts from Kitt Peak are presented for three times in the cycle.

As one might expect, at the approach to maximum the magnetic fields are concentrated at intermediate latitudes. The level of activity at this time is not high. It is interesting to note the zone within about 10° of the equator, which appears to be free of fields for most of the length of the synoptic chart. This results because of the fact that early in the cycle the active regions occur preferentially at high latitudes. It is not clear why the fields appear to expand mostly in the poleward direction.

As the cycle progresses the stronger magnetic fields are found at lower latitudes, and they occupy more area as the frequency of active regions increases. Near maximum, even the weak maximum of the last cycle (no. 20), the background fields become strong and generally fill up fully a large portion of the solar surface with fields in excess of 5 G. The post-maximum phase is characterized by the appearance of active regions at very low latitudes, a diminution of the level of activity, and the appearance of very large-scale magnetic structures. The synoptic charts close to minimum show very large-scale, weak, and diffuse patterns (Bumba & Howard 1965).

Bumba (1976b), using recent data, has shown that the large-scale patterns are quite similar for the years following maximum in cycles no. 19 and 20. In particular the role of the two polarities is very much the same.

The Variation of the Fields Through the Cycle

Stenflo (1972) and Yoshimura (1976) have examined digitized Mount Wilson magnetograph data over intervals of a cycle or more. Stenflo plotted a series of four-rotation average synoptic charts that showed in a simple form the features described above.

Yoshimura (1976) averaged a large amount of the same data. He divided the data into poloidal and toroidal components. Only the longitudinal fields were observed, but he averaged the fields in longitude to form the poloidal component, and he subtracted this field at each longitude from the measured field, and called the absolute value of this quantity the toroidal field.

The poloidal field evolution is quite interesting. As the cycle begins, two branches of the poloidal field appear in the mid-latitudes in each hemisphere; one propagates toward the equator and the other toward the poles. This behavior was predicted by Yoshimura's (1975) theoretical model of the solar cycle which is driven by the dynamo action of global-scale convection. The toroidal component of the field

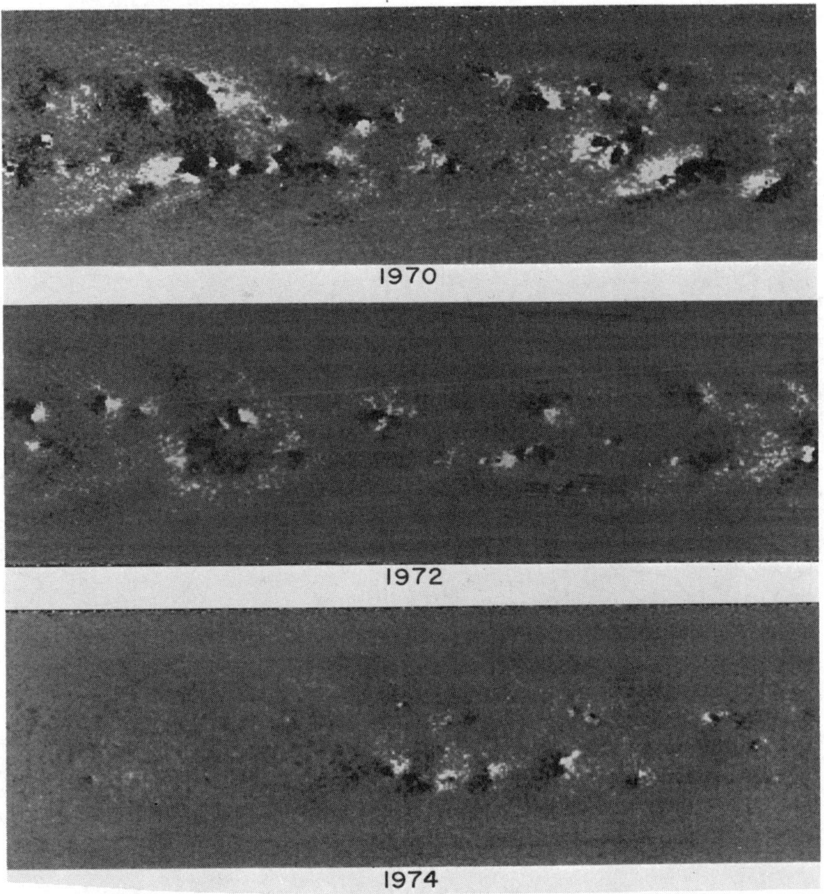

Figure 6 Kitt Peak magnetic synoptic charts for 3 Carrington rotations, 1558 (February–March 1970), 1589 (June 1972), and 1610 (January 1974). In the first two charts, black represents positive magnetic fields and white represents negative magnetic fields. In the third chart the sense is reversed. These represent a time near solar activity maximum (1558), the decline from maximum (1589), and the approach to minimum (1610). (Courtesy of Dr. J. W. Harvey.)

behaved somewhat like the butterfly diagram. Figure 7 shows the variation during the cycle of both the toroidal and poloidal components.

Variations in Magnetic Flux

An inspection of the magnetic flux values separately for each polarity (Howard 1974c) over the interval 1967–73 shows that an increase in the low-latitude flux around activity maximum roughly parallels the sunspot curve. The polar field polarity reversal in each hemisphere is seen to be the result of a wave of magnetic

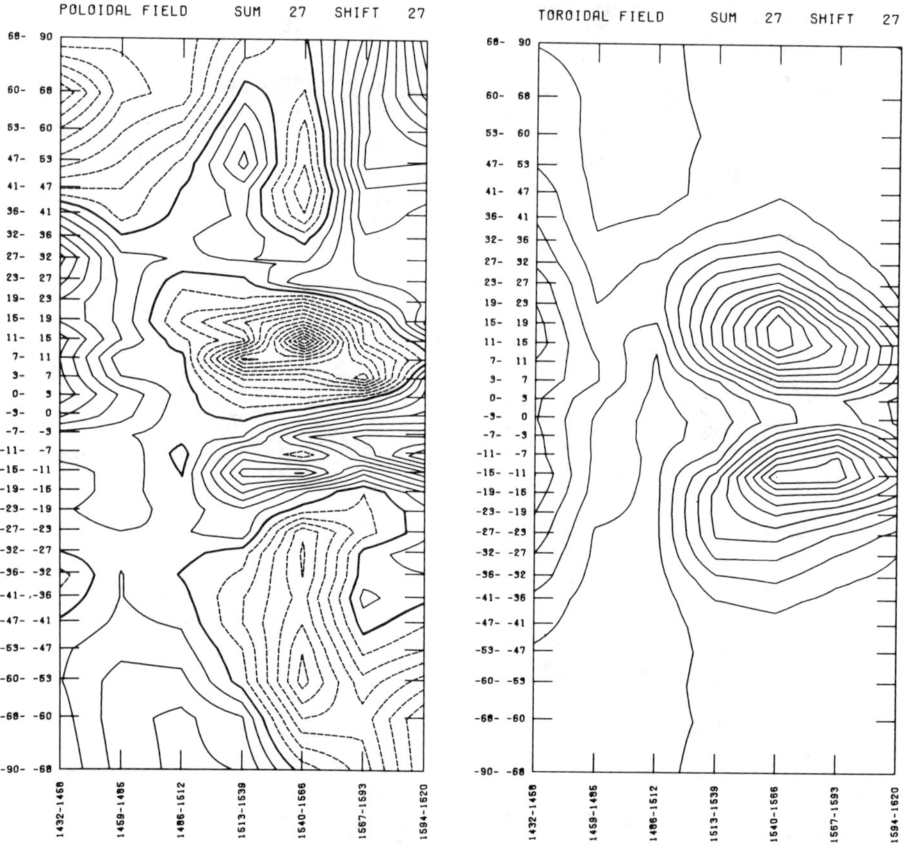

Figure 7 Evolutionary pattern of poloidal (left) and toroidal (right) magnetic fields on the solar surface. The ordinate is latitude, and the abscissa for each plot represents solar rotations 1432–1620 (1960 through 1974). The toroidal field behaves in much the same manner as the butterfly diagram, i.e. the latitudes of activity decrease as time progresses within the cycle. The contour interval is 0.1 G for these averaged data, and the maximum values are 6.9 G for the toroidal fields and 1.5 G for the poloidal fields. (From Yoshimura 1976).

flux of the appropriate polarity, which requires about one year to move from 40° latitude to the pole. Figure 8 shows the flux situation in the northern hemisphere in this interval.

Among the conclusions in this study was the fact that about 95% of the total magnetic flux of the Sun (F^T) is confined to latitudes below 40° in both hemispheres. The flux above 60° represents less than 2% of the total flux. The poleward flux

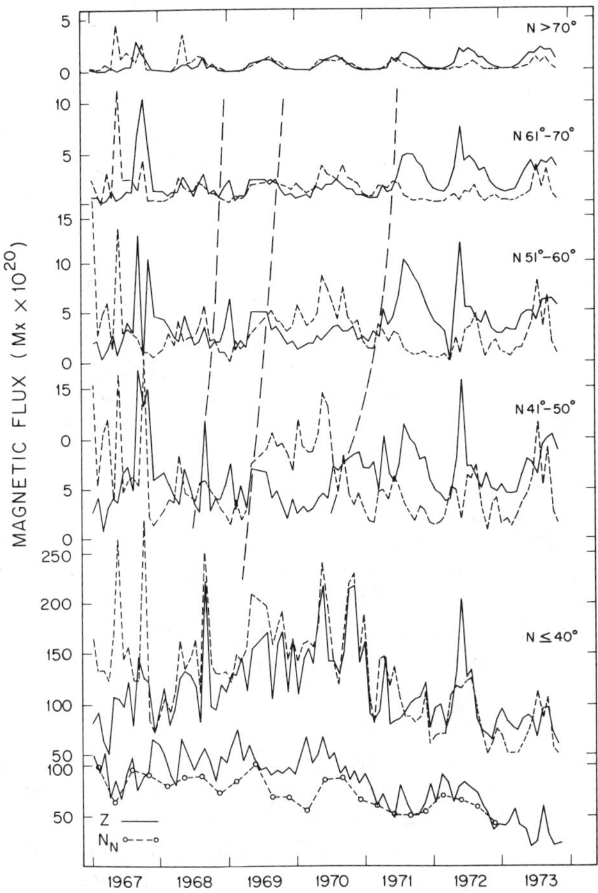

Figure 8 Positive polarity magnetic flux (solid lines) and negative polarity magnetic flux (dashed lines) in various latitude zones as a function of time in the northern hemisphere of the Sun. Each point represents the daily average over one Carrington rotation. The lowest solid curve is the Zurich full-disk Wolf (sunspot) number, and the lowest dashed curve represents the number of spot groups per quarter in the northern hemisphere. The vertical dashed lines represent three poleward magnetic flux migrations (positive, negative, positive). (From Howard 1974c.)

migrations, which were responsible for the polar field reversals, were accompanied by a poleward migration of filaments (Waldmeier 1973), which are known to separate large-scale magnetic fields of opposite polarities. The filament migration for cycle no. 10 was somewhat anomalous.

Recent work indicates (Howard 1977b) that the total flux at latitudes above 40° has, in the interval 1973–75, increased by a factor of about 2.5; no such increase has appeared in the low-latitude flux. This may be associated with changes in the latitude gradient of differential rotation at high latitudes during the approach to activity minimum (Howard 1976).

The Overall Pattern and its Variations

Svalgaard et al. (1974) have discussed a phenomenological model of the interplay between the very large-scale sector pattern of solar magnetism and the polar field. Referring to their Figure 1, the global-scale north-south neutral line takes on roughly an "S" shape, stretching from the northern to the southern hemisphere. This neutral line is overlaid in the corona by a current sheet that is tilted with respect to the central meridian. The current sheet is visible as a helmet streamer. The locations of helmet streamers in recent years agree with the predictions of this model. Also, the field distribution on the Sun shows such a shape (Svalgaard et al. 1977). The model presents a very much simplified arrangement of large-scale fields, but it predicts well the large-scale characteristics of the coronal and interplanetary magnetic field and the locations of large, long-lived coronal streamers.

Hansen & Hansen (1975, 1977) have used the differential rotation of large-scale surface magnetic features to explain evolutionary and sudden changes in the major structural components of the corona. Magnetic mergers and the large-scale reconnection of magnetic field lines are presumed to occur. No striking changes in the surface field patterns are required to produce these variations in the coronal structure.

There is no doubt that both these models are important in determining the broad characteristics of the corona and interplanetary magnetic fields.

The Magnetic Field of the Sun as a Star

A good correlation has been established (Severny et al. 1970) between the polarities of the magnetic field measured in integrated sunlight and the interplanetary magnetic field displaced by $4\frac{1}{2}$ days to account for the spiral pattern of the interplanetary fields. A recent study (Scherrer et al. 1977) compares the Crimean and Mount Wilson mean field observations over a period of 5 years or so. Because of the weak fields involved and the difficulty of obtaining good integrated sunlight in a large solar telescope, a perfect agreement can scarcely be expected. The cross-correlation coefficient between the two observations is 0.4 and peaks at a 12-hour lag, which represents the longitude difference between the two observatories.

The good agreement between the rather regular sector structure of the interplanetary magnetic fields and the large-scale solar fields is a bit puzzling. The solar sector pattern (Wilcox & Howard 1968, Wilcox & Svalgaard 1974) may represent a different and separate magnetic structure from that which results from the decay of active regions (Wilcox 1971, Svalgaard et al. 1977).

The Long View

Eddy (1976) has recently provided convincing evidence that the solar cycle has not continued without interruption since the first telescopic observations of sunspots. In the last half of the seventeenth century and early in the eighteenth century (the Maunder Minimum) there was no evidence of a cycle because the level of solar activity was practically zero.

It is not at all clear what the large-scale magnetic fields of the Sun looked like during this interval, although we could surmise that they resembled the last minimum but were two or three orders of magnitude weaker. Nevertheless, a successful theory of the solar activity cycle must be able to explain how it is possible for a gap of many cycles to occur, and then for there to be a long series of cycles such as we have seen since the end of the Maunder Minimum.

MAGNETIC FIELDS IN THE CORONA

It is not practical to measure magnetic fields in the corona; the spectrum lines are much too weak. One may infer magnetic-field structure from the shapes of coronal features seen at the limb or in X-rays and, alternatively, from extrapolations of the measured surface magnetic fields, using generally some approximation to the true coronal situation (Altschuler & Newkirk 1969, Schatten et al. 1969).

Expansion of the Surface Fields

The problem of calculating the magnetic-field configuration in the corona, given the surface field distribution, consists of finding a solution of Laplace's equation which satisfies the boundary conditions. Schatten et al. (1969) have used a Green's function to find a solution. Schatten (1968) was able to predict in a satisfactory way the appearance of the corona at the time of a solar eclipse, although such comparisons are inevitably somewhat subjective.

By far the most prolific coronal field calculators are Altschuler, Newkirk, and their collaborators. They solve Laplace's equation with a Legendre polynomial expansion, which is equivalent to an expansion of the magnetic potential in spherical harmonics. Newkirk et al. (1973) have published a microfilm atlas of coronal magnetic fields from which Figure 9 is one illustration.

The spherical harmonic solution represents a potential field. It should be remembered that transient or rapidly growing or decaying coronal features will probably not be well represented by this technique.

An interesting recent result of the potential field calculations during the time of the Skylab X-ray experiment is that a coronal hole is the locus of open field lines (Levine et al. 1977). Also, coronal holes appear to be the origin of high-speed solar wind streams (Krieger et al. 1973) and geomagnetic activity (Sheeley et al. 1976). A coronal hole may be seen in the X-ray observations, in many far ultraviolet lines, or in λ 10830 observations from the ground.

X-ray Coronal Loops

Soft X-ray photographs made by the AS & E S-054 experiment during the Skylab mission (Vaiana et al. 1974, Chase et al. 1977) show many loops connecting various features on the solar surface, some of which cross the equator.

These loops represent some of the magnetic-field lines in the corona. Figure 10 illustrates some of the loops. Sheeley et al. (1975) have shown that a newly emerging region can become interconnected with old fields in the neighborhood. Recent studies (Švestka et al. 1977, Howard & Švestka, 1977) have pointed out a number of characteristics of coronal loops, such as: (*a*) magnetic-field variations can cause transient brightenings and sharpenings of loops; (*b*) large loops connecting newly formed regions can form through reconnection of existing field lines; (*c*) a visible loop connection between two active regions does not result in the influence of the activity in one region on the other; (*d*) probably the most common way that loops interconnecting two active regions appear is by the brightening of a pre-existing field connection; (*e*) the basic magnetic interconnections between active regions, once born, live at least as long as both the interconnecting regions exist as distinct magnetic features; (*f*) individual loops, on the other hand, are visible for only short

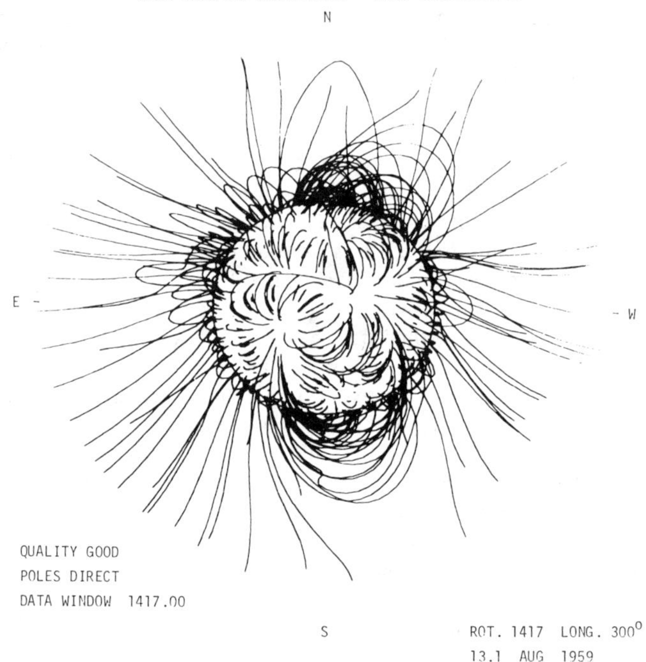

Figure 9 Magnetic field lines from a potential field calculation for 1.03 R to 2.5 R. This view is at longitude 300° in Carrington rotation 1417 (13.1 August 1959).

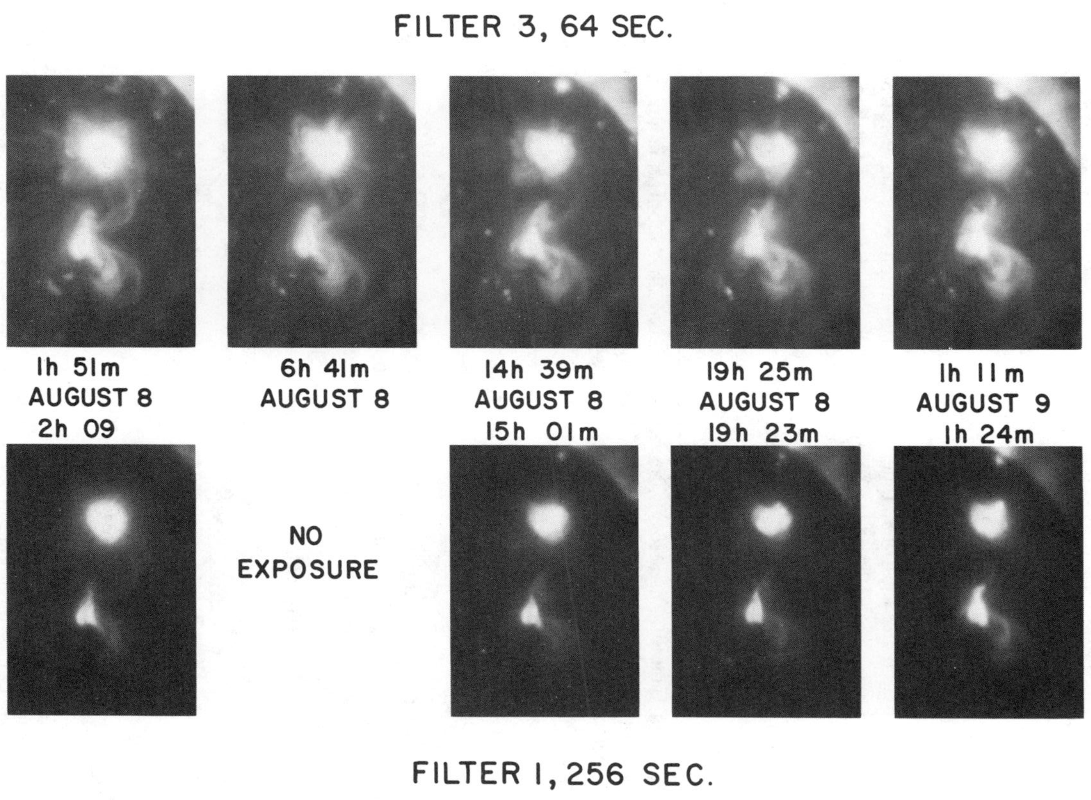

Figure 10 A.S. & E. Skylab X-ray photographs of the Sun on the dates indicated in 1973. Filter 1 has a passband of 2–17 Å, and Filter 3 has a passband of 2–32 Å and 44–54 Å. The loops connecting the two regions cross the solar equator. Note that the loops can change appearance in a matter of hours. (From Švestka et al. 1976.)

times—generally less than one day; (*g*) newly emerging magnetic flux tends to make some pre-existing field connections visible as loops; (*h*) the variability in the shapes of loops that are sometimes seen can at times be related to changes in photospheric magnetic field configurations; (*i*) a complex of activity (Bumba & Howard. 1965) appears to show remarkably ordered behavior, such as simultaneous brightenings of loops connecting different regions, and a characteristic pattern of loops internal and external to the active regions.

CONCLUDING REMARKS

After more than two decades of magnetograph observations, we have learned a great deal about magnetic fields on the solar surface. Perhaps the most fundamental result from this work is that practically every solar feature that can be seen in the photosphere, chromosphere, corona, and beyond to interplanetary space owes its existence in one way or another to the presence of magnetic fields. The one notable exception to this rule is the photospheric granulation pattern, which may be purely hydrodynamic in origin.

Solar activity is, therefore, the result of the motions of magnetic fields in and around the Sun. The Sun, as a result, becomes an interesting laboratory for large-scale magnetohydrodynamic phenomena. The activity cycle becomes a magnetic variation, perhaps a self-regenerating dynamo (cf Weiss 1971).

A number of questions remain to be answered. Among these we should include the following:

1. Is the solar sector magnetism a separate structure from the active-region fields that are seen to weaken, expand, and form the large-scale patterns and polar fields?
2. Can the behavior of magnetic fields at the solar surface be explained using only the random walk resulting from supergranular motions combined with the shearing effects of differential rotation, or are there other factors that contribute?
3. Do the large-scale patterns of magnetic fields represent the manifestation at the solar surface of a global-scale convection?

Literature Cited

Altschuler, M. D., Newkirk, G. Jr. 1969. *Solar Phys.* 9:131

Altschuler, M. D., Trotter, D. E., Newkirk, G. Jr., Howard, R. 1974. *Solar Phys.* 39:3

Altschuler, M. D., Trotter, D. E., Newkirk, G. Jr., Howard, R. 1975. *Solar Phys.* 41:225

Ambrož, P., Bumba, V., Howard, R., Sykorá, J. 1971. *Solar Magnetic Fields,* ed. R. Howard, p. 696. IAU Symposium No. 43

Babcock, H. D. 1959. *Ap. J.* 130:364

Babcock, H. W. 1953. *Ap. J.* 118:387

Babcock, H. W., Babcock, H. D. 1955. *Ap. J.* 121:349

Bumba, V. 1970. *Solar Phys.* 14:80

Bumba, V. 1976a. *Bull. Astron. Inst. Czech.* 27:153

Bumba, V. 1976b. *Bull. Astron. Inst. Czech.* 27:74

Bumba, V., Howard, R. 1965. *Ap. J.* 141:1502

Bumba, V., Howard, R. 1969. *Solar Phys.* 7:28

Bumba, V., Howard, R., Martres, M. J., Soru-Iscovici, I. 1968. *Structure and Development of Solar Active Regions,* ed. K. O. Kiepenheuer, p. 13. IAU Symposium No. 35

Butler, C. P. 1924. *MNRAS* 84:134

Chase, R. C., Krieger, A. S., Švestka, Z.,

Vaiana, G. S. 1977. *Space Res.* 16. In press
Dodson, H. W., Hedeman, E. R. 1970. *Solar Phys.* 13:401
Eddy, J. A. 1976. *Science* 192:1189
Gillespie, B., Harvey, J., Livingston, W. 1973. *Ap. J. Lett.* 186:L85
Hale, G. E. 1908. *Ap. J.* 28:315
Hale, G. E. 1913. *Ap. J.* 38:27
Hale, G. E., Seares, F. H., Van Maanen, A., Ellerman, F. 1918. *Ap. J.* 47:206
Hansen, R. T., Hansen, S. F. 1977. *Solar Phys.* In press
Hansen, S. F., Hansen, R. T. 1975. *Solar Phys.* 44:503
Harvey, K. L., Martin, S. F. 1973. *Solar Phys.* 32:389
Harvey, K. L., Harvey, J. W., Martin, S. F. 1975. *Solar Phys.* 40:87
Howard, R. 1959. *Ap. J.* 130:193
Howard, R. 1965. *Stellar and Solar Magnetic Fields,* ed. R. Lüst, p. 129. *IAU Symposium No. 22*
Howard, R. 1972. *Solar Phys.* 25:5
Howard, R. 1974a. *Solar Phys.* 38:283
Howard, R. 1974b. *Solar Phys.* 39:275
Howard, R. 1974c. *Solar Phys.* 38:59
Howard, R. 1976. *Ap. J. Lett.* 210:L159
Howard, R. 1977a. *Solar Phys.* In press
Howard, R. 1977b. In preparation
Howard, R., Stenflo, J. O. 1972. *Solar Phys.* 22:402
Howard, R., Švestka, Z. 1977. *Solar Phys.* In press
Howard, R., Yoshimura, H. 1976. *Basic Mechanisms of Solar Activity,* eds. V. Bumba, J. Kleczek, p. 19. IAU Symposium No. 71
Krieger, A. S., Timothy, A. F., Roelof, E. C. 1973. *Solar Phys.* 29:505
Leighton, R. B. 1959. *Ap. J.* 130:366
Leighton, R. B., Noyes, R. W., Simon, G. W. 1962. *Ap. J.* 135:474
Levine, R. H., Altschuler, M. D., Harvey, J. W. 1977. *J. Geoph. Res.* In press
Newkirk, G. Jr., Trotter, D. E., Altschuler, M. D., Howard, R. 1973. *A Microfilm Atlas of Magnetic Fields in the Solar Corona.* NCAR Technical Note STR-85
Schatten, K. H. 1968. *Nature* 220:1211
Schatten, K. H., Wilcox, J. M., Ness, N. F. 1969. *Solar Phys.* 6:442
Schatten, K. H., Leighton, R. B., Howard, R., Wilcox, J. M. 1972. *Solar Phys.* 26:283
Scherrer, P. H., Kotov, V., Severny, A. B., Wilcox, J. M., Howard, R. 1977. *Solar Phys.* In press
Severny, A. B. 1971. *Solar Magnetic Fields,* ed. R. Howard, p. 675. IAU Symposium No. 73
Severny, A., Wilcox, J. M., Scherrer, P. H., Colburn, D. S. 1970. *Solar Phys.* 15:3
Sheeley, N. R. Jr. 1976. *J. Geoph. Res.* 81:3462
Sheeley, N. R. Jr., Bohlin, J. D., Brueckner, G. E., Purcell, J. D., Scherrer, V., Tousey, R. 1975. *Solar Phys.* 40:103
Sheeley, N. R. Jr., Harvey, J. W., Feldman, W. C. 1976. *Solar Phys.* 49:271
Simon, G. W., Weiss, N. O. 1968a. *Structure and Development of Solar Active Regions,* p. 108, ed. K. O. Kiepenheuer. IAU Symposium No. 35
Simon, G. W., Weiss, N. O. 1968b. *Z. Astrophys.* 69:435
Stenflo, J. O. 1970. *Solar Phys.* 14:263
Stenflo, J. O. 1972. *Solar Phys.* 23:307
Svalgaard, L., Wilcox, J. M. 1974. *Science* 126:51
Svalgaard, L., Wilcox, J. M., Duvall, T. L. 1974. *Solar Phys.* 37:157
Svalgaard, L., Wilcox, J. M., Scherrer, P. H., Howard, R. 1977. *Solar Phys.* In press
Švestka, Z., Krieger, A. S., Chase, R. C., Howard, R. 1977. *Solar Phys.* In press
Vaiana, G. S., Krieger, A. S., Petrasso, R., Silk, J. K., Timothy, A. F. 1974. *Instrumentation in Astronomy—II,* eds. L. Larmore, D. Crawford, Seminar Proceedings, Soc. Photo-Opt. Instrum. Eng. 44:185
Vaiana, G. S., Krieger, A. S., Timothy, A. F. 1973. *Solar Phys.* 32:81
Wagner, W. J., Gilliam, L. B. 1976. *Solar Phys.* 50:265
Waldmeier, M. 1973. *Solar Phys.* 28:389
Weiss, N. O. 1971. *Solar Magnetic Fields,* ed. R. Howard, p. 757. IAU Symposium No. 73
Wilcox, J. M. 1971. *Comments Astrophys. Space Sci.* 3:133
Wilcox, J. M., Howard, R. 1968. *Solar Phys.* 5:564
Wilcox, J. M., Svalgaard, L. 1974. *Solar Phys.* 34:461
Yoshimura, H. 1975. *Ap. J. Suppl.* 29:467
Yoshimura, H. 1976. *Solar Phys.* 47:581

Ann. Rev. Astron. Astrophys. 1977. 15: 175–96

THE INTERACTION OF SUPERNOVAE WITH THE INTERSTELLAR MEDIUM

×2112

Roger A. Chevalier
Kitt Peak National Observatory,[1] P.O. Box 26732, Tucson, Arizona 85726

1 INTRODUCTION

More than 40 years ago, Baade and Zwicky realized the large energy release in extragalactic supernovae and envisioned a very interesting process for the production of the energy (Baade & Zwicky 1934). They speculated that a stellar core could collapse to form a neutron star and that the resulting release of gravitational energy could power an explosion. The hypothesis of collapse resulting in explosion is now commonly accepted as a model for the supernova phenomenon. Energy can be stored in the rotation of the compact object, but at present there is no reason to believe that this energy is larger than that in the ejected material. This review concentrates on our theoretical understanding of the interaction of the energy release in the supernova explosion with the ISM (interstellar medium). Observations of individual SNRs (supernova remnants) are not comprehensively reviewed. Woltjer (1972) discussed the data available at that time. More recently, a review of observational aspects of SNRs with an emphasis on radio remnants has been given by Mills (1974), and X-ray observations of remnants have been described by Gorenstein & Tucker (1976).

The discussion here carries further the standard picture of SNR evolution as taking place in a number of phases (Spitzer 1968, Woltjer 1972). Initially there is the supernova itself, at which time about 10^{49} ergs of optical radiation are emitted; most of the energy is deposited in the kinetic energy of the ejecta. The first phase of SNR evolution is one of free expansion. After the ejecta have interacted with approximately their own mass in the ambient medium, the energy is transferred to the ISM, resulting in an adiabatic blast wave. This is the second phase, which ends when the cooling time behind the shock wave becomes shorter than the hydrodynamic time. The resulting formation of a dense shell is the third phase of evolution. Recently, investigators have examined in detail the transition between the phases, have checked on the effects of heat conduction, and have estimated the

[1] Operated by the Association of Universities for Research in Astronomy, Inc., under contract with the National Science Foundation.

effects of an inhomogeneous ISM. These developments and the various aspects of the ISM for which energy deposition by supernovae is important are discussed in this review.

2 THE SUPERNOVA EXPLOSION

In order to understand the interaction of supernovae with the surrounding medium, it is necessary to consider the form of energy release by the initial explosion. Ionizing radiation emitted at the time of the explosion can have a fairly direct effect on the surroundings, while the kinetic energy of the ejecta has a delayed effect.

Hydrodynamic models of supernova explosions show that although there is some release of ionizing energy in the supernova, it is a small fraction of the total kinetic energy (estimated to be 10^{51} ergs). Models of Type II supernovae (Colgate 1974a,b, Chevalier 1976a) indicate that there are ultraviolet and hard X-ray or γ-ray bursts when the shock wave reaches the stellar surface. The energy in these bursts may be in the range 10^{47}–10^{49} ergs. The prompt effects of Type I supernovae may be smaller than those of Type II supernovae because their initial radii may be smaller. Colgate (1974b) estimates that 10^{44} ergs of hard γ-ray radiation emerges at the time of shock break-out. However, models of Type I supernovae with extended envelopes have been constructed (Lasher 1975), and these would have burst properties similar to the Type II supernovae.

The most stringent observational limit on the emission of hard radiation from supernovae can be deduced from the X-ray and γ-ray background radiation. The radiated energy per supernova above 0.2 keV is found to be less than 6×10^{49} (τ/100 yr) ergs, where τ^{-1}, the supernova frequency per galaxy, is assumed to be the same for all galaxies (Silk 1973). Considering the theoretical estimates of radiated energy, it is possible that supernovae do contribute to the background radiation.

No definite limits for emission below 0.2 keV are available. Morrison and his colleagues (Morrison & Sartori 1969, Chiu, Morrison & Sartori 1975) have proposed models for the light from Type I supernovae which require the release of a large amount of ionizing radiation at the time of the explosion. The mechanism for producing 10^{52} ergs of radiated energy is not given. Colgate (1972) has investigated ways of converting the kinetic energy into ionizing radiation at early times and found that the efficiency is low (see also Kahn 1974b). Another problem with the Morrison theory is that the medium in which the light pulse is moving must be very uniform with radius to give a constant decay time in the light curve. As the medium is a result of stellar mass loss (Morrison & Sartori 1969), its density is likely to decrease with distance from the star. Kafatos & Morrison (1971) claim that the existence of the Gum Nebula around the Vela XYZ supernova remnant lends observational support to the ionizing flux conjecture. On the other hand, Beuermann (1973) found that the flux from the hot stars in the Vela OB association could provide the required ionization energy for the nebula; but this result is still controversial (Kafatos et al. 1975). The remnant of Tycho's supernova, which was a Type I event, has been searched for a fossil H II region without success (van den Bergh et al. 1973).

Table 1 Properties of supernovae

	Type I	Type II
Ejected mass ($M_\odot$)	0.5	5
Mean velocity (km sec^{-1})	10 000	5000
Kinetic energy (erg)	5×10^{50}	1×10^{51}
Visual radiated energy (erg)	4×10^{49}	1×10^{49}
Ionizing radiated energy (erg)	10^{44} or 10^{48}–10^{49}	10^{48}–10^{49}
Frequency (yr^{-1})	1/60	1/40
Stellar population	old disc	young disc

Estimates of the basic properties of supernovae are summarized in Table 1. The ejected mass of Type I supernovae is highly uncertain, but is probably in the range 0.1 to 1.0 $M_\odot$. A lower limit to the mass of Type II supernovae can be deduced from the fact that they occur only in the spiral arms of spiral galaxies (Tinsley 1975, Maza & van den Bergh 1976). The actual ejected mass may be greater than that listed in the table. The mean velocity is derived on the hypothesis that supernovae eject shells (see the arguments for Type II supernovae in Chevalier 1976a). The velocities can be estimated from the photospheric radii at the time when the photosphere reaches its maximum extent, or from the velocities indicated by absorption lines at late times. Of course, there is material moving with far greater velocities but it is unclear whether this material contributes substantially to the kinetic energy, which is computed here from the given mass and mean velocity. The estimate of visual radiated energy is based on visual light curves and spectrophotometric temperatures, assuming a Hubble constant of 50 km sec^{-1} Mpc^{-1}, and it is probably accurate to a factor of 2 or 3. There may be a considerable intrinsic spread for Type II supernovae. The ionizing energy is based on uncertain theoretical predictions, as discussed above. The supernovae frequency in our galaxy and the stellar population of the progenitors are taken from Tammann (1974) and Tinsley (1975).

In conclusion, most of the observable supernova energy (i.e. excluding neutrinos and gravitational radiation) emerges in the form of the kinetic energy of the ejecta. The ejected material may be subject to a Rayleigh-Taylor instability at some phase of the explosion (Falk & Arnett 1973, Chevalier 1976a), so the ejecta may be clumpy. No detailed models of the outcome of the instability have been calculated. It is likely that part of the ejected material is in a relatively smooth shell and part is in clumps, but it is not known how the mass is divided between these two states.

3 THE EVOLUTION OF A SUPERNOVA REMNANT

3.1 *Transfer of Energy to the ISM*

In the last section we noted that the supernova energy is primarily deposited in the kinetic energy of the ejecta. In the hundreds to thousands of years following the explosion, this energy is transferred to the surrounding medium.

If the ejected shell is smooth, it sweeps up the surrounding medium as it expands. While the amount of swept-up gas is small, the expanding shell acts like an expanding piston with constant velocity. The interaction of such a piston with a uniform medium has been considered by Taylor (1946). For a $\gamma = 5/3$ gas, a shock wave forms at 1.1 times the piston radius and the ISM is compressed to 4 times the ambient density in the shock. As the amount of swept-up mass becomes comparable to the ejected mass, two effects become important. The first is a result of the low pressure in the ejected material which has been adiabatically expanding. The pressure difference between the shocked ISM and the ejecta drives a shock wave into the ejecta (Ardavan 1973, Kahn 1974a, McKee 1974, Mansfield & Salpeter 1974; see also Figure 1a). The converging shock wave may be unstable (Somon 1971), giving the shock front a corrugated appearance. The reverse shock wave is the beginning of the deceleration of the supernova ejecta, which leads to the second effect: the

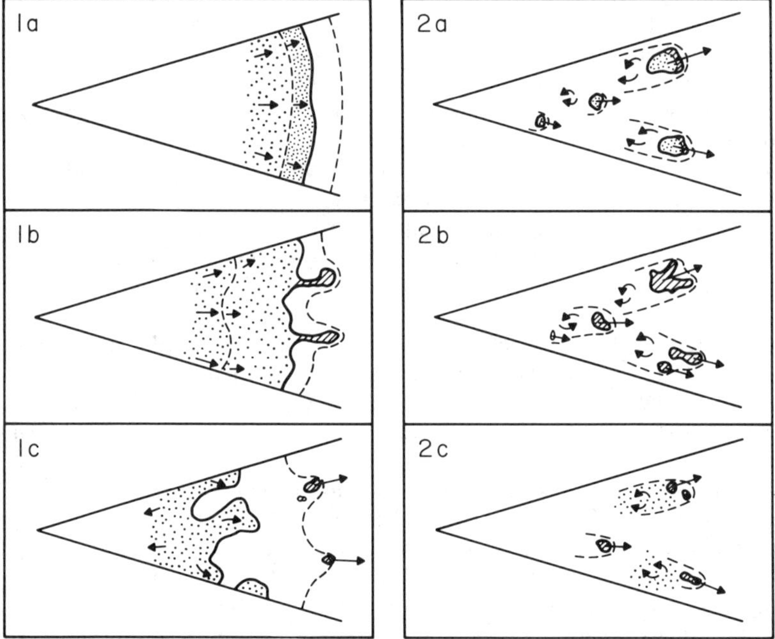

Figures 1 and 2 Schematic views of two models for the early evolution of SNRs. Dashed lines are shock fronts, solid lines are contact discontinuities, and the arrows indicate velocities. The supernova ejecta are stipled, the cool, compressed parts of the ejecta are hatched, and the ISM is plain. Figure 1 depicts the interaction of a smooth shell with the ISM, including the formation of a reverse shock wave and the development of the Rayleigh-Taylor instability (cf Gull 1975). Figure 2 shows the interaction of clumpy ejecta with the surrounding medium. The actual situation may be a combination of these two cases. See Section 3.1 for a detailed description.

Rayleigh-Taylor instability at the interface between the dense shell and the ambient medium. The deceleration of a dense gas by a low-density gas leads to the breakup of the dense gas as it is penetrated by the low-density gas (Figure 1b). Gurzadyan (1953, 1969) long ago pointed out that this instability would cause the breakup of expanding shells in the ISM. Gull (1973a, 1975) suggested that the instability in young SNRs generates turbulent motions that are important for the acceleration of relativistic particles (see Section 5.1). The fragments that form as a result of the instability are themselves subject to breakup by further instabilities (Figure 1c). Cowie (1975) pointed out that the breakup may be halted at a size determined by the action of viscosity.

The expansion of a uniform shell can lead to an interesting magnetic effect if the ambient field lines in the supernova environment are approximately straight. The tension in the swept-up magnetic field lines gives the dominant pressure close to the contact discontinuity between the expanding shell and the ISM (Kulsrud et al. 1965). The effect can be analyzed by estimating the swept-up magnetic flux in a plane perpendicular to the field lines. The magnetic pressure leads to a bulging out of the shock wave perpendicular to the field lines, although the deviation from sphericity is small. Kulsrud et al. further pointed out that the Rayleigh-Taylor instability at the contact discontinuity may cause the formation of magnetic filaments.

As mentioned in the previous section, it is possible that much of the ejecta is initially in clumps. Evidence for this phenomenon is found in the young SNR Cassiopeia A. Velocity studies of the fast-moving knots can be interpreted as undecelerated motion since the time of the explosion (van den Bergh & Dodd 1970). The overabundance of Ar, S, and O in these knots relative to H (Peimbert & van den Bergh 1971, Peimbert 1971) shows that the knots are supernova ejecta. While Cas A may have been a peculiar supernova (cf Chevalier 1976b), it does provide evidence for the early breakup of the supernova shell. The evolution of a knot depends on the internal velocity dispersion it receives at the time of formation. If the dispersion is significant, a shock wave similar to the reverse shock will move into the knot (Figure 2a). The deceleration would again be accompanied by instabilities (cf Blake 1972 and Figures 2b and 2c), and because the knot is surrounded by hot gas, heat conduction is expected to act on the knot (Chevalier 1975b). As the heated gas expands, it is more rapidly decelerated by the surrounding medium. Because the deceleration is larger for a knot of smaller radius on the assumption that the knot density is constant, the deceleration of a knot may proceed rapidly once it begins to break up.

3.2 *The Supernova Ejecta*

After the supernova ejecta have interacted with their own mass in the surrounding medium, significant deceleration is expected. In both types of evolution discussed above, the ejecta are likely to be at a fairly high temperature (greater than 10^6 K) at this time. In the reverse shock, the initial shock velocity is low and the gas density is high, so that cooling may occur behind the shock wave. However, the shock moves more rapidly as the density of expanding gas decreases, so that the shock wave soon heats the gas to a temperature at which its cooling time is long. Only a few percent of the ejected mass may be cool after the reverse shock has passed. It is unlikely that the Cas A knots formed in this way because of the large observed range in knot velocities (van den Bergh & Dodd 1970); only the outer, high-velocity part of the shell would be expected to cool in this model. In the case of initial knot ejection, the densities may be high enough so that a shock wave that brings a knot into pressure equilibrium is a cooling shock wave. Heating by conduction may bring the ejecta to X-ray emitting temperatures.

Once the supernova energy has been transferred to the ISM, the standard picture of SNRs describes the subsequent evolution by the self-similar adiabatic blast wave solution for a point explosion in a uniform medium (Sedov 1959). Of course, the supernova is not a point explosion and the supernova ejecta contribute to an enhanced density at the center of the explosion (Mansfield & Salpeter 1974). Chevalier (1975b) suggested that the enhanced X-ray emission observed at the center of the Cygnus Loop (Rappaport et al. 1973) can be explained as emission from the supernova ejecta. Rappaport et al. argued for a central point source but the more sensitive observations of Weisskopf et al. (1974) cast doubt on this interpretation.

As it is likely that the supernova ejecta are heated to X-ray emitting temperatures, observations of X-ray lines are of particular importance. Spatial resolution as well as spectral resolution will probably be important because regions with large over-abundances of heavy elements may only occupy a small fraction of the volume of the SNR. X-ray lines have now been definitely observed in Cas A and in Tycho (Davison et al. 1976, Pravdo et al. 1976), which confirms the thermal nature of these sources. The strengths of the lines, which are probably due to Fe XXV and Fe XXVI, do not indicate a large Fe overabundance. While the study of X-ray lines holds much promise, their interpretation must be undertaken with care because the hot gas is not likely to be in ionization equilibrium. If the departure from equilibrium occurs during heating (ionization), line strengths may be anomalously high (cf Kafatos & Tucker 1972 on Si lines), while if it occurs during cooling (recombination), the lines may be anomalously weak (cf Chevalier 1975b on O lines). Solar astronomers have already gained experience in this field from X-ray studies of the active regions (Culhane & Acton 1974). Another probable nonequilibrium situation involves the electron and ion temperatures behind the high-velocity shock waves that occur in young SNRs. This subject has not progressed beyond the discussion of Shklovsky (1968), who noted that Coulomb interactions are not sufficient to bring the electrons and ions into equilibrium, but that plasma instabilities in a collisionless shock may bring about equilibrium more rapidly (see also McKee 1974). Gorenstein et al. (1974) conjecture that the approach to equilibrium may be slow and that this may explain why the temperature derived from X-ray spectra (the electron temperature) is lower than the expected post-shock temperature in young remnants.

3.3 *The Effect of Heat Conduction*

Another assumption of the standard blast-wave solution is adiabaticity. If heat conduction is effective, deviations from the adiabatic solution are to be expected. Assuming the absence of a magnetic field and the applicability of the collisional heat conduction coefficient, Chevalier (1975a) showed that an electron conduction front can move out from the hot gas generated by a supernova more rapidly than a shock wave. The reason for this behavior is the high electron thermal velocity compared to the proton thermal velocity. A point is eventually reached where a transition to a shock wave is made. Cowie & McKee (1977) showed that the classical conduction coefficient can give a large overestimate of the conductive flux when the electron mean free path is larger than the electron temperature scale height (saturated conduction). They point out that the velocity of a thermal wave cannot greatly exceed twice the sound speed of the hot gas. They further conjecture that plasma instabilities do not allow a conduction front to propagate ahead of the shock wave.

Another effect of conduction is to smooth the interior temperature profile. Assuming infinite conductivity, Solinger et al. (1975) have investigated isothermal blast-wave solutions for supernova remnants. This solution would apply to a phase after thermal waves were important and before radiative cooling sets in. Solinger et al. show that the basic properties of SNRs (age, ambient density, and initial energy) derived from X-ray and optical observations (radius, X-ray temperature, and X-ray luminosity) are approximately the same for both the adiabatic and isothermal models. X-ray spectra to high energies will probably be most useful for discriminating among models. Chevalier (1975b) found that the high interior temperatures in the adiabatic models can give a significant deviation from a one-temperature bremsstrahlung spectrum on the assumption that the electrons and ions are in temperature equilibrium. Lerche & Vasyliunas (1976) have shown that the isothermal blast-wave solutions are unstable to radial perturbations. However, the one-dimensional computations of SNR evolution with heat conduction (Chevalier 1975a) should include these effects. The computations do not show any irregular behavior, but the temperature profile is never completely isothermal.

After shell formation, conduction results in a fairly smooth interior temperature distribution. This has two effects on X-ray observations of an old remnant: the X-rays are emitted closer to the dense shell and the characteristic X-ray temperature is higher than in the absence of conduction. Both of these effects go in the direction of bringing the dense shell model for the Cygnus Loop (Cox 1972c) into better agreement with the X-ray observations.

Conduction may also be important for the evaporation of clouds that are immersed in hot gas, as mentioned above. McKee & Cowie (1975) have analyzed the types of conduction fronts (or thermal waves) possible from considerations of the jump conditions at the front. They classify the fronts in a way analogous to the classification of ionization fronts. Cowie & McKee (1977) have further analyzed the evaporation of clouds by heat conduction and have derived rates of cloud mass loss for both classical and saturated conduction. The results show that cloud evaporation may be important in the evolution of SNRs if magnetic fields do not impede conduction.

3.4 *The Radiative Shock Phase*

As the SNR shock wave weakens (possibly because it enters a high-density region), radiative cooling causes the gas temperature to drop behind the shock wave.

Pikel'ner (1954) realized that the cooling radiation might have a characteristic spectrum, and he calculated approximate line intensities for comparison with observations of the Cygnus Loop. A more thorough calculation of the emitted radiation was undertaken by Cox (1972a), who included radiative transfer through the shocked gas and the effects of a magnetic field. Spectrophotometry of old remnants (cf Miller 1974, Osterbrock & Dufour 1973) has given strong support to the hypothesis that a radiating shock wave is being observed, although there are differences in detail. A new set of shock calculations has been completed by Raymond (1976), who improved the calculation by starting with the ionization balance of the pre-shock gas, rather than assuming ionization equilibrium at the post-shock temperature, and by separately following the flow of neutral and ionized species in cases where they do not behave as a single fluid. He also increased the parameter space covered by the calculations. It may now be possible to understand some of the details of observed shock spectra; for example, Raymond suggests that differences in spectra taken in different parts of the Cygnus Loop (Miller 1974) have to do with the pre-shock ionization state.

Shock spectra are a potentially powerful means of investigating old SNRs and the ISM. Characteristic line ratios can yield a good estimate of the shock velocity. The pre-shock density can be estimated either from the absolute flux in the Balmer lines or from density-sensitive line ratios if pressure equilibrium in the post-shock gas is assumed. The difficulty in applying the last method is shown by Miller's (1974) observations of the Cygnus Loop, where the density derived from the [S II] doublet ratio is 5 to 7 times higher than that derived from the [O II] doublet ratio, although the temperatures of the emitting regions are presumably similar. Inaccuracies in the atomic physics data may provide an explanation. Shock spectra may yield information on the pre-shock magnetic field; Raymond (1976) notes that the lines of neutral and singly ionized species are weak if the magnetic field is strong. Finally, it is possible to estimate element abundances. The results on the Cygnus Loop indicate that abundances of certain elements are higher than in a standard grain depletion model, perhaps implying that some grains are destroyed by the shock wave (Raymond 1976).

If the shock front around most of the SNR becomes radiative, the remnant is encompassed by a dense shell. Subsequent to the analytic work of Cox (1972b) on the formation of a dense shell, numerical calculations were performed that followed the shell formation stage in detail (Rosenberg & Scheuer 1973, Chevalier 1974, Mansfield & Salpeter 1974, Straka 1974, Falle 1975). These spherically symmetric calculations show that the gas recombines behind the shock wave to form a cool neutral shell. There have been neutral hydrogen radio observations of some SNRs that in fact show a supersonic, expanding shell; they are HB 21 (Assousa & Erkes 1973) and S 147 (Assousa, Balick & Erkes 1973). A subsonic expanding shell has been observed around W 44 with a radius many times that of the nonthermal emission (Knapp & Kerr 1974). Cornett & Hardee (1975) have interpreted this expansion as being due to preheating of the region around the SNR by either ionizing radiation or cosmic rays. Heiles (1976) has found large expanding shells in his survey of 21-cm emission. The momentum carried by the shells is consistent with a supernova origin.

The dense shell evolves primarily by momentum conservation. While there is hot gas in the interior of the shell, it loses most of its energy to radiation at the inner boundary of the shell. In fact, a shock wave can form at the inner edge of the shell as hot gas cools and accretes onto the shell (Chevalier 1974, Falle 1975). The momentum of this gas is added to the shell.

A substantial amount of the energy radiated behind the shock wave may be emitted in the infrared (Cox 1972a, Chevalier 1974). Radiation in fine-structure transition lines and in radiatively heated grains is expected. It has also been suggested that young remnants may have a large infrared flux from collisionally heated grains (Silk & Burke 1974). This process is of particular importance for supernovae that occur in dense clouds (densities greater than 10^2 cm^{-3}). Despite these predictions, the infrared stands out as a wavelength band in which SNRs have not been extensively observed. Even upper limits on the infrared fluxes of remnants may be of interest.

McCray, Stein & Kafatos (1975) have pointed out that narrow density fluctuations in the ambient medium are magnified in the cooling region behind the shock wave. This effect is only marginally observable; the cooling region in the Cygnus Loop subtends an angle of less than 1″ arc (Cox 1972a). However, this instability may be important in the interpretation of shock spectra. Also important is the frequently observed chaotic structure of the shock front, to be discussed in the next section.

3.5 *The Irregular Structure of Old Remnants*

Because supernova remnants are observed to have a highly irregular structure (cf van den Bergh et al. 1973), one-dimensional calculations cannot adequately describe them. The expanding neutral hydrogen shells mentioned above are by no means smooth or complete. A possible cause for the irregular structure was found by Chevalier & Theys (1975), who investigated the passage of a cooling shock wave over a density fluctuation using a two-dimensional hydrodynamic code. The part of the shock front passing over the region of increased density was retarded. If the flow was adiabatic, the shock front tended to straighten out after passing fluctuation; but with cooling included, the shock front became more irregular with time. In this sense, a cooling shock wave is unstable. While the instability is probably related to the Rayleigh-Taylor instability, the initial conditions are considerably more complicated. The numerical calculations can be understood as follows (see Figure 3). After passing the density fluctuation, the shocked gas is channeled toward the lagging part of the shock front (Figure 3b). In the case of adiabatic flow, this material is of high pressure and it re-expands. However, with radiative cooling, a long-lived clump can form because it is no longer a high-pressure region. The clump has a larger radial momentum per unit area than the rest of the shell, so that it eventually moves ahead of the rest of the shell (Figure 3c). Now, further clumps form around the initial one and the growth of irregular structure continues. This theory does not clearly predict the arclike structure observed in old remnants such as S 147. However, magnetic stresses, which were not included in the numerical calculations, may be the dominant pressure in the dense shell. As one part of the shell moves ahead, magnetic stresses may create the arclike structure that is observed.

The inhomogeneous nature of the interstellar medium undoubtedly has a profound effect on SNR evolution. The effect described in the previous paragraph depends on small density enhancements; much larger enhancements, interstellar clouds, also exist in the medium into which the supernova shock expands. The interaction of supernova blast waves with clouds has been investigated by Sgro (1972, 1975), McKee & Cowie (1975), and Bychkov & Pikel'ner (1975). The numerical calculations of Sgro show that the transmitted shock wave may be a cooling shock if the cloud density is greater than a critical value that depends on the shock velocity, ambient density, and thickness of the cloud. A lower density cloud will remain hot for some time, emitting X-rays. Sgro has applied these results to the X-ray emission from Cas A. For this model to be tenable, the mass in quasi-stationary floculli that do not cool must be considerably greater than the mass in those that do. McKee & Cowie (1975) describe the interaction of a blast wave with a cloud as follows (see Figure 4). After the shock front has reached the cloud boundary, there is a reflected shock and a transmitted shock into the cloud (Figure 4a). The pressure between the two shocks can be several times the initial post-shock pressure. If the flow past

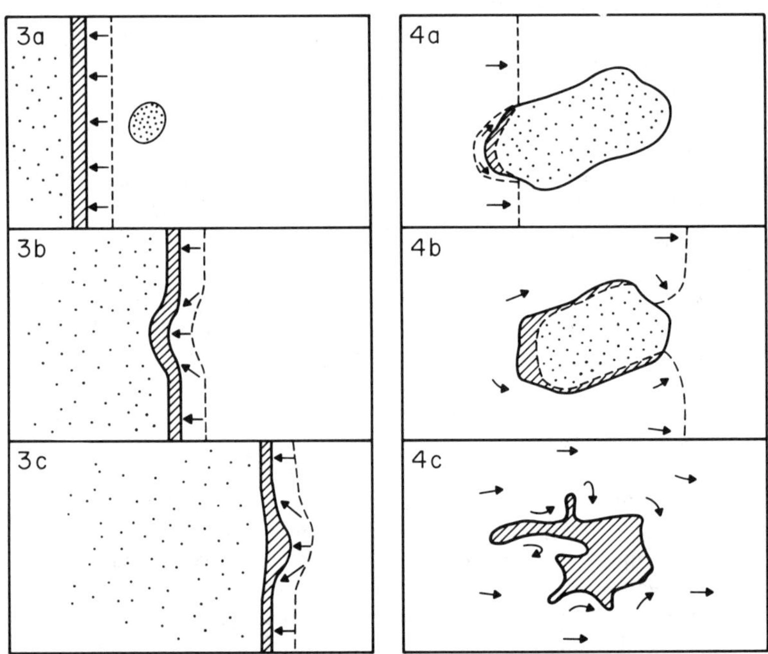

Figures 3 and 4 Schematic views of the interaction of shock waves with density inhomogeneities. Dashed lines are shock fronts, solid lines are the boundaries between high and low density regions, and hatched areas are cool, compressed regions. Arrows indicate velocities; in Figure 3, the velocities are relative to the dense shell. Figure 3 shows the interaction of a radiative shock wave with a density fluctuation, while Figure 4 depicts the interaction of a blast wave with an interstellar cloud. See Section 3.5 for details.

the cloud is supersonic, the reflected shock is maintained; if not, it disappears. The shock velocity in the cloud is slow, so the blast wave rapidly envelops the cloud. The high-pressure gas can drive secondary shocks into the sides and back of the cloud (Figure 4b). These general features are also found in the numerical computations of Sgro (1972, 1975) and Woodward (1976). Woodward computed the continued flow past the cloud, which results in the growth of instabilities (Figure 4c).

Bychkov (1973), Sgro, and McKee and Cowie interpret the optical emission from the Cas A quasi-stationary floculli as being due to shocked clouds. McKee and Cowie and Bychkov & Pikel'ner (1975) give quantitative support for the hypothesis that the X-ray and optical observations of the Cygnus Loop can be reconciled in terms of a cloud interaction model (cf Woltjer 1972). The X-rays are from the blast wave in the low-density medium while the optical filaments are places where clouds are being shocked. Further observational results relevant to this subject have been presented by Kirshner & Taylor (1976), who find Hα emitting gas in the Cygnus Loop with velocities up to 270 km sec^{-1}. These observations present a further puzzle in that gas with such a high velocity would be expected to have a much smaller emission measure than that observed; further observations to confirm these results would be useful. This study brought out the fact that the distance to the Cygnus Loop, which is based on proper motion and radial velocity studies of the optical filaments, is not as well known as is commonly assumed.

McKee and Cowie conjecture that the irregular structure of old remnants can be understood solely on the basis of interactions with clouds. However, as they point out, the curvature of observed SNR filaments is opposite to that expected in such a model. It seems more likely that the structure is due to the action of a Rayleigh-Taylor type instability, as discussed above. This hypothesis is supported by photographs of explosions in the earth's atmosphere. The interface between the debris products and the surrounding atmosphere is Rayleigh-Taylor unstable and usually shows a structure similar to that of old SNRs.

In addition to the instability discussed above involving radiative shock waves, there is an adiabatic instability involving shocked clouds. Richtmyer (1960) showed that the boundary between high and low density regions is unstable after being shocked. He numerically modeled the adiabatic flow resulting from the incidence of a shock wave on the corrugated boundary of a high density region. Because of the adiabatic assumption, both the transmitted and reflected shock waves straightened out with time; yet the contact discontinuity was Rayleigh-Taylor unstable and the corrugation became more prominent (see also Woodward 1976). In an astrophysical context, in addition to this instability, the cooling shock wave instability would occur.

The large compressive forces that can occur in SNRs suggest the possibility of star formation in the dense shell. Spherically symmetric models indicate that the magnetic pressure may prevent star formation (Chevalier 1974). However, the above-mentioned factors can result in flow along magnetic field lines and the channeling of gas into clumps. Intersecting shock waves can result in a pressure which is several times that resulting from a single shock. Thus, if star formation can occur on a timescale less than 10^6 years, it is reasonable to expect young stars to be associated with an old SNR. Berkhuijsen (1974) has noted that the Origem Loop may be an old SNR with radius 60 pc. There appear to be H II and early-type stars associated with the Loop, which she suggests have formed in the shell. Once star formation has occurred, the stars decouple from the shell and are no longer decelerated. The velocity of the stars associated with the Origem Loop are consistent with this expectation. Sancisi (1974) has found expanding neutral H shells connected with two OB associations. The idea that the expansion of OB associations might be related to star formation in expanding supernova shells goes back to Öpik (1953). Sancisi's results can be interpreted as support for this hypothesis. Ögelman & Maran (1976) have speculated that as the massive stars formed in a SNR shell themselves become supernovae, a cascade process is initiated.

4 COLLECTIVE EFFECTS OF SUPERNOVAE ON THE ISM

4.1 *Ionization Energy*

As discussed in Section 2, supernova explosions themselves are unlikely to be significant sources of ionizing radiation. However, the medium that is shocked by the supernova motions is sufficiently hot to emit ionizing radiation. The ionizing luminosity has been calculated in the one-dimensional numerical models (Chevalier 1974, Mansfield & Salpeter 1974). The results show that a significant amount (about 30%) of the supernova energy may be emitted as ionizing radiation. However, as the total energy of a supernova is about 10^{51} ergs, this energy in ionizing radiation is considerably less than that produced by a hot star during its lifetime. As the birthrate of early-type stars is similar to the birthrate of supernovae, SNRs are probably only minor contributors to the total ionization energy of the interstellar medium. However, the ionizing radiation from SNRs does have the property of extending further into the soft X-ray range than does the stellar radiation. This radiation is capable of partially ionizing an extended volume, giving a warm, partially ionized component of the ISM (Salpeter 1976).

4.2 *Thermal Energy*

Supernovae may contribute to the heating of the ISM in a rather special way: the high velocities that occur in supernovae can generate exceptionally high temperatures in the gas. Cox & Smith (1974) suggested that this gas might occupy a substantial fraction of the volume of the interstellar medium. They pointed out that the fraction of the supernova energy lost to radiative cooling is small if the ambient density is small. Thus, if the fractional volume of hot gas is above a critical limit, supernova remnants overlap and a young remnant may expand into a low density medium, losing little energy to radiation. If the fractional volume never reaches the critical value, only 10% or less of the ISM may be in the hot phase and SNRs remain distinct. Detailed simulations of the growth of overlapping SNRs by Smith (1977) indicates that at least 30% of the volume of a spiral arm near the galactic plane can be maintained in the hot phase. He considered the fact that when a young SNR encounters an old one, it must break through a dense shell in order to join up with the old one. Rayleigh-Taylor instabilities facilitate this process.

Observational data bearing on the hot phase of the ISM are the soft X-ray background (Williamson et al. 1974) and interstellar O VI absorption lines (Jenkins & Meloy 1974). The strong O VI lines in the spectra of stars behind the Vela SNR (Jenkins et al. 1976) provide evidence that at least in this case, the O VI absorbing gas is associated with a SNR. The observations further show that the correlation of O VI column density with stellar distance is poor, implying that the O VI absorbing gas is not uniformly distributed through the ISM. The gas temperature required to give the soft X-rays is somewhat higher than that of the O VI absorbing gas (cf Shapiro & Field 1976). In an analysis of the soft X-ray observations of the Wisconsin group, Cox (1976) found that a one-temperature model did not give an adequate fit. In a two-temperature model, the lower temperature may be consistent with that of the O VI absorbing gas, a temperature in the range $3\text{-}8 \times 10^5$K. A region with this temperature might occur at the interface between the higher temperature X-ray emitting gas and cool gas. Conductive energy transfer would occur at the interface.

A model of the ISM in which the evaporation of clouds by conduction plays an important role is that of McKee & Ostriker (1975). In their model, most of the volume of the ISM is made up of the hot gas from SNRs, although there are also cool clouds and a warm ($T \sim 10^4$K) component. The evaporation of clouds by a SNR significantly adds to its interior mass. There is a balance achieved between the formation of clouds in the late stages of SNR evolution and the evaporation of clouds by the hot gas. Rough agreement with the soft X-ray flux is obtained. In this picture, SNRs only enter the dense shell phase at a very large radius ($R \sim 170$ pc) as they are expanding in a low density medium. If it can be definitely shown that a relatively small remnant, like S 147, is in the dense shell phase, it would provide evidence against this model. Also, heat conduction is not the only mechanism for dissipating clouds. The radiation from an early-type star can ionize and heat nearby clouds, creating a region of near uniform density surrounding the star (Elmergreen 1976). The medium produced in this way is similar to that used in numerical simulations of SNR evolution. If most of the volume of the ISM is in the hot phase, as conjectured by McKee and Ostriker, there may be implications for the density wave theory of spiral structure. Galactic shocks would not form in the hot gas so narrow regions of star formation may not occur. A calculation of the propagation of a spiral density wave in a medium with various volume fractions of hot gas would be of some interest.

Jones (1975) has suggested that information on the fractional volume of the hot component can be deduced from statistics of SNR diameters derived from radio observations. He concludes that 80% of the ISM is in the hot phase. However, the observational data are subject to selection effects that are difficult to account for. For example, it is unclear how readily a large diameter SNR can be distinguished from the nonthermal radio background; the lack of such remnants forms a crucial part of Jones' argument. Also, as discussed in Section 5.2, the theoretical basis for deriving diameters from radio surface brightness measurements is poorly developed. It appears unlikely that useful information on the hot component can be deduced from radio observations at the present time.

The density gradient out of the plane of the Galaxy can have an important effect on SNRs. First, a strong explosion can break out of the plane of the Galaxy (Chevalier

& Gardner 1974) if it does not first reach pressure equilibrium with its surroundings. The SNR shock wave accelerates away from the plane because of the rapidly decreasing density. Another effect is that hot gas produced by a supernova is buoyant even after it has reached pressure equilibrium with its surroundings (Jones 1973, Chevalier & Gardner 1974). Thus hot gas produced by supernovae can contribute to a hot galactic corona. The idea of a hot corona was first discussed by Spitzer (1956), who suggested that the high temperature could be maintained by the motions of stellar ejecta.

A galactic fountain effect has been discussed by Shapiro & Field (1976). In this model, the hot gas radiatively cools as it rises out of the plane before reaching a height of 1 kpc. The cool, dense gas then falls back to the plane, perhaps giving rise to the high-velocity clouds. However, they have neglected the heat input of Type I supernovae, which probably have a scale height in the range 500–1000 pc. These may be able to maintain the gas at high temperature. In the absence of cooling, the scale height of 10^6K gas would be about 7500 pc; thus, the height of the hot gas above the plane may ultimately be determined by that of Type I supernovae.

If more hot gas is produced than can be held by the gravitational field of the galaxy, there is a flow of hot gas out of the galaxy. The properties of galactic winds in the spherically symmetric case have been discussed by Mathews & Baker (1971). The winds can occur even in the presence of radiative cooling; the cool material falls to the center of the galaxy. Winds driven by supernovae can be of particular importance in the early evolution of galaxies (Larson 1974). Certain characteristics of low-mass elliptical galaxies, such as their lack of central concentration, can be understood given the hypothesis that supernovae drove the ISM out of the young galaxies.

4.3 *Kinetic Energy*

Spitzer (1968) showed that supernovae can be important contributors to the kinetic energy of interstellar clouds; the energy is dissipated in cloud collisions. Recent calculations of SNR evolution have not substantially changed his estimates. Spitzer found that 3% of the initial supernova energy is in cloud motions when the motions have slowed to the mean velocity of interstellar clouds. Chevalier (1974) obtained an efficiency of 4% for an ambient density of 1 H atom cm^{-3} and an efficiency of 8% for a density of 0.01 H atom cm^{-3}. If the density is high, less energy is available for cloud motions because of radiative losses. Salpeter (1976) estimated a 10% efficiency and found that the kinetic energy supplied by supernovae may be larger than that required to maintain cloud motions. He conjectured that the excess energy may go into heating a neutral component of the ISM by cloud motions. Considering the large uncertainties in the input parameters, it can only be concluded that the amount of kinetic energy required for cloud motions is consistent with a supernova origin. Detailed characteristics of the clouds must be examined in order to make progress on this problem.

Observational data is available on the distribution of cloud velocities for the intermediate and high-velocity clouds. Siluk & Silk (1974) have compared this data with what would be expected on the hypothesis of the supernova origin of cloud motions. They assumed that interstellar clouds are to be identified with the dense shell that forms late in the evolution of a SNR and found that the supernova hypothesis is compatible with the data.

5 RELATIVISTIC PARTICLES

5.1 *Young Supernova Remnants*

Relativistic electrons can be studied by means of the nonthermal radio synchrotron radiation they emit. Young supernova remnants fall into two classes with regard to their radio properties. The first class is characterized by a centrally peaked, amorphous region of radiation and a flat spectral index ($\alpha = 0.0 - 0.25$ where α is defined by flux $\propto \nu^{-\alpha}$). The prototype of this class is the Crab Nebula; 3C 58 is another example (Wilson & Weiler 1976). The second class is characterized by an apparent annular region of emission and a steeper spectrum (α in the range 0.4–0.8). Examples are Cas A, Tycho's SNR, Kepler's SNR, and the remnant of SN 1006. The energy content in relativistic electrons appears to be similar for the two types of remnants (Woltjer 1972).

Interaction with the ISM may have only a small effect on the properties of the first class of remnants; these remnants are probably dominated by a central pulsar that supples the energy for the nebula. However, the ISM may play a role in confining the relativistic gas produced by the pulsar (Max et al. 1976).

The large particle energies in the second type of remnant imply that recent particle acceleration has taken place. The apparent annular structure of the remnants suggests that acceleration is occurring in the outer parts. A plausible hypothesis is that there is a turbulent acceleration mechanism associated with the interaction of the supernova ejecta with the ISM.

Gull (1973a, 1975) has suggested that turbulence is generated by the Rayleigh-Taylor instability which occurs when the ejected supernovae shell interacts with the interstellar medium (see Section 3.1). Using a numerical model, he has calculated the amount of energy in the convective-like motions as a function of time. The basic parameters are the mass of the ejected shell, its velocity, and the density of the ambient medium. The total energy in turbulent motions does not exceed 1% of the supernova energy. Gull assumes that equipartition is reached between the turbulent motions, the magnetic field, and the relativistic electrons. The model has been used to derive parameters for Cas A and the remnant of Tycho's supernova (Gull 1973b).

Gull derives an age of 200 years for Cas A; his model requires considerable deceleration to have occurred. However, the observations of the fast-moving knots are consistent with no deceleration and indicate an age of about 310 years (van den Bergh & Dodd 1970). As mentioned earlier, it is likely that the knots formed during or soon after the supernova explosion. Therefore, another model of turbulent particle acceleration has been developed for this remnant (Scott & Chevalier 1975; Chevalier, Robertson & Scott 1976). In this model, the knots formed during the explosion and have been moving at near constant velocity since that time. As they move through the surrounding medium, they create turbulent wakes (Figure 2). The knots lose mass as they move because of instabilities and heat conduction. It is conjectured that many have already been evaporated at this time.

Fermi-type acceleration in a turbulent medium appears to be adequate to explain the observations of Cas A (Chevalier, Robertson & Scott 1976). The Fermi gains are offset by adiabatic expansion losses; because the time scales for these two processes are similar, the correct spectral index is obtained. However, there is no underlying physical reason for the similarity in time scales so the model does not explain the small range of spectral indices observed in SNRs. Chevalier et al. solved for the diffusion of particles in energy space with time and were able to fit the following observed properties of Cas A: the spectral index, the curvature of the spectrum, the rate of flux decrease, and the rate of spectral flattening. The model is oversimplified and is probably not unique, but it does show that turbulent acceleration mechanisms hold promise for explaining the properties of young SNRs.

The recent map of Cas A with the Cambridge 5-km telescope shows a wealth of detail in the remnant (Bell, Gull & Kenderdine 1975). The radio remnant does appear to have a well-defined edge, which Bell et al. suggest is the perimeter of the shock front. A sharp edge is also found in the remnant of Tycho's supernova (Duin & Strom 1975). These observations raise the question of the propagation of relativistic particles inside a young supernova remnant. If propagation is slow, e.g., limited to the Alfvén velocity, particle acceleration in the shock wave may be required. The Cas A map shows evidence for cellular structure, as does a high-resolution polarization map of Tycho's SNR (Duin & Strom 1975). These features may be giving information on turbulent processes. More direct information on turbulence can be obtained from observations of changes in the remnants. In comparing the 5-km Cambridge map of Cas A with a previous map by Rosenberg (1970), Bell found that the radio peaks move with random velocities of several thousand km sec^{-1}. The mean outward velocity of the radio peaks is considerably less than the velocities of the fast-moving optical knots at the same distance from the center of expansion. The generation of turbulence by the Rayleigh-Taylor mechanism may no longer be effective for the slow-moving material, so the turbulence may be generated by the fast-moving knots traveling through a shell (A. R. Bell, personal communication).

5.2 *Old Supernova Remnants*

The mechanisms described above involve the generation of turbulence by the interaction of the supernova ejecta with the ISM. After this phase, when most of the supernova energy has been transferred to the ISM, it is expected that a blast wave is driven into the ISM. If the gas acted like a fluid and the ambient medium were uniform, the shock front would be stable and the flow would be laminar. Further particle acceleration would not occur and adiabatic expansion losses of the particles would be the dominant effect. However, the gas is actually a collisionless plasma and turbulence probably plays a role in the shock transition. Also, as the shock front passes irregularities in the ambient medium, turbulence may be generated in the post-shock flow (S. F. Gull, personal communication). These possibilities have not been thoroughly investigated; it is possible that particle acceleration occurs in the blast wave phase. Such a theory may be required to explain the radio radiation from Tycho's remnant, which is probably in the blast wave stage of evolution (van den Bergh 1971).

The radio flux is likely to increase when the remnant, or parts of the remnant, enters the cooling phase and a dense shell forms. The compression of ambient magnetic fields and relativistic electrons can give rise to enhanced synchrotron radiation (van der Laan 1962). Old remnants that emit radio radiation primarily by this mechanism constitute the third class of radio supernova remnants. Examples are the Cygnus Loop, S 147, and IC 443. This class of remnants appears to be observationally distinguishable from the second class in that predominantly tangential magnetic field configurations are found as opposed to radial fields in the second class of remnants. Recent data that tend to confirm the basic radiation mechanism are the high-resolution Westerbork observations of IC 443 (Duin & van der Laan 1975). The radio emission is seen to follow closely the optical emission, indicating that the emission is from the compressed regions behind the shock wave.

The Σ-D plot is used to describe the relation between surface brightness and diameter of SNRs (cf Woltjer 1972). Considering that there are at least three classes of radio SNRs, it is surprising that any relation exists. Besides, the radio flux from the last two classes of remnants, and possibly the first as well, is probably dependent on the density of the ambient medium. In view of the lack of theoretical understanding of the Σ-D relation, it should be used with care in deriving distances of SNRs. For example, remnants in the late stage of blast-wave evolution and remnants in a low-density medium may be anomalously weak radio emitters. Clark & Caswell (1976) have compiled a catalog with all SNRs down to a uniform level of surface brightness. The numbers they find for the frequency and energy of events giving rise to typical radio remnants are in fair agreement with those found from supernova explosions themselves (Section 2).

5.3 *Cosmic Rays*

Turbulent acceleration mechanisms in young remnants are expected to yield relativistic protons and heavy nuclei as well as electrons. Young remnants are thus candidates for the acceleration sites of the cosmic rays observed at earth. The spectral index of cosmic rays is similar to that deduced for the relativistic electrons in SNRs, although the remnants generally have a somewhat flatter spectrum. The composition of cosmic rays can also give information on their origin. The analysis of Hainebach, Norman & Schramm (1976) shows that about equal amounts of supernova ejecta and interstellar material are required to produce the cosmic ray composition. This composition results naturally from the kind of acceleration discussed above. Most of the particle acceleration occurs when the ejecta have swept up their own mass in surrounding material because the turbulent energy is greatest at that time. Assuming that the ejecta are mixed in with the turbulent gas, the correct cosmic ray composition results. It should be noted that because the particle gyroradii cannot be larger than the size of the accelerating turbulent clouds, proton energies above a few times 10^{15} eV are not created by this mechanism.

Another factor to be considered is whether supernovae can provide the energy of observed cosmic rays. The turbulent acceleration mechanism can probably yield at most a few percent of the total supernova energy, that is, a few $\times 10^{49}$ ergs. If Cas A has 100 times as much energy in relativistic protons as relativistic electrons and the particles are in energy equipartition with the magnetic field, this is approximately the particle energy that would be deduced from the synchrotron emission. This is also the energy per supernova required for cosmic rays based on the mass traversed by cosmic rays, the supernova rate, and the mass of gas in the galaxy (cf Lingenfelter 1973). However, the cosmic rays experience adiabatic losses while their energy density is larger than the mean galactic energy density (Wentzel 1973, Kulsrud & Zweibel 1975). This results from a hydromagnetic instability that is excited when relativistic particles stream along magnetic field lines. The cosmic ray energy is transferred to Alfvén waves that dissipate the energy in the thermal gas. This mechanism can reduce the cosmic ray energy produced in a remnant like Cas A by more than an order of magnitude. Yet if a significant fraction of supernovae occur in a low density medium, the adiabatic loss problem may be alleviated. It is also possible that a more complex model of cosmic rays in the Galaxy is required, taking into account inhomogeneities within the Galaxy. The long lifetime of cosmic rays found by Garcia-Munoz et al. (1975) may indicate that the required energy per supernova has been overestimated.

In addition to initially producing cosmic rays, SNRs can increase the energy of ambient cosmic rays by the betatron effect during the dense shell phase (Chevalier 1977). Cosmic rays may initially be reflected by the SNR shock, but it is assumed that they are eventually swept up by the shell because of the tangled nature of the ambient magnetic field. The energy increase can be more than 0.1% of the supernova energy if some particles escape from the dense shell or if there is moderate pitch angle scattering. Regardless of any energy increase, the cosmic ray energy content of the dense shell may be 10^{48}–10^{49} ergs. These particles can result in an observable flux of γ-rays through pion decays. Thus, γ-ray observations of old remnants may only yield information on swept-up interstellar cosmic rays rather than on recently accelerated particles. As the resolution of γ-ray experiments improves, observations of young remnants such as Cas A and Tycho's remnant should indicate their relativistic proton content. Because these would be particles produced in the remnant, we will then be in a better position to decide whether young remnants can supply the bulk of galactic cosmic rays.

6 CONCLUSIONS AND FUTURE PROSPECTS

Supernovae are likely to be important sources of energy in the ISM. In particular, they may give the kinetic energy of cloud motions, create the hottest ($T \approx 10^6$K) component of the ISM, give soft X-rays that partially ionize a component of the ISM, and accelerate the bulk of cosmic rays. Individual supernova remnants act as interstellar laboratories in which a wide range of physical phenomena may be observed. Examples are the action of heat conduction, turbulent amplification of magnetic fields, turbulent acceleration of relativistic particles, and the growth of Rayleigh-Taylor type instabilities. Future observations relevant to the interaction of supernovae with the ISM hold a great deal of promise. Possible observations in each band of the electromagnetic spectrum will be reviewed in order of increasing frequency.

Radio observations of young SNRs will be of particular interest. Temporal differences in the structure and spectrum of Cas A have already been found. Recently, Cas A has shown a flux increase at low frequencies, as opposed to a continued decrease at higher frequencies (Erickson & Perley 1975, Read 1977). This may be due to a sudden injection of relativistic particles or to an increase in the magnetic field if the electron spectrum steepens at low energy. As information on turbulent processes in the remnant may be obtained, monitoring of the spectrum should be continued. With the turbulent acceleration hypothesis, there are many acceleration sites throughout the remnant and there is no reason to expect each one to produce particles with the same spectral index. Detailed spectral index maps may give information on the diffusion of particles in the remnant. In this regard, a low-frequency interferometric map of Cas A would show whether the observed flux increase is localized. These types of observations may eventually be extended to fainter sources, like the remnant of Tycho's supernova.

Infrared studies of SNRs have not yet yielded significant results, although, as discussed in Section 3.4, radiation from cooling shock waves and collisionally heated grains in hot gas may be expected. Optical spectra of old remnants act as probes of the surrounding ISM (Section 3.4). Further spectrophotometry of old remnants will be valuable as improved theoretical calculations of shock wave spectra become available. The hot gas in remnants can be investigated in the coronal lines and, possibly, Hα (Kirshner & Taylor 1976). The Fe XIV line at $\lambda 5303$ is the most suitable coronal line for detailed study (cf Woodgate et al. 1974), and it may become possible to map older remnants in this line. The Ca XV line at $\lambda 5694$ is emitted by higher temperature gas, but is expected to be considerably fainter than the Fe XIV line. The future coronal line studies will supplement X-ray studies of the hot gas.

The energies to which particles can be accelerated by a turbulent mechanism in a remnant like Cas A is an important problem. Efforts should be made to detect nonthermal radiation from Cas A at infrared and optical wavelengths. This emission could be identified as such by comparison with the high-resolution radio maps of Cas A and by its polarization properties.

Lines from a wide range of ionized species are available in the ultraviolet. The Copernicus satellite has been used to observe stars behind the Vela remnant (Jenkins et al. 1976). Studies of high-velocity features in the ultraviolet may indicate whether they can be identified as SNR shells. If so, this would provide further evidence for the hypothesis that supernovae are important contributors of kinetic energy to the ISM. Imaging of SNRs in the ultraviolet is just beginning (Carruthers & Page 1976). It provides the opportunity to observe gas that is somewhat cooler than the X-ray emitting gas and thus may yield information on the transition regions between hot gas and cool clouds.

Future X-ray experiments promise improved spatial and spectral resolution. The source of the softer X-rays from the young remnant Cas A for which there are presently several theories will become clear. Information on the interaction of hot gas with clouds will be obtained, and it will be interesting to see whether soft X-rays become more pronounced near clouds. High spectral resolution observations of young remnants will be of particular interest if the hypothesis that the supernova ejecta are heated to X-ray emitting temperatures is correct. We will then have information on the initial process of mixing of supernova ejecta into the ISM. As discussed in Section 5.3, γ-ray observations will play a crucial role in determining whether SNRs produce the bulk of galactic cosmic rays.

There are two broad areas of theoretical research in this field that will probably command a great deal of attention in the future. The first is the plasma physics of supernova remnants. The structure of the collisionless shock waves in young remnants is poorly understood. This is a topic where experience from interplanetary shock waves, especially the earth's bow shock, may yield important insights. Turbulent processes leading to the amplification of magnetic fields and particle acceleration require further exploration. In these areas, the theory must be closely tied to observational results, as any general understanding of these processes is lacking. Finally, questions concerning the diffusion of relativistic particles through and out of a supernova remnant require further attention.

The second area is the role played by the hot gas phase in a complete picture of the ISM. It now seems likely that supernovae are capable of contributing hot gas to a galactic corona, as first envisioned by Spitzer (1956). There may be a continual recycling process through the corona; this process may be important for the structure of the galactic magnetic field, and it may play a role in a galactic dynamo. Further problems involve the importance of the hot gas for the propagation of cosmic rays and its interaction with the pressure perturbations generated by spiral density waves.

I am grateful to many colleagues for discussions of the material in this review and for sending preprints. I am particularly indebted to Don Cox for his comments on the manuscript. The hospitality of the Institute of Astronomy, Cambridge, where much of this review was written, is also acknowledged.

Literature Cited

Ardavan, H. 1973. *Ap. J.* 184:435

Assousa, G. E., Balick, B., Erkes, J. W. 1973. *Bull. Am. Astron. Soc.* 5:410

Assousa, G. E., Erkes, J. W. 1973. *Astron. J.* 78:885

Baade, W., Zwicky, F. 1934. *Phys. Rev.* 45: 138

Bell, A. R., Gull, S. F., Kenderdine, S. 1975. *Nature* 257:463

Berkhuijsen, E. M. 1974. *Astron. Astrophys.* 35:429

Beuermann, K. P. 1973. *Astrophys. Space Sci.* 20:27

Blake, G. M. 1972. *MNRAS* 156:67

Bychkov, K. V. 1973. *Astron. Zh.* 50:907 (*Sov. Astron.-AJ* 17:577)

Bychkov, K. V., Pikel'ner, S. B. 1975. *Pis'ma Astron. Zh.* 1:29 (*Sov. Astron. Lett.* 1:14)

Carruthers, G. R., Page, T. 1976. *Ap. J.* 205: 397

Chevalier, R. A. 1974. *Ap. J.* 188:501

Chevalier, R. A. 1975a. *Ap. J.* 198:355

Chevalier, R. A. 1975b. *Ap. J.* 200:698

Chevalier, R. A. 1976a. *Ap. J.* 207:872

Chevalier, R. A. 1976b. *Ap. J.* 208:826

Chevalier, R. A. 1977. *Ap. J.* 213:52

Chevalier, R. A., Gardner, J. 1974. *Ap. J.* 192: 457

Chevalier, R. A., Robertson, J. W., Scott, J. S. 1976. *Ap. J.* 207:450

Chevalier, R. A., Theys, J. C. 1975. *Ap. J.* 195:53

Chiu, B. C., Morrison, P., Sartori, L. 1975. *Ap. J.* 198:617

Clark, D. H., Caswell, J. L. 1976. *MNRAS* 174:267

Colgate, S. A. 1972. *Ap. J.* 174:377
Colgate, S. A. 1974a. *Ap. J.* 187:321
Colgate, S. A. 1974b. *Ap. J.* 187:333
Cornett, R. H., Hardee, P. E. 1975. *Astron. Astrophys.* 38:157
Cowie, L. L. 1975. *MNRAS* 173:429
Cowie, L. L., McKee, C. F. 1977. *Ap. J.* 211:135
Cox, D. P. 1972a. *Ap. J.* 178:143
Cox, D. P. 1972b. *Ap. J.* 178:159
Cox, D. P. 1972c. *Ap. J.* 178:169
Cox, D. P. 1976. Paper presented at IAU Gen. Assem., Grenoble
Cox, D. P., Smith, B. W. 1974. *Ap. J.* (*Lett.*) 189:L105
Culhane, J. L., Acton, L. W. 1974. *Ann. Rev. Astron. Astrophys.* 12:357
Davison, P. J. N., Culhane, J. L., Mitchell, R. J. 1976. *Ap. J.* (*Lett.*) 206:L37
Duin, R. M., Strom, R. G. 1975. *Astron. Astrophys.* 39:33
Duin, R. M., van der Laan, H. 1975. *Astron. Astrophys.* 40:111
Elmergreen, B. G. 1976. *Ap. J.* 205:405
Erickson, W. C., Perley, R. A. 1975. *Ap. J.* (*Lett.*) 200:L83
Falk, S. W., Arnett, W. D. 1973. *Ap. J.* (*Lett.*) 180:L65
Falle, S. A. E. G. 1975. *MNRAS* 172:55
Garcia-Munoz, M., Mason, G. M., Simpson, J. A. 1975. *Ap. J.* (*Lett.*) 201:L141
Gorenstein, P., Harnden, F. R. Jr., Tucker, W. H. 1974. *Ap. J.* 192:661
Gorenstein, P., Tucker, W. H. 1976. *Ann. Rev. Astron. Astrophys.* 14:373
Gull, S. F. 1973a. *MNRAS* 161:47
Gull, S. F. 1973b. *MNRAS* 162:135
Gull, S. F. 1975. *MNRAS* 171:263
Gurzadyan, G. A. 1953. *Izv. Akad. Nauk. Arm. SSR, Ser. Fiz.-Mat. Nauk.* 5, No. 2
Gurzadyan, G. A. 1969. *Planetary Nebulae.* New York: Gordon & Breach. 314 pp.
Hainebach, K. L., Norman, E. B., Schramm, D. N. 1976. *Ap. J.* 203:245
Heiles, C. 1976. *Ap. J.* (*Lett.*) 208:L137
Jenkins, E. B., Meloy, D. A. 1974. *Ap. J.* (*Lett.*) 193:L121
Jenkins, E. B., Silk, J., Wallerstein, G. 1976. *Ap. J.* Suppl. 32:681
Jones, E. M. 1973. *Ap. J.* 182:559
Jones, E. M. 1975. *Ap. J.* 201:377
Kafatos, M., Brandt, J. C., Maran, S. P., Stecher, T. P. 1975. *Bull. Am. Astron. Soc.* 7:400
Kafatos, M., Morrison, P. 1971. *Ap. J.* 168:195
Kafatos, M., Tucker, W. 1972. *Ap. J.* 175:837
Kahn, F. D. 1974a. In *The Interstellar Medium,* ed. K. Pinkau, p. 235. Dordrecht: Reidel
Kahn, F. D. 1974b. In *IAU Symposium No. 60,* eds. F. J. Kerr, S. C. Simonson, p. 329. Dordrecht: Reidel
Kirshner, R. P., Taylor, K. 1976. *Ap. J.* (*Lett.*) 208:L83
Knapp, G. R., Kerr, F. J. 1974. *Astron. Astrophys.* 33:463
Kulsrud, R. M., Bernstein, I. B., Kruskal, M., Fanucci, J., Ness, N. 1965. *Ap. J.* 142:491
Kulsrud, R. M., Zweibel, E. 1975. *Munich Cosmic Ray Conf., Pap. OG9 1–12*
Larson, R. B. 1974. *MNRAS* 169:229
Lasher, G. 1975. *Ap. J.* 201:194
Lerche, I., Vasyliunas, V. M. 1976. *Ap. J.* 210:85
Lingenfelter, R. E. 1973. *Astrophys. Space Sci.* 24:83
Mansfield, V. N., Salpeter, E. E. 1974. *Ap. J.* 190:305
Mathews, W. G., Baker, J. C. 1971. *Ap. J.* 170:241
Max, C., Arons, J., Blandford, R. 1976. Paper presented at the Berkeley ASP meeting
Maza, J., van den Bergh, S. 1976. *Ap. J.* 204:519
McCray, R., Stein, R. F., Kafatos, M. 1975. *Ap. J.* 196:565
McKee, C. F. 1974. *Ap. J.* 188:335
McKee, C. F., Cowie, L. L. 1975. *Ap. J.* 195:715
McKee, C. F., Ostriker, J. P. 1975. *Bull. Am. Astron. Soc.* 7:419
Miller, J. S. 1974. *Ap. J.* 189:239
Mills, B. Y. 1974. In *IAU Symposium No. 60,* eds. F. J. Kerr, S. C. Simonson, p. 311. Dordrecht: Reidel
Morrison, P., Sartori, L. 1969. *Ap. J.* 158:541
Ögelman, H. B., Maran, S. P. 1976. *Ap. J.* 209:124
Öpik, E. J. 1953. *Ir. Astron. J.* 2:219
Osterbrock, D. E., Dufour, R. J. 1973. *Ap. J.* 185:441
Peimbert, M. 1971. *Ap. J.* 170:261
Peimbert, M., van den Bergh, S. 1971. *Ap. J.* 167:223
Pikel'ner, S. B. 1954. *Izv. Krym. Astrofiz. Obs.* 12:93
Pravdo, S. H., Becker, R. H., Boldt, E. A., Holt, S. S., Rothschild, R. E., Serlemitsos, P. J., Swank, J. H. 1976. *Ap. J.* (*Lett.*) 206:L41
Rappaport, S., Cash, W., Doxsey, R., Moore, G., Borken, R. 1973. *Ap. J.* (*Lett.*) 186:L115
Raymond, J. C. 1976. PhD thesis. Univ. Wisconsin
Read, P. L. 1977. *MNRAS* 178:259
Richtmyer, R. D. 1960. *Commun. Pure Appl. Math.* 13:297

Rosenberg, I. 1970. *MNRAS* 151:109
Rosenberg, I., Scheuer, P. A. G. 1973. *MNRAS* 161:27
Salpeter, E. E. 1976. *Ap. J.* 206:673
Sancisi, R. 1974. In *IAU Symposium No. 60,* eds. F. J. Kerr, S. C. Simonson, p. 115. Dordrecht: Reidel
Scott, J. S., Chevalier, R. A. 1975. *Ap. J.* (*Lett.*) 197:L5
Sedov, L. 1959. *Similarity and Dimensional Methods in Mechanics.* New York: Academic
Sgro, A. 1972. PhD thesis. Columbia Univ.
Sgro, A. 1975. *Ap. J.* 197:621
Shapiro, P. R., Field, G. B. 1976. *Ap. J.* 205: 762
Shklovsky, I. 1968. *Supernovae.* New York: Wiley
Silk, J. 1973. *Ann. Rev. Astron. Astrophys.* 11:269
Silk, J., Burke, J. R. 1974. *Ap. J.* 190:11
Siluk, R. S., Silk, J. 1974. *Ap. J.* 192:51
Smith, B. W. 1977. *Ap. J.* 211:404
Solinger, A., Rappaport, S., Buff, J. 1975. *Ap. J.* 201:381
Somon, J. P. 1971. In *Physics of High Energy Density,* eds. P. Caldriola, H. Knoepfel, p. 189. New York: Academic
Spitzer, L. Jr. 1956. *Ap. J.* 124:20
Spitzer, L. Jr. 1968. *Diffuse Matter in Space.* New York: Interscience
Straka, W. C. 1974. *Ap. J.* 190:59
Tammann, G. A. 1974. In *Supernovae and Supernova Remnants,* ed. C. B. Cosmovici, p. 155. Dordrecht: Reidel
Taylor, G. I. 1946. *Proc. R. Soc. London Ser.* A 186:273
Tinsley, B. M. 1975. *Publ. Astron. Soc. Pac.* 87:837
van den Bergh, S. 1971. *Ap. J.* 168:37
van den Bergh, S., Dodd, W. W. 1970. *Ap. J.* 162:485
van den Bergh, S., Marscher, A. P., Terzian, Y. 1973. *Ap. J. Suppl.* 26:19
van der Laan, H. 1962. *MNRAS* 124:125
Weisskopf, M. C., Helava, H., Wolff, R. S. 1974. *Ap. J.* (*Lett.*) 194:L71
Wentzel, D. G. 1973. *Astrophys. Space Sci.* 23:417
Williamson, F. O., Sanders, W. T., Kraushaar, W. L., McCammon, D., Borken, R., Bunner, A. N. 1974. *Ap. J.* (*Lett.*) 193:L133
Wilson, A. S., Weiler, K. W. 1976. *Astron. Astrophys.* 49:357
Woltjer, L. 1972. *Ann. Rev. Astron. Astrophys.* 10:129
Woodgate, B. E., Stockman, H. S., Angel, J. R. P., Kirshner, R. P. 1974. *Ap. J.* (*Lett.*) 188:L79
Woodward, P. R. 1976. *Ap. J.* 207:484

Ann. Rev. Astron. Astrophys. 1977. 15: 197–234

TRANSITION PROBABILITY DATA FOR MOLECULES OF ASTROPHYSICAL INTEREST

R. W. Nicholls
Centre for Research in Experimental Space Science, York University, Downsview, Ontario M3J 1P3, Canada

1 INTRODUCTION

1.1 *General Comments*

This article is a review of transition probability data of molecular spectra, principally of diatomic molecular spectra that are of astrophysical importance. This topic is viewed in the context of contemporary spectroscopic research, which is advancing on two major fronts: (*a*) wavelength studies, from which atomic and molecular structure constants may be derived from the inferred energy-level separations, and (*b*) intensity studies, from which transition probability data and physico-chemical conditions in the light source or absorbing layer may be inferred. Both of these activities find many applications in astronomy and astrophysics. The difference in outlook between them are related more to the experimental methods of each than to fundamental differences between them. The energy levels, whose locations are determined in (*a*), provide a set of eigenvalues E_i against which models of atomic and molecular structure can be tested. These models provide a set of eigenfunctions ψ_i, one for each E_i. The eigenfunctions determine the transition-strength matrix elements S_{ij} through

$$S_{ij} = \left| \int \psi_i^* M \psi_j \, d\beta \right|^2, \tag{1}$$

which determine the relative intensities of the radiative transitions between i and j. M is the multipole moment and $d\beta$ is the element of configuration space. The transition probability data (such as S_{ij} itself, the Einstein A and B coefficients, the oscillator strengths, and absorption coefficients, etc) are all products of S_{ij} and various powers of the transition frequency ν_{ij}.

In spite of this strong conceptual link between the two major areas of spectroscopic research activity, there has not until recently been a great amount of cross-fertilization between them, except perhaps in astrophysical circumstances where both are needed. The overwhelming emphasis on wavelength spectroscopy has to a large extent been due to the exquisite precision it provides to the measurement of structure constants of atoms and molecules, against which theoretical models can be tested. In spite of the increased use of modern methods, spectral intensities cannot be measured nearly as accurately as line locations. Nevertheless, the diagnostic potential of the results of intensity spectroscopy has assured its continuing importance in astronomy and astrophysics.

1.2 *Atomic and Molecular Spectra*

The crucial need for definitive oscillator strength data for astrophysically important atomic lines and multiplets is so well understood by astronomers as to need no further emphasis here. Such data are provided by many diverse experimental methods including emission and absorption, "hook method," beam foil, atomic beam, and shock-tube spectroscopy, as well as by quantum considerations of atomic and molecular models. Various government laboratories, principally the U.S. National Bureau of Standards, have provided the very valuable services of compilation and assessment of available atomic transition probability data. This work augments the similar service that they have provided for many years on many aspects of wavelenth spectroscopy. In spite of such fine efforts, much more needs to be done even on the spectra of astrophysically important atoms and ions. For example, by no means all of the FeI lines yet have firm transition assignments (Weeks & Simpson 1967), and there still exists doubt as to the accuracy of some of the oscillator strengths (Blackwell 1975).

The situation for molecular spectra is far worse both for wavelength identifications and for intensity and transition probability data. This is due in part to the far greater richness of lines exhibited by molecular spectra, and in part to the fact that the number of lines emitted or absorbed in a system depends in a very sensitive manner on the excitation conditions (e.g. temperature) of the source, because of the small separations between molecular energy levels.

As is well known, the wavelength λ and frequency ν of a molecular line are related to changes in electronic, vibrational, and rotational energies by

$$\frac{hc}{\lambda} = h\nu = \Delta E_e + \Delta E_{vib} + \Delta E_{rot}. \tag{2}$$

The change ΔE_e in electronic energy locates the whole band system in the spectrum. It is a measure of the gross separation of the two potentials concerned. The change ΔE_{vib} in vibrational energy locates the position of the band in the band system. The change ΔE_{rot} in the rotational energy locates the line in the band. Astronomically interesting band systems occur between the vacuum UV ($<2000\ A$) and the near infrared ($>2\ \mu$). The vibration rotation ($\Delta E_e = 0$) molecular spectrum occurs in the near infrared. Most astronomically important cases are observed in absorption, often from the lowest vibrational level of the ground electronic state between 1 and 10 μ. Pure rotation spectra ($\Delta E_e = \Delta v = 0$) occur at higher wavelengths, including the microwave region, particularly if transitions between sublevels are included. There is great current astronomical interest in this region.

Transition probability parameters specific to the electronic, vibrational, and rotational components of molecular spectra are required for diagnostic applications in astronomy.

The existence of local thermodynamic equilibrium in many astronomical bodies that exhibit molecular emission and absorption spectra often gives rise to spectra that are somewhat similar in appearance to those of laboratory origin. Such spectra offer good diagnostic potential if sufficient reliable transition probability data are available on them. In certain cases however, unusual physical conditions (e.g. very low temperatures, the Swings effect in the fluorescent excitation of cometary spectra) give rise to molecular spectral features of unusual appearance, on which diagnosis may still be performed, but with great care.

In spite of many years of research on molecular spectra and the great wealth of lines in each spectrum, by no means all lines of all bands have yet been identified or analyzed. The contemporary literature shows that existing analyses are of somewhat variable quality insofar as molecular structure constants that they provide are concerned. Often such constants are not of high enough quality to synthesize high-resolution spectra with confidence on a computer.

Research on intensities of molecular spectra has two main components, namely (*a*) experimental intensity measurements in emission or absorption, and (*b*) theoretical studies (including ab initio quantum calculations) of molecular potentials, wavefunctions, and derived quantities of them such as Franck-Condon and Hönl-London factors, *r*-centroids, and transition moments. The sparseness of hard data in this field is far more extreme than the sparseness of molecular structure data. Firm transition probability data are thus available for relatively few astronomical molecular features, and far more laboratory work, specifically in support of astronomy, is needed.

2 MOLECULAR SPECIES OF ASTROPHYSICAL IMPORTANCE

2.1 *Molecular Species*

The most common diatomic and polyatomic molecular species whose emission and absorption spectra are astrophysically important are of course compounds of the most abundant elements: C, H, O, N, and their isotopes and ions. In addition, molecules containing other abundant elements, such as are found for example in the Sun and chondritic meteorites (Ross & Aller 1976), are also often found.

For a molecular species to form and remain stable against dissociating influences in an astronomical environment, the temperature must be sufficiently low and other energetic interactions must be sufficiently mild that the probability of breaking a chemical bond, once formed, is low. Thermochemical methods can be used to discuss such considerations quantitatively (Tsuji 1964, Sauval 1976). Such methods are routinely used in quantitative shock-tube spectroscopy (Arnold & Nicholls 1972, 1973b, Cooper & Nicholls 1975a, 1976). They are extremely sensitive to the choice of dissociation energy used in each case.

Table 1 displays astronomical locations from which molecular spectra are

Table 1 Astronomical locations of molecular spectra

Region	Excitation Processes or Absorption
Planetary (including terrestrial) Atmospheres	Cold Gas absorption Scattering Emission from solar wind bombardment (aurorae) Fluorescence Chemiluminescence (airglow)
Comets	Fluorescent Excitation (head) Photoionization, Fluorescence (tail)
Cooler stellar Atmospheres (including the Sun and Sunspots)	Hot Gas absorption, sometimes hot gas emission
Interstellar clouds	Cold Gas absorption and emission

commonly observed and the physical processes related to their observation. Temperatures in these locations lie between a few degrees (K) and a few thousand degrees (K).

While the identity of some molecular species in astronomical objects of the solar system, in stars, and in interstellar space has been firmly established for some years, for others even the identification is not yet firm. Thus, identification is the first area of uncertainty in the study of astronomical molecular spectra. Objects exhibiting simple spectra that are easily compared with laboratory spectra allow firm identifications to be made. It is probably easier to make firm identifications of an emission spectrum than an absorption spectrum, for there are so many more spectral features in the former than the latter to assist the identification. Some molecular spectra (e.g. CO^+ Comet Tail, C_3 bands, H_2O^+ bands) were first observed in astronomical light sources (comets) before being studied in the laboratory. The absorption spectra of hot molecular gases such as occur in some stellar envelopes are not often studied in the laboratory except in some shock-tube experiments. The identification of the CO Fourth Positive band system in absorption in the solar spectrum by a comparison between the absorption spectrum of a shock-heated CO sample in the laboratory and the UV spectrum of the Sun is one good example of this (Goldberg, Parkinson & Reeves 1965).

Sources in which a large number of atomic and molecular features are interlaced are the most difficult from which to make firm identifications of molecular features, unless they dominate one region of the spectrum and can easily be distinguished from atomic lines. Broida & Moore (1956) and Moore & Broida (1956, 1959a,b) have discussed the statistical problems of making molecular identifications in the spectra of stellar envelopes. They have also made a critical assessment of the certain presence, the probable presence, and probable absence of specific molecular features in stellar spectra. Moore (1966) has augmented these comments for the spectra of late-type stars, which are the most common stellar objects to exhibit molecular features in absorption. The Sun exhibits somewhat different molecular spectra from the disc and from sunspots, which are regions of different local temperatures (Sotirovski 1971).

Swings (1951, 1958, 1966) and Herzberg (1965, 1976) have made important and authoritative assessments of the presence of molecular species in cosmic sources. Most stellar molecular spectra are absorption features. There are however a very few examples of emission bands: the AlO emissions of Mira Ceti (Keenan, Deutsch & Garrison 1969), the diffuse emissions (inverse predissociation of AlH) in χ Cygni (Merrill 1953, Herbig 1956), and the CN bands in R Coronae Borealis and in the solar atmosphere (Swings 1966).

It is not the purpose of this article to discuss in detail the identifications of specific molecular species (and their spectra). This is discussed in the original literature. Rather it is emphasized that not all identifications are as firm as others and not all are easily made. For example, the tracings of the spectrum of 19 Piscium, and related discussion of Wojslaw & Peery (1976), indicate a very strong likelihood of the presence in the spectrum of this star of the lines of one, two, or three bands of the resonance (A-X) transitions of the CuH, ZnH, GeH, and SnH molecules.

Table 2 is a list of important certain and probable diatomic and small polyatomic species of astrophysical interest classified with respect to the astronomical location of occurrence. Only molecules for which observed spectra are reported in the literature have been included in the list. Molecules that have been proposed on general physical grounds but whose spectra have not yet been observed have not been included. Neither have the somewhat larger polyatomic molecules of interstellar space, lists of which are dated almost as soon as they are published, and most of which are detected by radioastronomical rather than by optical methods. Table 2 was compiled from the review articles cited above, a number of original articles, and the works of Aller (1963) and Allen (1973). The importance of certain metal oxides and hydrides in the spectrum of the cooler stars is noted. Many of the entries in Table 2 arise from ground-based observations in the visible and red regions of the spectrum. With the recent advent of astronomical observations from space it is to be anticipated that more molecular features in the vacuum UV region of the spectrum will be reported in the next few years. In Section 4 of this article, which reviews the available transition probability data, we do not restrict the discussion solely to the visible band systems of the species of Table 2, but include all band systems (including those in the vacuum UV) for which transition probability data are available.

The situation for molecular constituents of planetary atmospheres is still evolving, particularly with respect to the minor constituents, which in the case of terrestrial atmosphere play important roles in the spectra of the aurorae and the airglow, and also in bulk absorption through long atmospheric paths. The increase in interest in infrared astronomical spectra will probably give rise to reports of new polyatomic molecular constituents over the next few years.

2.2 *Availability of Spectroscopic Data*

There is a mistaken yet commonly held belief among some users of spectroscopic methods and data that all necessary analyses of important spectra have been

Table 2 Molecular species and their astronomical locations

	PLANETS							COMETS		STARS						INTER-STELLAR MEDIUM
Molecule	Venus	Earth	Mars	Jupiter	Saturn	Uranus	Neptune	Head	Tail	Sun or Sunspots	G-K	M	S	Carbon	Variable	
AlH										*		*				
AlO										*		*			*	
BH										*						
BO										*		*				
CaH										*		*				
CaO															*	
ClO		*														
CrO												*				
C_2								*		*		*		*	*	
CH								*		*	*	*		*	*	*
CH^+									*							*
CN								*		*	*	*		*	*	*
CO	*	*	*					*		*						*
CO^+	*	*	*						*							
CS																*
CuH														*		
FeH										*						
GeH														*		
H_2		*		*	*	*	*									*
HD																*
LaO													*			
MgF										*						
MgH										*		*				
MgO										*		*				
N_2	*	*	*													
N_2^+	*	*							*							
NH								*		*						
NO		*														
NO^+		*														
O_2	*	*								*						

O_2^+		*													
OH		*						*		*					*
OH^+									*						
SO															*
ScO										*				*	
SiF											*				
SiH											*				
SiH^+										*					*
SiN										*	*	*			*
SiO											*				*
SuH													*		
SrF										*					
TiH										*					
TiO											*				
VO											*				
YO											*	*			
ZnH													*		
ZrF														*	
ZrO											*	*		*	
C_3								*					*		
CH_4		*		*	*	*	*								
CO_2	*	*	*												
CO_2^+	*	*	*						*						
HCN								*							*
H_2O	*	*	*	*				*							*
H_2O^+									*						
H_2S															*
NH_2								*							
NH_3				*	*										*
NO_2		*													
N_2O		*													
OCS															*
O_3		*													
SO_2		*													
SiC_2													*		

made, that all of the resulting basic spectroscopic data are in tables, and that the only current requirement is the tidying up of details. This is very far from the truth. Indeed as long ago as 1962, in an assessment of the existing literature of diatomic molecular spectroscopy, Grinfeld (1962) pointed out that of a possible 5000 diatomic species, 400 had been examined spectroscopically in some fashion. Of these 15% had been rotationally analyzed to some extent and 30% had been vibrationally analyzed to some extent. The quality and provenance of the analyses are very variable, having been carried out by many workers over some decades. Thus structure constants in the literature have to be scrutinized very carefully indeed before full confidence may be placed on them for specific applications. Albritton et al. (1973a,b,c,d) and Zare et al. (1973) have pointed out the poor quality of some analyses, and the need to use not only methods strictly based on quantum mechanics, but also statistically reliable data analyses in the reduction of the observational data of many spectra. This has been confirmed by Cann & Nicholls (1974) and by Creek & Nicholls (1975) for certain atmospheric molecules. It is interesting to note that in many cases researches related to the development and use of molecular lasers, as well as to the needs of astronomy and astrophysics and aeronomy, have stimulated extensive critical assessments of the state of knowledge of specific molecular transitions. For example, in three recent volumes, Suchard (1975a,b) and Suchard & Melzer (1976) have made a massive assessment of the extent and quality of data that exist on rotational and vibrational analyses of spectra, dissociation energies, and transition probabilities, and of chemical kinetic data for 264 heteronuclear and 102 homonuclear molecules which may have important applications for the development of molecular lasers. The tables of this compilation are eloquent evidence of the fragmentary nature of existing reliable spectroscopic data on wavelengths and transition probabilities of diatomic molecules. Section 4 emphasizes this for the molecules of Table 2.

There are a number of often-used reference sources of diatomic molecular data that have been produced over the past quarter century. The tables of Herzberg's (1950) classic work are soon to be updated by Huber & Herzberg (1977). The compilations of Rosen (1951, 1952, 1970) and Barrow (1973) are invaluable. Rosen's (1970) work has been the springboard for the recent works of Suchard (1975a,b) and Suchard & Melzer (1976). Dissociation energies have been compiled by Gaydon (1968). Identification of molecular spectra has been supported by the works of Pearse & Gaydon (1976), Rosen (1952), Wallace (1962a,b), Krupenie (1966, 1972), Tilford & Simmons (1972), Lovas & Krupenie (1974) and by the identification atlases of molecular spectra from the author's laboratory (Tyte & Nicholls, 1964a,b, 1965, Hébert, Innanen & Nicholls 1967, Tyte, Innanen & Nicholls 1967, Degen et al. 1968, Harrington et al. 1970, Brocklehurst et al. 1971, 1972, Seel, Hébert & Nicholls 1977a,b).

The indispensible Berkeley Newsletter regularly issued by Phillips & Davis (1960–present) is the most useful continuing current bibliography of the literature of molecular spectroscopy. It is the source for the computer-based compilation of Suarez (1974) and has been used extensively in the preparation of this review.

3 TRANSITION PROBABILITY PARAMETERS OF MOLECULAR SPECTROSCOPY

3.1 *General Concepts*

In order to place in context Section 4, which is on existing transition probability data of diatomic astrophysical molecules, we make here a didactic review of the numerous types of transition probability parameters associated with molecular spectra.

Many of the parameters were first introduced as phenomenological constants of proportionality, which eventually were later given quantum meaning. Thus some parameters are derived from experimental measurements, others are derived from theory, and yet others are derived from both considerations. Einstein A and B coefficients, linear and mass absorption coefficients, absorption cross sections, band strengths, Hönl-London factors, Franck-Condon factors (and when necessary r-centroids), electronic transition moments, band strengths, oscillator strengths, and lifetimes are typical transition probability parameters.

3.1.1 PHENOMENOLOGICAL CONCEPTS The Einstein A and B coefficients were respectively introduced as unimolecular and bimolecular rate constants of the spontaneous emission and stimulated emission and absorption "reactions," in both of which the photon is a reactant. In the reactions below X is the general molecule, U and L indicate upper and lower states, and N is number density. That is,

Spontaneous Emission $\quad X^*(N_U) \xrightarrow{A_{UL}} X(N_L) + h\nu \qquad$ (3a)

Stimulated Emission and Absorption

$$h\nu + X^*(N_U) \xrightarrow{B_{UL}} X(N_L) + 2h\nu \tag{3b}$$

$$h\nu + X(N_L) \xrightarrow{B_{LU}} X(N_U). \tag{3c}$$

The kinetic equation for (3a) is

$$-\frac{dN_U}{dt} = N_U \sum_L A_{UL}. \tag{4a}$$

The summation takes account of all radiative channels $U \to L$. Thus

$$N_U = N_U^{(0)} \exp\left[-t/T_U\right]. \tag{4b}$$

Here

$$T_U = 1/\sum_L A_{UL} \tag{4c}$$

is the radiative lifetime T_U of the upper state, measurements on which are sometimes used to infer data on A_{UL}.

Similarly the kinetic equations for (3b) and (3c) are

$$-\frac{dN_U}{dt} = B_{UL} N_U \rho(\nu), \tag{5a}$$

$$-\frac{dN_L}{dt} = B_{LU} N_L \rho(\nu), \tag{5b}$$

where $\rho(\nu)$ is radiation density. In astronomical circumstances it is sometimes common to define B with respect to radiation flux rather than density (Aller 1963).

In emission spectra, the intensity I_{UL} or radiant power in the $U-L$ transition is

$$I_{UL} = KN_U A_{UL} h\nu_{UL}. \tag{6}$$

This is the fundamental equation of optically thin emission spectroscopy. K includes geometry and units. This equation allows A_{UL} to be inferred from measurements of I_{UL} provided N_U is known. The accurate determination of such populations is one major difficulty of the emission method and the need for heterochromatic detector calibration is another.

In absorption spectra, the measured differential extinction $-dI/I$ of an absorbing beam in a path length dx is

$$-dI/I = k_\nu\, dx = N_L B_{LU}\, dx/c. \tag{7}$$

Here k_ν (cm^{-1}) is the phenomenological linear (or volume) absorption coefficient, which is related to other quantities through

$$k_\nu = N_L \alpha_\nu = \rho K_\nu = \frac{1}{\Lambda_\nu} = \frac{\tau_\nu}{x}. \tag{8}$$

Here α_ν is the atomic or molecular absorption cross section (cm^2) and K_ν is the mass absorption coefficient ($cm^2\ gm^{-1}$) for an absorber of density ρ. Λ_ν is the photon mean free path (cm) and τ_ν is the optical depth.

Equation (7) integrates to

$$\ln(I_0/I) = \tau_\nu, \tag{9a}$$

which is Beer's law for optically thin ($\tau < 1$; $\Lambda > x$) absorption. It is sometimes recast, often in chemical circumstances, in terms of decadic logarithms

$$\log_{10}(I_0/I) = \varepsilon_\nu C x, \tag{9b}$$

where C is the absorbant concentration in moles cm^{-3}, x is in cm, and ε is the molar decadic extinction coefficient. Molar decadic extinctions in $cm^2\ mole^{-1}$ units, where C is in mole cm^{-3}, can be converted to cross sections (cm^2) by multiplication by 3.82×10^{-24}

Equations (9a) and (9b) are the basis of much absorption spectroscopy, where detailed knowledge of the wavelength response of linear detectors is not needed.

Oscillator strength and equivalent width are useful concepts in absorption spectroscopy. The oscillator strength f associated with a specific range of integration is defined through

$$\int \alpha_\nu\, d\nu = \frac{\Pi e^2 f}{mc} = \Pi c r_0 f, \tag{10}$$

where r_0 is the classical radius of the electron. The oscillator strength is the fraction of the classical electron associated with the transition.

The equivalent width W of a spectral feature is defined through

$$W_\nu = \int (I_0 - I_\nu)/I_0 \, d\nu = Nx \int \alpha_\nu \, d\nu = Nx\Pi c r_0 f. \tag{11a}$$

In wavelength terms we have

$$W_\lambda = Nx \int \alpha_\lambda \, d\lambda = \frac{N\lambda_0^2}{c} \int \alpha_\nu \, d\nu = \frac{W_\nu \lambda_0^2}{c} = Nx\lambda_0^2 r_0 f. \tag{11b}$$

The oscillator strength and equivalent width can thus be measured from an absorption tracing. In interpreting absorption spectra, the finite resolving power of the instrument has to be taken into account. If optically thick absorption spectra have to be treated, the equation of radiative transfer has to be solved. One solution, often used in quantitative shock-tube spectroscopy, for a plane parallel LTE slab is

$$I_\nu = B_\nu(T)[1 - \exp k_\nu(1 - \exp hc/\lambda kT)x] + I_0 \exp(-k_\nu x), \tag{12}$$

where B_ν is the Plankian. Numerous intensity measurements of astrophysically important diatomics by shock-tube radiometry have been based on this equation.

3.1.2 QUANTUM CONCEPTS The above phenomenological constants of absorption and emission spectroscopy may all be recast in quantum terms. The more common ones are all products of low powers of frequency and the transition strength element S_{UL} (Nicholls 1969a, Armstrong & Nicholls 1972), where

$$S_{UL} = \left| \int \psi_U^* M \psi_L \, d\beta \right|^2. \tag{13}$$

ψ_U and ψ_L are the wavefunctions of the upper and lower states respectively, M is the electric dipole moment, and $d\beta$ is the element of configuration space.

The Einstein A and B coefficients and the oscillator strength are then defined:

$$g_U A_{UL} = \frac{64\Pi^4 \nu_{UL}^3 S_{UL}}{3hc^3}, \tag{14a}$$

$$g_L B_{UL} = g_L B_{LU} = \frac{8\Pi^2 S_{UL}}{3h^2}, \tag{14b}$$

$$|gf| = |g_U f_{UL}| = |g_L f_{LU}| = \frac{8\Pi^2 \nu_{UL} S_{UL}}{3he^2}, \tag{14c}$$

and the constants have been placed in convenient numerical form by Allen (1973). Experimentally S is usually determined through applications of Equations (4c), (6), (9a), and (14). Theoretically S is usually determined from Equation (13) by ab initio quantum calculations.

3.2 *Diatomic Molecular Transition Probability Concepts*

Equations (14) and (15) are modified to take account of the electronic, vibrational, and rotational contributions from a diatomic molecule by factoring S_{UL}, within the constraints of the Born-Oppenheimer approximation (Born & Oppenheimer 1927). The individual components of such a transition have been illustrated by Whiting & Nicholls (1974). In most astronomical applications all Zeeman and nuclear hyperfine components are merged.

Thus the transition strength $S_{Lv''J''}^{Uv'J'}$ of a molecular line is

$$S_{L,v'',J''}^{U,v',J'} = S_{Lv''}^{Uv'} S_{J''\Lambda''}^{J'\Lambda'}. \tag{15}$$

$S_{Lv''}^{Uv'}$, which is factored into two terms below, is the band strength defined by

$$S_{Lv''}^{Uv'} = \left| \int \psi_{v'} R_e \psi_{v''}\, dr \right|^2. \tag{16}$$

$S_{J''\Lambda''}^{J'\Lambda'}$ is often called the Hönl-London, or line-strength factor. It determines the distribution of intensity line by line in the band. Λ is the component of electronic angular momentum or the internuclear axis (Herzberg 1950).

$\psi_{v'}$ and $\psi_{v''}$ are vibrational wavefunctions of the levels v' and v'' and r is internuclear separation. $R_e(r)$ is called the electronic transition moment defined through

$$R_e(r) = \int \psi_{eU}^{*} M_e \psi_{eL}\, d\gamma, \tag{17a}$$

where the ψ_e's are electronic wavefunctions and M_e is the dipole moment. Strictly (Whiting & Nicholls 1974) $R_e(r)$ does not occur alone, but as the summation of squares of components of the transition, that is $\sum |R_e|^2$. In this review the earlier notations, found in many of the references, are used interchangeably with the recommended correct notation. The electronic transition moment (and its square), depending as they do on electric dipole moment, are r-dependent quantities.

The empirical finite expansion

$$R_e(r) = \sum a_n r^n \tag{17b}$$

over the range of r encountered in a band system is often found in the literature. Equations (16) and (17b) thus suggest the need for integrals

$$H_{v'v''}^{(n)} = \int \psi_{v'} \psi_{v''} r^n\, dr, \tag{18}$$

through which the n'th degree r-centroid $\overline{r^{(n)}}$ and the first-degree r-centroid $\bar{r}_{v'v''}$ may be defined as

$$\overline{r_{v'v''}^{(n)}} \doteq H_{v'v''}^{(n)} / H_{v'v''}^{(0)}, \tag{19a}$$

$$\bar{r}_{v'v''} = H_{v'v''}^{(1)} / H_{v'v''}^{(0)}. \tag{19b}$$

Numerical and analytic examination of the properties of these quantities, which depend sensitively upon the relative phase of the wavefunctions involved, has demonstrated that for many bands of many systems the r-centroid approximation (Nicholls 1974)

$$\overline{r_{v'v''}^{(n)}} \sim (\bar{r}_{v'v''})^n \tag{20}$$

holds to within a few percent for small (<3) values of n.

Equations (16), (17b), and (20) imply

$$S_{Lv''}^{Uv'} = R_e^2(\bar{r}_{v'v''}) \cdot q_{v'v''} = \Sigma \,|\, R_e^2(\bar{r}) \,|\, q_{v'v''}, \tag{21a}$$

where

$$q_{v'v''} = \left| \int \psi_{v'} \psi_{v''} \, dr \right|^2 \tag{21b}$$

is the Franck-Condon factor of the (v', v'') band.

Equation (21a) is thus the decomposition of the band strength into electronic and vibrational components, which allows Equation (15) to be written as the product of components, one for each of the energy sinks of the molecule.

Thus

$$S_{Lv''J''}^{Uv'J'} = \Sigma \,|\, R_e^2(\bar{r}_{v'v''}) \,| \cdot q_{v'v''} \cdot S_{J''\Lambda''}^{J'\Lambda'}. \tag{22}$$

The molecular versions of Equations (14) are then

$$d_U(2J'+1)A_{Lv''J''}^{Uv'J'} = \frac{64\Pi^4}{3hc^3} (\nu_{Lv''J''}^{Uv'J'})^3 \, \Sigma \,|\, R_e^2(\bar{r}_{v'v''}) \,|\, q_{v'v''} S_{J''\Lambda''}^{J'\Lambda'}, \tag{23a}$$

$$d_U(2J'+1)B_{Lv''J''}^{Uv'J'} = d_L(2J''+1)B_{Uv'J'}^{Lv''J''} = \frac{8\Pi^2}{3h^2} \Sigma \,|\, R_e^2(\bar{r}_{v'v''}) \,| \times q_{v'v''} S_{J''\Lambda''}^{J'\Lambda'}, \tag{23b}$$

$$d_U(2J'+1)f_{Lv''J''}^{Uv'J'} = d_L(2J''+1)f_{Uv'J'}^{Lv''J''} = \frac{8\Pi^2 m}{3he^2} \nu_{Lv''J''}^{Uv'J'} \, \Sigma \,|\, R_e^2(\bar{r}_{v'v''}) \,| \times q_{v'v''} S_{J''\Lambda''}^{J'\Lambda'}, \tag{23c}$$

where d_U and d_L are the electronic degeneracies.

Equations (23) have been used as the springboard for realistic line-by-line computer simulation in laboratory atmospheric and astronomical circumstances. Often in such applications it is necessary to write the linear absorption coefficient at ν:

$$k_\nu = \frac{8\Pi^3}{3hc} \frac{N_{v''}}{d_L(2J''+1)} \nu [\Sigma \,|\, R_e^2(\bar{r}_{v'v''}) \,|] \left(\sum_{\text{bands}} q_{v'v''} S_{J'J''} \right) b_\nu. \tag{24}$$

Here the second summation is over all lines of all bands that contribute to the absorption at ν, and b_ν is the line-shape function. $S_{J'J''}$ is the Hönl-London factor. In optically thick cases this is the value of k_ν inserted in Equation (12).

It is often possible (Armstrong & Nicholls 1972) to make a formal summation of each of Equations (23) over the line structure of a band. The resulting set of equations for an integrated band are

$$d_U A_{Lv''}^{Uv'} = \frac{64\Pi^4}{3hc^3} \nu_{v'v''}^3 \, \Sigma \,|\, R_e^2(\bar{r}_{v'v''}) \,|\, q_{v'v''}, \tag{25a}$$

$$d_U B_{Lv''}^{Uv'} = d_L B_{Uv'}^{Lv''} = \frac{8\Pi^2}{3h^2} \Sigma \,|\, R_e^2(\bar{r}_{v'v''}) \,|\, q_{v'v''}, \tag{25b}$$

$$d_U f_{Lv''}^{Uv'} = d_L f_{Uv'}^{Lv''} = \frac{8\Pi^2 m}{3he^2} \nu_{v'v''} \, \Sigma \,|\, R_e^2(\bar{r}_{v'v''}) \,|\, q_{v'v''}, \tag{25c}$$

$$\int_{\text{band}} k_\nu \, d\nu = \frac{8\Pi^3}{3hc} \frac{N_{Lv''}}{d_L} \nu_{v'v''} \Sigma \, | \, R_e^2(\bar{r}_{v'v''}) \, | \, q_{v'v''}. \tag{25d}$$

Equations of this nature are often used to interpret experimental measurements on bands as a whole in terms of transition probability data. With regard to Equations (25) it is often convenient to discuss the integrated absorption coefficient of a band in terms of band oscillator strength, or equivalent width. In an optically thin situation the equivalent width of the band is the sum of the equivalent widths of all of its lines.

A further level of summation over the whole band system is sometimes attempted, which leads to a poorly defined concept of the electronic oscillator strength f_{LU} of the whole system. Thus we have

$$d_L f_{LU} = d_L \sum_{v'} f_{v'v''} = \frac{8\Pi^2 m}{3he^2} \sum_{v'} \nu_{v'v''} [\Sigma \, | \, R_e^2(\bar{r}_{v'v''}) \, |] q_{v'v''}. \tag{26a}$$

If an average frequency $\bar{\nu}$ can be defined for the system, and if the electronic transition moment square is a constant, and accepting (see Equation 30) that $\sum_{v'} q_{v'v''} = 1$,

$$d_L f_{LU} \sim \frac{8\Pi^2 m}{3he^2} \bar{\nu} \Sigma \, | \, R_e^2 \, |. \tag{26b}$$

Thus, from Equation (25c),

$$\frac{f_{LU}}{f_{v'v''}} = \frac{\bar{\nu} \Sigma \, | \, R_e^2 \, |}{\nu_{v'v''} \Sigma [R_e^2(\bar{r}_{v'v''})] q_{v'v''}}, \tag{26c}$$

from which the gross approximation

$$f_{LU} = f_{v'v''} / q_{v'v''} \tag{26d}$$

is obtained, and sometimes used. It is strongly recommended that this concept not be used.

Finally, with regard to the inference of transition probability data from vibrational lifetimes, Equations (4c) and (25a) imply

$$T_{v'} = \frac{1}{\sum_{v''} A_{v'v''}} = \frac{3hc^3 \, d_U}{64\Pi^4} \frac{1}{\sum_{v''} [\Sigma \, | \, R_e^2(\bar{r}_{v'v''}) \, |] \cdot q_{v'v''} \nu_{v'v''}^3}. \tag{27}$$

Unless there is a dominant contribution to $\sum_{v''}$ it is difficult to obtain band strength data from such measurements.

3.3 *Molecular Transition Probability Parameters: Theory and Experiment*

The preceding sections indicate the diverse transition probability data of molecular spectra. Central to them all is the transition strength matrix element S, to which they are nearly all linearly related through the fact that the transition probability parameter involves $\nu^p S$ where $p = 0$, 1, 3, or 4. Tatum (1967), in an important review, has discussed with great clarity ambiguities and problems of definition that arise when using such data.

Experimental and theoretical methods are used separately and together to determine the data. Experimental intensity measurements or lifetimes are interpreted with the aid of theoretically calculated Franck-Condon and Hönl-London factors, to evaluate which precise experimental wavelength data are needed. We discuss these concepts in the following sections.

3.3.1 THEORY

3.3.1.1 *Hönl-London Factors* The Hönl-London or intrinsic line-strength factor determines, apart from the relative population term, the relative intensity distribution line-by-line within a band. In a formal sense, the Hönl-London factor can be defined through

$$S_{J'J''} = \left| \int \psi_{J'\Lambda'M'} \,|\, D(\theta\kappa\phi) \,|\, \psi_{J''\Lambda''M''} \, d\tau \right|^2, \tag{28}$$

where the ψ's are molecular symmetric top rotational wavefunctions and D is the dyadic that transforms molecular coordinates to laboratory coordinates.

While these quantities are simple in concept (Nicholls 1969a, Armstrong & Nicholls 1972), there has until recently been confusion and ambiguity connected with their definition and normalization. Tatum (1967) has discussed these problems in detail. Hougen (1970) has reviewed the strict quantum mechanical bases of the factors. Kovács (1969) has also discussed them and compiled voluminous tables for a wide variety of molecular coupling cases and branches. Whiting (1972, 1973) has applied these ideas to the development of a computer program for numerical evaluation of factors for any coupling case. Whiting et al. (1973) have used this program to check the tables in Kovács' (1969) book and have published extensive corrections. Whiting & Nicholls (1974) have reviewed the complete theoretical bases of such calculations.

Hönl-London factors are usually simple quotients of polynomial functions of J and Λ. In band systems of complex rotational structure and intermediate coupling cases, it is extremely easy to make errors. The reader is referred to Whiting's papers above for a reliable assessment of the field.

3.3.1.2 *Franck-Condon Factors, r-Centroids and Molecular Potentials* Franck-Condon factors $q_{v'v''}$ and r-centroids $\bar{r}_{v'v''}$ are defined in Equations (21b) and (19b) through elementary integrals that involve vibrational wavefunctions on which they depend in a sensitive manner. Franck-Condon factors are often the only data readily available for interpretation of astronomical molecular spectra. The vibrational wavefunctions are the solution of the one-dimensional Schrödinger equation of the molecular vibrational oscillator (sometimes with rotation included) appropriate to the specific molecular potential U(r) in question. Thus the potential has to be uniquely defined. There have been many empirical analytical forms (Harmonic Oscillator, Morse, etc) of molecular potential proposed, each of which has some degree of realism (Nicholls 1969a). There have also been many calculations of "Morse" Franck-Condon factors. Empirical potentials are not always satisfactory.

In recent years therefore it has been common to use a number of methods to determine the realistic molecular potential numerically from wavelength measurements by use of the methods of Rydberg, Klein, Rees, and Dunham. The principle of the method is to determine the turning points $r_{\min}$ and $r_{\max}$ of the molecular oscillator at each vibrational level v. The method is based on a WKB approximation and has been described in some detail by Spindler (1965), Jarmain (1971, 1972), Coxon (1971), and LeRoy (1973). The great advantage (and possible limitation) of this method is that experimental measurements of energy eigenvalues are the basic input data but are not always of high quality. Jarmain (1963) has suggested two sensitive checks on the adequacy of the wavefunctions that result from numerical solution of the Schrödinger equation from these eigenvalues. They are

(*a*) to compare the input rotational constant B_v with its theoretical value $h/8\Pi^2 c \int [\psi_v(r)]^2 \, dr$;

(*b*) to check the orthogonality of the wavefunctions through the calculation of a "noise factor" $\left| \int \psi_{v_1} \psi_{v_2} \, dr \right|^2$, which should be zero.

A number of computer programs are now available that use measured eigenvalues as input data to calculate (*a*) the potentials, (*b*) the wavefunctions, and (*c*) the Franck-Condon factors and r-centroids. Such calculations are now relatively routine. The adequacy of each has to be assessed critically against the quality of the experimental energy-level data on which it is based. Reference is made in the next section to data available for astrophysically important band systems. In the author's own laboratory over 10^5 such calculations have been carried out.

Most Franck-Condon factors are calculated for the rotationless molecule. This is an idealization, for the potential U(r) in the Schrödinger equation should be augmented by a rotation energy term $(h^2/8\Pi^2 \mu r^2) J(J+1)$. While for relatively heavy (large μ) molecules under modest thermal (low J) conditions, the effects of this term are often not too severe (Hubisz 1968), for astrophysically important light molecules such as hydrides, the effect cannot be entirely neglected, as pointed out by Haycock (1963), LeRoy & Vrscay (1975), and Bell, Branch & Upson (1976).

One natural extension of the Franck-Condon factor concept is from bound-bound transitions to bound-free transitions, particularly for absorption into the photodissociation continuum. The Franck-Condon density (Jarmain & Nicholls 1964) is thereby defined for these transitions as

$$q_{v'',\nu} = \left| \int \psi_{v''} \psi_{\nu} \, dr \right|^2, \tag{29}$$

where ψ_ν is the wavefunction in the continuum.

Orthogonality considerations indicate that the sum rule

$$\sum_{v'} q_{v'v''} + \int q_{v''\nu} \, d\nu = 1 \tag{30}$$

holds for Franck-Condon factors and densities. The variation of Franck-Condon density with frequency correlates well with absorption (and photoionization) cross sections.

The profile of three-dimensional Franck-Condon factor surfaces, which represent $q_{v'v''}$ as a function of v' and v'', give a good qualitative indication of the distribution of oscillator strength over a band system. Nicholls (1964a, 1965a, 1973a) has discussed the systematics of these surfaces and their use in interpolation.

The r-centroid concept found its place in molecular intensity spectroscopy through such considerations as Equations (19a,b), (20), and (22). The r-centroid exhibits a quasi-monotonic function of wavelength or frequency that allows plots to be made as $f(\lambda)$ or $f(\nu)$ instead of $f(\bar{r})$ or cases where $\bar{r}$ is not known. This and a number of its other properties have been reviewed elsewhere (Nicholls 1965b, 1969a). Before the advent of computers a number of approximate analytic methods were used to calculate r-centroids (Nicholls & Jarmain 1956). Contemporary methods involve the direct evaluation through Equation (19b), together with a test of the method through calculation of the figure of merit

$$Y^{(n)}_{v'v''} = \frac{\overline{r^{(n)}_{v'v''}}}{(\bar{r}_{v'v''})^n} = \frac{H^{(n)}[H^{(0)}]^{n-1}}{[H^{(1)}]^n}. \tag{31}$$

In most cases ($n \leqq 3$), $Y^{(n)}$ is within a few percent of 1, which confirms the applicability of the approximation. Its mathematical bases have been studied (Drake & Nicholls 1969, Nicholls 1972, 1973b, 1976b).

Franck-Condon factors and r-centroids are related to the expectation values of r, $\langle r_v \rangle$ in a given vibrational level v through (Nicholls & Jarmain 1955)

$$\langle r_{v'} \rangle = \sum_{v''} \bar{r}_{v'v''} q_{v'v''}, \tag{32a}$$

$$\langle r_{v''} \rangle = \sum_{v'} \bar{r}_{v'v''} q_{v'v''}. \tag{32b}$$

Further, a matrix multiplication property of the vibrational overlap integrals exists through (Nicholls 1968)

$$(l, m) = \sum_n (l, n)(n, m), \tag{33a}$$

where

$$(a, b) = \int \psi_a(r)\psi_b(r)\, dr. \tag{33b}$$

3.3.1.3 *Electronic Transition Moments* The electronic transition moment defined in Equation (17a) was, until recently, nearly always inferred from spectroscopic intensity measurements. For example, the integrated intensity in emission of a band is

$$I_{v'v''} = KN_{v'} \Sigma \left| R_e^2(\bar{r}_{v'v''}) \right| \nu^4_{v'v''} q_{v'v''}. \tag{34}$$

Thus a plot of $(I\lambda^4/q)_{v'v''}$ vs $\bar{r}_{v'v''}$ for bands originating from a common upper level v' delineates the variation of electronic transition moment squared with internuclear separation. This has been the basis for experimental work in many laboratories (Nicholls 1974).

However, the reliability of ab initio calculations of transition probability data, including electronic transition moments, for spin-allowed transitions has improved greatly in the last few years. Such calculations require very high quality wavefunctions for both the excited and ground electronic states. In principle a very large number ($\sim 10^4$) of electronic configurations must be included to give correct electronic energies and wavefunctions. Das & Wahl (1972) have shown how by use of the optimized valence configuration (OVC) such calculations can be made. The OVC method uses only those configurations needed to describe the dissociation of the molecule into the correct atomic fragments, together with those that take account of most of the electronic binding (Billingsley 1975). Experience has shown that ~ 50 configurations are normally adequate to describe the principal electronic features of the molecule. E. E. Whiting and J. O. Arnold of the NASA Ames Laboratory have had excellent success in recent years in calculating electronic transition moments of band systems of such astrophysical molecules as CN, ClO, and SiO, which agree with experiment to a few percent. Specific examples of their work, and of the work of Michels, Popkie, and Henneker, Yoshimine, McLean, and Lui, and Yoshimine, Green, and Thaddeus are cited in Section 4.

3.3.2 EXPERIMENTAL METHODS The principal aims of experimental methods of determination of transition probabilities should be the provision of band strength $S_{v'v''}$ and the related quantities $f_{v'v''}$, $A_{v'v''}$. As part of this endeavour workers often determine the electronic transition moment and its variation with internuclear separation, and then infer $S_{v'v''}$ from Equation (21a). Klemsdal (1973), Hefferlin (1975, 1976), Hefferlin et al. (1976), and Kuznetsova et al. (1974) have made recent reviews of such determinations.

About 50 band systems have so far been measured. Of those that have been measured, the data from different laboratories are not always in good agreement. The principal astrophysical transitions have been measured but as is seen in Section 4 many important ones have not yet been measured. Lack of better data often forces users to employ Franck-Condon factors or approximate broad average concepts such as the electronic oscillator strength f_{UL} of a band system (Equation 26d) to interpret astrophysical spectra.

3.3.2.1 *Emission Measurements* Emission spectra are normally richer in bands than (cool) absorption spectra, which often contain bands from one or two ($v'' = 0, 1$) progressions. Emission spectra normally contain bands for many v'' ($v' =$ const) progressions, depending upon method of excitation (discharge, arc, shock tube, flame, etc).

Most emission studies use Equations (6) and (34) (or one of their equivalents) from which to infer transition probability data. It is noted that the intensity is determined by the product of population and transition probability. Different emission sources maintain different relative population distributions $N_{v'}$. Thus while intensity distributions may easily be interpreted in relative terms, absolute information on $N_{v'}$ is required before emission intensities can be interpreted in terms of absolute transition probabilities. In many conventional excitation sources quasi-

Boltzman vibrational and rotational distributions are established, in which cases the relative population $N_{v'}$, $N_{J'}$ are easy to establish. It is not easy however to establish absolute number densities. In certain thermal sources (e.g. flames, shock tubes) absolute number densities relative to those of atoms can often be inferred from thermochemical considerations. Absolute molecular intensities can then be inferred from comparison with atomic lines of known oscillator strength. Thus emission intensity measurements are best used to determine relative transition probability data for many bands, which can then be reduced to absolute terms with respect to absolute oscillator strength measurements of either one or two absorption bands (see below) or atomic lines, or by reference to lifetime measurements.

Since the advent of photoelectric spectroscopy, intensity measurements with relatively high accuracy ($\sim 10\%$) have been possible. This has been greatly assisted by the availability of the methods of computer synthesis of spectra on a line-by-line basis, for comparison with measured spectra. Transition probabilities, temperatures (populations), line shapes, and other parameters of the synthetic spectra can be adjusted until agreement between the experimental and synthetic spectra is obtained. One specific application of this approach has been the use of shock-tube radiometry in which thermal excitation produces known number densities, so that radiation measurements in known wave bands can be interpreted by use of synthetic spectra to provide transition probability data.

One limitation of emission measurements is the need for measurement of wavelength dependence of detector response. This calls for great precision in absolute photometry. Many (particularly early) measurements in the literature are in question for this reason. Others are in doubt because band head maximum intensities were used as an index of integrated band-intensity. In yet others incomplete allowance has been made for the effect of overlap of adjacent band structure or continua.

Many early measurements were made using photographic photometry, which is bedevilled with nonlinearity of response of photographic emulsions. Some of this work used "eye estimates" of band (head) intensities as cited by Pearse & Gaydon (1976). Early photographic intensity measurements are probably only useful to demonstrate qualitative trends in which intensities follow Franck-Condon factors.

The simplest intensity measurements are interpreted in terms of optically thin situations. In some sources (e.g. flames and shockwaves) optically thick circumstances are obtained and have to be treated accordingly.

3.3.2.2 *Absorption Measurements* The principal advantage of absorption measurements for the determination of molecular transition probability parameters is that the ratio (I/I_0) of emergent (I) and incident (I_0) intensities at the absorption cell, rather than intensity alone, is interpreted, under optically thin conditions, at each frequency in terms of the product $N_L\alpha$ through Equations (8) and (9a). Alternatively, the equivalent width $W_\nu = \int (I_0 - I)/I_0 \, d\nu$ is interpreted under the same conditions as $N_L f$ through Equation (11a). Thus, the response does not have to be known absolutely, provided that the detector has a linear response at each wavelength. The method calls for access to a relatively flat continuum light source and a homogenous slab of absorbing gas flow whose thermal and pressure conditions are well enough known to determine N_L with some precision. Thus in principle absolute values of absorption cross sections α may be directly determined from I/I_0 for those transitions that are recorded.

Often a room temperature absorption cell is used, in which case N_L refers to the population of molecules in $v'' = 0$. Thus the absorption spectrum of only the $v'' = 0$ progression of bands is recorded. Even for higher temperature gases, very seldom are more than a few vibrational levels, e.g. $v'' = 0, 1, 2$, sufficiently excited to allow progressions of bands from them to appear in an absorption spectrum. Thus one disadvantage of the method is that relatively few of the bands are recorded by absorption methods. Emission methods that excite far more bands can be used for relative measurements on many bands. Absorption measurements can be used for absolute measurements on a few bands. If circumstances are such that some bands are common to each method, absorption measurements can be used to place relative emission data on an absolute scale.

One important modification of the absorption method is to make the absorption tube one arm of a Mach-Zehnder interferometer (Ladenburg & Bershader 1954), which is crossed with the spectrograph. In this way the dispersive qualities of the gas are indicated on the spectrum by fringes, oblique to the direction of the dispersion, which trace out the anomalous dispersion properties of the gas at each absorption feature where the fringe goes through a complete undulation cycle about the absorption feature. The positive and negative loops of the undulation are called "hooks," the square of the wavelength separation between which is proportional to the oscillator strength of the feature. This is the basis of the so-called "hook method" of oscillator strength measurement of Rozhdestvenskii (1912) [see Marlow (1967) and Huber (1971)], which has long been used to measure atomic oscillator strengths. In recent years it has been applied to the measurement of oscillator strengths of molecular features and dispersive properties of molecular gases, first by Pery-Thorne and her collaborators at Imperial College, by the author and his colleagues at York University, and in the vacuum *UV* region of the spectrum by Parkinson and his colleagues at the Center for Astrophysics at Harvard. Reference to such measurements are made as appropriate in Section 4.

3.3.2.3 *Lifetime Measurements* It has been pointed out above that Equations (4) and (27) can in principle be used to interpret molecular lifetime measurements in terms of transition probability data provided that: (*a*) decay from the upper state is not compensated for in any way by cascading from higher states; (*b*) decay from the upper state occurs solely by radiative channels, unaugmented by nonradiative processes; and (*c*) one of the radiative channels is dominant, and it is for that one that transition probability data are inferred.

Some workers make direct applications of this method by measuring the lifetime for decay of radiation from pulsed discharges (Jeunehomme 1965 and references therein) or following pulsed laser pumping (Johnson, Capelle & Broida 1972). Others have measured the rate of decay of a spectral feature along the length of flowing molecular luminosity of known flow speed. Others have measured lifetimes by use of the methods of delayed coincidences, for example between a pulsed exciting electron beam and the resulting photons (Heron et al. 1954, 1956, Brannen et al. 1955, Bennett & Dalby 1959, 1960a,b, 1962, 1964, Schwenker 1965). Yet others, using the somewhat similar approach of the phase shift method, infer lifetimes from the phase shift between a sinusoidally modulated electron exciting beam and the resulting photon signal (Demtroder 1962, Fink & Welge 1964, 1968, Lawrence 1965, Hesser & Dressler 1966, Smith 1970a).

4 AVAILABLE DIATOMIC MOLECULAR TRANSITION PROBABILITY DATA FOR ASTRONOMICAL MOLECULES

Table 2 displays the principal diatomic and triatomic species of astrophysical interest and their common astronomical locations. This table was compiled from the sources cited in Section 2 and various reports at the IAU Grenoble Congress (1976). The species displayed have been variously observed at optical (*UV*, visible, *IR*) and radio frequencies. Because of space limitations of this review, this section is limited solely to optical transitions of the diatomic species of Table 2. Data relating to radio frequency transitions of astrophysically important molecular species have been the subjects of a number of recent reviews in the *Journal of Chemical and Physical Reference Data,* from staff of The National Bureau of Standards. References to these works are given in the recent report (Nicholls 1976a) compiled for Commission 14 of the IAU for the 16th General Assembly in Grenoble.

The specific transition by which a molecule manifests itself depends upon the molecule and the observing circumstances, location (ground based, or from a space platform), and equipment. In view of the potential manifestation of a species from observations on more than one of its characteristic transitions, in this section, as far as is possible, a review is made of all available transition probability data for each astronomical species.

In the following section, which contains entries for each of the diatomic species of Table 2, references are made to such available data as band strength $S_{v'v''}$, band oscillator strength $f_{v'v''}$, absorption coefficients, electronic transition moments, Hönl-London factors, Franck-Condon factors, and *r*-centroids. The Hönl-London factors can nearly always be obtained from Kovács (1969), Whiting (1972, 1973), Whiting et al. (1973), and Whiting & Nicholls (1974) and will not be cited in each case below.

AlH RKR Franck-Condon factors have been published for a few bands of the ($A^2\Pi - X^2\Pi$) system by Huron (1969).

AlO The main optical transitions of the molecule and data pertaining to them were reviewed by Tyte & Nicholls (1964a), Rosen (1970), and Suchard (1975a). The blue-green band system is the principal one for astronomical applications. It was designated $A^2\Sigma - X^2\Sigma$ until a low-lying $^2\Pi$ state (now designated $A^2\Pi$) was discussed by McDonald & Innes (1969). The blue-green system is now designated $B^2\Sigma - X^2\Sigma$.

Relative intensities of bands of the system were measured by Hébert & Tyte (1964) and Tyte & Hébert (1964) using a variety of light sources (including shock excitation) assumed to be optically thin. Linton & Nicholls (1969a) used shock-tube spectroscopy, interpreted by realistic line-by-line computer synthesis of spectra to take account of optical depths of lines, to measure relative band strengths.

Van Pee, Kineyko & Caruso (1970) report absolute intensity measurements and oscillator strength measurements of the band system excited in an O_2–C_2N_2 flame in which $(CH_3)_3Al$ was injected. No account was taken of the $A^2\Pi$ state. Johnson, Capelle & Broida (1972), using laser fluorescence methods, measured the lifetimes of the $v' = 0, 1, 2$ levels of $B^2\Sigma$ from which oscillator strengths of the (0, 0), (1, 0), and (2, 0) bands were inferred. These oscillator strengths are $\sim 10^{-2}$ and are an order of magnitude greater than those of Van Pee et al.

Michels (1972) made ab initio quantal calculations of the oscillator strengths in good agreement with the measurements of Johnson, Capelle & Broida (1972) and with those of Hébert & Tyte (1964). Yoshimine, McLean & Lui (1973) also made less accurate ab initio calculations that are somewhat larger than those of Michels. Hébert, Nicholls & Linton (1977) have recently placed the measurements of Linton and Nicholls on an absolute basis. Numerous workers have calculated Franck-Condon factors for this system using Morse wavefunctions (e.g. Nicholls 1962a, Linton & Nicholls 1969a) and wavefunctions derived from RKR potentials [Sharma (1967), McCallum, Nicholls & Jarmain (1970), Michels (1972), Liszt & Smith (1972)]. In general there is not a great variation between the Morse and the RKR Franck-Condon factors for observed bands of the blue-green system.

Sulzmann (1973, 1975) has provided relative and absolute data on the oscillator strengths of the infrared vibration-rotation bands of the $X^2\Sigma$ ground state.

Apart from the ab initio calculation (Yoshimine, McLean & Lui 1973) of band strengths of the infrared vibration-rotation bands of the $X^2\Sigma$ state, the $A^2\Pi$ state, and the $B^2\Sigma$ state and of the $A^2\Pi - X^2\Sigma$ band system no other reliable data on AlO exist.

BH Smith (1971) measured the lifetime of the $v' = 0$ level of the (principal) $A^2\Pi - X^2\Sigma$ transition from which an oscillator strength f_{00} equal to 3.5×10^{-2} was inferred. No laboratory intensity measurements appear to have been made. Nicholls, Fraser & Jarmain (1959) report Morse Franck-Condon factors for the $B^1\Sigma - A^1\Pi$ band system. Liszt & Smith (1972) report RKR Franck-Condon factors (~ 0.99) for the (0, 0), (1, 1), (2, 2) of the $A^1\Pi - X^1\Sigma$ transition.

BO Relative intensity measurements were made on the $\alpha(A^2\Pi - X^2\Sigma)$ and $\beta(B^2\Pi - X^2\Sigma)$ systems by Robinson & Nicholls (1960) and were interpreted in terms of electronic transition moment and band strengths. Morse Franck-Condon factors were provided for the α, β and intercombination $(B^2\Pi - A^2\Pi)$ systems by Nicholls, Fraser & Jarmain (1959) and Nicholls et al. (1960) and by Singh & Rai (1965). RKR Franck-Condon factors and r-centroids were calculated by McCallum, Nicholls & Jarmain (1970) and by Liszt & Smith (1971). Popkie & Henneker (1971) performed ab initio quantum calculations on the α-system. Average electronic transition moment and absorption oscillator strengths were recorded.

CaO No reliable intensity measurements appear to have been made. Morse Franck-Condon factors have been calculated by Mon (1966) for the infrared $(A^1\Sigma - X^1\Sigma)$ system. RKR Franck-Condon factors and r-centroids were calculated by Liszt & Smith (1971) for the blue $(B^1\Pi - X^1\Sigma)$ and $UV(C^1\Sigma - X^1\Sigma)$ systems.

ClO Absolute intensity measurements in absorption have been made by Mandelman & Nicholls (1977) on the $A^2\Pi - X^2\Pi$ band system, and in emission, using shock-tube radiometry by Cooper (1977). RKR Franck-Condon factors and r-centroids have been calculated by Coxon (1976) and by Danylewych, Mandelman & Nicholls (1976). Arnold, Whiting & Langhoff (1976) report calculations of an ab initio $\Sigma |R_e/a_0 e|^2$ of 1.75 in comparison to a measured value of 1.58 (Cooper 1977).

CrO Morse Franck-Condon factors have been published by Murthy & Nagaraj (1962).

C_2 Herzberg et al. (1969) recommend the new nomenclature now adopted for the electronic states of C_2

Cooper (1974) and Cooper & Nicholls (1975a, 1975b, 1976) report absolute transition probability data for seven band systems of C_2, namely the Ballik-Ramsay $(b^3\Sigma_g^- - a^3\Pi_u)$; the Swan $(d^3\Pi_g - a^3\Pi_u)$; the Fox-Herzberg $(e^3\Pi_g - a^3\Pi_u)$; the Phillips $(A^1\Pi_u - X^1\Sigma_g^+)$; the Deslandres d'Azambuja $(C^1\Pi_g - A^1\Pi_u)$; the Mulliken $(D^1\Sigma_u^+ - X^1\Sigma_g^+)$ and the Freymark $(E^1\Sigma_g^+ - A^1\Pi_u)$ systems. No data are available for the Messerle-Krauss $(C'^1\Pi - A^1\Pi_u)$ system. They used the method of shock tube radiometry and interpreted intensity measurements in terms of $\Sigma |R_e^2(\bar{r}_{v'v''})|$, $S_{v'v''}$, $A_{v'v''}$, and $f_{v'v''}$. Realistic RKR Franck-Condon factors of McCallum, Nicholls & Jarmain (1970) and Hönl-London factors of Schadee (1964) were used. Arnold (1968) used a similar method to study the C_2 Swan system. Danylewych (1971) and Danylewych & Nicholls (1974a,b) made extensive photoelectric relative intensity measurements in emission on many bands of the C_2 Swan system and interpreted them by use of computer synthesis of realistic spectra to provide tables of $f_{v'v''}$, $S_{v'v''}$, and $A_{v'v''}$. The work was placed on an absolute basis by comparison with an absolute value of f_{00} derived from the phase shift lifetime measurements of Fink & Welge (1967) and Smith (1969a). Smith (1969a) also reports f-values for the Mulliken system of C_2 derived from his lifetime measurement.

Roux, Cerny & D'Incan (1976) report oscillator strengths for the Phillips system from intensity measurements in the infrared on an oxyacetylene flame. Grevesse & Sauval (1973) have made assessments of laboratory oscillator strength data for the (0, 0) band of the C_2 Swan system, and of other astrophysically important systems, for comparison with oscillator strengths implied from equivalent width measurements on solar molecular lines.

CH Bennett & Dalby (1960b), Fink & Welge (1967), and Hesser & Lutz (1970) have measured vibrational lifetimes of the $B^2\Sigma^-$ state of CH and from them have inferred oscillator strengths for a few bands. Elander & Smith (1973a) and Brooks & Smith (1974) have made high-resolution lifetime measurements on individual rotational lines of bands of the $C^2\Sigma^+$ and $B^2\Sigma^-$ states, from which inferences on predissociation are made. Clerc & Schmidt (1972) have measured radiative lifetimes for the $A^2\Delta$ state and Anderson, Peacher & Wilcox (1975) have made similar measurements on the $B^2\Sigma^-$ state. McCallum, Nicholls & Jarmain (1970) and Liszt & Smith (1972) provide RKR Franck-Condon factors for band systems of CH.

Grevesse & Sauval (1973) made a critical review of laboratory measurements of f_{00} for the $(A^1\Delta - X^1\Pi)$ and $(B^2\Sigma^- - X^2\Pi)$ band systems to compare with their measurements of equivalent width measurement on lines in the solar photospheric spectrum.

CH^+ Hesser & Lutz (1970) and Smith (1971) measured lifetimes of vibrational levels of the $A^1\Pi$ state of CH^+ from which f_{00} values for the $(A\text{-}X)$ system were inferred. Brooks & Smith (1975) and Anderson, Peacher & Wilcox (1975) made similar measurements in individual lines of $(A–X)$ and $(B–X)$ bands. Yoshimine, Green & Thadeus (1973) made ab initio quantum calculations of f_{00} for the $A^1\Pi - X^1\Sigma^+$ system. Liszt & Smith (1972) have calculated RKR Franck-Condon factors for CH^+ systems. Green, Hornstein & Bender (1973), using ab initio calculation of potentials for the A and X state, also calculated RKR Franck-Condon factors that were not in agreement with those of Liszt and Smith.

CN The CN Red $(A^2\Pi - X^2\Sigma^+)$ and Violet $(B^2\Sigma - X^2\Sigma)$ band systems are of great astrophysical interest and many data related to them are collected in the Identification Atlases of Brocklehurst et al. (1971, 1972).

Dixon & Nicholls (1958) made relative emission intensity measurements across the whole CN Red system, from which the approximate constancy of the electronic transition moment was inferred and from which other relative transition probability data could be derived. Wentink, Isaacson & Morreal (1964) and Jeunehomme (1965) made lifetime measurements on levels of the $A^2\Pi$ state using the laser blow-off and decay of discharge methods respectively, which were not in good agreement. Arnold (1972) and Arnold & Nicholls (1972, 1973a) used an absolute shock-tube radiometry method to determine the electronic transition moment of the system. They confirmed that it was essentially constant and provided a table of data for $S_{v'v''}$, $f_{v'v''}$, and $A_{v'v''}$ for the principal bands of the system. They also made a critical review of other measurements. Danylewych & Nicholls (1977) have completed high-resolution emission intensity measurements on the CN Red system, using the synthetic spectrum method to interpret the results in terms of transition moments $f_{v'v''}$, $A_{v'v''}$, and $S_{v'v''}$.

Intensities of the CN Violet system have been studied for many years. Ornstein & Brinkmann (1931) made relative photographic measurements that were interpreted in terms of transition probability data by Nicholls (1956). White (1938, 1940) measured oscillator strengths and lifetimes optically. The first accurate lifetimes were measured by Bennett & Dalby (1962), followed by Moore & Robinson (1968) and Liszt & Hesser (1970). Luk & Bersohn's (1973) lifetime measurements agree with those of Liszt & Hesser (1970) and with laser fluorescence measurements of lifetime by Jackson (1974). However, the measurements of the lifetimes of the

$A^2\Pi(v = 0)$ and $B^2\Sigma(v = 0)$ states of CN from an analysis of the zero electric field linewidth of the level anti-crossing spectrum by Cook & Levy (1972) appear to be discrepant with other measurements. A number of shock-tube radiometer measurements of oscillator strength of the CN Violet system have been made, the last of which is that of Arnold & Nicholls (1973a,b), who also give an extensive comparative review of past measurements and report measurements on the dissociation energy of the ground ($X^2\Sigma$) state.

Engelman & Rouse (1975) report a recent high-resolution measurement of f_{00} for CN Violet using absorption measurements in thermally dissociated C_2N_2. Their measurements, which like those of shock-tube work depend on thermochemical constants (which themselves are not always well known), are a factor of 4.8 less than the average of other measurements.

Danylewych & Nicholls (1977) have completed high-resolution emission intensity measurements on the CN Violet system, using the synthetic spectrum method to interpret the results in terms of transition moments, $f_{v'v''}$, $A_{v'v''}$, and $S_{v'v''}$.

Morse Franck-Condon factors for CN systems have been available for many years (Nicholls, Fraser & Jarmain 1959, Nicholls 1964b, Halmann & Laulicht 1966), and realistic RKR values have also been available for some time (Spindler 1965, McCallum, Nicholls & Jarmain 1970, Schoonveld 1972). Grevesse & Sauval (1973) review oscillator strength data for the CN Red and Violet systems for comparison with their equivalent width measurements on CN lines in the solar spectrum. Whiting et al. (1976) report calculations of an ab initio value of $\Sigma|R_e|^2$ (in $e^2a_0^2$ units) of 0.89 in comparison with a measured value of 0.90 (Arnold & Nicholls 1973b).

CO Carbon monoxide is one of the most widely studied diatomic molecules and is one of the richest in electronic states and band systems. Thus, in this review, great selectivity of citation of references has been followed because of space limitations. Many of the data related to its spectra have been compiled by Krupenie (1966). Suchard (1975a) also provides an extensive reference list. The three spectral features of greatest astrophysical interest relate to the solar spectrum.

They are the vibration-rotation spectrum of the ground ($X^1\Sigma$) state (Goldberg et al. 1952a,b, Goldberg 1955), the Fourth Positive band system ($A^1\Pi - X^1\Sigma$) (Goldberg, Parkinson & Reeves 1965), and possibly the Cameron ($a^3\Pi - X^1\Sigma$) system, all of which can be detected in absorption from $X^1\Sigma$.

Laboratory measurements of oscillator strengths of the vibration-rotation spectrum have been measured by Penner & Weber (1951a,b,c,d), Roux, Effantin & D'Incan (1972), Carance & Verges (1975), and Moskalenko (1975). A theory of vibrational intensities in diatomic infrared transitions with application to CO has been developed by Herman & Shuler (1953, 1954) and extended by Herman, Rothery & Rubin (1958).

Relative intensity measurements were made on the CO Angstrom and Third Positive systems by Robinson & Nicholls (1958). The intensity distribution of the Fourth Positive system was discussed by Nicholls (1960) and a constant electronic transition moment inferred. McEwen (1965) measured the relative intensity distribution of the Fourth Positive system in emission, and James (1971a) and Rich (1968), using shock-tube methods, measured absolute intensities and oscillator strengths of the Fourth Positive bands. Pilling, Bass & Braun (1971), Meyer & Lassettre (1971), and Lassettre & Skerbele (1971) made oscillator strength measurements on Fourth Positive bands.

Absolute oscillator measurements have been made on the CO Cameron system by Fairbairn (1970), Hasson & Nicholls (1971), James (1971a,b), and Lawrence & Seitel (1973). Slanger & Black (1972, 1976) have measured relative transition moments for the $(d^3\Sigma - a^3\Pi)$ and $(e^3\Sigma - a^3\Pi)$ systems.

Numerous measurements have been made of lifetimes of CO states, typical of which are those of Bennett & Dalby (1960a), Hesser (1968), and Fink & Welge (1968). Measurements on $A^1\Pi$ lifetimes have been made by Wells & Gissler (1970), Imhof & Read (1971a), and Chervenak & Anderson (1971); on $B^1\Sigma^+$ lifetimes by Rogers & Anderson (1970b); on $C^1\Sigma$ lifetimes by Le Calve et al. (1970); on $a^3\Pi$ lifetimes by Slanger & Black (1971), Borst & Zipf (1971), Johnson & Van Dyck (1972), and Wicke & Tempra (1975); and $b^3\Sigma^+$ lifetimes by Rogers & Anderson (1970a) and Smith, Imhoff & Read (1973).

A good review of vacuum *UV* absorption cross section of CO has been given by Hudson (1971). Recent measurements have been made in this region by Lee, Carlson & Judge (1975), Watson (1975), and Davenport (1976).

Realistic RKR Franck-Condon factors and *r*-centroids (McCallum, Nicholls & Jarmain 1970) and Morse Franck-Condon factors and *r*-centroids (Nicholls 1962b) are available for many CO band systems (Krupenie 1966).

CO^+ The principal band systems of CO^+ are the First Negative $(B^2\Sigma^+ - X^2\Sigma^+)$, Comet-Tail $(A^2\Pi - X^2\Sigma^+)$ and Baldet-Johnson $(B^2\Sigma - A^2\Pi)$. The molecule and its band systems have many similarities with N_2^+.

Relative emission intensity measurements were made in the Comet-Tail system by Robinson & Nicholls (1960). The transition moments of all three systems have been studied by Judge & Lee (1972). Electronic transition moments of the Comet-Tail system have been studied by Maier & Holland (1972) and by Jain (1972). Isaacson, Marram & Wentink (1967) discuss the electronic transition moment of the First Negative system.

Lifetime measurements on vibrational levels of the *A* and *B* states from which oscillator strengths have been inferred have been made by Hesser (1968), Fink & Welge (1968), Fowler (1969), Anderson, Sutherland & Frey (1972) and Smith, Read & Imhoff (1975). Realistic RKR Franck-Condon factors and *r*-centroids (McCallum, Nicholls & Jarmain 1970) and Morse Franck-Condon factors and *r*-centroids (Nicholls 1962c) are available for the three band systems.

CS Smith (1969b) and Silvers, Bergman & Klemperer (1970) have measured lifetimes of the $A^1\Pi$ and $a^3\Pi$ states and have inferred oscillator strengths from them for the $A^1\Pi - X^1\Sigma^+$ and $a^3\Pi - X^1\Sigma^+$ transitions.

Felenbok (1965) has calculated Franck-Condon factors and *r*-centroids for the $A^1\Pi - X^1\Sigma^+$ system.

H_2 Hydrogen exhibits many band systems: 31 singlet systems and 18 singlet systems. Of these only the seven that connect with the ground ($X^1\Sigma_g^+$) state and lie in the vacuum *UV* are astrophysically important. The two most important systems of these are the Lyman ($B^1\Sigma_u^+ - X^1\Sigma_g^+$) and Werner ($C^1\Pi - X^1\Sigma_g^+$) systems. Absorption oscillator strengths for these systems have been measured by Haddad et al. (1968), Fabian & Lewis (1974) and Lewis (1974a,b).

The electronic transition moment of the Lyman band system has been measured by Browne (1969) and Marchetti & LaPaglia (1971). Allison & Dalgarno (1969, 1970) have calculated band oscillator strengths and transition probabilities for Lyman and Werner bands. Lifetimes have been measured for the $C^3\Pi_u$ state (Johnson 1972), for the $a^3\Sigma_g^+$ state (Imhoff & Read 1971b), and for the $a^3\Sigma_g^+$ state by Smith & Chevalier (1972). Franck-Condon factors have been calculated by Nicholls (1965c) and Spindler (1969a,b,c).

HD Allison & Dalgarno (1970) have calculated band oscillator strengths and transition probabilities for Lyman and Werner bands. McKellar (1973) has measured intensities of the (3, 0) and (4, 0) vibration-rotation bands.

LaO Of the five band systems that connect with the ground state, two, the Red ($A^2\Pi - X^2\Sigma^+$) and the Green ($B^2\Sigma^+ - X^2\Sigma^+$), are the most important in stellar spectra.

Murthy & Murthy (1969, 1972) have measured relative intensities of 40 bands of the Green system by photographic photometry, interpreted the integrated intensities in terms of variation of the electronic transition moment with internuclear separation, and thereby derived a table of relative band strengths. They also calculated a set of Morse Franck-Condon factors and *r*-centroids, as did Ortenberg, Glasko & Dimitriev (1964).

Dongre (1974) has also evaluated electronic transition moments for this system, and Maranan & Suarez (1974) have calculated Franck-Condon factors and *r*-centroids for the Green System of $La^{16}O$ and $La^{18}O$, in which account is taken of vibration-rotation interaction.

MgF No experimental intensity measurements are available. Rao & Laksham (1970a,b), Singh, Shukla & Maheshwari (1969), and Maheshwari, Singh & Shukla (1968) have published tables of Franck-Condon factors and *r*-centroids for the $A^2\Pi - X^2\Sigma^+$, $B^2\Sigma \rightarrow X^2\Sigma^+$, and $C^2\Sigma - X^2\Sigma$ band systems.

MgH McGregor (1975, 1977) has used shock-tube radiometry to determine the constancy of the electronic transition moment and absolute oscillator strength values for the (0, 0), (1, 1), (2, 2), and (3, 3) bands of the $A^2\Pi - X^2\Sigma$ system of MgH as 0.0037, 0.0033, 0.0029, and 0.0025 respectively. He also calculated Morse Franck-Condon factors that are in good agreement with those of McCallum, Nicholls & Jarmain (1970) and of Popkie (1971). Schadee (1964) and Mallia (1968) inferred f_{00} from solar spectra, based on values of the dissociation energy that have now been corrected. Main, Carlson, and Dupuis made estimates of ($f_{00}+f_{11}+f_{22}$) from measurements on a rocket exhaust. Their assumed dissociation energy values have now been corrected.

MgO The two band systems of astronomical interest are the Green ($B^1\Sigma^+ - X^1\Sigma^+$) and Violet ($E^1\Sigma - X^1\Sigma$) systems. Main, Carlson & Dupuis (1967) report oscillator strength measurements on these systems.

Main & Schadee (1969) report oscillator strengths of these systems. Franck-Condon factors have been calculated by McCallum, Nicholls & Jarmain (1970) and Nicholls (1962a).

$N_2, N_2^+, NO, NO^+, O_2, O_2^+$ The emission and absorption spectra of these molecules are of great importance and can be applied to such atmospheric phenomena as the aurora, the airglow, and atmospheric absorption. Many band systems are involved and an assessment of their transition probability parameters has been the subject of a number of reviews related to aeronomy. In view of the length limitations of this article only references to these books and reviews are given here: Chamberlain (1961), Goody (1964), Nicholls (1964c, 1969b), Hudson (1971, 1974), Banks & Kockarts (1973), and Jones (1974).

NH Lifetimes and transition probabilities of the $A^3\Pi$, $c^1\Pi$, and $d^1\Sigma$ states of NH have been measured by Smith (1969c) and Smith & Liszt (1971), who also publish Franck-Condon factors. Oscillator strengths have been measured by Harrington, Modica & Libby (1966). Lifetimes of levels of the $b^1\Sigma^+$ state have been measured by Gelernt, Filseth & Carrington (1975). Smith, Brzozowski & Erman (1976) made high-resolution lifetime studies on lines originating on $A^3\Pi(v=0,1)$ and $c^1\Pi(v=0,1)$.

OH This molecule has been studied for many years, and the spectrophotometric atlas of Bass & Broida (1953) and the analysis by Dieke & Crosswhite (1962) are most helpful aids for work with the $A^2\Sigma^+ - X^2\Pi_i$ Violet system. Early intensity measurements by Dieke & Crosswhite (1949) were interpreted by Nicholls (1956, 1964d) in terms of interim absolute transition probability data. Learner (1962) also measured intensities and took account of the vibration-rotation interaction. Crosley & Lengel (1975) have discussed relative transition probabilities and the electronic transition moment.

Oscillator strengths of bands of the Violet system have been measured by Oldenberg & Rieke (1938, 1939), Dyne (1958), Carrington (1959), and Lapp (1961); by Anketell (1967), Anketell & Learner (1967a,b), and Anketell & Pery-Thorne (1967); and by Rouse & Engelman (1973).

Lifetime measurements have been made for levels of the $A^2\Pi$ state and other states. They include Bennett & Dalby (1964), Smith (1970b), De Zafra, Marshall & Metcalf (1971), Elmergreen & Smith (1972), Sutherland & Anderson (1973), Becker & Haaks (1973), Becker, Haaks & Tatarczyk (1974), Hoggan & Davis (1974), Brophy, Silver & Kinsey (1974), German (1975), Smith & Stella (1975), and Wilcox, Anderson & Peacher (1975).

Morse Franck-Condon factors were calculated by Nicholls, Fraser & Jarmain (1959). The effect of vibration-rotation interaction on these was studied by Learner (1962) and by Hubisz (1968).

The infrared vibration-rotation spectrum of OH is of some importance. Garvin (1959), Garvin, Broida & Kostkowski (1960, 1962), D'Incan, Effantin & Roux (1971), and Roux, D'Incan & Cerny (1973) have made intensity measurements on these bands from which oscillator strengths have been inferred.

Heaps & Herzberg (1952), Herman & Shuler (1953, 1954), and Herman, Rothery & Rubin (1958) have calculated band strengths for the vibration-rotation spectrum.

OH^+ Brzozowski et al. (1974) have measured lifetimes of OH^+ states.

SO Hébert & Hodder (1974) made relative photoelectric intensity measurements on 28 bands of the $B^3\Sigma^- - X^3\Sigma^-$ region and determined the variation of electronic transition moment with internuclear separation. They placed their results on an absolute basis with respect to the lifetime measurements of Smith (1969b), and have published a table of $S_{v'v''}$, $A_{v'v''}$ and $f_{v'v''}$, $q_{v'v''}$. They used the realistic Franck-Condon factors of McCallum, Nicholls & Jarmain (1970).

Smith & Liszt (1971) published Franck-Condon factors and oscillator strengths for the (A–X) system as well as the lifetimes of $A^3\Pi$ levels, oscillator strengths, and Franck-Condon factors for the $A^3\Pi - X^3\Sigma^-$ band systems (Smith 1972).

ScO Shin (1976) and Shin & Nicholls (1977), using absolute shock-tube radiometry, have measured the variation of electronic transition moment with internuclear separation and have produced tables of $S_{v'v''}$, $A_{v'v''}$, $f_{v'v''}$, and $q_{v'v''}$ for the Blue-Green system ($B^2\Sigma - X^2\Sigma$). They used the realistic Franck-Condon factors of McCallum, Nicholls & Jarmain (1970). Ortenberg, Glasko & Dimitriev (1964) published Morse Franck-Condon factors for the ($A^2\Pi - X^2\Sigma$) and ($B^2\Sigma - X^2\Sigma$) systems.

SiF Kuz'menko, Smirnov & Kuzyakov (1970a,b) have measured intensities of the $A^2\Sigma^+ - X^2\Pi$ and $B^2\Sigma^+ - X^2\Pi$ band systems and thereby inferred the variation of electronic transition moment with internuclear separation. Wentink & Spindler (1970) published tables of realistic Franck-Condon factors and r-centroids for a number of systems of SiF as have Singh & Maheshware (1969), Mohanty & Singh (1969), and Sankaranayanan & Narayanan (1966).

SiH Radiative lifetimes of the $A^2\Delta$ state were measured by Smith (1969c). Franck-Condon factors and $A_{v'v''}$ values for the ($A^2\Delta - X^2\Pi$) system were published by Liszt & Smith (1971). Grevesse & Sauval (1971) have inferred oscillator f_{00} values for this system from measurements of equivalent widths of solar lines. Schadee (1964) and Lambert & Mallia (1970) have made similar inferences.

SiH^+ Grevesse & Sauval (1971) have inferred f_{00} and f_{01} values for the ($A^1\Pi - X^1\Sigma^+$) system from equivalent width measurements on solar lines. Liszt & Smith (1972) have calculated realistic Franck-Condon factors for the system.

SiN Stevens & Ferguson (1963) have measured relative band intensities for the $B^2\Sigma - X^2\Sigma$ system and have inferred electronic transition moment variation and relative transition probabilities from their measurements. They provide Morse Franck-Condon factors and r-centroids. Gohel & Shah (1975) have calculated realistic RKR Franck-Condon factors and r-centroids.

SiO Elander & Smith (1973b) measured radiative lifetimes of a number of levels of the $E^1\Sigma^+$ state from which $f_{v'0}$, $A_{v'0}$, and $R_e(\bar{r}_{v'0})$ were derived. They also provide Franck-Condon factors for bands of the $v'' = 0$ progression. Langhoff & Arnold (1977), using ab initio quantum methods, evaluated $\Sigma |R_e/a_0 e|^2$ as 0.75 compared to Elander and Smith's measurement of 0.65.

Smith & Liszt (1972) measured the radiative lifetimes of a number of levels of the $A^1\Pi$ state and thereby calculate oscillator strengths for bands of the $v' = 0$ progression of the $A^1\Pi - X^1\Sigma$ system. Liszt & Smith (1972) have calculated realistic RKR Franck-Condon factors for the $A^1\Pi - X^1\Sigma$ system.

Ruskin (1970), Czernichowski & Zyrnicki (1970) and Main, Marsell & Hooker (1968) report transition probability measurements for bands of the $A^1\Sigma - X^1\Sigma$ system.

Nicholls (1962a) and McGregor, Nicholls & Jarmain (1961) report Morse Franck-Condon factors for the $A^1\Pi - X^1\Sigma$ system. Hedelund & Lambert (1972) report transition probabilities for vibration-rotation bands of the $X^1\Sigma$ state.

TiO Zyrnicki (1975) has reported transition probabilities for the (0, 0) and (1, 1) bands of the $\beta(c'\phi - a'\Delta)$ system. Price, Sulzmann & Penner (1971, 1974) have measured band oscillator strengths for the $\alpha(C^3\Delta - X^3\Delta)$ and $\gamma(A^3\phi - X^3\Delta)$ systems. Fairbairn, Wolnik & Berthel (1974) have measured oscillator strengths for the $\alpha(C^3\Delta - X^3\Delta)$ system. Dube (1972) has reported on Einstein coefficients and oscillator strengths for the $\beta(c'\phi - a'\Delta)$ system. Linton & Nicholls (1969b, 1970) have measured intensities of the α and β systems and report Franck-Condon factors for the β system. McCallum, Nicholls & Jarmain (1970), Ortenberg & Glasko (1962) and Fraser, Jarmain & Nicholls (1953) report Franck-Condon factors for TiO band systems.

VO Harrington (1970) made absolute intensity measurements on the VO ($A^4\Sigma^- - X^4\Sigma^-$) Yellow Green system using the methods of shock-tube radiometry. This was reported on by Nicholls (1971). This work is summarized in the identification atlas of Harrington et al. (1970). McCallum, Nicholls & Jarmain (1970) have calculated realistic RKR Franck-Condon factors and r-centroids. Ortenberg & Glasko (1962) have calculated Morse Franck-Condon factors and r-centroids for VO band systems.

YO Dube (1972) and Dube, Rai & Singh (1972) have studied the variation of electronic transition moment with internuclear separation, and report $S_{v'v''}$, $f_{v'v''}$, and $A_{v'v''}$ for bands of the $B^2\Sigma - X^2\Sigma$ system. Murthy & Murthy (1967) and Ortenberg, Glasco & Dimitriev (1964) have calculated Franck-Condon factors and r-centroids for the $B^2\Sigma - X^2\Sigma$ and $A^2\Pi - X^2\Sigma^+$ band systems.

ZrO Realistic RKR Franck-Condon factors and *r*-centroids for ZrO band systems have been calculated by McCallum, Nicholls & Jarmain (1970) and Liszt & Smith (1971). Morse Franck-Condon factors and *r*-centroids for ZrO band systems have been calculated by Nicholls & Tyte (1967) and Singh & Pathak (1966, 1967).

5 CONCLUSION

Comparison between the list of astrophysical molecules of Table 2 and the individual entries of Section 4 indicates, with the exception of Hönl-London factors, complete lack of data for CuH, FeH, GeH, SnF, TiH, ZrH, and ZrF. In some other cases those transition probability data that are available are sparse and of variable quality. ZrO is a striking example of a most important astrophysical molecule for which the only intensity data that appear to exist are Morse Franck-Condon factors.

Nevertheless, it is clear from Section 4 that a very large number of varied data do exist, too large to be considered individually or critically in a review of this length. Workers interested in using specific data to which reference is made in Section 4 should assess their usefulness and probable reliability within the context of the comments of Section 3. A critical review of the data situation for a specific molecule or group of related molecules could easily occupy a separate review article. This is particularly the case of the atmospheric molecules O_2, O_2^+, N_2, N_2^+, NO, and NO^+, for which a number of reviews have been made for workers in aeronomy. The data available for any individual molecule can change markedly over a period of a few years and thus regular assessments of transition probabilities available for astrophysical molecules should be made.

Length limitations have precluded a review of transition probabilities of the small polyatomic molecules of Table 2. In general little is known of their emission spectra, and the absorption coefficients are the data most often needed. Hudson (1971, 1974) and Lee et al. (1973) have provided absorption coefficient data for some of them.

ACKNOWLEDGMENTS

It is a great pleasure to acknowledge valuable advice and comments received during the preparation of this review from Dr. J. B. Tatum of the University of Victoria, Drs. E. E. Whiting and J. O. Arnold (NASA Ames Research Center), Dr. E. M. Reeves (Center for Astrophysics, Harvard College Observatory), Dr. R. Hefferlin (Southern Missionary College), and Dr. W. H. Smith (Princeton University Observatory).

Literature Cited

Albritton, D. L., Harrop, W. J., Schmeltekopf, A. L., Zare, R. N. 1973a. *J. Mol. Spectrosc.* 46:25

Albritton, D. L., Harrop, W. J., Schmeltekopf, A. L., Zare, R. N. 1973b. *J. Mol. Spectrosc.* 46:67

Albritton, D. L., Harrop, W. J., Schmeltekopf, A. L., Zare, R. N. 1973c. *J. Mol. Spectrosc.* 46:89

Albritton, D. L., Harrop, W. J., Schmeltekopf, A. L., Zare, R. N. 1973d. *J. Mol. Spectrosc.* 46:103

Allen, C. W. 1973. *Astrophysical Quantities.* London: Athlone Press. 3rd ed.
Aller, L. H. 1963. *Astrophysics.* New York: Ronald Press. 2nd ed.
Allison, A. C., Dalgarno, A. 1969. *J. Quant. Spectrosc. Radiat. Transfer* 9:1543
Allison, A. C., Dalgarno, A. 1970. *At. Data* 1:289
Anderson, R., Sutherland, R., Frey, N. 1972. *J. Opt. Soc. Am.* 62:1127
Anderson, R. A., Peacher, J., Wilcox, D. M. 1975. *J. Chem. Phys.* 63:5287
Anketell, J. 1967. PhD thesis. Univ. London
Anketell, J., Learner, R. C. M. 1967a. *Proc. R. Soc. London Ser. A* 301:355
Anketell, J., Learner, R. C. M. 1967b. *Proc. R. Soc. London Ser. A* 301:355
Anketell, J., Pery-Thorne, A. 1967. *Proc. R. Soc. London Ser. A* 301:343
Armstrong, B. H., Nicholls, R. W. 1972. Emission Absorption and Transfer of Radiation in Heated Atmospheres. Oxford: Pergamon
Arnold, J. O. 1968. *J. Quant. Spectrosc. Radiat. Transfer* 8:1781
Arnold, J. O. 1972. PhD thesis. York Univ. Ann Arbor: Univ. Microfilms
Arnold, J. O., Nicholls, R. W. 1972. *J. Quant. Spectrosc. Radiat. Transfer* 12:1435
Arnold, J. O., Nicholls, R. W. 1973a. Recent Developments in Shock Tube Research, *Proc. 9th Int. Shock Tube Symp., Stanford Univ.,* ed. D. Bershader, W. Griffith, p. 340
Arnold, J. O., Nicholls, R. W. 1973b. *J. Quant. Spectrosc. Radiat. Transfer* 13:115
Arnold, J. O., Whiting, E. E., Langhoff, S. R. 1977. *J. Chem. Phys.* In press
Banks, P. M., Kockarts, G. 1973. *Aeronomy, Parts A & B.* New York: Academic
Barrow, R. F. 1973. *Diatomic Molecules, A critical bibliography of spectroscopic data*
Bass, A. M., Broida, H. P. 1953. *Natl. Bur. Stand. (U.S.) Circ. 541*
Becker, K. H., Haaks, D. 1973. *Z. Naturforsch. Teil A* 28:249
Becker, K. H., Haaks, D., Tatarczyk, B. 1974. *Chem. Phys. Lett.* 25:564
Bell, R. A., Branch, D., Upson, W. L. 1976. *J. Quant. Spectrosc. Radiat. Transfer* 16:177
Bennett, R. G., Dalby, F. W. 1959. *J. Chem. Phys.* 31:434
Bennett, R. G., Dalby, F. W. 1960a. *J. Chem. Phys.* 32:1111
Bennett, R. G., Dalby, F. W. 1960b. *J. Chem. Phys.* 32:1716
Bennett, R. G., Dalby, F. W. 1962. *J. Chem. Phys.* 36:399
Bennett, R. G., Dalby, F. W. 1964. *J. Chem. Phys.* 40:1414
Billingsley, F. P. 1975. *J. Chem. Phys.* 62:864
Blackwell, D. E. 1975. *Q. J. R. Astron. Soc.* 16:361
Born, M., Oppenheimer, J. R. 1927. *Ann. Phys.* 84:457
Borst, W. L., Zipf, E. C. 1971. *Phys. Rev. A* 3:979
Brannen, E., Hunt, F. R., Adlington, R. H., Nicholls, R. W. 1955. *Nature* 175:810
Brocklehurst, B., Seel, R. M., Hébert, G. R., Nicholls, R. W. 1971. *Identification Atlas of Molecular Spectra,* Vol. 8, *The CN Red System,* Cent. Res. Exp. Space Sci., York Univ.
Brocklehurst, B., Seel, R. M., Hébert, G. R., Nicholls, R. W. 1972. *Identification Atlas of Molecular Spectra,* Vol. 9, *The CN Violet System,* Cent. Res. Exp. Space Sci., York Univ.
Broida, H. P., Moore, C. E. 1956. *Mem. Soc. R. Sci. Liège,* 4th Ser. 18:216
Brooks, N. H., Smith, W. H. 1974. *Ap. J.* 194:513
Brooks, N. H., Smith, W. H. 1975. *Ap. J.* 196:307
Brophy, J. H., Silver, J. A., Kinsey, J. L. 1974. *Chem. Phys. Lett.* 28:418
Browne, J. C. 1969. *Ap. J.* 156:397
Brzozowski, J. 1974. *Phys. Scr.* 10:241
Cann, M. W. P., Nicholls, R. W. 1974. *Synthetic spectrum calculations with atmospheric applications, Spectrosc. Rep. 6,* Cent. Res. Exp. Space Sci., York Univ.
Carance, M., Verges, J. 1975. *J. Phys. B.* 8:3001
Carrington, T. 1959. *J. Chem. Phys.* 31:1243
Chamberlain, J. W. 1961. *Physics of Aurora and Airglow.* New York: Academic
Chervenak, J. G., Anderson, R. A. 1971. *J. Opt. Soc. Am.* 61:952
Clerc, M., Schmidt, M. 1972. *Faraday Discuss. Chem. Soc.* 53:217
Cook, T. J., Levy, D. H. 1972. *J. Chem. Phys.* 57:5059
Cooper, D. M. 1974. PhD thesis. York Univ. Ann Arbor: Univ. Microfilms
Cooper, D. M. 1977. *J. Quant. Spectrosc. Radiat. Transfer* 17:543
Cooper, D. M., Nicholls, R. W. 1975a. *J. Quant. Spectrosc. Radiat. Transfer* 15:139
Cooper, D. M., Nicholls, R. W. 1975b. *Modern developments in shock-tube research, Proc. 10th Int. Shock Tube Symp., Kyoto,* ed. G. Kamimoto, p. 696
Cooper, D. M., Nicholls, R. W. 1976. *Spectrosc. Lett.* 9:139
Coxon, J. A. 1971. *J. Quant. Spectrosc. Radiat. Transfer* 11:443
Coxon, J. A. 1976. Unpublished work
Creek, D. M., Nicholls, R. W. 1975. *Proc. R. Soc. London Ser. A* 341:517
Crosley, D. R., Lengel, R. K. 1975. *J. Quant.*

Spectrosc. Radiat. Transfer 15:579
Czernichowski, A., Zyrnicki, W. 1970. *Acta Phys. Pol.* 37A:865
Danylewych, L. L. 1971. PhD thesis. York Univ. Ann Arbor: Univ. Microfilms
Danylewych, L. L., Nicholls, R. W. 1974a. *Proc. R. Soc. London Ser. A* 339:197
Danylewych, L. L., Nicholls, R. W. 1974b. *Proc. R. Soc. London Ser. A* 339:213
Danylewych, L. L., Nicholls, R. W. 1977. *Proc. R. Soc. London.* Submitted for publication
Danylewych, L. L., Mandelman, M., Nicholls, R. W. 1976. Unpublished work
Das, G., Wahl, A. C. 1972. *J. Chem. Phys.* 56:1796
Davenport, J. W. 1976. *Phys. Rev. Lett.* 36: 945
Degen, V., Innanen, S. E. H., Hébert, G. R., Nicholls, R. W. 1968. *Identification Atlas of Molecular Spectra,* Vol. 6, *The O_2 Herzberg System,* Cent. Res. Exp. Space Sci., York Univ.
Demtroder, W. 1962. *Z. Phys.* 166:42
De Zafra, R. L., Marshall, A., Metcalf, H. 1971. *Phys. Rev. A* 3:1557
Dieke, G. H., Crosswhite, H. M. 1949. *Q. Rep. Oct. 1–Dec. 31, Contract NOrd 8036 JHB-3, Problem A*
Dieke, G. H., Crosswhite, H. M. 1962. *J. Quart. Spectrosc. Radiat. Transfer* 2:97
D'Incan, J., Effantin, C., Roux, F. 1971. *J. Quant. Spectrosc. Radiat. Transfer* 11: 1215
Dixon, R. N., Nicholls, R. W. 1958. *Can. J. Phys.* 36:127
Dongre, M. B. 1974. *Ind. J. Phys.* 48:289
Drake, J., Nicholls, R. W. 1969. *Chem. Phys. Lett.* 3:457
Dube, P. S. 1972. *Ind. J. Pure Appl. Phys.* 10:167
Dube, P. S., Rai, D. K., Singh, N. L. 1972. *Ind. J. Pure Appl. Phys.* 10:87
Dyne, P. J. 1958. *J. Chem. Phys.* 28:999
Elander, N., Smith, W. H. 1973a. *Ap. J.* 184: 663
Elander, N., Smith, W. H. 1973b. *Ap. J.* 184: 311
Elmergreen, B. G., Smith, W. H. 1972. *Ap. J.* 178:557
Engelman, R., Rouse, P. E. 1975. *J. Quant. Spectrosc. Radiat. Transfer* 15:831
Fabian, W., Lewis, B. R. 1974. *J. Quant. Spectrosc. Radiat. Transfer* 14:523
Fairbairn, A. R. 1970. *J. Quant. Radiat. Transfer* 10:1321
Fairbairn, A. R., Wolnik, S. J., Berthel, R. O. 1974. *Ap. J.* 193:217
Felenbok, P. 1965. *Proc. Phys. Soc.* 86:676
Fink, von E. H., Welge, K. H. 1964. *Z. Naturforsch. Teil A* 19:1193
Fink, von E. H., Welge, K. H. 1967. *J. Chem. Phys.* 46:4315
Fink, von E. H., Welge, K. H. 1968. *Z. Naturforsch. Teil A* 23:358
Fowler, R. G. 1969. *J. Chem. Phys.* 50:4133
Fraser, P. A., Jarmain, W. R., Nicholls, R. W. 1953. *Ap. J.* 118:228
Garvin, D. 1959. *J. Am. Chem. Soc.* 81:3173
Garvin, D., Broida, H. P., Kostkowski, H. J. 1960. *J. Chem. Phys.* 32:880
Garvin, D., Broida, H. P., Kostkowski, H. J. 1962. *J. Chem. Phys.* 37:193
Gaydon, A. G. 1968. *Dissociation Energies and Spectra of Diatomic Molecules.* London: Chapman & Hall. 3rd ed.
Gelernt, B., Filseth, S., Carrington, T. 1975. *Chem. Phys. Lett.* 36:238
German, K. R. 1975. *J. Chem. Phys.* 62:2584
Gohel, V. B., Shah, M. P. 1975. *Ind. J. Pure Appl. Phys.* 13:162
Goldberg, L. 1955. *Am. J. Phys.* 23:203
Goldberg, L., McMath, R. R., Mohler, O. C., Pierce, A. K. 1952a. *Phys. Rev.* 85:140
Goldberg, L., McMath, R. R., Mohler, O. C., Pierce, A. K. 1952b. *Phys. Rev.* 85:481
Goldberg, L., Parkinson, W. H., Reeves, E. M. 1965. *Ap. J.* 141:1293
Goody, R. M. 1964. *Atmospheric Radiation.* Oxford: Oxford Univ. Press
Green, S., Hornstein, S., Bender, C. F. 1973. *Ap. J.* 179:671
Grevesse, N., Sauval, A. J. 1971. *J. Quant. Spectrosc. Radiat. Transfer* 11:65
Grevesse, N., Sauval, A. J. 1973. *Astron. Astrophys.* 27:29
Grinfeld, R. 1962. *Proc. Colloq. Spectrosc. Int. 10th,* ed. E. R. Lippincott, M. Margoshes, p. 597. Washington D.C.: Spartan Books
Haddad, G. N., Lokan, K., Farmer, A. J. D., Carver, G. H. 1968. *J. Quant. Spectrosc. Radiat. Transfer* 8:1193
Halmann, M., Laulicht, I. 1966. *Ap. J. Suppl.* 12:307
Harrington, J. A. 1970. PhD thesis. York Univ.
Harrington, J. A., Modica, A. P., Libby, D. R. 1966. *J. Quant. Spectrosc. Radiat. Transfer* 6:799
Harrington, J. A., Seel, R. M., Hébert, G. R., Nicholls, R. W. 1970. *Identification Atlas of Molecular Spectra,* Vol. 7, *The VO Yellow-Green and Red Systems,* Cent. Res. Exp. Space Sci., York Univ.
Hasson, V., Nicholls, R. W. 1971. *J. Phys. B* 4:681
Haycock, S. C. 1963. MSc thesis. Univ. Western Ontario, London, Ontario
Heaps, H. S., Herzberg, G. 1952. *Z. Phys.* 133:48
Hébert, G. R., Hodder, R. V. 1974. *J. Phys.*

B 7:2244
Hébert, G. R., Innanen, S. E. H., Nicholls, R. W. 1967. *Identification Atlas of Molecular Spectra,* Vol. 4, *The O_2 Schumann-Runge System,* Cent. Res. Exp. Space Sci., York Univ.
Hébert, G. R., Nicholls, R. W., Linton, C. 1977. *J. Quant. Spectrosc. Radiat. Transfer.* Submitted for publication
Hébert, G. R., Tyte, D. C. 1964. *Proc. Phys. Soc.* 83:629
Hedelund, J., Lambert, D. L. 1972. *Astrophys. Lett.* 11:71
Hefferlin, R. 1975. *J. Quant. Spectrosc. Radiat. Transfer* 15:925
Hefferlin, R. 1976. *J. Quant. Spectrosc. Radiat. Transfer* 16:1101
Hefferlin, R., Mashburn, J., Flechas, J., Main, R. P. 1976. *J. Tenn. Acad. Sci.* 51:100
Herbig, G. 1956. *Publ. Astron. Soc. Pac.* 68: 204
Herman, R. C., Rothery, R. W., Rubin, R. J. 1958. *J. Mol. Spectrosc.* 2:369
Herman, R. C., Shuler, K. E. 1953. *J. Chem. Phys.* 21:373
Herman, R. C., Shuler, K. E. 1954. *J. Chem. Phys.* 22:481
Heron, S., McWhirter, R. W. P., Rhoderick, E. H. 1954. *Nature* 174:564
Heron, S., McWhirter, R. W. P., Rhoderick, E. H. 1956. *Proc. R. Soc. London Ser. A* 234:565
Herzberg, G. 1950. *Spectra of diatomic molecules.* London: Van Nostrand
Herzberg, G. 1965. *J. Opt. Soc. Am.* 55:229
Herzberg, G. 1976. *Mem. Soc. R. Sci. Liège, 6th Ser.* 9:115
Herzberg, G., Lagerqvist, A., Malmberg, C. 1969. *Can. J. Phys.* 47:2735
Hesser, J. E. 1968. *J. Chem. Phys.* 48:2518
Hesser, J. E., Dressler, K. 1966. *J. Chem. Phys.* 45:3419
Hesser, J. E., Lutz, B. L. 1970. *Ap. J.* 159: 703
Hoggan, T., Davis, B. D. 1974. *Chem. Phys. Lett.* 29:555
Hougen, J. 1970. *Natl. Bur. Standards (U.S.) Monogr.* 115
Huber, K., Herzberg, G. 1977. *Spectroscopic Constants of Diatomic Molecules.* Princeton: Van Nostrand-Reinhold. In press
Huber, M. C. E. 1971. *Modern Optical Methods in Gas Dynamics,* ed. D. Dosanjh, p. 85. New York: Plenum
Hubisz, J. 1968. PhD thesis. York Univ. Ann Arbor: Univ. Microfilms
Hudson, R. D. 1971. *Rev. Geophys. Space Phys.* 9:305
Hudson, R. D. 1974. *Can. J. Chem.* 52:1465
Huron, B. 1969. *Physica* 41:58
Imhof, R. E., Read, F. H. 1971a. *Chem. Phys. Lett.* 11:326
Imhof, R. E., Read, F. H. 1971b. *J. Phys. B* 4:1063
Isaacson, L., Marram, E. P., Wentink, T. 1967. *J. Quant. Spectrosc. Radiat. Transfer* 7:691
Jackson, W. M. 1974. *J. Chem. Phys.* 61:4177
Jain, D. C. 1972. *J. Phys. B* 5:199
James, T. C. 1971a. *J. Mol. Spectrosc.* 40: 545
James, T. C. 1971b. *J. Chem. Phys.* 55:4118
Jarmain, W. R. 1963. *Can. J. Phys.* 41:1926
Jarmain, W. R. 1971. *J. Quant. Spectrosc. Radiat. Transfer* 11:421
Jarmain, W. R. 1972. *J. Quant. Spectrosc. Radiat. Transfer* 12:603
Jarmain, W. R., Nicholls, R. W. 1964. *Proc. Phys. Soc. A* 84:417
Jeunehomme, M. 1965. *J. Chem. Phys.* 42: 4086
Johnson, C. E. 1972. *Phys. Rev. A* 5:1026
Johnson, C. E., Van Dyck, R. S. 1972. *J. Chem. Phys.* 56:1506
Johnson, S. E., Capelle, G., Broida, H. P. 1972. *J. Chem. Phys.* 56:663
Jones, A. V. 1974. *Aurora.* Dordrecht: Reidel
Judge, D. L., Lee, L. C. 1972. *J. Chem. Phys.* 57:455
Keenan, P. C., Deutsch, A. J., Garrison, R. F. 1969. *Ap. J.* 158:261
Klemsdal, H. 1973. *J. Quant. Spectrosc. Radiat. Transfer* 13:517
Kovács, I. 1969. *Rotational Structure in the Spectra of Diatomic Molecules.* London: Adam Hilger
Krupenie, P. H. 1966. *Natl. Stand. Ref. Data Ser. 5,* Natl. Bur. Stand.
Krupenie, P. H. 1972. *J. Phys. Chem. Ref. Data 1, No. 2*
Kuz'menko, N. E., Smirnov, A. D., Kuzyakov, Y. Y. 1970a. *Vestn. Mosk. Univ. Khim.* 11:357
Kuz'menko, N. E., Smirnov, A. D., Kuzyakov, Y. Y. 1970b. *Vestn. Mosk. Univ. Khim.* 11:478
Kuznetsova, L. A., Kuz'menko, N. E., Kuzyakov, Y. Y., Plastinin, Y. A. 1974. *Sov. Phys.—Usp.* 17:405; *Usp. Fiz. Nauk.* 113:285
Ladenberg, R., Bershader, D. 1954. *Physical Measurements in Gas Dynamics and Combustion,* Vol. IX, p. 46. Princeton Univ. Press
Lambert, D. L., Mallia, E. A. 1970. *MNRAS* 148:313
Langhoff, S. R., Arnold, J. O. 1977. Private communication, in preparation for publication
Lapp, M. 1961. *J. Quant. Spectrosc. Radiat. Transfer* 1:30
Lassettre, E. N., Skerbele, A. 1971. *J. Chem.*

Phys. 54: 1597
Lawrence, G. M. 1965. *J. Quant. Spectrosc. Radiat. Transfer* 5: 359
Lawrence, G. M., Seitel, S. C. 1973. *J. Quant. Spectrosc. Radiat. Transfer* 13: 713
Learner, R. C. M. 1962. *Proc. R. Soc. London Ser. A* 269: 311
Le Calve, J., Bourene, M., Schmidt, M., Cleve, M. 1970. *J. Phys.* 30: 807
Lee, L. C., Carlson, R. W., Judge, D. L. 1975. *Mol. Phys.* 30: 1941
Lee, L. C., Carlson, R. W., Judge, D. L., Ogawa, M. 1973. *J. Quant. Spectrosc. Radiat. Transfer* 13: 1023
LeRoy, R. J. 1973. *Energy Levels of a Diatomic Near Dissociation. Molecular Spectroscopy, Spec. Period. Rep.*, Vol. 1, Chap. 3. London: Chemical Society
LeRoy, R. J., Vrscay, E. R. 1975. *Can. J. Phys.* 53: 1560
Lewis, B. R. 1974a. *J. Quant. Spectrosc. Radiat. Transfer* 14: 537
Lewis, B. R. 1974b. *J. Quant. Spectrosc. Radiat. Transfer* 14: 722
Linton, C., Nicholls, R. W. 1969a. *J. Quant. Spectrosc. Radiat. Transfer* 9: 1
Linton, C., Nicholls, R. W. 1969b. *J. Phys. B* 2: 490
Linton, C., Nicholls, R. W. 1970. *J. Quant. Spectrosc. Radiat. Transfer* 10: 311
Liszt, H. S., Hesser, J. E. 1970. *Ap. J.* 159: 1101
Liszt, H. S., Smith, W. H. 1971. *J. Quant. Spectrosc. Radiat. Transfer* 11: 1043
Liszt, H. S., Smith, W. H. 1972. *J. Quant. Spectrosc. Radiat. Transfer* 12: 947
Lovas, F. J., Krupenie, P. H. 1974. *J. Phys. Chem. Ref. Data* 3: 245
Luk, C. K., Bersohn, C. K. 1973. *J. Chem. Phys.* 58: 2153
Maheshwari, R. C., Singh, I. D., Shukla, M. M. 1968. *J. Phys. B* 2: 993
Maier, B., Holland, R. F. 1972. *J. Phys. B* 5: L118
Main, R. P., Carlson, D. J., Dupuis, R. A. 1967. *J. Quant. Spectrosc. Radiat. Transfer* 7: 805
Main, R. P., Marsell, A. L., Hooker, W. S. 1968. *J. Quant. Spectrosc. Radiat. Transfer* 8: 1527
Main, R. P., Schadee, A. 1969. *J. Quant. Spectrosc. Radiat. Transfer* 9: 713
Mallia, E. A. 1968. *Solar Phys.* 5: 281
Mandelman, M., Nicholls, R. W. 1977. *J. Quant. Spectrosc. Radiat. Transfer.* 17: 483
Maranan, J., Suarez, C. B. 1974. *Spectrosc. Lett.* 7: 303
Marchetti, M. A., LaPaglia, S. R. 1971. *J. Chem. Phys.* 55: 1655
Marlow, W. C. 1967. *Appl. Opt.* 6: 1715
McCallum, J. C., Nicholls, R. W., Jarmain, W. R. 1970. *Franck-Condon factors and Related Quantities for Diatomic Molecular Spectra, Spectrosc. Rep. No. 1,* Cent. Res. Exp. Space Sci., York Univ.
McDonald, J. K., Innes, K. K. 1969. *J. Mol. Spectrosc.* 32: 501
McEwen, D. J. 1965. PhD thesis. Univ. Western Ontario. Ann Arbor: Univ. Microfilms
McGregor, A. T. 1977. In process of publication
McGregor, A. T. 1975. PhD thesis. Univ. Western Ontario, London, Ontario. Ann Arbor: Univ. Microfilms
McGregor, A. T., Nicholls, R. W., Jarmain, W. R. 1961. *Can. J. Phys.* 39: 1215
McKellar, A. R. W. 1973. *Ap. J.* 185: L53
Merrill, P. W. 1953. *Ap. J.* 118: 453
Meyer, V. D., Lassettre, E. N. 1971. *J. Chem. Phys.* 54: 1608
Michels, H. H. 1972. *J. Chem. Phys.* 56: 665
Mohanty, B. S., Singh, O. N. 1969. *Ind. J. Pure Appl. Phys.* 7: 109
Mon, J. P. 1966. *C. R. Acad. Sci.* 262: 1276
Moore, C. E. 1966. *Future Spectroscopy of Late Type Stars, Colloq. Late Type Stars,* ed. M. Hack, p. 15. Univ. Trieste Obs.
Moore, C. E., Broida, H. P. 1956. *Mem. Soc. R. Sci. Liège, 4th Ser.* 18: 252
Moore, C. E., Broida, H. P. 1959a. *J. Res. Natl. Bur. Stand. Sect. A* 63: 19
Moore, C. E., Broida, H. P. 1959b. *J. Res. Natl. Bur. Stand. Sect. A* 63: 279
Moore, J. H., Robinson, D. W. 1968. *J. Chem. Phys.* 48: 4870
Moskalenko, N. I. 1975. *Opt. Spectrosc. USSR* 38: 382
Murthy, N. S., Nagaraj, S. 1962. *Proc. Phys. Soc. London* 84: 827
Murthy, N. S., Murthy, B. N. 1967. *Proc. Phys. Soc. London* 90: 881
Murthy, N. S., Murthy, B. N. 1969. *Nature* 223: 181
Murthy, N. S., Murthy, B. N. 1972. *J. Phys. B* 5: 714
Nicholls, R. W. 1956. *Proc. Phys. Soc. London Sect. A* 69: 741
Nicholls, R. W. 1960. *Nature* 186: 958
Nicholls, R. W. 1962a. *J. Res. Natl. Bur. Stand. Sect. A* 66: 227
Nicholls, R. W. 1962b. *J. Quant. Spectrosc. Radiat. Transfer* 2: 433
Nicholls, R. W. 1962c. *Can. J. Phys.* 40: 1772
Nicholls, R. W. 1964a. *Nature* 204: 373
Nicholls, R. W. 1964b. *J. Res. Nat. Bur. Stand. Sect. A* 68: 75
Nicholls, R. W. 1964c. *Ann. Geophys.* 20: 144
Nicholls, R. W. 1964d. *Supersonic Flow, Chemical Processes and Radiative Transfer,* ed. D. B. Olfe, V. Zakkay, p. 413.

Oxford: Pergamon
Nicholls, R. W. 1965a. *J. Quant. Spectrosc. Radiat. Transfer* 5: 647
Nicholls, R. W. 1965b. *Proc. Phys. Soc. London Sect. A* 85: 159
Nicholls, R. W. 1965c. *Ap. J.* 141: 819
Nicholls, R. W. 1968. *Nature* 219: 151
Nicholls, R. W. 1969a. *Electronic Spectra of Diatomic Molecules. Physical Chemistry,* ed. H. Eyring, D. Henderson, W. Jost, Vol. 3, Chapter 6. New York: Academic
Nicholls, R. W. 1969b. *Can. J. Chem.* 47: 1847
Nicholls, R. W. 1971. *Modern Optical Methods in Gas Dynamic Research,* ed. D. S. Dosanjh, p. 1. New York: Plenum
Nicholls, R. W. 1972. *Chem. Phys. Lett.* 17: 252
Nicholls, R. W. 1973a. *J. Quant. Spectrosc. Radiat. Transfer* 13: 1059
Nicholls, R. W. 1973b. *Chem. Phys. Lett.* 20: 261
Nicholls, R. W. 1974. *J. Quant. Spectrosc. Radiat. Transfer* 14: 233
Nicholls, R. W. 1976a. *Report of Committee 5, Molecular Spectra, of Commission 14 in Transactions of International Astronomical Union XVI.* Netherlands: Reidel
Nicholls, R. W. 1976b. *Spectrosc. Lett.* 9: 23
Nicholls, R. W., Fraser, P. A., Jarmain, W. R. 1959. *Combust. Flame* 3: 13
Nicholls, R. W., Fraser, P. A., Jarmain, W. R., McEachran, R. P. 1960. *Ap. J.* 131: 399
Nicholls, R. W., Jarmain, W. R. 1955. *J. Chem. Phys.* 23: 1561
Nicholls, R. W., Jarmain, W. R. 1956. *Proc. Phys. Soc. London Ser. A* 69: 713
Nicholls, R. W., Tyte, D. C. 1967. *Proc. Phys. Soc.* 91: 489
Oldenberg, O., Rieke, F. F. 1938. *J. Chem. Phys.* 6: 439
Oldenberg, O., Rieke, F. F. 1939. *J. Chem. Phys.* 7: 485
Ornstein, L. S., Brinkman, W. 1931. *Proc. Acad. Sci. Amsterdam* 34: 498
Ortenberg, F. S., Glasco, V. B. 1962. *Astron. J. USSR* 39: 921
Ortenberg, F. S., Glasco, V. B., Dmitriev, A. I. 1964. *Sov. Astron. AJ* 8: 258
Pearse, R. W. B., Gaydon, A. G. 1976. *The Identification of Molecular Spectra.* London: Chapman & Hall. 4th ed.
Penner, S. S., Weber, D. 1951a. *J. Chem. Phys.* 19: 807
Penner, S. S., Weber, D. 1951b. *J. Chem. Phys.* 19: 817
Penner, S. S., Weber, D. 1951c. *J. Chem. Phys.* 19: 1351
Penner, S. S., Weber, D. 1951d. *J. Chem. Phys.* 19: 1361
Phillips, J. G., Davis, S. 1960–present. *Berkeley Mol. Spectrosc.* Newsl., Univ. Calif., Berkeley
Pilling, M. J., Bass, A. M., Braun, W. 1971. *J. Quant. Spectrosc. Radiat. Transfer* 11: 1593
Popkie, H. E. 1971. *J. Chem. Phys.* 54: 4597
Popkie, H. E., Henneker, W. H. 1971. *J. Chem. Phys.* 55: 617
Price, M. L., Sulzmann, K., Penner, S. S. 1971. *J. Quant. Spectrosc. Radiat. Transfer* 11: 427
Price, M. L., Sulzmann, K., Penner, S. S. 1974. *J. Quant. Spectrosc. Radiat. Transfer* 14: 1273
Rao, T. V. R., Laksham, S. V. J. 1970a. *Physica* 46: 609
Rao, T. V. R., Laksham, S. V. J. 1970b. *J. Quant. Spectrosc. Radiat. Transfer* 10: 945
Rich, J. C. 1968. *Ap. J.* 153: 327
Robinson, D., Nicholls, R. W. 1958. *Proc. Phys. Soc. London Ser. A* 71: 957
Robinson, D., Nicholls, R. W. 1960. *Proc. Phys. Soc. London Ser. A* 75: 817
Rogers, J., Anderson, R. 1970a. *J. Quant. Spectrosc. Radiat. Transfer* 10: 515
Rogers, J., Anderson, R. 1970b. *J. Opt. Soc. Am.* 60: 278
Rosen, B. 1951. *Données Spectroscopiques Concernant les Molécules Diatomiques.* Paris: Herman et Cie
Rosen, B. 1952. *Atlas des longeurs d'onde charactéristiques des bandes d'émission et d'absorption des molécules diatomiques.* Paris: Herman et Cie
Rosen, B. 1970. *Spectroscopic data relative to diatomic molecules.* Oxford: Pergamon
Ross, J. E., Aller, L. H. 1976. *Science* 191: 1223
Rouse, P. E., Engelman, R. 1973. *J. Quant. Spectrosc. Radiat. Transfer* 13: 1503
Roux, F., Cerny, D., D'Incan, J. 1976. *Ap. J.* 204: 940
Roux, F., D'Incan, J., Cerny, D. 1973. *Ap. J.* 186: 1141
Roux, F., Effantin, C., D'Incan, J. 1972. *J. Quant. Spectrosc. Radiat. Transfer* 12: 97
Rozhdestvenskii, D. S. 1912. *Ann. Phys.* 39: 307
Ruskin, A. D. 1970. *Vestn. Mosk. Univ. Khim.* 11: 526
Sankaranayanan, S., Narayanan, P. S. 1966. *Proc. Natl. Inst. Soc.* 32: 56
Sauval, A. J. 1976. *List of thermochemically expected molecules in stellar atmospheres,* circulated at IAU Gen. Ass., Grenoble, 1976
Schadee, A. 1964. *Bull. Astron. Inst. Neth.* 17: 311
Schoonveld, L. 1972. *J. Quant. Spectrosc. Radiat. Transfer* 12: 1139
Schwenker, R. P. 1965. *J. Chem. Phys.* 42:

1895
Seel, R. M., Hébert, G. R., Nicholls, R. W. 1977a. *Identification Atlas of Molecular Spectra 10,* NO_2, Cent. Res. Exp. Space Sci., York Univ. In press
Seel, R. M., Hébert, G. R., Nicholls, R. W. 1977b. *Identification Atlas of Molecular Spectra 11,* SO_2, Cent. Res. Exp. Space Sci., York Univ. In press
Sharma, A. 1967. *J. Quant. Spectrosc. Radiat. Transfer* 7:283
Shin, J. B. 1976. PhD thesis. York Univ. Ann Arbor: Univ. Microfilms
Shin, J. B., Nicholls, R. W. 1977. *Proc. 11th Shock Tube Symp., Seattle*
Silvers, S. J., Bergeman, T. H., Klemperer, W. 1970. *J. Chem. Phys.* 52:4385
Singh, I. D., Maheshwari, R. C. 1969. *Ind. J. Pure Appl. Phys.* 7:708
Singh, P. D., Pathak, A. N. 1966. *Proc. Phys. Soc.* 90:543
Singh, P. D., Pathak, A. N. 1967. *Proc. Phys. Soc.* 91:497
Singh, R. B., Rai, D. K. 1965. *J. Quant. Spectrosc. Radiat. Transfer* 5:723
Singh, I. D., Shukla, M. M., Maheshwari, R. C. 1969. *J. Quant. Spectrosc. Radiat. Transfer* 1:533
Slanger, T. G., Black, G. 1971. *J. Chem. Phys.* 55:2164
Slanger, T. G., Black, G. 1972. *J. Phys. B* 5:1988
Slanger, T. G., Black, G. 1976. *J. Chem. Phys.* 64:219
Smith, A. J., Imhof, R. E., Read, F. H. 1973. *J. Phys. B* 6:1333
Smith, A. J., Read, F. H., Imhof, R. E. 1975. *J. Phys. B* 8:2869
Smith, W. H. 1969a. *Ap. J.* 156:791
Smith, W. H. 1969b. *J. Quant. Spectrosc. Radiat. Transfer* 9:1191
Smith, W. H. 1969c. *J. Chem. Phys.* 51:520
Smith, W. H. 1970a. *Nucl. Instrum. Meth.* 90:115
Smith, W. H. 1970b. *J. Chem. Phys.* 53:792
Smith, W. H. 1971. *J. Chem. Phys.* 54:1384
Smith, W. H. 1972. *Ap. J.* 176:265
Smith, W. H., Brzozowski, J., Erman, P. 1976. *J. Chem. Phys.* 64:4628
Smith, W. H., Chevalier, R. 1972. *Ap. J.* 177:835
Smith, W. H., Liszt, H. S. 1971. *J. Quant. Spectrosc. Radiat. Transfer* 11:45
Smith, W. H., Liszt, H. S. 1972. *J. Quant. Spectrosc. Radiat. Transfer* 12:505
Smith, W. H., Stella, G. 1975. *J. Chem. Phys.* 63:2395
Sotirovski, P. 1971. *Astron. Astrophys. Suppl.* 6:85
Spindler, R. J. 1965. *J. Quant. Spectrosc. Radiat. Transfer* 5:165
Spindler, R. J. 1969a. *J. Quant. Spectrosc. Radiat. Transfer* 9:597
Spindler, R. J. 1969b. *J. Quant. Spectrosc. Radiat. Transfer* 9:627
Spindler, R. J. 1969c. *J. Quant. Spectrosc. Radiat. Transfer* 9:1041
Stevens, A. E., Ferguson, H. I. S. 1963. *Can. J. Phys.* 41:240
Suarez, C. B. 1974. *Can. J. Phys.* 52:921
Suchard, S. N. 1975a. *Spectroscopic Data,* Vol. 1, *Heteronuclear Diatomic Molecules,* Part A. New York: Plenum
Suchard, S. N. 1975b. *Spectroscopic Data,* Vol. 1, *Heteronuclear Diatomic Molecules,* Part B. New York: Plenum
Suchard, S. N., Melzer, J. E. 1976. *Spectroscopic Data,* Vol. 2, *Homonuclear Diatomic Molecules.* New York: Plenum
Sulzmann, K. 1973. *J. Quant. Spectrosc. Radiat. Transfer* 13:931
Sulzmann, K. 1975. *J. Quant. Spectrosc. Radiat. Transfer* 15:313
Sutherland, R. A., Anderson, R. A. 1973. *J. Chem. Phys.* 58:1226
Swings, P. 1951. *Astrophysics,* ed. J. A. Hynek, Chap. 4, p. 145. New York: McGraw Hill
Swings, P. 1958. *Handb. Phys.* 50, *Astrophysics* I, p. 109. Berlin: Springer
Swings, P. 1966. *Proc. Colloq. Late Type Stars,* ed. M. Hack, p. 11. Univ. Trieste Obs.
Tatum, J. B. 1967. *Ap. J. Suppl. No. 124* 14:21
Tilford, S. G., Simmons, J. D. 1972. *J. Phys. Chem. Ref. Data 1, No. 1*
Tsuji, T. 1964. *Ann. Tokyo Astron. Obs. II* IX:1
Tyte, D. C., Hébert, G. R. 1964. *Proc. Phys. Soc.* 84:830
Tyte, D. C., Innanen, S. E. H., Nicholls, R. W. 1967. *Identification Atlas of Molecular Spectra 5, The* C_2 *Swan System,* Cent. Res. Exp. Space Sci., York Univ.
Tyte, D. C., Nicholls, R. W. 1964a. *Identification Atlas of Molecular Spectra,* Vol. 1, *The A1O Blue-Green System.* Mol. Excitation Group, Univ. Western Ontario
Tyte, D. C., Nicholls, R. W. 1964b. *Identification Atlas of Molecular Spectra,* Vol. 2, *The* N_2 *Second Positive System.* Mol. Excitation Group, Univ. Western Ontario
Tyte, D. C., Nicholls, R. W. 1965. *Identification Atlas of Molecular Spectra,* Vol. 3, *The* N_2^+ *First Negative System.* Mol. Excitation Group, Univ. Western Ontario
Van Pee, M., Kineyko, W. R., Caruso, R. 1970. *Combust. Flame* 14:381
Wallace, L. V. 1962a. *Ap. J. Suppl.* 6:445
Wallace, L. V. 1962b. *Ap. J. Suppl.* 7:165
Watson, W. S. 1975. *Planet. Space Sci.* 23:

384
Weeks, D. M., Simpson, E. A. 1967. *Sci. Rep. XIX,* Harvard College Obs.
Wells, W. C., Gissler, R. C. 1970. *Phys. Rev. Lett.* 24:705
Wentink, T., Isaacson, L., Morreal, J. 1964. *J. Chem. Phys.* 41:278
Wentink, T., Spindler, R. J. 1970. *J. Quant. Spectrosc. Radiat. Transfer* 10:609
White, J. U. 1938. *J. Chem. Phys.* 6:294
White, J. U. 1940. *J. Chem. Phys.* 8:459
Whiting, E. E. 1972. PhD thesis. York Univ., Toronto
Whiting, E. E. 1973. *NASA Tech. Note D-7268*
Whiting, E. E., Arnold, J. O., Garber, J. A., Burnett, K. 1976. Private Communication, in preparation for publication
Whiting, E. E., Nicholls, R. W. 1974. *Ap. J. Suppl. No. 235* 27:1
Whiting, E. E., Paterson, J. A., Kovács, I., Nicholls, R. W. 1973. *J. Mol. Spectrosc.* 47:84
Wicke, V. G., Tempra, W. 1975. *J. Chem. Phys.* 63:3756
Wilcox, D., Anderson, R., Peacher, J. 1975. *J. Opt. Soc. Am.* 65:1368
Wojslaw, R. S., Peery, B. F. 1976. *Ap. J. Suppl.* 31:75
Yoshimine, M., Green, S., Thaddeus, P. 1973. *Ap. J.* 183:899
Yoshimine, M., McLean, A. D., Liu, B. 1973. *J. Chem. Phys.* 58:4412
Zare, R. N., Schmeltekopf, A. L., Harrop, W. J., Albritton, D. L. 1973. *J. Mol. Spectrosc.* 46:37
Zyrnicki, W. 1975. *J. Quant. Spectrosc. Radiat. Transfer* 15:575

Ann. Rev. Astron. Astrophys. 1977. 15: 235–66

RECENT THEORIES OF GALAXY FORMATION

×2114

J. Richard Gott, III
Department of Astrophysical Sciences, Princeton University, Princeton, New Jersey 08540

1 INTRODUCTION

Galaxy formation is an essential part of any cosmological theory. This field has experienced rapid progress along a number of lines in recent years, a process punctuated by many interesting and lively debates. There have been several excellent reviews of various aspects of this subject (Jones 1976, Field 1975, Silk 1974, Rees 1971). Cosmological turbulence theories are not discussed as these have been reviewed extensively by Jones (1976), nor is spiral density wave theory discussed as this is well summarized in *La Dynamique Des Galaxies Spirales*, Proceedings of CNRS International Colloquium No. 241 (Paris, 1975) and by Toomre in the present volume. We first review the observational data, then describe how galaxies may originate in a standard big-bang model, and finally consider several detailed theories of formation. As a participant in this field, I apologize if my own biases are occasionally apparent.

2 OBSERVATIONS OF GALAXIES

Galaxies represent large-amplitude inhomogeneities in the universe. The mean density within 10 kpc of the center of our galaxy is $\sim 2 \times 10^{-24}$ g cm^{-3}, while the mean density in the universe probably lies between 5×10^{-30} and 5×10^{-31} g cm^{-3}.

Hubble (1926) found that galaxies are divided into distinct morphological types: ellipticals, spirals, S0's, and irregulars.

Elliptical galaxies are spheroidal systems of stars with smooth elliptical isophotes. Ellipticities $\varepsilon = (a-b)/a$ (where a and b are the semi-major and semi-minor axes respectively) range from 0 to 0.7 (i.e. E0–E7). E0–E1 galaxies are the most commonly observed types, those of higher ellipticity being less frequent (Sandage, Freeman & Stokes 1970). Deducing the distribution of true ellipticities requires consideration of random projection effects as well as observational errors (Hubble 1926, Sandage, Freeman & Stokes 1970, Thuan & Gott 1977). One problem is that ellipticity as a function of radius is usually not constant (Wilson 1974), and ellipticals are typed according to their maximum ellipticity. Another problem is that ellipticals

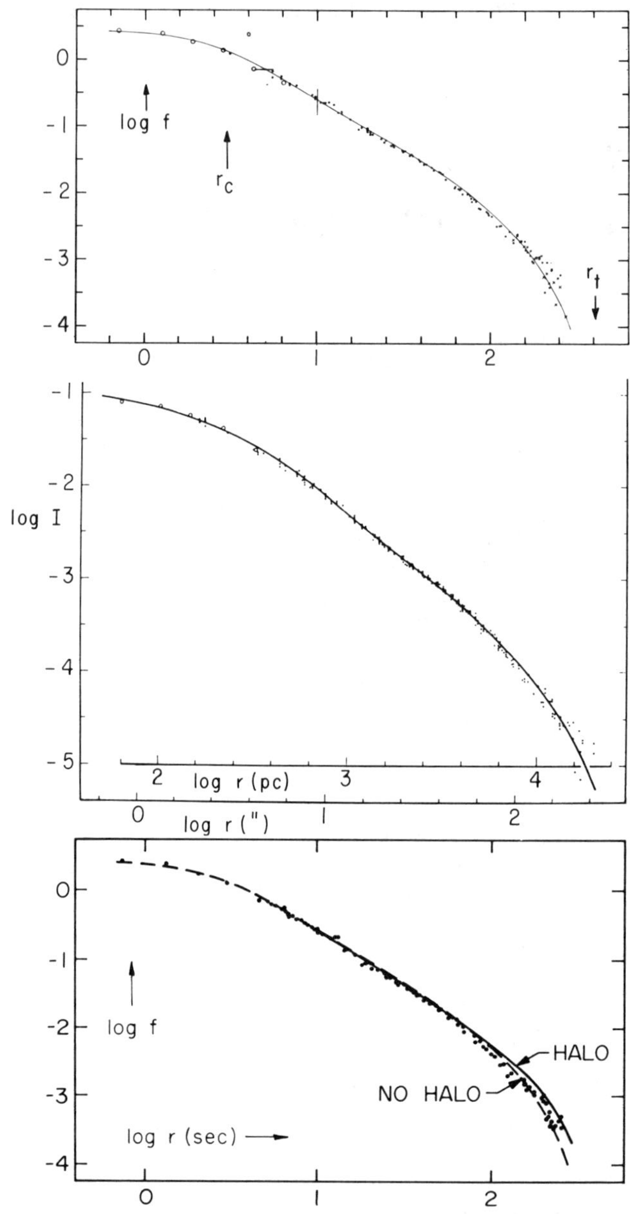

Figure 1 Miller & Prendergast's (1962) observations of surface brightness as a function of radius in the E0 galaxy NGC 3379 (data points), compared with three theoretical models (smooth curves). The same observational data is plotted three times (the middle graph

may be triaxial ellipsoids rather than oblate spheroids as is usually supposed. Surface brightness distributions as a function of radius in elliptical galaxies are remarkably similar and are approximated well by Hubble's (1930) law:

$$I(r) = I_0/(1+r/a)^2, \tag{1}$$

where I_0 is the central surface brightness and a is a characteristic radius. This law implies that $\rho \propto r^{-3}$ in the envelope and thus predicts a divergent total luminosity as $r \to \infty$. Real galaxies fall below Hubble's Law at large radii (Wilson 1974, Oemler 1974). Another equally good approximate fitting function which produces a finite total luminosity is deVaucouleurs' (1959) law (see also deVaucouleurs & deVaucouleurs 1964).

$$I(r) = I_0 \exp\left[-7.669\,(r/R_e)^{1/4}\right], \tag{2}$$

where R_e is the radius that contains half the luminosity. The dynamical models that fit the data best over their entire range are those of King (1966), based on a Maxwellian velocity distribution with a tidal energy-cutoff. A King model fitted to the data for NGC 3379 from Miller & Prendergast (1962) is shown in Figure 1. The fit is outstandingly good when one considers that only one adjustable fitting parameter ($r_{\rm core}/r_{\rm tidal}$) is used. Using deVaucouleur's Law, Fish (1964) claimed that for ellipticals $L_{\rm tot} \propto R_e^2$; however this remains a controversial result because of photometric errors and selection effects (cf Disney 1976). Faber & Jackson (1976) found that the core luminosity of an elliptical, its King core radius, and its central velocity dispersion are related:

$$L_c \propto r_c, \tag{3}$$

$$L_c \propto V^4. \tag{4}$$

---

merely has a different vertical scaling). The top curve shows a King (1966) model (cf Section 2) based on a simple isothermal Maxwellian velocity distribution with an upper tidal energy cutoff. In addition to simple horizontal and vertical scaling this model has one free-fitting parameter, the ratio of the core radius r_c to the tidal cutoff radius r_t. The values of r_c and r_t are indicated by the arrows. The middle curve is a Larson (1974a) model (cf Section 5) based on the combined effects of gas dissipation and star formation. This model has three fitting parameters in addition to simple scaling. The bottom curve is a Gunn (1976) heavy-halo model (cf Section 6). The curve marked "no halo" is a simple King model exactly like the top curve. The curve marked "halo" is Gunn's two-component King model containing an observed component (plotted here) and an invisible heavy-halo component approximating $\rho \propto r^{-2}$ out to the tidal cutoff radius. At small r the "halo" and "no halo" curves are nearly identical, so only one is plotted. Gunn's halo model has a total mass-to-light ratio six times larger than the Larson or King models. It is distinguishable from a simple King model in that the velocity dispersions of the observed stars vary more slowly with radius. If the central velocity dispersion of the observed stars is $\langle V_c^2\rangle^{1/2}$, then at $r = 100''$ in the above model the velocity dispersion of the observed stars is $0.94\,\langle V_c^2\rangle^{1/2}$, while the velocity dispersion of stars in the King model at $r = 100''$ is $0.73\,\langle V_c^2\rangle^{1/2}$. Gunn's model has three-fitting parameters in addition to simple scaling, but must produce the correct $\rho \propto r^{-2}$ heavy-halo envelope in addition to fitting the observed stellar distribution.

Elliptical galaxies typically show rotation curves when spectra are taken along the major axis, and it is generally supposed that the nonzero ellipticities of elliptical galaxies are due to rotation (but cf Section 6). Normally, ellipticals contain no significant amounts of gas and show no obvious new star formation; however metal-abundance gradients are observed between the nuclei and envelopes (Spinrad et al. 1972). Strom et al. (1977) argue that the U-R color is a good metal-abundance indicator: a redder color suggests higher metal abundance. They find color gradients extending far out into the envelopes of NGC 4486 (E1–E2), NGC 2768 (E5), and NGC 3115 (E7–S0). Miller & Prendergast (1962) find the nucleus of NGC 3379 to be redder than the envelope with the latter having a constant B-V color as a function of radius. Many workers, from Baum (1959) to Faber (1973), have found that colors and absolute magnitudes of ellipticals are well correlated. The brightest are the reddest, implying that the most massive ellipticals are the most metal-rich. While typical ellipticals contain no significant amounts of gas, J. S. Gallagher, G. R. Knapp, S. M. Faber, and B. Balick (private communication) find some ellipticals with appreciable amounts of neutral hydrogen gas. It is not clear whether such gas lies in disks or is a diffuse component.

The other great classes of galaxies are the spirals and S0's. These have both spheroidal and disk components. The spheroidal components resemble elliptical galaxies (deVaucouleurs 1974), while the disks are quite flat ($\varepsilon > 0.8$) and are supported by rotation. S0 disks typically have no new star formation and no gas, while spiral disks have both. The surface brightness distributions of the disks may be approximated by:

$$I(r) = I_0 \exp(-r/r_d), \tag{5}$$

where I_0 is the central surface brightness and r_d is a characteristic scale (deVaucouleurs 1959, Freeman 1970). Freeman (1970) finds $I_0 \sim 21.65$ mag arc sec^{-2} for the disk component independent of absolute B magnitude for a large class of spirals and S0's that contain pure exponential disks plus spheroidal components. Kormandy (1976) has argued that the small observed scatter in the extrapolated central surface brightness of the disk may result from confusion between the disk and spheroidal components, and that some of the disks may in fact have central surface brightnesses considerably lower than 21.65 mag arc sec^{-2} (see also Disney 1976).

Hubble (1926) classified 3% of the galaxies in his sample as irregular. These are typically low-luminosity companion galaxies containing gas and having an irregular shape. The Large Magellanic Cloud is an example.

Ostriker & Peebles (1973) have argued on the basis of their own and earlier numerical work and on the basis of theoretical results in Kalnajs (1972) that the classical model of our galaxy (a relatively cold self-gravitating disk held up by rotation) is violently unstable to bar-like modes. They begin with a disk model having a reasonable rotation curve and random velocity dispersions comparable to those in the Galaxy [and satisfying Toomre's (1964) criterion for radial stability]. The model is violently unstable to the growth of a bar and relaxes to a more centrally condensed distribution with large random velocities. At the end, $T_{rot} \sim$

$0.14|W|$ (virial equilibrium implies $T_{rot} + T_{random} = 0.5|W|$). As suggested by Kalnajs, relatively cold disks could be stabilized by adding a spheroidal halo with mass comparable to that of the disk. Ostriker & Peeble's combined halo-disk models with $T_{rot} < 0.14|W|$ were stable. These results suggest that approximately half the mass in the galaxy interior to the Sun may be in a heavy spheroidal halo. The bulk of the halo mass must come from low-luminosity stars (or black holes). Roberts & Whitehurst (1975) (see Figure 2) have shown that the rotation curve of M31 remains flat out to 30 kpc, implying $M(<r) \propto r$. This can be explained if most of the mass is in an invisible heavy halo with an envelope like an isothermal sphere $\rho_h \propto r^{-2}$. If the halo extends to 100 kpc, a mass of $10^{12}\ M_\odot$ for M31 is deduced in agreement with timing arguments for the dynamics of Local Group (Gunn 1974). Galaxy M/L ratios of ~ 10 are deduced from rotation curve studies at $r \sim 10$ kpc; if heavy halos extend to 100 kpc (Ostriker, Peebles & Yahil 1974), then we can explain total M/L values of ~ 100 obtained from virial analyses of binaries, groups, and clusters (cf Turner 1976, Gott & Turner 1977, Oemler 1974).

Schechter (1976) has shown that the luminosity function of galaxies in rich clusters is approximated well by

$$\phi(L)\,dL = \phi^* \left(\frac{L}{L^*}\right)^{-\alpha} \exp(-L/L^*)\,dL, \tag{6}$$

where $\alpha = 1.25$, ϕ^* is a constant, and L^* is the luminosity corresponding to a $B(0)$ absolute magnitude of -20.6, consistent with previous studies by Abell (1962). Turner & Gott (1976) find that the luminosity function in small groups is well

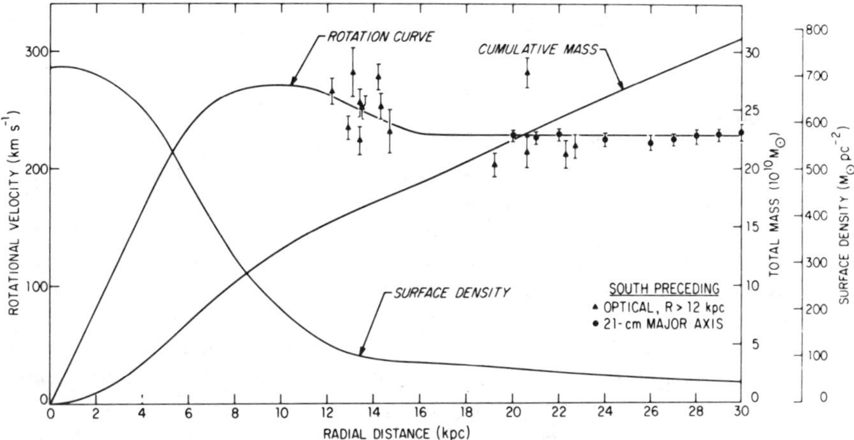

Figure 2 Roberts & Whitehurst's (1975) 21-cm observations show the rotation curve of M31 to be flat at large distances. The surface density and cumulative mass curves are for a highly flattened mass distribution; however the increase in cumulative mass is a general feature. This additional mass must have a high mass-to-light ratio since the surface brightness beyond $r = 20$ kpc is quite small. These observations support the heavy-halo picture of Ostriker, Peebles & Yahil (1974).

approximated by the above formula with $\alpha = 1.0$, $L^* = 3.4 \times 10^{10} L_\odot$ corresponding to a Zwicky absolute magnitude of -20.85. The two results differ by less than 2σ. Shapiro's (1971) data on galaxies in the local supercluster is consistent with the existence of a common luminosity function for spiral, S0, and elliptical galaxies.

The high-star formation rate expected in young galaxies suggests the possibility that they might be detectable at large redshift (cf Partridge & Peebles 1967a,b). Unsuccessful searches for such primeval galaxies have been made (Partridge 1974, Davis & Wilkinson 1974). Meier (1976a) has shown using Larson-type models (Larson 1974a) that primeval galaxies of 10^{11} $M_\odot$ might be detectable ($m < 21$) if $\Omega_0 > 1$, while if $\Omega_0 \sim 0.1$ (Ω_0 defined in Section 3) they should be undetectable because they would be expected to form at higher redshift (cf also Kaufman 1975, 1977). Meier (1976b) points out that OH 471 ($z = 3.40$) and 4C 05.34 ($z = 2.877$) have spectra similar to that expected for primeval galaxies, but notes that these objects are brighter than expected and have greater line widths. Tinsley (1976) has suggested that counts of galaxies to faint magnitudes can put interesting constraints on early galactic evolution.

For additional observational material on galaxies consult the recent reviews of deVaucouleurs (1974) and Freeman (1975).

3 FORMATION OF GALAXIES FROM SMALL INHOMOGENEITIES IN THE EARLY UNIVERSE

As we have noted, galaxies represent large density inhomogeneities in the universe. The standard idea is that galaxies and galaxy clusters arise via gravitational instability from small perturbations present at recombination (cf Rees 1971, Peebles 1974, Field 1975).

For a review of standard big-bang cosmology see Peebles (1971a), Weinberg (1972), and Rees (1971). The standard Friedmann cosmological models with no cosmological constant are determined by two parameters: H_0 and Ω_0. The present rate of expansion of the universe is $\mathbf{V} = H_0\mathbf{r}$, where H_0 is Hubble's constant; for convenience we take $H_0 = 50$ km sec^{-1} Mpc^{-1} (Sandage & Tammann 1975). If the present mean density of the universe is ρ_0, then $\Omega_0 = 8\pi G\rho_0/3H_0^2$. If $\Omega_0 > 1$ the universe will eventually recollapse, while $\Omega_0 \leqq 1$ implies expansion will continue forever. We concentrate on two models: $\Omega_0 = 1$ (Einstein-deSitter case) and a standard open model $\Omega_0 = 0.1$. The latter is claimed to have the advantage that it can simultaneously explain the mass-to-light ratios of galaxies, the age of the elements and globular cluster stars, and the cosmic deuterium abundance (Gott, Gunn, Schramm & Tinsley 1974). There are two key cosmological epochs: recombination and the epoch of equal matter and radiation density. Recombination occurs when the radiation temperature drops to $\sim$3000 K and the protons and electrons combine to form hydrogen. Given the present radiation temperature of 2.7 K this occurs at a redshift $(1+z_{\rm rec}) \sim 1000$ for all values of Ω_0 between 0.1 and 1. For $\Omega_0 = 1$, the epoch of equal matter and radiation density occurs at $(1+z_{\rm eq}) \sim 10^4$, while for $\Omega_0 = 0.1$ it occurs at approximately $(1+z_{\rm eq}) \sim 1000$.

For $\Omega_0 = 1$ the expansion of the universe is given by: $R = (1+z)^{-1} \propto t^{1/2}$ for

$0 < R < 10^{-4}$ (radiation dominated) and $R \propto t^{2/3}$ (matter dominated) for $10^{-4} < R$ [i.e. from $(1+z) = 10^4$ up through the present epoch, $R = 1$]. In the case $\Omega_0 = 1$ the mass within the horizon M_H (= the mass of observable universe) is proportional to the epoch t since $\rho = 3H^2/8\pi G$ at all times, and $\frac{1}{2} < Ht < \frac{2}{3}$, giving $M_H \cong \rho(ct)^3 \propto t$. At $(1+z_{\rm eq})$, $M_H = 10^{17}\ M_\odot$. Fluctuations on the mass scales of galaxies and clusters of galaxies thus come within the horizon during the radiation-dominated era and are frozen at approximately their initial amplitudes until recombination because the radiation and matter are locked together by Thomson drag, and the radiation fluid is quite stiff [Jeans mass $\sim M_H$ prior to $(1+z_{\rm eq})$ (cf Rees 1971)]. At recombination the Jeans mass drops to $4 \times 10^5\ M_\odot$ (cf Field 1975), and on all scales larger than this, density fluctuations in the linear regime can grow according to the well-known gravitational instability growing mode

$$\frac{\delta\rho}{\rho} \propto R \propto t^{2/3} \tag{7}$$

from recombination until the present epoch (cf Rees 1971, Field 1975). Thus, for galactic mass scales 10^{11}–$10^{12}\ M_\odot$, growth by a factor of 1000 is possible for perturbations in the linear regime. Once a perturbation has grown to the point where $\delta\rho/\rho \sim 1$, nonlinear effects accelerate the enhancement process. However, density fluctuations at recombination greater than 0.1% are needed to produce galaxies by the present epoch in an $\Omega_0 = 1$ cosmology. Furthermore, there is no known way to grow these finite perturbations from infinitesimal ones prior to recombination. Thus, as is discussed below, the perturbations must be primordial (Zeldovich 1972). For $10^{-4} < R < 10^{-3}$, $M_{\rm Jeans} \sim {\rm const} \sim 10^{17}\ M_\odot$ so fluctuations on mass scales larger than $10^{17}\ M_\odot$ can begin to grow at the rate given by Equation 7 at $R = 10^{-4}$. Thus perturbations on scales slightly larger than $10^{17}\ M_\odot$ can grow by a factor of 10 relative to those slightly below (Peebles & Yu 1970).

For the model with $\Omega_0 = 0.1$ the results are similar. No fluctuations can grow before $(1+z_{\rm eq}) \sim (1+z_{\rm rec}) \sim 10^3$, both because of Thomson drag and because small matter fluctuations do not grow in a universe dynamically dominated by radiation (Meszaros 1974). In general, perturbations in the matter cannot grow prior to $(1+z_{\rm eq})$ even if this is later than recombination. For the $\Omega_0 = 0.1$ universe we have $R \propto t^{1/2}$ for $0 < R < 10^{-3}$ (radiation dominated), $R \propto t^{2/3}$ for $10^{-3} < R < 10^{-1}$ (matter dominated, approximates Einstein-deSitter case), and $R \propto t$ for $10^{-1} < R$ [i.e. from $(1+z_{\rm op}) = \Omega_0^{-1} = 10$ up through the present epoch]. The last phase is that in which the density has dropped significantly below the critical value $(3H^2/8\pi G)$ so that the matter no longer significantly decelerates the Hubble expansion, and a linear expansion law prevails. At recombination $M_{\rm Jeans} \sim 10^6\ M_\odot$, $M_H \sim 10^{19}\ M_\odot$, so perturbations on mass scales in this range are free to grow according to Equation 7 for $10^{-3} < R < 10^{-1}$, but for $10^{-1} < R < 1$ we have

$$\frac{\delta\rho}{\rho} \sim {\rm const} \tag{8}$$

for perturbations in the linear regime. During this period the matter is dynamically insignificant; small density fluctuations do not perturb the mean Hubble flow and do not grow relative to the background. Thus in the $\Omega_0 = 0.1$ cosmology, growth by a factor of 100 is possible in the linear regime, requiring perturbations $\gtrsim 1\%$ at recombination for galaxy formation.

Notice that for the gravitational instability picture to work we must have $\Omega_0 > 0.01$ because at this limiting value we have $(1+z_{eq}) = (1+z_{op}) = \Omega_0^{-1} = 10^2$. Perturbations cannot grow prior to $(1+z_{eq})$ because the universe is radiation dominated, and cannot grow subsequent to this because the universe is so open that it has already entered its linear expansion phase.

Zeldovich (1972) has argued that the proper place to discuss primordial density fluctuations is when they first come within the horizon. After perturbations come within the horizon we can calculate their behavior. Density fluctuations on the scale of the horizon tell how well the different parts of the universe are "sewn" together, how well matched different parts of the universe are prior to their coming in causal contact. One may express any arbitrary primordial density fluctuation spectrum in terms of the amplitude of fluctuations coming within the horizon as a function of the mass within the horizon

$$\left(\frac{\delta\rho}{\rho}\right)_H = f(M_H). \tag{9}$$

For the phases of interest $M_H \propto t$ where t is the age of the universe. Zeldovich has argued that if nature is not capricious this spectrum should have the simplest possible form. In particular it should have no preferred mass scales; otherwise the universe would show a preferred mass scale that we would never have any way of predicting from astrophysics. [Indeed, as we shall see, covariance function studies by Peebles (1974) suggest that there are no intrinsic scales in galaxy clustering.] If there are to be no preferred mass scales f must be a power law: $f \propto M_H^x$. If x is positive the universe becomes more and more clumpy with time. Eventually $(\delta\rho/\rho)_H \sim 1$ and the universe becomes non-Friedmannian. Since the universe is so uniform today [$(\delta\rho/\rho)_H < 10^{-3}$ from isotropy of the 2.7 K radiation], it is hard to believe that x is positive. On the other hand, if x is negative, the universe would have been very non-Friedmannian at early epochs. The Friedmann model gives the right cosmological helium and deuterium abundances and these elements were formed when the mass within the horizon was only a few hundred solar masses, suggesting that the universe was approximately Friedmannian at this stage. Thus x cannot be very negative. Zeldovich proposes that $f = \kappa \sim 10^{-4}$. This is the simpliest possible primordial density fluctuation spectrum. Zeldovich suggests that $\kappa \sim 10^{-4}$ is simply one of the dimensionless constants of the universe, like the number of photons per baryon (10^{8-9}), and tells how well the universe approximates a Friedmann model. Indeed it has been suggested that there is a connection between the number of photons per baryon and κ (Zeldovich 1972, Gott & Rees 1975). So the Zeldovich hypothesis is

$$\left(\frac{\delta\rho}{\rho}\right)_H = \kappa \sim 10^{-4}. \tag{10}$$

There are two major types of initial density fluctuations: adiabatic and isothermal. Adiabatic perturbations conserve exactly the entropy per baryon and are like sound waves with equal photon and baryon fluctuations. Isothermal fluctuations do not conserve entropy per baryon; they represent baryon fluctuations in a uniform photon bath. The fate of adiabatic perturbations obeying the Zeldovich hypothesis has been studied by Doroshkevich, Sunyaev & Zeldovich (1974). Adiabatic perturbations enter the horizon at an amplitude of 10^{-4} and do not grow prior to recombination. (We have noted an exception to this for $M > 10^{17}\ M_{\odot}$ in the $\Omega_0 = 1$ cosmology.) Silk (1974) finds that adiabatic fluctuations on scales smaller than a characteristic damping mass M_D are damped due to photon viscosity; i.e. the excess photons can random-walk out of the perturbation on a Hubble expansion time scale. For $\Omega_0 = 1$, $M_D \sim 10^{12}\ M_{\odot}$; for $\Omega = 0.1$, $M_D \sim 10^{14}\ M_{\odot}$. At recombination one finds no significant fluctuations on scales smaller than M_D, but on all mass scales larger than M_D fluctuations are of the order of 10^{-4}. These sound waves involve peculiar velocities $\delta v \sim c(\delta\rho/\rho) \sim 10^{-4}\ c$. At recombination the restoring force due to the radiation suddenly disappears; regions that happened to be contracting will contract by an amount $\delta r \sim \delta v \cdot t_{\text{rec}}$ where t_{rec} is the Hubble expansion time scale at recombination. The velocity perturbations thus lead to baryon density fluctuations $(\delta\rho/\rho)_{\text{baryon}} \sim 3(\delta r/r) \sim 3\,\delta v t_{\text{rec}}/r \propto r^{-1}$. Since $M \propto r^3$ this leads to a baryon density fluctuation spectrum:

$$\left(\frac{\delta\rho}{\rho}\right)_{\text{baryon}} \propto M^{-1/3}. \tag{11}$$

See Figure 3, where these results are displayed in schematic form including the effects of additional growth for $M > 10^{17}\ M_{\odot}$ in the $\Omega_0 = 1$ cosmology.

The fate of isothermal fluctuations obeying the Zeldovich hypothesis has been studied by Gott & Rees (1975). We have $(\delta\rho/\rho)_{\text{total}} \sim 10^{-4}$ on the scale of the horizon at all times. Now for isothermal perturbations

$$\left(\frac{\delta\rho}{\rho}\right)_{\text{total}} = \left(\frac{\delta\rho}{\rho}\right)_{\text{baryon}} \cdot \frac{\rho_{\text{baryon}}}{\rho_{\text{total}}}. \tag{12}$$

Since the universe is radiation-dominated for the epochs of interest $\rho_{\text{total}} \propto R^{-4}$, $\rho_{\text{baryon}} \propto R^{-3}$, so $\rho_{\text{baryon}}/\rho_{\text{total}} \propto R$ where R is the expansion factor. Recall that $M_H \propto t$, $R \propto t^{1/2}$, so $M_{H,\text{total}} \propto R^2$ and $M_{H,\text{baryons}} \propto R^3$. Expressing the density fluctuations in terms of the mass scale in baryons we have

$$\left(\frac{\delta\rho}{\rho}\right)_{\text{baryon}} \propto M^{-1/3}. \tag{13}$$

These baryon fluctuations are not subject to damping and cannot grow because they are locked to the radiation via Thomson scattering, so they remain at this amplitude until recombination (cf Figure 3). For $\kappa = 10^{-4}$ (as illustrated in Figure 3) $(\delta\rho/\rho)_{\text{baryon}} \sim 1$ at scales of $10^5\ M_{\odot}$ ($\Omega_0 = 1$) or $10^7\ M_{\odot}$ ($\Omega_0 = 0.1$). Since the density of baryons can never be negative, at smaller scales equal positive and negative total-density fluctuations of amplitude $\kappa = 10^{-4}$ cannot be produced by baryon fluctuations alone. One speculates that on all smaller scales adiabatic fluctuations are dominant and that baryon fluctuations are of order unity (Gott & Rees 1975). Note that baryon fluctuations of this order would not interfere with cosmological helium and deuterium production (Epstein & Petrosian 1975). General perturbations can include both isothermal and adiabatic components.

As illustrated in Figure 3, Zeldovich's favored value of $\kappa = 10^{-4}$ produces isothermal fluctuations on galactic mass scales of a few percent for $\Omega_0 = 0.1$ and somewhat less than a percent for $\Omega_0 = 1$, sufficient to produce galaxies by the present epoch. The value of κ can be estimated by matching galaxy collapse times

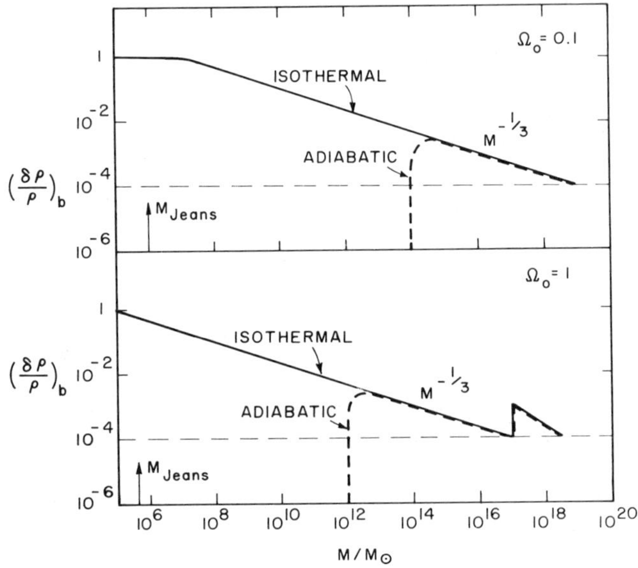

Figure 3 Schematic diagram of the expected baryon density fluctuation spectra just after recombination for Zeldovich's constant density fluctuation hypothesis. For purposes of illustration we have adopted $(\delta\rho/\rho)_H = \kappa = 10^{-4}$, and have considered two cosmological models $\Omega_0 = 0.1$ and $\Omega_0 = 1$. Adiabatic perturbations initially have amplitudes of 10^{-4} (light dashes), but perturbations on mass scales smaller than 10^{14} $M_\odot$ ($\Omega_0 = 0.1$) or 10^{12} $M_\odot$ ($\Omega_0 = 1$) are damped by photon viscosity. After recombination an enhancement process described in the text boosts the amplitudes to give the $(\delta\rho/\rho)_b \propto M^{-1/3}$ dependence shown above (heavy dashes). Isothermal baryon density fluctuation amplitudes at recombination as a function of mass scale are shown by the solid lines. Again we assume $(\delta\rho/\rho)_H = 10^{-4}$. On mass scales of interest for galaxy formation isothermal fluctuations give $(\delta\rho/\rho)_b \propto M^{-1/3}$. The feature at 10^{17} $M_\odot$ for both adiabatic and isothermal fluctuations in the $\Omega_0 = 1$ cosmology is due to growth of fluctuations on scales larger than 10^{17} $M_\odot$ between $(1+z_{eq})$, and recombination as described in the text. The curves stop at the mass within the horizon at recombination $M_H \sim 10^{19}$ $M_\odot$ ($\Omega_0 = 0.1$), $M_H \sim 3 \times 10^{18}$ $M_\odot$ ($\Omega_0 = 1$) where $(\delta\rho/\rho)_b \sim 10^{-4}$. The Jeans mass at recombination is shown by an arrow for each model.

(cf Equation 21) or by matching the observed amplitude of the covariance function of galaxies (cf Equation 14). Estimates of κ by Gott & Rees (1975) using the latter technique give values of 1.5×10^{-4} and 2.2×10^{-4} for $\Omega_0 = 1$ and $\Omega_0 = 0.1$ respectively. This model predicts fluctuations in the microwave background of $\delta T/T \sim \kappa \sim 10^{-4}$ on angular scales $>2°$ ($\Omega_0 = 1$), $>6°$ ($\Omega_0 = 0.1$), provided that the last scattering surface is at $z \sim 1000$ (Gott & Rees 1975). This is compatible with current limits on microwave background isotropy.

Doroshkevich, Sunyaev & Zeldovich (DSZ) (1974) favor pure adiabatic fluctuations in an $\Omega_0 = 0.1$ cosmology. Such an initial density fluctuation spectrum can lead to formation of bound protoclusters of galaxies of mass $10^{14}\ M_\odot$. Upon collapse of such a protocluster, shock fronts would form giving rise to dense pancakes of matter that could fragment into galaxies via the usual Jeans instability. Production of bound protoclusters of $10^{14}\ M_\odot$ requires $(\delta\rho/\rho) > 10^{-2}$ on this mass scale (cf Equation 19) and therefore $\kappa > 2.2 \times 10^{-4}$ (see Figure 3). However, collapse to a pancake in one direction may occur while a protocluster is still expanding in the other two directions (Sunyaev & Zeldovich 1972), so the protoclusters can be slightly unbound and $\kappa = 10^{-4}$ is acceptable. The adiabatic scheme applied to the $\Omega_0 = 1$ model could produce galaxies of $10^{12}\ M_\odot$ directly.

Peebles (1974) has argued that galaxies form first and clusters of galaxies form later by statistical clustering of the galaxies. Indeed, Peebles & Dicke (1968) propose that globular clusters, corresponding to the Jeans mass at recombination, were the first objects formed in the universe and that galaxies are aggregates of such primordial objects. This picture implies an isothermal density fluctuation spectrum that extends down to scales of the Jeans mass at recombination (cf Figure 3). Peebles cites as evidence for this picture the fact that the covariance function of galaxies (Totsuji & Kihara 1969, Peebles 1974) has a power law form with no intrinsic scales. The covariance function $\xi(r)$ is defined as the excess probability of finding a galaxy at a distance r from a random galaxy:

$$dP = n[1+\xi(r)]\,dV, \tag{14}$$

where n is the average number density of galaxies. Peebles finds

$$\xi(r) \sim 68\, r_{\mathrm{Mpc}}^{-1.77}. \tag{15}$$

If the DSZ scheme were correct one might have expected to find a strong feature in the covariance function on a scale of $10^{14}\ M_\odot$ which is not seen.

Isothermal density fluctuations carry no velocity perturbations, so the initial Hubble flow is uniform on galactic mass scales, and galaxies must gain their angular momentum via mutual tidal interactions. This effect is greatest at the point when the density perturbations on galactic mass scales have grown to be of order unity at redshifts $z \sim 10$–30. Since $(\delta\rho/\rho)$ is of order unity on galactic scales at this epoch, the protogalaxies (of mass M and size r) are quite irregular in shape having random quadrupole moments of order $Q \sim Mr^2$. The separations between protogalaxies are of order r so the tidal torque exerted on a protogalaxy by neighboring protogalaxies is of order $\tau \sim GMQ/r^3 \sim GM^2/r$. As the universe expands the protogalaxies move away from each other and the torques go down, so the effective time scale over which the torques act is the Hubble expansion time scale $t \sim (G\rho)^{-1/2} \sim G^{-1/2}M^{-1/2}r^{3/2}$. The angular momentum gained is $J \sim \tau t \sim G^{1/2}M^{3/2}r^{1/2}$. This amount of angular momentum is just sufficient to hold a protogalaxy up in centrifugal equilibrium: $V_{\rm cir}^2 \sim GMr^{-1}$, $J_{\rm eq} \sim MVr \sim G^{1/2}M^{3/2}r^{1/2}$. Thus, tidal interactions can transfer a dynamically significant amount of angular momentum. The spin angular momentum thus generated is equal and opposite to the generated orbital angular momentum of neighboring protogalaxies (Jones 1976). Detailed calculations by Peebles (1969, 1971b) and Thuan & Gott (1977) suggest that this process is just marginally sufficient to explain galaxy angular momenta.

One way to check the initial density fluctuation picture is by examining two statistics of galaxy clustering, the covariance and multiplicity functions.

The covariance function suggests that the spectrum at recombination is of a general power law form:

$$\left(\frac{\delta\rho}{\rho}\right)_{\rm b} \propto M^{-1/2-n/6}, \tag{16}$$

where n is a free parameter: $n = 0$ is a Poisson spectrum while $n = -1$ is the density fluctuation spectrum predicted by Equation 13. The three-point correlation function (Peebles & Groth 1975) suggests that galaxies are in a hierarchical clustering pattern with clumps within clumps within clumps. So $\xi(r) \approx \rho$ is the average density of clumps of size r. Thus we can deduce the average density of a clump today as a function of its mass. But we can also integrate a cosmological model to find how a small density perturbation at recombination evolves by the present epoch, i.e. we can find $\rho_{\rm present}$ as a function of $(\delta\rho/\rho)_{\rm initial}$. Given the initial density fluctuation spectrum, we can calculate the expected present covariance function and compare with observations. Rough estimates by Peebles (1974) suggest that an $n = 0$ initial spectrum is appropriate, while more extended calculations of a similar nature by Gott & Rees (1975) suggest that an $n = -1$ spectrum gives a better fit to the overall $\xi(r)$ curve than the $n = 0$ spectrum for both the $\Omega_0 = 1$ and the $\Omega_0 = 0.1$ model. Miyoshi & Kihara (1975) also find from N-body simulations for the $\Omega_0 = 1$ case that $n < 0$ is required.

An independent way to estimate the initial density fluctuation spectrum is to examine the multiplicity function of galaxies. By this we mean the luminosity function of groups of galaxies picked as satisfying a specific volume density criterion ($\rho_{\rm group} > \rho'$). Such a set of groups can be constructed for a large region of space by drawing the largest possible sphere containing a density $> \rho'$, which becomes the largest group; then one finds the second largest sphere containing a density $> \rho'$ and not overlapping the first, which becomes the second group, and so forth. The smallest spheres enclose only one galaxy each, i.e. single galaxies. Press & Schechter (1974) have developed a theory for such a catalogue of groups. The predictions are:

$$\eta_g(M)\,dM = (2\pi)^{-1/2}\frac{\bar{\rho}}{M_c}\left(1+\frac{n}{3}\right)\left(\frac{M}{M_c}\right)^{-3/2+n/6}\exp\left[-\frac{1}{2}\left(\frac{M}{M_c}\right)^{1+n/3}\right]\frac{dM}{M_c}, \tag{17}$$

where $\eta_g(M)\,dM$ is the number density of groups in mass range dM about M,

$\bar{\rho}$ is the average density of the universe, M_c is the characteristic mass scale such that on that scale the rms density fluctuation amplitude is just sufficient to produce a group of present density ρ', and n is the coefficient of the initial density fluctuation spectrum in Equation (16). As Schechter (1975) has pointed out, this formula depends only on the shape of the initial density fluctuation spectrum and not at all on Ω_0; it is entirely an initial value problem. This functional form is reminiscent of the galaxy luminosity function, Equation (6). Indeed if one believed that in galaxies we are simply seeing perturbations that had collapse times shorter than a critical value T_c (i.e. were greater than a critical density) (cf Equations 21, 22), then we would expect the form of the galaxy luminosity function and the multiplicity function to be identical. However, for $n \leq 0$ the slope of the theoretical multiplicity function at the faint end is far steeper than that of the galaxy luminosity function. Tidal stripping of galaxies (Richstone 1976) offers a possible explanation; brighter galaxies are stripped first, sharpening the cutoff at the bright end and flattening the galaxy luminosity function at the faint end. Written in integral form the above equation gives the fraction of mass in the universe in groups of a given density smaller than mass M. The flatter the initial density fluctuation spectrum the broader the range of galaxy cluster sizes at a given density. Gott & Turner (1977b) find that $\sim 50\%$ of the luminosity density of the universe is contributed by groups as small or smaller than the Local Group (i.e. $\leq L^*$), while $\sim 1\%$ of the luminosity of the universe is in clusters as large or larger than Coma ($\geq 200\ L^*$). Fitting the Press and Schechter forms to their group catalogue data they estimate $n = -1.3 \pm 0.3$, with different techniques for analyzing the multiplicity function yielding values of n between -0.9 and -1.4. These results are consistent with the $n = -1$ spectrum expected for isothermal fluctuations obeying the Zeldovich constant density fluctuation hypothesis (Equation 13).

4 THE COLLAPSE PICTURE OF GALAXY FORMATION

For perturbations of galactic size we may assume that growth begins at $(1+z_{\text{rec}}) \sim 10^3$. The Hubble expansion rate at this epoch is

$$H_{\text{rec}}^2 \sim \Omega_0 H_0^2 (1+z_{\text{rec}})^3. \tag{18}$$

Defining $\rho_{\text{cr}} = 3H_{\text{rec}}^2/8\pi G$ one finds (cf Sunyaev 1971, Gunn & Gott 1972)

$$\rho_{\text{cr}} - \bar{\rho} \approx \rho_{\text{cr}} \frac{1-\Omega_0}{\Omega_0(1+z_{\text{rec}})}, \tag{19}$$

where $\bar{\rho}$ is the mean density of the universe at recombination. Assume that a perturbation originally participates in the mean Hubble flow. To be bound, a perturbation must have a mean density $\rho > \rho_{\text{cr}}$. In the $\Omega_0 = 1$ cosmology all positive density fluctuations $(\delta\rho/\rho) > 0$ are bound, while in the $\Omega_0 = 0.1$ cosmology, density fluctuations must exceed a critical value $(\delta\rho/\rho) > (\rho_{\text{cr}} - \bar{\rho})/\bar{\rho} \approx 10^{-2}$ in order to be bound. Any bound perturbation will expand for a while but eventually will halt its expansion at a maximum radius R_0 and begin to collapse under its own gravitational attraction. Let T_c be the age of the universe when the galaxy completes its recollapse. The galaxy reaches its maximum radius R_0 at $t = \frac{1}{2}T_c$. One can calculate T_c in terms of the original density ρ in the perturbation at recombination, $(\delta\rho/\rho) = (\rho - \bar{\rho}/\bar{\rho})$:

$$T_c \approx \frac{\pi}{H_{\rm rec}}\left(\frac{\rho_{\rm cr}}{\rho - \rho_{\rm cr}}\right)^{3/2} \tag{20}$$

For perturbations of interest (where $\delta\rho/\rho > \rho_{\rm cr} - \bar{\rho}/\bar{\rho}$ and $\rho_{\rm cr} \sim \bar{\rho}$), this can be approximated by:

$$T_c \approx \frac{\pi}{H_0\Omega_0^{1/2}(1+z_{\rm rec})^{3/2}}\left(\frac{\delta\rho}{\rho}\right)^{-3/2}. \tag{21}$$

Thus, the bigger the initial density fluctuation the shorter the collapse time. We also have the relation

$$T_c = \frac{\pi}{(2)^{1/2}}(R_0^3/GM_0)^{1/2}, \tag{22}$$

where M_0 is the mass of the galaxy. At the point of maximum expansion the kinetic energy T is low and the total energy is $E = W_0 = -\frac{3}{5}GM_0^2/R_0$. If the gas comprising the protogalaxy has formed stars by $t = T_c$ so there is a dissipationless collapse (as proposed in some models for elliptical galaxies), the galaxy will reach virial equilibrium $2T + W = 0$, $E = W_0 = T + W = \frac{1}{2}W = -\frac{3}{10}GM_0^2/R_{\rm eq}$ with an equilibrium radius $R_{\rm eq} \sim \frac{1}{2}R_0$. Of course the final galaxy will be centrally condensed but we may define the mean harmonic radius $R_H = |GM_0^2W^{-1}|$ in the usual way so that $V^2 = GM_0/R_H$ where V is the velocity dispersion of the stars. If we define the crossing time of the system

$$\Delta t = \left(\frac{3}{5}\right)^{3/2} R_H V^{-1}, \tag{23}$$

then

$$T_c = 2\pi\,\Delta t. \tag{24}$$

If dissipation has been important T_c will be longer than this. Picking characteristic numbers $V \sim 250$ km sec^{-1} and $R_H \sim 10$ kpc, we find $T_c \sim 10^8$ yr, which requires that $\delta\rho/\rho = 0.07$ ($\Omega_0 = 1$), $\delta\rho/\rho = 0.14$ ($\Omega_0 = 0.1$) at recombination.

The idea that galaxies form from a collapse process received support from the classic work of Lynden-Bell (1967) on violent relaxation. Prior to this, one of the great mysteries of stellar dynamics had been how elliptical galaxies achieved such apparently relaxed stellar distributions. The two-body relaxation time for a system of N stars in a self-gravitating system is of the order

$$T_R \sim \frac{N}{2\ln N}\Delta t. \tag{25}$$

For additional information see Chandrasekhar (1960), Spitzer & Thuan (1972), Gott (1973). T_R is the time scale over which stars have their individual energies per unit mass ($E_S = \frac{1}{2}V_S^2 + \Phi_S$) altered by an amount comparable with their own magnitude, $\Delta E_S/E_S \sim 1$. For $N = 10^{11}$ stars and $\Delta t \sim 2 \times 10^7$ yr, T_R is 6 orders of magnitude longer than the age of the universe so no significant relaxation can occur. Lynden-Bell pointed out that in a collapsing system where the potential is changing, stars do not move in energy conserving orbits, but

$$\frac{DE_S}{Dt} = \frac{\partial \Phi}{\partial t}, \tag{26}$$

where Φ is the changing gravitational potential. At the beginning of the collapse a star's energy is of the order $E_S \sim \Phi_S \sim GM/R_0$. During the collapse Φ changes by an amount comparable with its original value, so during this time a star's energy changes by an amount

$$\frac{\Delta E_S}{E_S} \sim \frac{\Delta \Phi}{\Phi} \sim 1, \tag{27}$$

a relaxation effect equal to one whole relaxation time's worth of two-body relaxation. (Unlike two-body relaxation, violent relaxation produces velocity dispersions that are independent of stellar mass.) Numerical calculations (Peebles 1970, Bovier & Janin 1970, Gott 1973) show that by $t \sim \frac{3}{2}T_c$, phase mixing and violent relaxation lead to an equilibrium state in virial equilibrium. After this, $\Phi \sim$ const so that violent relaxation turns off. Since the violent relaxation process is necessarily incomplete, numerical studies are needed to delineate what final states may be achieved. For rotating stellar systems of finite extent the maximally relaxed state is one of solid-body rotation with isothermal, isotropic random velocities. Numerical experiments for axisymmetric systems of 2000 point masses (Gott 1973) indicate that violent relaxation is sufficient to achieve these conditions in the central regions but that distributions that are anisotropic and differentially rotating prevail in the envelopes of the galaxies. Violent relaxation appears sufficient to produce approximately Maxwellian velocity dispersions in the central regions, relaxed elliptical isophotes, and central "isothermal" cores as in the King models. One problem is that simple models involving the collapse of a sphere of stars (Peebles 1970, Gott 1973) do not show envelopes as extended as those of observed elliptical galaxies. The simulations have envelopes with $\rho \propto r^{-4}$ instead of $\rho \propto r^{-3}$ as required by Hubble's Law (but cf Sections 5 and 6).

Eggen, Lynden-Bell & Sandage (1962), and Sandage, Freeman & Stokes (1970) have utilized these ideas to explain the structure of our galaxy. The relaxed distribution of stars in the spheroidal component argues for its formation via collapse and violent relaxation. The stars in the halo with orbits of large eccentricity are metal-poor, while the stars in the disk with circular orbits are all metal-rich, indicating that the stars in the disk were formed after the halo stars. The halo stars thus represent those stars formed prior to the final collapse of the galaxy ($t < T_c$). Any gas left over at $t = T_c$ quickly dissipated its energy and formed a disk, and stars formed subsequently from gas in the disk. Sandage, Freeman & Stokes (1970) propose that angular momentum is a key factor in determining the star-formation rate, noting that the halo has a lower angular momentum per unit mass than the disk. This simple dynamical scenario has been further investigated and extended recently by Gott & Thuan (1976). Numerical models using both gas and stars show that the gas and stars move together on free-fall trajectories during the collapse, but when the point of maximum collapse is reached (a pancake configuration for a rotating system), the stars already formed cross through the plane to form a relaxed spheroidal component, while the gas immediately dissipates its energy and forms a thin disk. There is a crisp division between spheroidal and disk components, as occurs in nature. DeVaucouleurs (1974) has long argued that all galaxies can be decomposed into disk and spheroidal components and that the spheroidal components taken separately resemble elliptical galaxies in every respect. Indeed, Ostriker (1975) has cited several examples including M31, where the spheroidal component separately follows perfectly Faber and Jackson's $L \propto V^4$ and $L \propto r_c$ curves for ellipticals.

Once a flat disk of gas has formed, relaxation processes in the disk can cause it to relax to a centrally condensed exponential form (cf Equation 5). Numerical calculations by Hohl (1971) start with the surface-density distribution of a flattened Maclaurin spheroid in solid-body rotation with random velocities satisfying Toomre's (1964) criterion. It is unstable to bar-like modes and quickly evolves to a more centrally condensed state with differential rotation and relatively large random velocities. The final density distribution gives a high-density central core plus an outer disk that closely approximates the deVaucouleurs-Freeman exponential form.

Thus we have a basic scheme for galaxy formation and can now explore differing detailed models within this general framework as well as opposing ideas.

5 DISSIPATION MODELS FOR ELLIPTICALS

Larson (1974a, 1974b, 1975) has made the most detailed models for elliptical galaxies to date. He begins by considering spherical systems. He lets most but not all stars form by $t = T_c$, the remaining mass being in gas which undergoes a slow dissipation. Stars continue to form as the gas settles toward the center. Larson finds it is best to have the gas dissipation time scale, star formation time scale, and dynamical time scale about equal: $\tau_{\rm diss} \sim \tau_{\rm star} \sim \tau_{\rm dyn}$. This is plausible; if $\tau_{\rm diss} \sim \tau_{\rm star}$ for a spherical system we might not be surprised if an equal number of stars were left in each logarithmic interval of radius as the gas settles to the center, giving a Hubble's law envelope $\rho \propto r^{-3}$. For this scenario to work $\tau_{\rm diss} \gtrsim \tau_{\rm dyn}$, because if $\tau_{\rm diss} \ll \tau_{\rm dyn}$ the gas cools quickly and undergoes a free-fall collapse to the center. In a spherical system this would cause all the gas to end up in an extremely dense nucleus. (In a rotating system it would end up in a flat disk.) Larson is able to achieve extraordinarily good fits to the observed surface brightness distributions. In Figure 1 is shown the fit of Larson's model to the observations of NGC 3379. While the detailed fit of this model has been achieved by adjusting three parameters ($\tau_{\rm star}/\tau_{\rm dyn}$, $\tau_{\rm diss}/\tau_{\rm dyn}$, and the velocity dispersion at $t = \frac{1}{2}T_c$) as opposed to only one free parameter for the King model, it does show that a consistent model is possible. The Larson models naturally predict radial metal-abundance gradients in E0 galaxies. The gas becomes more metal-enriched with time as it settles toward the center owing to dissipation, so that later-generation, metal-rich stars are formed close to the center. The Larson models are compared with the observations in Figure 4. There is good qualitative agreement. Some Larson models produce metal-abundance gradients in the nucleus only while others predict gradients extending into the envelope, as found by Strom et al. (1977). In the Larson models the gas eventually reaches such a low density that supernovae in the galaxy can heat it to the point where it escapes from the system. Any further gas ejected by dying stars will then be continuously swept out of the system by supernovae (Mathews & Baker 1971). These models thus give a qualitative explanation of the color–absolute-magnitude relations of Faber (1973). In a more massive galaxy with a higher escape velocity the remaining gas fraction can become lower before ejection occurs. Massive galaxies should retain their gas longer, form more metal-rich stars, and be redder than smaller galaxies (Larson 1974b).

When Larson (1975) applies his models to rotating systems he finds that the isophotes become far too flat ($\varepsilon > 0.9$) as one approaches the center (the observations of elliptical galaxies show typically $\varepsilon \sim$ const or $\varepsilon \to 0$ as $r \to 0$). The rotating gas cloud simply settles into a thin disk as it dissipates energy. Larson (1975) is able to solve this problem by allowing for a gas viscosity. In a self-gravitating, differentially rotating system the effect of viscosity is to take angular momentum from the central regions and deposit it in the outer regions allowing the central regions to further contract (cf Lynden-Bell & Pringle 1974). Larson finds that the dissipation time scale must be approximately equal to the time scale for viscous rearrangement of the angular momentum: $\tau_{\text{diss}} \sim \tau_{\text{vis}}$. This way the gas at the center loses its angular momentum at the same rate as it loses energy, maintaining the same ellipticity as it contracts.

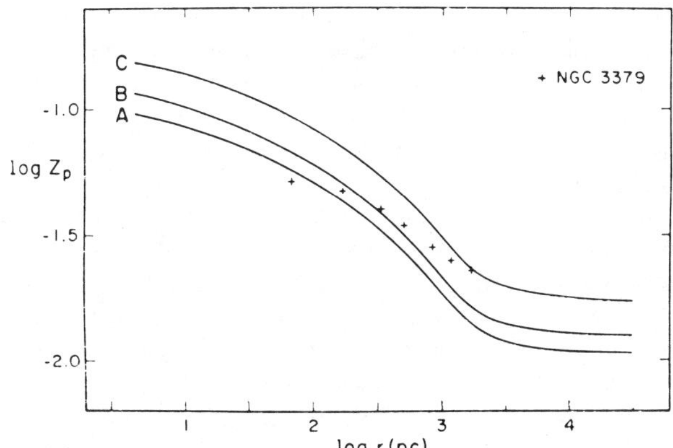

Figure 4 Stellar metal abundance averaged along the line of sight through Larson's (1974a) models A, B, C, plotted versus projected distance from the center. Model B is compared with surface photometry of NGC 3379 in Figure 1. Models A and C have slightly different parameters. The + symbols are derived from observations of NGC 3379 (Spinrad et al. 1972) by assuming that a "super metal rich" star has $Z = 0.06$ and a "normal" star has $Z = 0.015$ (Larson 1974a).

The Larson models require unusually slow dissipation and large viscosity. The bremsstrahlung cooling time for a cloud of gas in virial equilibrium with radius R is (cf Gott & Thuan 1976, Ostriker & Rees 1976)

$$\tau_{\rm cool} = 2.2 \times 10^{14}\, s \left(\frac{10^{11}\, M_\odot}{M}\right)^{1/2} \left(\frac{R}{10\ \rm kpc}\right)^{5/2}. \tag{28}$$

With $\tau_{\rm cool} = \tau_{\rm diss}$, $\tau_{\rm dyn} \sim \frac{1}{2} T_c$

$$\frac{\tau_{\rm diss}}{\tau_{\rm dyn}} = \left(\frac{R}{73\ \rm kpc}\right). \tag{29}$$

With $R \sim 10$ kpc, $\tau_{\rm diss}$ is much too short for the Larson models to work. Also in the interstellar medium today $\tau_{\rm diss} \sim 10^7$ yr $\ll \tau_{\rm dyn}$ because of cloud-cloud collisions (Spitzer 1968). Likewise, the viscosity required by Larson cannot be due to turbulent viscosity in a diffuse gas since the requirement $\tau_{\rm vis} \sim \tau_{\rm diss}$ would imply an effective Reynolds number $\mathbf{R} \sim 3$ (Gott & Thuan 1976), whereas the lowest Reynolds number observed in *any* turbulent flow is $\mathbf{R} \sim 30$ and for more relevant flows between rotating cylinders $\mathbf{R} \sim 10^4$, implying $\tau_{\rm vis} \gg \tau_{\rm diss}$.

The best possibility is that the gas consists of discrete clouds with mean free paths λ between collisions. Let R be the radius of the system and $V = (GM/R)^{1/2}$ be the velocity dispersion of the clouds. Then $\tau_{\rm dyn} = R/V$; $\tau_{\rm diss} = \tau_{\rm dyn} \cdot (\lambda/R)$; $\tau_{\rm vis} = \tau_{\rm dyn} \cdot (R/\lambda)^2$ for $\lambda < R$ and $\tau_{\rm vis} = \tau_{\rm dyn} \cdot (\lambda/R)$ for $\lambda \geqq R$ (Gott & Thuan 1976). If $\lambda \sim R$ one has Larson's parameters exactly: $\tau_{\rm diss} = \tau_{\rm vis} = \tau_{\rm dyn}$. If $\lambda > R$ one gets a Larson-type evolution on a time scale $\tau_{\rm diss}$. If $\lambda < R$ the gas quickly dissipates its random energies and undergoes a free-fall collapse, forming a flat disk with viscosity being unimportant. This is a scenario for the formation of a spiral galaxy (Gott & Thuan 1976).

One potential problem with the details of the Larson rotating models is that the viscosity coefficients are calculated via the Navier Stokes equations by extrapolating the short mean free path formula to the $\lambda \sim R$ regime. Larson (1975) therefore derives a second viscosity formula valid for the long mean free path limit. This formula appears to simulate the effects of violent relaxation but the viscosity goes to zero when the system reaches equilibrium. This second formula does not produce nicely elliptical isophotes as does the first formula. J. Binney (private communication) has pointed out that treatment of viscosity in a long mean free path system is complicated (cf Biermann 1951), and in some cases the viscosity can even be negative. Although heuristic arguments suggest that such extreme situations might not arise in real galaxies, there is a general uncertainty as to how to treat the viscosity properly.

Even with viscosity included, Larson's models show strong metal-abundance gradients perpendicular to the plane. If (U-R) colors are a good metallicity indicator (Strom et al. 1977), then the isochromes for an elliptical galaxy should be flatter than the isophotes. For example, Larson's (1975) Model I at 5 to 10 kpc from the center has isophotes with a flattening of $\varepsilon \approx 0.4$ and isochromes with a flattening $\varepsilon \approx 0.75$. Data on U-R colors for the E5 galaxy NGC 2768 and for the E1–E2 galaxy NGC 4486 by Strom (private communication) show the isophotes and isochromes to be of approximately equal flattening. The galaxy NGC 3115 (Strom et al. 1977) is more difficult to analyze since it also includes a disk of nearly constant color, which by itself tends to make the isochromes very flat. However, by fitting the spheroidal component separately, Strom et al. find isophote ellipticities of $\varepsilon = 0.64, 0.57, 0.46$ and isochrome ellipticities of $\varepsilon = 0.74, 0.60, 0.49$ at distances of $40''$, $80''$, $160''$ respectively along the minor axis ($40'' = 1.6$ kpc). The Larson models predict more dramatic differences between isophotes and isochromes than are observed. Whether this constitutes a serious difficulty for this class of models remains to be seen.

The Larson models are not directly extendable to production of spiral and S0 galaxies. Larson (1976) finds that in order to produce systems with appreciable disk components the star formation rate must dramatically drop after the collapse and the viscosity must become negligible.

6 DISSIPATIONLESS COLLAPSE MODELS FOR ELLIPTICAL GALAXIES

Dissipationless collapse models can give a natural account of the dynamical properties of elliptical galaxies (and spheroidal components of spiral galaxies) relying on stellar dynamics alone. As described earlier the dissipationless collapse models of Gott (1973) give smooth elliptical isophotes like those observed. Furthermore, the observed ellipticities of elliptical galaxies are given a simple explanation. A nonrotating galaxy at the point of maximum expansion ($t = \frac{1}{2}T_c$) will have a characteristic radius R_0; if stars are formed prior to $t = T_c$ the collapse will be dissipationless, and when the galaxy reaches virial equilibrium it will have a characteristic size $R \sim \frac{1}{2}R_0$ (cf Section 4). Consider a spherical protogalaxy with an amount of angular momentum near the maximum allowed by the tidal angular momentum picture (cf Section 3), i.e. it has acquired by $t = \frac{1}{2}T_c$ an angular velocity $\Omega = (GM_0/R_0^3)^{1/2}$ sufficient to support it in the equatorial plane. Such a protogalaxy will not collapse significantly in the equatorial plane as it is already supported in that direction by rotation giving $R_{\text{equator}} \sim R_0$, but rotation does not inhibit collapse perpendicular to the plane so that collapse should proceed in the polar direction just as in the spherical case, giving $R_{\text{polar}} \sim \frac{1}{2}R_0$. This produces an E5 galaxy. These simple virial theorem arguments suggest that elliptical galaxies should have flattenings up to about 2 to 1 as is observed. Thuan & Gott (1975) have recently made these arguments somewhat more quantitative using Maclaurin spheroid models. With an approximately spherical protogalaxy, they find $\varepsilon < 0.71$ for the equilibrium galaxy regardless of the initial angular momentum. The agreement between this and the observed maximum ellipticity of $\varepsilon = 0.7$ is quite suggestive even though the models are admittedly rather crude.

The following relations from Gott & Thuan (1976) are useful. For a dissipationless collapse where the original protogalaxy is a nonrotating sphere of size R_0 at $t = \frac{1}{2}T_c$ and relaxes to an equilibrium configuration obeying deVaucouleurs Law,

$$R_e = 0.275R_0, \tag{30}$$

where R_e is the radius containing half the light as in Equation (2). For ellipticals in the same absolute magnitude range as our own galaxy ($M_B = -20$, $L \sim 1.5 \times 10^{10} L_\odot$, $M \sim 1.5 \times 10^{11} M_\odot$ for the optically visible portion), one finds $R_e \sim 2.5$ kpc, $R_0 \sim 9$ kpc and therefore $T_c \sim 7 \times 10^7$ yr. For spiral galaxies we have the relation

$$r_d = 0.382 t_i R_0, \tag{31}$$

where $t_i = |T_{\rm rot}/W|$ at $t = \frac{1}{2}T_c$ and r_d is the deVaucouleur-Freeman disk radius defined in Equation (5), and we have used conservation of angular momentum. Monte-Carlo simulations of tidal torque generated by the tidal interaction picture suggest $(t_i)_{\rm median} \sim 0.075$. With no dissipation this amount of angular momentum would produce an E3 galaxy. However, if gaseous dissipation occurs and a disk is formed $r_d = 0.03 R_0$ for this case. For galaxies in the luminosity range of our galaxy typically $r_d \sim 3$ kpc (Freeman 1970). This gives a typical value of $R_0 \sim 100$ kpc and with a disk mass $M \sim 1.5 \times 10^{11}$ $M_\odot$ gives $T_c \sim 2.7 \times 10^9$ yr. This value for R_0 is consistent with the arguments of Eggen, Lynden-Bell & Sandage (1962). Any heavy halo components should lie outside and not greatly affect the collapse dynamics of the inner parts.

If the star formation rate per unit volume is proportional to ρ^2 (Schmidt 1959), then $\tau_{\rm star} \propto \rho^{-1}$. Now the collapse time of a galaxy is related to its mean density during the expansion and collapse phase $T_c \propto (G\rho)^{-1/2}$, so $\tau_{\rm star}/T_c \propto \rho^{-1/2} \propto T_c$ and ellipticals with their characteristically shorter collapse times should have more complete star formation by $t = T_c$ than the spirals which have lower densities and longer collapse times. In this picture (Gott & Thuan 1976), the protogalaxies with the highest initial density perturbations at recombination (and therefore the shortest collapse times) will become ellipticals, while those with lower initial density perturbations become spiral galaxies.

If a region has large positive density fluctuations so as to produce elliptical galaxies, then it will also have a high mean density and will become a dense cluster of galaxies. Gott and Thuan point out that this can explain why the frequency of elliptical galaxies is highest in the cores of rich clusters of galaxies (cf Oemler 1974).

In every protogalaxy perturbation the excess density is expected to be the highest at the center. So we would expect the central region to have a short collapse time and efficient star formation. It should relax to form a spheroidal system of stars. The outer regions will complete their collapse later and should have a good deal of left-over gas, which will then settle into a disk. Thus, in agreement with the observations, we expect the spheroidal component to resemble a normal elliptical galaxy. The observed spheroidal component is expected to have less angular momentum per unit mass than the disk because with its shorter collapse time it would have a smaller quadrupole moment at $t \sim \frac{1}{2}T_c$ and will gain less angular momentum from tidal interactions (Section 3). If, as Ostriker & Peebles (1973) argue, a substantial fraction of the mass is in the elliptical component, then it is clear that the universe efficiently produces elliptical galaxies with the presence or absence of later infall gas determining whether or not they will become spirals. Ostriker & Thuan (1975) have models where the majority of gas in the disk is not primordial at all, but is second-generation metal-enriched gas from dying stars in the spheroidal component. Such models can explain why there are so few metal-poor stars in the disk without relying on changes with time in the stellar birthrate as a function of mass.

J. E. Gunn (private communication) has pointed out how a dissipationless model can explain the Faber & Jackson (1976) relations for the cores of ellipticals. Since $L_c \propto r_c$, $L_c \propto V_c^4$ we have $M_c \propto L_c^{3/2}$. With a dissipationless collapse $r_c \propto R_0$, $T_c \propto (M_c R_0^{-3})^{-1/2}$ and using Equation (21) we find an initial $(\delta\rho/\rho)_c \propto M_c^{-1/3}$, the same relationship between initial density fluctuations and mass scales expected from Equation (13).

We recall that simple dissipationless models of the collapse of a sphere of stars produce an envelope with $\rho \propto r^{-4}$ rather than $\rho \propto r^{-3}$ as required by Hubble's Law. This problem can be solved by cosmological infall (Gott 1975). Consider a perturbation with size R_i, mass M_i, and excess density $(\delta\rho/\rho)_i$, and let the density outside this perturbation be the average cosmological density. The central region will collapse at time $t = T_c$, but material at a larger radius will still be bound to it and suffer infall at a later time. [Material at radius r initially will have an excess density interior to its radius of $(\delta\rho/\rho) = (\delta\rho/\rho)_i \cdot (r/R_i)^{-3}$.] If the mass of the galaxy is taken to include all mass that has crossed through the center, then for $t > T_c$

$$\frac{M(t)}{M_i} = 1 + \frac{(t/T_c)^{2/3} - 1}{1 + (1 - \Omega_0)(H_0 t/\pi\Omega_0)^{2/3}} \tag{32}$$

(Gunn & Gott 1972). For $T_c < t \ll H_0^{-1}$ one finds

$$M(t) \sim M_i \cdot (t/T_c)^{2/3} \tag{33}$$

independent of Ω_0. This infall material is less tightly bound than the original perturbation proper and will produce an extended envelope. If each infalling shell were artificially halted at its radius of maximum expansion and held there it would produce an envelope with

$$\rho \propto r^{-2.25}. \tag{34}$$

(Gott 1975). This is certainly an upper limit to the extent of the envelope. The shape of the envelope will depend on the details of violent relaxation. For example, if each shell of stars relaxes to a $\rho \propto r^{-3}$ configuration, then a number of r^{-3} distributions with different scale factors are added and the total envelope will still have $\rho \propto r^{-3}$. A numerical experiment with these infall conditions using 1000 stars (100 in the central perturbation) gives an envelope with $\rho \propto r^{-2.8}$ in good agreement with Hubble's Law (Gott 1975). An axisymmetric model using 2000 point masses and including both infall and tidal torque by neighboring protogalaxies produces a model that gives a good fit to the surface brightness distribution of NGC 4697 (Figure 5). The model fits the data very well without the use of any free-fitting parameters. The overall slope in the envelope is set by the infall, giving good agreement with Hubble's Law; the cutoff at large radii is due to tidal stealing by neighboring protogalaxies. The dissipationless models naturally produce a slow variation of ellipticity with radius.

Gunn (1976) has argued that if one waits long enough violent relaxation becomes less and less important so that eventually one reaches the limiting relation $\rho \propto r^{-2.25}$ far from the center. In this regime each star falls into an average equilibrium radius that is a constant factor smaller than its radius at maximum expansion. This is the type of mass distribution expected for heavy halos, i.e. $\rho \propto r^{-2}$. One could have the following scenario. The central perturbation proper collapses and relaxes leaving an envelope with $\rho \propto r^{-4}$. Gott's numerical calculations show that the $\rho \propto r^{-4}$ distribution of the stars from the central perturbation is not disturbed by the infall stars. Since the density in the central region is high throughout the collapse it should have a high rate of star formation and reprocessing. Thus, these stars might be expected to be metal rich. Relatively metal-poor infall stars add a factor of 10 in mass and with violent relaxation important settle into an

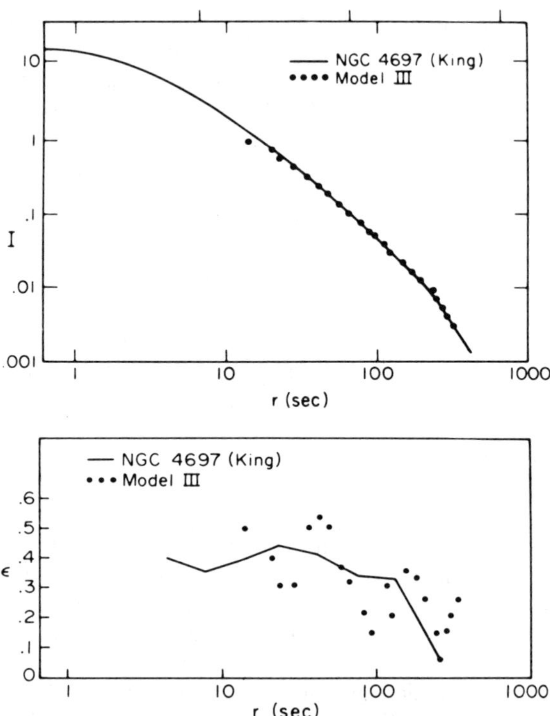

Figure 5 Comparison of Gott's (1975) dissipationless collapse model with King's photometry of the E5 galaxy NGC 4697. The observations are indicated by solid lines while the model results (Model III) are indicated by dots, the upper graph giving surface brightness as a function of radius and the lower graph giving ellipticity as a function of radius. While the model's ellipticity data show considerable statistical scatter, in the mean they also closely parallel the NGC 4697 data. Aside from simple scaling the model uses no free-fitting parameters.

extended envelope with $\rho \propto r^{-3}$ in agreement with Hubble's Law. One could have a metal-abundance gradient extending far out into the envelope with the average metal abundance falling off as rapidly as r^{-1}. Further out, as still another factor of 10 in mass is added by infall, the violent relaxation becomes smaller and the envelope falls off as $\rho \propto r^{-2.25}$. This outer envelope must be composed primarily of low mass, high M/L stars or black holes. (From earlier arguments we might have expected the outer envelope to be left as gas, so this is a somewhat ad hoc part of the model.)

The Gunn model has a halo with $\rho \propto r^{-2.25}$, based on infall with the assumption that beyond the central perturbation the density was equal to the average density. This is the situation expected for a Poisson $n = 0$ spectrum, where the distribution of point masses is uncorrelated. The $n = -1$ spectrum of Section 3 has more power at large scales. Point masses in such a distribution would have a covariance function $\xi(r) \propto r^{-2}$ (Gott & Rees 1975, Aarseth, Gott & Turner 1977). Thus, instead of having $(\delta\rho/\rho) = (\delta\rho/\rho)_i \cdot (r_i/R_i)^{-3}$ as in the $n = 0$ case we have $(\delta\rho/\rho) = (\delta\rho/\rho)_i \cdot (r_i/R_i)^{-2}$ for the $n = -1$ case leading to

$$\rho \propto r^{-2} \tag{35}$$

for the final heavy-halo envelope, just as suggested by the observational data (cf Section 2), although perhaps not discernibly different from Gunn's result.

Gunn (1976) has made a two-component, self-gravitating King model to illustrate his heavy-halo theory. The model is determined by three parameters: the depth of the central potential well, the ratio of central densities of the observed stellar component and the invisible heavy-halo component, and the ratio of the central velocity dispersions of the two components. The fit of Gunn's observed stellar component to the NGC 3379 data is shown in Figure 1. The observed and halo components have similar tidal limiting radii, but the halo component has an envelope that follows $\rho \propto r^{-2}$ accurately up to that point while the observed stars have a more rapid density fall-off. The halo dominates the mass in the outer regions giving this model a total mass-to-light ratio six times larger than a simple King model. The Gunn model has the great advantage that it gives flat rotation curves like M31 and a total $M/L \sim 100$ in agreement with numerous studies of groups and clusters of galaxies. (Gunn 1974, Ostriker, Peebles & Yahil 1974, Oemler 1974, Turner 1976, Gott & Turner 1977a). The Gunn model can be distinguished observationally from a simple King model in that the velocity dispersion in a simple King model becomes smaller as one goes away from the nucleus, whereas in the Gunn model the velocity dispersion of the observed stars remains virtually constant (cf Figure 1). Observations of this sort are just now becoming feasible.

Consider the following simple heavy-halo model for M31. Roberts & Whitehurst (1975) find the rotation curve of M31 to be constant at large distances: $V_{\text{rot}} = 228$ km sec^{-1}. This tells us directly the gravitational potential

$$\frac{d\Phi}{dr} = \frac{V_{\text{rot}}^2}{r}, \tag{36}$$

and from $\nabla^2\Phi = 4\pi G\rho$ we find that the halo mass distribution in the envelope is given by $\rho_H \propto r^{-2}$, like the envelope of a self-gravitating isothermal sphere. Assuming that the velocity distribution of the heavy-halo stars is isothermal and isotropic with rotation unimportant, the stationary collisionless Boltzmann equation (cf Mihalas & Routly 1968) gives

$$-\frac{d\Phi}{dr} = \frac{\langle V^2 \rangle}{3r} \frac{d \ln \rho}{d \ln r}. \tag{37}$$

Combining (37) with (36) gives the velocity dispersion of the halo stars: $\langle V_h^2 \rangle^{1/2} =$ 279 km sec^{-1}. If the observed stars and globular clusters in the spheroidal component have an approximately isothermal isotropic distribution and sit in the potential well produced by the heavy-halo stars, then to satisfy Equation (37) and have an envelope like Hubble's Law $\rho \propto r^{-3}$, they must have a velocity dispersion $\langle V_{\text{ob}}^2 \rangle^{1/2} = 228$ km sec^{-1}. This corresponds to a line-of-sight V_{rms} of 132 km sec^{-1}. The ratio $\langle V_h^2 \rangle = 1.5 \langle V_{\text{ob}}^2 \rangle$ deduced from this simple model agrees exactly with the ratio found empirically by Gunn (1976) from his two component King models. The average line-of-sight velocity dispersion for all globular clusters in M31 listed by Hartwick & Sargent (1974) is $V_{\text{rms}} = 124 \pm 7$ km sec^{-}1. Their data shows a tendency for the velocity dispersion to decrease with increasing distance from the center. The velocity dispersion of stars in the nucleus has been estimated to be 130 ± 20 km sec^{-1} by Morton & Elmergreen (1976). However Schechter (private communication) finds 160 km sec^{-1} and Faber & Jackson (1976) find 180 km sec^{-1}. The nucleus of M31 is quite dense and may be a separate dynamical entity (cf Tremaine et al. 1975). It might not be surprising if it had a larger velocity dispersion than the overall average for the bulge. Faber & Jackson (1976) find lower velocity dispersions ($\leqq 150$ km sec^{-1}) just 30 pc from the center. All in all, the data seem encouraging.

While Gott's model of NGC 4697 provides a good description of the surface brightness distribution it predicts a higher rotation curve than is observed. The central line-of-sight velocity dispersion in this galaxy was measured by King & Minkowski (1966) to be 310 km sec^{-1}. Normalized to this value the model would have predicted a peak in the observed rotation curve of ~ 270 km sec^{-1}, while Bertola (1972) observes a peak value in the rotation curve of only ~ 65 km sec^{-1}. Gott (1975) has pointed out that this would not only be a problem for his model but for any model where the ellipticity is produced by rotation in the usual way. (An ad hoc solution would be to take an existing rotating model and simply reverse approximately half the star orbits, but this seems implausible physically.) Preliminary inspections of spectra of this galaxy by G. Illingworth (private communication) have tended to confirm the original estimates. Binney (1976a) has noted that the original velocity dispersion is probably not too high since the Faber & Jackson (1976) relations suggest a velocity dispersion of 270 km sec^{-1} for a galaxy of this absolute magnitude. King (1976) has recently announced that there is a whole class of elliptical galaxies with reasonable ellipticity and yet very low rotation curves. Typically the ellipticity of these galaxies is low in their central regions but increases with radius from the center.

Binney (1976a) has come up with a beautiful explanation for systems of this type. Motivated by the Zeldovich pancake idea Binney starts with 400 stars in a (non-rotating) flat distribution ($\varepsilon = 0.95$) and lets them collapse and relax. The violent relaxation in the central regions is sufficiently efficient to produce an isotropic velocity distribution and nearly round isophotes. However, in the outer regions the violent relaxation is not as efficient and the equilibrium configuration is left with an anisotropic velocity dispersion (largest dispersion in the galactic plane) and elliptical outer isophotes ($\varepsilon = 0.5$). Thus, Binney's model can produce elliptical galaxies with little or no rotation and predicts that the ellipticity should typically increase outwards as found by King.

The apparent major axis of the ellipsoidal component in M31 is tilted 10° relative to the apparent major axis of the disk. Lindblad (1956) has suggested that this can be explained most easily if the ellipsoidal component is in the shape of a bar rotating in the plane of the disk. The least bar-like model satisfying the observations is a rotating triaxial ellipsoid with axis ratios 1:1.37:2.16 (Stark 1977). This leads to the suggestion that many if not all ellipticals may be triaxial. If a rotating bar configuration is left after violent relaxation it could maintain itself indefinitely (rotation and pattern speed need not be similar). Numerical simulations of flat stellar disks (Hohl 1971) produce rotating bars (axis ratios 2:3), persisting to the end of the calculations (6 rotations). Miller (1976) has performed full 3-D calculations of the collapse and relaxation of a stellar system using 115,000 mass points and a fast Fourier transform technique, which calculates the potential on a grid $64 \times 64 \times 64$. Miller starts with a sphere of uniform density in solid-body rotation, initially just holding itself up by rotation in its equatorial plane. After a complex collapse and relaxation process involving a number of transient features the system settles into a rotating triaxial configuration. A key observational test of the triaxial hypothesis would be measurement of rotation curves on the apparent minor axes of elliptical galaxies; a nonzero result would indicate a nonaxisymmetric galaxy.

A crucial problem the pure stellar dynamical models must solve is the observed metal-abundance gradients in elliptical galaxies. One possibility outlined above is that the stars from the central perturbation are metal enriched relative to the infall stars. This could produce metal-abundance gradients extending far out into the envelope as observed by Strom et al. (1977). Another promising mechanism is dynamical friction. Tremaine, Ostriker & Spitzer (1975) have recently proposed that nuclei of galaxies may be formed by globular clusters that have spiraled to the center by dynamical friction. Dynamical friction steals energy from a massive orbiting body as the system tries to establish equipartition of energy (Chandrasekhar 1943). The characteristic time scale for a globular cluster of mass $M_{\rm cl}$ to spiral in is of the order

$$T_{\rm sp} \sim \frac{M_{\rm gal}/M_{\rm cl}}{2 \ln (M_{\rm gal}/M_{\rm cl})} \Delta t. \tag{38}$$

It would be just equal to the two-body relaxation time of the galaxy if it were composed entirely of globular clusters of mass $M_{\rm cl}$ instead of stars (cf Equation 25).

Two-body relaxation involves a statistical random walk in velocity due to two-body encounters and a Chandrasekhar drag term. If the system consists of equal point masses the drag term and the relaxation term are of comparable magnitude. For a body $M_{cl} \gg M_{star}$ only the drag term is important. The drag term on a body depends only on the mean density and velocity dispersion of the stars comprising the mass of the galaxy and not on their masses, so the spiral time above is independent of M_{star} as long as $M_{cl} \gg M_{star}$. For typical values $M_{gal} \sim 10^{11}\ M_{\odot}$, $\Delta t \sim 2 \times 10^{7}$ yr, and $M_{cl} \sim 2 \times 10^{6}\ M_{\odot}$ one obtains $T_{sp} \sim 5 \times 10^{10}$ yr, of the order of the age of the universe. Detailed calculations (Tremaine, Ostriker & Spitzer 1975) indicate that $\sim$25 globular clusters should have spiraled to the center of M31 by now, and the observed nucleus has a mass of just this order. Also, the expected depletion of globular clusters in the central region is seen. Another interesting case is M32 (Tremaine 1976). It is tidally limited by M31 so that any globular cluster associated with M32 is required to be inside the tidal limiting radius of 2.2 kpc. It is calculated that a globular cluster within this radius should have spiraled into the center by now. Indeed, M32 has a rather massive nucleus and no globular clusters. With an isothermal density fluctuation spectrum at recombination (Section 3), we expect protogalaxies to be composed of a hierarchy of subunits of different masses down to globular cluster size. These primordial clusters will spiral toward the center until they are tidally disrupted upon reaching a region of the galaxy where the mean density interior to that radius is as great as the internal density of the cluster. If denser clusters have more reprocessing and higher metallicity, the spiraling in and disruption process will leave the galaxy with a metallicity gradient with the metal-rich stars in the central regions. Since dynamical friction decreases a cluster's angular momentum at the same time as it decreases its energy, one would expect this effect to produce elliptical galaxies with isochromes and isophotes of approximately equal flattening, in agreement with observations by Strom (private communication) and Strom et al. (1977) (cf Section 5). The theory that metallicity gradients arise from metallicity differences between stars from the dense central regions of a protogalaxy and the infall stars would also predict isochromes and isophotes of equal flattening. On the other hand, Binney (1976b) has shown that in an elliptical galaxy with an anisotropic velocity dispersion (as in his model), dynamical friction acts to pull massive clusters toward the plane. Thus in this case it is even possible to produce an elliptical galaxy with isochromes flatter than isophotes by the action of stellar dynamics alone. A study of metallicity gradients in ellipticals with low rotation curves would be especially interesting.

One final piece of evidence favoring the dissipationless collapse picture is that velocity dispersions in galaxies and groups of galaxies are quite similar (Einasto et al. 1976). Velocity dispersions in galaxies are typically a couple of hundred km sec^{-1}, while 63% of the groups of galaxies in Gott & Turner's (1977) sample have velocity dispersions $\leqq 300$ km sec^{-1}. If dissipation were important for the bulk of the mass in galaxies then they could in principle have much larger velocity dispersions. As it is, galaxies (with their heavy halos) seem to constitute a smooth continuation of the hierarchical gravitational clustering seen on larger scales (cf Equation 14 and Fall 1975).

7 OTHER MODELS

Toomre & Toomre (1972) have produced computer simulations of tidally interacting galaxies (e.g. NGC 4676 and NGC 4038/9) that duplicate the observed galaxy photographs in such detail that there is little doubt that the tidal interaction model is correct. About 10 out of the $\sim$4000 NGC galaxies can be classed as violently interacting systems. Lynds & Toomre (1976) have shown that ring galaxies can be produced by a "direct hit" of two galaxies. As the intervening galaxy nucleus passes through the center of a disk galaxy it perturbs the orbits in the disk setting up a ring-shaped density wave in the disk. This explains why ring galaxies always have companions near the minor axis of the ring and more spectacularly explains the occurence of double-ring galaxies (where two disk galaxies pass through each other). The impact parameters for production of ring galaxies are more precise than for interacting galaxies, and they estimate that ring galaxies should occur about 1/300 as often. Also, the ring phase should last a shorter time. Indeed, the redshifts of the nearest ring galaxies are about ten times larger than those of the nearest interacting galaxies, suggesting a volume density 10^{-3} as great. In another 10^9 yrs the interacting galaxies in the NGC sample will have coalesced due to dynamical friction and violent relaxation to leave spheroidal systems of stars. Over the age of the universe (2×10^{10} yr) Toomre & Toomre (1972) expect $\sim$500 galaxy collisions in the NGC sample since the collision rate should have been higher in the past. Thus they propose that the ($\sim$400) elliptical galaxies in the NGC survey might simply be debris from galaxy collisions. This could explain why elliptical galaxies are seen so often in rich clusters. Numerical simulations show that if two elliptical galaxies with deVaucouleur radii R_{eo} are dropped toward each other from infinity they will coalesce via violent relaxation to form a homologous system with deVaucouleur radius $R_e = 2R_{eo}$ (Toomre 1974). The final system has the same total binding energy as the two original galaxies. Thus Equations 30 and 22 can underestimate the collapse time of an elliptical galaxy since they ignore the binding energy in subclumps. This lessens the amplitude of density fluctuations required to produce ellipticals. The longer collapse times also ease the angular momentum problem for ellipticals since an elliptical can have a large quadrupole moment at $t = \frac{1}{2}T_c$ for gaining angular momentum through tidal interactions and yet have a large binding energy. Toomre further proposes that gas is everywhere very dissipative, always forming disks, and that stars form only in such high density gas disks. When star formation is complete one has disks of stars (tiny S0 galaxies) which can then collide to form ellipticals. Such gas disks would ease the initial fluctuation and angular momentum constraints still further since these gas disks can be a tenth the size of the collapsing clumps from which they originated (Section 6). Gold (1976) has proposed similar ideas.

Elliptical galaxies have a covariance function of about twice the amplitude of that possessed by spirals (Davis & Geller 1976). Inspection of the form of the three-point correlation function of Peebles & Groth (1975) shows that a tight binary galaxy has twice the covariance function amplitude of an average galaxy, so if the binary coalesces into an elliptical it would give the correct result. On the other hand S0 galaxies also have a covariance function about twice the amplitude of that for spirals, which would not be predicted. The Gott & Thuan (1976) picture that ellipticals via their shorter collapse times should start clustering earlier would likewise predict a higher covariance function for elliptical galaxies only. Perhaps a simpler explanation lies in the fact that elliptical and S0 galaxies have about twice the total mass-to-light ratio of spirals (~ 120 versus ~ 65) as suggested by studies of binary galaxies (Turner 1976) and groups (Gott & Turner 1977a). There is evidence (Shapiro 1971) that ellipticals, S0's, and spirals have similar luminosity functions. Thus typical elliptical and S0 galaxies in a magnitude-limited survey will be twice as massive as corresponding spirals. Again, relying on the three-point correlation-function results, we notice that a tight binary is dynamically indistinguishable from a single galaxy of twice the mass, so we might expect elliptical and S0 galaxies to be statistically more clustered by the desired amount.

The most difficult observation for the Toomre theory to explain would seem to be the color versus absolute magnitude relations of Faber (1973). A random collision process would tend to wash out any such tight correlations. Also, the disks of S0 galaxies are metal enriched compared with their spheroidal envelopes (cf Strom et al. 1977), and it is not clear that collisions between such galaxies will produce ellipticals with the correct metal abundances. Whether or not these constitute serious problems for the Toomre theory remains to be seen. Metal-abundance gradients can be produced by dynamical friction as in Section 7. The Toomre picture of ellipticals (and spheroidal components of spirals) being built from smaller pieces meshes with the picture of Peebles & Dicke (1968) and an initial isothermal density fluctuation spectrum (cf Section 3).

While Toomre's collision mechanism might not explain every spheroidal system (in particular, dwarf ellipticals and globular clusters), it might explain a number of them, especially those in the centers of rich clusters like Coma. Oemler's (1976) data on cD galaxies suggests that the envelopes of these galaxies consist of debris from galactic collisions. Gunn & Gott (1972) found that spiral galaxies falling into the centers of rich clusters like Coma should have their gas stripped out by ram pressure from the hot 10^8 K intracluster gas detected in X-ray studies. The stripped spirals would look like S0's, which explains why there are no spirals seen in the center of the Coma cluster, only S0's and ellipticals. Gisler (1976) has investigated a similar stripping mechanism for ellipticals. If ellipticals complete their star formation by $t = T_c$, then the gas production rate due to the death of stars decreases with time (approximately proportional to t^{-1} for $t > 2T_c$) (Gisler 1976). He estimates that a 10^{10} $L_{\odot}$ elliptical should currently ($t = 2 \times 10^{10}$ yr) have a gas production rate of ~ 0.07 $M_{\odot}$ yr^{-1}. If the intergalactic ram pressure is sufficient, the gas will be swept out as it is produced; but if the gas production is high enough, gas can accumulate in the elliptical, perhaps powering a radio source or quasar in the nucleus. In the past, ellipticals had higher gas production rates and were harder to strip, so a greater fraction would retain gas and the fraction having violent activity would be greater. This can explain qualitatively the density evolution law for quasars (Gisler 1976).

Hickson, Richstone & Turner (1977) have investigated galaxy types as a function of group crossing time for dense groups. In groups where many collisions are expected the number of spiral galaxies drops. However, the implied collision efficiency appears to be less than the geometrical cross section. It is thus surprising that some of these systems have two or more remaining spirals.

Ostriker & Rees (1977) and Silk (1977) have developed a different galaxy formation picture utilizing the fact that for galaxies larger than 10^{12} $M_{\odot}$ the cooling time and the dynamical time are equal for a system of radius $r \sim 73$ kpc (cf Equation 29) independent of mass. A protogalaxy of 10^{12} $M_{\odot}$ may have a long collapse time with $R_0 > 73$ kpc; when it collapses it will heat up to the appropriate virial temperature and begin to cool quasi-statically until it reaches the critical radius of 73 kpc. It then cools quickly to 10^4 K and begins a free fall collapse. One can imagine star formation taking place at this point leaving a dissipationless collapse of stars as in Section 6. In support of this theory we note that M31, which has a mass of 10^{12} $M_{\odot}$ [from local group timing arguments (Gunn 1974)], has a halo with a velocity dispersion of 279 km sec^{-1} (cf Section 6) giving it a binding energy equivalent to a 10^{12} $M_{\odot}$ sphere of radius $r \sim 76$ kpc, just as predicted. Like the Toomre model this model allows longer collapse times, easing angular momentum and initial fluctuation amplitude requirements. This model would work most naturally for an adiabatic fluctuation spectrum in an $\Omega_0 = 1$ cosmology, where fluctuations on scales smaller than 10^{12} $M_{\odot}$ are damped. With an isothermal spectrum the smaller subclumps ($r < 73$ kpc) would be expected to cool and form stars before the galaxy as a whole collapsed unless star formation were somehow impeded.

8 CONCLUSION

The field of galaxy formation has experienced rapid development in recent years. Many important new ideas have emerged: violent relaxation, dynamical friction, heavy halos, cosmological infall, and angular momentum via tidal interactions. With the discovery of the microwave background radiation attention has focused on the standard hot big-bang Friedmann cosmologies. One can now ask what are natural initial conditions for these cosmological models and explore how galaxies might originate from small density fluctuations at recombination via gravitational instability. Also, we have seen the advent of direct numerical simulations of galaxy formation which allow the detailed predictions of different models to be tested.

One of the greatest problems in the field today is that various theoretical models have enough legitimate free parameters (unknown star formation rates, gas properties, initial conditions, etc.) so that they can be made to fit the observations very well. Figure 1 gives an example of this. Three models with rather different physical assumptions all give outstandingly good fits to the same data. Because of this situation, it has been difficult to find conclusive observational tests of the competing theories. Sometimes we may learn more when the models fail in some dramatic way than when they succeed. For example, the failure of my model to account for the low rotation curve of NGC 4697 prompted Binney to investigate a new and more promising type of model.

On the theoretical side I suspect the most important advances in the next few years will come from new ideas and from the utilization of additional qualitative clues about galaxy formation. As I have emphasized here, I think there are important new clues in the observed clustering properties of galaxies and in the relation between the multiplicity function of galaxies and the galaxy luminosity function. On the observational side the most promising areas would seem to be 1. measurements of minor axis rotation curves for ellipticals, 2. velocity dispersions and rotation curves away from the nucleus in ellipticals, 3. studies of metallicity gradients in ellipticals, S0's, and spirals, 4. accurate photometry of large numbers of galaxies of all types, 5. virial mass determinations and other heavy-halo tests, 6. systematic studies of galaxy morphology as a function of environment, and 7. searches for primeval galaxies at large redshift. I am confident that galaxy formation will remain an area of lively debate and progress for the foreseeable future.

ACKNOWLEDGMENTS

It is a pleasure to thank James E. Gunn and Stephen E. Strom for permission to present some of their results prior to publication, and James Binney, Jeremiah P. Ostriker, Edwin L. Turner, and Joseph Silk for their comments on the manuscript.

Literature Cited

Aarseth, S. J., Gott, J. R., Turner, E. L. 1977. In preparation
Abell, G. O. 1962. In *Problems of Extragalactic Research*, ed. G. C. McVittie. New York: MacMillan. 213 pp.
Baum, W. A. 1959. *Publ. Astron. Soc. Pac.* 71:106
Bertola, F. 1972. *The Rotation of Galaxies, Atti Riunione Soc. Astron. Ital. 15th, Bologna, 1971*
Biermann, L. 1951. *Z. Astrophys.* 28:304
Binney, J. 1976a. *MNRAS* 177:19
Binney, J. 1976b. *MNRAS*. Submitted for publication
Bovier, P., Janin, G. 1970. *Astron. Astrophys.* 5:127
Chandrasekhar, S. 1943. *Ap. J.* 97:255
Chandrasekhar, S. 1960. *Principles of Stellar Dynamics.* New York: Dover
Davis, M., Geller, M. J. 1976. *Ap. J.* 208:13
Davis, M., Wilkinson, D. T. 1974. *Ap. J.* 192: 251
deVaucouleurs, G. 1959. *Handb. Phys.* 53: 311
deVaucouleurs, G. 1974. *The Formation and Dynamics of Galaxies, IAU Symp. No. 58*, ed. J. R. Shakeshaft, pp. 1–53. Dordrecht: Reidel
deVaucouleurs, G., deVaucouleurs, A. 1964. *Reference Catalogue of Bright Galaxies.* Austin: Univ. Texas Press
Disney, M. J. 1976. *Nature* 263:573
Doroshkevich, A. G., Sunyaev, R. A., Zeldovich, Y. B. 1974. *Confrontation of Cosmological Theories and Observational Data*, ed. M. S. Longair. Holland: Reidel. 213 pp.
Eggen, O. J., Lynden-Bell, D., Sandage, A. 1962. *Ap. J.* 136:748
Einasto, J., Joeveer, M., Kaasik, A., Vennik, J. 1976. *Targu Astron. Obs. Preprint.* 8
Epstein, R. I., Petrosian, V. 1975. *Ap. J.* 197:281
Faber, S. M. 1973. *Ap. J.* 179:731
Faber, S. M., Jackson, R. E. 1976. *Ap. J.* 204:668
Fall, S. M. 1975. *MNRAS* 172:23P
Field, G. B. 1975. In *Stars and Stellar Systems*, ed. A. Sandage, M. Sandage, J. Kristian, 9:359–407. Chicago: Univ. Chicago Press
Fish, R. A. 1964. *Ap. J.* 139:284
Freeman, K. C. 1970. *Ap. J.* 160:811
Freeman, K. C. 1975. See Field 1975, pp. 409–507
Gisler, G. R. 1976. *Astron. Astrophys.* 51:137
Gold, T. 1976. *IAU Symp. Galaxy Formation, Grenoble*
Gott, J. R. 1973. *Ap. J.* 186:481
Gott, J. R. 1975. *Ap. J.* 201:296
Gott, J. R., Gunn, J. E., Schramm, D. N., Tinsley, B. M. 1974. *Ap. J.* 194:543
Gott, J. R., Rees, M. J. 1975. *Astron. Astrophys.* 45:365
Gott, J. R., Thuan, T. X. 1976. *Ap. J.* 204: 649

Gott, J. R., Turner, E. L. 1977a. *Ap. J.* 213: 309
Gott, J. R., Turner, E. L. 1977b. *Ap. J.* 216: In press
Gunn, J. E. 1974. *Comments Astrophys. Space Phys.* 6:7
Gunn, J. E. 1976. In preparation
Gunn, J. E., Gott, J. R. 1972. *Ap. J.* 176:1
Hartwick, F. D. A., Sargent, W. L. W. 1974. *Ap. J.* 190:283
Hickson, P., Richstone, D. O., Turner, E. L. 1977. *Ap. J.* 213:323
Hohl, F. 1971. *Ap. J.* 168:343
Hubble, E. P. 1926. *Ap. J.* 64:321
Hubble, E. P. 1930. *Ap. J.* 71:231
Jones, B. J. T. 1976. *Rev. Mod. Phys.* 48: 107
Kalnajs, A. J. 1972. *Ap. J.* 175:63
Kaufman, M. 1975. *Astrophys. Space Sci.* 33:265
Kaufman, M. 1977. *Astrophys. Space Sci.* In press
King, I. R. 1966. *Astron. J.* 71:64
King, I. R. 1976. *IAU Symp. Galaxy Formation, Grenoble*
King, I. R., Minkowski, R. 1966. *Ap. J.* 143: 1002
Kormendy, J. 1976. PhD thesis. Calif. Inst. Technol.
Larson, R. B. 1974a. *MNRAS* 166:585
Larson, R. B. 1974b. *MNRAS* 169:229
Larson, R. B. 1975. *MNRAS* 173:671
Larson, R. B. 1976. *MNRAS* 176:31
Lindblad, B. 1956. *Stockholm Obs. Ann.* 19, No. 2
Lynden-Bell, D. 1967. *MNRAS* 136:101
Lynden-Bell, D., Pringle, J. E. 1974. *MNRAS* 168:603
Lynds, R., Toomre, A. 1976. *Ap. J.* 209:382
Mathews, W. G., Baker, J. C. 1971. *Ap. J.* 170:241
Meier, D. L. 1976a. *Ap. J.* 207:343
Meier, D. L. 1976b. *Ap. J.* 203:L103
Meszaros, P. 1974. *Astron. Astrophys.* 37: 225
Mihalas, D., Routly, P. McR. 1968. *Galactic Astronomy*. San Francisco: Freeman
Miller, R. H. 1976. *IAU Symp. Galaxy Formation, Grenoble*
Miller, R. H., Prendergast, K. H. 1962. *Ap. J.* 136:713
Miyoshi, K., Kihara, T. 1975. *Publ. Astron. Soc. Jpn* 27:333
Morton, D. C., Elmergreen, B. G. 1976. *Ap. J.* 205:63
Oemler, A. 1974. PhD thesis. Calif. Inst. Technol.
Oemler, A. 1976. *Ap. J.* 209:693
Ostriker, J. P. 1975. *Conf. Galaxy Formation, Cambridge*
Ostriker, J. P., Peebles, P. J. E. 1973. *Ap. J.* 186:467
Ostriker, J. P., Peebles, P. J. E., Yahil, A. 1974. *Ap. J.* 193:L1
Ostriker, J. P., Rees, M. J. 1977. *MNRAS*. Submitted for publication
Ostriker, J. P., Thuan, T. X. 1975. *Ap. J.* 202:353
Partridge, R. B. 1974. *Ap. J.* 192:241
Partridge, R. B., Peebles, P. J. E. 1967a. *Ap. J.* 147:868
Partridge, R. B., Peebles, P. J. E. 1967b. *Ap. J.* 148:377
Peebles, P. J. E. 1969. *Ap. J.* 155:393
Peebles, P. J. E. 1970. *Astron. J.* 75:13
Peebles, P. J. E. 1971a. *Physical Cosmology*. Princeton: Princeton Univ. Press
Peebles, P. J. E. 1971b. *Astron. Astrophys.* 11:377
Peebles, P. J. E. 1974. *Ap. J.* 189:L51
Peebles, P. J. E., Dicke, R. H. 1968. *Ap. J.* 154:891
Peebles, P. J. E., Groth, E. J. 1975. *Ap. J. (Suppl.)* 196:1
Peebles, P. J. E., Yu, J. T. 1970. *Ap. J.* 162: 815
Press, W. H., Schechter, P. 1974. *Ap. J.* 187: 425
Rees, M. J. 1971. *Italian Physical Society: Proceedings of the International School of Physics. "Enrico Fermi", Course 47: General Relativity and Cosmology*, ed. B. K. Sachs. New York: Academic
Richstone, D. O. 1976. *Ap. J.* 204:642
Roberts, M. S., Whitehurst, R. N. 1975. *Ap. J.* 201:327
Sandage, A., Freeman, K. C., Stokes, N. R. 1970. *Ap. J.* 160:83
Sandage, A., Tammann, G. A. 1975. *Ap. J.* 197:265
Schechter, P. 1975. PhD thesis. Calif. Inst. Technol.
Schechter, P. 1976. *Ap. J.* 203:297
Schmidt, M. 1959. *Ap. J.* 129:243
Shapiro, S. L. 1971. *Ap. J.* 188:233
Silk, J. 1974. In *Confrontation of Cosmological Theories with Observational Data*, ed. M. S. Longair. Holland: Reidel. 175 pp.
Silk, J. 1977. *Ap. J.* 211:638
Spinrad, H., Smith, H. E., Taylor, D. J. 1972. *Ap. J.* 175:649
Spitzer, L. 1968. *Diffuse Matter in Space*. New York: Wiley Interscience
Spitzer, L., Thuan, T. X. 1972. *Ap. J.* 175: 31
Stark, A. A. 1977. *Ap. J.* 213:368
Strom, K. M., Strom, S. E., Jensen, E. B., Moller, J., Thompson, L. A., Thuan, T. X. 1977. *Ap. J.* 212:335
Sunyaev, R. A. 1971. *Astron. Astrophys.* 12: 190

Sunyaev, R. A., Zeldovich, Y. B. 1972. *Astron. Astrophys.* 20:189
Thuan, T. X., Gott, J. R. 1975. *Nature* 257:774
Thuan, T. X., Gott, J. R. 1977. *Cal. Tech. Orange Aid Preprint 457*
Tinsley, B. M. 1976. *IAU Symp. Galaxy Formation, Grenoble*
Toomre, A. 1964. *Ap. J.* 139:1217
Toomre, A. 1974. See deVaucouleurs 1974, pp. 347–65
Toomre, A., Toomre, J. 1972. *Ap. J.* 178:623
Totsuji, M., Kihara, T. 1969. *Publ. Astron. Soc. Jpn.* 21:221
Tremaine, S. D. 1976. *Ap. J.* 203:345
Tremaine, S. D., Ostriker, J. P., Spitzer, L. 1975. *Ap. J.* 196:407
Turner, E. L. 1976. *Ap. J.* 208:304
Turner, E. L., Gott, J. R. 1976. *Ap. J.* 209:6
Weinberg, S. 1972. *Gravitation and Cosmology.* New York: Wiley Interscience
Wilson, C. P. 1974. PhD thesis. Univ. Calif., Berkeley
Zeldovich, Y. B. 1972. *MNRAS* 160:1P

Ann. Rev. Astron. Astrophys. 1977. 15: 267–93

FORMATION AND DESTRUCTION OF DUST GRAINS

E. E. Salpeter
Departments of Physics and Astronomy, Cornell University, Ithaca, N.Y. 14853

1 AN OVERVIEW

Since the classic book by Spitzer (1968) and the review by Lynds & Wickramasinghe (1968) two excellent reviews on interstellar dust grains have appeared recently (Wickramasinghe & Nandy 1972, Aannestad & Purcell 1973) as well as the proceedings of three conferences (Greenberg & Van de Hulst 1973, Field & Cameron 1975, Wickramasinghe & Morgan 1976). Because of this excellent coverage I give references mainly to very recent papers.

Interstellar dust has been known or suspected for almost half a century (Russell 1928). Over most of this period a chief puzzle was the question of how and where dust grains form. This is still a key question, but less of a puzzle, and the destruction of grains has become equally important. This change in emphasis is mainly due to three developments in observational astronomy: (*a*) Microwave and *UV* spectroscopy has shown that "molecular clouds" (H_2 plus a large variety of complex molecules; see Watson 1977) are fairly common in the interstellar medium (Burton 1976); (*b*) *UV* absorption line studies, especially from the Copernicus satellite (Spitzer & Jenkins 1975), have shown that elements which can form refractory solid grains are highly depleted in most of the interstellar gas; and (*c*) infrared emission features in the spectra of cool but luminous stars have shown the presence of silicate particles (and probably of SiC and graphite) in or above the atmospheres of such stars (Woolf 1973). The relevance of these developments can be summarized as follows: (*c*) provides direct observational evidence for one environment where solid particles are formed out of gas that was originally free of solids. Cool stars that have a *small* outflow rate (so that the primary spectrum of the star is not masked strongly) are of particular diagnostic importance for theories of nucleation and growth of grains (Sections 3.1, 3.2, 5.1, and 5.2). (*b*) suggests that interstellar gas containing dust grains but few condensable atoms is the rule rather than the exception. Why some atoms are more depleted than others then becomes an interesting problem (Field 1974), and we have to consider both the sputtering of atoms off grain surfaces (Sections 3.4, 6.1, 6.2) into the interstellar gas and the

Table 1 Turnover rates, in units of $(10^{10}\ \text{yr})^{-1}$, for formation and destruction

Grain-material			Rate
Core only	Formation	Cool stars; P.N.; novae	1 or 2
Both	Destruction	Protostellar nebulae and star formation	4 to 10
At least core	Formation	(Protostellar nebulae and star formation)	0.1 to 6
Core (and mantle)	Destruction	Sputtering in SNR	~10
Mantle only	Destruction	Sputtering in cloud collisions	20 to 100
Mantle (and core)	Formation and reprocessing	Molecular clouds and H II regions	200 to 10^4

deposition of gas atoms onto grain surfaces (Sections 3.3, 6.3). Finally, (*a*) points to complex situations where dust grains and molecules already coexist in some kind of symbiotic relationship.

The "cosmic" abundance of elements making highly refractory grains (mainly Mg, Si, and Fe) is about a factor of 10 lower than the abundance of C, N, and O, which form (in the presence of hydrogen) relatively volatile ices. It is customary to discuss composite grains with a small refractory (silicate and iron) core and a larger more volatile mantle of "dirty ices" (Whipple 1950). Because of the great difference in both the condensation temperature and in the resistance to destruction by sputtering we shall consider separately the life-cycles for refractory grain-cores and for ice-mantles in Table 1. In reality one should consider a third category of grain material—"oily plastics," say—which is made by the polymerizing action of stellar *UV* and cosmic rays on "dirty ices." This third category might be quite important but little is known about it at present (see Section 6.3).

The main "punchline" of this review is Table 1, which gives rates for various processes based on the papers reviewed in the text but ultimately requiring subjective order-of-magnitude estimates. Unfortunately our theoretical and observational understanding of processes is inversely correlated with their numerical importance: Circumstellar grain formation (line 1, Table 1) is well documented and understood, but the reprocessing in and near molecular clouds (line 6, Table 1) is not. The rate of a "process' in Table 1 is defined as the ratio of M_{pr}, the total mass of grain-containing interstellar gas "processed" per unit time, to the present total mass of interstellar gas M_{gas}. Assuming a galactic disk of radius 15 kpc and including molecular hydrogen (Scoville & Solomon 1975, Burton & Gordon 1976), we shall adopt $M_{gas} = 4 \times 10^9\ M_\odot$ (see Section 4.1). Line 1 of Table 1 is discussed in Section 5.1, lines 2 and 3 in Section 5.3, lines 4 and 5 in Sections 3.4, 6.1, and 6.2, and line 6 in Sections 6.3 and 6.4.

2 ELEMENT ABUNDANCES

2.1 *Solar and "Cosmic" Abundances*

Fortunately, there is relatively little controversy about the solar system abundances of the more common elements that are of interest for grain formation: In many cases, independent abundance values or ratios for the solar photosphere, solar corona, and meteorites agree and there have been few major changes (except a slight abundance decrease for Ar) from the earlier compilation by Cameron (1973a) to the most recent one by Ross & Aller (1976). The recent values for abundance f (by number, relative to 10^{12} for hydrogen) are given in the first line of Table 2 for a few common elements. They should be accurate to within better than a factor of two (except for some lingering doubt on the important iron-abundance).

We are interested in "cosmic" abundances, or at least in *total* abundances in the interstellar medium in our Galaxy (within a few kpc of the Sun) summed over both gases and solids. We simply *assume* the solar values for this purpose. Theoretical and observational arguments for this assumption, as well as the doubts, are given in the detailed review by Trimble (1975). It is gratifying, for instance, that abundance ratios for most common elements (excluding H and He) obtained from galactic cosmic rays (Shapiro & Silberberg 1974) agree within a factor of 4 with the solar values. Abundances in O and B stars are comparable to solar values. There is considerable data on abundances from emission-line studies of H II regions (Peimbert & Costero 1969, Osterbrock 1974), especially of the Orion Nebula. Abundances (relative to H) in Orion of C, N, O, Ne, S, and Ar are close to the solar values. On the other hand, the abundance of Fe in the ionized gas in Orion is about a factor of 20 smaller (Olthof & Pottash 1975) than in the Sun. Since we have theoretical and observational reasons (Sections 3.4 and 6.3) to believe that iron and silicate grain-cores can survive in H II regions, we take the underabundance of iron in the gas phase to indicate condensation on dust grains rather than deviations from "cosmic" abundances. The same should apply to Mg and Si in Orion, but no accurate abundances are available as yet.

Abundance data from emission lines in the ionized gas of some planetary nebulae is also available (Aller & Czyzak 1968, Osterbrock 1974). Elements like N and O have abundances close to solar values in planetary nebulae, but the iron abundance in the gas phase is again lower than in the Sun by a factor of about 30 (Shields 1975) in NGC7027. This is again not surprising, since some refractory dust grains are expected in planetary nebulae; but it is then disturbing that Ca seems *not* to

Table 2 Solar abundance $f_{\odot}(A)$ (by number) of element A, normalized to $\log_{10} f(H) = 12$ and "typical" interstellar abundance values $f_{rs}(A)$ in the line of sight towards reddened stars

	C	N	O	Ne	Na	Mg	Al		
$\log_{10} f_{\odot}$	8.6	8.0	8.8	7.6	6.3	7.6	6.5		
$\log_{10} f_{\odot}/f_{rs}$	0.8	1.0	0.7	—	0.8	1.4	3.2		
	Si	**P**	**S**	**Ar**	**K**	**Ca**	**Ti**	**Fe**	**Ni**
$\log_{10} f_{\odot}$	7.6	5.5	7.2	6.0	5.1	6.3	5.0	7.5	6.3
$\log_{10} f_{\odot}/f_{rs}$	1.8	0.9	0.7	0.5	1.0	3.5	2.8	2.2	2.3

be underabundant (Aller & Czyzak 1968), since calcium minerals are even more refractory than iron.

Some fears remain that solar abundances (relative to H) may deviate from "cosmic" abundances in some systematic way, e.g. because of contamination of outer layers of the Sun by infalling debris (Joss 1974), selective formation of stars in metal-rich regions (Talbot & Arnett 1973), or other evolutionary effects (Audouze & Tinsley 1976). We provisionally disregard these fears and assume that the abundances from Ross & Aller (1976) in Table 2 are representative to within factors of about 2 (or, at worst, 4).

2.2 *Absorption Lines in the Interstellar Medium (ISM)*

Only about 10% of the mass of the ISM resides in H II regions, so we are mainly (but not exclusively) interested in H I regions where hydrogen is mainly neutral. H I regions are sometimes subdivided into three components, (*a*) dense, cool clouds ($n \gtrsim 10^3$ H atoms cm^{-3}, $T < 50$ K) where the hydrogen is mainly H_2 and complex molecules are present; (*b*) "standard interstellar clouds" (density $n \sim 1$ to 100 cm^{-3}, temperature $T \sim 50$ to 100 K) where the hydrogen is mainly neutral atomic; and (*c*) the "intercloud medium" with $n \sim 0.2$ cm^{-3} and temperature a few thousand degrees (see Spitzer 1968, Field, Goldsmith & Habing 1969). There is of course a continuous gradation between "standard" and "dense" clouds, and evidence is accumulating (see, e.g., Salpeter 1976, Dickey, Salpeter & Terzian 1977) that there is also an overlap between "standard clouds" and the "intercloud medium." Interstellar absorption lines, discussed below, for a "highly reddened star" should arise mainly in "dense, cool clouds" (*a*), and lines for an "unreddened" star should arise mainly in the intercloud medium (*c*), but in both cases there are likely to be contributions from intermediate types of regions.

Ground-based studies have been augmented greatly by interstellar absorption-line measurements in the *UV* down to wavelength $\lambda \sim 1000$ Å. These abundance measurements are described in the recent review by Spitzer & Jenkins (1975). The radial velocity distribution of the interstellar gas along the line of sight to a particular star is not Maxwellian but is affected by the systematic velocities of various cloud components. This complicates the curve-of-growth analysis and lowers the accuracy of the derived abundances, but in principle allows separate abundance determinations for the different velocity components. The data on reddened stars comes mainly from ζ Oph (Morton 1975) and from o Per (Snow 1976), two fairly similar stars (09.5 and B1) at similar distances (190 and ~300 pc). In the bottom line in Table 2 each abundance f (by number, relative to atomic and molecular hydrogen combined), divided by the solar abundance $f_{\odot}$, is an average over the range of possible values for both ζ Oph and o Per [note that the solar abundances of Ross & Aller (1976) are used in Table 2, which leads to slightly lower values of $f(Ar)/f_{\odot}(Ar)$ than in Morton and Snow's papers]. The range of uncertainty (measurement error and difference between ζ Oph and o Per) is indicated roughly by the height of the rectangles in Figure 1, where essentially the same data are plotted against a condensation parameter T_{cond} to be discussed in Section 3.1. The most striking features of the data are that the depletion factor $f/f_{\odot}$ can reach extremely large values and, as pointed out by Field (1974), that this factor correlates well with T_{cond}.

For stars with little or no reddening, attempts have been made to separate abundance values for the various velocity component. For the "low velocity" components ($v_{\text{rad}} \lesssim 20$ km s^{-1}, typical of the bulk of the ISM) the situation is as follows: Data are available for λ Sco, a star with a moderate amount of reddening, $E_{B-V} = 0.03$ (York 1975), and for a few unreddened stars including γ^2 Vel (Grewing et al. 1973), ζ Pup (deBoer & Pottash 1973), MuCol, and HD-50896 (Shull & York 1977). The H I regions in the line of sight to unreddened stars generally show smaller depletion factors $f_{\odot}/f$ than those for reddened stars (with intermediate values for λ Sco). "Typical" values are denoted by crosses in Figure 1, indicating that elements like C and S show little or no depletion, but elements like Mg, Ca, and Fe are definitely underabundant (there is considerable variability and the exact location of the crosses in Figure 1 is not significant).

High-velocity gas ($|V_r| \sim 50$ to 200 km s^{-1}) has been observed in the line of sight to a number of unreddened stars, such as HD-50896 (Shull 1977), and two stars behind the Vela supernova remnant (Jenkins, Silk & Wallerstein 1977), as well as MuCol and HD-28497. In many, if not all, cases even elements such as Mg, Ca, and Fe are *not* depleted relative to the solar values! From the location of the stars involved it is clear that one is *not* dealing with the *very distant* high-velocity clouds

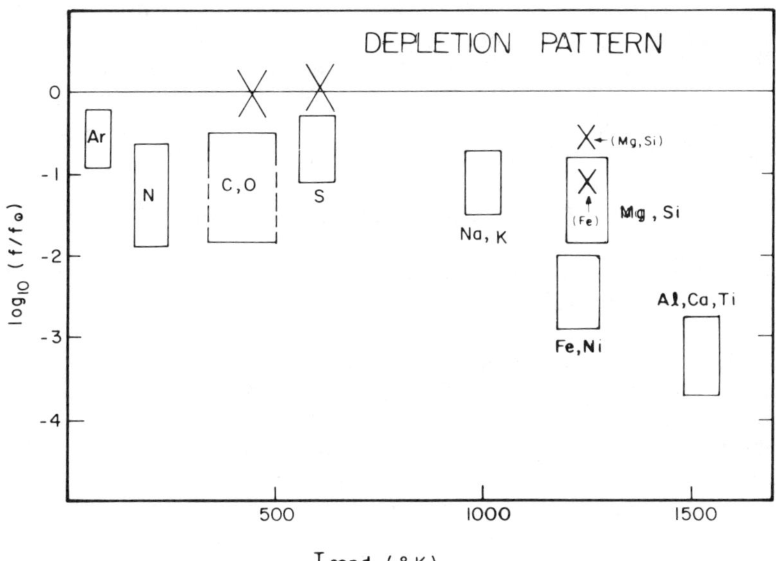

Figure 1 A schematic plot of the depletion factor ($f/f_{\odot}$) for an element plotted against the condensation temperature T_{cond} of minerals that first condense out most of that element. Rectangles refer to lines of sight towards highly reddened stars, crosses towards unreddened stars (T_{cond} for C, O refers to a hypothetical polymerized solid, not to ices).

sometimes found in 21-cm studies (Verschuur 1975), but most probably with interstellar gas that had been shock-accelerated by a blast wave from a supernova remnant that destroyed the dust grains (see Section 6.1 and the current review by Chevalier 1977).

In connection with this high-velocity gas and the destruction of grains it is interesting to note that regions containing highly ionized species such as O VI have also been found (Jenkins & Meloy 1974). These regions must be very hot ($T > 10^5$ K) and may have been heated by supernova blast waves. The abundance ratio of Na I to Ca II in the interstellar gas also varies with the velocity of the gas (Routley & Spitzer 1952, Siluk & Silk 1974). Although other explanations are possible (Pottash 1972), this variation is probably also connected with the depletion history of the total calcium abundance (Hobbs 1975, Jura 1975, 1976).

3 CHEMICAL PHYSICS OF GRAIN FORMATION AND DESTRUCTION

3.1 *Thermochemistry*

We return later to condensation processes in instellar space, where conditions are violently *non*-equilibrium, but consider first the simpler process of condensing solids out of a cooling gas mass that originally was so hot as to contain no solids. Section 3.2 summarizes classical nucleation theory and Section 3.3 discusses some complications, but we first consider simple thermodynamic equilibrium. Besides specifying the temperature T and total pressure P of a gas, we also have to specify all the element abundances. For outflow from the atmospheres of *normal,* cool stars and for protostellar nebulae we simply assume "cosmic," i.e. solar, abundances. For outflow from stars the relevant pressures range from $\sim(10 \text{ to } 10^{-4})$ dyn cm^{-2}, for protostellar nebulae somewhat higher; the relevant temperatures range from 2000 K downwards (see Figures 2 and 3).

Thermochemical calculations basically rely on tables of free energies for many gas-phase atoms and molecules and many minerals, such as the compilation by JANAF (1971). Molecular abundances have been discussed by Tsuji (1973) and by Johnson, Beebe & Sneden (1975). An important feature is the very great stability of gaseous CO at temperatures of one to two thousand degrees. As a consequence, whichever of C and O has the smaller abundance (by number) is effectively bottled up in CO and not available for making solids. For normal cosmic abundances we therefore expect no carbon compounds to be formed (above 1000 K). For the remaining of the most common elements we find that, depending on pressure, H is mainly atomic to the left of the dotted line in Figure 3 and mainly molecular to the right; O is mainly in the form of H_2O (below $\sim$2500 K); N is in the form of N_2, Si is in the form of SiO; and most metals, including Mg, Ca, and Fe, are mainly monatomic.

Condensation curves for solids have been reviewed by a number of authors (Lord 1965, Donn et al. 1968, Anders 1971, Lewis 1972, Grossman & Larimer 1974, Barshay & Lewis (1976), and a few such curves are given (solid curves) in Figure 2. Unlike the equilibrium between different gas-phase molecules, which changes continuously with temperature, each such curve represents a sharp phase transition. For instance, at any point in the pressure-temperature plane above (and to the left of) the curve labelled Mg_2SiO_4 only gaseous Mg, SiO, and H_2O are present; if the curve is crossed at condensation temperature $T_{cond}(P)$, forsterite (a Mg_2SiO_4-mineral) beings to form. At any point below and to the right of the curve, enough forsterite forms to keep the partial pressures of Mg, SiO, and H_2O equal to their values on the curve (for the same temperature). The few minerals (out of many) displayed in Figure 2 represent the first mineral to condense out for each common element (at perfect thermal equilibrium) as temperature is lowered (tungsten is not a common element, but is shown since it forms one of the most refractory solids). It is important to note that Al + Ca + Ti are about ten times less abundant than Mg + Si or Fe but form solids first (i.e. at higher temperature), such as corundum (Al_2O_3), perovskite ($CaTiO_3$), and gehlenite ($Ca_2Al_2SiO_7$). Next to form are the more abundant magnesium silicates (forsterite, Mg_2SiO_4, condenses first over most of the pressure range shown but at the lowest pressures enstatite, $MgSiO_3$, would condense first) and metallic iron; sodium and potassium silicates (e.g. feldspars, also containing Al) form at somewhat lower temperatures.

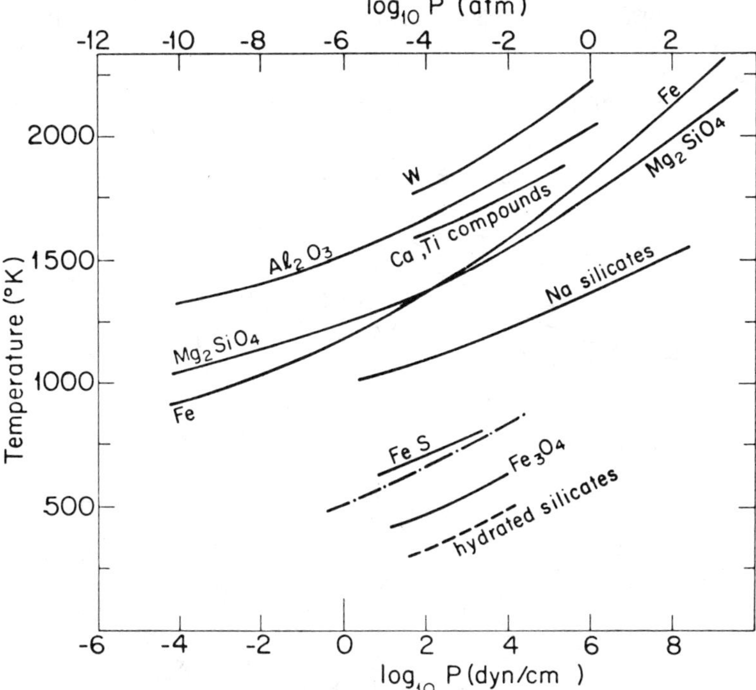

Figure 2 Plots of condensation temperature versus total gas pressure *P* for various minerals, assuming normal cosmic abundances. The dash-dot curve indicates conditions where half of the CO is destroyed by interaction with H_2.

The uncertainties in the thermochemical data are small enough so that the derived values for $T_{cond}(P)$ should be accurate to better than 10%. In some pressure ranges the condensation order of forsterite, enstatite, and iron might be interchanged, but otherwise there is little controversy above ~ 1000 K. At lower temperatures many equilibrium curves are also well known (such as the ones for troilite, FeS, and magnetite, Fe_3O_4, in Figure 2), but it is controversial whether equilibrium can be reached in practice, since molecular rearrangements are required and reactions are slow at low temperatures. This is particularly true for hydrated silicates, which are stable only at fairly low temperatures, and for the dash-dot curve in Figure 2: Below this curve the equilibrium balance shifts from $(CO + 3H_2)$ toward $(CH_4 + H_2O)$, but reactions between CO and H_2 are notoriously slow in the absence of catalysts (see Sections 3.3, 4.2).

Because of the great stability of gaseous CO, the thermochemistry is different for the outflow from stars of anomalous composition where the carbon abundance (by number) exceeds the oxygen abundance. In all such "carbon-excess" stars solid oxides and silicates cannot be formed, but the details still depend on whether (*a*) the star is "genuinely carbon-rich," i.e. the abundance-ratios C/H and C/N exceed solar values, or (*b*) the star is actually "oxygen-poor" but C/H and C/N are less than solar (Fujita 1970). To furnish large amounts of graphite particles for the ISM type (*a*) is of course much more efficient if it should be a common occurrence. Helium burning plus deep mixing can bring excess C^{12} to a star's surface (Sackmann, Smith & Despain 1974), and there is some evidence from helium stars (Dinger 1970)

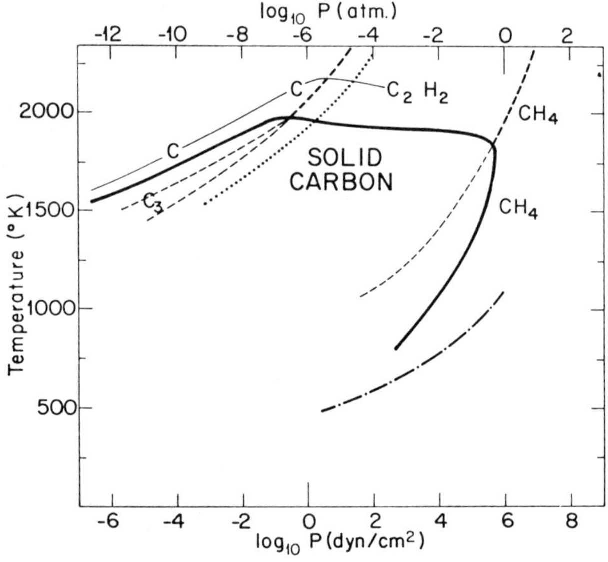

Figure 3 A temperature-pressure phase-diagram for carbon-rich material (excess carbon $\sim 10^{-3.5} \times$ hydrogen). The thick solid curve encloses the region where solid carbon (graphite) is stable.

that mixing plus mass loss had taken place. Unfortunately, neither theory nor observation gives reliable estimates on how common carbon-rich stars of type (*a*) are. Figure 3 gives some relevant thermochemical data for an intermediate case where the abundance of excess carbon (by number) is $f_c \sim 10^{-3.5} f_H$.

By far the most abundant solid to form at equilibrium is graphite (Donn et al. 1968, Fix 1969a,b, 1970) (but amorphous solid carbon, even with some impurities, would give essentially the same condensation curve). The abundant gas-phase species depend strongly on the total pressure: To the left of the dotted curve in Figure 3 hydrogen is mainly atomic, to the right mainly molecular; the three *thin* dashed lines indicate which carbon molecules would be most important if there were no solids, namely C at lowest pressure, C_3, C_2H_2 next, and CH_4 at highest pressures. The thick solid curve encloses the region of stability for solid-carbon. At pressures P of $\sim$(1 to 10^4) dyn cm^{-2}, the relevant gas-phase species is acetylene, C_2H_2, and the condensation temperature is almost independent of pressure (Salpeter 1973, 1974a). This is an illustration of Lechatelier's principle, since the reaction $C_2H_2 \rightarrow$ graphite $+ H_2$ leaves the number of gas-phase molecules unchanged. For the assumed composition the gas-phase abundance of HCN is always less than 10% of C_2H_2. Only for an extreme type (*b*) composition, if $f_c < 0.05\ f_N \sim 10^{-4.5}\ f_H$, is the dominant gas-phase molecule HCN; the *thin* solid curve is the condensation curve for solid carbon for extreme type (*a*) composition, $f_c \sim 10^{-2.5}\ f_H$. At low pressures, C (plus some C_3) is the abundant gas-phase species.

Solid metallic iron will form at lower temperatures than graphite (just as in Figure 2 for normal abundance); silicon is monatomic in the carbon-rich gas phase and some solid SiC will form, but also only at lower temperatures than graphite. As the temperature is lowered below T_{cond} for graphite (at fixed P, say), the (saturation vapor) partial pressures of *all* carbon-containing gas-phase molecules decrease drastically and are essentially negligible for $T \sim$ (700 to 1200) K. At lower temperatures still, the equilibrium again shifts towards hydrocarbons (but it is unlikely that equilibrium can be attained in practice, see Section 3.3). The dash-dot curve in Figure 3 again indicates the (hypothetical) equilibrium shift from $(CO + 3H_2)$ toward $(CH_3 + H_2O)$.

3.2 *Homogeneous Nucleation Theory*

We summarize here the theory of homogenous nucleation, reviewed in detail by Hirth & Pound (1963), Feder et al. (1966), and Abraham (1974). This theory deals with originally gaseous material that is *almost* in complete thermal equilibrium at temperature T and pressure P, but T is lowered at a very slow, predetermined rate. Consider first the specially simple case where we have only one kind of gas-phase monomer (atom or molecule) A with concentration c_1, and a macroscopic grain (or drop) is simply a complex (or cluster) A_N in the limit of large values of N. At *complete* thermal equilibrium, for finite N, the law of mass action for the reaction $(A_N + A \rightleftarrows A_{N+1})$ gives for the cluster concentrations c_N and c_{N+1} (expressed in special units) the relation

$$c_{N+1}/c_N = c_1 \exp\left[-\Delta F_N(T)/kT\right], \tag{3.1}$$

where ΔF is a free-energy difference. In this language, a point T_{cond} on the condensation curve—equilibrium between a macroscopic solid and vapor—means a point in the (c_1, T) plane where Equation (3.1) gives $c_{N+1}/c_N = 1$ in the limit of $N \to \infty$. The condensation curve $T_{\text{cond}}(c_1)$ and a cooling-track $\{T(t), c_1(t)\}$ as a function of time t are shown schematically in Figure 4*a* and 4*b*, with the two curves crossing at time zero and temperature T_0. Define units of concentration such that $\Delta F_N(T)$ for $N \to \infty$ is independent of T in the immediate vicinity of T_0 along the cooling track. Let B be this value of $\Delta F(T_0)$ and define a quantity D such that

$$-\Delta F_N(T) = B - N^{-1/3} D_N(T). \tag{3.2}$$

The first term B on the right-hand side of Equation (3.2) is essentially the binding energy per particle in a macroscopic grain, and the second term is the difference in surface energy for the $(N+1)$ cluster and N cluster. For large spherical liquid drops the surface energy is proportional to $N^{2/3}$ and, in the limit of large N, D_N becomes independent of N and is related to the surface tension of the liquid. *Classical* nucleation theory simply makes the drastic approximation of replacing D_N for all N by this limiting constant D_∞, even though D_N is required mainly for N in the vicinity of some "critical cluster size" N_{crit}, which is not very large (see Equation 3.3). Replacing D_N by a constant D, but leaving open how to determine its value, we can summarize the results of classical nucleation theory in an approximate manner (see Draine & Salpeter 1977, who correct some errors in Salpeter 1974a):

We have two moderately large dimensionless ratios, $(B/kT_0) \sim 15$ to 50 and $(B/D) \sim 4$ to 10, and a third *very* large dimensionless parameter η which depends on the dynamics. Let t_0 be the "cooling time," i.e. the period during which T decreases below T_0 by a fractional amount (kT_0/B), and let η be the number of "sticking collisions" with monomers which an individual surface site on a grain experiences during time t_0 (see Equation 5.1). As shown schematically in Figure 4*b*, the actual temperature T drops below the condensation temperature for a time

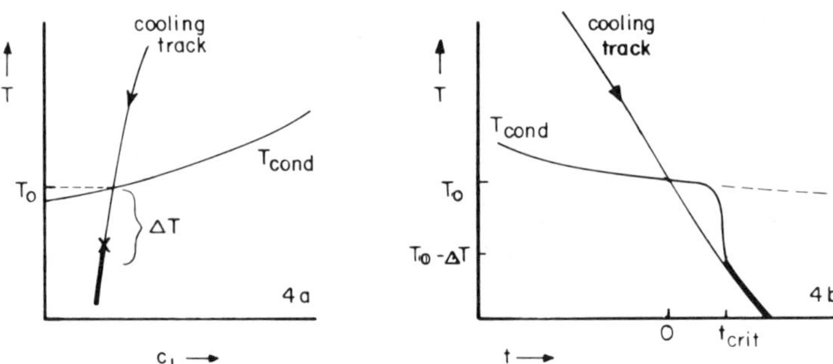

Figure 4 A schematic cooling track leading to condensation. (*a*) Temperature T plotted against monomer concentration c_1. T_0 is the condensation temperature T_{cond} for thermal equilibrium, $T_0 \times \delta \equiv \Delta T$ is the amount of supercooling required for copious condensation (thick line). (*b*) The cooling track plotted against time; t_{crit} is the time delay required for supercooling by ΔT.

period t_{cr} when it reaches a value $T_{cr} = T_0(1-\delta)$, after which most of the gas-phase monomers are rapidly converted to the solid (and the condensation temperature follows the actual temperature closely). In the critical time period the cluster size N is in the vicinity of a "critical size N_{cr}" which, together with the critical temperature T_{cr} and the critical delay time t_{cr}, is given by

$$T_{cr} \equiv T_0(1-\delta),\ \delta \equiv (\ln \eta)^{-1/2}(D^3/8B^2kT_0)^{1/2};$$
$$t_{cr} \equiv (\ln \eta)^{-1/2}(D/2kT_0)^{3/2}t_0; \qquad (3.3)$$
$$N_{cr} \equiv (\ln \eta)^{3/2}(8kT_0/D)^{3/2}.$$

The typical number N_{fin} of molecules in a "final grain" (after most of the gas-solid conversion) is given to rough order of magnitude by

$$N_{fin} \sim (\eta/\ln \eta)^3. \qquad (3.4)$$

For a gas-phase species with abundance f we write $\eta \equiv f\eta_0$, with η_0 independent of f. The abundance (by number relative to H) f_{fin} of grains is f/N_{fin} or

$$f_{fin} \sim f^{-2}(\ln \eta/\eta_0)^3, \qquad (3.5)$$

i.e. the *number* density of grains decreases with increasing element abundance!

The literature on nucleation theory is mainly concerned with liquid drops made of saturated stable molecules where the bulk surface tension (and therefore D_∞) is known accurately, and many controversial papers (see, for instance, Katz 1970 and Abraham 1974) attempt to calculate $(D_{N_{crit}} - D_\infty)$ and to calculate the nucleation rate at a predetermined temperature. Some papers on stellar-grain formation (see Tabak et al. 1975, Walker 1975) apply these "corrections to classical theory" to the formation of graphite (or silicate) grains, which are held together by strong valence forces. Such applications are formally correct but not really practical since the bulk surface tension is poorly (if at all) known for refractory solids, and the extrapolation downwards to N_{cr} is dubious. In one case, the formation of graphite from monatomic C (with $B \sim 45\ kT_0 \sim 7$ eV), direct data on free energies for C_N molecules is available for N up to 9 (Palmer & Shelef 1968), which gives $D \sim 0.9$ eV $\sim 0.12B$, and Equation (3.3) gives $\delta \approx 0.1\,(\ln \eta)^{-1/2}$. The situation is more complicated for silicates and oxides (Donn 1976) where the gas-phase molecules are not the monomer for the solid, such as Al and H_2O in the gas building up $(Al_2O_3)_N$ in corundum: The condensation curves and B ($\sim 60\ kT_0$, referred to Al) are still accurately known, but no free-energy data are available at the moment even for Al_2O_3 in the gas. Nevertheless, from the free-energy differences of Al, AlO, and Al_2O_2 (see JANAF 1971) I estimate $D/B \sim 0.15$ to 0.25 for corundum and $\delta \sim 0.2\,(\ln \eta)^{-1/2}$. This value of δ could be in error by a factor of 2, but for large values of η the uncertainty in $T_{cr}/T_0 \equiv 1-\delta$ is still fairly small.

3.3 *Surface Nucleation and Reprocessing of Grains*

We are most interested in the condensation along a cooling track of the most abundant solids, e.g. magnesium silicates for normal chemical abundances. If homogeneous nucleation applied, the final size and (number) abundance of Mg_2SiO_4 grains would be given by Equations (3.4) and (3.5) with $f \sim 10^{-4.5}$. However, very small corundum (Al_2O_3) grains will condense first (if η_0 is sufficiently large), and Equation (3.5) with $f \sim 10^{-5.5}$ gives a *larger* grain abundance. Consequently, the magnesium silicates will *not* form by homogeneous nucleation but will grow as mantles around the corundum grains. Most of the mass of the final grains is Mg_2SiO_4 but the final number of grains is determined by f and η for the earlier condensation of Al_2O_3.

It would be important to know if pure minerals or amorphous mixtures are formed and whether further chemical reprocessing takes place. Unfortunately, no quantitative calculations and only few experiments (Meyer 1971, Dorfeld & Hudson 1973) have been carried out to date and one can only state trends: If η is very large (many collisions) and if the gas density is high (hence high condensation temperatures and great thermal mobilities), then pure minerals can form or recrystallize; e.g. magnesium-rich silicate grains would form separately from pure metallic iron grains for normal abundances (or pure graphite grains for carbon-rich atmospheres). In practice, η and the density may not be sufficiently great, and amorphous grain mixtures of oxides, silicates, and iron may form (or amorphous carbon with a small admixture of hydrocarbons included for carbon-rich atmospheres). This tendency towards "dirty grains" would also be enhanced by deviations from thermal equilibrium between grains, gas, and radiation (Arrhenius & De 1973, Czyzak & Santiago 1973, Donn 1976). The possibility of forming "low-temperature" minerals such as magnetite (Shapiro 1975) and hydrated phyllosilicates (Zaikowski, Knacke & Porco 1975) is somewhat paradoxical: If η and gas density ρ are *extremely* high, such minerals may be reprocessed at "near-equilibrium" during subsequent cooling; if η and ρ are very small and deviations from equilibrium are violent, then these minerals (and almost everything else) might form right at the beginning; but for intermediate values of η and ρ only pure silicates and iron form at the beginning and the subsequent cooling is too rapid for any reprocessing. It is also not clear whether the abundant CO molecules in the gas can be reprocessed into complex molecules and solids by surface catalysis during the subsequent cooling below $\sim$500 K (Anders, Hayatsu & Studier 1974).

3.4 *Grain-Grain Collisions, Sputtering, and UV Effects*

When two grains collide with relative velocity ΔV, four different outcomes are possible—elastic scattering, coalescence, evaporation, and shattering. There is a copious literature on collisions between liquid droplets (see, for instance, Mason 1971), but only few experiments have been carried out for collisons between two solid particles of comparable size (Kerridge & Vedder 1972, Vedder & Mandeville 1974). Order-of-magnitude estimates will be found in theoretical papers on grain destruction and coalescence in protostellar nebulae (Carusi et al. 1975, Cameron 1973b, 1975, Burke & Silk 1976), circumstellar outflows (Gilman 1972, Salpeter 1974a, Hayashi & Nakagawa 1975), interstellar shocks (Aannestad 1973, Spitzer 1976), and supernova shocks (Barlow & Silk 1977). Coalescence is more likely for $\Delta V < 1$ km sec^{-1}, and shattering and/or evaporation is more likely for $\Delta V > 20$ km sec^{-1} (for refractory grains like silicates or graphite); quantitative details depend not only on ΔV and binding energy of the solid, but also on temperature and the shape of the grain.

The sputtering of a surface atom or molecule on a grain by the impact of a fast gas atom is better understood: There is considerable experimental data (see, for instance, Laegrid & Wehner 1961, Stuart & Wehner 1962), and semi-empirical formulae have been given by the theoretical papers mentioned above and by Draine (1977). The sputtering yield depends mainly on the ratio $[M_p M_T (M_p + M_T)^{-2} E_p B_T^{-1}]$, where M_p and E_p are the mass and kinetic energy of the (gas-atom) projectile, M_T and B are the mass and binding energy of a surface atom of the grain. For a gas with cosmic abundances, He is the most efficient projectile for sputtering grains. For a silicate (or graphite) grain, velocities of order 200 km sec^{-1} relative to the gas are required for appreciable sputtering efficiency (see Figure 7).

The most obvious possibly destructive effect of radiation on a grain is heating and the consequent evaporation. Calculations of grain temperatures (Werner & Salpeter 1969, Kudo 1974, Greenberg & Hong 1974, Purcell 1976) depend on the optical properties of the solid (see, for instance, Bussoletti & Zambetta 1976), but thermal evaporation is likely to be unimportant even in H II regions except close to the exciting star. For a foreign atom or molecule adsorbed on a grain surface (even if it is strongly chemisorbed), the efficiency of photodesorption by *UV* photons of energy $\sim$10 eV is quite high (Watson & Salpeter 1972, Greenberg & van de Hulst 1973), but these same photons (or even more energetic ones) do very little damage to atoms in a good crystal. Consequently, a fully grown grain can survive the radiation field even in an H II region, whereas the build-up of molecules to form a new grain-mantle is prevented even by the milder radiation field in typical H I regions.

4 OPTICAL PROPERTIES OF GRAINS

4.1 *Interstellar Observations*

Interstellar extinction (and, to some extent, scattering and polarization) in the visible, *UV*, near- and mid-infrared has been studied extensively and reviewed by Aannestad & Purcell (1973) and Spitzer & Jenkins (1975). Figure 4 gives (schematically) three "typical" curves of extinction efficiency Q plotted against wavelength λ, normalized to unity for the efficiency Q_V in the visible ($\lambda \sim 5500$ Å). The most striking features are the "hump" near $\lambda = 2200$ Å (Stecher 1965), which is contributed mainly by absorption (Lillie & Witt 1976), and the continuing rise in Q all the way to $\lambda \sim 1000$ Å. As indicated by Figure 4, the normalized extinction in the far *UV* varies appreciably from one case to another, and the peak of the "hump" varies somewhat but not its position and shape (Savage 1975). With $A(\lambda)$ the extinction in magnitudes, much data is available on A_V and $E_{B-V} \equiv A_B - A_V$. The ratio $R \equiv A_V/E_{B-V}$ is about 3 to 3.5 and varies less than the *UV* extinction with the length and direction of the path over which it is measured (Hackwell & Gehrz 1974, Sandage 1975, Schultz & Wiemer 1975). This constancy of R suggests that the grains responsible for extinction in the visible vary rather little in size from one region to another. However, R is considerably larger in dense, dark clouds (Vrba et al. 1975), especially in the Orion Nebula, indicating larger grains.

There is now considerable data on the ratio of neutral hydrogen column density N_H to reddening E_{B-V} for various path lengths, both from globular clusters and galaxies outside the galactic disk (Knapp & Kerr 1974, Heiles 1976) and from stars close to the galactic plane (Bohlin 1975). If molecular hydrogen is included in the column density, the mean ratio N_H/E_{B-V} is about 5×10^{21} (protons cm^{-2}) mag^{-1}. In the galactic plane the corresponding average absolute densities (Burton & Gordon 1976) are $N_H \sim 3 \times 10^{21}$ (protons cm^{-2}) kpc^{-1}, $n_H \sim 1$ proton cm^{-3}, $E_{B-V} \sim 0.6$ mag kpc^{-1} and $A_V \sim 2$ mag kpc^{-1}. Very dense and opaque clouds (consisting mainly of *molecular* hydrogen), which occupy a small fraction of the volume but contribute about half the total mass of the interstellar medium, are included in these recent estimates. The corresponding estimate for the total mass of the interstellar gas (including H_2 and He) is about $4 \times 10^9\ M_\odot$ (assuming a galactic disk of radius 15 kpc).

For wavelengths in the mid-infrared there is a striking feature in the interstellar extinction curve, a peak centered at $\lambda \approx 9.7\ \mu m$ (for shape see Figure 5*a*) with a weaker feature (see insert to Figure 4) centered near $\lambda \sim 20\ \mu m$ (Stein & Gillett 1969, Woolf 1973, Aitken & Jones 1974, Gillett et al. 1975a). The 9.7 μm "silicate feature" is quite common, but an appreciable fraction of the absorbing material can be in the outer layers of the continuum source (Kwan & Scoville 1976); the extinction ratio $A(9.7\ \mu m)/A_V$ for the general interstellar medium is therefore difficult to determine (Persson, Frogel & Aaronson 1976), but is probably in the range 0.04 to 0.1.

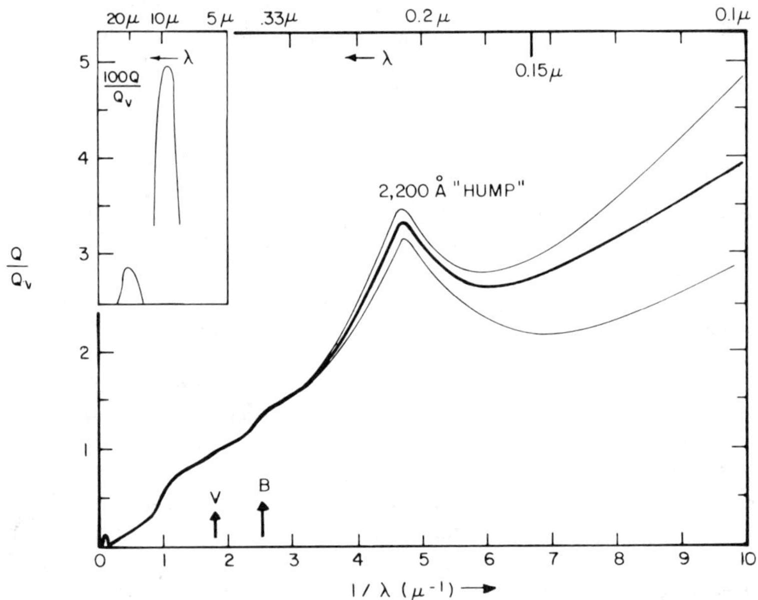

Figure 5 The extinction efficiency Q, normalized to the efficiency Q_V in the visible, plotted against wavelength λ.

Water-ice has a narrow absorption feature near $\lambda = 3.08\ \mu m$, whose position and shape are not affected very strongly by temperature (Knacke, Cudaback & Gaustad 1969) or by admixture with methane and ammonia ices (Hunter & Donn 1971). This absorption feature is found in the interstellar extinction for some lines of sight but (*a*) only if dense, molecular clouds are present and (*b*) even for such a molecular cloud the peak absorption at 3.08 μm is only comparable with that in the vicinity of 9.7 μm (Merrill, Russell & Soifer 1976).

An emission feature centered on $\lambda \sim 11\ \mu m$ (see Figure 5*b*), found in the spectra of cool carbon stars, is discussed in Section 5.1. The diffuse interstellar bands are not discussed here (see e.g. Herbig 1975, Greenberg & Hong 1976) since their relationship to interstellar grains is still not clear, nor are measurements of infrared polarization (Dyck & Beichman 1974, Capps & Knacke 1976) or other indicators of grain alignment.

4.2 *Grain Models*

The optical data on interstellar extinction and polarization are complex enough, so that no model for grains of a single size and composition can fit the data. Crudely speaking, at least three different types of grains seem to be required to fit three features whose strengths seem to vary separately from region to region: (*a*) the data in the visible, (*b*) the "hump" near $\lambda \sim 2200$ Å and (*c*) the rise in extinction towards $\lambda \sim 1000$ Å. Once three or more grain types are allowed in a grain model, the model is by no means unique. One class of models (Gilra 1971) uses SiC grains (radius $\sim 0.1\ \mu m$) for much of the optical behavior in the visible, small graphite grains (radius $\sim 0.02\ \mu m$) for the "hump" near 2200 Å, and silicate grains (radius $\sim 0.05\ \mu m$) for the extinction for $\lambda \sim$ (1800 to 1200) Å. Another class of models (see Greenberg & Hong 1974) uses "core-mantle" grains for the behavior in the visible, with the mantles consisting of H_2O, NH_3, and CH_4 ices as suggested by van de Hulst (1948). In spite of the general controversy, a few points seem fairly clear, especially when data on circumstellar grains are also considered (see Section 5.1):

The $\lambda \sim 9.7\ \mu m$ feature is almost certainly due to some form of silicate grains. The exact type of silicate mineral is not clear, partly because quantitative details of the spectra depend on particle size and temperature (Day, Steyer & Huffman 1974, Day 1976). The total amount of silicates required to fit the 9.7 μm data is not known accurately, but is quite compatible with the cosmic abundance of Mg and Si in the interstellar medium going mainly into such grains (Gillett et al. 1975b). It is possible that much of the interstellar silicate material has been reprocessed into hydrated silicates, as are found in carbonaceous chondrite meteorites (Zaikowski, Knacke & Porco 1975, Zaikowski & Knacke 1975, Penman 1976). Wickramasinghe (1975) has pointed out that polyoxymethylene (polymerized formaldehyde) has a similar absorption feature near $\lambda \sim 10\ \mu m$, but this must be at least partly coincidental because the feature is also seen in the outflow from oxygen-rich stars where polymerized formaldehyde could not have formed (Section 5.1).

The "hump" near $\lambda \sim 2200$ Å may be fit best by graphite (Gilra 1971) or some other kind of very small, carbon-rich particle (Donn 1968). Alternative models using quartz or silicates seem to give only a moderate fit to the extinction data and scattering albedo (Egan & Hilgeman 1975). It is clear that some *very* small grains (radius $\lesssim 0.02\ \mu$m) are required to fit the extinction rise towards $\lambda \sim 1000$ Å, but the chemical composition is still fairly arbitrary.

Ironically, the oldest data—the visible extinction—are at the moment the most controversial: SiC-grains (in addition to graphite and silicates) can fit the data in principle, but the *amount* of silicon required seems to be more than cosmic abundances allow (and mass requirements can be stated in fairly general terms in terms of the Kramers-Krönig relation; e.g. Purcell 1969, Caroff et al. 1973). Since the abundances of C, N, and O exceed those of Si by factors of more than 10, there is no such embarrassment for grains with ice mantles of the appropriate size (Aannestad 1975, Dempsey & Wickramasinghe 1975); the cosmic abundances even allow excess C, N, and O for very small or very large grains that are optically inefficient (Greenberg 1974). As mentioned, water-ice is found in molecular clouds but in very small amounts compared with the cosmic abundance of O (only about one tenth of the mass in silicates when the optical depths are comparable; see Merrill, Russell & Soifer 1976). I favor the (also controversial) explanation that much of the ices has been polymerized by the action of stellar *UV* (Section 6.3).

5 CIRCUMSTELLAR DUST

5.1 *Red Giants, Supergiants, and Infrared Stars*

Circumstellar dust grains (Woolf & Ney 1969, Gilman 1969, Woolf 1973, 1975) are involved in mass loss from evolved stars. The death rate of stars with original main sequence mass $\sim$(1.2 to 4) $M_\odot$ and final white dwarf mass $< 1\ M_\odot$ is about 2 yr^{-1} (Weidemann 1967). With $\sim 1\ M_\odot$ per star returned to the interstellar medium (and assuming a galactic disk of radius 15 kpc), the rate of mass return (gas plus dust) is $\sim 1.5\ M_\odot\ \text{yr}^{-1} \sim M_{\text{gas}}/3 \times 10^9$ yr, where $M_{\text{gas}} \approx 4 \times 10^9\ M_\odot$ is the present mass of gas in the galactic disk. Some fraction of the mass loss from a star is likely to occur when the effective surface temperature $T_{\text{eff}} > 2500$ K and its mass-flow rate $\phi < 5 \times 10^{-9}\ M_\odot\ \text{yr}^{-1}$ (Cohen 1976), in which case grains probably do not form in the outflow. However, a large fraction of the total mass loss is likely to occur over short periods when the star becomes an extremely cool *M* giant or supergiant or when it sheds a planetary nebula (Section 5.2). Assuming that one or two thirds of the total mass loss is accompanied by grain formation leads to the first row of Table 1.

For cool oxygen-rich stars (atmospheres with normal cosmic abundance), there is an abundance of data for the $\lambda \sim 9.7\ \mu$m "silicate feature" seen in emission. The strength of this feature increases with the bolometric luminosity L_{bol} of the star, especially when $L_{\text{bol}} \gtrsim 10^4\ L_\odot$ (Cohen & Gaustad 1973), and increases very drastically when T_{eff} becomes *less* than $\sim$2500 K (Dyck, Lockwood & Capps 1974). Apart from the silicate feature, there is also evidence for attenuation of the primary starlight and re-emission in the continuum in the infrared, which also suggests a circumstellar shell. The color-temperature of the re-emission decreases (outer radius of the shell increases) with the optical depth τ of the shell, and the strength of the silicate emission feature increases first with increasing τ, reaches a maximum, decreases, and changes sign, i.e. for very large τ (and continuum color temperature below $\sim$400 K) *silicate absorption* increases with increasing τ (Merrill & Stein 1976a,b). This change of sign is predicted by theoretical models for circumstellar dust envelopes (Jones & Merrill 1976); it is analogous to Fraunhofer lines showing deep absorption for an optically thick atmosphere if the continuum emission peaks at longer wavelength than the spectral line. Unlike stellar atmospheres, these circumstellar dust-shells have a large ratio of outer radius r_2 to inner radius r_1: For one star at least (IRC+10011), occultation measurements (Zappala et al. 1974) indicate (*a*) a stellar radius of $R_* \sim 10^{14}$ cm, (*b*) some re-emission at $\lambda \sim 2.2\ \mu$m from an inner radius of order $r_1 \sim 5\times10^{14}$ cm, and (*c*) re-emission at longer wavelengths from larger radii, up to $r_2 \sim 5\times10^{15}$ cm for $\lambda \sim 20\ \mu$m.

Rates of mass loss ϕ in the range of $(10^{-8}$ to $10^{-5})\ M_\odot\ \mathrm{yr}^{-1}$ have been measured directly (Sanner 1976), using spectral lines of *gas* atoms, and $\phi \sim 10^{-6}\ M_\odot\ \mathrm{yr}^{-1}$ seems to correspond to moderate silicate emission strength in the sequence of Merrill and Stein. OH and H_2O masers connected with late *M*-type Mira variables and supergiants suggest values of ϕ well above $10^{-5}\ M_\odot\ \mathrm{yr}^{-1}$ at terminal flow speeds $V_\infty \sim$ (10 to 20) km sec^{-1}; detailed theoretical models for the flow subsequent to the sonic point have been constructed by Goldreich & Scoville (1976). The possibility that dust grains form in a cool stellar atmosphere and radiation pressure on the grains drive out both gas and dust was already suggested by Hoyle & Wickramasinghe (1962) and by Donn & Stecher (1965). Theoretical calculations are relatively simple if $T_{\mathrm{eff}} \lesssim 2000$ K so that grains already form close to the star's photosphere where flowspeeds are very subsonic. Calculations are again relatively simple if $T_{\mathrm{eff}} \gg 2000$ K, if there is mass loss due to some cause unconnected with grains such as a nova explosion (Section 5.2), and if grains form only long after passing the sonic point. The most typical realistic situation seems to be an intermediate one where T_{eff} is larger than 2000 K, but not by very much, and grains form beyond the sonic point but only a few stellar radii out. It is not clear how the grains induce the flow in such cases, but the flow is in any case highly sporadic and far from spherical symmetry (Forrest, Gillett & Stein 1972), which may be caused by giant convection cells below the photosphere (Schwarzschild 1975) and accentuated by grain-driven instabilities (Salpeter 1974b).

For purposes of nucleation and grain growth, the parameter η in Equation (3.4) is important, evaluated at the inner radius r_1 where grains are just beginning to form. With $s \sim 0.5$ a sticking coefficient, ϕ the mass flow rate in $M_\odot\ \mathrm{yr}^{-1}$, and V_f the flow velocity near r_1, we have (Draine & Salpeter 1977)

$$\eta \sim \left(\frac{s}{0.5}\right)\left(\frac{10\ \mathrm{km\ sec}^{-1}}{V_f}\right)^2\left(\frac{5\times10^{14}\ \mathrm{cm}}{r_1}\right) f\phi\times4\times10^9. \tag{5.1}$$

For the nucleation of Mg_2SiO_4 with $f \sim 10^{-4.5}$ and for $r_1 \sim 5\times10^{14}$ cm we have $\eta \sim 10^5\ \phi$. For $\phi < 10^{-5}$ this expression would fail to give any grain formation, whereas stars even with $T_{\mathrm{eff}} > 2500$ K and $\phi \sim (10^{-7}$ to $10^{-6})$ seem to manage the growth of grains. This discrepancy may be connected with the sporadic, "clumpy" nature of the outflow, so that the density in the "clumps" (to which η

is proportional) might be larger than the average by a factor of 100, say. Nevertheless, η is likely not to be enormous and the final grain size (Equation 3.4) should be fairly small. As discussed in Section 3.1, temperatures $T \lesssim 500$ K and very many collisions are required to hydrate silicates and to make organic macromolecules from CO. By the time T has dropped below 500 K, η has decreased further and there simply is no time for such reactions, at least for $\phi \lesssim 10^{-4}$. In these mass flows, at least, the "silicate feature" near 9.7 μm must indeed be due to refractory silicates.

For carbon-rich cool giants and supergiants, the silicate feature is missing (as it should be) and is replaced by some black-body emission with $T \lesssim 1000$ K in the mid-infrared plus the emission feature sketched in Figure 6*b* (Merrill & Stein 1976a,b). The black-body continuum is probably due to graphite, and SiC has been suggested for the 11 μm feature. The question has been raised (Tarafdar & Wickramasinghe 1975) whether all this mid-infrared emission could instead be due to hydrocarbon molecules in the gas phase (which are fairly abundant for $T \sim 2000$ K). To show up in emission rather than absorption, the material must be far above the stellar surface and therefore cooler ($T \sim 1000$ K, say) and at low density; thermochemistry (Section 3.1, Figure 3) then shows that hydrocarbons in the gas phase have negligible abundances because of the low density (abundances drop even more once graphite forms).

As mentioned in Section 3.1, carbon-rich stars are less common than oxygen-rich ones by a factor of possibly 10. If most of these "carbon-rich" stars are only oxygen-poor, the abundance of excess carbon may be only a few times the silicon abundance and the mass production of graphite grains would be only about one third of silicate production. However, the average excess carbon abundance might be ten times larger, say, giving a graphite/silicate production ratio of about three.

5.2 *Novae and Planetary Nebulae*

A prerequisite for a nova outburst in a hot white dwarf is the presence of a fairly close (evolved) companion star that provides nuclear fuel for a thermonuclear

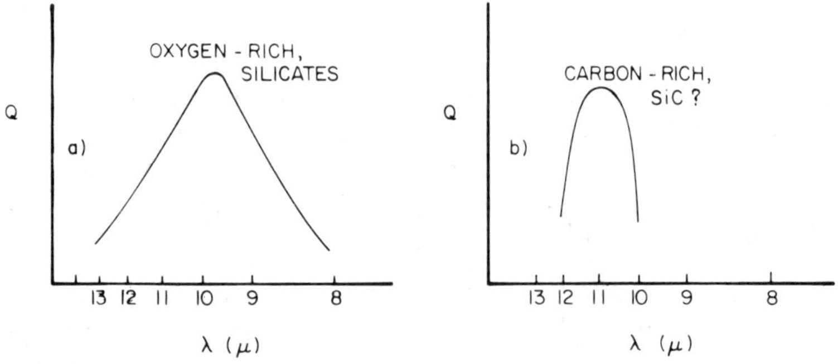

Figure 6 The qualitative shape of the extinction efficiency Q wavelength λ curve in the "mid-infrared," (*a*) for oxygen-rich material, (*b*) for carbon-rich material.

runaway. Since only a fraction of all stars are in close binaries, the total mass injected into the interstellar medium by novae must be much less than that from planetary nebulae and supergiants. Nevertheless, novae provide an excellent test case for models on grain production, partly because it is clear that the mass outflow is not *caused* by grains (but by a nuclear explosion) and partly because only *some* novae show evidence of grain formation, so typical conditions there must be borderline for being able to nucleate grain formation.

A favorable case is Nova FH Ser 1970, which was observed for about 100 days in the visible, the UV (Gallagher & Code 1974), and the infrared (Hyland & Neugebauer 1970, Geisel, Kleinmann & Low 1970). Models for the thermonuclear runaway (Sparks, Starrfield & Truran 1976) indicate that the combined abundances of C, N, and O must be larger by factors of order 10, and there is also some observational evidence for some overabundance (Antipova 1969). Grains formed in the outflow should have interesting isotopic composition anomalies (Clayton & Hoyle 1976). It is not clear at the moment if the outflow is carbon-rich or oxygen-rich, but models of grain formation have been constructed for both cases (Clayton & Wickramasinghe 1976, Yamamoto & Nishida 1977). With $L_{\rm bol} \sim 3 \times 10^4\ L_\odot$ and $T_{\rm eff} \gtrsim 10^4$ K, grains acquire a somewhat larger temperature than the "energy density temperature"; they first form at a rather large distance, $r_1 \sim 7 \times 10^{14}$ cm, which is reached after about 40 days at typical flow speeds of $V_{\rm fl} \sim 1500$ km sec^{-1}. Mass flow rates are of order $\phi \sim 10^{-3}\ M_\odot$ yr^{-1} (say $2 \times 10^{-4}\ M_\odot$ over the flow period, ~2 months). Equation (5.1) then gives $\eta \sim 200\ f$. If carbon is 10 times more abundant than in the Sun, then $f \sim 10^{-2.4}$ and $\eta \sim 1$. In reality the carbon might be overabundant by an even larger factor and/or "clumping" could raise the effective value of η somewhat, as evidenced by the fact that the infrared emission from Nova FH Ser shows the presence of grains after a month or so (but it is also reasonable that η may be less than unity in other novae that do not form grains).

As mentioned in Section 5.1, material is returned to the interstellar medium from evolved cool giants and supergiants at a rate of $\sim 1\ M_\odot$ yr^{-1}, partly in the form of planetary nebulae (Bonilha & Terzian 1976). An appreciable fraction of 1 $M_\odot$ is ejected in a planetary nebula with speeds of $V_f \sim$ (10 to 20) km sec^{-1} over a time period of order 10^3 years (Härm & Schwarzschild 1975). The parameter η for grain formation (Equation 5.1) is of order $\eta \sim f\phi \times 4 \times 10^9 \sim f \times 10^{-6}$, and abundances in planetary nebulae are probably close to the cosmic values and $\eta \sim 10^{0.5}$ for the formation of Al_2O_3, for instance. The outflow is again likely to be "clumpy," which increases the effective value of η, and there is little doubt theoretically that grains should be able to form.

There is also little doubt observationally that planetary nebulae contain grains at temperature $T \sim 200$ K, as evidenced by emission near $\lambda \sim 10\ \mu$m and 18 μm (Cohen & Barlow 1974, Grasdalen 1977). For at least one planetary nebula, far infrared emission has been measured out to $\lambda \sim 300\ \mu$m (Telesco & Harper 1976), and the total mass of grains required (some at $T \lesssim 100$ K) is compatible with all of the Si, say, in the nebula being in the form of grains. The infrared spectra are complicated by gas-based features and the type of grain is not known observationally. The exact location of the grains is also not known directly, but some indirect dynamical arguments (Bohuski & Smith 1974, Ferch & Salpeter 1975) favor (at least some) grains mixed with the ionized gas.

Supernova explosions return matter to the interstellar medium at a rate about 20 times less than supergiants and planetary nebula (van den Bergh & Maza 1976, Chevalier 1977). However, the supernova ejecta are enriched in the medium and heavy elements and could be an important source of condensable material. It is conceivable that grains could form in the outflow, but the environment is probably too hostile (Simons & William 1975, Falk & Scalo 1975, Clayton 1975).

5.3 *Protostellar Nebulae and Star Formation*

Interstellar matter is used up to form new stars at a rate only slightly greater than the return rate from old stars to the interstellar medium (Audouze & Tinsley 1976), about 2 $M_{\odot}$ yr^{-1}. There is no doubt that all grains that are incorporated into stars are destroyed, giving the minimum destruction rate in the second line of Table 1. However, the process of star formation is very complex (Larson 1973, Kondo 1975, Westbrook & Tarter 1975, Cameron & Truran 1977), and it is not clear how much *additional* interstellar material per star came close enough to the young star to be heated sufficiently for the grains to evaporate. Estimates for the additional heated material have ranged up to slightly more than the final mass of the star, giving the larger value for the destruction rate in the second line of Table 1.

The additional heated material is eventually expelled from the protostellar nebula and new grains may form in this material. Such grain production in very young stars was suggested already by Herbig (1963, 1969, 1970), but the production rate is not clear: The most conservative estimate of processing about a Jupiter mass of total gas per star formation gives a negligible rate. The ad hoc assumption of processing about a hundred times this amount ($\sim 0.1\ M_{\odot}$) to give a Jupiter mass of condensable material (Mendis & Wickramasinghe 1975) still produces less new grain material than is swallowed up in stars, but the production rate (third line in Table 1) could be as large as the difference of the two destruction rates in the second line in Table 1. Apart from newly formed grains, the formation of a cocoon from grains originally present in the interstellar material was suggested by Davidson & Harwit (1967). Observational data is discussed by Strom, Strom & Grasdalen (1975), and one possible scenario for reprocessing grains is given in detail by Burke & Silk (1976). Deviations from thermal equilibrium are more severe when grains form in a protostellar environment (see, for instance, Alfvén & Arrhenius 1973, 1974) than in the outflow from cool giants.

6 THE INTERSTELLAR MEDIUM

6.1 *Supernova Remnants*

As discussed in Section 3.4, interstellar grains are not easily destroyed by *UV* radiation, but are vulnerable to sputtering by high-speed atoms. For refractory grains (silicates, iron, graphite), velocities of ~ 200 km sec^{-1} are required, which are attained in blast-wave shockfronts when a supernova remnant progresses through the interstellar medium. Estimates of the total energy in a supernova blast wave ($E_0 \sim 10^{51}$ erg) vary somewhat (Chevalier 1977), as do estimates for the shock-speed V_s required to sputter away most of the mass of a typical interstellar grain (Barlow & Silk 1977, Draine 1977), but for magnesium silicate a typical value is $V_s \sim 180$ km sec^{-1}. For $E_0 \sim 5 \times 10^{50}$ erg the blast wave can sputter grains in a mass of interstellar gas of $\sim 300\ M_\odot$ (Chevalier 1977), the rate of supernova occurrence in the galactic disk is $\sim$(1/40 yr) (van den Bergh & Maza 1976), and (with $M_{gas} \sim 4 \times 10^9\ M_\odot$) the destruction rate is $\sim 1.5 \times 10^{-9}$ yr^{-1} (see Figure 7).

It is gratifying that measurements of element depletion corroborate the effects of supernova remnants at least qualitatively: As discussed in Section 2.2, some high-velocity clouds associated with the Vela supernova remnant show no depletion of Mg and Fe relative to solar abundances (Jenkins, Silk & Wallerstein 1976, 1977). Stellar winds from OB supergiants (Hutchings 1976) also eject material at speeds comparable to supernova ejecta, which could also initiate grain sputtering. Rate estimates have not yet been made, but this effect might increase appreciably the rate in the fourth line of Table 1.

6.2 *Cloud-Cloud Collisions*

The concept of an "interstellar cloud" (Routley & Spitzer 1952), a density enhancement with a fairly well-defined space velocity V, has stood the test of time (although it must not be taken too literally, see Heiles 1974). For cloud velocities with a

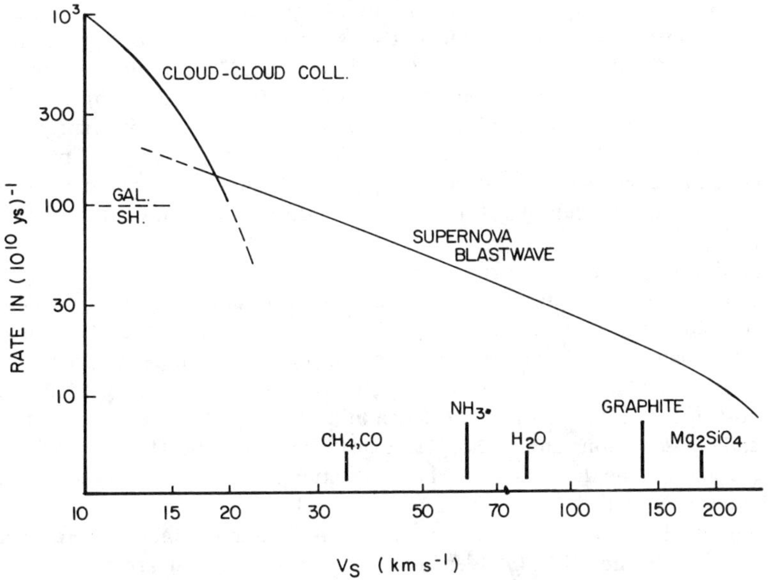

Figure 7 An order-of-magnitude estimate for the rate at which interstellar material is engulfed by a shock wave of shock speed V_s, caused either by a cloud-cloud collision or a supernova blast wave. The required values of V_s for destruction of a mineral by sputtering are indicated by vertical lines.

radial component up to ~20 km sec^{-1} there is good statistical data (Mast & Goldstein 1970, Falgarone & Lequeux 1973). One can estimate probabilities for cloud-cloud collisions with different relative velocities, even though estimates of "collision cross sections" are very dubious (Heiles 1974) and the statistics are complicated by an anticorrelation between optical depth and velocity of a cloud (Dickey, Salpeter & Terzian 1977). Cloud-cloud collisions lead to interstellar shock waves, and crude estimates for the rate of occurrence as a function of shock speed V_s (Draine 1977) are shown in Figure 7. The separation of this figure into two curves is artificial, since supernova blast waves probably contribute the high-velocity "tail" of the cloud-velocity distribution.

The vertical marks in Figure 7 show critical shock speeds for grain destruction due to gas-atom sputtering (Aannestad 1973, Draine 1977). Grains can be destroyed by grain-grain collisions (Oort & van de Hulst 1946) at much smaller relative speeds, but such collisions are less frequent and it is not clear at the moment (Spitzer 1976) by how much the destruction rates are enhanced due to this effect. Line five of Table 1 omits this enhancement but compensates by using the theoretical "supernova" curve in Figure 7 (which gives a larger rate than suggested by extrapolating the low-velocity data upwards).

6.3 *Molecular Clouds and H II Regions*

Dark, dense molecular clouds and H II regions are opposite extremes regarding the flux of *UV* photons, but they occur close to each other in larger "molecular cloud complexes" where some young, hot stars have already formed and produced an H II region but most of the gas mass is neutral. There is now appreciable observational data on infrared emission (as well as optical and radio) from such H II regions (Petrosian 1973, Grasdalen 1974, Gillett et al. 1975a, Balick 1976, Pipher et al. 1976). Unique interpretations are difficult because of the proximity of contrasting regions, but it is clear that (*a*) *much* dust is associated with H II regions in some way and (*b*) at least *some* of the dust is physically mixed with the ionized gas.

Quantitatively there is controversy: The authors quoted above interpret their data in terms of only small amounts of dust (below cosmic abundance by factors of 5 to 200) mixed with the ionized gas. Panagia (1975) and Natta & Panagia (1976) interpret the same observational data in terms of a more detailed model, which includes a "hole" near the exciting star and two kinds of grains but which has most of the grains in the ionized regions. Regarding the life history of dust grains, the observations on planetary nebulae (Section 5.2) and H II regions in any case corroborate the theoretical conclusion (Section 3.4) that refractory grain-cores are *not* destroyed when they enter an ordinary H II region.

The molecular cloud observations most relevant for grain formation were already mentioned in Section 4.1: (*a*) Elements like Mg, Ca, and Si are very strongly depleted from the gas phase; (*b*) the abundant elements C, N, and O are significantly depleted from the gas phase; (*c*) water-ice is found in only small amounts (Merrill, Russell & Soifer 1976) in molecular clouds (and not at all outside these clouds); and (*d*) the total/selective extinction ratio R is unusually large in particularly dark and dense regions, such as the center of the Rho Ophinchi cloud complex (Carrasco, Strom & Strom 1973, Vrba et al. 1975), indicating grains of larger size than usual. The presence of larger grains is correlated with enhanced depletion (Crutcher 1976) of Na (the absence of similar enhancement for K and Ca may be due to grain-charge effects).

It is gratifying from a theoretical point of view that (*a*) the presence of complex molecules in the gas phase, (*b*) large (and abundant) dust grains and (*c*) depletion of atoms in the gas phase are all found in particularly dark and dense regions. Because of the high density of gas and dust, gas atoms form molecules readily (see, for instance, Watson & Salpeter 1972, Watson 1977, Allen & Robinson 1977), and both atoms and molecules stick to grain surfaces readily; because of the low flux of energetic photons there is little photodesorption of atoms or molecules from grain surfaces (Greenberg 1973) and little photodissociation of molecules. The formation of "dirty ice" grain-mantles from molecules sticking to grain-cores leads naturally to larger grains, but there are two complications: (*a*) larger grains could also be produced by coagulation of smaller ones in regions of particularly high density (Cameron 1973a,b, Lefèvre 1974, Simons & Williams 1975) with turbulence greatly enhancing the rate of coalescence (Cameron 1973a,b, Scalo 1977), and (*b*) little of the grain-mantles can be in the form of water-ice, which would be the most abundant material made from fully saturated molecules.

A likely explanation for the low abundance of recognizable water-ice comes from the effect of *UV* photons. Although such photons are rare (but not absent, see Whitworth 1975) in the dark cloud, they are common in the nearby H II regions, and the cloud "turns inside out" often enough so that volatile solids are reprocessed. There is even a possibility of explosive radicals building up (Greenberg 1976), but it is more likely that *UV* will partly detach smaller molecules but will partly polymerize the material in the grain-mantle. In other words, only some photon events in a grain-mantle lead to a sturdier solid, but if the cycling from *UV* to dense cloud (where the grain-mantle is laid down again) occurs often enough, the end result will be some "oily plastic" that is more resistant to *UV* and to sputtering than ices.

We can estimate the turnover rate as follows: A moderately dark cloud (Lynds 1962) implies a high abundance of H_2 (and CO) and vice versa (Hollenbach, Werner & Salpeter 1971). About half of all the interstellar gas is in such molecular clouds at any one time (Scoville & Solomon 1975, Burton & Gordon 1976). The lifetime of a cloud can be estimated as 10^6 to 10^7 years from different arguments: (*a*) the travel time at turbulent velocities of a few km sec^{-1} over cloud sizes of ~ 10 pc and (*b*) detailed estimates relating to the H-H_2 balance (Barlow & Silk 1976, Allen & Robinson 1976, Hill & Hollenbach 1976). An atom or grain spends about half its life in and out of molecular clouds (not always the same cloud); it remains only a few times 10^6 years in each state but this is sufficient time for molecule formation and deposition of grain-mantles in a dense cloud (and for *UV* processing out of the cloud)—hence line six of Table 1.

6.4 *A Second Overview*

As Table 1 shows, the processes in and near molecular clouds are numerically the most important, although not understood in detail: It is clear that most condensable atoms and molecules are deposited on grain mantles in the interior of a dark molecular cloud, but it is less clear (*a*) whether metal and Si atoms can migrate through the mantle to join the refractory grain-core; (*b*) how important coagulation (grain-coalescence) is inside such clouds; (*c*) how polymerization of grain-mantles compares with sputtering once grains are exposed to stellar *UV* and to interstellar shocks initiated by cloud-cloud collisions; (*d*) how strongly grain buildup is affected by grain-charge and the ionization stage of gas atoms (Snow 1975, Crutcher 1976); and (*e*) how closely the resistance of a grain material to sputtering is correlated to its condensation temperature T_{cond}. These questions are particularly poignant in connection with the enormous gas-phase depletion factors (Figure 1, Table 2) for Al and Ca: Grains containing these elements have particularly large values of T_{cond}, and it seems attractive to invoke condensation under conditions near thermal equilibrium (Field 1974). Such condensation "under clean conditions," lines 1 and 3 in Table 1, is certainly important, but the rate is only comparable with the average rate for complete destruction of even the most refractory grain, line 4 in Table 1. Potentially the largest rates are in line 6 of Table 1, and it is my guess that complicated mineralogy will be needed to understand Figure 1 fully: Assume, for instance, that a particular atom sticks well to most grain surfaces and protects itself from sputtering in H II regions and in cloud collisions by rapidly migrating inward from a grain surface. Such an atom would be highly depleted in the gas phase, since it is released only in rare supernova remnant episodes and withdrawn much more often when it wanders into a dark cloud.

Supernova remnants are important both in breaking up pre-existing interstellar dust-grains and in supplying new gas which is "nuclearly" enriched in elements from C to Fe. Here too puzzles remain: do the newly manufactured atoms (*a*) form grains on their own, (*b*) mingle with the newly sputtered material and then form grains, or (*c*) wait to be deposited on ordinary interstellar grains later on? Plenty is left for the next review on this topic!

I am indebted to B. Draine, B. Lynds, and L. Spitzer for useful discussions and to NSF Grant AST 75-21153 for partial support.

Literature Cited

Aannestad, P. A. 1973. *Ap. J. Suppl.* Ser. 25: 223

Aannestad, P. A. 1975. *Ap. J.* 200:30

Aannestad, P. A., Purcell, E. M. 1973. *Ann. Rev. Astron. Astrophys.* 11:309

Abraham, F. F. 1974. *Homogeneous Nucleation Theory.* New York: Academic

Aitken, D. K., Jones, B. 1974. *MNRAS* 167: 11P

Alfvén, H., Arrhenius, G. 1973. *Astrophys. Space Sci.* 21:117

Alfvén, H., Arrhenius, G. 1974. *Astrophys. Space Sci.* 29:63

Allen, M., Robinson, G. W. 1976. *Ap. J.* 207:745

Allen, M., Robinson, G. W. 1977. *Ap. J.* 212: 319

Aller, L., Czyzak, S. 1968. *Planetary Nebulae, IAU Symp. No. 34,* ed. D. Osterbrock, C. R. O'Dell. New York: Springer

Anders, E. 1971. *Ann. Rev. Astron. Astrophys.* 9:1

Anders, E., Hayatsu, R., Studier, M. H. 1974. *Ap. J.* 192:L101

Antipova, L. I. 1969. *Soviet Astr.-A.J.* 12:288

Arrhenius, G., De, B. R. 1973. *Meteoritics* 8:297

Audouze, J., Tinsley, B. M. 1976. *Ann. Rev. Astron. Astrophys.* 14:43

Balick, B. 1976. *Ap. J.* 208:75

Barlow, M. J., Silk, J. 1976. *Ap. J.* 207:131

Barlow, M. J., Silk, J. 1977. *Ap. J.* 211:L83

Barshay, S. S., Lewis, J. S. 1976. *Ann. Rev. Astron. Astrophys.* 14:81

Bohlin, R. C. 1975. *Ap. J.* 200:402

Bohuski, T. J., Smith, M. G. 1974. *Ap. J.* 193:197

Bonilha, J. R. M., Terzian, Y. 1976. *Bull. Am. Astron. Soc.* 8:371

Burke, J. R., Silk, J. 1976. *Ap. J.* 210:341

Burton, W. B. 1976. *Ann. Rev. Astron. Astrophys.* 14:275

Burton, W. B., Gordon, M. A. 1976. *Ap. J.* 207:L189

Bussoletti, E., Zambetta, A. M. 1976. *Astron. Astrophys. Suppl.* 25:549

Cameron, A. G. W. 1973a. *Space Sci. Rev.* 15:121

Cameron, A. G. W. 1973b. *Icarus* 18:407

Cameron, A. G. W. 1975. *Icarus* 24:128

Cameron, A. G. W., Truran, J. W. 1977. *Icarus* 30:447

Capps, R. W., Knacke, R. F. 1976. *Ap. J.* 210:76

Caroff, L. J., Petrosian, V., Salpeter, E., Wagoner, R., Werner, M. 1973. *MNRAS* 164:295

Carrasco, L., Strom, S. E., Strom, K. M. 1973. *Ap. J.* 182:95

Carusi, A., Coradini, A., Federico, C., Fulchignoni, M., Magni, G. 1975. *Astrophys. Space Sci.* 33:369

Chevalier, R. A. 1977. *Ann. Rev. Astron. Astrophys.* 15:175

Clayton, D. D. 1975. *Ap. J.* 199:765

Clayton, D. D., Hoyle, F. 1976. *Ap. J.* 203:490

Clayton, D. D., Wickramasinghe, N. C. 1976. *Astrophys. Space Sci.* 42:463

Cohen, J. G. 1976. *Ap. J.* 203:L127

Cohen, M., Barlow, M. J. 1974. *Ap. J.* 193:401

Cohen, M., Gaustad, J. E. 1973. *Ap. J.* 186:L131

Crutcher, R. M. 1976. *Ap. J.* 208:382

Czyzak, S. J., Santiago, J. J. 1973. *Astrophys. Space Sci.* 23:443

Davidson, K., Harwit, M. 1967. *Ap. J.* 148:443

Day, K. L. 1976. *Ap. J.* 203:L99

Day, K. L., Steyer, T. R., Huffman, D. R. 1974. *Ap. J.* 191:415

deBoer, K. S., Pottash, S. R. 1973. *Astron. Astrophys.* 28:155

Dempsey, M. J., Wickramasinghe, N. C. 1975. *Astrophys. Space Sci.* 34:185

Dickey, J., Salpeter, E., Terzian, Y. 1977. *Ap. J.* 211:L77

Dinger, A. S. 1970. *Astrophys. Space Sci.* 6:118

Donn, B. 1968. *Ap. J.* 152:L129

Donn, B. 1976. *Mem. Soc. R. Sci. Liège* (6 Ser.) 9:499

Donn, B., Stecher, T. P. 1965. *Ap. J.* 142:1681

Donn, B., Wickramasinghe, N., Hudson, J., Stecher, T. 1968. *Ap. J.* 153:451

Dorfeld, W. G., Hudson, J. B. 1973. *Ap. J.* 186:715

Draine, B. T. 1977. PhD thesis. Cornell Univ., New York

Draine, B. T., Salpeter, E. E. 1977. *J. Chem. Phys.*

Dyck, H. M., Beichman, C. A. 1974. *Ap. J.* 194:57

Dyck, H. M., Lockwood, G. W., Capps, R. W. 1974. *Ap. J.* 189:89

Egan, W. G., Hilgeman, T. 1975. *Astron. J.* 80:587

Falgarone, E., Lequeux, J. 1973. *Astron. Astrophys.* 25:253

Falk, S. W., Scalo, J. M. 1975. *Ap. J.* 202:690

Feder, J., Russell, K., Lothe, J., Pound, G. M. 1966. *Adv. Phys.* 15:111

Ferch, R., Salpeter, E. E. 1975. *Ap. J.* 202:195

Field, G. 1974. *Ap. J.* 187:453

Field, G., Cameron, A. G. W., eds. 1975. *The Dusty Universe.* New York: Neale Watson. 323 pp.

Field, G., Goldsmith, D. W., Habing, H. J. 1969. *Ap. J.* 155:L149

Fix, J. D. 1969a. *MNRAS* 146:37

Fix, J. D. 1969b. *MNRAS* 146:51

Fix, J. D. 1970. *Ap. J.* 161:359

Forrest, W. J., Gillett, F. C., Stein, W. A. 1972. *Ap. J.* 178:L129

Fujita, Y. 1970. *Spectra and Atmospheric Structure in Cool Stars.* Baltimore: University Park Press

Gallagher, J. S., Code, A. D. 1974. *Ap. J.* 189:303

Geisel, S. L., Kleinmann, D. E., Low, F. J. 1970. *Ap. J.* 161:L101

Gillett, F. C., Forrest, W. J., Merrill, K. M., Capps, R. W., Soifer, B. T. 1975a. *Ap. J.* 200:609

Gillett, F. C., Jones, T. W., Merrill, K. M., Stein, W. A. 1975b. *Astron. Astrophys.* 45:77

Gilman, R. C. 1969. *Ap. J.* 155:L185

Gilman, R. C. 1972. *Ap. J.* 178:423

Gilra, D. P. 1971. *Nature* 229:237

Goldreich, P., Scoville, N. 1976. *Ap. J.* 205: 144
Grasdalen, G. L. 1974. *Ap. J.* 193: 373
Grasdalen, G. L. 1977. *Ap. J.* In press
Greenberg, J. M. 1974. *Ap. J.* 189: L81
Greenberg, J. M. 1976. *Astrophys. Space Sci.* 39: 9
Greenberg, J. M., Hong, S. S. 1974. In *Galactic Radio Astronomy, IAU Symp. No. 60*, ed. F. Kerr, R. Simonson. Dordrecht: Reidel
Greenberg, J. M., Hong, S. S. 1976. *Astrophys. Space Sci.* 39: 31
Greenberg, J. M., van de Hulst, H. C., eds. 1973. *Interstellar Dust and Related Topics.* Dordrecht: IAU
Grewing, M., Lamers, H. J., Walmsley, C. M., Wulf-Mathies, C. 1973. *Astron. Astrophys.* 27: 115
Grossman, L., Larimer, J. W. 1974. *Rev. Geophys. Space Phys.* 12: 71
Hackwell, J. A., Gehrz, R. D. 1974. *Ap. J.* 194: 49
Härm, R., Schwarzschild, M. 1975. *Ap. J.* 200: 324
Hayashi, C., Nakagawa, Y. 1975. *Prog. Theor. Phys.* 54: 93
Herbig, G. H. 1963. *Ap. J.* 137: 200
Herbig, G. H. 1969. *Lick Obs. Contr.* No. 302, p. 13
Herbig, G. H. 1970. *Mem. Soc. Roy. Sci. Liège* (Ser. 5) 19: 13
Herbig, G. H. 1975. *Ap. J.* 196: 129
Heiles, C. 1974. In *Galactic Radio Astronomy*, ed. F. Kerr, R. Simonson. Reidel: Dordrecht
Heiles, C. 1976. *Ap. J.* 204: 379
Hill, J. K., Hollenbach, D. J. 1976. *Ap. J.* 209: 445
Hirth, J. P., Pound, G. M. 1963. *Condensation and Evaporation.* New York: McMillan
Hobbs, L. M. 1975. *Ap. J.* 202: 628
Hollenbach, D., Werner, M. W., Salpeter, E. E. 1971. *Ap. J.* 163: 165
Hoyle, F., Wickramasinghe, N. C. 1962. *MNRAS* 124: 417
Hunter, C. E., Donn, B. 1971. *Ap. J.* 167: 71
Hutchings, J. B. 1976. *Ap. J.* 204: L99
Hyland, A. R., Neugebauer, G. 1970. *Ap. J.* 160: L177
JANAF 1971. *Thermochemical Tables*, 2nd ed. (NSRDS-NGS37)
Jenkins, E. B., Meloy, D. A. 1974. *Ap. J.* 193: L121
Jenkins, E. B., Silk, J., Wallerstein, G. 1976. *Ap. J.* 209: L87
Jenkins, E. B., Silk, J., Wallerstein, G. 1977. *Ap. J. Suppl.* 32: 681
Johnson, H. R., Beebe, R. F., Sneden, C. 1975. *Ap. J. Suppl.* 29: 123
Jones, T. W., Merrill, K. M. 1976. *Ap. J.* 209: 509
Joss, P. 1974. *Ap. J.* 191: 771
Jura, M. 1975. *Ap. J.* 200: 415
Jura, M. 1976. *Ap. J.* 206: 691
Katz, J. L. 1970. *J. Chem. Phys.* 52: 4733
Kerridge, J. F., Vedder, J. F. 1972. *Science* 177: 161
Knacke, R. F., Cudaback, D. D., Gaustad, J. E. 1969. *Ap. J.* 158: 151
Knapp, G. R., Kerr, F. J. 1974. *Astron. Astrophys.* 35: 361
Kondo, A. 1975. *Publ. Astron. Soc. Jpn* 27: 215
Kudo, A. 1974. *Sci. Rep. Tohoku Univ.* Ser. I, 57: 97
Kwan, J., Scoville, N. 1976. *Ap. J.* 209: 102
Laegrid, N., Wehner, G. K. 1961. *J. Appl. Phys.* 32: 365
Larson, R. B. 1973. *Ann. Rev. Astron. Astrophys.* 11: 219
Lefèvre, J. 1974. *Astron. Astrophys.* 37: 17
Lewis, J. S. 1972. *Icarus* 16: 241
Lillie, C. F., Witt, A. N. 1976. *Ap. J.* 208: 64
Lord, H. C. 1965. *Icarus* 4: 279
Lynds, B. T. 1962. *Ap. J. Suppl.* 7: 1
Lynds, B. T., Wickramasinghe, N. C. 1968. *Ann. Rev. Astron. Astrophys.* 6: 215
Mason, B. J. 1971. *The Physics of Clouds.* Oxford: Clarendon
Mast, J., Goldstein, S. J. 1970. *Ap. J.* 159: 319
Mendis, D. A., Wickramasinghe, N. C. 1975. *Astrophys. Space Sci.* 38: L13
Merrill, K. M., Russell, R. W., Soifer, B. T. 1976. *Ap. J.* 207: 763
Merrill, K. M., Stein, W. A. 1976a. *Publ. Astron. Soc. Pac.* 88: 285
Merrill, K. M., Stein, W. A. 1976b. *Publ. Astron. Soc. Pac.* 88: 294
Meyer, C. 1971. *Geochim. Cosmochim. Acta* 35: 551
Morton, D. C. 1975. *Ap. J.* 197: 85
Natta, A., Panagia, N. 1976. *Astron. Astrophys.* 50: 191
Olthof, H., Pottash, S. R. 1975. *Astron. Astrophys.* 43: 291
Oort, J. H., van de Hulst, H. C. 1946. *Bull. Astron. Netherlands* 10: 187 (No. 376)
Osterbrock, D. E. 1974. *Astrophysics of Gaseous Nebulae.* San Francisco: Freeman. 251 pp.
Palmer, H., Shelef, M. 1968. *Chem. Phys. of Carbon* 4: 85
Panagia, N. 1975. *Astron. Astrophys.* 42: 139
Peimbert, M., Costero, E. 1969. *Bol. Obs. Tonantzintla Tacubaya* No. 31,3
Penman, J. M. 1976. *MNRAS* 175: 149
Persson, S. E., Frogel, J. A., Aaronson, M. 1976. *Ap. J.* 208: 747

Petrosian, V. 1973. In *Interstellar Dust and Related Topics,* ed. J. Greenberg, H. van de Hulst. Dordrecht: IAU
Pipher, J. L., Sharpless, S., Savedoff, M., Kerridge, S., Krassner, J., Shurmann, S., Soifer, B., Merrill, K. M. 1976. *Astron. Astrophys.* 51:255
Pottash, S. R. 1972. *Astron. Astrophys.* 20: 245
Purcell, E. M. 1969. *Ap. J.* 158:133
Purcell, E. M. 1976. *Ap. J.* 206:685
Ross, J. E., Aller, L. H. 1976. *Science* 191: 1223
Routly, P. M., Spitzer, L. 1952. *Ap. J.* 115: 227
Russell, H. N. 1928. *Ap. J.* 67:83
Sackmann, I. J., Smith, R. L., Despain, K. H. 1974. *Ap. J.* 187:555
Salpeter, E. E. 1973. *J. Chem. Phys.* 58:4331
Salpeter, E. E. 1974a. *Ap. J.* 193:579
Salpeter, E. E. 1974b. *Ap. J.* 193:585
Salpeter, E. E. 1976. *Ap. J.* 206:673
Sandage, A. R. 1975. *Publ. Astron. Soc. Pac.* 87:853
Sanner, F. 1976. *Ap. J.* 204:L41
Savage, B. D. 1975. *Ap. J.* 199:92
Scalo, J. M. 1977. *Astron. Astrophys.* 55: 253
Schultz, G. V., Wiemer, W. 1975. *Astron. Astrophys.* 43:133
Schwarzschild, M. 1975. *Ap. J.* 195:137
Scoville, N. Z., Solomon, P. M. 1975. *Ap. J.* 199:L105
Shapiro, M. M., Silberberg, R. 1974. *Phil. Trans. R. Soc.* A277:319
Shapiro, P. 1975. *Ap. J.* 201:151
Shields, G. A. 1975. *Ap. J.* 195:475
Shull, J. M. 1977. *Ap. J.* 212:102
Shull, J. M., York, D. J. 1977. *Ap. J.* 211: 803
Siluk, R. S., Silk, J. 1974. *Ap. J.* 192:51
Simons, S., Williams, I. P. 1975. *Astrophys. Space Sci.* 32:493
Snow, T. P. 1975. *Ap. J.* 202:L87
Snow, T. P. 1976. *Ap. J.* 204:759
Sparks, W. M., Starrfield, S., Truran, J. W. 1976. *Ap. J.* 208:819
Spitzer, L. 1968. *Diffuse Matter in Space.* New York: Interscience. 262 pp.
Spitzer, L., Jenkins, E. B. 1975. *Ann. Rev. Astron. Astrophys.* 13:133
Spitzer, L. 1976. *Comments. Astron. Astrophys.* 6:177
Stecher, T. P. 1965. *Ap. J.* 142:1683
Stein, W. A., Gillett, F. C. 1969. *Ap. J.* 189: L27
Strom, S. E., Strom, K. M., Grasdalen, G. L. 1975. *Ann. Rev. Astron. Astrophys.* 13:187
Stuart, R. V., Wehner, G. K. 1962. *J. Appl. Phys.* 33:2345
Tabak, R. G., Hirth, J. P., Meyrick, G., Roark, T. P. 1975. *Ap. J.* 196:457
Talbot, R. J., Arnett, W. D. 1973. *Ap. J.* 186:69
Tarafdar, S. P., Wickramasinghe, N. C. 1975. *Astrophys. Space Sci.* 35:L41
Telesco, C. M., Harper, D. A. 1977. *Ap. J.* 211:475
Trimble, V. 1975. *Rev. Mod. Phys.* 47:877
Tsuji, T. 1973. *Astron. Astrophys.* 23:411
van de Hulst, H. C. 1948. *Ned. Tijdschr. Natuurkunde* 10:251
van den Bergh, S., Maza, J. 1976. *Ap. J.* 519:20
Vedder, J. F., Mandeville, J. C. 1974. *J. Geophys. Res.* 79:3247
Verschuur, G. L. 1975. *Ann. Rev. Astron. Astrophys.* 13:257
Vrba, F. J., Strom, K. M., Strom, S. E., Grasdalen, G. L. 1975. *Ap. J.* 197:77
Walker, G. H. 1975. *Astrophys. Lett.* 16:115
Watson, W. D. 1977. *Rev. Mod. Phys.* 48: 513
Watson, W. D., Salpeter, E. E. 1972. *Ap. J.* 174:321
Weidemann, V. 1967. *Z. Astrophys.* 67:286
Werner, M. W., Salpeter, E. E. 1969. *MNRAS* 145:249
Westbrook, R., Tarter, C. B. 1975. *Ap. J.* 200:48
Whipple, F. L. 1950. *Ap. J.* 111:375
Whitworth, A. P. 1975. *Astrophys. Space Sci.* 34:155
Wickramasinghe, N. C. 1975. *MNRAS* 170: 11P
Wickramasinghe, N. C., Morgan, D. J., eds. 1976. *Solid State Astrophysics.* Dordrecht: Reidel. 314 pp.
Wickramasinghe, N. C., Nandy, K. 1972. *Rep. Prog. Phys.* 35:157
Woolf, N. J. 1973. In *Interstellar Dust and Related Topics,* ed. J. Greenberg, H. van de Hulst. Dordrecht: IAU
Woolf, N. J. 1975. In *The Dusty Universe,* ed. G. Field, A. Cameron. New York: Neale Watson
Woolf, N. J., Ney, E. P. 1969. *Ap. J.* 155: L181
Yamamoto, T., Nishida, S. 1977. Preprint
York, D. J. 1975. *Ap. J.* 196:L103
Zaikowski, A., Knacke, R. F. 1975. *Astrophys. Space Sci.* 37:3
Zaikowski, A., Knacke, R. F., Porco, C. C. 1975. *Astrophys. Space Sci.* 35:97
Zappala, R. R., Becklin, E. E., Matthews, K., Neugebauer, G. 1974. *Ap. J.* 192:109

Ann. Rev. Astron. Astrophys. 1977. 15: 295–362

THE GALACTIC CENTER

J. H. Oort
Sterrewacht, Leiden, the Netherlands

INTRODUCTION AND OVERVIEW

This article is partly a review and partly a research paper. The discussion and conclusions can be summarized as follows:

Gravitational field and mass distribution Within $R = 750$ pc this is determined from the rotation of the H I nuclear disk (Table 1). Emission at 2.2 μ from the main stellar population has been used, in combination with data on M31, to estimate the mass distribution in the inner 50 pc. Observations of the Ne II emission at 12.8 μ yield information on the mass inside 0.5 pc. An ultracompact radio source of radius smaller than ~ 10 A.U. is presumably the actual center of the Galaxy. It may have a mass in the order of five million solar masses, and there is a suspicion that it may contain the "engine" responsible for the many expulsion phenomena observed throughout the central region.

Table 1 Mass within R, potential, circular velocity, time of revolution, and density

R (pc)	$M(R)$ (10^6 $M_\odot$)	$\Phi(0)-\Phi(R)$ (10^4 km^2 sec^{-2})	Θ_c (km sec^{-1})	T (10^6 yr)	$\rho(R)$ ($M_\odot$ pc^{-3})
0.1	0.3[a]	0.64	114	0.006	3×10^7
1.0	4.4[a]	4.4	138	0.045	4×10^5
5	30	8.2	161	0.20	2×10^4
10	70	10.0	174	0.36	7×10^3
20	160	12.2	186	0.67	1.9×10^3
50	490	15.6	206	1.5	
100	920	18.6	200	3.1	1.1×10^2
200	2500	21.9	233	5.4	
500	8300	27.8	268	12	6
1000	16400	32.8	266	24	

[a] Recent dynamical data have shown that the actual mass within $R = 0.4$ pc is about three times higher than the mass of 1.5×10^6 $M_\odot$ interpolated from the table (Section 6.3.2).

Expulsive phenomena Four different regimes of expulsive phenomena have been established; presumably all four are causally connected with the nucleus:

1. Arm-like H I features extending to a few kpc. Their H I masses range up to two million solar masses. The radial motions are generally between 100 and 200 km sec^{-1} (Table 2). The two largest, the 3-kpc arm and the expanding arm at $+135$ km sec^{-1}, may be of a different nature, and they may be parts of resonance orbits instead of results of expulsion. Many expanding features move at an angle with the galactic plane. The origin of the expulsion is unknown. (Figures 11–15).
2. Massive complexes of molecular clouds, within 300 pc from the center. Their relatively slow motions indicate that they must have been expelled. Direct evidence for this is shown by a ring of molecular clouds at $R \sim 190$ pc expanding at an average velocity of 150 km sec^{-1}. One of the giant molecule agglomerations, the "+40 km sec^{-1} cloud," may be very close to the center and of quite recent origin. The longitude distribution of the bulk of the molecules is extremely anisotropic: by far the greater part lies at positive longitudes and has positive velocities. The total mass in the molecular complexes is in the order of $10^8\ M_\odot$ (Figures 19, 21, 26).
3. Ionized gas within $R \sim 50$ pc. This also occurs in extended agglomerations, including the "Arc," about 15′ from the center (Figure 31). Again, they have small velocities, from which it may be inferred that they come from nuclear expulsion.
4. High-velocity streams within 1 pc from the center. Velocities determined by the Ne II line at 12.8 μ give evidence of large radial (as well as transverse) motions (Figure 34).

Table 2 H I features with large radial motions at $l = 0°$

	M_{HI} ($10^6\ M_\odot$)	V_E (km sec^{-1})	Limits of l (°)	(°)	$\langle b \rangle$ (°)	$\langle z \rangle$ (pc)	R (kpc)
3-kpc arm	36[a]	−53	338	6	+0.1	0	4.0
The expanding arm at +135 km sec^{-1} = van der Kruit I	>7[b]	+137	355	22	+0.4	+85	2.2
van der Kruit VII	2	+100	355	0	+1.0	+183	0.9
van der Kruit XII	1.4	−99	358	9	−2.5	−330	2.4
Rougoor 2 = Cohen II	1.2	+73	357	~20	0.0	0	4.0
van der Kruit X	0.07	−162	358	5	−3.0	−436	1.7
feature E	0.06	−170	0		−1.5	−260	?
Cugnon's object[c] = van der Kruit VIII	2:	high	349	354	+3	+500	2?
Mirabel & Turner's object[d]	2	high	351	359	−14	−2400	2.4

[a] Integrated over a full quadrant.

[b] Integrated only over the longitude range 355° to 10°.

[c] This large object has been observed down to 349° longitude where it has its maximum intensity; a possible farther extension is quite uncertain. Its radial velocity of +50 km sec^{-1} is strongly forbidden (Cugnon 1968). As it does not extend to $l = 0°$, its association with the center is uncertain; it *might* be a relatively nearby high velocity complex.

[d] An elongated structure extending from $l = 359°$, $b = -7^{\circ}.5$ to $l = 351°$, $b = -20°$, pointing away from the center, and having a forbidden radial velocity of +44 km sec^{-1} (Mirabel & Turner 1972). Because of its appreciable distance from the center its allocation to the central region remains uncertain.

The infrared core The same 1-pc region contains a remarkable concentration of discrete sources of far-infrared radiation emitted by dust that is heated partly by the smoothly distributed old stars and partly by young O stars. It is optically thin (Figures 35–38).

Need for a massive black hole? Nothing is known about the mechanism of expulsion, nor of the way in which the nucleus is replenished with interstellar gas on a sufficiently short time scale. Something out of the ordinary appears to be required.

1 POSITION AND DISTANCE OF THE NUCLEUS

The concentration of near-infrared radiation and, in particular, the still unresolved core of the radio source Sagittarius A West define the direction to the galactic center to within about $0''.1$. Becklin & Neugebauer's (1975) high resolution map at 2.2 μ shows the centroid to lie near their source 16, at $\alpha = 17^h42^m29^s.3 \pm 0^s.15$, $\delta = -28°59'18'' \pm 3''$ (1950). The position of the unresolved core of Sgr A West is $\alpha = 17^h42^m29^s.291 \pm 0^s.005$, $\delta = -28°59'17''.6 \pm 0''.1$ (1950) (Ekers et al. 1975). The two positions agree exactly. The galactic co-ordinates are $l = -3'.34$, $b = -2'.75$.

At present the most reliable way to determine the distance to the center is by observations of the RR Lyrae variables. These are strongly concentrated toward the center. The dispersion of their mean absolute magnitudes is small, of the order of ± 0.2 mag. However, there is still a rather serious uncertainty, estimated to be about ± 0.15 mag, in the calibration of the absolute magnitude.

From the data obtained by Plaut in an extensive survey of the RR Lyrae variables with the 48-inch Palomar Schmidt in some fields near the center, Oort & Plaut (1975) derived a distance of 8.7 kpc for the center. The estimated mean square error was ± 0.6 kpc, the largest contribution to the error coming from the uncertainty in the mean absolute magnitude, the second largest from the rough way in which the absorptions in the various fields were determined.

To avoid possible confusion in comparisons with other investigators the data and discussions in the present article have all been based on the standard distance of 10 kpc.

2 THE GRAVITATIONAL FIELD AND THE NUCLEAR DISK

Information on the mass density in the central region can be obtained in three ways: (*a*) from the distribution of radiation in the near infrared, (*b*) from the rotation velocity of gas concentrated in the galactic plane, and (*c*) from the density distribution and velocity dispersion of Population II objects.

2.1 *Near-infrared Radiation*

Because of the heavy obscuration in the galactic layer our knowledge of the central region is confined to what we can learn from observations in the infrared, and at mm and radio wavelengths.

Only a limited range of the infrared spectrum, from about 1.5 to 3.5 μ, can be

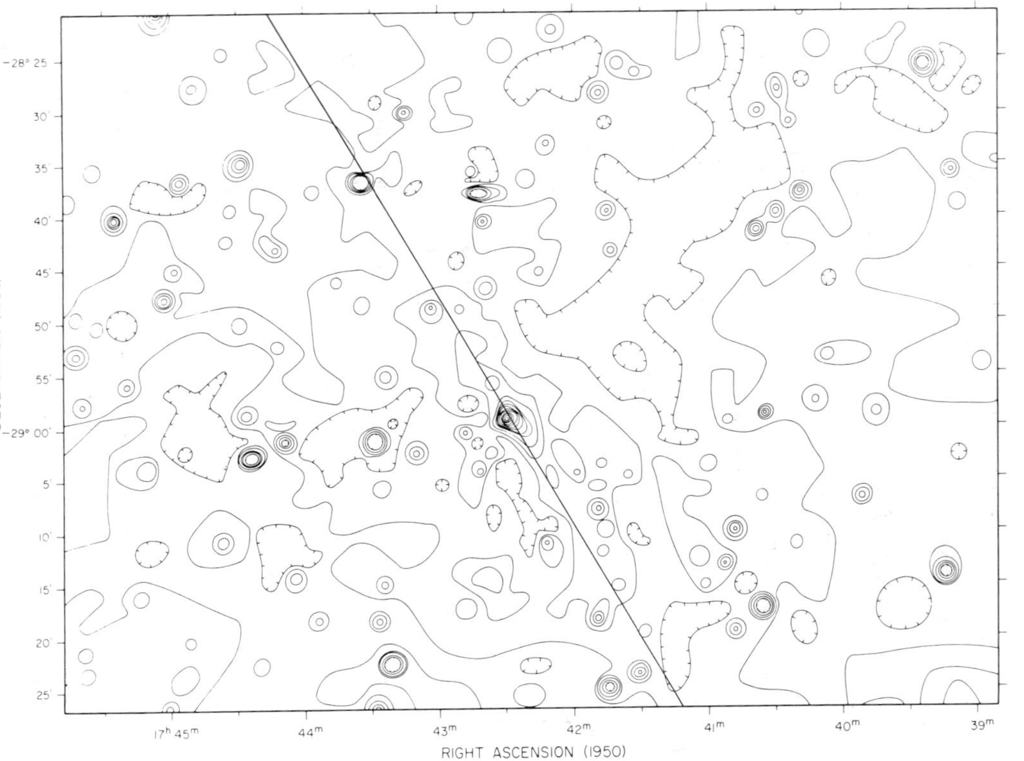

Figure 1 A 2.2-μ contour map of the central 1° of the galactic center made with an angular resolution of 1.'2. The contour levels correspond to 1.7 Jy in a 1.'2 aperture. The hatched areas correspond to point sources whose flux density is greater than 6.5 Jy. The zero level of the map was taken as the average flux 1° east and west of the galactic center; the ticked contours show the direction of decreasing flux. The map was constructed from right ascension drift scans on the 1-m telescope at Las Campanas, Chile; the present data were obtained with only a single measuring beam (Becklin, Neugebauer & Early 1977). The straight line is the true galactic equator ($b = -2.'75$).

used for exploring the mass distribution. At shorter wavelengths the absorption becomes prohibitive, while at much longer wavelengths the radiation comes mainly from interstellar dust and gas. Around 2 μ, however, the radiation is largely stellar.

Figure 1 shows the distribution of the surface brightness at 2.2 μ in a $1^\circ \times 1^\circ$ region around the center, as measured by Becklin, Neugebauer & Early (1977). It shows a strong concentration towards the center of the picture, which coincides with what on other grounds is considered to be the actual center of the Galaxy, and a gradual falloff with the distance from this center. The distribution is clearly elongated, with an axial ratio of about 0.4; the elongation is along the galactic equator. The patchiness of the picture is probably partly due to unevenness in the interstellar extinction, which totals about 3 magnitudes at 2.2 μ, and partly to individual stars of very high intrinsic brightness in the central region. The absorption cannot be reduced much further by observing at still longer wavelengths because, as Becklin & Neugebauer (1969) have clearly shown, nonstellar radiation begins to preponderate beyond about 3.5 μ.

The assumption that the radiation around 2 μ is stellar radiation receives support from observations of the center of the Andromeda nebula, where the optical absorption is relatively small, so that the *optical* spectrum can be studied along with the infrared emission. Sandage, Becklin & Neugebauer (1969) have shown that the distribution of the surface brightness is practically identical for various optical and near-infrared wavelengths, including 2 μ. We know from the spectrum that the optical radiation comes from stars, probably principally K giants. The brightness at 2.2 μ corresponds roughly to what would be expected from such stars, and it is

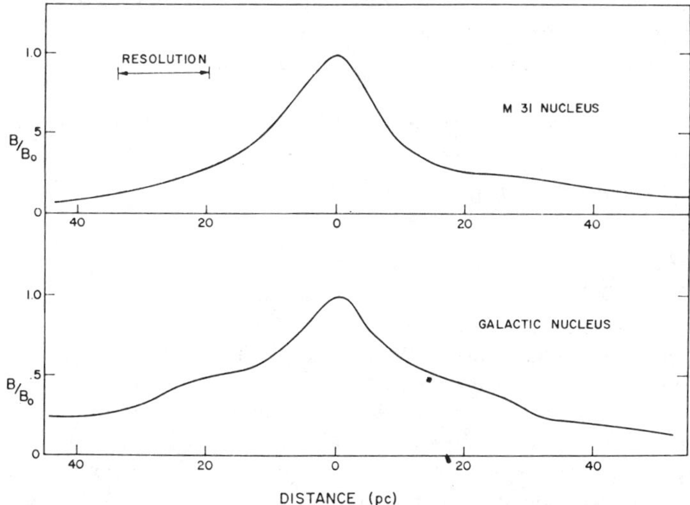

Figure 2 Comparison of distribution of 2.2-μ brightness around the galactic center with that around the nucleus of M31 at the same spatial resolution (Becklin & Neugebauer 1968).

therefore plausible to assume as a working hypothesis that the 2.2-μ radiation gives a measure of the star density. We shall assume that the same applies to our own Galaxy.

The two galaxies resemble each other in the distribution as well as in the absolute intensity of the 2.2-μ radiation (cf Figure 2). However, the light distribution in the innermost part of M31 is less flattened than the 2.2-μ radiation in the Galaxy, and the absolute value of the surface brightness seems to be 2.4 times fainter in M31.

2.2 *Mass-to-Light Ratio. Mass Distribution from 2.2-μ Radiation*

The only way in which we can determine the ratio of the mass to infrared emission is through a comparison with the Andromeda nebula, in which one can probably see the true light distribution down into the nucleus. Mass-to-light ratios (M/L) for its nuclear region have been estimated from various spectral data. These indicate that the bulk of the light is contributed by late-type giants. Population models constructed by Faber give $M/L_V = 15$ (cf van den Bergh 1975). An independent value can be derived from measurements of the velocity dispersion and the distribution of the light density in the nucleus. Applying the virial theorem to these data, Morton & Thuan (1973) obtained $M/L_V \sim 13$. It may be noted that a very similar ratio has been found for the main body of M31 from the rotation curve.

As we know the ratio of the 2.2 μ to the visual brightness we can in this way find the ratio of mass density to that of the infrared emission.

In order to find the mass distribution in our own Galaxy from the measurements at 2.2 μ we can assume as a working hypothesis that the stellar composition of the nuclear regions is the same in the two galaxies. We then have all the data that are needed to obtain the mass density distribution. The resulting gravitational field can finally be confronted with that derived from the rotation of the H I nuclear disk, which gives an entirely independent and essential check on the various assumptions made.

On the basis of Becklin & Neugebauer's 1968 data the following representation of the mass density has been found:

$$\rho(\alpha) = 7.6 \times 10^5 \alpha^{-1.8} M_\odot \, \mathrm{pc}^{-3}, \tag{1}$$

where

$$\alpha^2 = \varpi^2 + z^2(a/c)^2,$$

ϖ being the distance from the rotation axis, z that from the galactic plane, both in pc, and c/a the axial ratio of the surface brightness distribution. The above expression has been taken from an article by Sanders & Lowinger (1972). They used a somewhat more roundabout way than that sketched above to convert the 2.2-μ densities into mass densities. The axial ratio was taken to be 0.4.

Infrared intensities are available only for the region within roughly 50 pc from the center. In order to extend the data to larger distances, which is necessary if one wants to compare them with rotation velocities in the central disk, one must once more recur to a comparison with M31 and make use of the light distribution in its central region.

In view of the similarity of the 2.2-μ radiation in the innermost regions of M31 and the Galaxy, and in view of the fact that the light distributions in the central bulges of different galaxies appear to be practically identical, it is reasonable to assume that the similarity will extend beyond 50 pc. From data collected by Kinman (1965), Ruiz (1976) has derived space densities in M31 up to about 500 pc from the center. Between 10 and 500 pc the density can be well represented by $\rho \propto a^{-1.9}$, if a is the distance from the center measured in the equatorial plane and the axial ratio is assumed to be 1.45. This is practically the same power law as in (1), and supports the assumption made by Sanders & Lowinger (1972) that the mass density in the Galaxy can be represented by formula (1) up to at least 500 pc.

The following section shows that the agreement between the results obtained in this way and the rotation of the central disk is satisfactory.

2.3 *The Interstellar Gas*

The interstellar hydrogen in the central region has been extensively investigated in various ways. It is partly in atomic form, in which it can be observed in the 21-cm line. A large fraction is molecular and is concentrated in dense clouds; the H_2 molecules in these clouds have not been observed directly, but only through the emission or absorption lines of other molecules (CO, OH, H_2CO) that always occur side by side with H_2.

Finally, there are numerous H II regions in the central area. These are observed in the continuous radio spectrum as well as in recombination lines.

The contribution of the gas to the total mass density in the central region is negligible, but the *motion* of the gas gives important information on this total density. The clearest data, so far, come from the 21-cm line observations.

2.4 *The Rotating Nuclear Disk*

In regions where the gas is moving in approximately circular orbits, measurements of its velocity yield directly the gravitational force. Pressure gradients are almost always negligible. In the central region, up to about 4 kpc from the center, large deviations from circular motion occur, so that the velocity of the gas cannot give unambiguous information on the gravitational field. The deviations are at least partly due to gas expulsion from the nucleus. Between 1 and 4 kpc they may also be caused by a deviation of the gravitational field from circular symmetry. We return to these problems in Sections 3 and 4.

For the present context it is important to note that there is a region, between roughly 50 and 750 pc from the center, where the atomic hydrogen appears to move in practically circular orbits, and thus permits the determination of the gravitational field.

Figures 3 and 4, taken from an early investigation of the central region by Rougoor (1964), give a survey of the H I density in the galactic plane, as a function of longitude and velocity. The low-velocity parts, which contain mainly gas outside the central region, have been left blank. A more recent, higher-sensitivity map made with the Dwingeloo telescope has been published by Burton (1970).

The figures give an impression of the complexity of the phenomena in the central region. Most striking is the asymmetry. If the motions were governed principally by the galactic rotation the upper right-hand quadrant should have been a mirror image of the lower left-hand quadrant, and similarly for the lower right-hand and upper left-hand quadrants. Evidently the reality is totally different.

Particularly intriguing are some features with large *radial* motions relative to the center. The most important of these are as follows:

1. The intense ridge at negative velocity is seen in absorption at $l = 0°$. It moves away from the center at a velocity of 53 km $\sec^{-1}$. Figure 4 indicates that it becomes "tangential" around $-22°$ longitude. At the time of its discovery in 1957, when the distance R_0 from the Sun to the center was assumed to be 8.2 kpc, the longitude 22° corresponded to $R = 3$ kpc. The feature was therefore called the 3-kpc arm.
2. Around $l = 0°$ relatively strong emission is seen at velocities up to nearly $+200$ km $\sec^{-1}$. The gas at these velocities does not absorb the radiation of Sgr A; it must therefore lie behind it and move away from the center. The stronger part has been called "the expanding arm at $+135$ km $\sec^{-1}$" (Rougoor 1964) or "feature I" (van der Kruit 1970, Cohen 1975). Figure 3 shows that the high positive velocities of the feature extend to $-5°$ longitude. At negative longitudes the radial velocity due to rotation should be negative; the hydrogen

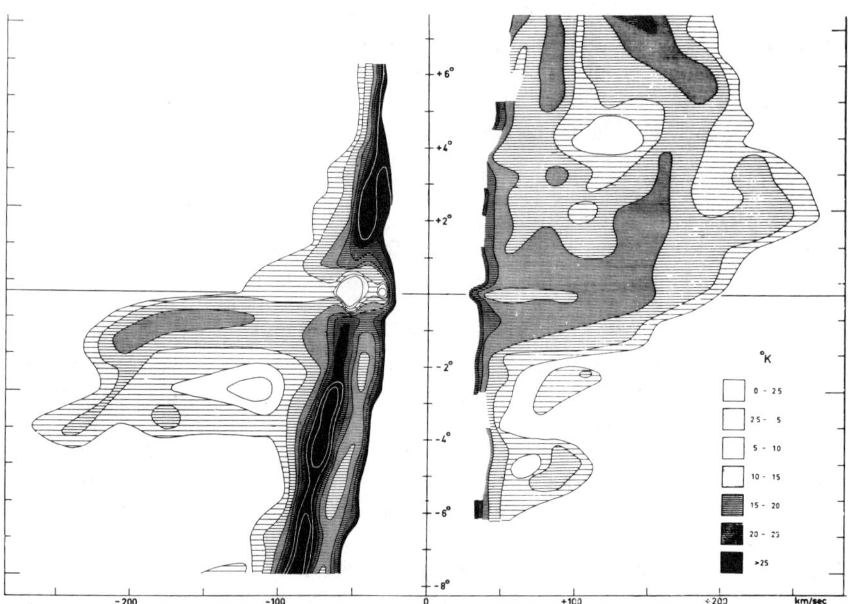

Figure 3 Velocity-longitude contours of atomic hydrogen in the galactic plane between $-8°$ and $+8°$ longitude, observed with the Dwingeloo radiotelescope with a half-power beam of 34′ (Rougoor 1964).

in this part of feature I must therefore have a very large radial component directed away from the center.

A striking feature in Figures 3 and 4 is the wing of high negative velocities between $l = -4°$ and $l = 0°$. I have added Figure 4 to emphasize the uniqueness of this feature. It was first discovered by Rougoor and Oort in 1959, who interpreted it as part of a rotating nuclear disk on the basis of two independent arguments (Rougoor & Oort 1960). The first was the sudden cutoff at the exact longitude of the center, and the lack of any clear sign of expansional radial motions. The second was that the rotation velocities that followed from this interpretation agreed with the circular velocities to be expected if the distribution of the mass near the galactic center were the same as in M31. The mass density in M31 was calculated from the brightness distribution on the assumption that the value of M/L in its central region was the same as that for the main body of the nebula.

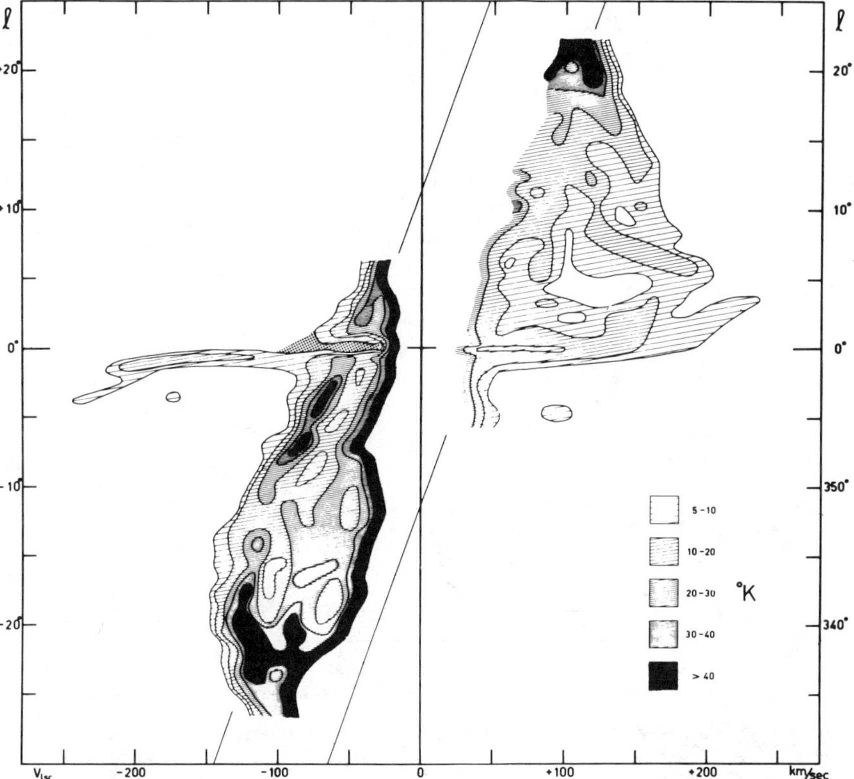

Figure 4 Extension of Figure 3 to greater distances from the center, indicating the uniqueness of the wing ascribed to the negative-longitude half of the rotating nuclear disk (Rougoor 1964).

A complete rotating disk should show a similar *positive* velocity wing at positive longitudes. However, on the positive velocity side the phenomena are muddled by high-velocity expanding gas, so that the other side of the supposed disk does not show up so convincingly. However, one does find a fair symmetry of the two sides if one confines oneself to the highest velocities, as in Figure 5, which gives a comparison of the column densities of H I atoms with velocities above 175 km sec^{-1}.

For somewhat lower velocities there are clearly asymmetries in the part within $l = \pm 1^{\circ}.5$, as is seen most clearly in Figure 6, for instance in the absence of a counterpart at positive longitudes of the ridge *d*, which is so striking on the negative side. Sanders, Wrixon & Mebold (1977), to whom Figure 6 is due, interpret this feature as a small, steeply inclined arm spiraling into the nucleus. The innermost part of the proposed rotating disk seems to be seriously disturbed on the positive longitude side. *Outside* 1°.5 longitude the disk may well be symmetrical: the hydrogen with velocities from +200 to +250 km sec^{-1} between +1°.5 and +4°.0 longitude corresponds exactly with the gas at −200 to −250 km sec^{-1}

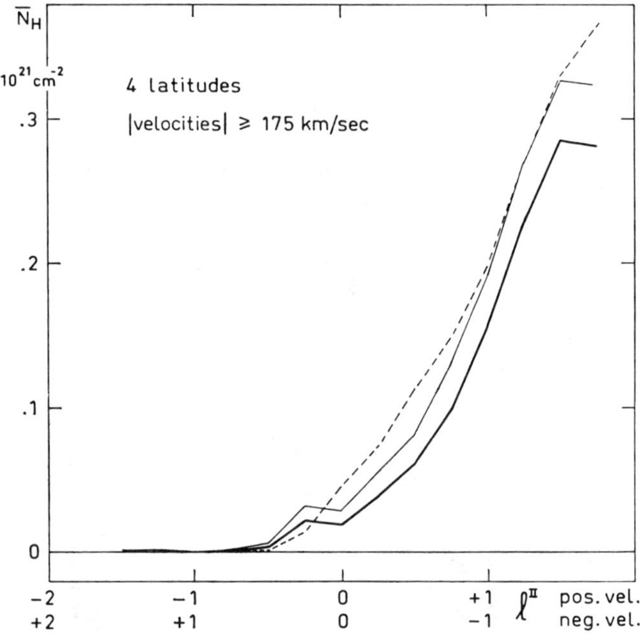

Figure 5 Variation with longitude of the number of hydrogen atoms with velocities above 175 km sec^{-1}. N_H has been averaged over latitudes b^{II} = +0°.69, +0°.44, −0°.56, and −0°.81. The heavy line is for the positive velocities above +175 km sec^{-1}, the thin line is for the positive velocities above +170 km sec^{-1}. This latter line coincides fairly well with the dashed curve for negative velocities. The longitude scale for the positive velocities has been reversed (Rougoor 1964).

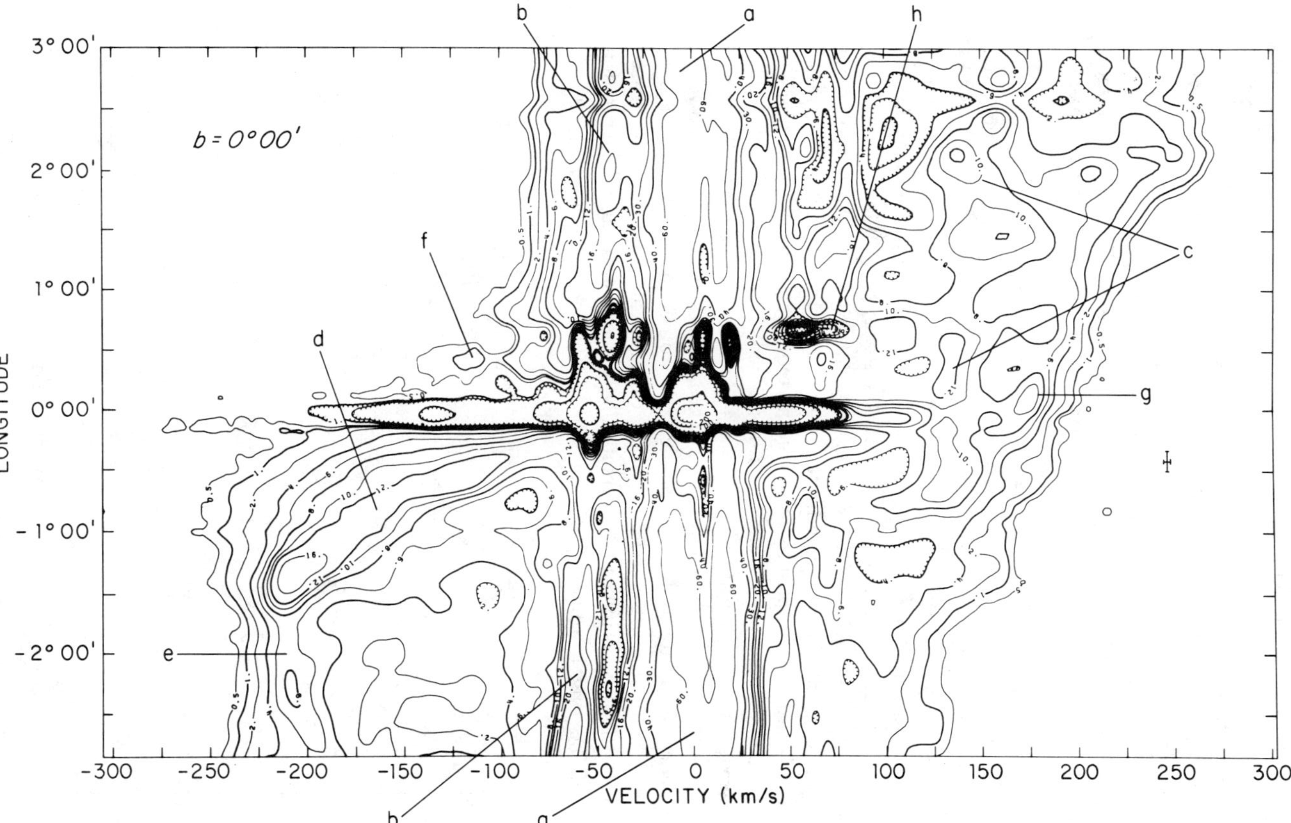

Figure 6 Velocity-longitude contours of H I in the galactic plane within 3° of the center observed with the Effelsberg radiotelescope. Beamwidth 9′ (Sanders, Wrixon & Mebold 1977).

between $-1^\circ\!.5$ and $-4^\circ\!.0$ longitude. Moreover, there is evidence for a similar cut-off at $l = 4^\circ\!.0$ on both sides of the center.

From a combination of the measured rotation in the central disk and data on the light distribution in M31 Rougoor & Oort (1960) derived a rotation curve for the central region down to quite small distances from the center (Figure 7). Assuming that pressure effects are small compared to gravitation they used these data to compute the mass distribution shown in Table 1. The table is intended only to give a general idea of the masses and the gravitational field. For this rough approximation the field was assumed to be spherically symmetrical.

The most extensive comparison of the rotation curve predicted from the light distribution in M31 and the 2.2-μ radiation in our Galaxy with the measured rotations has been made by Sanders & Wrixon (1973). They computed 21-cm line contour-velocity diagrams starting from this predicted rotation curve and assuming various density distributions of the gas. They compared these with a map for the inner $\pm 3^\circ$ longitude constructed from observations with the 140-foot Green Bank telescope, with a resolution of 21′ (Wrixon & Sanders 1973). The authors found that the predicted rotation curve (indicated by crosses in Figure 7), which is almost identical with the curve adopted by Rougoor and Oort, could give a good representation of the longitude-velocity diagram if combined with a suitable distribution of the gas density. The density distribution adopted by Sanders and Wrixon is shown by the solid drawn curve in Figure 8. The actual distribution may, however, differ considerably from this curve (cf Sanders, Wrixon & Mebold 1977). Observations extending to larger distances suggest the existence of a thin ring of increased density at $R = 0.7$ kpc, as indicated schematically by the dashed curve. The rotation velocity of the ring is roughly 240 km sec^{-1} (cf Rougoor & Oort 1960 and Burton 1974).

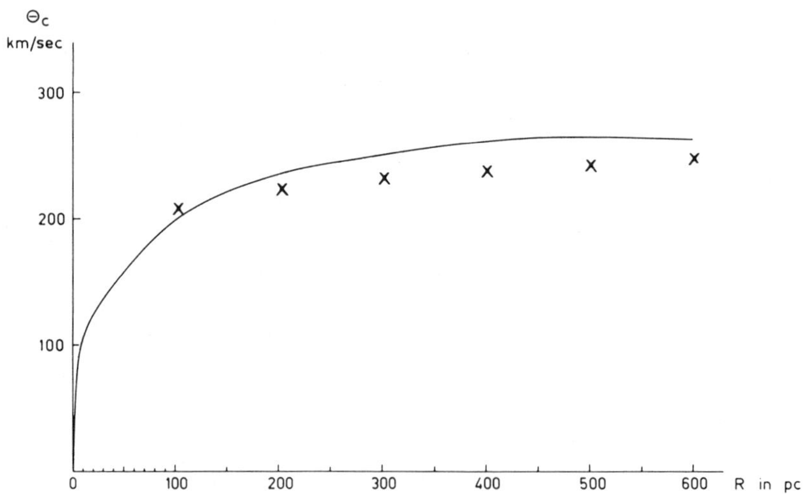

Figure 7 Circular velocity in the central region. The curve is from Rougoor & Oort (1960), the crosses are from Sanders & Wrixon (1973).

From the angle subtended in latitude the thickness of the disk between half-density surfaces is found to be 50 pc for l between $-0\overset{\circ}{.}3$ and $-1\overset{\circ}{.}5$ (with R between 50 and 250 pc) and 80 pc between $l = -1\overset{\circ}{.}5$ and $-3\overset{\circ}{.}0$ (with R between 250 and 500 pc). In the ring it rises to about 200 pc, comparable to the thickness of the H I layer in the outer regions of the Galaxy, up to about 10 kpc. Sanders and Wrixon point out that velocity dispersions greater than 100 km sec^{-1} would be necessary to maintain a disk of such thickness. However, the observed line profiles appear to exclude random velocities of such magnitude, and the authors suggest that the observed thickness may be due to the action of magnetic fields. The amount of atomic hydrogen in the nuclear disk is roughly $4 \times 10^6\ M_\odot$; the *total* gaseous mass contained in it must be at least ten times, and possibly as much as fifty times higher, the additional mass being concentrated in dense molecular clouds (cf Section 5.3). The ring between $R \sim 700$ and ~ 900 pc is inclined to the plane at an angle of 8°, having an average latitude of $+0\overset{\circ}{.}9$ at $l = -5°$ and $-0\overset{\circ}{.}5$ at $l = +5°$. The small disk *may* have a similar inclination.

Summarizing this section we conclude that there is fair evidence that the wing of high negative velocity hydrogen at negative longitudes should be interpreted as being due to a rapidly rotating disk rather than to gas expelled from the center.

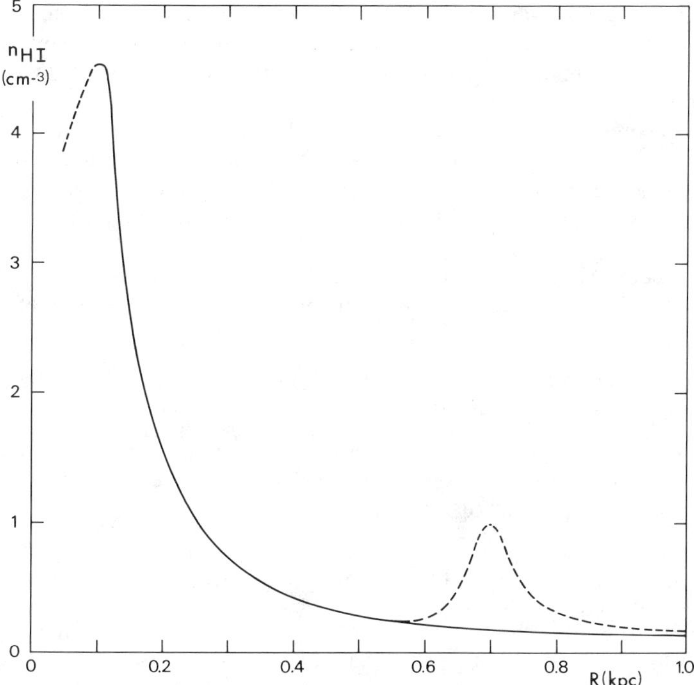

Figure 8 Density of atomic hydrogen in the galactic plane in the model used by Sanders & Wrixon (1973). The dashed secondary hump indicates schematically the ring structure suggested by Rougoor & Oort (1960). Ordinates are numbers of atoms per cm^3.

This interpretation is supported by (*a*) the wing's sharp cutoff at precisely the longitude of the center, and (*b*) the fact that the rotation curve derived from the velocities in the wing corresponds in shape as well as absolute value with what is expected from the 2.2-μ emission in our Galaxy and from the light distribution in the central region of M31. The evidence for the sharp cutoff between $-0^\circ.5$ and 0° longitude has been particularly strengthened by the recent very high resolution map made with the Effelsberg telescope by Sanders, Wrixon & Mebold (1977) (Figure 6). These observations may make it possible to find the rotation of the disk from direct measurements down to $R \sim 50$ pc.

Although the model of the rotating disk thus appears plausible, the evidence is not entirely conclusive. In particular, the accumulating evidence for the existence of numerous "expanding" features in the central region must keep us alive to the possibility of an alternative interpretation of the high-velocity wing.

It therefore remains desirable to confirm the gravitational field derived from the disk by data of a different nature, such as can in principle be derived from objects belonging to an old population.[1]

2.5 *Upper Limit to a Compact Mass at the Center*

The existence of various massive expanding features around the center, which is discussed in the following sections, has led to the idea that the Galaxy might contain a massive compact nucleus of unknown nature, possibly a black hole. Such a nucleus might be invisible at optical and infrared wavelengths, and be discoverable only by its gravitational field.

Up to the present the only *direct* source of information on the gravitational field has been the rotation of the nuclear disk, and this could not be measured closer than 50 to 100 pc from the center. The uncertainty at that distance must be at least 10%, corresponding with about $10^8\ M_\odot$. This means that we can add an invisible mass of up to $10^8\ M_\odot$ at the center without coming into conflict with observations. A specific effort to get as reliable an upper limit as possible to this hypothetical mass has been made by Sanders & Wrixon (1973). This led to the value of $10^8\ M_\odot$ mentioned.[1]

2.6 *Possibility of Determining the Gravitational Field from the Density and Velocity Distribution of Population II Objects*

Stars belonging to Population II, which is characterized by a strong concentration to the center and high random velocities, are ideal for an alternative determination of the gravitational field in the central region.

[1] Since this chapter was written the reviewer received extremely important data throwing new light on this problem. Members of the Berkeley team [Wollman et al. (1976), and in particular Wollman (1976)] have determined motions of the ionized gas within 1 pc from the center by observations of the Ne II line at 12.8 μ. These data permit a direct estimate of the mass inside ~ 0.4 pc, namely $\sim 5 \times 10^6\ M_\odot$ (cf Section 6.3.2). If there is a black hole at the center this gives an indication of its mass. If there is no black hole it provides a check on the density distribution in Table 1. Interpolated for $R = 0.4$, this table gives $1.5 \times 10^6\ M_\odot$, three times lower than the mass found from the Ne II motions.

Two types of Population II objects can be observed close enough to the center to be of interest, namely planetary nebulae and type IIb OH masers.

Extensive surveys of planetary nebulae by Minkowski (cf Minkowski 1965 for a general description) have clearly brought out their strong concentration towards the center, where, in contrast to their disk distribution through most of the Galaxy, they appear to be distributed nearly isotropically. This is confirmed by the radial velocities, which are known for about 60 nebulae less than 5° from the center. These show no sign of galactic rotation. The distribution is Gaussian, with a dispersion of 125 km sec^{-1}. Although the strong emission lines have enabled investigators to discover these nebulae up to large distances and through considerable amounts of absorption, the absorption close to the center is so strong that between −5° and +5° longitude relatively few have been found closer than 2° from the galactic equator, in the region that is of main interest for the gravitational field. The planetaries in this obscured region *are* discoverable at radio wavelengths, and a search is being made. But owing to their faintness, progress is slow. To obtain a significant accuracy in the determination of the field a considerable extension of the radial velocity data would also be needed.

If the velocity distribution is Maxwellian the gravitational potential Φ can be determined from the following relation

$$\Phi(0) - \Phi(R) = \frac{\log \nu(R) - \log \nu(0)}{\text{Mod disp}^2\, v}, \tag{2}$$

where ν is the space density and v the radial velocity.

For investigating the possibility of a massive nucleus it is essential to obtain observations on objects very close to the center. The minimum distance that can be reached is determined mainly by the space density of the objects used.

An interesting development that promises to provide data of the same kind as are being sought in the planetaries is the recent discovery by B. Baud, H. Habing and others (private communication) that type IIb OH masers in the central region probably share the dynamical characteristics of planetary nebulae, namely a strong concentration towards the center combined with a very large velocity dispersion. These Population II masers can be identified without ambiguity, and their observation provides at once their radial velocity. However, their space density may be too small for a significant test of the field of force near the center.

3 EXPANDING H I FEATURES

Outside the nuclear disk and ring the hydrogen density drops to low values. At our side of the center it only becomes appreciable again when we reach the 3-kpc arm.

This well-defined expanding feature, described on p. 302, has been observed over a galactocentric sector of about 90°. The H I contained in this sector has a mass of $3.6 \times 10^7\ M_\odot$ (Cohen & Davies 1976). The *total* mass is higher because of the presence of H_2, and because some 50% in weight should be added for the helium. From CO observations Bania (1977) has estimated that the H_2 mass is about half that of H I. This leads to a total gaseous mass of about $8 \times 10^7\ M_\odot$ for the observed quadrant of the 3-kpc arm.

It is interesting that the proportion of molecules seems to be much lower in the 3-kpc arm than in the undisturbed medium at similar distances from the center. From Gordon & Burton's (1976) analysis of the CO distribution in the Galaxy the ratio between the column density of H_2 and H I in the general medium around $R = 3$ kpc may be estimated as 3.6 in weight, as against 1.5 in the 3-kpc arm itself.

The second massive feature with a large radial motion, Rougoor's "expanding arm at +135 km sec^{-1}," which is situated on the opposite side of the center, is much less regular than the 3-kpc arm. Its H I mass may be estimated at between 1.0 and 1.5 × 10^7 $M_{\odot}$, one third, or one half, of that of the 3-kpc arm. Besides these two principal structures there are in the central region a number of smaller features that appear to be similarly moving away from the center. These are discussed in the next section.

Figure 9 gives a sketch of the possible situation of the 3-kpc arm, the expanding arm at +135 km sec^{-1}, and the nuclear disk and ring.

Have the radial motions of the 3-kpc arm and the other features been caused by

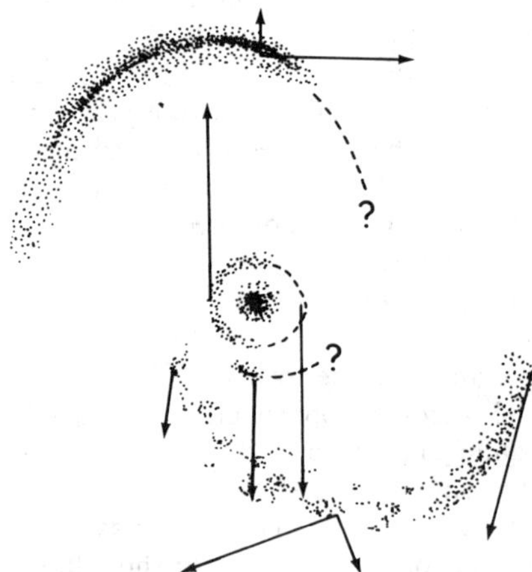

Figure 9 Sketch of the possible situation and motion of the 3-kpc arm, the +135 km sec^{-1} expanding arm, and the nuclear disk and ring (Rougoor & Oort 1960).

expulsion of gas from the galactic center, or are they connected with gas accumulated in the region of the inner Lindblad resonance, where highly excentric quasi-steady orbits may exist? The two large expanding arms may lie in such a resonance region. Attempts to represent the general motions within $R \simeq 5$ kpc by a resonance model, or by "dispersion orbits," have been made by Shane (1972) and Simonson & Mader (1973). They have not produced convincing evidence. On the theoretical side no gas-dynamical model has yet been given that bears resemblance to the observed motions. The question therefore remains unanswered.

One difficulty facing all models that ascribe the deviations from circular motion to the gravitational field is the lack of symmetry apparent in the great difference between the expansion velocity of the expanding arm at $+135$ km sec^{-1} and that of the 3-kpc arm, and the equally striking disparity between the entire velocity longitude contours for positive and negative velocities, as shown by Figures 3 and 4. That there is no counterpart of the 3-kpc arm behind the center is shown particularly clearly by CO observations (Bania 1977). Further discussion about the nature of the noncircular motions is postponed to Sections 4 and 5.8.

An apparently very serious obstacle for a gravitational field interpretation is the existence of features outside the galactic plane. The first evidence that H I exists at appreciable latitudes and has motions in a "forbidden" direction (that is, in a direction opposite to that which would correspond to the normal rotation) was found in 1966 by Shane (cf Oort 1968). It was followed by a systematic search by van der Kruit (1970). He observed several features lying distinctly outside the galactic plane and having "forbidden" velocities. At negative longitudes they were found to be situated above the galactic plane, at positive longitudes below it. His conclusion was that the data suggested an ejection of clouds from the nucleus in two roughly opposite directions at a large angle with the plane. The total H I mass of the expanding features lying outside the plane must be about 4×10^6 $M_{\odot}$ (Cohen & Davies 1976).

Van der Kruit's survey was extended with considerably greater sensitivity to higher negative latitudes by Sanders, Wrixon & Penzias (1972), who used a horn antenna with a beam width of 2°. But by far the most detailed survey was made by Cohen at Jodrell Bank (Cohen 1975, Cohen & Davies 1976). Several new features were found, and already-known features were outlined in much greater detail. A sample of the Jodrell Bank latitude-velocity maps is shown in Figure 10.

Most of the features supposed to be connected with the galactic center are situated within the limits of 5° in l and b of the Jodrell Bank survey or at least pass through this region. Exceptions are the two objects in the last lines of Table 2. One other feature has been found within 10° from the center. It was discovered by Shane as an isolated cloud with a geometrical mean radius of 1° at $l = 8°$, $b = -4°$, and a velocity of -213 km sec^{-1} relative to the LSR (Saraber & Shane 1974). The H I mass is 3×10^4 $M_{\odot}$ if it lies at the distance of the center. According to the authors it may have been ejected from the galactic nucleus. An unpublished sensitive survey between 349° and 12° longitude, $-10°$ to $+10°$ latitude, where v is between -500 and $+500$ km sec^{-1}, has revealed no other new object of similar or larger size. I am indebted to Dr. Burton for this information.

It is evident that structure and motions in the central region are very complex. Table 2 summarizes data on the more important features having motions deviating strongly from circular orbits. They are arranged in order of their H I mass, shown

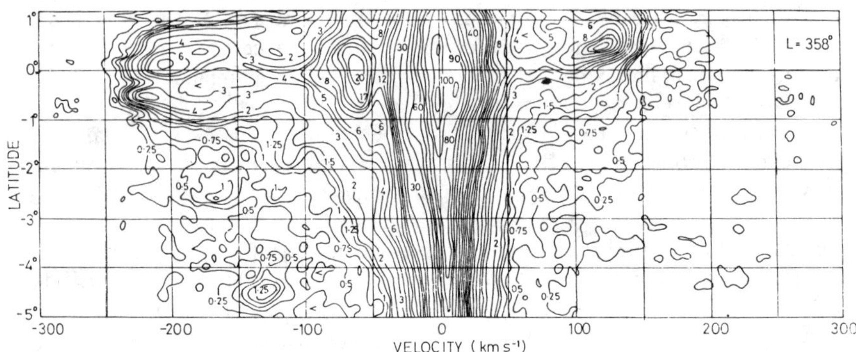

Figure 10 Latitude-velocity contours of H I at $l = 358°$ taken with the Mark IA radio telescope (Cohen 1975). Resolution $13' \times 7.3$ km sec^{-1}. The contours are for antenna temperatures of 0.25, 0.5, 0.75, 1, 1.5, 2, 2.5, 3, 4 K. They should be multiplied by 1.56 to convert to brightness temperatures. Velocities are relative to the LSR. The intense feature on the right is the $+135$ km sec^{-1} expanding arm, the 3-kpc arm is seen around $b = 0°$, $v = -60$. The powerful feature near $v = -200$ is due to the nuclear disk. The small concentration at $b = -4\overset{\circ}{.}4$, $v = -130$ is feature J1 (see text).

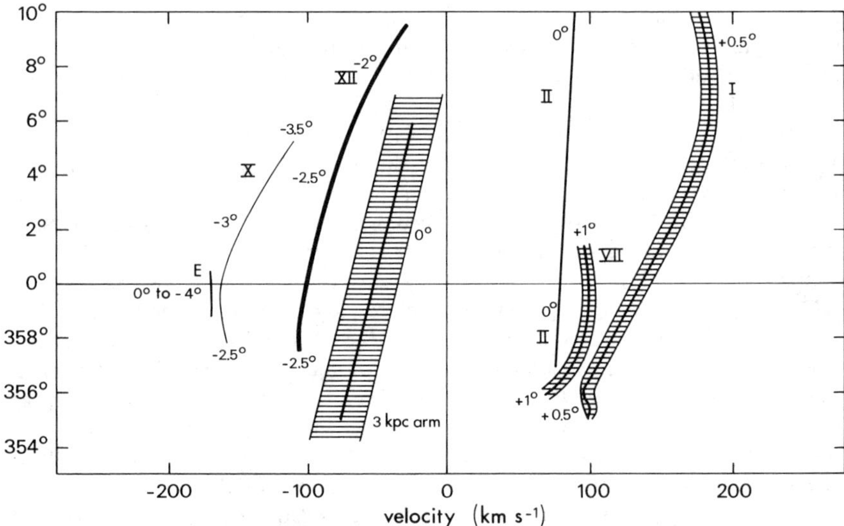

Figure 11 The principal "expanding" features listed in Table 2. Roman numbers refer to the designations by Rougoor, van der Kruit, and Cohen. Small arabic figures indicate average galactic latitudes. Width of the shading or thickness of the curve is roughly proportional to the H I mass per unit interval of l.

in the second column. V_E gives the radial component of the feature's motion at 0° longitude. Negative signs indicate structures on our side, positive signs those on the far side of the center. The next column shows the range in longitude over which the feature is observed. The next to last column gives the mean distance from the galactic plane. All data were taken from the article by Cohen & Davies (1976). For smaller features, not listed in Table 2, reference should be made to Cohen (1975) and Cohen & Davies (1976).

In the last two lines data are given for two features, Cugnon's and Mirabel & Turner's objects (Cugnon 1968, Mirabel & Turner 1972), situated outside the region surveyed by Cohen. Their relation with the galactic center is somewhat uncertain; the second feature is elongated in a radial direction relative to the center.

The situation of the various features in the l/v plane is shown in Figure 11. Figures 12 and 13 show the distribution on the sky of the column density in features XII and VII + J2 respectively.

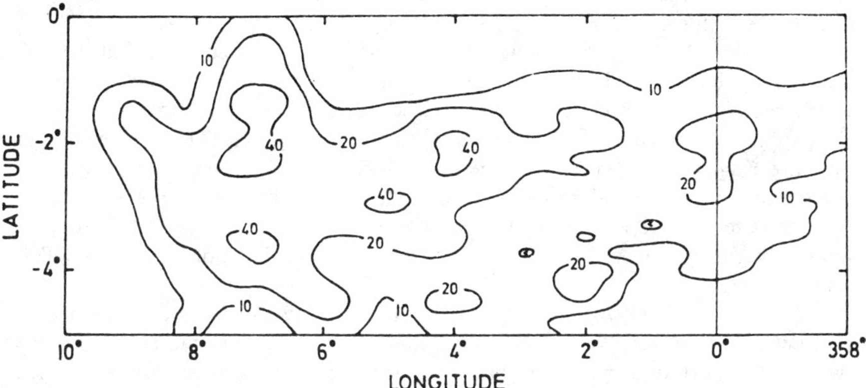

Figure 12 H I column density of van der Kruit's feature XII in units of 10^{19} atoms per cm^2.

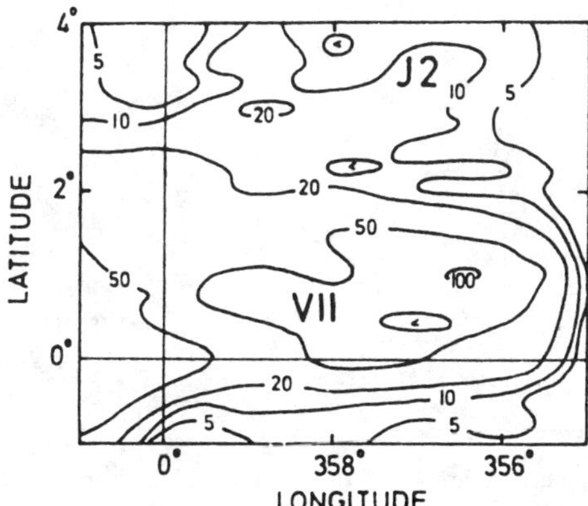

Figure 13 H I column density in units of 10^{19} atoms per cm^2 of features VII and J2.

A general characteristic of most features is that they extend over considerable stretches in longitude. They are apparently parts of rings or arms rather than clouds or jets.

However, the smaller features, E, J1, and J2, may be of a somewhat different nature. E was found by Sanders, Wrixon & Penzias (1972) at $l = 0°$ at a velocity of -170 km sec^{-1}, with a half width of 40–50 km sec^{-1}. It sticks out almost perpendicular to the plane, extending down to $b = -4°$. The authors suggested that it "might be associated with continual high angle expulsion of mass from the nucleus." J1, found by Cohen, is a rather compact cloud centered at $l = 357.°8$, $b = -4.°5$, with a half-width of about 1° in each coordinate and a velocity of -130 km sec^{-1}. It has faint extensions in l, to about 2° on either side (cf Figure 10, and Cohen's Figure 7). The H I mass is 5×10^4 $M_\odot$ if it lies at the distance of the center. J2, similarly found by Cohen, extends from $l \sim 356.°5$ to $\sim 359.°5$ and has a latitude of $+3.°0$. It has a strongly forbidden velocity of $+100$ km sec^{-1}. It is adjacent to, and possibly connected with, feature VII (see Figure 13). Like E it makes a very large angle with the galactic circle. Its H I mass is 3×10^4 $M_\odot$.

In all cases where a feature includes the direction of the center, so that we can see whether or not it absorbs the radiation from Sgr A and thus determine on which side of the center it is situated, the motions are directed away from the center. This applies to four of the six major features listed in Table 2. The two remaining ones are Cugnon's object and van der Kruit's feature XII. If we adopt expulsion as the general cause for all anomalous motions in the central region it is probable that the latter feature is at present likewise moving away from the center. For if it were falling back it would be hard to understand why it has not previously fallen into the disk.

Cohen & Davies (1976) have tried to locate in space all features identified in the central region, in a somewhat similar manner as was done in Figure 9 for the two most important arms. Many have not been observed at their tangent points,

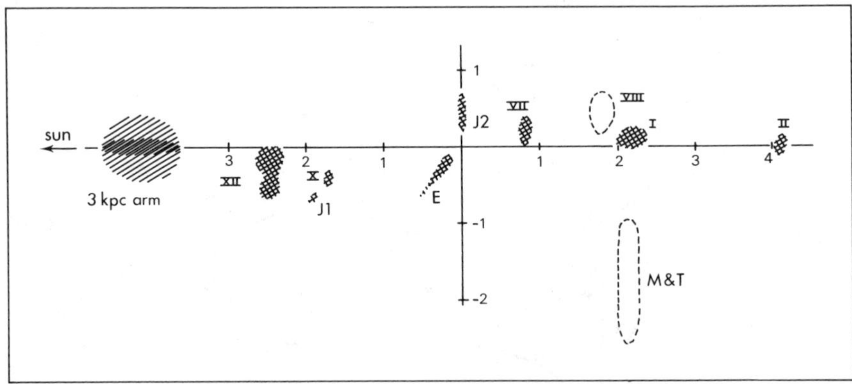

Figure 14 Cross cut perpendicular to the galactic plane through $l = 0°$, schematically indicating the situation of various "expanding" features. Distance scales are in kpc.

so that there is no direct way of estimating their distance from the center. Cohen and Davies determined this distance from the way in which the radial velocity varied with longitude. They assumed that each feature was rotating as well as expanding and that it was lying on a circle. The radius of the circle determined in this way is given under R in the last column of the table. These distances must be considered as quite uncertain, mainly because of the assumption of circular shape. Most features might well be parts of considerably inclined spirals.

It is seen from Table 2 that the majority of the high-velocity arms lie at considerable distance from the galactic plane. This is further illustrated in Figure 14, which gives a schematic cross section at $l = 0°$. The features shown are those in Table 2 plus J1 and J2 mentioned in the text. Cugnon's arm and Mirabel and Turner's feature, which do not extend to $l = 0°$, and whose pertinence to the central region is uncertain, are indicated by dotted contours. It should be stressed again that the distances from the center are mostly uncertain. The distances z from the plane are, however, well determined.

There is reason to believe that at least part of the gas complexes considered owe their large expansional motions and their high z values to ejection of massive clouds from a small nuclear region. Rather direct evidence in support of this thesis is shown by features E and J2, which lie close to the center and point away from it under large angles with the galactic plane.

No recent data on K_z are available, but we can get a sufficient impression of the velocities required to attain the observed distances from the plane from the data in Table 1. Spherical symmetry was assumed in this table; in the actual case the potential differences in the z direction will be higher than in the spherical case for the same distance from the center, so that the velocities cited below, which were computed on the assumption that $\Phi(0)-\Phi(z)$ was equal to $\Phi(0)-\Phi(R)$, will be too low.

From the values of the potential given in Table 1 we find that if there is no friction a body thrown out from the center needs a velocity of 610 km sec^{-1} to reach a maximum distance of 100 pc and 740 km sec^{-1} to get to 500 pc. It is this order of velocity that must have been imparted to gas complexes containing in some cases more than a million solar masses of H I.

We return to this problem and that of the arms in the plane in Section 4 and Section 5.8.

3.1 *Evidence for an Inclination of the Layer of the Expanding H I Complexes*

We have already mentioned that van der Kruit's observations (van der Kruit 1970) indicated clearly that the high-velocity gas near the center is tilted with respect to the galactic equator. At longitudes close to zero the tilt of the layer of forbidden-velocity H I may be quite steep. It is interesting to note that the continuum radiation at a wavelength of 20 cm shows two ridges about 250 pc long and steeply inclined to the galactic equator in the same sense: towards positive b at negative l and towards negative b at positive l (Kerr & Sinclair 1966).

Cohen & Davies' (1976) data fully confirm the tilt. The authors conclude that the

H I within 2.5 kpc of the center forms an inclined layer, with a pole in the direction $l = 124° \pm 28°$, $b = 81° \pm 4°$.

The nuclear disk, with inclusion of its ring, appears to be similarly tilted at an angle of about 6° (R. D. Davies, private communication).

A similar tilt has been observed by Kerr (1968) in the H I, with highest velocities in the *permitted* direction. Integrating in each point over the highest 70 km sec^{-1} he finds column densities as plotted in Figure 15. The tilt appears to extend to $l = 345°$ and $l = 8°$ on the two sides of the center, with average z values of $+100$ and -60 pc respectively. Near the center the angle is about 8°. Kerr suggests that the gas concerned forms a central bar structure which at its outer end might connect up with the 3-kpc arm. However, the evidence for a bar is not convincing.

4 ORIGIN OF THE RADIAL MOTIONS AND THE DEVIATIONS FROM THE GALACTIC PLANE

4.1 *Introduction*

In the twenty years that have elapsed since the discovery of the 3-kpc arm no essential progress has been made toward answering the central question of whether its large deviation from circular motion is connected with a large-scale deviation from axial symmetry in the Galaxy's gravitational field ("the field model") or whether it is due to expulsion of gas from the nucleus ("the ejection model").

Great progress has, however, been made on the *observational* side, in three main directions: (*a*) the identification of other "expanding arms"; (*b*) the discovery in the central region of H I features at considerable distance from the galactic plane,

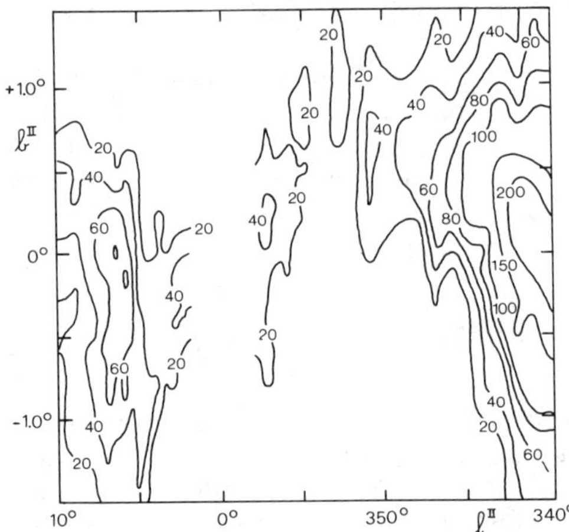

Figure 15 Integrated brightness temperature for atomic hydrogen of highest velocities. For $l > 0°$ the interval over which b^{II} has been integrated extends over the highest 70 km sec^{-1}, for $l < 0°$ it extends also over 70 km sec^{-1}, the lower limit being the lowest velocity observed at each position (Kerr 1968). Note the expanded latitude scale.

some of them quite large and massive; (*c*) the observation of numerous dense molecular clouds in the nuclear disk, mostly with considerable velocity components in radial direction relative to the center.

For these molecular clouds an expulsive origin seems indicated, notwithstanding the severe difficulty presented by their large masses. Similarly, for the jet-like H I features E and J2, which seem to point away from the center and extend to about 700 pc, an ejective origin is strongly suggested. But there are no such directly suggestive circumstances for the bulk of the H I features.

In connection with the confrontation of the two hypotheses two things should be stressed. The first is that violent activity leading to the expulsion of discrete masses of gas at high velocity is apparently of rather common occurrence in the nuclei of massive galaxies. It may be estimated that at least 10% of the spiral galaxies with masses of the order of that of our Galaxy are in a very active Seyfert stage, in which large discrete "clouds" are being continually thrown out from the nucleus at velocities of the order of 1000 km sec^{-1}. Unfortunately we have no sufficient information concerning the total masses of gas that are expelled during a Seyfert stage, so that we cannot make a quantitative confrontation with the phenomena observed in our Galaxy. That in some cases the expelled mass must be quite large is shown by the filaments around NGC 1275.

It is quite possible that the Seyfert stage is intermittent, and that most large spirals pass through active periods that in total might make up on the order of a tenth of their lives.

Another type, or possibly, another stage, of activity is observed in NGC 4258. The strong synchrotron emission of this spiral and especially the remarkable structure of the synchrotron arms has been interpreted as being due to the ejection of clouds with a combined mass of about 10^8 $M_\odot$ and velocities of the order of 1000 km sec^{-1} into two opposite cones at a small angle with the equatorial plane (cf van der Kruit, Oort & Mathewson 1972, van der Kruit 1974, van Albada & Shane 1975, de Bruyn 1977). The case of NGC 4258, which does not seem to be unique, indicates that anisotropic expulsions involving masses of the same order as those of the largest expanding H I features in the Galaxy *can* occur in ordinary spirals. It should be mentioned that at present the nucleus of our Galaxy is in a quiescent stage, showing no resemblance to a Seyfert nucleus. It is interesting that the nucleus of NGC 4258, notwithstanding the enormous activity it must have had in the recent past, shows only moderate activity at the present time.

There is a third nearby galaxy that displays still another type of gigantic nuclear activity, namely NGC 5128. Plasma clouds thrown out in directions roughly perpendicular to the equatorial plane have created a giant radio galaxy. The total energy content in the form of relativistic particles and magnetic fields is at least 2×10^{59} erg.

The second thing to be stressed is that to be acceptable, a theory of the expanding features must also account for the conservation of the nuclear disk.

For the discussion of "field" models a detailed study of the central regions of barred spirals would be desirable. This would, however, go beyond the scope of the present review. But attention may be drawn to the possible existence of mini-bars in the centers of spirals that are not classified as barred systems. Such small bars have been suggested as mechanisms to activate spiral waves. Attention should also be drawn to the fact that in several of the nearby spiral galaxies small dust arms are seen to spiral in close to the nucleus (cf Sandage 1961), which might indicate a nonaxisymmetric field.

4.2 *Some Expulsion Models*

An attempt to explain the outward motion of the 3-kpc arm and the +135 km sec^{-1} expanding arm was made by van der Kruit (1971). His observations of H I outside the galactic plane having suggested that this gas had been expelled from the nuclear region, it appeared natural to inquire whether the large expanding arms *in* the plane could not likewise have received their outward momentum from gas expelled at an angle with the plane, thereby avoiding the destruction of the nuclear disk. He supposed that the ejection took place in two opposite directions, rotating fast around the axis of the Galaxy so that all galactocentric longitudes were covered during the period of expulsion. The ejection should have consisted of a large number of small clouds and was presumably triggered in a small core. Before leaving the nuclear disk the clouds would have swept up disk gas and acquired some angular momentum. Van der Kruit assumed that the clouds would have emerged from the disk around $R = 100$ pc, and that beyond this distance they moved under the sole influence of the gravitational field until coming down again into the gas layer at $R \sim 3$ kpc. They were then rapidly braked by friction, transferring their momentum to the gas in the layer.

Figure 16 gives an impression of the orbits followed by clouds starting with a velocity of 625 km sec^{-1} at $R = 0.1$ kpc under various angles i with the galactic plane. The forces were computed from the Schmidt model, amended in the central region by adding an inhomogeneous central spheroid in order to represent the rotation of the nuclear disk.

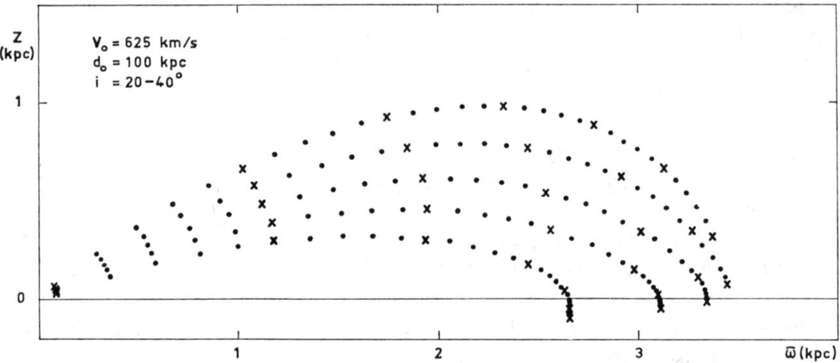

Figure 16 Orbits of gas clouds for an expulsion velocity of 625 km sec^{-1} and inclinations $i = 20$, 25, 30, 35, and 40°. Time interval between dots 5×10^5 yr, between crosses 2.5×10^6 yr (van der Kruit 1971).

According to van der Kruit's calculation the outward motion of the 3-kpc arm may have been caused by an ejection some twelve million years ago of gas clouds with a total mass between five and ten million solar masses, which would have started with a velocity of 600–650 km sec^{-1} at $R = 100$ pc under an angle of 25°–30° with the galactic plane. Similar numbers would be required to produce the motion of the +135 km sec^{-1} expanding arm.

The ejection of the massive features observed *outside* the galactic plane was thought to have occurred some six million years ago.

A different attempt to explain the motion of the 3-kpc arm by an ejection process was made by Sanders & Prendergast (1974). They considered the gas-dynamic effect of an isotropic explosion on the normal gas layer and found that a radially oscillating ring of cold gas is produced. In order to get such a ring at the location of the 3-kpc arm the explosion of a central mass of the order of 10^8 $M_\odot$ with a total explosion energy in the order of 10^{58} ergs is required. The arm would oscillate back and forth for a considerable time, estimated as perhaps 10^8 years. During such a period the nuclear disk, which was destroyed by the explosion, might have been restored by the normal mass loss of evolving stars. The principal difficulty is the size of the initial mass which has to be exploded.

4.3 *The Gravitational Field Hypothesis*

We are confronted with two problems: (*a*) the radial motions, and (*b*) the systematic deviations from the galactic plane.

While the first might be attributed to a systematic deviation from axial symmetry in the mass distribution, the second requires that the mass distribution in the center is inclined to the galactic plane. Without such a tilted distribution the field model could hardly produce the Z velocities of the order of 500 km sec^{-1} required to take hydrogen masses up to the observed distances of ~300 pc from the galactic plane.

In the central region the only important contributors to the mass density are old Population II stars with high random velocities. It is for these that one would have to assume a tilted distribution.

That such a situation should not be dismissed a priori is indicated by two phenomena: in the first place by the large angles that are often seen between the minor axes of elliptical radio galaxies and their radio axes, and in the second place by the structure of the core of the Andromeda nebula. The circumstance that in double radio galaxies, which also have components close to the nucleus, the latter lie generally in the same direction as the outer components suggests that this direction may be parallel to the rotation axis of the compact nucleus from which the radio emission presumably originates, and that therefore this axis often differs considerably from that of the E galaxy as a whole. As the nucleus is likely to be generically related to the galaxy's inner region its tilt may be an indication that the rotation axis of this inner mass may likewise make an angle with the axis of the main mass of the galaxy.

The stratoscope II observations of M31 (Light, Danielson & Schwarzschild 1974) suggest that its distinctly flattened nucleus is somewhat tilted relative to the major axis of the main system. The nucleus has a radius of roughly 3 pc. Because of its importance for the present discussion I reproduce in Figure 17 the contour map given by Light et al. for the central 3″ (or 9 pc) of M31, in which I have indicated the direction of the major axis of the entire galaxy. From ellipses fitted to the contours the authors find that on the average these have a roughly 25° larger position angle. A disturbing feature is the somewhat asymmetrical position of the brightest region; this may indicate that the contours are affected by absorption effects.

A somewhat smaller tilt in the same direction had previously been observed by H. M. Johnson (1961). Lindblad (1956) has pointed to a similar small deviation in the isophotes between 300 and 1500 pc radius which he, however, interpreted as being due to a central bar lying in the equatorial plane.

The color and spectrum of the nucleus is identical with that of the surrounding "nuclear bulge." Both consist mainly of later-type stars, and may be assumed to reflect the mass distribution.

If the tilt is confirmed by observations at longer wavelengths it would indicate that the plane of symmetry near the center of a galaxy can be appreciably tipped with respect to the general equatorial plane. This would evidently be a discovery of fundamental significance.

It is possible that the H I features outside the galactic plane in the center of our own Galaxy indicate that *its* central plane of symmetry is similarly tilted, at an angle of perhaps 10°. In that case part of the anomalous features could be naturally understood on the basis of the field hypothesis. There would, of course, remain the need for a nonaxially symmetric mass distribution to explain the large radial motions.

A considerable amount of work has been done to explain the observed motions by dynamical models without the aid of an explosion (see, for instance, Sanders &

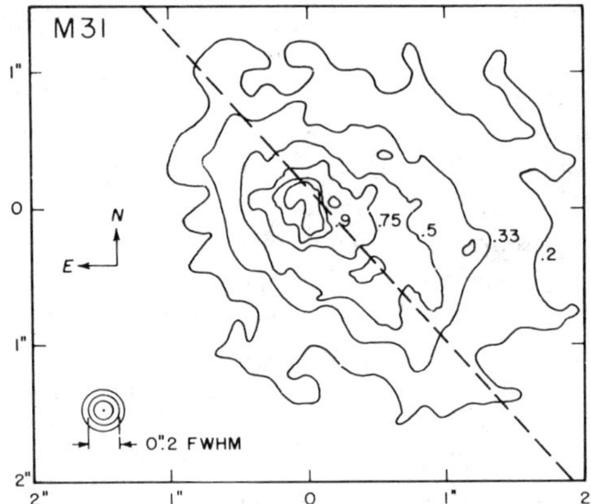

Figure 17 Contour map of the nucleus of M31. The levels are labeled by the fraction of peak intensity. (Light, Danielson & Schwarzschild 1974.)

Huntley 1976), but it has not yet proved possible to construct a plausible model that reproduces the observed structures and motions in a convincing way.

Peters (1975) has tried to represent the H I velocity longitude diagram for the central region by gas moving in concentric elliptical orbits with axial ratios of about 0.5. He suggested that such orbits might be sustained by a bar-like perturbation, but no quantitative dynamical foundation was given.

4.4 *Ejection or Gravitational Field?*

We close this section by summarizing the main difficulties encountered by either model.

4.4.1 PROBLEMS WITH THE EXPULSION HYPOTHESIS The most serious problem is the vastness of the mass that has to be expelled. Van der Kruit estimated that a quantity of H I of between 0.5 and $1.0 \times 10^7\ M_\odot$ would have had to be expelled to give the 3-kpc arm its outward velocity. Cohen's high-sensitivity survey has increased the H I mass of the 3-kpc arm by a factor of 1.8. If one includes the hydrogen in molecular form as well as the helium the total gaseous mass becomes a factor of 4 higher than that considered by van der Kruit. The resulting amount of gas to be expelled would then become approximately $3 \times 10^7\ M_\odot$. This is for a collimated ejection; an isotropic ejection would require some five times more. The ejection took place roughly ten million years ago. Averaged over such a period the mass loss from the region within $R = 50$ pc would be of the order of 10 $M_\odot$ per year. Such a quantity cannot be replenished by evolving stars. Normal evolution of the stars within $R = 100$ pc would yield no more than 0.0004 $M_\odot$ per year (cf van der Kruit, Oort & Mathewson 1972). Even including all stars within 500 pc would yield but 0.004 $M_\odot$, so that it would take 10^{10} years to produce the gas needed for the outward momentum of the 3-kpc arm to be caused by expulsion. The expulsion hypothesis can for these massive features only be maintained if the central region is replenished by infalling material. The rate of infall would then have to be at least two orders of magnitude higher than that indicated by the high velocity complexes in the vicinity of the Sun. Alternatively one might consider the possibility that there is a mechanism (provided perhaps by a massive black hole) that would disintegrate stars at a much faster rate than normal evolution (cf also p. 343).

The collimation assumed by van der Kruit forms another problem. It is possible, however, that the collimation is only apparent, and is in reality an effect of selection, i.e., the clouds expelled under small inclinations may be stopped in the disk and transformed into molecular clouds, while those with very large inclinations might not have become dense enough to be observable.

The nucleus today shows nothing like the degree of violence that should have existed in order to eject such high-velocity features as listed in Table 2. The turbulent ionized gas seen in the nucleus is negligible in quantity as well as velocity compared to what is needed for such ejections (see Section 6). For the expulsion theory to be acceptable one must suppose that a radical change in the activity of the nucleus has taken place in the past ten million years.

4.4.2 PROBLEMS FOR THE GRAVITATIONAL FIELD HYPOTHESIS It is difficult to see how a field due to probably well-mixed stars with high random velocities could produce jet-shaped features, like E and J2 of Table 2, pointing away from the center under a very large angle with the galactic plane. It seems equally difficult to construct a field that could have imparted to the numerous dense molecular clouds observed in the nuclear disk their considerable velocities without at the same time systematically affecting the H I in the disk. But in particular it is hard to conceive that the recently confirmed, almost complete ring of molecular clouds, which has a radius of 190 pc and is expanding at a velocity of 150 km sec^{-1} (cf Section 5.5), was produced by an asymmetry in the field.

Whether it is possible or impossible for the field model to explain the large H I features outside the plane depends entirely on whether or not the central mass distribution can be tilted. And even if it *is* tilted there would still remain, beside E and J2 already mentioned, a number of other features, for instance X and J5 of Cohen's survey, which would lie so far outside the tilted plane as to require some ejection or infall mechanism to explain them.

4.4.3 CONCLUSION The molecular clouds in the nuclear disk and some of the highly inclined H I features near the center have probably been formed by ejection. For the large H I features at 1 kpc and farther we do not know. For these we may have to revert to the hypothesis of a tilted bar structure. For both hypotheses the actual driving mechanism has until today remained hidden. The problem is reconsidered in Section 5.8 in connection with the molecular clouds.

4.5 *The Disk's Magnetic Field*

The large differential rotation in the nuclear disk tends to increase the magnetic field strength on a short time scale. It is conceivable that it would become sufficiently strong to affect appreciably the motion of the gas. The problem has been discussed by Sanders & Wrixon (1973) in connection with the thickness of the disk. They noted that a velocity dispersion of about 100 km sec^{-1} would be needed to maintain the observed z distribution of the H I. So high a dispersion seems incompatible with the structure observed in the plane. According to the authors the largest dispersion permitted by these latter observations is 24 km sec^{-1}. They concluded that either cosmic-ray pressure or magnetic pressure contribute primarily to the support of the disk. If it is magnetic pressure a field on the order of 20 μG would be required. Sanders and Wrixon suggest that by the wrapping in the disk the field strength is increased to a "quasi-static equilibrium value, where further increase is prevented by loops of force rising out of the plane and dragging some neutral gas along. This might also account for the high-velocity features outside of the plane near the center."

A limiting factor is the brightness temperature of the nonthermal radio emission from the disk region. If the density of cosmic ray electrons in this region is comparable to that in the solar neighborhood the maximum permissible field strength would be on the order of 10^{-5} G. Such a strength would be quite insufficient to throw out massive gas streams at velocities of several hundred km sec^{-1}, as would be required for most of the features listed in Table 2.

5 MOLECULAR CLOUDS AND TOTAL GAS DENSITY

5.1 *Introduction*

Inward from about 4 kpc the interstellar density decreases considerably and remains low until we reach the edge of the nuclear disk at $R \sim 750$ pc. This low-density region is the domain of the "expanding" H I features discussed in Section 3. In the nuclear disk the "expanding" regime changes again into a more orderly state in which the H I motions seem to be practically circular. But at a few hundred parsecs from the center a third radical change occurs. The density increases strongly, and the hydrogen is almost entirely in the molecular state and is mostly concentrated in massive and dense conglomerations whose motions deviate widely from the rotation of the nuclear disk described in Section 2. The conglomerations are further characterized by very large internal velocity dispersions. They lie close to the galactic plane and are quite thin in the z-direction (20 or 30 pc).

The knowledge of this inner region has come mainly from observations of interstellar molecules, especially hydroxyl, formaldehyde, and carbon monoxide. The most compelling evidence on the great abundance of molecular hydrogen and on the large masses of the nuclear aggregates has been provided by CO.

5.2 *Total Gas Density Derived from Carbon Monoxide Observations*

CO is observed in emission in the $J = 1 \rightarrow 0$ rotational transition at a wavelength of 2.6 mm. It occurs largely in dense cool clouds where it is shielded from dissociation by ultraviolet radiation, and it is excited through collisions. In order to populate the upper level of the 2.6-mm transition with a Boltzmann distribution corresponding to the excitation temperatures of about 15 K found from the observed emission profiles, particle densities of the order of 10^3 cm^{-3} are required. As the density of hydrogen atoms is far too small the only particles that can be sufficiently numerous to excite the CO are hydrogen molecules.

Minimum cloud masses have been estimated from these considerations. Such estimates are, however, quite uncertain, because the cloud agglomerations considered are probably very inhomogeneous, and it is impossible to determine "filling factors."

The mass estimates given in the following are based on the intensities of the CO emission. If a feature is optically thin in the CO line and the distribution over the levels concerned is "thermalized," the total number of CO molecules per cm^2 can be directly computed from the measured surface brightness. A practical difficulty is that all the more important features in the central region have a large optical thickness in the ^{12}CO line, and that at the time of writing of this report measurements of the fainter ^{13}CO line are still scarce: they are limited to the Sgr B2 complex (cf Section 5.7) and some small other regions. Solomon et al. (1972) as well as Liszt et al. (1975) measured ^{12}CO to ^{13}CO ratios in a small part of the +40 km sec^{-1} feature near the galactic center. For the integrated profile the latter found an average ratio of about 7. The former observed considerable variation of the ratio with velocity, with extreme values of ~ 5 to ~ 18. As the terrestrial abundance ratio of the two carbon isotopes is 89, these measurements may give an impression of the degree of saturation of the ^{12}CO line. However, other investigators have argued that the clouds might *not* always be optically thick (cf Leung & Liszt 1976). A summarizing discussion of the problem has been given by Gordon & Burton (1976).

Most observations have been made in the ^{12}CO line. In their large survey of CO in the Galaxy Gordon & Burton (1976) use the following semi-empirical formula to derive the column density of CO per cm^2

$$N(CO) = 3.43 \times 10^{14} \frac{^{12}CO}{^{13}CO} \int T^*(^{12}CO)\, dv = 1.6 \times 10^{16} \int T^*(^{12}CO)\, dv, \qquad (3)$$

where T^* is the antenna temperature corrected for antenna and atmospheric losses, and v is in km sec^{-1}. For the abundance ratio $^{12}CO/^{13}CO$ they adopt a value of 40 suggested by Wannier et al. (1976).

In order to find total hydrogen masses we must know two other things, namely, 1. the relative number of the carbon atoms that have been condensed onto solid particles, and 2. the abundance ratio of H to C.

1. From a comparison of the CO emission and the optical extinction of a dust cloud, Milman (1975) has found an upper limit of 0.1 for the fraction of carbon in the form of CO, while on theoretical grounds Leung & Liszt (1976) also found 0.1. In most mass estimates a 10% value has been adopted, but this should be considered with very great reserve. Gordon & Burton also use this fraction. Assuming a carbon abundance of 1:3000 they thus arrive at a ratio $n(CO):n(H_2)$ of 6×10^{-5}. The same ratio was used by Bania (1977) in his investigation of the molecular clouds near the center.

The uncertainty in the fraction of the C that has not condensed onto solid particles introduces a great uncertainty in the mass estimates of the molecular clouds. In particular it is questionable whether the fraction derived from observations in the neighborhood of the Sun is also applicable to the cloud complexes near the center, which have formed and evolved under rather different conditions. There are, indeed, indications that the absorption in the central region is considerably less than what would be expected from the high column densities of gas derived by Bania on the assumption that only 10% of the carbon atoms would be in CO. Using formula (3) and the ratio $n(CO)/n(H_2)$ given above, we find from Figure 21 in the central 2 or 3° a column density of 8×10^{23} hydrogen atoms per cm^2. If the dust-to-gas ratio, or the ratio between the number of H atoms and the visual extinction, were the same as that near the Sun, this column density would correspond to a visual extinction of roughly 300^m, roughly half of which might lie in front of the center. From photometric measures at various wavelengths in the infrared Willner (PhD thesis, 1976) quotes a visual extinction of about 28^m in front of the central sources. The general extinction in the galactic disk is probably between 10 and 15^m, leaving only about 15^m for the extinction in the dense central region, which differs by a factor of ten from the above estimate from the CO observations. This may indicate that in the region of the molecular clouds near the center a much larger fraction of carbon is in gaseous state than the 10% found from observations near the Sun.

An independent estimate of the gas-to-dust ratio may be obtained from measurements of the surface brightness in the mm range, such as have been made by Westbrook et al. (1976) in Sgr B2. From the flux at 1 mm combined with the temperature determined from the far-infrared spectrum they derive the column density of the dust particles in g cm^{-3}. As, with normal abundances, about 1/1000 of the hydrogen can condense onto solid particles, an estimate of the mass of the solid particles will yield also an estimate of the total gaseous mass. In this way the authors found a hydrogen mass of 10^5 $M_\odot$ for the central 1′ of Sgr B2. A comparison with the core mass of 6×10^5 estimated by Scoville, Solomon & Penzias (1975) from CO observations indicates again that the assumption that only 10% of the carbon is in CO, on which the latter estimate is based, may not apply in the central region.

2. An analogous problem is presented by the general abundance of C and O. All published mass estimates rest on the assumption that the abundances are the same as those in the Sun and its surroundings, namely $n(\mathrm{H})/n(\mathrm{C}) \sim 3 \times 10^3$. However, since there is considerable evidence that the abundance of at least some heavier elements, such as O and N, is an order of magnitude higher in the nuclear region of spirals than in the outer parts, the published mass estimates of molecular features in the central region may also on this account be too high.

In view of the above arguments it seems that recent mass estimates of molecular clouds in the central region may be an order of magnitude too high. In the following parts of this section two mass estimates for the hydrogen will therefore be given: (*a*) that of the authors, and (*r*) a ten times lower value. In the reviewer's opinion the second may be closer to the truth.

Notwithstanding these various uncertainties, important information can be obtained on the general molecular density in the Galaxy, and in particular on the very high density in the nuclear region. The most complete general survey to date is that of Gordon & Burton (1976), who observed from $l = 10°$ to $36°$ with a sampling interval of $0.^\circ2$, and over some samples at higher longitudes. They found that the fraction of molecular hydrogen increases gradually towards the center. While in the surroundings of the Sun the masses of atomic and molecular hydrogen in a column perpendicular to the galactic plane are roughly the same, between 2 and 6 kpc from the center the mass of atomic hydrogen diminishes to about one fifth of that of the molecular component. In the nuclear disk it is still smaller, while in the innermost two or three hundred parsecs it seems to become almost negligible.

The total gas density has a maximum between $R \sim 4$ and ~ 7 kpc; it drops sharply inside 4.0 kpc, and between 2 and 4 kpc the density per column perpendicular to the galactic plane is only about four tenths of that between 4 and 6 kpc. The entire region between roughly 1 and 4 kpc is remarkably devoid of gas. It is not known what causes this. It may be that the gas has been used up by more intensive star formation which might, for instance, be caused by the increase of the angular

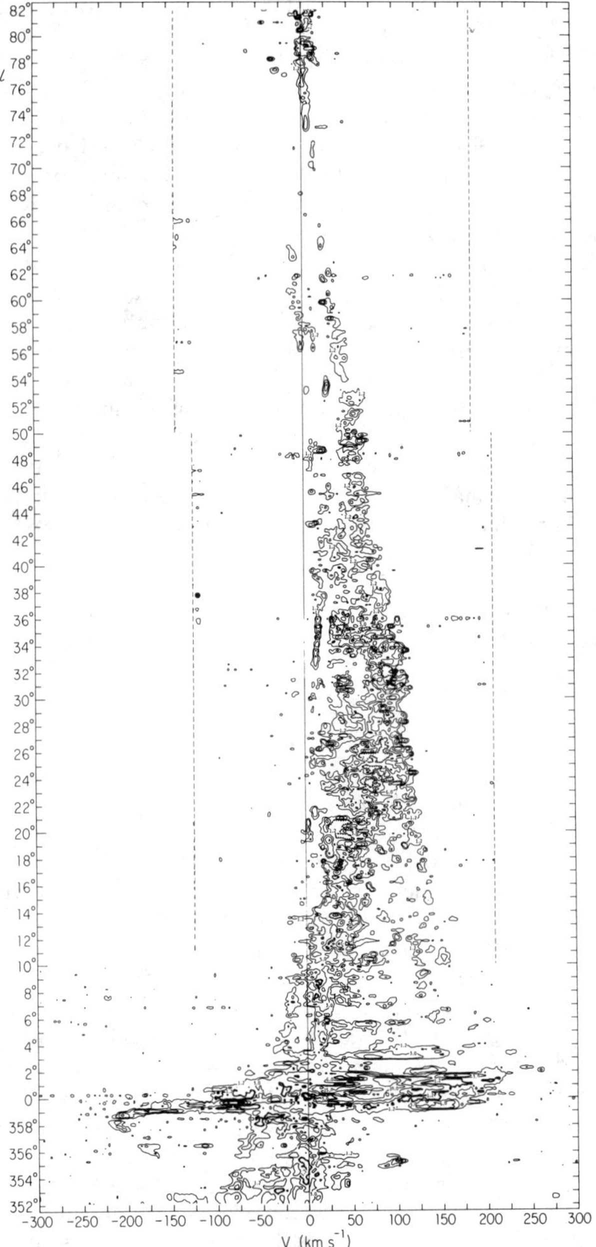

Figure 18 Distribution of CO in longitude and velocity at $b = 0°$ from $-8°$ to $+82°$ longitude, from observations with the 11 m NRAO telescope on Kitt Peak. Beam width 65″. (Burton & Gordon 1977 for $l > 10°$; Bania 1977 for $l \leqq 10°$.)

velocity towards the center and the consequent greater frequency with which the gas meets the spiral pattern (cf Oort 1974a; Segalovitz 1975); or, alternatively, the gas may have been swept away by expanding arms.

The gradual increase in molecular density from $l = 85°$ to $l = 21°$ and the sharp rise within 4° from the center may be illustrated by Figure 18, which gives the ^{12}CO intensity as a function of v and l. An instructive complement to this figure is shown in Figure 1 of Gordon & Burton (1976), which gives an l/v contour diagram between $l = 36°$ and 10°, beautifully illustrating the cloudy structure of the CO, as well as the remarkably sharp velocity cutoff around 110 km sec^{-1} between $l = 36°$ and 21°.

Two things stand out in Figure 18 as well as in Gordon & Burton's Figure 1: the high intensity and very large velocity spread between $l = +4°$ and $-4°$ and the absence of velocities larger than about +60 km sec^{-1} between $l = 21°$ and 4°. This absence reflects the general lack of gas between $R = 800$ and 3500 pc; the lower velocity CO observed in this longitude range lies outside 4 kpc. The large velocity dispersion at $|l| < 4°$ is due in part to the rotation of the nuclear disk, and for an equally important part to features with large noncircular motions. The latter are similar to what has been found in H I. However, the observations of the molecules have made it possible to study these motions in greater detail and nearer to the center.

5.3 *Distribution and Motions of Molecules in the Nuclear Disk. CO Observations*

The first somewhat complete survey of CO emission from the central region was made by Scoville, Solomon & Jefferts (1974), who gave an instructive longitude velocity contour diagram at $b = 0°$. Recently these observations have been greatly extended by Bania (1977) and by Liszt et al. (1977), with a much finer grid and at various latitudes. The observations at $b = 0$ are illustrated in Figures 19 and 20. In the part within $|l| = 1°$, from Liszt et al., the latitude is $-2'45''$ and the longitudes are counted from the galactic center position at $l = -3'20''$, $b = -2'45''$. Observations were also made at other latitudes, up to 12′ (35 pc). No important new features appeared in these, but they indicate the thinness of the CO layer: the half-width of the main part appears to be between 30 and 40 pc.

Although there is some CO emission over the entire extent of the H I nuclear disk, the high-intensity radiation is limited to the interval from roughly $l = -1.^\circ5$ to $+2.^\circ0$. The strength of this concentration is illustrated in Figure 21, showing the intensity integrated over all velocities as a function of l. Tentatively we infer from these data that the molecular clouds are almost wholly concentrated in a region within $R \simeq 250$ pc. From this figure Bania estimated the total H_2 mass within $R = 300$ pc at (*a*) $3.5 \times 10^8 M_\odot$; the alternative value (*r*) is $3.5 \times 10^7 M_\odot$ (see p. 325).

The broad strip of low-velocity emission in Figure 19 is from gas outside the central region. At $v \sim -50$ km sec^{-1} the 3-kpc arm is visible in emission as well as absorption. Absorption is also observed in the low-velocity gas. Evidently there must be clouds of extremely low temperature.

The strongest emission is very asymmetrical relative to the center, by far the greatest part lying at positive longitudes. Within this large region of high intensity two concentrations may be noted: one around 0′ longitude and one around 45′. Both extend over about 15′ (40 pc) in longitude as well as in latitude, and comprise a wide range of velocities, from about 0 to 80 km sec^{-1} for the first, and 40 to 100 for the second. The first has been referred to as the +40 km sec^{-1} cloud, or as the Sgr A complex; it may, or may not, be connected with the nuclear source of the Galaxy. The second contains the unique H II region Sgr B2; we shall therefore refer to it as the Sgr B2 complex. It is possible that it has a weak counterpart on the opposite side of the center, around $l = -30'$, $v = -100$ km sec^{-1}. Around $v = +25$ and $v = +90$ there appear to be "bridges" between the Sgr A and Sgr B2 complexes.

All of this gas has velocities in the direction of galactic rotation, but its average radial velocity is much smaller than would be expected on the basis of the rotation curve discussed in Section 2.4 for clouds with circular motion around the center. If we accept the rotation curve derived from the H I observations, we must conclude that the CO clouds have very large systematic deviations from circular motion. The low radial velocity cannot be explained by situating them far from the "sub-

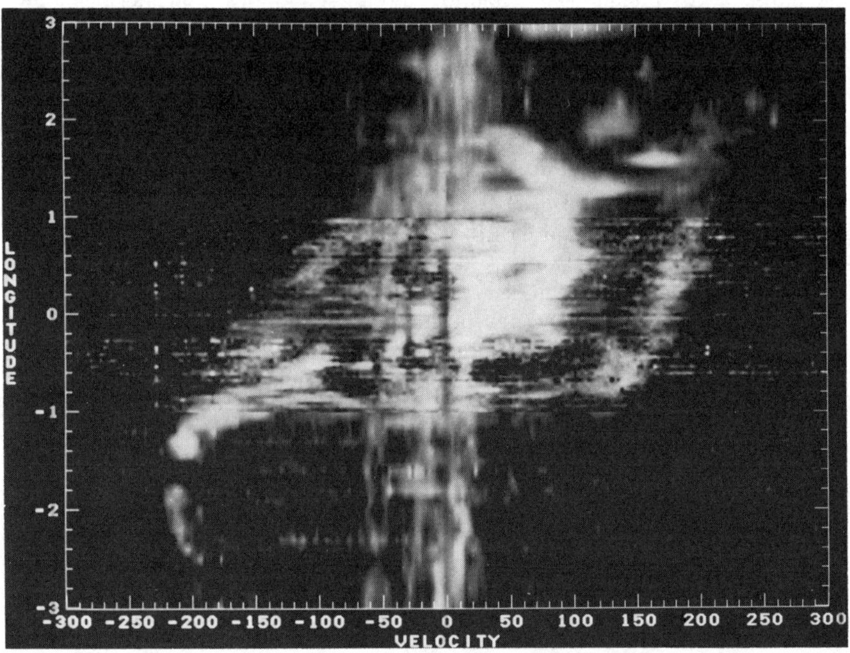

Figure 19 CO distribution in l and v at $b = 0°$ between $-3°$ and $+3°$ longitude, from observations with the 11 m NRAO telescope on Kitt Peak. Beam width 65″. Spacings 12′ between $|l| = 2°$ and 3°, 6′ between 1° and 2°, 2′ for $|l-3'|$ between 10′ and 60′, and 1′ for $|l-3'| < 10'$. The part between 359° and 1° is from a study by Liszt et al. (1977). (Bania 1977).

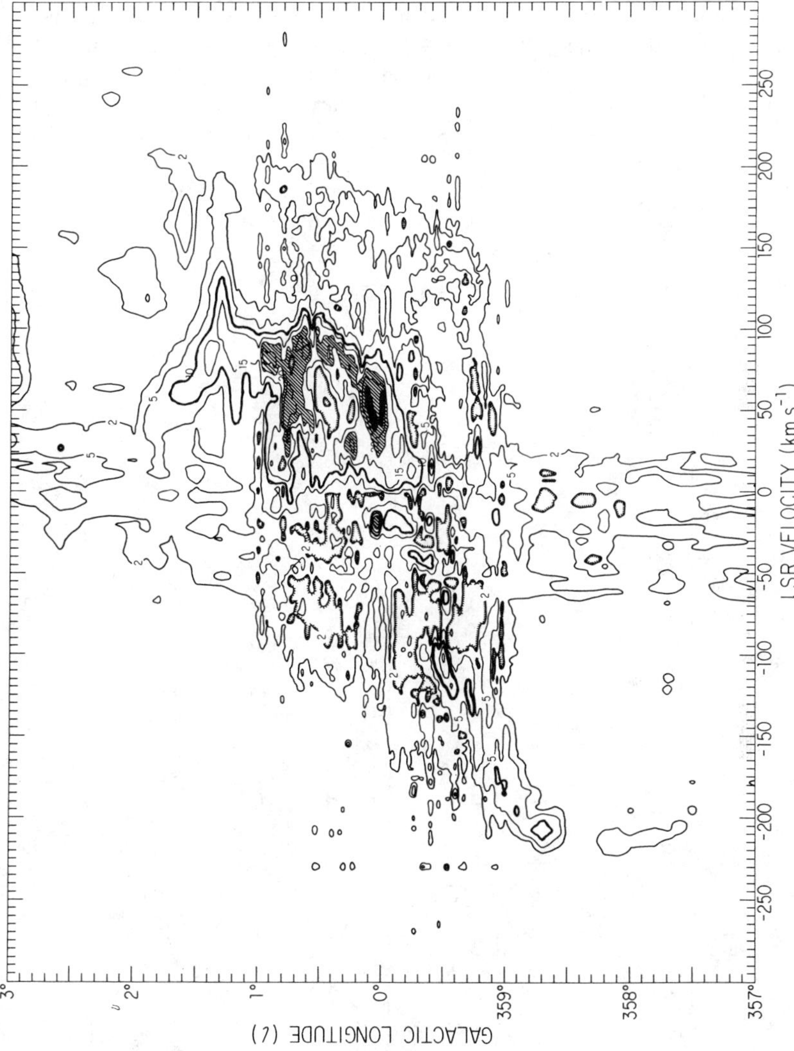

Figure 20 Longitude velocity contour map of ^{12}CO emission at $b = 0$ from the same data as Figure 19. Contour levels are drawn at $T^* = 2, 5, 10, 15, 20, 25, 30$ K. Heavy line: 10 K contour; hatched areas are above 20 K; black is above 30 K.

central points" because, as we have argued above, they must in all likelihood lie within about 250 pc from the center.

The most remarkable features in Figure 19 are the almost straight ribbons at large negative as well as large positive velocities, extending from $l = -50'$ to $l = +60'$. At $l = 0$ their velocities are -125 and $+165$ km sec^{-1}, respectively. From evidence to be discussed below it is clear that the -125 km sec^{-1} feature lies on our side of the center, while the $+165$ km sec^{-1} ribbon lies behind it. Both features are thus moving away from the center. Their slopes in the v/l diagrams indicate that they have also transverse velocities. These are in the direction of the galactic rotation, but lower by factors of ~ 4 and ~ 8 respectively than the circular velocities at $R \sim 200$ pc.

A feature at $l \sim 3\overset{\circ}{.}2$, $v \sim 50$ to 150 km sec^{-1} merits special mention because of its high intensity and remarkably large spread in velocity. It might be connected

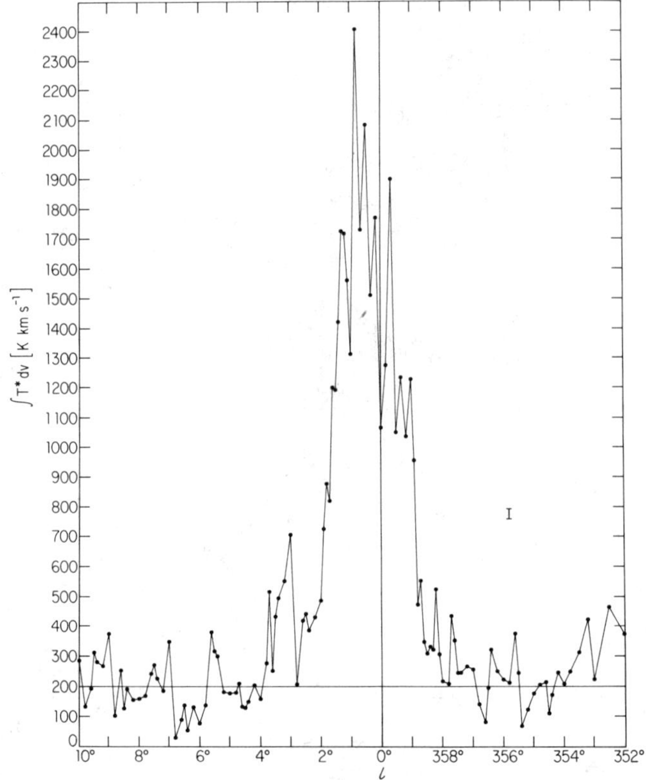

Figure 21 CO emission at $b = 0'$ integrated over all velocities as a function of galactic longitude. Ordinates are $\int T^* dv$, T^* being proportional to the antenna temperature (see text) (Bania 1977).

with the large 100-μ infrared source at the same position. Half of it is visible at the upper edge of Figure 19.

5.4 *Distribution and Motions of Molecules in the Nuclear Disk. OH and H_2CO Observations*

Figure 22 shows the absorption spectrum of Sgr A in OH (full drawn curve) and H I (dashed curve) (Bolton et al. 1964). There are two extremely deep and wide absorption bands in the OH radiation, centered on the velocities of the -125 km sec^{-1} feature and the $+40$ km sec^{-1} complex discussed above in connection with the CO observations. The fact that they are here seen in absorption proves that they must lie in front of Sgr A, or at least of an important part of Sgr A. Their great widths in velocity are probably due to the fact that they consist of various fairly large clouds or sheets with different velocities rather than to the velocity dispersion of a swarm of compact cloudlets. For, as Scoville, Solomon & Penzias (1975) have remarked in their discussion of Sgr B2, the great depths of the absorption lines are hard to explain unless the "clouds" would at each velocity cover the entire

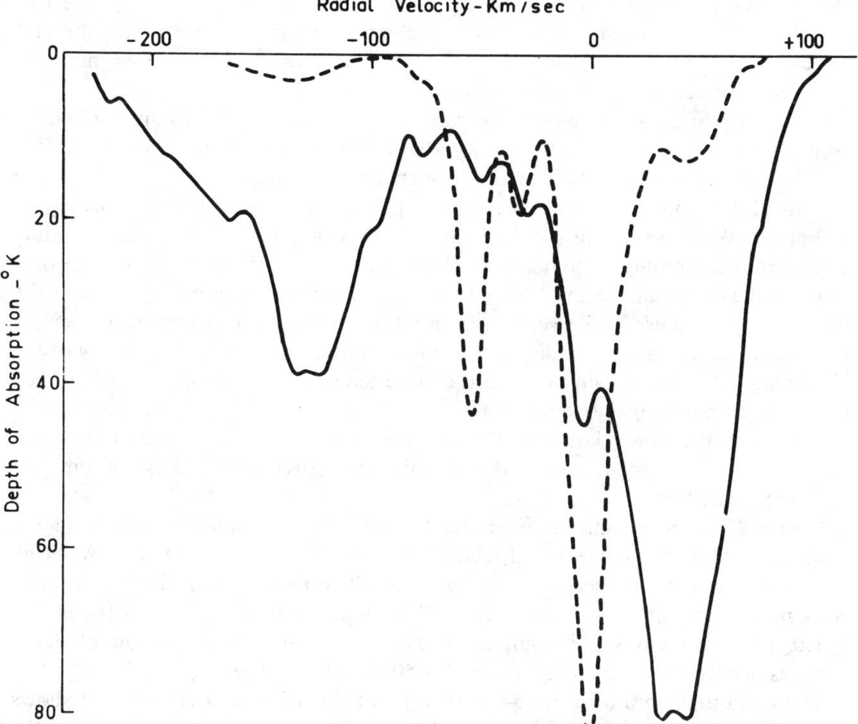

Figure 22 Absorption profile of Sgr A observed with the Parkes telescope (Bolton et al. 1964). Full curve: OH; dashed curve: H I.

continuum source. To effect this with randomly moving small clouds would require such large numbers that they would collide on a time scale short compared to the presumable ages of the features concerned. The wide dispersion has been observed equally in H I and formaldehyde in the $+40$ km sec^{-1} feature. From the extensive observations of H I, OH, and H_2CO made by Sandqvist (1970) with the 40-m Green Bank telescope at high-velocity resolution, the half power width of the $+40$ km sec^{-1} feature may be estimated at 50 km sec^{-1}. The -130 km sec^{-1} band in OH gives 48 km sec^{-1}.

The great difference between the dashed and continuous curves in Figure 22 illustrates once more the enormous concentration of molecules relative to H I in these features close to the center. There is relatively little OH absorption at zero velocity, illustrating on one hand that the molecules in the 10-kpc path between the Sun and the center are relatively scarce and on the other hand that the molecular gas *within* the central region is largely, if not entirely, concentrated in complexes with large radial motions.

It is also interesting to compare the conditions in the 3-kpc arm at -53 km sec^{-1} with those in the two arms in the nuclear region. Figure 22 shows that in the former the opacity in OH is considerably smaller than that in H I. In the two nuclear arms the optical depth of OH is clearly very much *larger* than the H I optical depth, so that the abundance of H I relative to that of OH must be much smaller than in the 3-kpc arm.

The first and most extensive observations of molecular clouds in the central region were of hydroxyl (Bolton et al. 1964; Robinson & McGee 1970; McGee et al. 1970). They were made in its absorption line at a wavelength of 18 cm. Strong absorption was found between $-1°$ and $+2°$ longitude. According to Robinson (1974), weaker absorption can be traced to $l \simeq 4°$, but notwithstanding considerable searching none has been found between $-1.°5$ and $-3°$. The absorptions as measured in the brightness temperature depend on the brightness of the background. Because of the very large variation of this background in the central region they do not give anything like a representative picture of the quantity of the absorbing gas. To a certain extent this can be remedied by plotting what has been called the apparent opacity defined as $\tau_a = -T_{line}/T_c$, where T_{line} is the temperature measured in the line relative to that outside, and T_c is the temperature of the source behind the cloud, supposed to be high compared with the excitation temperature of the line.

Figure 23 gives a contour diagram at $b = 0°$ of the apparent opacity of the 1667 MHz OH line as measured recently in Jodrell Bank (Cohen & Few 1976). The 1665 line was also measured, and gives a full confirmation of the 1667 results. Striking features are the row of "clouds" at high negative velocity between $l \sim \pm 1.°0$, the $+40$ km sec^{-1} complex at $l = 0°$, and the concentration of dense features between $l \sim 0.°5$ and $2°$ and $v \sim -50$ to $+100$ km sec^{-1}.

The contour diagram of the 6-cm absorption line of H_2CO (Figure 24) shows the same features, and in addition, the 3-kpc arm. The observations, which give the apparent opacity at $b = -12'$, are due to Scoville & Solomon (1973). The beam width is $6.'6$.

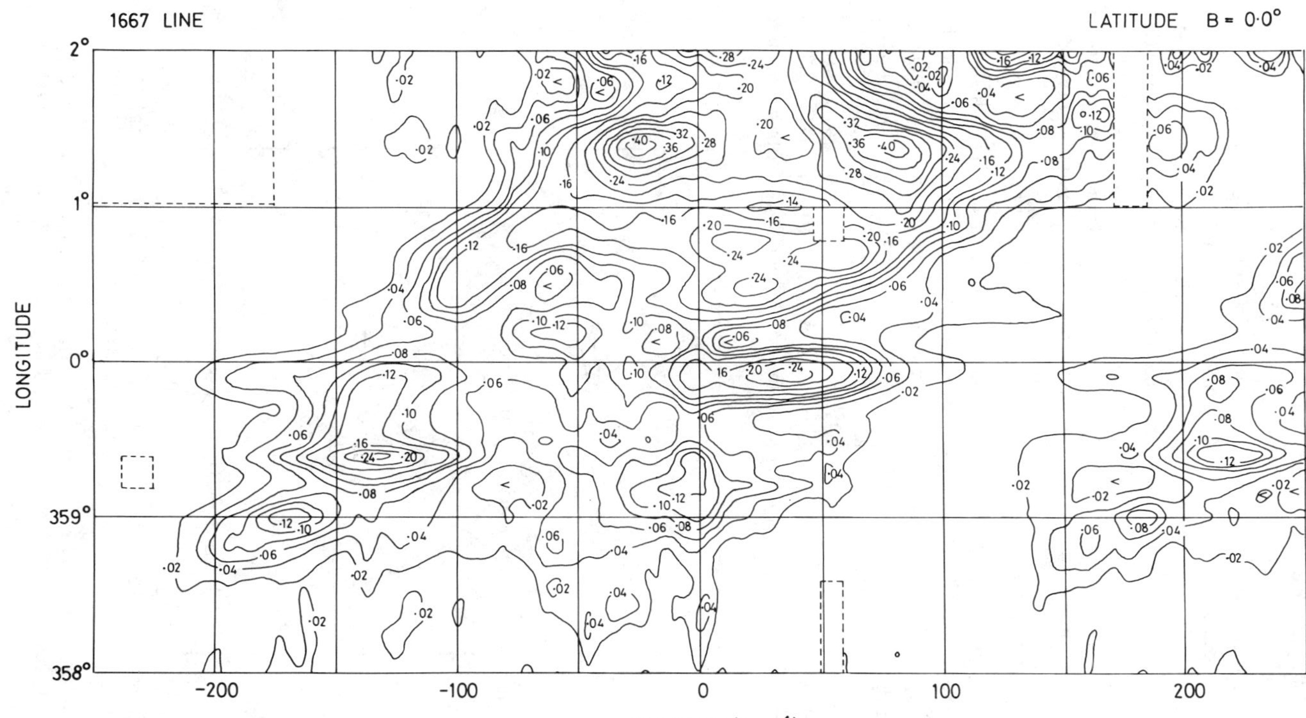

Figure 23 Longitude velocity map of apparent opacity of the 1667 MHz line of OH at $b = 0\overset{\circ}{.}0$ from observations with the Jodrell Bank Mark I telescope. The wing on the right-hand side is due to the 1665 MHz line; the real velocities for this wing are 350 km sec^{-1} lower than indicated at the bottom. The half power beam width is 11′, the velocity resolution 6.1 km sec^{-1} (Cohen & Few 1976).

The string of negative-velocity clouds seen in both pictures coincides essentially with the negative-velocity feature of the CO emission. The fact that it is seen in absorption against the continuous radiation of the central radio sources over the entire longitude range from 359°.0 to 1°.4 shows that it lies in front of these, and is moving outward. In Figure 23, which extends to sufficiently high positive velocities, no trace is seen of the +165 km sec^{-1} feature. This holds for both the 1667 MHz and the 1665 MHz line: within 1° from the center there is no absorption above +100 km sec^{-1}. It is thus evident that the +165 km sec^{-1} feature lies behind all the brighter continuum sources, including Sgr B2, and that it is moving outward.

5.5 *The Expanding Molecular Ring*

It has been suggested by various authors (Kaifu, Kato & Iguchi 1972 from OH observations, Scoville 1972 from H_2CO data) that the string of negative-velocity clouds would form part of an expanding ring around the center. The suggested locus of such a ring is sketched in the full-drawn curve in Figure 24. The ring would have a radius of about 250 pc, an expansion velocity of ~140 km sec^{-1}, and a rotation of 50 km sec^{-1}. The velocity half-width is estimated at 41 km sec^{-1} by Cohen & Few (1976), who give a column density of 8×10^{15} T_s OH molecules per cm^2 (where T_s is the excitation temperature in °K, which is probably around

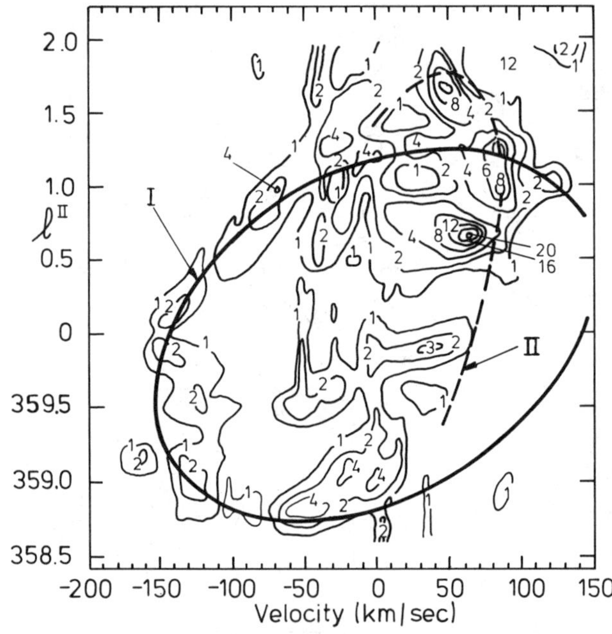

Figure 24 Contours of apparent H_2CO absorption averaged for $b = -2'$ and $-12'$. Beam width 6'.6, velocity resolution 1.6 km sec^{-1}, 6' grid in l (Scoville & Solomon 1973). Full curve: locus of expanding molecular ring as suggested by Scoville (1972).

15 K). The mighty Sgr B2 complex might belong to the "ring." But its most convincing part is the string of negative velocities. In the OH and H_2CO data there is no good evidence for the existence of the opposite half. The half that *is* observed is interesting enough, in showing that the mechanism which has produced these apparently "expanding" cloud complexes must have worked in a coordinated manner over a sector of about 180°. It is therefore unlikely that it would be connected with supernova explosions.

That there is no evidence in the OH and H_2CO absorption for the existence of the opposite half may be due to the fact that this would lie behind the majority of the continuum sources. The earlier CO observations did not show clear evidence for either the front or the rear side of the hypothetical ring, no doubt partly due to the fact that the front side is centered about 6′ below the galactic plane. However, as described in Section 5.3, the recent more complete observations by Liszt et al. at $b = -2'45''$ (Figure 19) show clearly the more or less continuous ribbon at negative velocities as well as a very similar ribbon at high positive velocities. It is tempting to believe that these two features have resulted from one explosive event, and show the "ring" in a somewhat more complete form. Bania (1977) has approximated the features by a ring of radius 190 pc, expanding at an average velocity of 150 km sec^{-1} and rotating with a velocity of 65 km sec^{-1}. The observed emission is somewhat asymmetrical, the positive velocity side having a 20 km sec^{-1} higher expansion velocity than the average of 150 km sec^{-1}, while the negative velocity side has a 20 km sec^{-1} lower expansion velocity than the average, but such asymmetries are to be expected in expulsive phenomena, especially since there must have been considerable interaction with ambient gas. In this case they are all the more probable in view of the glaring asymmetry in longitude of the denser CO at smaller velocity. Because of its greater completeness Bania's model of the ring appears preferable to that derived earlier from the OH and H_2CO observations.

Atomic hydrogen in the ring has been observed by Cohen (1977). From a comparison of the observed profile with emission profiles he found an average excitation temperature $T_s = 92$ K. The cool molecular clouds are therefore embedded in a much hotter H I medium. This medium has a thickness in the z direction of about twice that of the molecules.

In the same article Cohen has given data illustrating the systematic increase of the ratio of the optical thickness of OH and H_2CO to that of H I with decreasing distance from the center. With optical thicknesses integrated over all velocities the ratio changes by more than a hundred between $R = 10$ kpc and the nuclear region.

5.6 *The +40 km sec^{-1} Feature*

Of the two wide absorption bands in the spectrum of Sgr A, one has been interpreted as being caused by the expanding ring discussed in the preceding section. The other, though resembling the first in its width (both have a half-width of roughly 50 km sec^{-1}), appears to be connected with a rather different object. Contrary to the ring the +40 km sec^{-1} feature is strongly concentrated in a small region, near the direction of the center. As may be seen in Figure 20 it contains

Table 3 Observations of the +40 km sec^{-1} feature near Sagittarius A

Gas	Authors	Region l	Region b	Method	Beam
HI	Sandqvist (1970, 1973, 1974)	6′ diam.		lunar occ.	40″ × 70″
HI	Schwarz, Shaver & Ekers (1977)	8′ diam.		ap. synth.	25″ × 120″
OH	Sandqvist (1970, 1973, 1974)	6′ diam.		lunar occ.	40″ × 70″
OH	Bieging (1976)	10′ diam.		ap. synth.	3′.25
H_2CO	Sandqvist (1970, 1973, 1974)	6′ diam.		lunar occ.	40″ × 70″
H_2CO	Gardner & Whiteoak (1970)	359° to 1°	0′ to −12′	pencil beam	4′
H_2CO	Scoville, Solomon & Thaddeus (1972)	359°.4 to 2°.2	−2′	pencil beam	6′
H_2CO	Scoville & Solomon (1973)	358°.5 to 1°.0	−2′, −12′	pencil beam	6′
H_2CO	Fomalont & Weliachew (1973)	5′ diam.		ap. synth.	40″ × 120″
H_2CO	Whiteoak, Rogstad & Lockhart (1974)	4′ diam.		ap. synth.	20″ × 40″
CO	Solomon et al. (1972)	359°.7 to 2°.8	−2′	1′ to 2′ grid	1′
CO	Sanders & Wrixon (1974)	−10′ to +10′	0′	1′ grid	1′
CO	Liszt, Sanders & Burton (1975)	−4′ to 0′	−5′ to −2′	1′ grid	1′
CO	Liszt et al. (1977)	−60′ to +60′	−7′ to +1′	1′ to 5′ grid	1′
CO	Bania (1977)	352° to +10° [a]	0′	6′ grid	1′
HCN	Fukui et al. (1977)	~ −12′ to +24′	~ −10′ to +8′	2′ grid	2′

[a] For the part between 359° and 1° the data were taken from Liszt et al. (1977). The 6′ grid quoted is for $|l|$ between 1°.0 and 2°.0; for the remainder of the survey it was 12′.

the strongest CO emission observed anywhere in the central region. It has therefore been generally thought that it might be connected with the nucleus, and has consequently been the subject of numerous investigations, an overview of which is given in Table 3.

The center of the feature, as inferred from Figure 20, lies practically at $l = 0'$. At $+40$ km sec^{-1} it has a half-width of about $15'$ (45 pc) in l. It extends from $\sim +25$ to $\sim +75$ km sec^{-1} in velocity. Between $+70$ and $+100$ km sec^{-1} a sort of bridge connects it with the other main concentration of CO, which lies around Sgr B2, at $l \sim 0^\circ.7$.

A rough outline of the "cloud" can also be obtained from the absorption data, cf Cohen & Few (1976) (Figure 23), Sandqvist (1974), and Bieging (1976) (Figure 25). From the latter we find the OH extending over an area of half-widths of about $12'$ in l and $6'$ in b, centered at $l \sim -3'$, $b \sim -6'$; the longitude coincides with that of the galactic center, the latitude is $3'$ below it.

All maps that have sufficient resolution show that there is considerable internal structure. This is seen very clearly in the high resolution charts of Whiteoak, Rogstadt & Lockhart (1974). These authors pointed out that the most extensive absorption is found in the vicinity of the eastern component of the central radio source (Sgr A East), while no absorption is seen against Sgr A West (which is supposed to be the real nucleus of the Galaxy; see Section 6). This has been taken as an indication that the main part lies behind the nucleus, and may have been expelled, as tentatively sketched in Figure 26, in which it has been supposed that

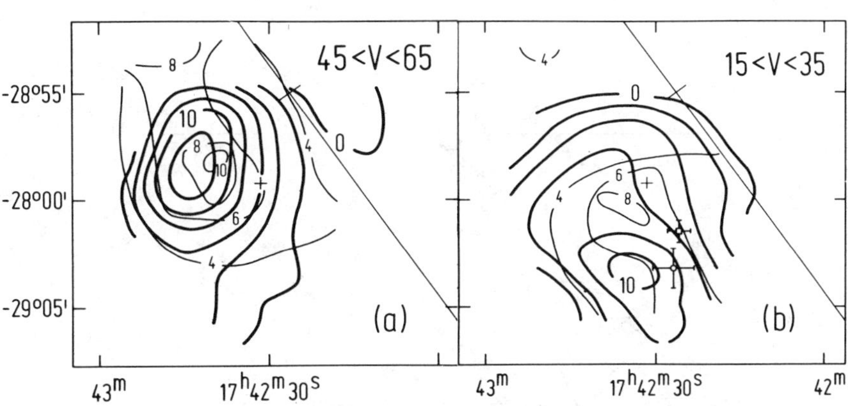

Figure 25 (*a*) Comparison of OH and CO distributions in the direction of Sgr A, integrated over the velocity range 45 to 65 km sec^{-1}. *Bold contours:* $N(\mathrm{OH})T_s$ in units of 3.2×10^{14} cm^{-2} K^{-1}. *Light contours:* ^{12}CO peak line antenna temperature in K, from Solomon et al. (1972). The cross marks the 1.67 GHz continuum peak. The straight line is the galactic equator, on which zero longitude has been marked. (*b*) As for Figure 25a, for the velocity range 15 to 35 km sec^{-1}. Positions, with error bars, of peak emission from 2 mm H_2CO and 12 mm NH_3 lines are indicated by the open square and circle, respectively (Bieging 1976).

Sgr A East lies behind the center. It should be emphasized, however, that the distribution of the H_2CO as indicated by Whiteoak et al. is quite irregular, and shows large variation in optical depth and in velocity over the 4′ field they surveyed. The small absorption in the direction of the Western component might well be accidental, and cannot therefore be taken as a proof that the object lies *behind* Sgr A West.

The alternative is that the +40 km sec^{-1} gas lies in front of both central sources and is moving towards them. Some investigators have suggested that it is part of an extended positive velocity mass covering a large sector of the central region, and that the coincidence of the dense concentration with the position of the central source is purely accidental. However, such a model is contradicted by what we know about the gas between +70 and +100 km sec^{-1} that forms the bridge with Sgr B2, and by the high-velocity part of Sgr B2 itself. This gas, which the CO observations show to be quite dense, is all but invisible in the OH and H_2CO absorption maps; it must therefore be situated behind all the continuum sources in this sector, and must be moving away from the central region, just like the ring at +165 km sec^{-1}.

Considering the evidence presently available it appears probable that the +40 km sec^{-1} cloud is actually related with the center and that it may have been expelled. The fact that no specific peculiarities are observed at the center itself may be because the molecules have formed only at some distance from the explosive source.

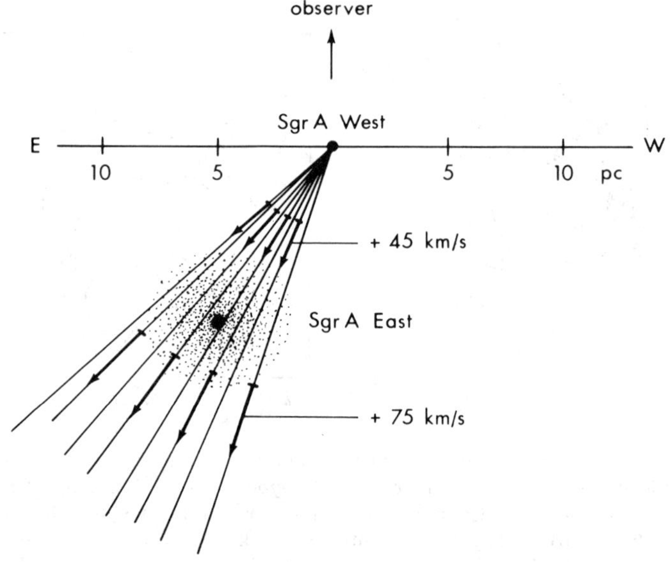

Figure 26 Possible situation of the gas showing CO emission between +25 and +90 km sec^{-1}, and seen in absorption in OH and H_2CO between +30 and +70 km sec^{-1} (Oort 1974b).

We now return to the structural and dynamical properties of the feature. The structures found by Whiteoak et al. have typical dimensions of $2' \times 1'$. Liszt, Sanders & Burton (1975) found striking differences in intensity and velocity structure on scales of 1′ in *l* and *b*. The velocities are high on the Eastern and higher longitude side, and low in the opposite part. A similar tendency may be seen in Figure 25. This may indicate that the feature is rotating, in the same sense as the galactic rotation, but more slowly than the circular velocities in the central part of the nuclear disk (Table 1). Alternatively, the velocity differences might have been caused by asymmetries in the expulsion process.

Interesting new observations have been made in the hydrogen cyanide emission line at 3.4 mm by Fukui et al. (1977) at the Tokyo Observatory. The contours of maximum antenna temperatures in Figure 27 show an elongated structure, with the half-intensity contour extending over roughly 16′ in longitude and about 5′ in latitude. The authors point to the fan shape of the outer contours; they believe that their observations support the ejection hypothesis.

All anomalously moving molecular features found in the nuclear region can also be observed in H I. The +40 km sec^{-1} cloud has been seen by several observers in the 21-cm absorption spectrum of Sgr A. An accurate profile has been given by Sandqvist (1970). In a recent investigation Schwarz, Shaver & Ekers (1977) have derived a column density of about 1.3×10^{22} cm^{-2} for H I. Cohen (1977) found

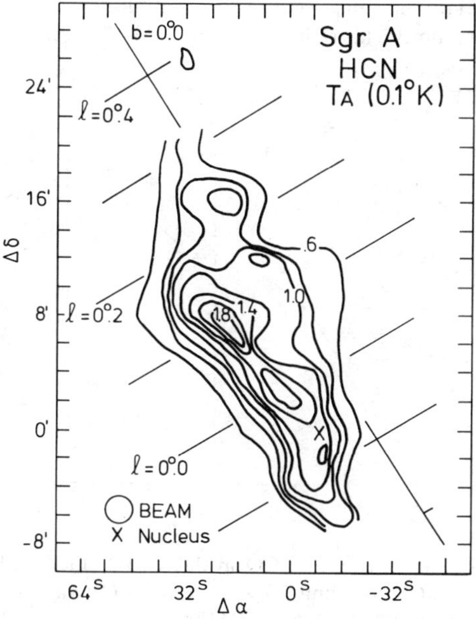

Figure 27 Contours of maximum antenna temperature of HCN emission from the +40 km sec^{-1} cloud as a function of α and δ measured at the Tokyo Astronomical Observatory (Fukui et al. 1977).

the spin temperature of the ambient H I to be 110 K, and the ratio of OH to H I column densities as 4×10^{-5} (if the temperature of the molecules is supposed to be 20 K). Again, the molecular clouds appear to be floating in a much hotter surrounding medium.

So far as I am aware no observable quantity of ionized hydrogen has with certainty been identified with the +40 km sec^{-1} cloud. It differs in this respect from many other dense CO concentrations, and in particular from the nearby Sgr B2 complex.

A rough estimate of the mass of the cloud can be made from the contour map of Figure 20. Assuming that the thickness is half that of the dimensions in the plane, and adding 50% helium, we find the mass within the 20 K contour to be (*a*) $4 \times 10^6\ M_\odot$; the alternative value (*r*) is $4 \times 10^5\ M_\odot$ (see p. 325). If we include the surrounding region up to $\Delta l = 10'$, the mass becomes a factor of 10 higher.

5.7 *Sagittarius B2*

Next to the galactic nucleus itself and the +40 km sec^{-1} cloud, the most striking object in the nuclear region is Sgr B2. It is a strong radio source, containing at least seven compact H II regions that are intrinsically as bright as the Orion nebula (Martin & Downes 1972). It is a unique object in the Galaxy, a uniqueness that may be connected with a possible origin from explosive activity in the nucleus. It has become especially famous because the combination of large mass with high density has made it one of the richest hunting grounds for new species of inter-

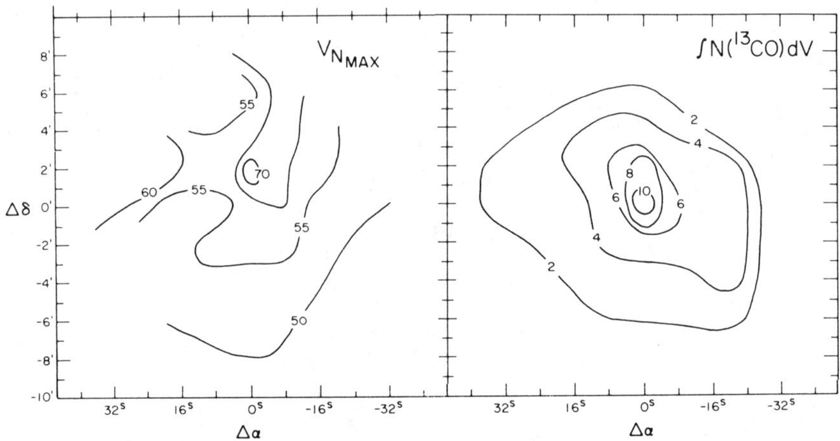

Figure 28 Right: Column density of ^{13}CO in the Sgr B2 cloud with velocities between +30 and +75 km sec^{-1}. Contour unit 5.7×10^{16} cm^{-2}. Origin of coordinates is the OH maser at the center of the cloud at $\alpha_{1950} = 17^h44^m11^s$, $\delta_{1950} = -28°22'30''$. This lies 1^s east and 15″ south of the 2-cm continuum peak. Left: Velocity variation over the cloud. Numbers give the velocity in km sec^{-1} at the maximum column density (Scoville, Solomon & Penzias 1975).

stellar molecules. In Figure 20 the Sgr B2 region stands out as an elongated patch of strong CO emission centered around 40′ longitude and extending from +45 to +100 km sec^{-1} in velocity. The region has been the subject of numerous studies, on interstellar molecules, on the H II regions in its densest part, and on the radiation in the far infrared. It is impossible in the scope of this article to present an adequate discussion. I therefore limit myself to a summary of the properties that appear to be the most relevant in connection with the general structure and dynamics of the nuclear region as a whole. Most of the data cited are from a recent article by Scoville, Solomon & Penzias (1975), in which references to other investigations may be found.

Sgr B2 is an agglomerate of gas and dust clouds; the best-defined part has a diameter of roughly 10′, or 30 pc. Scoville et al. made extensive measurements of ^{12}CO, ^{13}CO, and CS emission over the cloud with a beam of about 1′; they also observed the 2-cm K-doublet of H_2CO in absorption. Because the ^{13}CO line is generally optically thin they could derive CO column densities and determine the structure and space density in the cloud. The column densities are shown in Figure 28, which illustrates the more or less isolated character of the cloud and its strong central concentration. The velocity structure is complicated. At each position the dispersion is large, averaging about 15 km sec^{-1}. Differences in systematic velocity across the cloud are of the same order as the "dispersions" found in a column through the cloud. These observations give "strong evidence that structured flows in the cloud are at least as large as random motions" (quoted from the article by Scoville et al.).

From their ^{13}CO intensities the authors derived a mass (*a*) of $3 \times 10^6\ M_\odot$ for the 10′ diameter cloud, with about 20% in the core; the alternative mass (*r*) is $3 \times 10^5\ M_\odot$.

They point out that their mass is just about equal to the mass that according to the virial theorem could keep the cloud together in the face of its internal motions. However, as the observed velocity distribution over the face of the cloud shows no sign of its having been kept together by its own gravitation, I believe that it is more likely that the agreement of the two mass determinations is fortuitous, and that the velocity differences are not governed by the cloud's mass.

We return to this problem as well as to that of the motion of the entire feature in Section 5.8.

Scoville et al. also discussed the relation between gas and dust in Sgr B2 and came to the conclusion that the two constituents may well be in approximate thermal equilibrium, at a temperature between 20 and 30 K.

Detailed maps of the distribution and temperature of the *dust* have been published by Harvey, Campbell & Hoffmann (1976) at 53, 100, and 175 μ, and by Westbrook et al. (1976) at 1 mm wavelength. From the surface brightness at 1 mm the latter derived a mass of $10^5\ M_\odot$ for the central 1′.

5.8 *Masses and Dynamics*

If we accept the gravitational field derived from the H I nuclear disk (Table 1), the dynamics of most molecular clouds in the nuclear region must be far from equilibrium conditions. For, as pointed out in Section 5.3, their velocities are generally much lower than the rotational velocities inferred for the H I disk. The bulk of the molecular clouds must therefore have been expelled, presumably from the nucleus.

Fairly direct evidence for such expulsion is shown by the expanding ring and the $+40$ km sec^{-1} cloud.

In order to reach the "ring" at $R = 190$ pc with a radial velocity component $\Pi = 150$ km sec^{-1}, an object moving unimpeded through the interstellar gas would have to start from the nucleus at a velocity of 692 km sec^{-1}. This is derived from Table 1. Should there be any invisible point mass in the center the initial velocity should of course be higher.

Such a picture of clouds expelled at high velocity and decelerated solely by the gravitational field cannot, however, be realistic. Even if compact clouds of sufficient density to move freely through the surrounding medium could have been formed—which is not impossible—no plausible expulsion model can be devised in which all *observed* velocities are nearly an order of magnitude smaller than the required ejection velocities.

We are thus led to the conclusion that the dynamics of the expelled molecular clouds are largely determined by forces other than gravitation. Their motions have presumably been braked by interstellar "friction," and they must therefore consist for an important part of swept-up gas. That the molecular features contain much swept-up gas is confirmed by the fact that they possess angular momentum, although less than what would correspond to the circular velocities listed in Table 1. It is evident that if the friction is taken into account the initial velocities of the clouds which caused the formation of the "ring" must have been considerably higher than the 692 km sec^{-1} estimated above.

The acceleration of the large molecular complexes may have been a gradual process, due to the outward pressure exerted on existing disk gas by the collisions of many small, presumably ionized, clouds ejected at high velocity. A mechanism of this kind is directly suggested by what is observed in Seyfert galaxies.

The *main mass* of the CO may be ascribed to another eruptive period in which the accelerated clouds did not move out as far as the ring, and may at present be near their maximum distance. Liszt et al. (1977) have drawn attention to a pattern extending from about $\Delta l = -50'$, $v = -160$ km sec^{-1}, through $\Delta l = -30'$, $v = -100$, across $v = 0$ at $\Delta l = 0$ to $\Delta l = +30'$, $v = +90$, which looks like a disk rotating at a constant angular velocity. This may well be part of a very anisotropic ring, rotating, at $|l| \sim 30'$, at roughly half the circular velocity given in Table 1.

The $+40$ km sec^{-1} complex, extending from the nucleus to about $R = 50$ pc, may be the most recent formation, perhaps no more than half a million years old. The apparent absence of H II regions in this exceptionally dense structure might be considered as a confirmation of its youth.

The time scales for the other features are uncertain. If the outer ring is on its first journey outward it would be about one million years old. But if, as is quite conceivable, it has moved back and forth, it could be five to ten times older. The age of the remainder of the CO is hard to estimate, but it is unlikely to be much greater than the time of revolution—at this distance about three million years—because otherwise it could hardly have remained so far from an equilibrium stage as it evidently is.

As in the case of the expanding H I features the question arises how the H I nuclear disk can have survived these eruptions. On the positive longitude side it may indeed have been swept away from the region inside 200 pc where the dense CO is observed (cf Section 2.4). On the other side, where the rotating disk *is* observed down to $R \sim 50$ pc, there is little CO, except for the expanding ring. This latter has apparently been ejected at an angle with the plane, for it is inclined at some 10°.

An intriguing problem is presented by the large *masses* of the molecular features and the rate of ejection needed to form them. The total hydrogen mass within 300 pc is estimated to be between 3×10^7 and 3×10^8 $M_\odot$ (Section 5.3); with 50% helium mass added the total gaseous mass becomes (*a*) 5×10^7, (*r*) 5×10^8. From integrations of Figure 20 the mass of the expanding ring is roughly estimated to be between 3×10^6 and 3×10^7 $M_\odot$, and that of the +40 km sec^{-1} cloud insofar as it lies within the $T^* = 20$ contour to be between 10^6 and 10^7 $M_\odot$.

Even if only 10% of the mass consisted of gas expelled from the nucleus, the rest having been swept up, and even if we suppose that the expulsion of this mass of gas would occur only once in 10^7 years, the nucleus would still have to expel on the order of a solar mass per year in high-velocity clouds; and this is certainly an underestimate, because it does not take account of the gas expelled at an angle with the disk.

The gas produced by evolving stars in a core of 1-pc diameter is many orders of magnitude less than what is required (cf Section 4.4). Additional gas might be produced by stellar collisions or tidal disruptions by a massive central black hole (cf Salpeter 1964, Lynden-Bell & Rees 1971, Frank & Rees 1976, and references given therein), but it is highly unlikely that this could produce the amount needed. Nor can infalling gas from the halo or from outside the Galaxy supply more than a minor fraction of what is needed.

The only way out that I see is a *circulation* of the clouds. As we have seen, the molecular agglomerates have transverse velocities considerably smaller than the circular velocities. They will therefore fall back towards the center. The minimum distances R they can reach are mainly determined by their angular momenta; they may average a few tens of parsecs. The clouds cannot get into the core proper unless they get rid of their angular momenta. It is not at all clear how this could be accomplished. Magnetic fields cannot be of sufficient strength. Though the fast differential rotation—revolution periods ranging from ~40,000 years at $R = 1$ pc to ~400,000 years at $R = 10$ pc—is likely to set up strong turbulence, it is doubtful whether this would suffice to effect the rapid redistribution of angular momentum required.

An attractive alternative idea, suggested in a discussion with E. N. Parker and H. G. van Bueren, is that the gas would intermittently contract into a dense ring at $R \sim 20$ pc and be expelled by an outburst of *radiation* and relativistic particles

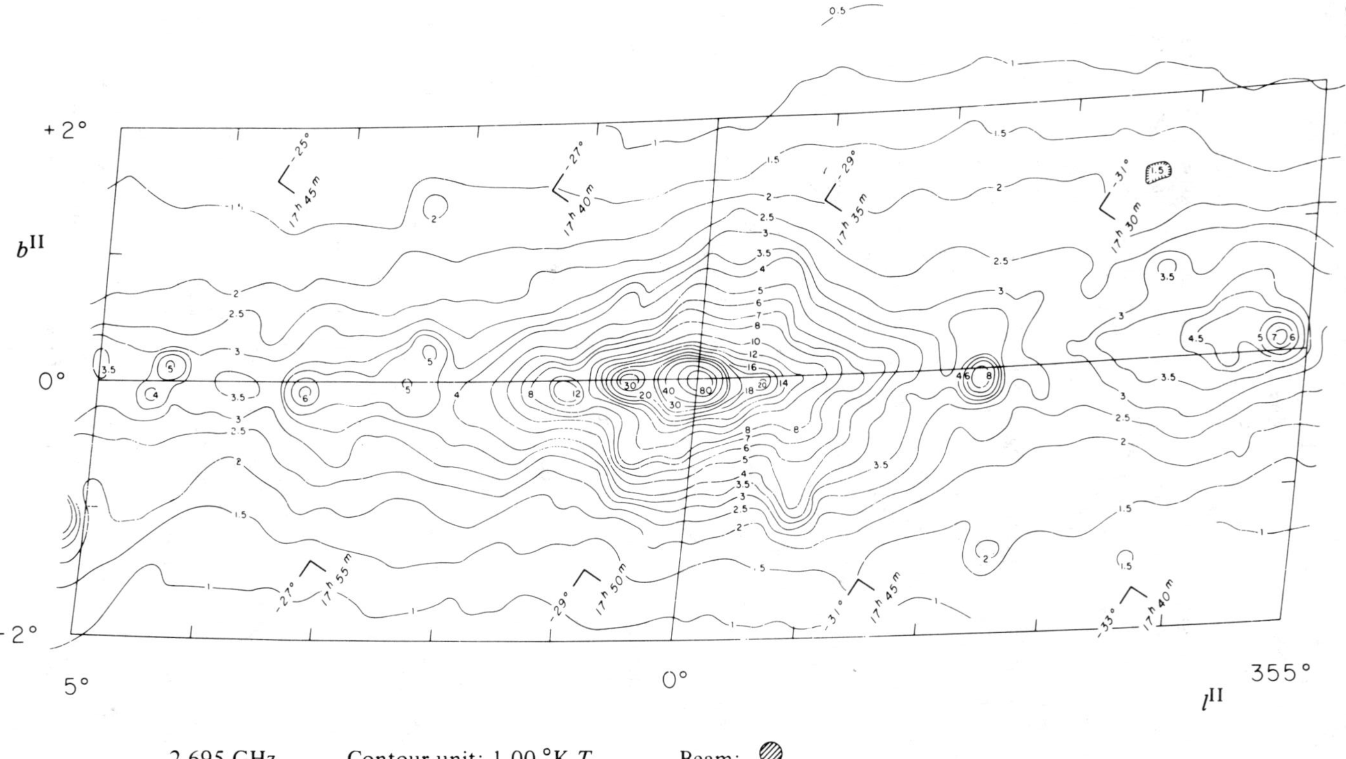

Figure 29 The galactic center in nonthermal radiation. From a survey at 11-cm wavelength made with the 140-ft antenna of the National Radio Astronomy Observatory with a beam width of about 11′. At 11 cm the nonthermal radiation dominates (Altenhoff et al. 1970).

from the nucleus. The energy required for the outward impulse (10^{55} or 10^{56} ergs) and the intervals of the order of 10^6 years would be comparable to what is observed in some radio galaxies.

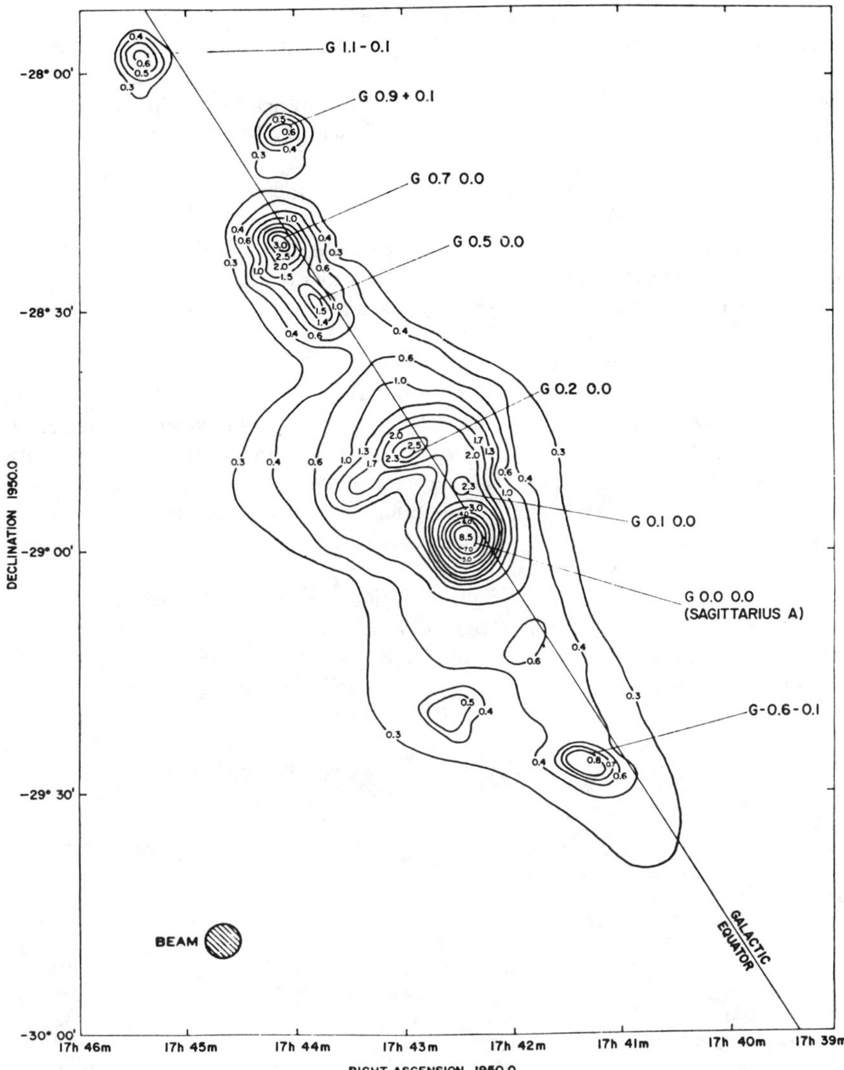

Figure 30 Distribution of thermal radiation around the galactic center. From a survey at 3.75 cm by Downes & Maxwell (1966) with the 120-ft Haystack antenna; beam 4′.2; contours represent antenna temperature.

6 EMISSION AT RADIO FREQUENCIES AND IN THE 12.8-μ NEON II LINE

The central region emits a mixture of thermal and nonthermal radio waves. Figures 29 and 30 show the general distribution of the two kinds of radiation. The first, at 11 cm, gives a good impression of the general distribution of the synchrotron radiation outside the galactic plane. At the lowest latitudes it is mixed with thermal emission. For $|b| < 0^\circ.25$ a purer picture of the nonthermal component may be obtained from a map at 75-cm wavelength made with the Molonglo Cross telescope (Little 1974).

Figure 30, at 3.75-cm wavelength, shows mainly the thermal emission. The thermal character is indicated by the spectral index as well as by the observation of recombination lines. Moreover, at longer wavelengths, the central body of ionized gas can be seen in absorption against the nonthermal radiation (cf Mills 1956). The observations with Mills' cross at λ 3.5 m show a striking dip extending to approximately 1° from the center.

The outer diameters are roughly 5° × 2° (900 × 350 pc) for the nonthermal, and $1^\circ.5 \times 0^\circ.5$ (260 × 90 pc) for the thermal radiation. Both are strongly concentrated towards the center. In both domains the contours have significant irregularities, on scales of 20′ and less. In addition there are a number of discrete sources; the strongest have been marked by their galactic coordinates in degrees preceded by G, which is the usual way of designating sources and concentrations in the central region. A few are fore- or background objects, but all bright sources with $|l| < 2^\circ.5$, with the possible exception of G 0.87+0.10, lie in the central domain. They are listed in Table 4, the data of which are from an article by Mezger, Churchwell & Pauls (1974). For Sgr A, which will be extensively discussed in Section 6.3, only the nonthermal component is given. The electron density for a homogeneous source is designated by n_e, E is the emission measure, and L_c the number of Lyman quanta required for the ionization.

Table 4 Radio sources near the galactic center (Mezger, Churchwell & Pauls 1974)

Source	$S_{5\,\mathrm{GHz}}$ (Jy)	α	V_{lsr} (km sec^{-1})	n_e (cm^{-3})	E (10^5 pc cm^{-6})	L_c (10^{49} sec^{-1})
G 357.66−0.09	15.7	−0.4				
G −0.53−0.06	15.4	−0.7				
G −0.05−0.05 (Sgr A)	190	−0.7				
G 0.07+0.01	80	−0.1	−41	105	2.8	71
G 0.18−0.03	157	−0.1	−26	76	2.3	139
G 0.53−0.03	36	−0.2	+46	96	1.9	31
G 0.67−0.02 (Sgr B2)	48	−0.2	+63	201	5.6	42
G 0.87+0.10[a]	8	+0.2				
G 1.15−0.06	10	0.0	−22	98	1.3	9
Region $1^\circ.5 \times 0^\circ.5$[b]	410			16[b]		360

[a] Uncertain whether this source belongs to the central region.
[b] Half widths 61′ × 20′ in $l \times b$; a homogeneous spheroid of this size would produce the observed radiation if $n_e = 16$.

All central region sources lie in an extremely narrow layer, between $-0.^{\circ}09$ and $+0.^{\circ}01$ latitude. Their average distance from the true plane of symmetry, at $b = -3'$, is $\pm 1.'4$, or 4 pc.

6.1 *Thermal Emission*

The thermal radiation comes in part from giant H II regions. These lie all at positive longitudes, thus showing the same asymmetry in distribution as CO. The strongest appear to be directly related with molecular clouds: Sgr A is, as we have seen, probably connected with the mighty +40 km sec^{-1} complex, while Sgr B2 coincides in position and velocity with the second densest agglomeration of clouds. Details on the structure of the six brightest sources may be found in an extensive review by Downes (1974).

The emission profiles of the sources, and of the central area in general, discern themselves from H II regions in other parts of the Galaxy by the generally high velocities of the various components, and especially by their widths. From measurements in the region within $1.^{\circ}5$ longitude, Pauls & Mezger (1975) find a median half-width of 54 km sec^{-1} for the components considered to lie spatially near the center, as against 20 km sec^{-1} for foreground or background components.

The emission within $\sim 20'$ from the center has a peculiar structure, as indicated in Figure 31 (Pauls et al. 1976). Beside the compact central source Sgr A we note, firstly, the striking asymmetry in longitude already mentioned, and, secondly, a structure which the Bonn group has called the Arc, formed by a string of extended "clouds." The combined flux density of these clouds at 5 MHz is about 240 Jy according to Mezger (1974). Pauls et al. have measured recombination line velocities at 15 points in various parts of the field. Along the western branch of the Arc the radial velocities range from -48 km sec^{-1} around the longitude of the center to about -20 at the sharp bend in the north. In the northern arm the velocities are small. At first sight it seems surprising that these velocities are an order of magnitude smaller than the rotation velocities indicated by the H I nuclear disk and the mass model discussed in Section 2, or than the motions to be expected if the virial theorem would be applicable. We are again tempted to question whether we have not been radically mistaken in our interpretation of the H I motions as those of a rotating disk. It appears to me that the difficulties one is faced with if one attempts to interpret the H I in a different manner are so great that this alternative can practically be eliminated, and that we must therefore accept that the H II velocities deviate widely from those corresponding to the virial theorem. This is not so unlikely as it may have seemed at first, because already the peculiar shape of the Arc by itself indicates that it is due to expulsion. In features resulting from a gradual expulsion the lowest velocities required to reach a certain distance will strongly preponderate and may not give any indication of the gravitational field.

In Figure 31 a small extension of the Sgr A source toward greater longitude may be noted. In the tip of this protuberance a velocity of +45 km sec^{-1} was found, deviating rather strongly from the velocities in neighboring regions. It might be connected with the +40 km sec^{-1} molecular cloud.

It is possible that the Arc clouds are not smooth structures, but are made up of swarms of smaller components. Recent Westerbork observations of some of the

Arc "clouds" at wavelengths of 21 and 6 cm have indicated that they indeed contain a considerable number of compact H II regions, which seem to lie preferentially near the Arc's outer edge. I am indebted to J. Wouterloot in Leiden for this provisional account of his observations.

According to Mezger, Churchwell & Pauls (1974) the discrete thermal sources listed in Table 4 would together require 34 O6 supergiants to provide for their ionization, while 43 more are required to ionize the extended H II regions. This shows that star formation in the region must be taking place at a rapid rate. The abundance of the large and dense molecular clouds fully supports this.

There is of course the possibility that the ionization is partly due to collisions among the various fast moving features.

6.2 *Nonthermal Emission*

It is unknown whether the relativistic electrons responsible for the radiation have been supplied by the galactic nucleus or by supernovae. In view of the fast rate of

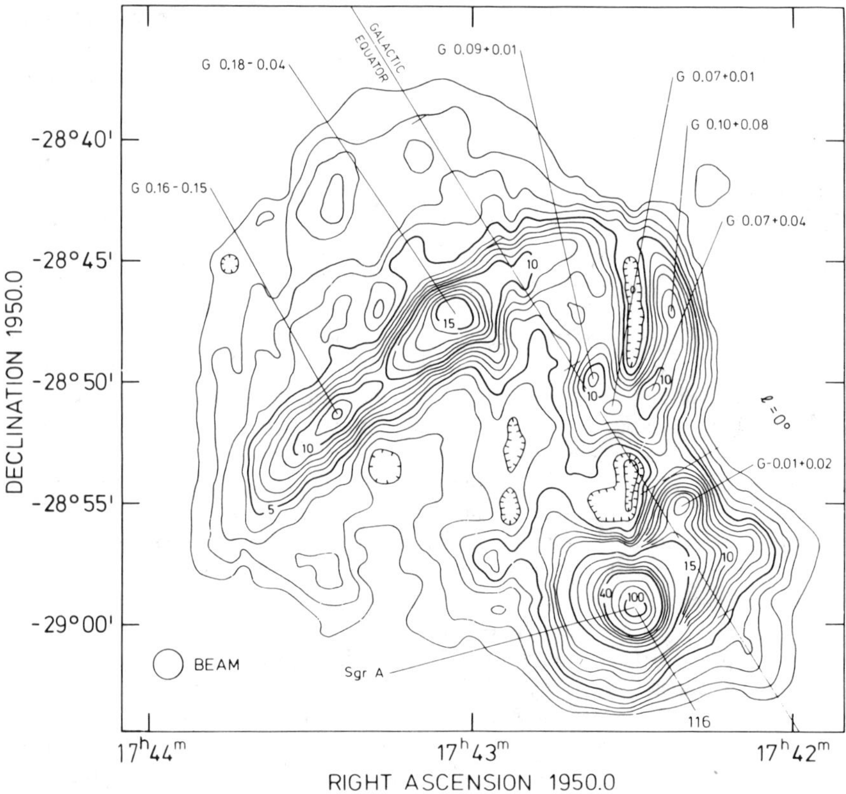

Figure 31 Map of thermal radiation between −10′ and +20′ longitude showing Sgr A and the Arc structure. This 10.7-GHz (2.8-cm) map (beam width 1′.3; contour unit 0.285 K) was made with the 100-m Effelsberg telescope (Pauls et al. 1976).

star formation the frequency of supernovae is probably considerably higher than elsewhere in the Galaxy.

A number of discrete nonthermal sources are known in the central region, three of which are listed in Table 4. The one connected with Sgr A is further discussed in the following section.

A transient radio source, presumably identical with a transient X-ray source, was found by Davies et al. (1976) at 53″ from the center, in position angle 223°. It was found on March 30, 1975, when it had a flux density of 0.48 Jy at 0.96 GHz. It was invisible ($S < 0.05$ Jy) on 6 other days in the period 1974/75.

A sketch of various features in the disk's central part is given in Figure 32.

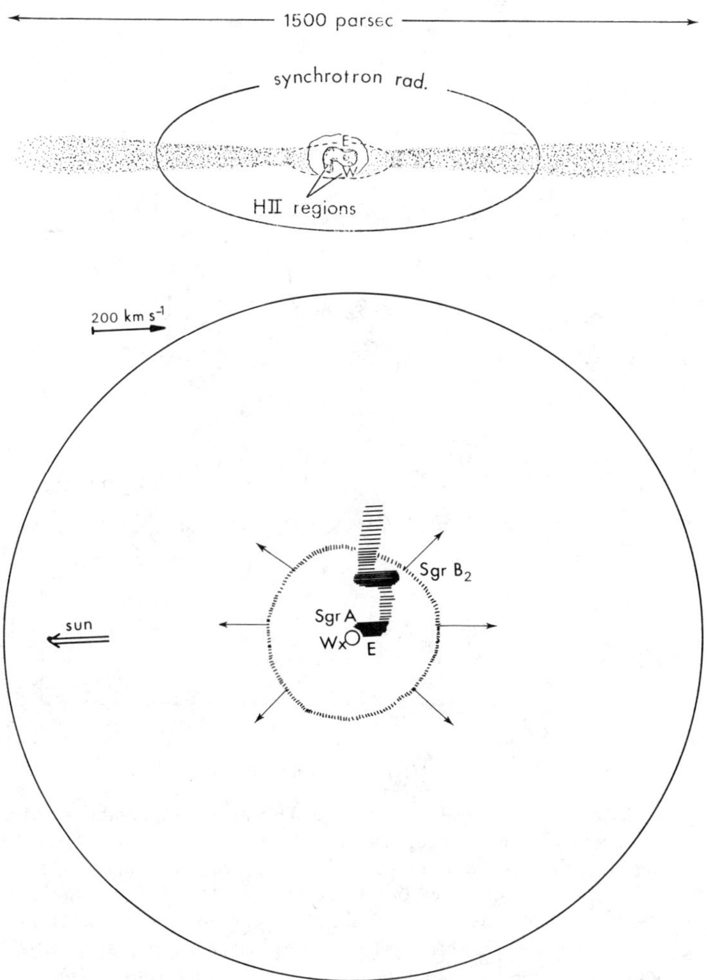

Figure 32 Sketch of the situation of the molecular clouds and other nuclear features relative to the H I disk.

6.3 *Sagittarius A*

As may be seen in Figures 29 and 30 the thermal as well as the synchrotron emission increase gradually with decreasing distance from the center. A new sharp increase sets in around $R = 2'$. Figure 33 (Ekers et al. 1975) shows a map of this inner region, which is called Sgr A. The map is on a 5 times larger scale than that of Figure 31, which itself is on a 3 times larger scale than Figure 30. All three are for cm wavelengths. Figure 33 was obtained from a combination of observations with the synthesis telescope at Westerbork and the Owens Valley radiotelescope. The angular resolution is $6''.3 \times 34''$ (or 0.3×1.7 pc). Many other maps exist,

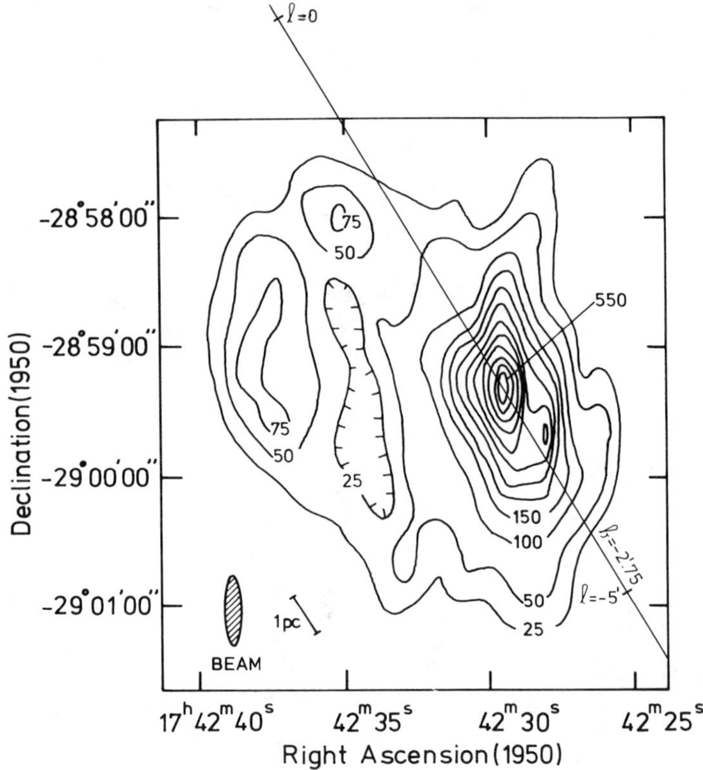

Figure 33 Full synthesis map of Sgr A at 5 GHz. This map, from combined Westerbork-Owens Valley Radio Synthesis Telescopes, is sometimes referred to as the "WORST" map. Half-power widths of synthesized beam $6''.3 \times 34''$ ($\alpha \times \delta$); contour unit 1.2 K in brightness temperature. The zero contour corresponds approximately to the 60-K contour in a 6-cm survey with the Parkes 64-m telescope by Whiteoak & Gardner (1973). The straight line is the latitude circle at $b = -2'.75$, which is probably the real galactic equator; the zero point of galactic longitude is marked; and the center of the concentrated source, which is presumably the actual center of the Galaxy, lies at $l = -3'.34$ (Ekers et al. 1975).

including high-resolution interferometer maps (Downes & Martin 1971, Whiteoak, Rogstadt & Lockhart 1974, Balick & Sanders 1974) or charts constructed from lunar occultation observations (Gopal-Krishna et al. 1972, Sandqvist 1974). The map in Figure 33 gives probably the best overall high-resolution picture.

It shows two major components, Sgr A West and Sgr A East, as well as structural details within these components. There is little doubt that both lie in the immediate vicinity of the galactic center. The eastern component has a nonthermal spectrum, as is shown by observations at longer wavelengths. It is probably a supernova remnant; in spectral index, diameter (about 7 pc), and luminosity it resembles common supernova remnants; there is even a hint of a shell structure. Its brightest part lies $1\overset{'}{.}8$ (5.2 pc) from the peak of Sgr A West.

Sgr A West has a flat spectrum between the wavelengths of 6 and 21 cm where Ekers et al. (1975) have determined its flux density. Its radiation at short wavelengths is probably mostly thermal, although at longer wavelengths it appears to emit also nonthermal radiation (cf Gopal-Krishna & Swarup 1976). Ekers et al. estimate the diameter to half power to be $1' x < 0\overset{'}{.}6$, with the long axis in position angle 32°; the flux density at 6 cm is 26 Jy. The structure differs from that of other giant H II regions in the Galaxy by its much smoother structure, which as the authors remark could arise if the exciting stars are more evenly distributed, or if there is only a single exciting source, which may be associated with the compact nucleus described in Section 6.3.1. The central H II is also unique by its large internal motions (Section 6.3.2).

Pauls et al. (1976) describe a "halo" of 3′ radius surrounding Sgr A East and West, which at 3 cm has a flux density of 70 Jy. About 60% of this would be nonthermal.

Sgr A West lies in the midst of a unique agglomeration of infrared sources covering an area of about 20″ (1 pc) diameter (cf Section 7, Figure 35). The agglomeration contains among others the sharp maximum of 2.2 μ radiation, which is supposed to be the dynamical center of the Galaxy. Radio observations at 8085 MHz show some fine structure of the same 5″ to 10″ scale as the infrared (Balick & Sanders 1974, cf the inserts in their Figures 3 and 5); the shoulder at ~25″ SW of the maximum in Figure 33 may be an outlying condensation of the same category. It lies in the general direction where at 53 microns a conspicuous maximum has been observed (cf Figure 38). The central component of the fine structure has an ultracompact core (Section 6.3.1), which is probably the actual nucleus of our Galaxy. This coincides with the 2.2 μ maximum within the errors of a few seconds of the 2.2 μ position (Ekers et al. 1975, Borgman, Koornneef & de Vries 1974, Becklin & Neugebauer 1975).

6.3.1 THE COMPACT NUCLEUS Sgr A West has an extremely compact core. This was first observed in 1971 by Ekers & Lynden-Bell; later observations by Balick & Brown (1974) showed that it was smaller than $0\overset{''}{.}1$ and had flux densities of 0.6 and 0.8 Jy at 11 and 3.8 cm respectively, corresponding to a brightness temperature in excess of 10^7 K. The 1950 position of this small source is $\alpha = 17^h42^m29\overset{s}{.}291 \pm 0\overset{s}{.}005$, $\delta = -28°59'17\overset{''}{.}6 \pm 0\overset{''}{.}1$, or $l = -3\overset{'}{.}34$, $b = -2\overset{'}{.}75$. The source was subsequently observed by Lo et al. (1975) with a medium-baseline interferometer at 3.7 cm. They found it to be smaller than about 0.″02, with a flux density of 0.6 Jy. More recent observations at a wavelength of 3.8 cm, with a much longer baseline, indicate an overall size of 0.″01 to 0.″02, with about 25% of the flux, or 0.2 Jy, in a core of about 0.″001 (private communication from Kellermann; cf Kellermann et al. 1977). The radio luminosity of the core is 10^{33} ergs sec^{-1}. The brightness temperature must be about 10^{10} K, and the radiation is probably nonthermal. The core diameter is smaller than 10 astronomical units, or 80 light-minutes. This is smaller than the limits measured in any other nucleus.

Davies, Walsh & Booth (1976) have suggested from their measurements at 0.408, 0.96, and 1.66 GHz that the observed diameter of the nuclear source would be mainly due to interstellar scattering. The new VLBI result by Kellermann et al. seems to show that the scattering is considerably less than proposed by Davies et al. However, judging from the scintillation observations of other distant galactic sources, the scattering towards the center is unlikely to be smaller than 0.″001 at 3.8 cm. The true diameter may therefore be appreciably smaller still.

The existence of such an extremely compact radio-emitting body in the galactic center is intriguing and leads naturally to the question of whether it might contain the engine providing the energy and mass required for the expulsion phenomena discussed in the preceding sections.

It may be mentioned in this connection that the Schwarzschild radius of a mass of $5 \times 10^6 M_\odot$, which may be the mass of the compact core (cf Section 6.3.2), is $\sim 1.5 \times 10^{12}$ cm, or 0.1 astronomical unit.

Several radio galaxies are known to contain cores of small dimensions, which are suspected to be the sources of the enormous energies poured into their radio lobes. The galactic nucleus is, however, 7 to 9 powers of ten weaker than the nuclei of strong radio galaxies and quasars, so that the confrontation may not be meaningful. It should be noted, however, that there is an enormous spread in the emission at radio frequencies, such as shown for instance by the large class of radio-quiet quasars, and by the very great differences in radio emission by the cores of common galaxies, as is indicated by a comparison of our Galaxy with the very similar nearby spirals M31 and M81. While no compact core is seen in M31, M81 has a nonthermal radio core that may be of similar dimension as that in the Galaxy, but has a 10^4 times higher radio power. Interferometer measures show that its diameter is less than 1300 au, or 7 light-days (Kellermann et al. 1976); intensity variations at cm waves indicate it to be smaller than one light-day.

6.3.2 HIGH-VELOCITY IONIZED GAS AROUND SGR A WEST AND THE MASS OF THE NUCLEUS Observations of the 109α hydrogen recombination line at 5010 MHz in Sgr A showed that this is exceptionally broad (Pauls, Mezger & Churchwell 1974). As observed with the Effelsberg telescope, with a beam width of ∼2.′5 or 7.5 pc, the line was found to have a half-width of ∼200 km sec^{-1} centered around zero velocity. The authors commented that such a width is almost certainly the result of systematic motions, and not of random turbulence.

New information on the nature of these motions has come from the observation of the Ne II fine-structure line at 12.8 μ by Townes and co-workers (Wollman et al. 1976, Wollman 1976). The published observations were made at the Tololo Observatory. Subsequently, data with much better sensitivity as well as better resolution were obtained by Wollman with the Lick 3-m telescope. The reviewer is greatly indebted to Dr. Wollman for putting these results at his disposal prior to publication. Because of their eminent interest, a full reproduction of his principal results is given in Figure 34. The resolution of these observations, which is 28 times better than that of the hydrogen recombination lines, made it possible not only to investigate the distribution of the ionized gas around Sgr A West, but also to get some insight in the internal motions.

From the data in Figure 34, combined with less sensitive observations in surrounding fields, it was found that the ionized gas is concentrated in an area of about 15″ (or 3/4 pc) radius, the center of which coincides with the compact radio source discussed in the preceding subsection and indicated by an asterisk in Figure 34. As described in Section 7 a unique concentration of discrete infrared sources has been found in just the same area. I therefore propose to refer to this region as the "infrared core." Figure 34 shows the positions at which the high-resolution Ne II line measurements were made, projected onto a 10μ continuum map.

Some of the profiles, in particular in the southwestern part, are extremely wide; in the area b' the emission seems to cover an interval of about 1000 km sec^{-1} (half-width $\sim$ 500 km sec^{-1}). The central velocity is seen to change systematically from about $+100$ km sec^{-1} at $\Delta l = +8''$ to -200 km sec^{-1} at $\Delta l = -7''.5$, Δl being measured from the longitude of the compact radio source. The shift is perpendicular to the rotation axis of the Galaxy and in the direction of the galactic rotation. If interpreted as rotation it indicates a rotation velocity of roughly 150 km sec^{-1} at $7''.8$, or 0.4 pc, from the center. The *circular* velocity at $R = 0.4$ pc is likely to be higher, for two reasons: firstly because the measures give the projection of the rotation on the line of sight, averaged over a column through the whole mass, and secondly because the large linewidths indicate that there must be large radial motions in addition to rotation. Therefore, the true "circular" velocity at $R = 0.4$ pc may well be between 200 and 250 km sec^{-1}, and possibly even higher. It should, however, be stressed that there remains a possibility that the observed systematic change of velocity is *not* due to rotation but to radial motions mimicking rotation around the galactic axis.

In a sense the ionized gas at 0.5 pc gives a repetition of what the H I nuclear disk gave at a thousand times larger distance.

In the case of a spherical mass distribution, or a point mass, the mass inside 0.4 pc would be between 4 and 6 × 10^6 $M_\odot$. The mass within the same radius as estimated from the distribution of the 2.2 μ radiation and the light distribution in M31 was 1.5 × 10^6 $M_\odot$ (interpolated from Table 1). It is therefore possible that practically the whole of the 4–6 × 10^6 $M_\odot$ deduced from the Ne II observations would be concentrated in the ultracompact radio source; it might be the mass of a black hole in the center. The kinematics of the nuclear complex are evidently complicated. Radial streamings of comparable velocity must be present beside the rotation; otherwise the profiles of fields c and b', both at negative longitude relative

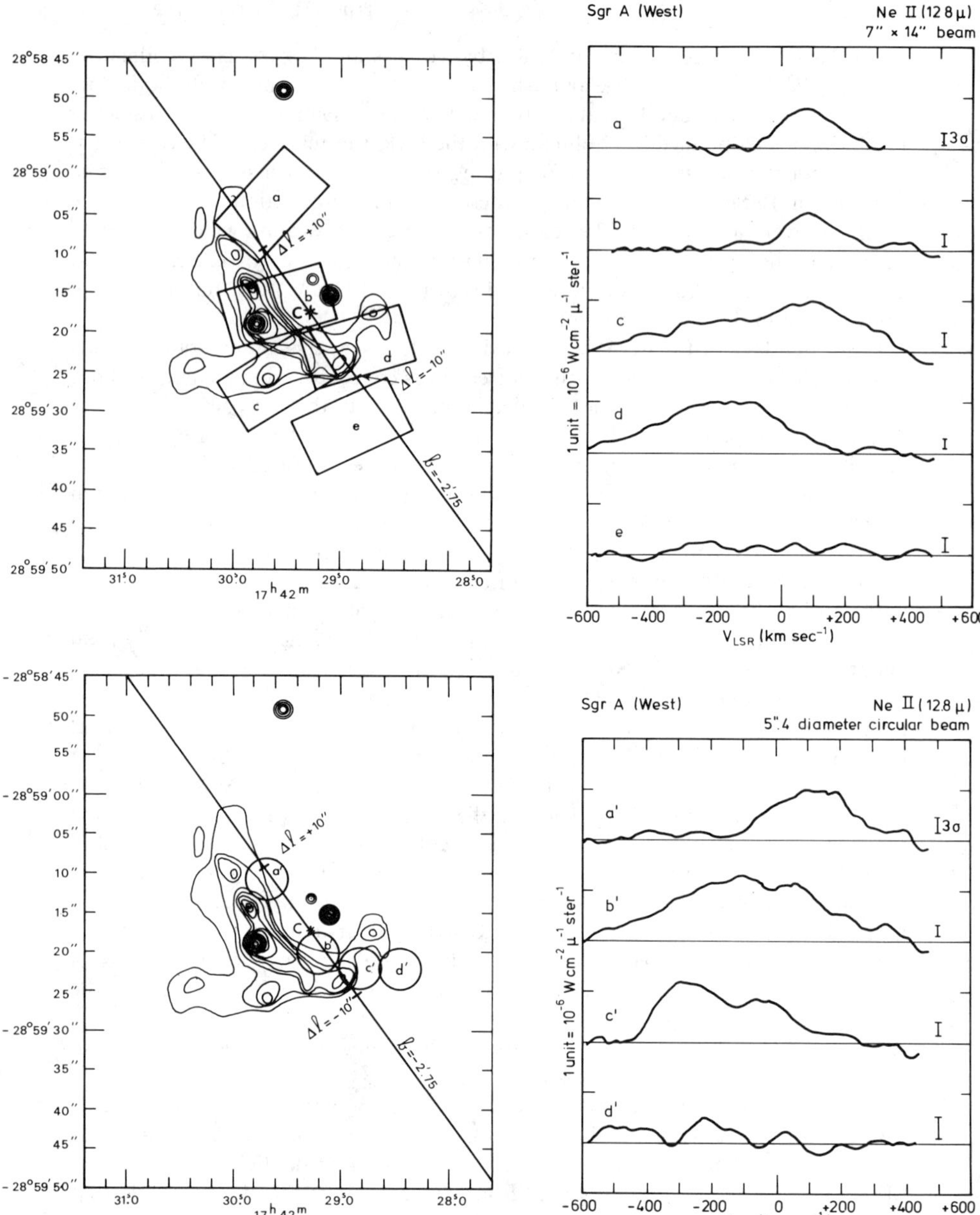

Figure 34 Profiles of the Neon II line at 12.8 μ. The regions measured are shown projected on a 10-μ continuum map. The straight line is the latitude circle at $b = -2'.75$, and the position of the galactic center is shown by an asterisk and a capital C (Wollman 1976, from observations with the Lick 3-m telescope).

to the center, could not have such long extensions to positive velocities. Nor could we otherwise have such large asymmetries.

The lifetime of these large-scale irregularities must be short, comparable with the time of revolution, which is roughly 15,000 years. To maintain the observed state of turmoil there should be major eruptions on time scales of at most 10,000 years, each imparting a kinetic energy of $10^{49}-10^{50}$ erg in the form of large-scale motions. This is of the order of what can be yielded by a type II supernova. But it is unlikely that supernovae would be sufficiently frequent.

It is more probable that the same unknown mechanism that is needed to expell the massive molecular features and the expanding H I arms is also responsible for the motions in this inner region.

Evidently, further observations of the Ne II line in the galactic nucleus are of extreme interest.

7 THE INFRARED CORE

Observations at infrared wavelengths have shown a remarkable concentration of objects in a region of 10 to 15″ radius around the center.

Radiation at 2.2 μ was first extensively observed by Becklin & Neugebauer (1968), who interpreted it as emission from stars at the center. They also observed at other wavelengths, and found relatively strong radiation at 10 and 20 μ concentrated in an area of about 1 pc diameter. Rieke & Low (1973), observing at various wavelengths between 3.5 and 20 μ, discovered that the area contained a number of discrete sources. Later observations with better resolving power (Borgman, Koornneef & de Vries 1974, Rieke, unpublished) yielded improved absolute positions of the Rieke & Low features, and brought out some more detail. The most detailed maps presently available are by Becklin & Neugebauer (1975). They are shown in Figures 35 (2.2 μ) and 36 (10 μ), the latter being from unpublished observations by Willner (Willner, Becklin & Neugebauer, private communication, Willner 1976). Individual sources in these maps are numbered, and will in the following be referred to as IRS 1, IRS 2, etc.

As suggested by Becklin & Neugebauer (1968), the near infrared (2–3 μ) emission comes probably mainly from the central concentration of the later-type stars of which the spheroidal bulges of galaxies are made; it contains an outstanding point-like object (IRS 7). As discussed in Section 2 these observations have given important information on the density distribution and total star density between about 0.2 and 50 pc from the center. An extension of the map to distances of about $0^\circ.5$ is shown in Figure 1. Source IRS 16 of Figure 35 coincides with the ultracompact central radio source; it provides an indication of the star density within 2″ of the actual center. The irregularity of the distribution in Figures 35 and 1 is mainly caused by the dominating role of a relatively small number of intrinsically very bright M-type stars, all situated in the region of the center.

The maps in Figures 35 and 36 are strikingly different. This is due to the fact that the emission at 2.2 μ is mainly direct stellar radiation, while that at 10 μ comes from solid particles, heated by stars, but radiating at a much lower temperature. The

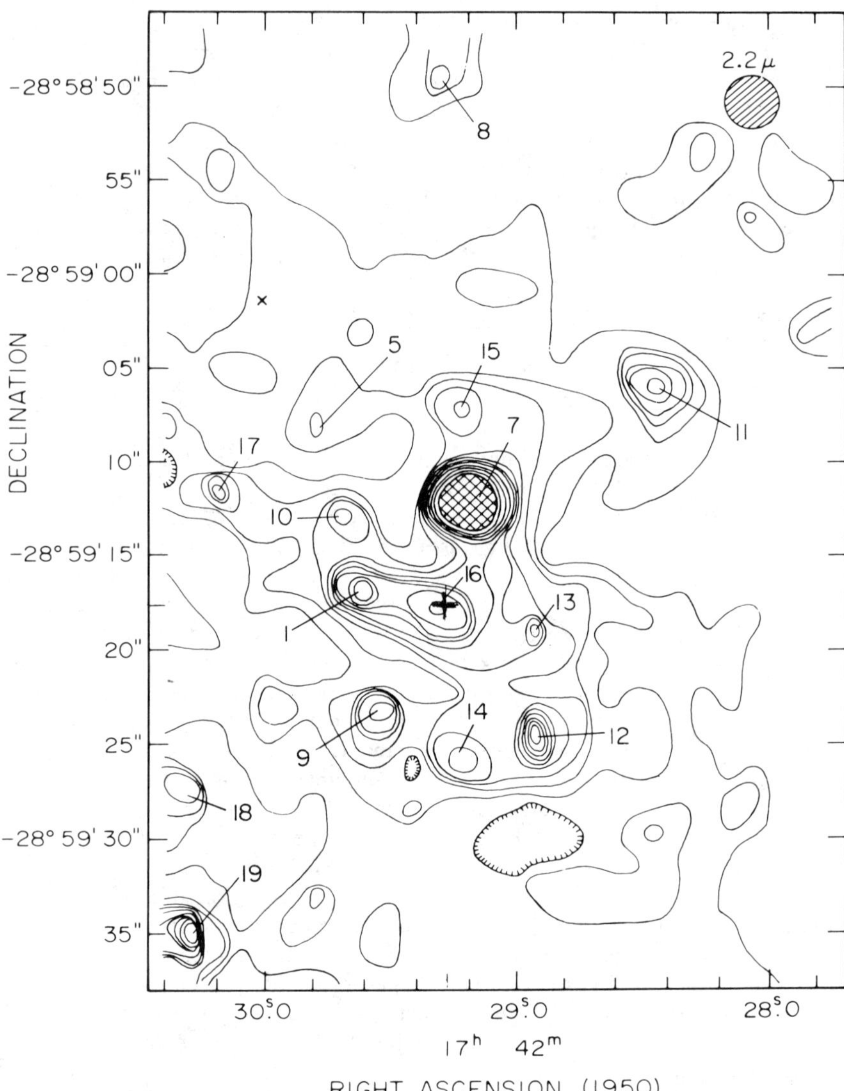

Figure 35 Map of the galactic center region at 2.2 μ, made with a 2″.5 diaphragm on the 5-m Hale telescope. The contour levels each correspond to 2.5×10^{-18} W m^{-2} Hz^{-1} sr^{-1}. The crosshatching corresponds to 35 contour levels. The cross marks the galactic center. Discrete sources are indicated by numbers which will be referred to in the text as IRS 1, etc. (Becklin & Neugebauer 1975).

dust is apparently rather evenly distributed over a sort of ridge that is seen in the 10-μ picture to extend from IRS 2 to a little North of IRS 5. It is heated partly by the Population II stars (whose strong concentration towards the center causes the dust temperature to increase with decreasing distance from the center) and partly by high-luminosity O stars. The condensations in Figure 36 are presumably the locations of such newly formed stars. By an extensive analysis of the spectra between

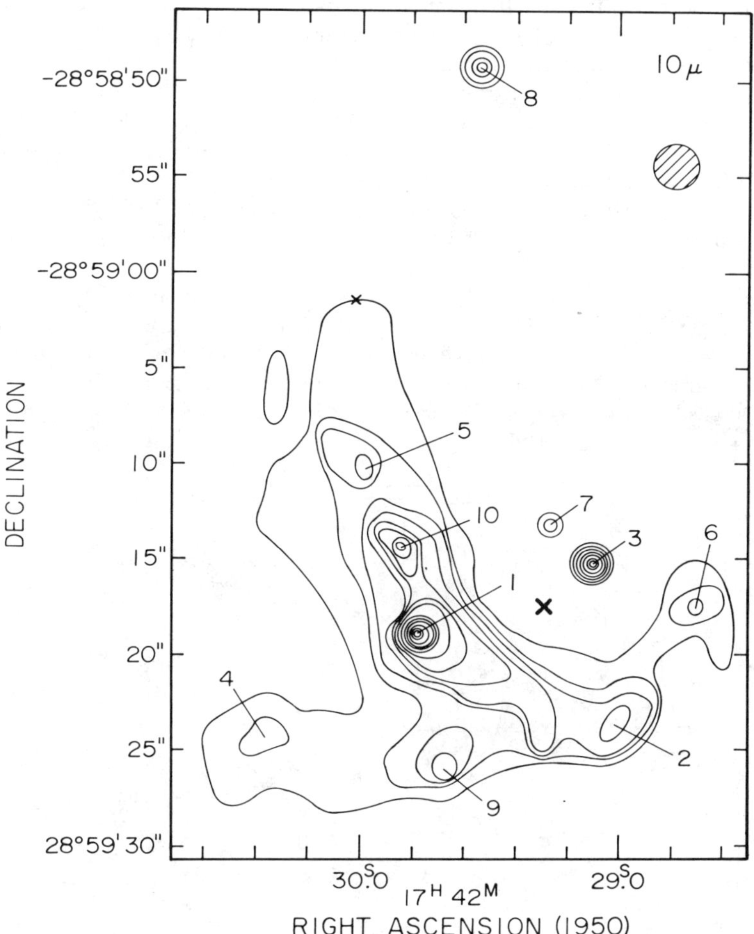

Figure 36 Map of the galactic center region at 10 μ, made with a 2".3 circular aperture on the 5-m Hale telescope. The aperture size is shown by the circle in the upper right-hand corner. The contour interval is 4×10^{-16} W m^{-2} Hz^{-1} sr^{-1}. The cross marks the galactic center (Willner 1976, Becklin et al. 1977).

1.2 and 12.5 μ, Willner (1976) has found that IRS 7, 11, 12, 16, and 19 of Figure 35 are stars or star clusters (IRS 16 is the star concentration at the center) with luminosities of the order of $10^4 L_\odot$. No. 7, with $10^5 L_\odot$, is probably an MIa supergiant. All except IRS 7 are invisible at 10 μ; cf Becklin et al. 1977.

The "ridge" which contains the dust that produces the emission around 10 μ and longer is at the same time a density-bounded concentration of ionized gas, as is shown by the high resolution Ne II observations discussed in Section 6.3.2. Willner has further argued that the 10-μ sources do not come from increased dust density but rather from increased local heating of the dust; the heated regions have a typical size of 0.1 pc, comparable to that of compact H II regions.

The extinction towards most of the sources appears to be moderately uniform. From their photometric measurements in the infrared Becklin, Neugebauer and Willner (Willner 1976) conclude that the visual extinction is about 28 magnitudes, and that most of it occurs in the Galaxy at large. The dust density in the central region itself appears to be relatively small in comparison with the high gas density. The visual extinction across the central 10 pc is estimated to be only about 3^m (Gatley et al. 1977).

Radiation in the *far* infrared (25–300 μ) was first extensively studied by Hoffmann, Frederick & Emery (1971) (cf also Low & Aumann 1970, Harper 1974). The early observations were made with large apertures, and did not therefore reveal any of the fine structure. A radical improvement has recently become possible by the use of NASA's Gerard P. Kuiper Airborne Infrared Observatory. With this equipment Gatley et al. (1977) mapped a region of 10′ diameter around the center in three

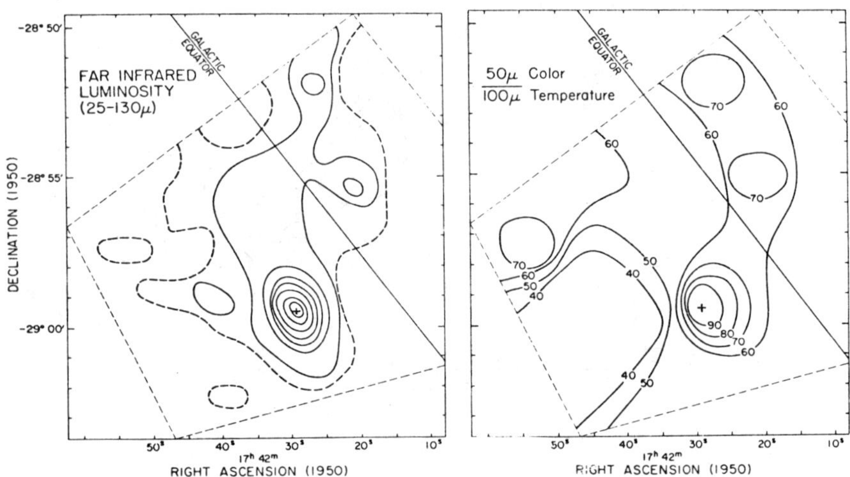

Figure 37 Map of the far-infrared luminosity of the galactic center region obtained by integration from 25- to 130-μ wavelength (left); the contour interval is equivalent to 1.0×10^{-10} W m^{-2} into a 1′ beam. The right-hand diagram gives the color temperatures (Gatley et al. 1977).

wavelength bands, at 30 μ, 50 μ, and 100 μ, with a resolution of $\sim 1'$. Figure 37a shows the distribution of the integrated luminosity between 25 and 130 μ; Figure 37b indicates the run of the color temperature obtained by fitting a Planck curve to their data. The far-infrared radiation is seen to be strongly concentrated to the same region as the near-infrared emission (the galactic center position is shown by a cross). The 25–130 μ luminosity of the central 1′ is 2.3×10^6 $L_\odot$, the color temperature is ~ 100 K, while the brightness temperature at 50 μ is ~ 40 K. From the comparison of these two temperatures the authors estimated that the optical depth $\tau_{50\mu} \sim 0.05$. The optical depth was found to be practically the same at almost all points, which indicates that the volume density of dust is fairly uniform over the whole region within about 15 pc from the center.

Observations with much higher resolving power with the same airborne observatory have been made and discussed by Harvey, Campbell & Hoffmann

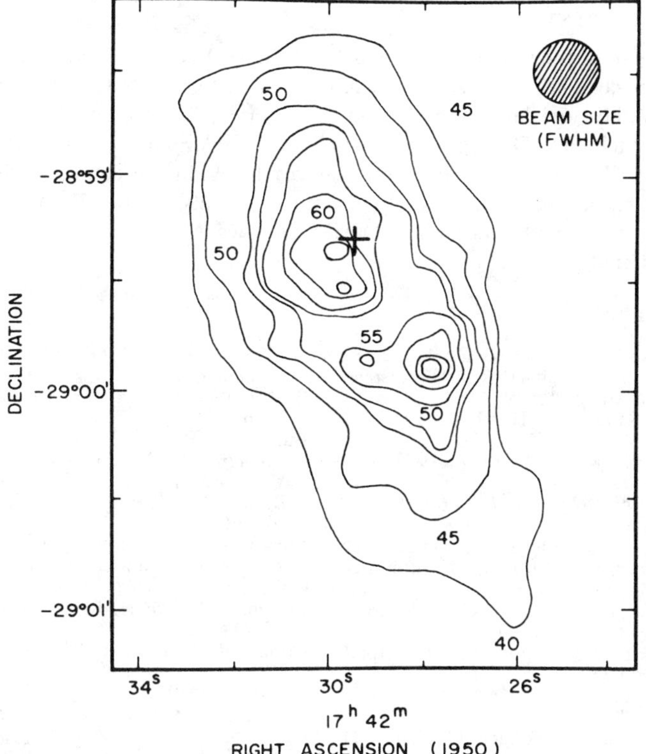

Figure 38 Map of the galactic center at 53 μ with a 17″ beam. The contour levels are at 5, 10, 15, 18, 21, 24, 27, and 30 $\times 10^{-16}$ W m^{-2} Hz^{-1} sr^{-1}. Dust temperatures are also indicated. The cross marks the position of the ultracompact radio source at the galactic center (Harvey, Campbell & Hoffmann 1976).

(1976). Their map at 53 μ, with 17″ resolution, is shown in Figure 38. It shows a strongly elongated structure, more or less (but not precisely) along the true galactic equator, in which there are two maxima, about 45″ (2 pc) apart; the NE maximum coincides with the galactic center (Sgr A West), the other with a *weak* 10 μ source. It may be noted that the "shoulder" shown in the 6-cm map (Figure 33) SW of the center lies in the same position angle, though at a somewhat smaller distance. The authors observed also at 100 and 175 μ, with a beam of $\sim$30″. Like Gatley et al. they find that the decrease in surface brightness with distance from the center (in this case between 0.7 and 2 pc) can be most readily explained by a temperature gradient, with little or no change in dust column density. The amount of dust required for the far-infrared emission is comparable to that required to explain the near-infrared and visual *extinction*. The optical depths they find from their model range from 0.1 at 53 μ to 0.01 at 175 μ. The paper contains an interesting graph of the energy distribution of Sgr A West between $\sim$10 and 400 μ.

ACKNOWLEDGMENT

I am grateful for the kindness with which colleagues have put unpublished material at my disposal. The article could not have been written in its present form but for the pre-publication data received from Drs. Bania, Burton, and Wollman. I am similarly indebted to Drs. Becklin, Cohen, Fukui, Kellermann, Lo, Sanders, Willner, and others for early information on new results. Valuable comments have been given by Drs. van Bueren, Dekker, Ekers, Mezger, Pauls, van den Bout, and van der Kruit. I want to thank Drs. Cohen, Davies, Downes, Mezger, and Townes for help of various kinds.

Literature Cited

Altenhoff, W. J., Downes, D., Goad, L., Maxwell, A., Rinehart, R. 1970. *Astron. Astrophys. Suppl.* 1:319
Balick, B., Sanders, R. H. 1974. *Ap. J.* 184: 415
Balick, B., Brown, R. L. 1974. *Ap. J.* 194: 265
Bania, T. M. 1977. *Ap. J.* Submitted for publication
Becklin, E. E., Matthews, K., Neugebauer, G., Willner, S. P. 1977. *Ap. J.* Submitted for publication
Becklin, E. E., Neugebauer, G. 1968. *Ap. J.* 151:145
Becklin, E. E., Neugebauer, G. 1969. *Ap. J. Lett.* 157:L31
Becklin, E. E., Neugebauer, G. 1975. *Ap. J. Lett.* 200:L71
Becklin, E. E., Neugebauer, G., Early, D. 1977. In preparation
Bieging, J. H. 1976. *Astron. Astrophys.* 51: 289
Bolton, J. G., Gardner, F. F., McGee, R. X., Robinson, B. J. 1964. *Nature* 204:30
Borgman, J., Koornneef, J., de Vries, M. 1974. *Proc. 8th Eslab Symp. H II Reg. Galactic Cent.*, p. 229
Burton, W. B. 1970. *Astron. Astrophys. Suppl.* 2:261
Burton, W. B. 1974. In *Galactic and Extra-Galactic Radio Astronomy*, ed. G. L. Verschuur, K. J. Kellermann, pp. 82–117. Berlin, Heidelberg, New York: Springer. 402 pp.
Burton, W. B., Gordon, M. A. 1977. *Astron. Astrophys.* Submitted for publication
Cohen, R. J. 1975. *MNRAS* 171:659
Cohen, R. J., Davies, R. D. 1976. *MNRAS* 175:1
Cohen, R. J., Few, R. W. 1976. *MNRAS* 176:495
Cohen, R. J. 1977. *MNRAS*. 178:547
Cugnon, P. 1968. *Bull. Astron. Inst. Neth.* 19:363
Davies, R. D., Walsh, D., Booth, R. S. 1976. *MNRAS* 177:319
de Bruyn, A. G. 1977. *Astron. Astrophys.* 58: 221
Downes, D., Maxwell, A. 1966. *Ap. J.* 146: 653

Downes, D., Martin, A. H. M. 1971. *Nature* 233:112

Downes, D. 1974. *Proc. 8th Eslab Symp. H II Reg. Galactic Cent.*, p. 247

Ekers, R. D., Lynden-Bell, D. 1971. *Astrophys. Lett.* 9:189

Ekers, R. D., Goss, W. M., Schwarz, U. J., Downes, D., Rogstad, D. H. 1975. *Astron. Astrophys.* 43:159

Fomalont, E. B., Weliachew, L. N. 1973. *Ap. J.* 181:781

Frank, J., Rees, M. J. 1976. *MNRAS* 176: 633

Fukui, Y., Iguchi, T., Karifu, N., Chicada, Y., Morimoto, M., Nagane, K., Miyazawa, K., Miyaji, T. 1977. *Publ. Astron. Soc. Jpn.* In press

Gardner, F. F., Whiteoak, J. B. 1970. *Astrophys. Lett.* 5:161

Gatley, I., Becklin, E. E., Werner, M. W., Wynn-Williams, C. G. 1977. Preprint

Gopal-Krishna, Swarup, G., Sarma, N. V. G., Hoshi, M. N. 1972. *Nature* 239:91

Gopal-Krishna, Swarup, G. 1976. *Astrophys. Lett.* 17:45

Gordon, M. A., Burton, W. B. 1976. *Ap. J.* 208:346

Harper, D. A. 1974. *Ap. J.* 192:557

Harvey, P. M., Campbell, M. F., Hoffmann, W. F. 1976. *Ap. J. Lett.* 205:L69

Hoffmann, W. F., Frederick, C. L., Emery, R. J. 1971. *Ap. J. Lett.* 164:L23

Johnson, H. M. 1961. *Ap. J.* 133:309

Kaifu, J., Kato, T., Iguchi, T. 1972. *Nature Phys. Sci.* 238:105 (*not Nature* 238:105, as has been erroneously cited by several authors)

Kellermann, K. I., Shaffer, D. B., Pauliny-Toth, I. I. K., Preuss, E., Witzel, A. 1976. *Ap. J. Lett.* 210:L121

Kellermann, K. I., Shaffer, D. B., Clark, B. G., Geldzahler, B. J. 1977. *Ap. J. Lett.* 214:L61

Kerr, F. J., Sinclair, M. W. 1966. *Nature* 212:166

Kerr, F. J. 1968. In *Radio Astronomy and the Galactic System, IAU Symp. No. 31*, p. 239, ed. H. van Woerden

Kinman, T. D. 1965. *Ap. J.* 142:1376

Leung, Chun Ming, Liszt, H. S. 1976. *Ap. J.* 208:732

Light, E. S., Danielson, R. E., Schwarzschild, M. 1974. *Ap. J.* 194:257

Lindblad, B. 1956. *Stockholm Obs. Ann.* 19: No. 2

Liszt, H. S., Sanders, R. H., Burton, W. B. 1975. *Ap. J.* 198:537

Liszt, H. S., Burton, W. B., Sanders, R. H., Scoville, N. Z. 1977. *Ap. J.* 213:38

Little, A. G. 1974. In *Galactic Radio Astronomy, IAU Symp. No. 60*, ed. F. J. Kerr, S. C. Simonson III, pp. 491–97

Lo, K. Y., Schilizzi, R. T., Cohen, M. H., Ross, H. N. 1975. *Ap. J. Lett.* 202:L63

Low, F. J., Aumann, H. H. 1970. *Ap. J. Lett.* 162:L79

Lynden-Bell, D., Rees, M. J. 1971. *MNRAS* 152:461

Martin, A. H. M., Downes, D. 1972. *Astrophys. Lett.* 11:219

McGee, R. X., Brooks, J. W., Sinclair, M. W., Batchelor, R. A. 1970. *Aust. J. Phys.* 23: 777

Mezger, P. G. 1974. *Proc. ESO/SRC/CERN Conf. Prog. New Large Telescopes*, p. 79

Mezger, P. G., Churchwell, E. B., Pauls, T. A. 1974. *Proc. Eur. Astron. Meet., 1st* 2:140

Mills, B. Y. 1956. *Observatory* 76:65

Milman, A. S. 1975. *Ap. J.* 202:673

Minkowski, R. 1965. *Stars and Stellar Systems*, eds. A. Blaauw, M. Schmidt, Vol. 5, p. 321. Univ. Chicago Press

Mirabel, I. F., Turner, K. C. 1972. *Bull. Am. Astr. Soc.* 4:413

Morton, D. C., Thuan, T. X. 1973. *Ap. J.* 180:705

Oort, J. H. 1968. In *Non-Stable Phenomena in Galaxies, Proc. IAU Symp. No. 29*, ed. Armenian Acad. Sci., pp. 41–44

Oort, J. H. 1974a. In *The Formation and Dynamics of Galaxies, IAU Symp. No. 58*, ed. J. R. Shakeshaft, pp. 378–82

Oort, J. H. 1974b. See Little 1974, pp. 539–47

Oort, J. H., Plaut, L. 1975. *Astron. Astrophys.* 41:71

Pauls, T., Mezger, P. G., Churchwell, E. 1974. *Astron. Astrophys.* 34:327

Pauls, T., Mezger, P. G. 1975. *Astron. Astrophys.* 44:259

Pauls, T., Downes, D., Mezger, P. G., Churchwell, E. 1976. *Astron. Astrophys.* 46:407

Peters, W. L. 1975. *Ap. J.* 195:617

Rieke, G. H., Low, F. J. 1973. *Ap. J.* 184:415

Robinson, B. J., McGee, R. X. 1970. *Aust. J. Phys.* 23:405

Robinson, B. J. 1974. See Little 1974, pp. 521–35

Rougoor, G. W., Oort, J. H. 1960. *Proc. Natl. Acad. Sci. USA* 46:1

Rougoor, G. W. 1964. *Bull. Astron. Inst. Neth.* 17:381

Ruiz, Maria Teresa. 1976. *Ap. J.* 207:382

Salpeter, E. E. 1964. *Ap. J.* 140:796

Sandage, A. R. 1961. *The Hubble Atlas of Galaxies, Carnegie Inst. Washington Publ. 618*, Plate 31

Sandage, A. R., Becklin, E. E., Neugebauer, G. 1969. *Ap. J.* 157:55

Sanders, R. H., Huntley, J. M. 1976. *Ap. J.* 209:53

Sanders, R. H., Lowinger, T. 1972. *Astron. J.* 77:292
Sanders, R. H., Prendergast, K. H. 1974. *Ap. J.* 188:489
Sanders, R. H., Wrixon, G. T. 1973. *Astron. Astrophys.* 26:365
Sanders, R. H., Wrixon, G. T. 1974. *Astron. Astrophys.* 33:9
Sanders, R. H., Wrixon, G. T., Mebold, U. 1977. To appear in *Astron. Astrophys.*
Sanders, R. H., Wrixon, G. T., Penzias, A. A. 1972. *Astron. Astrophys.* 16:322
Sandqvist, Aa. 1970. *Astron. J.* 75:135
Sandqvist, Aa. 1973. *Astron. Astrophys. Suppl.* 9:391
Sandqvist, Aa. 1974. *Astron. Astrophys.* 33:413
Saraber, M. J. M., Shane, W. W. 1974. *Astron. Astrophys.* 36:365
Schwarz, U. J., Shaver, P. A., Ekers, R. D. 1977. *Astron. Astrophys.* 54:863
Scoville, N. Z., Solomon, P. M., Thaddeus, P. 1972. *Ap. J.* 172:335
Scoville, N. Z. 1972. *Ap. J. Lett.* 175:L127
Scoville, N. Z., Solomon, P. M. 1973. *Ap. J.* 180:55
Scoville, N. Z., Solomon, P. M., Jefferts, K. B. 1974. *Ap. J. Lett.* 187:L63
Scoville, N. Z., Solomon, P. M., Penzias, A. A. 1975. *Ap. J.* 201:352
Segalovitz, A. 1975. *Astron. Astrophys.* 40:401
Shane, W. W. 1972. *Astron. Astrophys.* 16:118
Simonson, S. C. III, Mader, G. L. 1973. *Astron. Astrophys.* 27:337
Solomon, P. M., Scoville, N. Z., Jefferts, K. B., Penzias, A. A., Wilson, R. W. 1972. *Ap. J.* 178:125
van Albada, G. D., Shane, W. W. 1975. *Astron. Astrophys.* 42:433
van den Bergh, S. 1975. *Ann. Rev. Astron. Astrophys.* 13:217
van der Kruit, P. C. 1970. *Astron. Astrophys.* 4:462
van der Kruit, P. C. 1971. *Astron. Astrophys.* 13:405
van der Kruit, P. C. 1974. *Ap. J.* 192:1
van der Kruit, P. C., Oort, J. H., Mathewson, D. S. 1972. *Astron. Astrophys.* 21:169
Wannier, P. G., Penzias, A. A., Linke, R. A., Wilson, R. W. 1976. *Ap. J.* 204:26
Westbrook, W. E., Werner, M. W., Elias, J. H., Gezari, D. Y., Hausier, M. G., Lo, K. Y., Neugebauer, G. 1976. *Ap. J.* 209:94
Whiteoak, J. B., Gardner, F. F. 1973. *Astrophys. Lett.* 13:205
Whiteoak, J. B., Rogstad, D. H., Lockhart, I. A. 1974. *Astron. Astrophys.* 36:245
Willner, S. P. 1976. *Compact infrared sources: NGC 7538 and the galactic center.* PhD thesis. Calif. Inst. Technol. 56 pp.
Wollman, E. R., Geballe, T. R., Lacy, J. H., Townes, C. H., Rank, D. M. 1976. *Ap. J. Lett.* 205:L5
Wollman, E. R. 1976. *Ne II 12.8 μ emission from the galactic center and compact H II regions.* PhD thesis. Univ. Calif., Berkeley
Wrixon, G. T., Sanders, R. H. 1973. *Astron. Astrophys. Suppl.* 11:339

Ann. Rev. Astron. Astrophys. 1977. 15:363–87

MASS AND ENERGY FLOW IN THE SOLAR CHROMOSPHERE AND CORONA

George L. Withbroe and Robert W. Noyes
Center for Astrophysics, Harvard College Observatory and Smithsonian Astrophysical Observatory, Cambridge, Massachusetts 02138

OVERVIEW

Introduction

In the past few years there has been substantial progress in the development of an understanding of the mass and energy flow in the outer solar atmosphere. Much of this progress has resulted from observations made by rocket and satellite experiments that probe the chromosphere and corona using radiation emitted at *UV, XUV,* and X-ray wavelengths and measurements made by satellite experiments that sample the solar wind in situ. These data have demonstrated more clearly the fundamental role that magnetic fields play in defining the structure and the mass and energy flow in the chromosphere and corona and have resulted in the identification of coronal holes as the source of recurrent high-speed solar wind streams. There has also been substantial progress in the measurement and interpretation of parameters measured with ground-based instrumentation, particularly the photospheric velocity and magnetic fields. This has led to the discovery that a significant fraction, perhaps all, of the photospheric magnetic flux is concentrated in small spatial elements where the strength of the fields is of the order of a kilogauss. Analyses of photospheric motions have led to a better understanding of how much energy is represented by wave phenomena, energy that is potentially available for heating the upper atmospheric layers.

In this review we discuss some results of investigations into the mass and energy flow in the solar chromosphere and corona. The objective of these investigations is the development of a physical model that will not only account for the physical conditions in the outer atmosphere of our nearest stellar neighbor, the Sun, but also can be applied to the study of the chromospheres and coronae of other stars. The currently available physical models (e.g. Ulmschneider 1967, Kuperus 1969, deLoore 1970, Landini & Monsignori-Fossi 1973, Leibacher & Stein 1974, Hearn 1975) need to be improved before they can accurately predict conditions in the mechanically heated layers of stellar atmospheres. Because there is now empirical evidence for the presence of both chromospheres and coronae in stars other than the Sun (e.g. Jordan & Avrett 1973, Catura, Acton & Johnson 1975, Dupree 1975, Evans, Jordan & Wilson 1975, McClintock et al. 1975, Mewe et al. 1975, Rogerson & Lamers 1975, Dupree & Baliunas 1976), the development of reliable physical models for outer stellar atmospheres has become increasingly important. These models also have important implications concerning the prediction of stellar winds and mass loss from both early- and late-type stars.

Role of Magnetic Fields

Differences in the physical conditions (temperature, density, and mass and energy flow) in the chromospheric and coronal layers over different areas of the solar surface appear to be intimately related to the configuration and strength of the magnetic fields in these areas. The X-ray photograph in Figure 1 illustrates three major classes of structure: 1. active regions and coronal bright points, which are characterized by strong coronal magnetic fields with closed configurations, 2. coronal holes, which are associated with weak coronal magnetic fields having open configurations, and 3. quiet regions, which have weak coronal magnetic fields that appear to be closed on a large scale. The areas of intense X-ray emission in Figure 1 are active regions and bright points, the large low-emission area near the center of the photograph is a coronal hole, and the areas of intermediate intensity surrounding the coronal holes, bright points, and active regions are the so-called quiet regions. The evident role of magnetic fields in shaping the structures seen in Figure 1 must be due at least in part to the ability of magnetic fields to channel the flow of mass and energy in the corona. It may also be due in part to direct deposition of magnetic energy stored in the field.

The development of a reliable physical model for the mass and energy flow in the solar atmosphere depends critically on knowledge of the relevant heating and cooling mechanisms. One way of obtaining information about these mechanisms is through investigation of the physical conditions in solar structures characterized by different coronal magnetic field strengths and configurations. From these studies one can derive estimates of energy losses in each region and thereby obtain estimates of the required energy input. These estimates, combined with other information determined from empirical and theoretical studies, provide insights concerning possible heating mechanisms. In the following sections we briefly summarize the basic heating and cooling mechanisms and then discuss results of empirical and theoretical investigations into the physical conditions and the mass and energy flow in the solar atmosphere.

Mechanisms for Energy Loss

The three primary mechanisms of energy loss in the solar atmosphere are: 1. radiation, 2. thermal conduction, and 3. mass flow. In the chromosphere the principal source of energy loss is radiation, while in the corona all three sources are important because it loses energy by radiation, by conduction to the lower atmosphere, and by mass flowing outward as solar wind and downward into the chromosphere. The configuration of the magnetic field is an important factor in

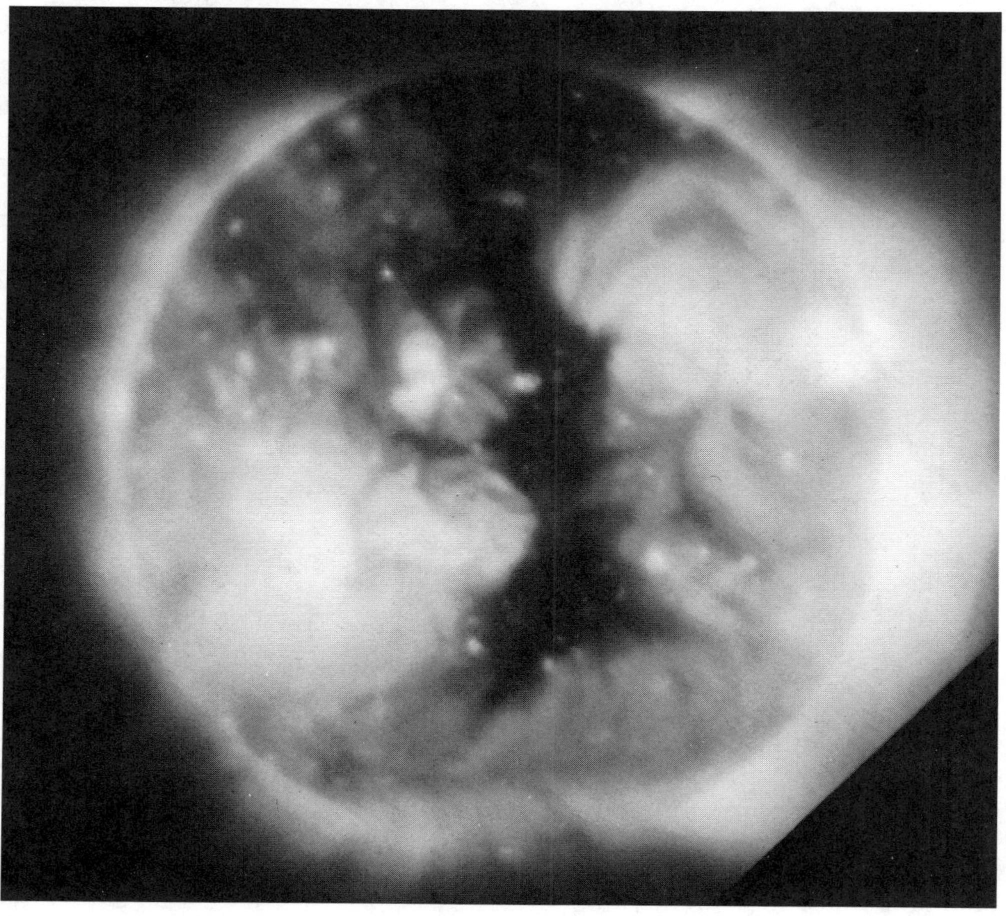

Figure 1 Soft X-ray photograph of corona on 1 June 1973 obtained by the Skylab experiment of American Science & Engineering. North is at the top, east on the left. (Photograph courtesy of American Science and Engineering/Harvard College Observatory.)

controlling energy losses. In the chromosphere and low corona (heights below one solar radius $R_{\odot}$), the flow of energy transported by thermal conduction and flow of plasma are constrained by the magnetic fields to move along field lines. In addition, if the magnetic configuration is closed (both ends of the field lines connected to the solar surface) and if the magnetic pressure exceeds the gas pressure, as is generally the case in the chromosphere and low corona, then the magnetic field prevents the outward flow of plasma or solar wind from transporting mass and energy away from the atmospheric layers contained within the closed configuration.

Mechanisms for Energy Input

There are four primary mechanisms of energy input to the chromosphere and corona: 1. dissipation of MHD or acoustic waves, 2. dissipation of energy stored in magnetic fields, 3. thermal conduction, and 4. mass flow. As is discussed later, dissipation of energy carried by waves generated in the convection zone is the most likely source of the energy heating the chromosphere. Energy transported downward from the corona by thermal conduction may also contribute to the heating of the upper chromosphere, particularly in active regions. In some regions the energy released by material falling into the chromosphere (e.g. from spicules, prominences, and cooling coronal material) may also be important. Although it has been believed traditionally that the corona is heated by dissipation of MHD or acoustic waves, magnetic heating is an alternative possibility that deserves more attention than it has received. Energy released by ohmic dissipation of currents or magnetic field annihilation may provide a substantial fraction, perhaps all, of the energy required to heat the coronal layers of active regions, bright points, and perhaps even quiet regions and coronal holes. Dissipation of magnetic energy may also contribute to the heating of chromospheric layers where strong fields are found, as in the chromospheric network and active regions. The magnetic field configuration plays an important role in energy input by channelling mass and energy (energy carried by thermal conduction and MHD waves) flow along the field lines. The magnetic configuration is also important in determining where dissipation of magnetic energy is feasible.

MASS AND ENERGY FLOW IN REGIONS WITH WEAK CORONAL MAGNETIC FIELDS

Chromosphere

INTRODUCTION Since most of the mechanical energy passing upward through the solar photosphere is deposited in the chromosphere, knowledge concerning the physical conditions and energy flow in the chromosphere is of overriding importance in determining how much mechanical energy must be transported from the solar interior through the photosphere to heat the outer atmospheric layers. The physical conditions in the chromosphere are complex. In spectral lines and continua formed in the upper photosphere and chromosphere one observes a network structure consisting of an irregular pattern of bright emission overlying the downflowing outer edges of the large-scale photospheric cells known as the supergranulation (e.g. Bray & Loughhead 1974, Athay 1976). Most of the magnetic flux in quiet regions is in the network and at photospheric levels is concentrated in small areas ($\lesssim 700$ km) where the fields are 1–2 kG (see Stenflo 1975, 1976, 1977, and references given therein). Extending above the chromosphere into the hotter transition region and low corona are dynamic inhomogeneities in the form of spicules that consist of relatively cool material ($T \approx 10^4$ K) moving at velocities of the order of 25 km sec^{-1} (e.g. Bray & Loughhead 1974, Athay 1976).

Information about the physical conditions in the chromosphere is provided by measurements made in optically thick spectral lines and continua, such as the Ca II H and K lines, Mg II H and K lines, hydrogen Lyman lines and continua, infrared continuum ($\lambda > 20$ μm), and UV continuum ($\lambda < 2000$ Å). Many of these spectral features are affected by departures from local thermodynamic equilibrium (LTE). Because of the difficulties inherent in non-LTE analyses of optically thick spectral lines and continua, existing models for the chromosphere are generally based on the assumption of a homogeneous, plane-parallel atmosphere that is in hydrostatic equilibrium. Although such models are too simple, they provide us with important insights regarding the physical conditions in the chromosphere, particularly the layers below 2000 km where the chromosphere appears to be as a first approximation, plane-parallel. Hence we describe the homogeneous models first and then discuss the role of inhomogeneities.

HOMOGENEOUS MODELS Several representative models for the photosphere and low chromosphere are presented in Figure 2. The models of Gingerich et al. (1971) and Vernazza, Avrett, & Loeser (1976) are based on non-LTE analyses of continuum measurements, while the model by Ayres & Linsky (1976) is based on a non-LTE analysis of the profiles of the Ca II and Mg II H and K lines. The model of Holweger & Müller (1974) is based on an LTE analysis of spectral lines and continua and therefore is unreliable for atmospheric layers above $\tau_{5000} = 10^{-3}$, where departures from LTE become important. The model by Kurucz (1974) is a theoretical line-blanketed model that assumes radiative-convective equilibrium based on mixing-length theory, LTE, and no mechanical heating. For further discussion of chromospheric models, see recent reviews by Athay (1976), Avrett (1977), and Linsky (1977).

Determination of the requirements for mechanical heating in the vicinity of the temperature minimum depends critically on a comparison of temperature differences between empirical models and theoretical models based on radiative equilibrium. At $\tau_{5000} = 10^{-4}$ an input of 2×10^5 erg cm^{-2} sec^{-1} is required to raise the temperature 100 K above the radiative equilibrium temperature, while at $\tau_{5000} = 10^{-3}$, 2×10^6 erg cm^{-2} sec^{-1} is required to raise the temperature 100 K (Athay 1976). Given the uncertainty in the radiative equilibrium model and the wide variation between the empirical models in the layers $10^{-3} > \tau_{5000} > 10^{-5}$, one cannot reliably estimate the heating requirements for this part of the atmosphere or determine the exact height at which mechanical heating becomes important. As a representative number for the radiative losses in the low chromosphere, the region just above the temperature minimum, we take Athay's estimate of 2×10^6

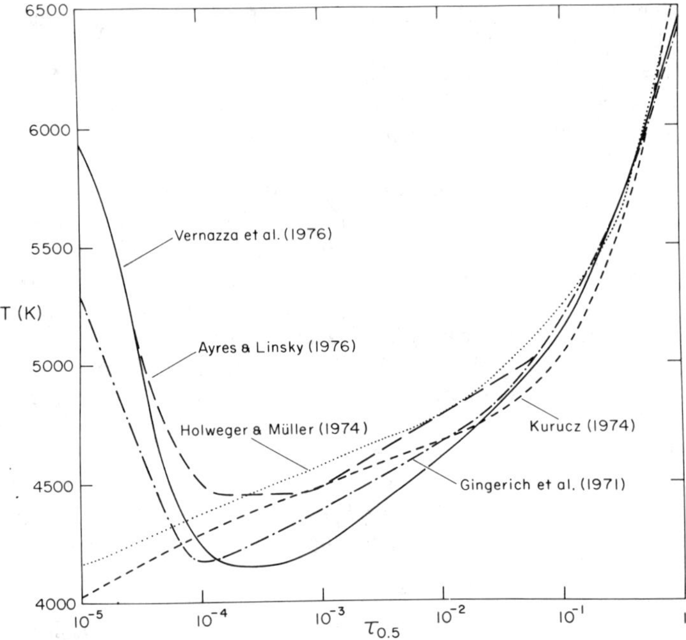

Figure 2 Temperature as a function of optical depth at 0.5 μm (5000 Å) in the upper photosphere and temperature-minimum region according to several recent models. (Figure courtesy of E. H. Avrett, Center for Astrophysics.)

erg cm^{-2} sec^{-1} (Athay 1976). For the middle chromosphere, which extends upward to the level where the temperature reaches 10^4 K, the radiative losses are also about 2×10^6 erg cm^{-2} sec^{-1} (Athay 1976) based on the measured intensities of spectral lines and continua formed there. The principal sources of radiative loss in the middle chromosphere are strong lines of Fe I, Na I, Ca II, Mg II, and the hydrogen Balmer lines and continua. In the upper chromosphere, $10^4 \lesssim T \lesssim 3 \times 10^4$ K, the principal source of radiative loss is hydrogen Lyman alpha emission, which yields a loss rate of about 3×10^5 erg cm^{-2} sec^{-1} (Athay 1976). Ulmschneider's (1974) estimates for the radiative losses in the low and middle chromosphere, 2.5 to 3.3×10^6 erg cm^{-2} sec^{-1}, are in good agreement with Athay's.

NETWORK Observations of the chromospheric network in various spectral lines and continua (e.g. Bray & Loughhead 1974, Athay 1976) indicate that at all chromospheric heights, from the temperature minimum upwards, the radiative losses in the network are larger than in the cells. For areas with sizes of a few times 10^3 km the network appears to radiate about two times more energy per unit area than the cells (based on results given by Skumanich, Smythe & Frazier 1975, Stenflo 1975, 1976, Athay 1976, Reeves 1976). Consequently, there must be more mechanical energy deposited per unit area in the chromospheric layers of the network than in the cells. The increased heating in the network is presumably related to the local concentration of magnetic fields there.

SPICULES These features, the major component of the chromospheric fine structure, are cylindrical structures with characteristic widths of 10^3 km and lengths of 10^4 km (e.g. Beckers 1972, Bray & Loughhead 1974, Athay 1976) and appear to be jets of material at chromospheric temperatures ($T \approx 10^4$ K) that extend from the chromospheric network upward into the hotter corona ($T \approx 10^6$ K). The mass and energy flow represented by spicules is a critical factor in the overall mass and energy balance of the chromosphere and corona. Spicules with densities of 1.7×10^{-13} g cm^{-3} moving upward at velocities of 25 km sec^{-1} carry an upward mass flux of about 4 to 8×10^{-9} g cm^{-2} sec^{-1} averaged over the surface of the Sun (since spicules cover 1–2% of the surface). Because the solar wind mass loss averaged over the surface is about 3×10^{-11} g cm^{-2} sec^{-1}, most of the spicular material must return to the chromosphere (cf Beckers 1972, Athay 1976).

If the energy required to create spicules is derived from a nonthermal source, such as energy stored in magnetic fields, spicules may provide a mechanism for converting this nonthermal energy into thermal energy. The energy required to raise chromospheric material to the heights reached by spicules is about 2×10^{24} ergs per spicule. If this energy is converted into thermal energy when the spicular material falls back into the chromosphere, the amount of energy released at the base of the spicule is about 10^6 erg cm^{-2} sec^{-1} (or 2×10^4 erg cm^{-2} sec^{-1} when averaged over the solar surface). Since this is comparable to the amount of energy radiated in the upper chromosphere at the base of spicules, falling spicular material may provide a significant fraction of the energy required to heat these areas. However, spicules represent only a small fraction of the total energy flow in the chromosphere, when averaged over the entire disk.

A variety of mechanisms for generating spicules have been proposed (see reviews by Beckers 1972, Bray & Loughhead 1974, Athay 1976, Stenflo 1976). Some models utilize energy stored in chromospheric magnetic fields to accelerate chromospheric material upward to form spicules, while others assume that spicules are driven by convective instabilities in the chromosphere or by shock waves propagating along magnetic field lines. At present there is no clear choice as to the best model.

CHROMOSPHERIC LAYERS OF CORONAL HOLES In most spectral lines and continua there is little difference in the appearance and intensity of the chromosphere observed in coronal holes and quiet regions (e.g. Munro & Withbroe 1972, Huber et al. 1974, Reeves 1976). An exception to this is provided by the helium lines and continua, whose intensities tend to be lower in coronal holes than in quiet regions by 40 to 50% (e.g. Munro & Withbroe 1972, Tousey et al. 1973, Huber et al. 1974, Bohlin, Sheeley & Tousey 1975, Bohlin 1977a,b). The behavior of the helium lines in coronal holes has been attributed to the effects of diffusion (Jordan 1975b, Milkey 1975, Shine, Gerola & Linsky 1975) and to the decreased coronal *XUV* radiation incident on the chromosphere, which affects the ionization of helium (Zirin 1975, Avrett, Vernazza & Linsky 1976).

The other major difference between the chromospheric layers of coronal holes and quiet regions is in the spicules. Spicules appear to extend farther above the surface in coronal holes than in quiet regions, as evidenced by observations indicating that spicules at the poles (where coronal holes are usually found) are taller than in equatorial regions (e.g. Beckers 1972, Bray & Loughhead 1974, Athay 1976). In addition, very large spicules, known as macrospicules (Bohlin et al. 1975, Withbroe et al. 1976) extend as high as 40,000 km above the limb in polar coronal holes. Because of the amount of energy required to produce these features, 3×10^{26} ergs for a large macrospicule (Withbroe et al. 1976), they are most likely generated by a mechanism that utilizes energy stored in magnetic fields to accelerate chromospheric material upward. The observation (Moore et al. 1976) that some macrospicules are associated with bright-point flares supports this hypothesis. If the same mechanism is responsible for the generation of both spicules and macrospicules, then a successful model must explain why macrospicules seem to occur predominantly in coronal holes (Bohlin et al. 1975).

Because there appears to be little difference between the brightness of the chromosphere as observed in spectral lines and continua radiating most of the energy from the chromospheric layers of coronal holes and quiet regions, the chromospheric radiative losses are approximately the same for the two types of regions. Consequently, the observations imply that approximately the same amount of mechanical energy is deposited in the chromospheric layers of these regions.

Transition Layer and Corona in Quiet Regions

MAGNETIC CONFIGURATION We define as "quiet regions" those areas of the solar atmosphere away from active regions, filaments, and coronal holes. Quiet areas of the chromosphere and corona overlie regions with mean weak photospheric magnetic fields. Evidence that these areas have magnetic configurations that are closed on a large scale (few tenths $R_{\odot}$) is provided by 1. calculations of coronal magnetic fields from measurements of photospheric fields (e.g. Levine et al. 1977) and 2. the appearance of the corona as observed in X-ray filtergrams (e.g. Vaiana, Krieger & Timothy 1973, Vaiana et al. 1973, Vaiana 1976). However, the evidence is not yet sufficiently reliable to rule out the possibility that some open field lines originate in quiet regions and that solar wind flows out from these areas. If quiet regions are partially open, then the mean energy loss due to the solar wind is less than or equal to a few times 10^4 erg cm^{-2} sec^{-1} (e.g. Kopp 1972), while the mass loss is less than or equal to 2×10^{-11} g cm^{-2} sec^{-1}. In coronal holes, as we shall see below, the mass and energy losses due to the solar wind are an order of magnitude larger. Figure 3 illustrates the variation with height of the temperature and density in typical quiet region in which horizontal structures are neglected.

In the transition layer and low corona the small-scale configuration of the field (few times 10^4 km) is important. Two types of structures are relevant: 1. small-scale, magnetically closed regions that are often associated with coronal bright

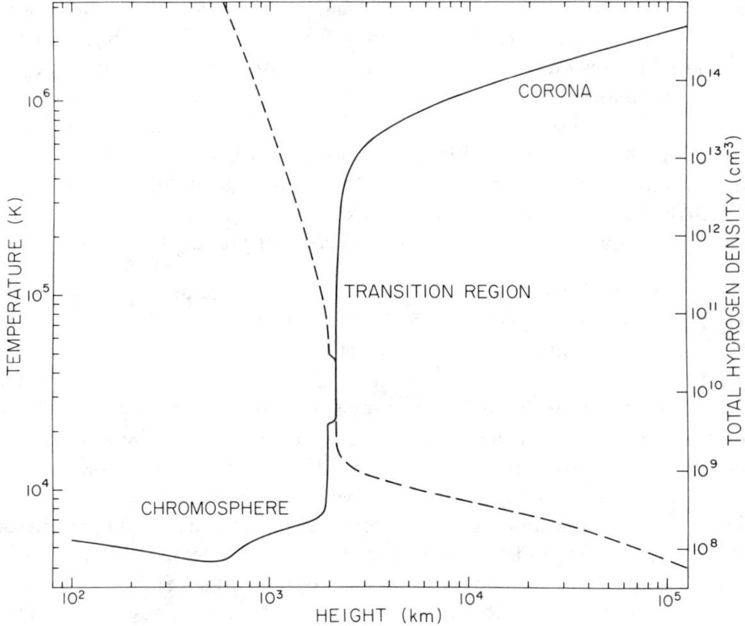

Figure 3 Variation with height of the temperature and density in the chromosphere, transition layer, and corona in a typical quiet region.

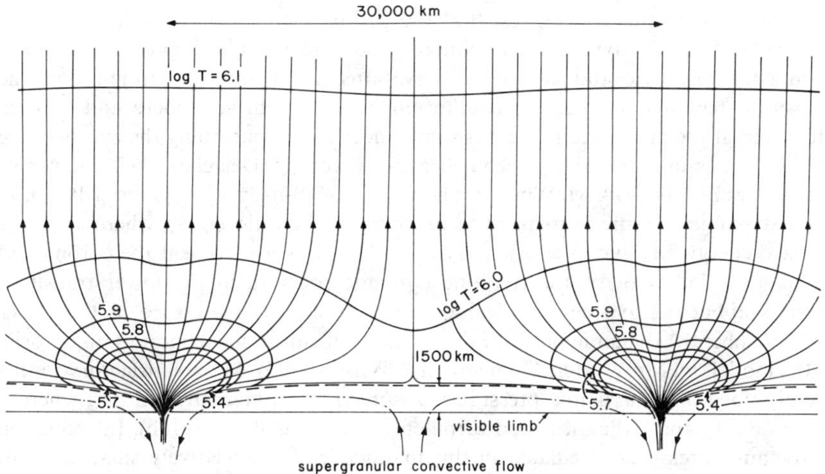

Figure 4 Magnetic field configuration associated with the network and temperature contours from Gabriel's (1976a) model for the transition region and low corona in a quiet region.

points and 2. regions where the magnetic field diverges from the chromospheric network, as shown schematically in Figure 4 from a model by Gabriel (1976a). In quiet regions, it appears that the field lines extending from the network, which appear to be radial in the idealized configuration shown in this figure, eventually return to the Sun and form a large-scale closed configuration, while in a coronal hole they remain open. The field lines channel energy carried by conduction from the corona into the chromospheric network (Kopp & Kuperus 1968, Kopp 1972, Gabriel 1976a,b). They also guide the flow of material between the chromosphere and corona (e.g. spicules) and control the direction of propagation of MHD waves.

EMPIRICAL TEMPERATURE DENSITY MODELS Empirical models for the transition layer and lower corona (heights $< 0.5\ R_{\odot}$) have been derived from measurements of XUV spectral line intensities and radio brightness temperatures (see reviews by Athay 1971, 1976, Noyes 1971, Jordan & Wilson 1971, Noyes & Withbroe 1972, Gabriel 1976b, Withbroe 1976). These models give the variation with height of the temperature and density for the atmospheric layers from which the XUV and radio emissions originate. From the temperature gradients, temperatures, and densities in these models one can evaluate the radiative and conductive energy losses as a function of height in the atmosphere through use of formulae relating the conductive flux to the temperature gradient (e.g. Spitzer 1962, Ulmschneider 1970) and radiative losses to the temperature and density (e.g. Cox & Tucker 1969, McWhirter, Thonemann & Wilson 1975, Raymond, Cox & Smith 1976). These models indicate that the radiative energy losses in the corona are typically 10^5 erg cm^{-2} sec^{-1} and that the mean conductive flux from the corona into the lower layers of the atmosphere is about 2×10^5 erg cm^{-2} sec^{-1} (see reviews by Gabriel 1976b, Withbroe 1976).

Most empirical models suggest that the gas pressure in the transition layer is between 0.1 and 0.2 dyn cm^{-2}, a value in good agreement with pressures inferred from chromospheric and coronal observations under the assumption that the transition region is in pressure equilibrium with the chromosphere and corona. However, at the present time there is some uncertainty concerning the effects of the dynamic pressure caused by turbulence or waves (cf Delache 1967, Flower & Pineau des Forets 1974, McWhirter, Thonemann & Wilson 1975, Gabriel 1976a,b). If the dynamic pressure is assumed to be given by the ratio ϕ_w/V_s, where ϕ_w is the upward energy flux due to waves carrying sufficient energy to heat the corona and V_s is the sound velocity, the resulting dynamic pressure in the lower transition layer is about 0.2 dyn cm^{-2}, a value comparable to the pressure in the upper chromosphere. A dynamic pressure this high is difficult to incorporate satisfactorily in the models (cf McWhirter, Thonemann & Wilson 1975). More satisfactory results are obtained if the dynamic pressure is assumed to be given by $1/2\ \rho v^2$, where ρ is the density and v the measured turbulent velocity (Gabriel 1976b). Introduction of dynamic pressures calculated in this manner leads to relatively small changes (less than a factor of 2) in the densities and temperature gradients in lower transition layer as compared to models where the dynamic pressure is not included.

ENERGY BALANCE MODELS Several semi-empirical models for the transition layer and lower corona have been derived by assuming energy balance (e.g. Moore & Fung 1972, Pneuman 1973, McWhirter, Thonemann & Wilson 1975, Gabriel 1976a,b, Kopp & Orrall 1976, Rosner & Vaiana 1977). These models are based on the assumption of net balance between the various energy sinks and sources in each atmospheric layer, that is

$$\nabla \cdot (\mathbf{F}_r + \mathbf{F}_c + \mathbf{F}_m + \mathbf{F}_g + \mathbf{F}_k + \mathbf{F}_e) = 0,$$

where $\mathbf{F}_r$ is the radiative flux, $\mathbf{F}_c$ is the conductive flux, $\mathbf{F}_m$ is the mechanical flux, $\mathbf{F}_g$ is the gravitational potential energy flux, $\mathbf{F}_k$ is the kinetic energy flux, and $\mathbf{F}_e$ is the enthalpy flux. On the basis of the studies performed to date it appears that the divergence of the downward conductive flux provides sufficient energy to balance the radiative losses in the upper transition region ($10^5 < T < 10^6$ K), but that mechanical heating is required below 10^5 K.

Figure 4 shows the magnetic geometry and temperature contours from an energy balance model developed by Gabriel (1976a). This model takes explicitly into account the influence of the channeling of conductive energy by the magnetic fields associated with the network and satisfactorily explains two of the chief characteristics of *XUV* observations made in lines formed in the transition region and lower corona: 1. that the network is bright in lines formed at temperatures between 10^4 and 10^6 K and concentrated in structures with widths of the order of 10 arc seconds and 2. that in lines formed above 10^6 K the quiet Sun is observed to be relatively uniform (e.g. Mariska & Withbroe 1975, Reeves 1976, Withbroe 1976).

Table 1 summarizes the various energy losses for the chromosphere, transition region, and corona in a typical quiet area.

Coronal Holes

MAGNETIC CONFIGURATION Coronal holes appear to be associated with weak-field magnetic structures that have an open diverging configuration. This is illustrated in Figure 5, which compares the location of coronal holes as shown by the contours in the lower portion of figure with the points of origin of magnetic field lines that are swept out by the solar wind (upper portion of the figure). The field lines were calculated by Levine, Altschuler & Harvey (1977) from potential theory using Kitt Peak measurements of photospheric magnetic fields. The contours of the coronal holes were determined from X-ray photographs obtained by the American Science and Engineering experiment on Skylab (Nolte et al. 1976b). As is readily apparent, the open-field lines originate in the coronal holes.

SOLAR WIND ASSOCIATION Because coronal holes appear to be associated with magnetic structures that have an open configuration, solar wind can flow out of these regions. In fact, there is now a substantial body of evidence that coronal holes are a major source, perhaps the primary source, of the solar wind, particularly the recurrent high-speed streams responsible for recurrent geomagnetic disturbances (e.g. Krieger, Timothy & Roelof 1973, Krieger et al. 1974, Neupert & Pizzo 1974,

Bell & Noci 1976, Nolte et al. 1976a, Sheeley, Harvey & Feldman 1976). These high-speed solar wind streams are a major source of energy loss in coronal holes, a factor that has important ramifications regarding the mass and energy flow all the way down to the chromosphere.

The total energy loss (consisting of the kinetic energy, potential energy, and

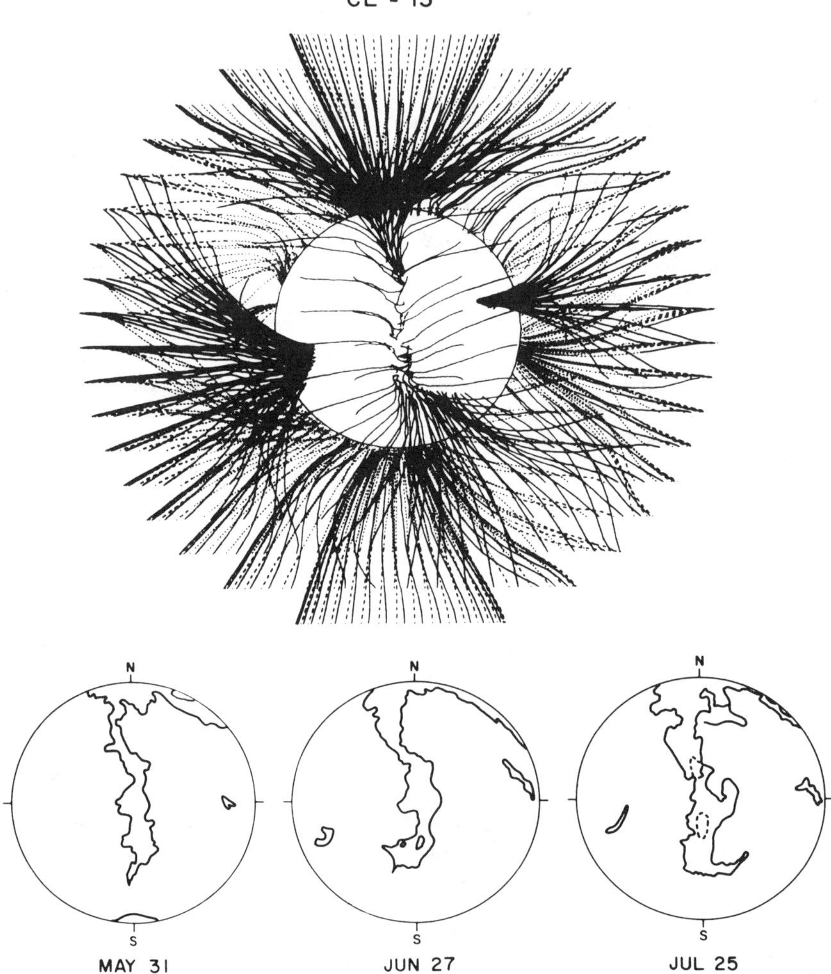

Figure 5 Comparison of calculated magnetic field configuration (Levine, Altschuler & Harvey 1977) for field lines swept out by the solar wind (upper part of the figure) and contour maps giving locations of coronal holes determined from X-ray photographs similar to that of Figure 1 (Nolte et al. 1976b). (Figure courtesy R. H. Levine, Center for Astrophysics.)

enthalpy) represented by a typical high-speed stream is of the order of 7×10^5 erg cm^{-2} sec^{-1}. The total mass loss is approximately 2×10^{-10} g cm^{-2} sec^{-1}. These estimates are based on an assumed proton density at the earth of 6 cm^{-3}, solar wind velocity of 650 km sec^{-1} (e.g. Hundhausen 1972), and an expansion factor of 6 (e.g. Bohlin 1977a,b, Zirker 1976, Munro & Jackson 1977), where the expansion factor is the ratio between the angular area covered by a high-speed stream measured at the earth and the angular area of the coronal hole from which the stream originates.

The chromospheric source of the mass lost in high-speed streams is not known at the present time. One possibility is spicules and macrospicules. As indicated earlier, the upward mass flux represented by spicules is of the order 4 to 8×10^{-9} g cm^{-2} sec^{-1} averaged over the surface of the Sun. Hence only a few percent of this mass needs to be deposited in the corona to replenish the mass lost in the solar wind.

EMPIRICAL MODELS The principal cause of the low coronal emission in coronal holes (e.g. Figure 1) is the low density in these regions, which is a factor of 2 to 3 lower than in normal quiet regions (e.g. Munro & Withbroe 1972, Huber et al. 1974, Vaiana 1976). A second difference between coronal holes and quiet regions is that the temperature gradient in the transition layer is less steep in coronal holes by a factor of about five and therefore the conductive flux from the corona to the transition layer is reduced by the same factor, to a value of about 6×10^4 erg cm^{-2} sec^{-1} (e.g. Munro & Withbroe 1972, Huber et al. 1974, Bohlin, Sheeley & Tousey 1975, Mariska 1976). This result has two important implications: 1. the conductive flux from the corona to the transition region is too low to balance the radiative losses in the transition region in the layers where $10^5 < T < 10^6$ K, thereby implying that mechanical heating must take place in the transition region in coronal holes and 2. the loss of energy from the corona by conduction to the lower atmosphere is no longer the dominant coronal cooling mechanism that it is in the quiet Sun.

Another important factor in the structure of coronal holes that appears to be related to the fact that they are sources of high-speed solar wind streams is the rapid increase in the angular area covered by a coronal hole with increasing distance from the solar surface. This rapid divergence with height has been demonstrated by calculations of the magnetic field configuration (e.g. Figure 5; Altschuler, Trotter & Orrall 1972, Levine, Altschuler & Harvey 1977, Levine et al. 1977) and by analysis of coronal observations made with X-ray and *XUV* telescopes and white light coronagraphs (e.g. Munro & Jackson 1977, Bohlin 1977b). Munro and Jackson find that at 3 $R_\odot$ the angular area of the polar hole they studied is seven times larger than at the surface. Similar expansion factors have been found for other coronal holes (e.g. Zirker 1976).

Table 1 compares some of the physical properties of coronal holes and quiet regions obtained from empirical models discussed above. As indicated earlier, there appears to be little difference in the chromospheric emission measured in quiet regions and coronal holes, so we have assumed that the chromospheric radiation losses in both regions are equal. This suggests that the requirement for mechanical

Table 1 Chromospheric and coronal energy losses

Parameter	Quiet Sun	Coronal hole	Active region
Transition layer pressure (dyn cm^{-2})	2×10^{-1}	7×10^{-2}	2
Coronal temperature (K, at $r \approx 1.1\ R_{\odot}$)	1.1 to 1.6×10^6	10^6	2.5×10^6
Coronal energy losses (erg cm^{-2} sec^{-1})			
Conduction flux F_c	2×10^5	6×10^4	10^5 to 10^7
Radiative flux F_r	10^5	10^4	5×10^6
Solar wind flux F_w	$\lesssim 5 \times 10^4$	7×10^5	$(<10^5)$
Total corona loss $F_c+F_r+F_w$	3×10^5	8×10^5	10^7
Chromospheric radiative losses (erg cm^{-2} sec^{-1})[a]			
Low chromosphere	2×10^6	2×10^6	$\gtrsim 10^7$
Middle chromosphere	2×10^6	2×10^6	10^7
Upper chromosphere	3×10^5	3×10^5	2×10^6
Total chromospheric loss	4×10^6	4×10^6	2×10^7
Solar wind mass loss (g cm^{-2} sec^{-1})	$\lesssim 2 \times 10^{-11}$	2×10^{-10}	$(<4 \times 10^{-11})$

[a] Based on Athay's (1976) estimates for the quiet Sun; see text.

heating in the chromospheric layers of the two types of regions is also the same. There are indications that the requirement for coronal heating is a factor of 2 to 3 larger in coronal holes than in quiet regions. However, given the uncertainties in the estimates, which could be in error by a factor of 2, it seems reasonable to adopt the hypothesis that the requirements for mechanical heating in the chromospheric and coronal layers of quiet regions and coronal holes are the same.

THEORETICAL MODELS FOR CORONAL HOLES A number of theoretical models for the transition layer and corona in open-field regions have been developed in recent years. Pneuman (1973) demonstrated that because of solar wind losses, open-field regions will have lower coronal temperatures and densities than closed-field regions. Kopp & Holzer (1976) studied the hydrodynamic properties of an expanding corona for flow tubes characterized by an angular cross section that increases with height low in the corona and then becomes constant at large distances above the solar surface. This is the configuration that is believed to occur in coronal holes because of the diverging configuration of the coronal magnetic field associated with these features. Kopp and Holzer find that if the rate of divergence is sufficiently large, the solar wind becomes supersonic at a critical point low in the corona in the region of high divergence. This behavior makes it possible to reconcile the low values of electron density in coronal holes with the large particle fluxes in the associated high-speed streams observed in the solar wind.

Several investigators (e.g. Gabriel 1976b, Kopp & Orrall 1976, McWhirter 1976, Rosner & Vaiana 1977, see also references given by Zirker 1976) have constructed models in which radiative, conductive, and solar wind losses were taken into account in the energy equation and where the mechanical heating was assumed to be due to waves or a source whose energy deposition decreased exponentially with height. These models account for many of the observed characteristics of coronal holes and suggest that direct momentum transfer from the mechanical flux to the plasma is a likely acceleration mechanism for the solar wind. Both the theoretical and empirical models (see Munro & Jackson 1977, Zirker 1976) indicate that the acceleration of the solar wind occurs near the Sun, within two to five solar radii.

MASS AND ENERGY FLOW IN REGIONS WITH STRONG CORONAL MAGNETIC FIELDS

Introduction

The fundamental structures in strong-field regions appear to be loops or arches connecting areas of opposite magnetic polarity. Figure 6 shows examples of active region loops as observed in spectroheliograms obtained in the lines Ne VII λ465 (mean temperature of formation $T_f = 6 \times 10^5$ K), Mg IX λ368 ($T_f = 10^6$), and Fe XV λ285 ($T_f = 2.5 \times 10^6$ K). A Kitt Peak magnetogram is also shown for comparison. In order to understand the mass and energy flow in active regions, we need to determine the physical conditions in these loop structures and how they change with time. If one can specify: 1. the radiative and conductive energy losses as a function of position in individual loops, 2. the mass flow, and 3. the rate of heating or cooling of material contained in the loops, then one can obtain constraints on possible heating mechanisms. Development of detailed models specifying all of these parameters is still in its infancy. However, sufficient work has been performed that we can get some insights regarding the mass and energy flow in strong-field regions.

Chromosphere

Empirical models for the upper photospheric and low chromospheric layers of active regions have been derived from center-to-limb observations of continua and lines in the visible part of the spectrum (e.g. Schmahl 1967, Chapman 1970, Stellmacher & Wiehr 1971, 1973, Stenflo 1975). These models indicate that the temperature minimum is located at a larger optical depth in active regions than in quiet regions, perhaps as low as $\tau_{5000} = 10^{-1}$ (as compared to $\tau_{5000} = 10^{-4}$ in quiet regions). In addition the minimum temperature is higher, perhaps as high as 5000 K. However, since the existing models are not in agreement on the height at which the temperature minimum occurs or on the temperature gradient in the vicinity of the temperature minimum, it is not possible to estimate reliably how much mechanical heating is required in the lower chromospheric layers of active regions. As indicated in the discussion of the quiet chromosphere, relatively small differences between the temperatures in radiative equilibrium and empirical models lead to very large differences in the heating requirements in the low chromosphere. The empirical models clearly indicate that mechanical heating becomes important at a lower level in the atmosphere and is more intense in active regions than in quiet regions (outside of areas of the network where the magnetic fields are strong).

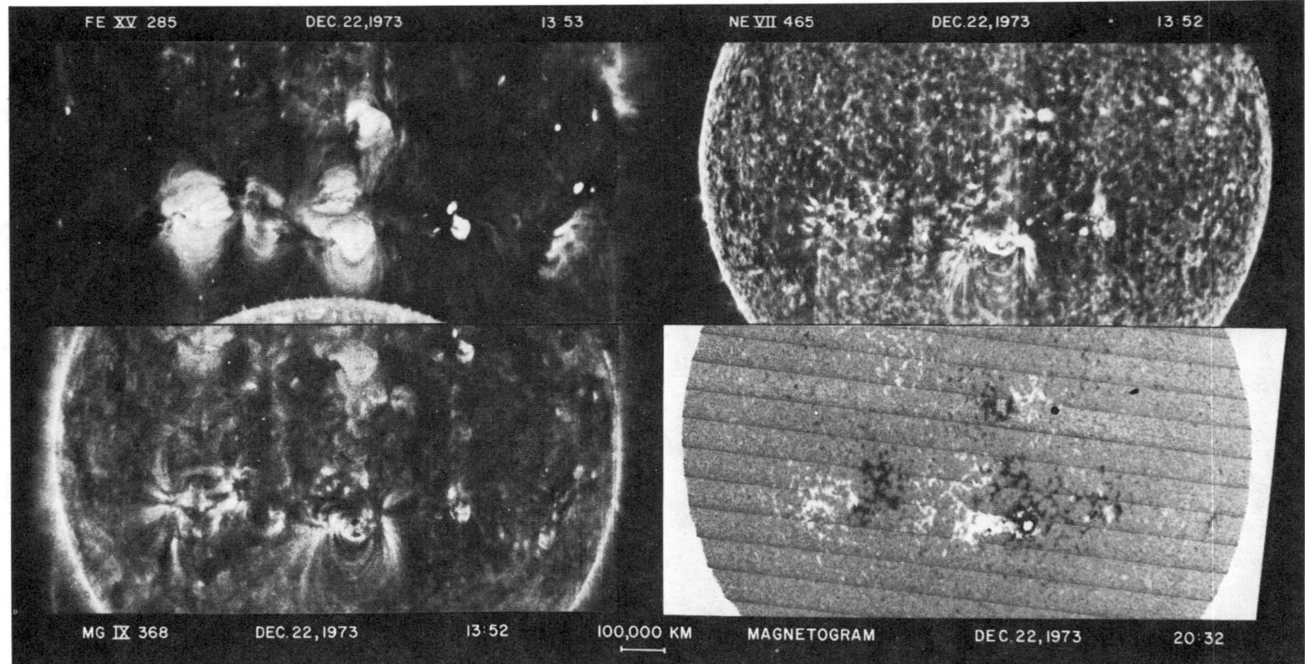

Figure 6 Spectroheliograms in several *XUV* spectral lines showing coronal loops associated with several active regions. The light and dark areas in the magnetogram are regions of enhanced magnetic fields of opposite polarity. The spectroheliograms are from the Skylab experiment of the Naval Research Laboratory (NRL) and the magnetogram is from the Kitt Peak National Observatory. (Figure courtesy of N. R. Sheeley, NRL, and KPNO.)

Given the temperature increases found by Chapman (1970), Stenflo (1975), and others referred to above, the heating requirements for the upper photosphere and low chromosphere in active regions could easily exceed 10^7 erg cm^{-2} sec^{-1}.

Shine & Linsky (1974) have developed a series of models for the chromospheric layers of active regions using measured profiles of the Ca II H, K, $\lambda 8498$, $\lambda 8542$, and $\lambda 8662$ lines and boundary conditions imposed by the photosphere and hydrogen Lyman continuum. They find that the increase in emission in the Ca II lines from quiet to active regions appears to be a consequence of an increased temperature gradient in the low chromosphere and a resulting increase in pressure and electron density at all chromospheric levels. The intensities of lines formed in the middle and upper chromosphere are observed to be a factor of 3 to 5 brighter in active regions than quiet regions (e.g. Noyes, Withbroe & Kirshner 1970, Skumanich, Smythe & Frazier 1975) suggesting that the radiative losses in these chromospheric layers are larger by a similar factor. This yields loss rates of about 10^7 erg cm^{-2} sec^{-1} and 1.5×10^6 erg cm^{-2} sec^{-1} respectively for the middle and upper chromospheric layers of active regions (see Table 1).

Transition Region and Corona

Early empirical models for the transition region and corona based on *XUV* measurements with low spatial resolution (e.g. Noyes, Withbroe & Kirshner 1970, Withbroe & Gurman 1973) indicated: 1. that the mean conductive flux from the corona to the chromosphere is about 5×10^6 erg cm^{-2} sec^{-1}, 2. that the pressure in the transition region and low corona is about one dyn cm^{-2}, and 3. that the mean coronal temperature is about 2.5×10^6 K. Given the complexity of the structures in active regions, the existence of the coronal loops containing material at temperatures ranging from 10^5 to 10^7 K, such models are overly simplistic. For example, loops that have one end terminating in a sunspot appear to have very low temperature gradients along the length of the loop, corresponding to downward conductive fluxes of the order 10^5 erg cm^{-2} sec^{-1} (e.g. Noyes 1974, Cheng & Moe 1977, Foukal 1976), while hot loops (loops filled with plasma having $T > 10^6$ K) can have conductive fluxes of the order of a few times 10^6 erg cm^{-2} sec^{-1} to 10^7 erg cm^{-2} sec^{-1} (e.g. Gabriel & Jordan 1975, Jordan 1975a, 1976, unpublished results from Harvard Skylab experiment). The gas pressure in these loops can range from 0.1 dyn cm^{-2} for some sunspot loops to several dyn cm^{-2} in hot X-ray emitting loops (e.g. Noyes 1974, Gabriel & Jordan 1975, Davis et al. 1975, Jordan 1975a, 1976, Landini et al 1975, Foukal 1975a, 1976, Neupert, Nakagawa & Rust 1975, Levine & Withbroe 1977). However, since most of the material contained in active-region loops is at a temperature of about 2.5×10^6 K and pressure $P \approx 2$ dyne cm^{-2}, one can estimate typical values for the coronal radiative loss rates, which appear to be of the same order of magnitude or larger than the conductive loss rates in these loops. Consequently, we obtain for active regions the parameters given in Table 1. A comparison of the data in this table indicates that the chromospheric and coronal energy losses in active regions exceed those in quiet regions and coronal holes by approximately one order of magnitude and therefore that the requirements for mechanical heating are an order of magnitude larger.

OBSERVATIONAL EVIDENCE FOR WAVE HEATING AND SYSTEMATIC MASS FLOWS

XUV Brightenings

One of the fundamental problems in solar physics is the nature of the coronal heating mechanism. Although it is generally believed that the corona is heated by dissipation of energy carried upward from the convection zone by waves, we still do not have unambiguous observational proof that this is the case. However, measurements of spectral line intensities and profiles have provided some important constraints. Vernazza et al. (1975) have studied the time variation of the intensities of *XUV* lines. Simultaneous observations made with lines formed in the chromosphere, transition region, and corona show aperiodic brightenings. Cross correlations of the data suggest that on the average there is a time delay of a few seconds between brightenings observed in the chromosphere or lower transition layer and brightenings observed in the upper transition layer. This suggests the possibility that the brightenings are caused by an upward propagating phenomena, possibly waves, because the time delays are consistent with a disturbance moving at the sound speed. However, the identification of this phenomena with waves is uncertain, and the analysis of the data has not yet resulted in an estimate of how much upward propagating energy flux could be associated with these brightenings. Except for chromospheric lines, there is no significant evidence for a 300-second periodicity in these brightenings.

Spectral Line Profiles

Other important clues regarding the mass and energy flow in the solar atmosphere are provided by measurements of spectral line profiles. Several recent reviews have summarized work in this area (e.g. Beckers & Canfield 1975, Deubner 1976b, White 1976, Withbroe 1976). The observed profiles of lines formed in the chromosphere, transition layer, and corona are broadened by nonthermal motions in the line-forming medium. Analyses of the observations indicate that in the atmospheric layers above the temperature minimum, the nonthermal velocities are isotropic and increase with height from a few kilometers per second in the low chromosphere to 25–30 km sec^{-1} at the level where the temperature reaches a few times 10^5 K in the transition region (e.g. Beckers & Canfield 1975, Boland et al. 1973, 1975, Moe & Nicolas 1977, Doschek et al. 1976, Feldman, Doschek & Tousey 1975a,b, Feldman, Doschek & Patterson 1976). Measurements of coronal forbidden lines in the visible (see reviews by Beckers & Canfield 1975, Athay 1976) and permitted and forbidden lines in the *UV* and *XUV* (Feldman & Behring 1974, Moe & Nicolas 1977) indicate that the nonthermal velocities in the corona are in the range 10–30 km sec^{-1}.

Boland et al. (1973, 1975), McWhirter, Thonemann & Wilson (1975), and Jordan (1976) suggest that this nonthermal broadening may be caused by acoustic or MHD waves responsible for heating the corona. On the basis of the calculations of Boland et al. it appears that the observed nonthermal spectral line broadening is consistent with a wave flux carrying energy of about 4×10^5 erg cm^{-2} sec^{-1}, a value adequate to balance the radiative and conductive losses from the corona in quiet regions and coronal holes. However, as Boland et al. point out, the nonthermal broadening may also be caused by nonpropagating waves.

Feldman, Doschek & Tousey (1975a,b) have compared *XUV* line profiles measured at the limb in a polar coronal hole and a quiet region. On the basis of these profiles it appears that the half-widths of lines formed in the transition region are the same in coronal holes and the quiet Sun. This indicates that the nonthermal broadening is the same in both types of regions and is evidence that the wave flux passing through the transition region is also the same—if we attribute the nonthermal broadening to propagating waves.

Measurements of line profiles also provide information about oscillatory motions and systematic flows. Since Beckers & Canfield (1975), Athay (1976), Deubner (1976b), White (1976), and others have reviewed this subject in some detail, we mention only a few recent results. Observations of *XUV* lines by rocket and satellite experiments (see White 1976) indicate that lines formed in the transition layer of quiet regions and coronal holes exhibit systematic redshifts corresponding to velocities of 2 to 15 km sec^{-1}. These flows appear to be predominately into regions of high magnetic field strength in the network. This behavior could result from cooling due to local thermal instabilities in coronal material that becomes denser than the surrounding coronal plasma and falls into the chromosphere (Foukal 1976, Pneuman & Kopp 1977). Measurements of Doppler shifts in coronal lines have provided evidence for systematic outflows of coronal material moving with velocities of 10 to 20 km sec^{-1} in some quiet regions and coronal holes (Cushman & Rense 1976, Brueckner 1976).

Periodic motions have been measured through the photosphere and chromosphere into the low transition layer (see reviews by Beckers & Canfield 1975, Athay 1976, Deubner 1976b, White 1976). The 300-second oscillation is present throughout this range of heights with shorter period oscillations also having been observed in the chromosphere and low transition layer. Near the temperature minimum Deubner (1976a,b) finds evidence for waves at virtually every period between 300 sec and 10 sec and a total mechanical flux of between 10^8 and 10^9 erg cm^{-2} sec^{-1}. Since the heating requirements for the chromospheric and coronal layers of quiet regions and coronal holes are only a few times 10^6 erg cm^{-2} sec^{-1} (cf Table 1), these waves contain sufficient energy to heat the outer atmosphere in these regions even if only a small fraction of the wave flux propagates into and is dissipated in these layers.

In active regions measurements of Doppler shifts and analyses of *XUV* observations of coronal structures indicate that there are systematic downflows of cool ($10^4 < T < 10^6$ K) plasma in loops whose footpoints are sunspots (Brueckner 1976, Foukal 1976, Levine & Withbroe 1977). Similar mass downflows have been observed in the visible part of the spectrum for many years (e.g. Tandberg-Hanssen 1967). In general it appears likely that the cool ($T < 10^6$ K) material observed in coronal loops associated with active regions is not in hydrostatic equilibrium, but flows along the magnetic field lines defining the loops. Most of this material appears to result from the condensation of hot ($T > 10^6$ K) coronal gas that becomes thermally unstable, cools, and falls into the chromosphere at either end of a magnetic loop. Some of the cool material in loops results from the injection of chromospheric matter into them by surges and small flares that are often observed to occur at the footpoints of loop structures. The problem of the interchange of material between the chromosphere and corona in active regions is one requiring much additional observational and theoretical work.

HEATING MECHANISMS

Chromosphere

In order to balance the energy losses in the chromosphere a mean energy input of about 4×10^6 erg cm^{-2} sec^{-1} is required in quiet regions and coronal holes and 2×10^7 erg cm^{-2} sec^{-1} is required in active regions (see Table 1). As discussed in the previous section, observations in spectral lines formed near the temperature minimum suggest that there is a large mechanical energy flux, of the order of 10^8 to 10^9 erg cm^{-2} sec^{-1}, carried by waves having periods between 10 and 300 sec. Even though radiation damping in the photosphere can remove a significant fraction of the wave energy, up to 90% (cf Leibacher & Stein 1974), an ample amount remains to balance the energy losses in the chromosphere and corona. Theoretical investigations of the dissipation of vertically propagating waves indicate that the low chromosphere can be heated by short-period waves (a few tens of seconds) that shock and deposit their energy in this layer (e.g. Ulmschneider 1971a,b, 1974, Leibacher & Stein 1974). Longer-period waves (e.g. 300 sec.) shock higher in the atmosphere and can heat the upper chromosphere and possibly the corona.

Hydromagnetic waves, whose heating increases rapidly with increasing field strength, most likely provide the increased energy required to heat the chromosphere in regions with strong magnetic fields, such as those found in active regions and the chromospheric network (e.g. Osterbrock 1961, Milkey 1970, Piddington 1973, Stenflo 1975, 1976, 1977, Uchida & Kaburaki 1974). The discovery that most of the magnetic flux in the photosphere is concentrated in kilogauss fields located in small spatial elements has led to the suggestion that the chromospheric heating in the network and active region plages is the same; the primary difference is in the number of magnetic elements concentrated in a given region (Stenflo 1975, 1976).

Although theoretical and observational studies indicate that wave heating is an attractive chromospheric heating mechanism, heating by dissipation of magnetic energy provides an alternative possibility. Because of the concentration of the magnetic flux into small spatial elements with high field strengths, there is a substantial amount of energy stored in these fields. If this energy is dissipated on a scale of 10 minutes, an amount of energy comparable to that radiated by the chromosphere will be released (Stenflo 1976). Stenflo argues that lifetimes of this length are consistent with the observations. However, more study is required before one can conclude whether or not magnetic heating provides a feasible alternative or important supplement to wave heating of the chromosphere.

Corona

WAVE HEATING The energy input required to balance coronal energy losses (see Table 1) could be provided either by dissipation of acoustic or hydromagnetic waves or by dissipation of magnetic energy. As indicated in reviews by Leibacher & Stein (1974), Athay (1976), and others, self-consistent shock-heated coronal models are much more difficult to construct than chromospheric models. The predicted coronal heating is very sensitive to all of the approximations and uncertainties in the models. Weak shock theory is not valid in the transition region and must be replaced by nonlinear calculations. The steep temperature gradient in the transition region and inhomogeneities in the upper chromosphere and corona lead to complicated reflection and refraction effects. These and other difficulties have slowed the development of a reliable theoretical model for coronal heating by waves. As indicated in the discussion of *XUV* brightenings and spectral line widths above, there is some evidence for upward propagating waves in the transition region and low corona. However, the evidence is not conclusive. Measurements of *XUV* or X-ray coronal line profiles made with high spatial, temporal, and spectral resolution could resolve this problem. In regions with strong magnetic fields, such as bright points and active regions, the increased coronal heating could be provided by hydromagnetic waves that are enhanced in regions with strong magnetic fields. Both magneto-acoustic and Alfvén waves are a possibility (e.g. Osterbrock 1961, Piddington 1973, Leibacher & Stein 1974, Uchida & Kaburaki 1974, Wentzel 1974, Jordan 1976).

MAGNETIC HEATING An alternative method of heating the corona, especially in strong field regions, is magnetic heating resulting from dissipation of currents or magnetic field annihilation. Of particular interest in this regard are the results of comparisons made between coronal structures observed in high-resolution X-ray and *XUV* pictures and magnetic field configurations calculated from measurements of photospheric magnetic fields. The calculated field configurations are based on the assumption that the coronal fields are either potential (current free) or force free ($\nabla \times \mathbf{B} = \alpha\mathbf{B}$), where α is a constant. On the basis of comparisons of calculated and observed coronal structures, Neupert, Nakagawa & Rust (1975), Foukal (1975b), Levine (1976), Krieger, deFeiter & Vaiana (1976), and others have concluded that the magnetic field configurations in some active regions require the presence of currents of sufficiently large magnitude to force the magnetic field structures to depart significantly from a potential configuration. These currents are a possible source of energy for balancing the radiative and conductive energy losses in the coronal layers of active regions. Two methods of releasing the energy stored in nonpotential magnetic fields are generally recognized: current dissipation and field reconnection.

Current dissipation Associated with a force-free nonpotential field is a current $\mathbf{J} = (c/4\pi)\nabla \times \mathbf{B} = \alpha\mathbf{B}$, which gives rise to heating at a rate $Q = J^2\eta$, where η is the resistivity. Classical resistivity, due to coulomb collisions between electrons and ions, is orders of magnitude too small in the corona to be important, but turbulent resistivity, where electrons interact with elements of the current-driven plasma turbulence, gives rise to significant heating. Tucker (1973) suggests that active region loops exist in a quasi-steady state where new current is generated by twisting of magnetic fields through photospheric motions; this current buildup is balanced by decay of currents through turbulent resistivity in the corona. In turn, the energy dissipated by the currents leaves the loops through conduction and radiation. Glencross (1975) has drawn similar conclusions from X-ray bright points, which he interprets as due to twisted braids of magnetic flux that differ from active regions (and perhaps all large-scale closed coronal structures) only in that they are compact and tightly twisted.

Field reconnection A somewhat different mechanism for magnetic dissipation has been suggested by Parker (1975), who calculates the stability of a magnetic flux tube as it rises through the photosphere to produce an X-ray bright-point loop. If the twist of the flux tube exceeds a modest amount, it becomes unstable and rapidly forms kinks in which oppositely directed field elements are pressed together. This leads to rapid field reconnection and energy release across the neutral sheet. Such behavior could be cause of bright-point flares, which are observed to exhibit many characteristics of active region flares. If this process occurs continuously, it could also produce quasi-steady heating of bright points or other magnetic flux tubes. Levine (1974) suggests that the entire corona is heated by reconnection through magnetic neutral sheets. However, in this case the neutral sheets would have to be finely distributed through the large-scale coronal structures, at a scale small compared to the resolution of Apollo Telescope Mount (ATM) instruments or of ground-based (e.g. Fe XIV 5303A) observations, in order to explain the apparent spatial smoothness and lack of rapid time variability of these structures. While dynamic microstructure in the field with continual formation of magnetic neutral sheets is not ruled out by the observations, it seems hard to reconcile with the observed rather simple structure of the field on observable spatial scales (Figure 1).

It may be that both current dissipation and magnetic reconnection play a role in the conversion of magnetic energy into heat in the corona, the former producing quasi-stable heating and the latter being associated with more dynamic field instabilities on all scales from bright points up to major flares.

SUMMARY

There has been substantial progress in the development of knowledge concerning the mass and energy flow in the solar chromosphere and corona. Because of the improved quality of observations acquired by ground-based and satellite instrumentation, our knowledge of energy-loss mechanisms and relative importance of these mechanisms in different regions of the solar atmosphere has increased significantly. This has led to improved estimates of requirements for energy input and the variations in these requirements between regions characterized by different magnetic field strengths and configurations.

The problem of energy input mechanisms, particularly for the corona, continues to be an area of uncertainty. It is interesting that in spite of the enormous amount of theoretical and observational work in recent years on the heating of the corona, we still do not have a definitive answer to the question of whether the heating is primarily through wave or magnetic energy deposition. Very often in past work it has been taken as an a priori assumption that coronal heating is by wave deposition. However, recent observations from space vehicles now argue that such an assumption must be viewed with caution:

1. A number of careful searches have failed to turn up conclusive evidence for significant propagation or deposition of shock wave energy in the corona.
2. The forms of X-ray and ultraviolet emission from the corona show magnetic fields to delineate regions of highest radiative loss. While not requiring that the primary energy source be magnetic field energy (since the field could simply channel the energy flow from other sources) this result is at least consistent with direct heating by magnetic fields.

The contributions of recent space data, especially ATM, have been fundamental in enlarging our viewpoint on solar heating. Surely the most significant contribution so far has been the revelation of the importance of open and closed magnetic structures and the controlling role they play in solar wind mass flow and the coronal energy balance. As the analysis of the rich store of space data continues, attention is being focussed on the detailed structure of individual magnetic features. It is not too much to expect that such studies will ultimately lead to a detailed understanding of specific coronal heating mechanisms, for example by showing if there is a relation between the magnitude of departure from a potential configuration in a given coronal structure and the energy dissipated in that structure. At that point extrapolation to heating of coronas of other stars will be on a far firmer footing.

ACKNOWLEDGMENTS

We are indebted to A. K. Dupree, P. V. Foukal, R. H. Levine, and R. Rosner for their helpful suggestions and comments. This work was supported in part by the National Aeronautics and Space Administration through contract NAS 5-3949 and the Air Force Geophysical Laboratory through contract F19628-76-C-0281.

Literature Cited

Altschuler, M. D., Trotter, D. E., Orrall, F. Q. 1972. *Solar Phys.* 26:354–65

Athay, R. G. 1971. In *Physics of The Solar Corona,* ed. C. J. Macris, pp. 36–65. Dordrecht, Holland: Reidel. 345 pp.

Athay, R. G. 1976. *The Solar Chromosphere and Corona: Quiet Sun.* Dordrecht, Holland: Reidel. 504 pp.

Avrett, E. H., Vernazza, J. E., Linsky, J. L. 1976. *Ap. J.* 207:L199–204

Avrett, E. H. 1977. In *The Solar Output and Its Variations,* ed. O. R. White, J. A. Eddy, D. F. Heath. In press

Ayres, T. R., Linsky, J. L. 1976. *Ap. J.* 205: 874–94

Beckers, J. M. 1972. *Ann. Rev. Astron. Ap.* 10:73–100

Beckers, J. M., Canfield, R. C. 1975. *Motions In The Solar Atmosphere,* AFCRL-TR-0592. 85 pp.

Bell, B., Noci, G. 1976. *J. Geophys. Res.* 81:4508–16

Bohlin, J. D. 1977a. *Solar Phys.* 51:377–98

Bohlin, J. D. 1977b. See Avrett 1977

Bohlin, J. D., Sheeley, N. R. Jr., Tousey, R. 1975. In *Space Research XV,* ed. A. C. Strickland, pp. 651–56. Berlin: Akademie-Verlag. 737 pp.

Bohlin, J. D., Vogel, S. N., Purcell, J. D., Sheeley, N. R. Jr., Tousey, R., VanHoosier, M. E. 1975. *Ap. J.* 197:L133–35

Boland, B. C., Dyer, E. P., Firth, J. C., Gabriel, A. H., Jones, B. B., Jordan, C., McWhirter, R. W. P., Monk, P., Turner,

R. F. 1975. *MNRAS* 171:697–724
Boland, B. C., Engstrom, S. F. T., Jones, B. B., Wilson, R. 1973. *Astron. Ap.* 22:161–69
Bray, R. J., Loughhead, R. E. 1974. *The Solar Chromosphere.* London: Chapman & Hall. 384 pp.
Brueckner, G. E. 1976. In *IAU Colloq. No. 36, The Mass and Energy Balance and Hydrodynamics of the Solar Chromosphere and Corona,* Nice, France. In press
Catura, R. C., Acton, L. W., Johnson, H. M. 1975. *Ap. J.* 196:L47–L49
Chapman, G. A. 1970. *Solar Phys.* 14:315–27
Cheng, C. C., Moe, O. K. 1977. *Solar Phys.* In press
Cox, D. P., Tucker, W. H. 1969. *Ap. J.* 157:1157–67
Cushman, G. W., Rense, W. A. 1976. *Ap. J.* 207:L61–62
Davis, J. M., Gerassimenko, M., Krieger, A. S., Vaiana, G. S. 1975. *Solar Phys.* 45:393–410
Delache, P. 1967. *Ann. Astrophys.* 30:827–60
deLoore, C. 1970. *Astrophys. Space Sci.* 6:60–100
Deubner, F. L. 1976a. *Astron. Astrophys.* 51:189–94
Deubner, F. L. 1976b. See Brueckner 1976
Doschek, G. A., Feldman, U., VanHoosier, M. E., Bartoe, J. D. F. 1976. *Ap. J. Suppl.* 31:417–43
Dupree, A. K. 1975. *Ap. J.* 200:L27–31
Dupree, A. K., Baliunas, S. 1976. *Bull. Am. Astron. Soc.* 8:397 (Abstr.)
Evans, R. G., Jordan, C., Wilson, R. 1975. *MNRAS* 172:585–602
Feldman, U., Behring, W. E. 1974. *Ap. J.* 189:L45–46
Feldman, U., Doschek, G. A., Patterson, N. P. 1976. *Ap. J.* 209:270–81
Feldman, U., Doschek, G. A., Tousey, R. 1975a. *Ap. J.* 202:L147–50
Feldman, U., Doschek, G. A., Tousey, R. 1975b. *Ap. J. Suppl.* 31:445–66
Flower, D. R., Pineau des Forets, G. 1974. *Astron. Ap.* 37:297–300
Foukal, P. V. 1975a. *Solar Phys.* 43:327–36
Foukal, P. V. 1975b. *Bull. Am. Astron. Soc.* 7:346 (Abstr.)
Foukal, P. V. 1976. *Ap. J.* 210:575–81
Gabriel, A. H. 1976a. *Philos. Trans. R. Soc. London Ser. A* 281:339–52
Gabriel, A. H. 1976b. See Brueckner 1976
Gabriel, A. H., Jordan, C. 1975. *MNRAS* 173:397–418
Gingerich, O., Noyes, R. W., Kalkofen, W., Cuny, Y. 1971. *Solar Phys.* 18:347–65
Glencross, W. M. 1975. *Ap. J.* 199:L53–56
Hearn, A. G. 1975. *Astron. Astrophys.* 40:355–64
Holweger, D. G., Müller, E. A. 1974. *Solar Phys.* 39:19–30
Huber, M. C., Foukal, P. V., Noyes, R. W., Reeves, E. M., Schmahl, E. J., Timothy, J. G., Vernazza, J. E., Withbroe, G. L. 1974. *Ap. J.* 194:L115–18
Hundhausen, A. J. 1972. *Coronal Expansion and Solar Wind.* Berlin: Springer. 238 pp.
Jordan, C. 1975a. In *Solar Gamma X-, and EUV Radiation,* ed. S. R. Kane, pp. 109–31. Dordrecht, Holland: Reidel. 439 pp.
Jordan, C. 1975b. *MNRAS* 170:429–40
Jordan, C. 1976. *Philos. Trans. R. Soc. London Ser. A* 281:391–404
Jordan, S. D., Avrett, E. H. 1973. *Stellar Chromospheres.* NASA SP-317. 318 pp.
Jordan, C., Wilson, R. 1971. See Athay 1971, pp. 219–36
Kopp, R. A. 1972. *Solar Phys.* 27:373–93
Kopp, R. A., Holzer, T. E. 1976. *Solar Phys.* 49:43–56
Kopp, R. A., Kuperus, M. 1968. *Solar Phys.* 4:212–23
Kopp, R. A., Orrall, F. Q. 1976. *Astron. Astrophys.* 53:363–75
Krieger, A. S., deFieter, L. D., Vaiana, G. S. 1976. *Solar Phys.* 47:117–26
Krieger, A. S., Timothy, A. F., Roelof, E. C. 1973. *Solar Phys.* 29:505–25
Krieger, A. S., Timothy, A. F., Vaiana, G. S., Lazarus, A. J., Sullivan, J. D. 1974. In *Solar Wind Three,* ed. C. T. Russell, pp. 132–39. Los Angeles: Inst. Geophys. Planet. Phys., Univ. Calif. 487 pp.
Kuperus, M. 1969. *Space Sci. Rev.* 9:713–39
Kurucz, R. L. 1974. *Solar Phys.* 34:17–23
Landini, M., Monsignori-Fossi, B. C. 1973. *Astron. Astrophys.* 25:9–16
Landini, M., Monsignori-Fossi, B. C., Krieger, A. S., Vaiana, G. S. 1975. *Solar Phys.* 44:69–82
Leibacher, J., Stein, R. F. 1974. *Ann. Rev. Astron. Ap.* 12:407–35
Levine, R. H. 1974. *Ap. J.* 190:457–66
Levine, R. H. 1976. *Solar Phys.* 46:159–70
Levine, R. H., Altschuler, M. D., Harvey, J. W. 1977. *Geophys. Res.* 82:1061–65
Levine, R. H., Altschuler, M. D., Harvey, J. W., Jackson, B. V. 1977. *Ap. J.* In press
Levine, R. H., Withbroe, G. L. 1977. *Solar Phys.* 51:83–101
Linsky, J. L. 1977. See Avrett 1977
Mariska, J. T. 1976. *Bull. Am. Astron. Soc.* 8:325 (Abstr.)
Mariska, J. T., Withbroe, G. L. 1975. *Solar Phys.* 44:55–68
McClintock, W., Linsky, J. L., Henry, R. C., Moos, H. W. 1975. *Ap. J.* 202:165–82
McWhirter, R. W. P. 1976. See Brueckner 1976

McWhirter, R. W. P., Thonemann, P. C., Wilson, R. 1975. *Astron. Astrophys.* 40:63–73
Mewe, R., Heise, J., Gronenschild, E. H. B. M., Brinkman, A. C., Schrijver, J., den Boggende, A. J. F. 1975. *Ap. J.* 202:L67–71
Milkey, R. W. 1970. *Solar Phys.* 14:62–76
Milkey, R. W. 1975. *Ap. J.* 199:L131–32
Moe, O., Nicolas, K. R. 1977. *Solar Phys.* In press
Moore, R. L., Fung, P. C. W. 1972. *Solar Phys.* 23:78–102
Moore, R. L., Tang, F., Bohlin, J. D., Golub, L. 1976. *Bull. Am. Astron. Soc.* 8:333 (Abstr.)
Munro, R. H., Jackson, B. V. 1977. *Ap. J.* 213:874–86
Munro, R. H., Withbroe, G. L. 1972. *Ap. J.* 176:511–20
Neupert, W. M., Nakagawa, Y., Rust, D. M. 1975. *Solar Phys.* 43:359–76
Neupert, W. M., Pizzo, V. 1974. *J. Geophys. Res.* 79:3701–9
Nolte, J. T., Krieger, A. S., Timothy, A. F., Gold, R. E., Roelof, E. C., Vaiana, G., Lazarus, A. J., Sullivan, J. D., McIntosh, P. S. 1976a. *Solar Phys.* 46:303–22
Nolte, J. T., Krieger, A. S., Timothy, A. F., Vaiana, G. S., Zombeck, M. V. 1976b. *Solar Phys.* 46:291–301
Noyes, R. W. 1971. See Athay 1971, pp. 192–218
Noyes, R. W. 1974. Presented at *IAU Colloq. No. 27, Ultraviolet and X-ray Spectroscopy of Astrophysical and Laboratory Plasmas,* Cambridge, Mass.
Noyes, R. W., Withbroe, G. L. 1972. *Space Sci. Rev.* 13:612–37
Noyes, R. W., Withbroe, G. L., Kirshner, R. P. 1970. *Solar Phys.* 11:388–98
Osterbrock, D. E. 1961. *Ap. J.* 134:347–88
Parker, E. N. 1975. *Ap. J.* 201:494–501
Piddington, J. H. 1973. *Solar Phys.* 33:363–73
Pneuman, G. W. 1973. *Solar Phys.* 28:247–62
Pneuman, G. W., Kopp, R. A. 1977. *Astron. Astrophys.* 55:305–6
Raymond, J. C., Cox, D. P., Smith, B. W. 1976. *Ap. J.* 204:290–92
Reeves, E. M. 1976. *Solar Phys.* 46:53–72
Rogerson, J. B. Jr., Lamers, H. J. G. L. M. 1975. *Nature* 256:190
Rosner, R., Vaiana, G. S. 1977. *Ap. J.* In press
Schmahl, G. 1967. *Z. Astrophys.* 66:81–117
Sheeley, N. R., Harvey, J. W., Feldman, W. C. 1976. *Solar Phys.* 49:271–78
Shine, R. A., Gerola, H., Linsky, J. L. 1975. *Ap. J.* 202:L101–5
Shine, R. A., Linsky, J. L. 1974. *Solar Phys.* 39:49–77
Skumanich, A., Smythe, C., Frazier, E. N. 1975. *Ap. J.* 200:747–64
Spitzer, L. 1962. *Physics of Fully Ionized Gases,* New York: Interscience. 107 pp.
Stellmacher, G., Wiehr, E. 1971. *Solar Phys.* 18:220–25
Stellmacher, G., Wiehr, E. 1973. *Astron. Astrophys.* 29:13–16
Stenflo, J. O. 1975. *Solar Phys.* 42:79–105
Stenflo, J. O. 1976. See Brueckner 1976
Stenflo, J. O. 1977. In *IAU Symposium No. 71, Basic Mechanisms of Solar Activity,* Prague, eds. V. Bumba, J. Kleczek. In press
Tandberg-Hanssen, E. 1967. *Solar Activity,* Waltham, Mass: Blaisdell. 464 pp.
Tousey, R., Bartoe, J. D. F., Bohlin, J. D., Brueckner, G. E., Purcell, J. D., Scherrer, V. E., Sheeley, N. R. Jr., Schumacher, R. J., VanHoosier, M. E. 1973. *Solar Phys.* 33:265–80
Tucker, W. H. 1973. *Ap. J.* 186:285–89
Uchida, Y., Kaburaki, O. 1974. *Solar Phys.* 35:451–66
Ulmschneider, P. 1967. *Z. Astrophys.* 67:193–218
Ulmschneider, P. 1970. *Astron. Astrophys.* 4:144–51
Ulmschneider, P. 1971a. *Astron. Astrophys.* 12:297–309
Ulmschneider, P. 1971b. *Astron. Astrophys.* 14:275–82
Ulmschneider, P. 1974. *Solar Phys.* 39:327–36
Vaiana, G. S. 1976. *Philos. Trans. R. Soc. London Ser. A* 281:365–74
Vaiana, G. S., Davis, J. M., Giacconi, R., Krieger, A. S., Silk, J. K., Timothy, A. F., Zombeck, M. V. 1973. *Ap. J.* 185:L47–51
Vaiana, G. S., Krieger, A. S., Timothy, A. F. 1973. *Solar Phys.* 32:81–116
Vernazza, J. E., Avrett, E. H., Loeser, R. 1976. *Ap. J. Suppl.* 30:1–60
Vernazza, J. E., Foukal, P. V., Huber, M. C. E., Noyes, R. W., Reeves, E. M., Schmahl, E. J., Timothy, J. G., Withbroe, G. L. 1975. *Ap. J.* 199:L123–26
Wentzel, D. G. 1974. *Solar Phys.* 39:129–40
White, O. R. 1976. See Breuckner 1976
Withbroe, G. L. 1976. See Brueckner 1976
Withbroe, G. L., Gurman, J. B. 1973. *Ap. J.* 183:279–89
Withbroe, G. L., Jaffe, D. T., Foukal, P. V., Huber, M. C. E., Noyes, R. W., Reeves, E. M., Schmahl, E. J., Timothy, J. G., Vernazza, J. E. 1976. *Ap. J.* 203:528–32
Zirin, H. 1975. *Ap. J.* 199:L63–66
Zirker, J. 1976. See Brueckner 1976

Ann. Rev. Astron. Astrophys. 1977. 15:389–436

JUPITER'S MAGNETOSPHERE ×2118

C. F. Kennel[1,2] *and F. V. Coroniti*[1,3]
Physics Department, University of California, Los Angeles, California 90024

INTRODUCTION

Jupiter's magnetosphere, discovered twenty years ago by radio astronomers, has recently been traversed by the Pioneers 10 and 11 spacecraft. It is presently the only object in the cosmos for which inferences drawn from remote astronomical observation can be compared with direct in situ measurement of its neutral atom, plasma, and magnetic field environment.[4] Because Jupiter's magnetosphere is a rotating magnetized source of radio emissions and relativistic particles, it may resemble astrophysical cosmic-ray and radio sources such as pulsars.

Pioneers 10 and 11 reached three important, unanticipated conclusions concerning Jupiter's magnetosphere. First, the magnetic field in the outer magnetosphere is radially extended into a highly time-variable disk-like configuration that differs fundamentally from Earth's outer magnetosphere. Jupiter's outer disk region, and the energetic particles confined within it, are modulated at Jupiter's 10-hr rotation period. Moreover, its entire outer magnetosphere appears to change drastically on time scales of a few days to a week, perhaps in response to variable solar-wind conditions. Secondly, in addition to its known modulation of the Jovian decametric radio bursts, Jupiter's satellite Io was found by Pioneers 10 and 11 to have two other extremely important effects on its magnetosphere. Io absorbs some radiation-belt particles and accelerates others, and most importantly, Io is a source of neutral atoms and by inference a heavy ion plasma that may significantly affect, and perhaps dominate, the hydromagnetic flow in Jupiter's magnetosphere. Thirdly, Jupiter's outer magnetosphere generates, or permits to escape, such intense fluxes of relativistic electrons that Jupiter is the dominant source of 1–30-MeV cosmic-ray electrons in the heliosphere.

More than twenty years ago, Burke & Franklin (1955) detected bursts of radiation from Jupiter in the decametric wavelength band. In 1956, Mayer et al. (1958a,b)

[1] Department of Physics, University of California, Los Angeles, California.

[2] Institute of Geophysics and Planetary Physics, University of California, Los Angeles, California.

[3] Department of Astronomy, University of California, Los Angeles, California.

[4] We expect the magnetospheres of Saturn and Uranus to join Jupiter in this class in the next decade. Radio emissions, similar to Jovian decametric emissions, have been recently detected from Saturn and, with less certainty, from Uranus.

detected 3.15-cm radiation from Jupiter, which corresponds to a blackbody disk temperature of about 150 K. In 1958, 10.3-cm measurements indicated a temperature of 650 K (Sloanaker 1959, McClain & Sloanaker 1959), which suggests that the decimetric radiation was nonthermal. Subsequent measurements at longer wavelengths confirmed this trend. The Earth's radiation belts were also discovered in 1958 (Van Allen 1959). In 1959, Field discussed several possible decimetric radiation mechanisms, one of which was synchrotron emission from a Jovian radiation belt (Field 1959). Drake & Hvatum (1959) and Roberts & Stanley (1959) also considered the synchrotron mechanism. In 1960, Radhakrishnan and Roberts found that the Jovian decimetric radiation was linearly polarized, confirming the synchrotron hypothesis (Radhakrishnan & Roberts 1960). Thus, by 1960, our understanding of the terrestrial and Jovian radiation belts was comparable: both contained fluxes of energetic electrons trapped in the dipolar regions of their magnetic fields.

During the 1960s, plasma measurements in the Earth's magnetosphere were extended to low energies, and its magnetic field was completely mapped. At the same time it became possible to delineate the basic processes by which turbulent plasma is dissipated in the magnetosphere, and the magnetosphere's response to changes in solar-wind boundary conditions was understood phenomenologically. By 1970, a fairly complete understanding of the processes controlling the terrestrial radiation belts had emerged. The Earth's radiation-belt particles, numerically and energetically a minor constituent of its magnetosphere, derive their energy by radially diffusing inward towards stronger magnetic fields and are subject to losses caused by pitch-angle scattering due to plasma turbulence. The radial diffusion is driven by fluctuations in solar-wind dynamic pressure and electric field. Throughout the 1960s, refined measurements of the Jovian synchrotron source were also made.

Terrestrial radiation-belt theory and Jovian source measurements were combined and examined together at a 1971 workshop sponsored by the Jet Propulsion Laboratory to evaluate the radiation hazard to the Pioneer spacecraft. The JPL workshop (Beck 1972) stimulated the production of several quantitative radiation-belt models based on physical principles. Their strategy was to predict the intensity and spatial distribution of the synchrotron emission by using radial diffusion to energize and transport electrons from Jupiter's outer magnetosphere to its synchrotron region near 2 Jovian radii (2 R_J), while accounting for foreseeable losses due to sweep-up by Jupiter's satellites and turbulent scattering from known instabilities. Prior to the first Pioneer 10 encounter, as much theory as possible had been developed given the remote data. These pre-encounter radiation-belt models stood the test of in situ measurements well (Coroniti 1975) and were the first success in comparative magnetosphere studies. The outlines of Jovian radiation-belt theory are now relatively clear. Energetic ions and electrons diffuse radially inward, thereby gaining energy, subject to losses from absorption by Jupiter's satellites and probably from scattering by microscopic plasma turbulence. Those electrons that diffuse as far inwards as two Jovian radii from Jupiter's center and thereby reach relativistic energies then lose energy by the synchotron process to the Jovian decimetric radio emissions. The radial diffusion is driven by fluctuating hydromagnetic electric fields coupled to turbulent upper atmospheric winds. For the Earth, on the other hand, the radial diffusion is driven by fluctuating hydromagnetic electric fields coupled to variations in the solar wind. In retrospect, our success in this area is due to the fact that the Jovian radiation belts are in the inner dipolar region of the magnetosphere, where it is most like the Earth's.

Radio astronomical methods were unable to detect most of the plasma and most of the energy in the Jovian magnetosphere. Since the properties of the solar wind at Jupiter could be estimated theoretically, and since the 40-MHz cutoff of decametric bursts suggested a 10-G surface magnetic field (assuming 40 MHz is the surface cyclotron frequency), one could attempt to scale the interaction of the solar wind with Jupiter's magnetosphere. Brice & Ioannidis (1970), in a landmark paper, estimated that Jupiter's magnetopause would stand 50 R_J upstream from Jupiter in a shocked solar wind, making it evident how much of Jupiter's magnetosphere had been missed. However, Jupiter's rapid rotation made it doubtful that scaling of the Earth's magnetospheric physics would suffice. Several authors (Ellis 1965, Melrose 1967, Gledhill 1967, Piddington 1967, Brice & Ioannidis 1970 Kennel 1973) observed that centrifugal forces should create an outer magnetosphere radically different from the Earth's. Just how different became clear when Pioneer 10 encountered the magnetosphere four days ahead of the schedule set by the 50 R_J magnetopause estimate.

Our modern understanding of the Jovian magnetosphere begins with Pioneers 10 and 11. It is not surprising that there is an extensive primary literature on their results. However, this literature is surprisingly well organized; papers from each experimental group are discussed in special issues of *Science* (Vols. 183, 188) and *Journal of Geophysical Research* (Vol. 79) devoted to Pioneers 10 and 11. Since this information will not be augmented until the Mariner-Jupiter-Saturn mission encounters Jupiter in 1979, it is not surprising that review articles have begun to appear. However, this literature is strikingly complete and unified. A symposium on the magnetospheres of Earth and Jupiter held at Frascati in May, 1974, led to a review volume (Formisano 1975) and a special issue of *Space Science Reviews* (Vol. 17). In May, 1975, after Pioneer 11, a symposium covering all aspects of the planet Jupiter, its satellites, and its magnetosphere was held at Tucson. The magnetospheric papers contributed to this meeting were published in a special issue of the *Journal of Geophysical Research* (Vol. 81). The forty-four remarkably detailed and complete review articles were published together in a single volume (Gehrels 1976).

This is a selective review of Jupiter's magnetosphere. We discuss neither the Jovian radiation belts, which have been reviewed experimentally in Gehrels (1976) and theoretically by Coroniti (1975), nor the Jovian radio emissions, which have been excellently reviewed by Berge & Gulkis (1976), Carr & Desch (1976), and R. A. Smith (1976). We concentrate on where, in our opinion, our uncertainties are greatest: the hydromagnetic structure of Jupiter's middle and outer magnetosphere. Secondly, since theory has not been as well reviewed as experiment, we emphasize theoretical questions, even though we are less secure in our judgements. Finally, given the wealth of detailed, accessible, and unified primary and review literature on the Pioneer 10 and 11 experimental results, we review them only briefly, emphasizing those that appear to have theoretical significance at this time. We can only apologize in advance for whatever subjectivity has been introduced by our choice of subjects and methods.

THE PIONEER 10 AND 11 MAGNETOSPHERIC EXPERIMENTS: AN OVERVIEW

On Board Experiments

Pioneer 10 carried six experiments of interest to magnetospheric physics, and Pioneer 11, seven. The emphasis was on energetic particles, as it had been in the infancy of the studies of the Earth's magnetosphere. Each spacecraft contained four separate energetic-particle detectors from the universities of Iowa, Chicago, and California (San Diego), and the University of New Hampshire Goddard Space Flight Center. These detectors measured fluxes, energy, and angular spectra of electrons and ions with energies exceeding 60 KeV. Both spacecraft carried a helium-vector magnetometer from the Jet Propulsion Laboratory that was capable of measuring fields < 1.4 G with an accuracy of $10^{-2}\ \gamma = 10^{-7}$ G. Pioneer 11 carried an additional Goddard flux-gate magnetometer devoted to measuring strong magnetic fields up to 10 G. Measurements of low-energy (0–60 KeV) plasma were not made, though some useful information has been derived from an Ames Research Center plasma detector designed for solar-wind studies. Low-frequency (1–100 KHz) electrostatic and electromagnetic plasma-wave turbulence, essential to understanding the terrestrial radiation belts, was not investigated. Despite Jupiter's honorable history as a radio source, radioastronomy packages were not flown on Pioneers 10 and 11.

Trajectories

A flyby of the Jovian magnetosphere, from first to last encounter of the bow shock standing upstream ahead of it, lasts several weeks. The encounter trajectories of Pioneers 10 and 11 are shown in Figures 1a and 1b, which have been taken from Simpson & McKibben (1976). It is convenient to break the two encounters down into four "passes"—inbound and outbound for Pioneers 10 and 11. Each one has proven different. They may be categorized Jovigraphically as follows:

P-10 Inbound: Low southerly Jovigraphic latitudes, near 1000 hr local time (1000 LT)
P-10 Outbound: Low northerly Jovigraphic latitudes, near the dawn meridian (0600 LT)
P-11 Inbound: Low southerly Jovigraphic latitudes near 1000 hr local time
P-11 Outbound: Middle Jovigraphic latitudes, nearly local noon (1200 LT).

The P-10 and P-11 inbound passes traversed the same regions of space, relative to Jupiter, albeit a year apart. Only P-11 outbound traversed reasonably high latitudes. There have been no passages into the all-important magnetic tail (2400 LT) or the local dusk (1800 LT) sectors. Comparison of local dawn and local dusk passages might reveal east-west asymmetries induced by Jupiter's rotation.

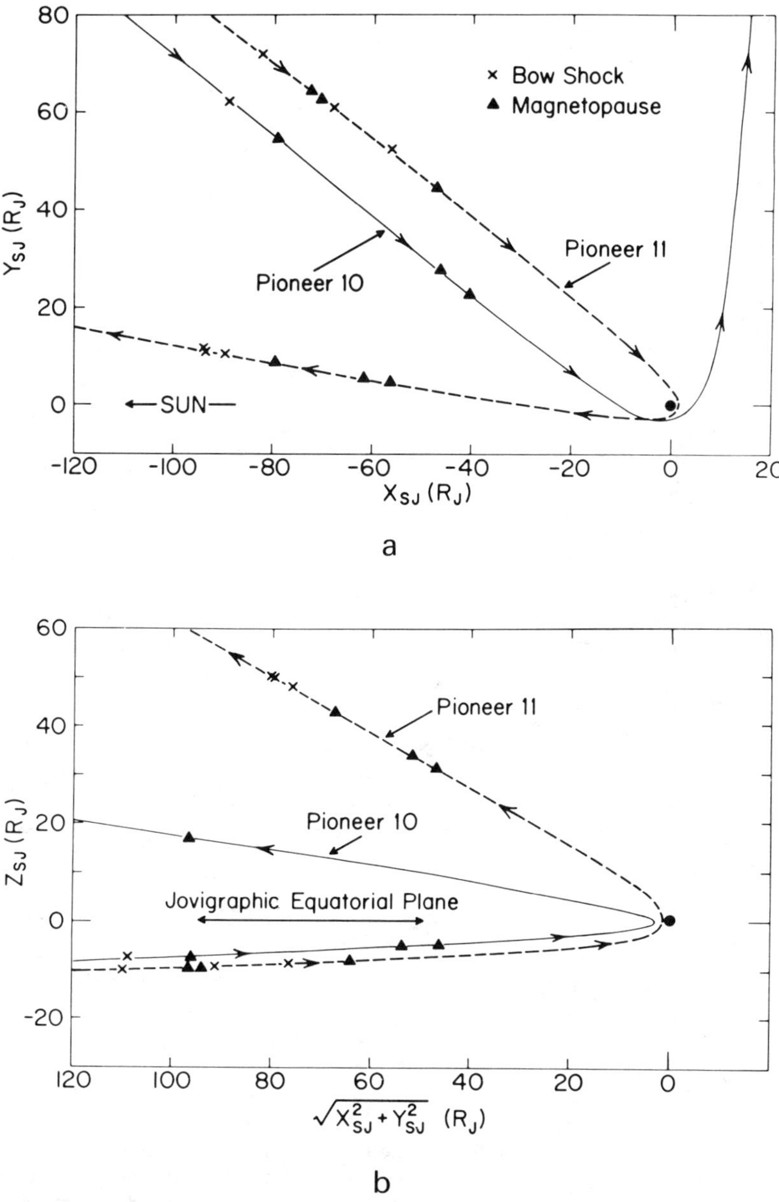

Figure 1 P-10 and 11 trajectories. X_{SJ} is in the ecliptic plane, positive radially away from the Sun along the Sun-Jupiter line; Z_{SJ} is positive toward the north pole and perpendicular to the ecliptic; Y_{SJ} completes the right-handed coordinate system. 1 Jovian radius = 1 R_J = 71,372 km (from Simpson & McKibben 1976).

Overview of Data

At the Earth, the magnetopause is the boundary between the shocked solar wind (the "magnetosheath") and the Earth's magnetic field. To a first approximation, the solar wind does not penetrate the magnetopause. Because the magnetosphere is a blunt obstacle standing in the hypersonic, hyper-Alfvénic solar wind, a collisionless bowshock stands upstream of the magnetopause. In Figures 1a and 1b, the positions of the analogous crossings of the Jovian bowshock and magnetopause are marked

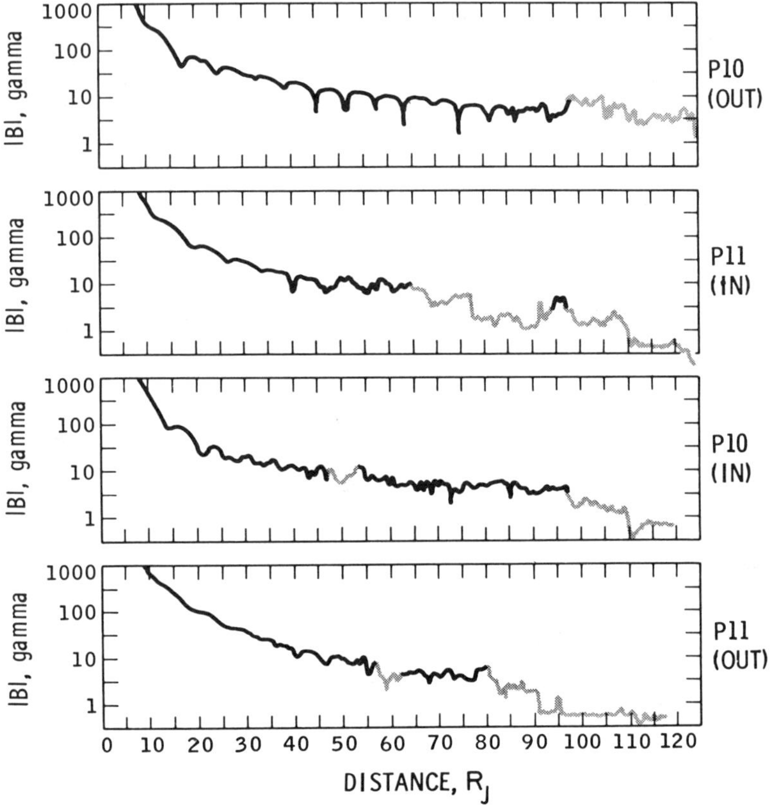

Figure 2 P-10 and 11 magnetic field magnitude. The magnetic field magnitude measured by the JPL magnetometer is plotted against radial distance from Jupiter's center in R_J. 1 gamma = 1 $\gamma = 10^{-5}$ G. When the spacecrafts were within (outside) the magnetosphere, the data are plotted as a solid (gray) line. P-11 (in), P-10 (in), and P-11 (out) all encountered a magnetosheath region inside a magnetospheric region, indicating that the magnetosphere was unsteady. P-10 (out) encountered the magnetosphere again between 130–150 R_J. The dips in the P-10 (out) pass correspond to current sheet crossings. The data have been ordered in local time, with predawn data in the top panel and prenoon data in the bottom panel (from Smith et al. 1976).

by crosses and triangles respectively. For P-10, these crossings were identified using the solar-wind plasma probe (Wolfe et al. 1974). Upon crossing the bowshock, the plasma probe detected thermalized ion fluxes, and upon crossing the magnetopause, it detected a complete dropout in ion flux. For P-11, the crossings were detected by characteristic signatures in the magnetic field data. There were multiple shock and magnetopause crossings on each of the passes.

Figure 2, taken from Smith et al. (1976), shows the magnetic field magnitude plotted against radial distance for the four passes. The portions within the magnetosphere are drawn in blackline and within the magnetosheath or solar wind in grayline. Smith et al. (1976) have argued that the field latitude δ best characterized the structure of the Jovian magnetic field. Figure 3 shows plots of δ versus distance

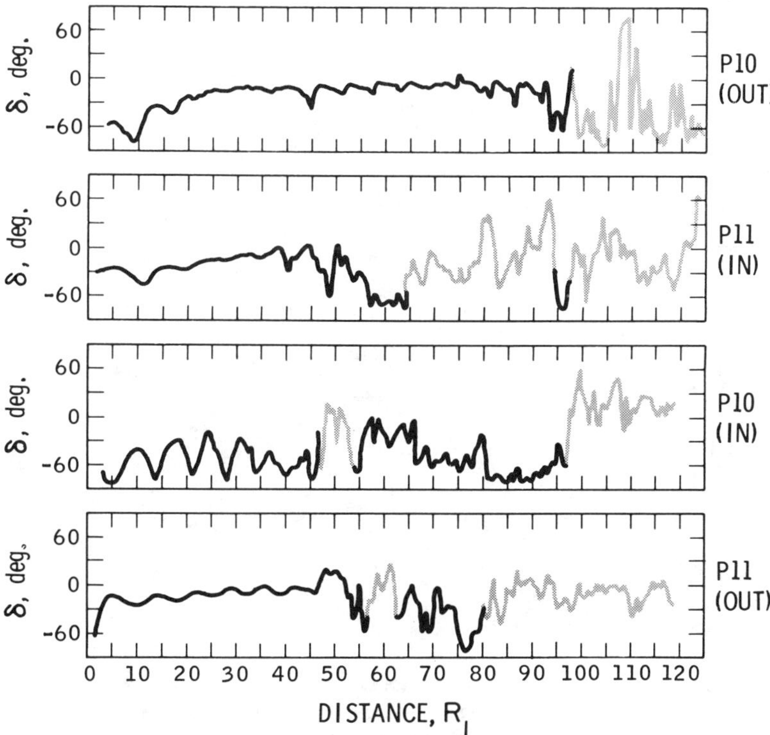

Figure 3 P-10 and 11 magnetic field longitude. δ is the Jovigraphic latitude towards which the magnetic field points, and $\delta < 0$ implies a southward directed field. These data are derived from hourly averages. Features to be discussed in the text include the following: On P-10 (out), the field was directed radially outward between 20–95 R_J. The smooth oscillations with a 10-hr period within 40 R_J (P-10 in), 45 R_J (P-11 out), and the one cycle visible within 20 R_J on P-10 (out) resemble those expected from Jupiter's rotating misaligned dipole field. Within but near the magnetopause, the field turns sharply southward at 95 R_J (P-10 out), 55 R_J (P-11 in), 80 R_J (P-10 in), and at 48 and 75 R_J on P-11 out (from Smith et al. 1976).

for the four passes; $\delta < 0$ denotes a magnetic field pointed in a southerly direction, $\delta > 0$ northerly, and $\delta \simeq 0$, radial field. Figure 4 shows a typical survey plot from one of the energetic particle detectors, the University of Chicago 6–30-MeV energetic-electron detector (McKibben & Simpson 1974). Both inbound and outbound Pioneer 10 passes are shown; magnetopause and bowshock crossings are indicated by MP and BS respectively.

The dominant first impression from the P-10 and P-11 data is of high time variability. On its inbound pass, P-10 first passed from the shocked magnetosheath into the magnetosphere at a distance of 98 R_J and then passed back into the magnetosheath at 55 R_J, reentering the magnetosphere finally at 48 R_J. This behaviour is typical. P-11 encountered three magnetopause crossings within 100 R_J on its inbound pass, and three on its outbound pass. Only on P-10 outbound did Jupiter's magnetosphere remain in the same state for a significant fraction of the pass. For this reason, P-10 outbound has attracted disproportionate attention. Yet even here, the magneto-

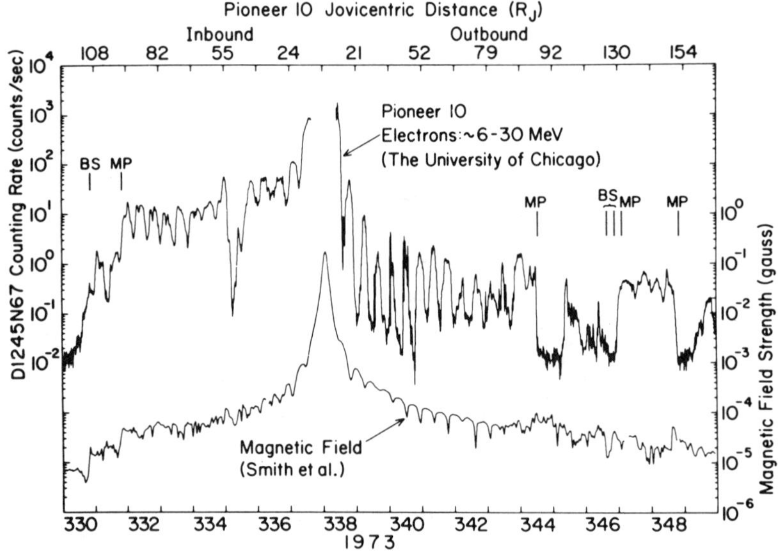

Figure 4 P-10 6–30-MeV electron fluxes and magnetic field. 6–30-MeV electron fluxes in counts/sec and the magnetic field magnitude are plotted against day of year 1973 (bottom) and radial distance (top). Features to be discussed include: 1. the steady increase of 6–30-MeV electron flux prior to the first magnetopause crossing (MP), which indicates that Jupiter is a source of cosmic-ray electrons to the solar wind; 2. the smooth increase in electron flux with decreasing radial distance within 20 R_J defining the radiation-belt region; 3. the pronounced sawtooth modulation with 10-hr period on the outbound pass associated with current sheet crossings. Note that when P-10 outbound reentered the magnetosphere between 130–150 R_J, the 6–30-MeV electron fluxes were comparable to the peaks found further in. Bow-shock crossings are indicated by BS (from McKibben & Simpson 1974).

sphere was reencountered between roughly 130–150 R_J. These facts alone make it evident that flyby data on Jupiter's magnetosphere is severely time-aliassed, despite the inordinately long duration of the encounters: the time scale for changes of magnetospheric state is comparable to the duration of each pass. Separation of spatial from temporal variations in the data requires the extended coverage provided by an orbiter.

The second important aspect of Jupiter's magnetosphere is the surprisingly great spatial scale of its temporal variations. Balancing solar-wind dynamic pressure with the magnetic pressure of Jupiter's vacuum dipole field places its magnetopause at about 40 R_J at the subsolar point, yet magnetopause crossings beyond 80 R_J were encountered on each pass. Since, in the vacuum dipole model, the magnetopause distance depends only upon the sixth root of the dynamic pressure, this discrepancy cannot be explained by an inordinately low dynamic pressure and is not consistent with the measured dynamic pressure. Moreover, the temporal variations in size of the magnetosphere are a factor of two—whereas 10% variations are typical for the Earth, and a factor of two is rare.

Thirdly, the magnetopause has never been encountered within the classical magnetopause position of 40 R_J though on P-10 and P-11 outbound, the innermost magnetopause crossings lie near the classical position. Thus, the total pressure within Jupiter's magnetosphere typically exceeds that of the vacuum dipole. Figure 2 makes this obvious; at the most distant magnetopause crossing on each of the passes, the magnetic field amplitude is roughly $10\ \gamma = 10^{-4}$ G, whereas the dipole field would be less than $1\ \gamma$. The magnetic pressure exceeds the dipole pressure by a factor of one hundred. Moreover, on P-10 outbound, the magnetosheath magnetic pressure exceeded that in the magnetosphere, which suggests that the internal particle and magnetic pressures were comparable within the magnetosphere.

Energetic particles serve as useful tracers of magnetospheric structure. Even casual inspection of Figure 4 reveals a marked difference between the smooth behavior of the energetic-electron fluxes within 20 R_J and the marked 10-hr variations in intensity in the outer magnetosphere. All the energetic experimental groups have remarked that the magnetopause is a true boundary, since the particle fluxes drop rapidly by at least an order of magnitude. This indicates that the particles do not freely penetrate the magnetopause. However, Jovian energetic particles do find their way out of its magnetosphere somewhere, somehow, since Jupiter is probably the source of most of the few MeV electrons in the heliosphere.

How should we proceed with a more detailed study of the Pioneer data? If the magnetosphere were steady, we could divide Jupiter's magnetosphere into several physically different spatial regions and study them separately. But given that Jupiter's magnetosphere might switch in time between several distinct states, we should consider each pass unique. In fact, we combine the two strategies. Despite the difficulties of separating space from time variations, the various experimental groups have all separated Jupiter's magnetosphere into three regions. One of the regions, the innermost one, is immediately obvious in Figure 4. Here the magnetic field is approximately dipolar, the energetic particle fluxes increase monotonically approaching Jupiter, and the magnetosphere is the most independent of time. This region was theoretically anticipated and subsequently confirmed by P-10 and P-11 to closely resemble a more energetic version of the terrestrial radiation belts.

The outer magnetosphere is clearly distinct from the radiation belts; it is also natural to have a transition region or "middle magnetosphere" separating the two. Since the magnetopause is time variable, the boundary separating the middle and outer magnetosphere should be fuzzy and time variable as well. It is typically placed between 20 and 40 R_J. In Section 3, we review in greater detail the observations of the middle and outer magnetosphere. Our understanding of the basic hydromagnetic processes occurring in the middle and outer magnetosphere is sufficiently weak that in our opinion the physics of the energetic particles there, however interesting, must take second position in our review; we will use energetic particles primarily as tracers of magnetospheric structure. In Section 4, we consider various hydromagnetic theories of Jupiter's magnetosphere.

PIONEER OBSERVATIONS OF JUPITER'S MIDDLE AND OUTER MAGNETOSPHERE

We now turn to observations made by Pioneers 10 and 11 of the middle ($12\ R_J \lesssim R \lesssim 40\ R_J$) and outer ($R > 40\ R_J$) magnetospheres. We review first the quasi-steady P-10 outbound pass and then P-11 outbound high-latitude passes and construct the conceptual models that could emerge from these passes alone. We then consider together the P-10 and P-11 inbound passes, which had very similar trajectories.

Pioneer 10 Outbound

Figure 5 is a different plot of the JPL magnetometer data (Smith et al. 1974a,b). Each point in the bottom inset represents a 1-hr average value of the latitude δ. The top insert represents δ_D, the latitude, had the field been a vacuum field. A 10-hr periodicity in δ_D due to the rotation of Jupiter's oblique dipole is apparent. In fact, about two cycles of this periodicity are apparent in the data, when P-10 was within 20 R_J. Thereafter, *the field became nearly radial* somewhat past 20 R_J. The solar-wind plasma probe detected magnetosheath flowing plasma for the first time at 98 R_J. The magnitude of the magnetic field (Figure 2) varied approximately as $1/R^3$ within 20 R_J and as $1/R^{1.7}$ beyond, so that the magnetic pressure exceeded the dipole pressure by several hundred at the magnetopause crossing at 98 R_J. Quantitative calculations (Roederer 1970) would place the classical magnetopause near 75 R_J on the dawn meridian. Smith et al. (1974a,b) also plotted the *longitude* towards which the field points. In the region between 20 and 90 R_J the field, while remaining very nearly in a meridian plane, was gradually swept back so that by 90 R_J it pointed 30° out of the meridian plane, lagging precise corotation. A theoretically more meaningful display of the same data is shown in Figure 6, taken from Goertz et al. (1976). Plotted is the ratio of the azimuthal (B_ϕ) to radial (B_ρ) magnetic field components, divided by radial distance, denoted by ρ, against radial distance between 0 and 80 R_J. B_ϕ is negative nearly everywhere (the few positive points are questionable), and $B_\phi/\rho B_\rho$ is weakly dependent on distance. Thus, the garden-hose angle increased almost linearly with distance.

The magnetopause at 98 R_J identified by the solar-wind plasma probe is not evident in the 1-hr averages of field latitude, magnitude, or longitude. (It is identifiable, with magnetic data with a higher time resolution.) However, at roughly

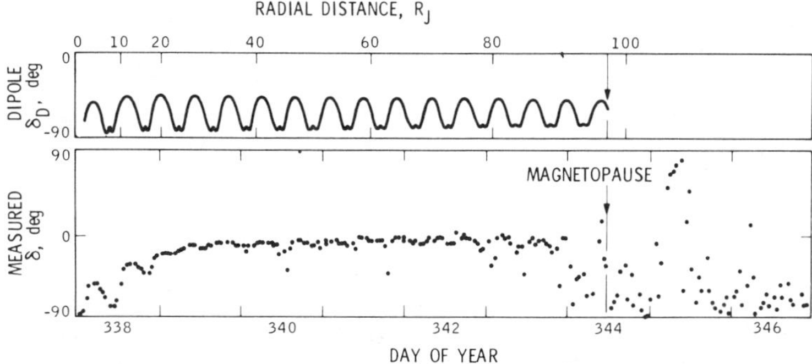

Figure 5 P-10 outbound magnetic longitudes. Shown in the bottom panel, point by point, is the hourly average magnetic longitude δ; δ_D, in the top panel, is the longitude for Jupiter's rotating misaligned dipole field. One cycle of the 10-hr modulation of δ_D appears within 20 R_J. Beyond 20 R_J, δ is very small, indicating a predominantly radial field. Hourly averages do not clearly reveal current sheet crossing; however, these appear as southward excursions represented by a few data points (e.g. near 45 R_J). Before the magnetopause was crossed, the field turned southward, and the hourly averages reveal a highly irregular behavior (from Smith et al. 1974a,b).

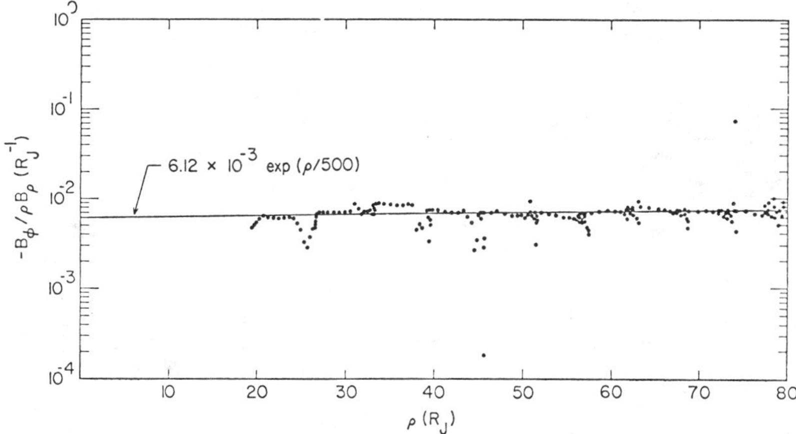

Figure 6 P-10 outbound gardenhose angle. The azimuthal field B_ϕ divided by the radial distance ρ and the radial field B_ρ is plotted versus distance in units of R_J^{-1}. The solid line is an empirical best fit. Deviations from the theoretical line occur at current sheet crossings. In a radially expanding super-Alfvénic wind, the radial flow speed $v_\rho \simeq |\Omega \rho B_\rho / B_\phi|$, where Ω is Jupiter's rotation frequency, 1.75×10^{-4} rad sec^{-1}. The above data imply v_r exceeds 10^3 km sec^{-1}; however, the data may not be usable in a velocity determination (from Goertz et al. 1976).

90 R_J, a transition to a strongly southward, much more unsteady field occurs. An apparently related change in behavior is noticeable in the 6–30-MeV electron fluxes (Figure 4), where the amplitude of the 10-hr flux modulations is sharply diminished beginning at 90 R_J. A similar change is even more apparent in the energetic electron data of Van Allen et al. (1974). Thus, there is evidence of a separate layer near the magnetopause where the magnetic field is southward and highly time variable.

Figure 4 reveals that the 6–30-MeV electrons are modulated by a factor of 100 between 20 and 90 R_J during the P-10 outbound pass. Ten-hr modulations of the energetic-particle fluxes were common on all the P-10 and P-11 passes, but they were most pronounced on P-10 outbound. Figure 7 shows similar data for the 1.1–2.15-MeV (upper curve) and 14.8–21.2-MeV (lower curve) proton fluxes (Trainor et al. 1974). The proton and electron modulations are in phase. According to Trainor et al. (1974), the peak 0.41–2-MeV electron fluxes were roughly 10^5 cm^{-2} sec^{-1}, and the peak 1.1—2.15-MeV proton fluxes were 10^4 cm^{-2} sec^{-1}, so that the peak energy densities of few MeV electrons and protons were comparable. The peaks in electron and proton fluxes are associated with partial or complete traversals of a current sheet-field reversal layer. Figure 8 (Smith et al. 1976) shows a current sheet traversal at about 75 R_J. While there is evidently considerable structure, the field magnitude generally diminishes and changes from radial outside to more southward inside. Of particular significance is the 180° jump in field longitude near 22 GMT; evidently the spacecraft passed between lobes of the magnetosphere and the magnetic field reversed direction. The peak energetic-particle fluxes coincide with sheet crossings; moreover, the electron energy spectra are consistently harder at the magnetic field minima. Figure 9 shows 0.5–1.8-MeV proton and 6–30-MeV electron fluxes with one-minute resolutions for a P-10 outbound sheet crossing at 40 R_J (Simpson et al. 1974a). The fluxes exhibit considerable fine structure on the minute time scale within the sheet but are much more slowly varying outside. Of particular significance is the

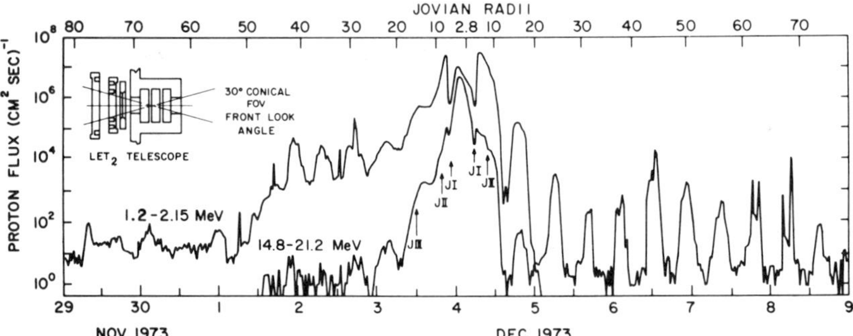

Figure 7 P-10 energetic proton fluxes. The fluxes of 1.2–2.15 MeV and 14.8–21.2 MeV protons are plotted against time below and radial distance above. Crossings of the orbits of Jupiter's satellites are indicated by JI, JII, etc. Within 15–20 R_J in the radiation-belt region, the proton fluxes increase smoothly with decreasing distance. Ten-hr sawtooth modulations of the 1.2–2.15-MeV proton fluxes are evident between 20–80 R_J. They are in phase with the electron modulations (from Trainor et al. 1974).

fact that electron and proton fluxes jump simultaneously in minutes. The proton flux exhibits strong fluctuating field-aligned anisotropies; the electrons do not. Simpson & McKibben (1976) make two important suggestions: first, the simultaneous electron and proton increases are evidence of local acceleration; second, the acceleration may be due to magnetic field reconnection in the sheet. Sakurai (1976) has estimated the reconnection rate to be sufficient to accelerate particles to MeV energies rapidly; however, a detailed theory of reconnection in the Jovian current sheet is still lacking. The rapid flux changes in Figure 9 suggest that the magnetic structure in Figure 8 may be temporal and not spatial.

Since the energetic-particle fluxes tend to be isotropic at the current sheet crossings, the particles may be contained on closed-field lines within the sheet. This is consistent with the tendency for the magnetic field to be southward within the sheet. On P-10 outbound, the particle fluxes were modulated by a factor of 100 at sheet crossings—much more than on the other passes—which suggests that the field lines bounding the current sheet may be open (Goertz et al. 1976). Goertz et al. (1976) emphasize that this reasoning is only applicable to P-10 outbound.

The global properties of the current sheet are difficult to ascertain with measurements from a single spacecraft. To know its thickness, for example, we

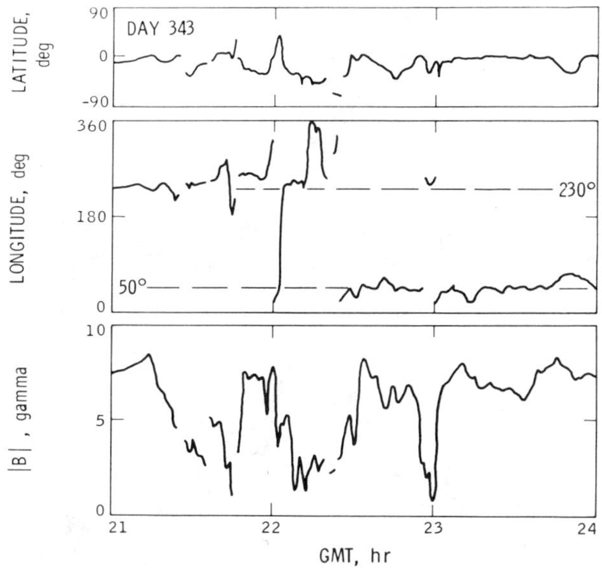

Figure 8 P-10 outbound current sheet crossing. Shown here are 4 hours of high time-resolution magnetic-field data for a current sheet crossing near 90 R_J at $5^h\ 30^m$ local time. The most important feature is the 180° reversal of the longitude towards which the magnetic field points near 22 GMT, which indicates that P-10 passed between lobes of the magnetosphere. While the magnitude fluctuates it diminishes to roughly 1 γ during the crossing. There is a tendency for the field to point more southward within the sheet than outside (from Smith et al. 1976).

must construct a model for its motions. Figure 10, taken from Goertz (1976b), sketches two possible models which, according to Goertz, cannot be resolved by present experimental data. The "hinged current sheet" model on the top is due to Smith et al. (1974a). Since the perturbation magnetic field (actual minus dipole) proved to be relatively invariant with respect to Jupiter's dipole axis, they argued that the current sheet is parallel to the rotation equator. They fix the hinge point between dipole and sheet at $\simeq 20\ R_J$ consistent with P-10 outbound observations. Goertz (1976b) argues that the current sheet should remain almost precisely in the magnetic equatorial plane. We defer further discussion of these models until Section 4.

The flapping-sheet model predicts that the maximum energetic-particle fluxes should have occurred when the latitude of the spacecraft was a minimum, i.e. when

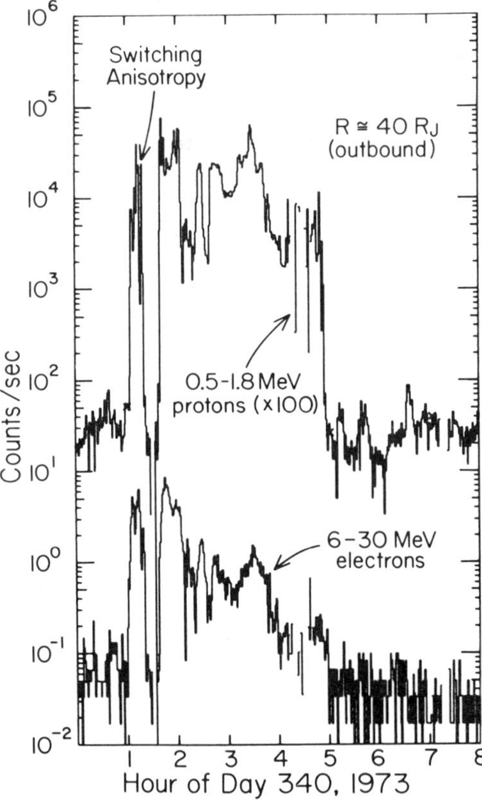

Figure 9 Rapid temporal particle fluctuations in a current sheet model. Shown here is high time-resolution 6–30-MeV and 0.5–1.8-MeV proton-flux data for a current sheet crossing near 40 R_J on P-10 outbound. Note that electrons and protons are accelerated *simultaneously* on minute time scales. Simpson & McKibben (1976) suggest that magnetic field reconnection within the sheet accelerates these particles (from Simpson et al. 1974a).

it was at a Jovian System III longitude of 42° on the outbound pass. The flux maxima lagged behind the 42° meridian slightly by an amount that increases monotonically with distance, which is reminiscent of the behavior of B_ϕ (Northrop et al. 1974). While this conceivably supports the sheet model for P-10 outbound, leads as well as lags were found on the other passes, so that the overall situation remains unclear.

In summary, from P-10 outbound alone, one would be tempted to construct the following phenomenological model of Jupiter's magnetosphere. The magnetosphere is divided into two lobes, northern and southern, separated by a thin current sheet that moves up and down as Jupiter rotates. The magnetic field is nearly radial near to but outside the sheet and is weakly swept back out of meridian planes into a garden-hose configuration, slightly lagging behind corotation. The magnetic field is on the average southward in the sheet, indicating that energetic particles are contained in it. However, field reconnection may occur there, accelerating particles to high energy and temporarily reversing the dominant southward direction of the sheet magnetic field. The plasma whose pressure is needed to balance the magnetic pressure on either side of the current sheet has not yet been detected, although the intense fluxes of MeV protons suggest that 10–100-KeV ions may be responsible. Next to the magnetopause, there is a thin (8 R_J radial thickness) layer where the current sheet is less prominent and the magnetic field is southward and irregular.

Pioneer 11 Outbound Pass

Pioneer 11 remained close to the local noon meridian plane, passing through the magnetosphere at Jovigraphic latitudes of 30–35°. We begin our examination of the

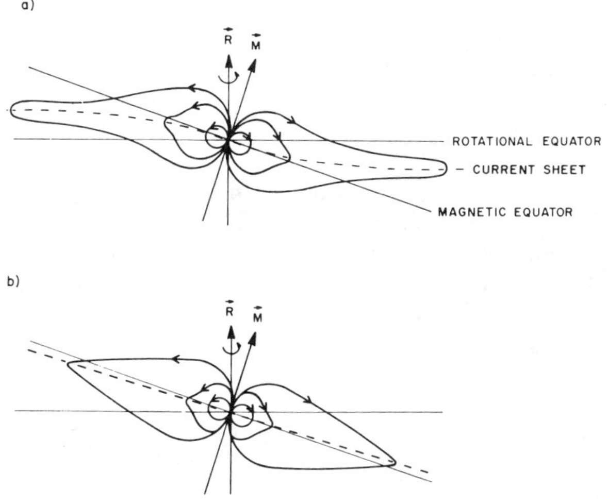

Figure 10 Two possible disk models. At the top is shown the "hinged current sheet" model of Smith et al. (1974a,b). At the bottom is Goertz's (1976b) model of a sheet localized to the magnetic equatorial plane. This figure is taken from Goertz (1976b).

magnetic field magnitude, Figure 2, and longitude, Figure 3, at a radial distance near 20 R_J where $X_{SJ} \simeq Z_{SJ} \simeq 15\, R_J$. As P-11 proceeded outward, the magnitude diminished smoothly, following an R^{-3} law much more closely than R^{-2}, until just before the first magnetopause crossing at 55 R_J, $X_{SJ} \simeq 45$ R_J, $Z_{SJ} \simeq 30$ R_J. (On P-10 outbound, the magnitude deviated from an R^{-3} law beginning at 20 R_J radial distance.) The P-11 outbound magnetic latitude, Figure 3, manifests weak 10-hr modulations between 10 and 50 R_J that are similar to those in the top inset of Figure 5 for the undistorted dipolar field. The magnetic longitude data, not shown here, show that the field was not swept back but remained in a meridian plane. Thus within 55 R_J P-11 outbound found a weakly distorted dipole field; the distance of its first magnetopause crossing was consistent with this fact. However, the magnetosphere was evidently not stationary, for Pioneer 11's passage into the magnetosheath at 55 R_J was transient, and it found itself back inside the magnetosphere again at 63 R_J. This time the magnetic field strength exceeded the dipolar field, pointed strongly southward, and was irregular. P-11 exited a final time into the magnetosheath at $R \simeq 80$ R_J, $X_{SJ} \simeq 65$ R_J, $Z_{SJ} \simeq 42$ R_J.

Figure 11 has been adapted by Kivelson (1976) from Trainor et al. (1975). The bottom curve shows the fluxes of 0.1–2-MeV electrons, the middle curve 1.2–2-MeV protons, and the top curve, 0.5–2.1-MeV protons, for both inbound and outbound P-11 passes; the outbound portion of the flyby begins December 3, 1974. Bowshock and magnetopause crossings are denoted by B and M, respectively. Following Smith et al. (1975), Kivelson (1976) studied the time variability of the magnetic field in the internal boundary layer next to the magnetopause. Within this layer, the field is highly turbulent, down to time scales as short as tens of seconds. These turbulent regions, shaded in Figure 11, lie near but within the magnetopause, and just as for P-10 outbound, the particle fluxes show little or no periodicity. A transition to the internal boundary layer is therefore marked by three characteristics: the hourly average field shifts southward and becomes more variable; magnetic turbulence is present; and the energetic particle fluxes are smoother and less distinctly modulated.

Two facts concerning the P-11 energetic-particle fluxes, Figure 11, stand out. The outbound high-latitude fluxes are comparable in magnitude with the inbound low-latitude fluxes. While the 10-hr modulations inbound might suggest disk confinement, 4 days later there were copious fluxes at latitudes above those expected from a disk model. Secondly, the electron flux exceeded the proton flux by somewhat more than an order of magnitude, again suggesting that the number densities of MeV electrons and protons were comparable.

In summary, if only the P-11 outbound data within 55 R_J were available, one would say that Jupiter's magnetosphere was Earth-like along the Jupiter-Sun line, with the magnetopause position fixed approximately by the balance of solar-wind dynamic pressure and undistorted dipole magnetic pressure, and with the entire magnetosphere filled with energetic particles out to the magnetopause, as at Earth. Why, then, a day after first passing the magnetopause, P-11 reencountered something resembling a high-latitude version of the turbulent internal boundary layer found next to the magnetopause on P-10 outbound remains a mystery.

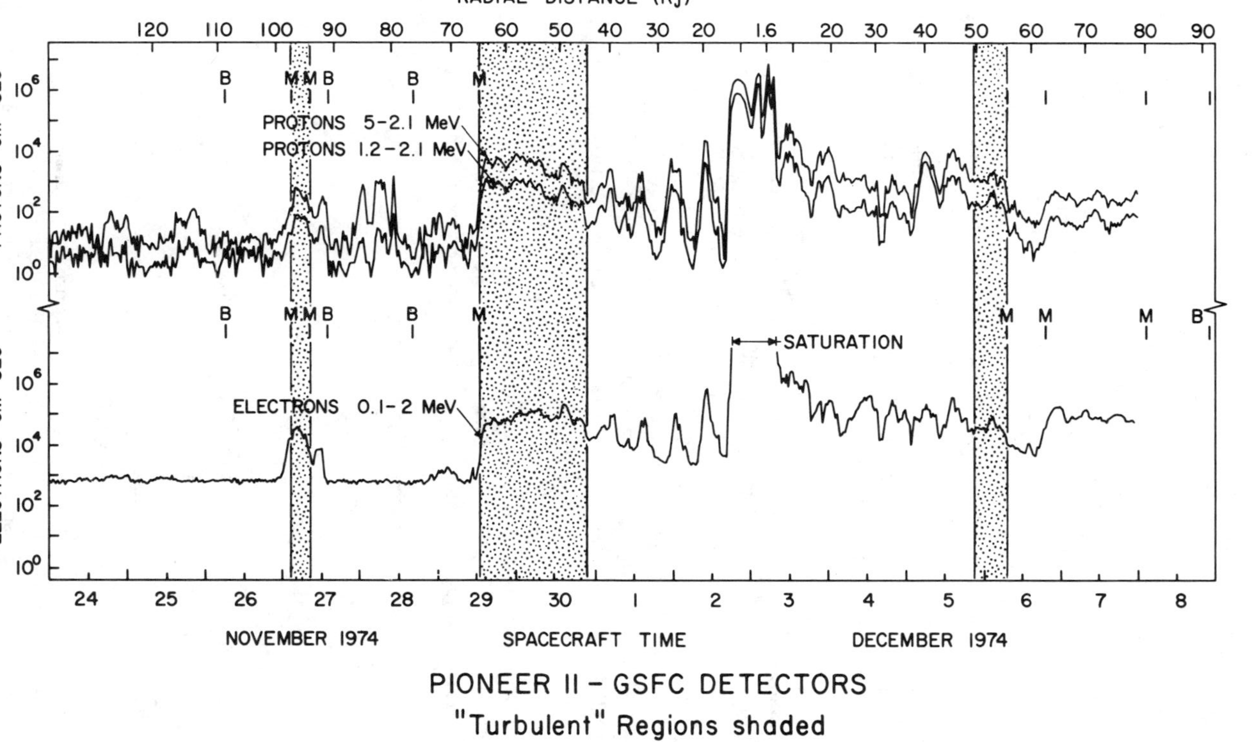

Figure 11 P-11 energetic electrons, protons, and turbulent regions. The fluxes of 0.1–2-MeV electrons, 1.2–2.1-MeV protons, and 0.5–2.1-MeV protons are plotted against time (lower) and radial distance (upper scale). Magnetopause and bowshock crossings are denoted by M and B, respectively. The shaded regions indicate where Kivelson (1976) has found magnetic turbulence in the JPL magnetometer data. Note the turbulent region within the magnetopause between 45–65 R_J on P-11 inbound; the energetic particles showed little or no 10-hr modulation in this turbulent boundary layer. Within the classical magnetopause radius at 45 R_J, electrons and protons showed pronounced 10 hr in-phase modulations reminiscent of P-10 outbound. P-11 outbound was at high latitudes, and the particle fluxes were weakly modulated at best; however, turbulence was encountered at high latitudes next to the magnetopause at 55 R_J. Note that the particle fluxes on P-11 outbound were comparable with the peak fluxes observed inbound; this seems inconsistent with particle confinement to a thin disk (from Kivelson 1976).

Pioneers 10 and 11 Inbound

Looking at the magnetic field data alone, Figures 2 and 3, a case could be made that P-10 inbound encountered an approximately Earth-like magnetosphere from its last magnetopause crossing at $R \simeq 45\ R_J$ inward. The magnetic latitude δ showed periodic modulations approximately compatible with those of the spinning oblique vacuum dipole, with the magnetic field magnitude exceeding dipolar by about factor 2, consistent with the dipolar pressure balance model, and following the dipolar spatial variation closely. Since the field near the Jovigraphic equator pointed dominantly southward during this portion of the pass, Smith et al. (1974a,b) felt that there was little evidence for current sheet crossings in this region. Certainly, the 10-hr variations in particle flux were much less pronounced than on P-10 outbound.

Trainor et al. (1974) have published the proton-flux energy spectra shown in Figure 12 for this quiet portion of the P-10 inbound pass. Near the magnetopause, the proton energy spectrum is quite steep, varying as E^{-3}, and there is no evidence of a turnover. However, if the E^{-3} spectrum were to prevail below several tens of KeV, the plasma $\beta = 8\ \pi NT/B^2$ would exceed unity, which is contradicted by the magnetic field measurements. Very likely, the plasma β is near unity at the magnetopause, a common occurrence near Earth's. magnetopause. The relatively small velocities of $\sim$ MeV protons, together with a knowledge of the proton energy spectrum, permit a test of whether or not the protons corotate with Jupiter. Trainor et al. (1974) fit their angular distributions with a function of the form $J(\theta) = A_0 + A_1 \sin(\theta - \theta_1) + A_2 \sin 2(\theta - \theta_2)$, where θ_1 is the angle between their detector look direction and the normal to the ecliptic plane. Their results for A_1/A_0 and θ_1 are shown in Figures 13a and b, respectively. While A_1/A_0 varies significantly from the value expected on the basis of corotation beyond about 25 R_J, it appears that within 25 R_J, the particles corotate, coming from $\theta_1 = 90°$ as expected. The region 25–45 R_J is evidently a transition region, with a hint that

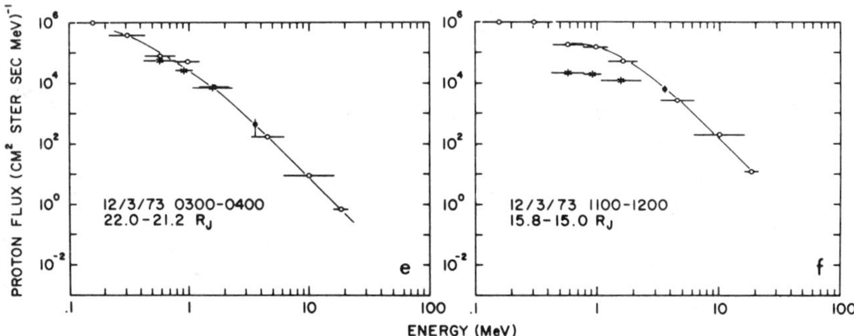

Figure 12 P-10 outbound proton energy spectra. Shown here are proton energy spectra between 100 KeV and 20 MeV. The spectra fall off as $1/E^3$. The spectra are quite similar at larger radial distances. If the $1/E^3$ spectrum extends down to, say, 10 KeV at larger radial distances, the proton pressure could be comparable to the magnetic pressure (from Trainor et al. 1974).

corotation may be preserved. Interpretation of such plots becomes ambiguous if the particle fluxes have significant spatial gradients.

Despite the fact that they passed through approximately the same region of space relative to Jupiter, P-10 and 11 inbound encountered significantly different magnetospheres. P-11 inbound encountered the bowshock at 110 R_J and crossed the magnetopause twice at 97 and 92 R_J, which suggested that its position was oscillatory. Then it crossed the shock again at 92 R_J to remain in the solar wind, until the shock was again encountered at 78 R_J. The final durable crossing of the magnetopause occurred at 65 R_J. According to Figure 11, the spacecraft remained in the turbulent internal boundary layer until roughly the classical magnetopause position of 45 R_J. Within 45 R_J the behavior was much more like P-10 outbound than P-10 inbound. Strong modulations of the energetic particles associated with a current sheet were found. Rapid variations in intensity that were simultaneous for both electrons and protons, albeit less pronounced than on P-10 outbound, were again found, and again suggested local particle acceleration (Simpson & McKibben 1976). Within about 50 R_J, the field was swept back slightly out of the meridian planes, again reminiscent of P-10 outbound.

The evidence (Figures 13a,b) for corotation found by Trainor et al. (1974) on P-10 inbound corresponds to a time when the magnetosphere was dipolar within 50 R_J. Sentman & Van Allen (1976) performed a similar analysis for their 0.6–3.4-MeV proton data on P-11 inbound and found that corotation prevailed within the last durable magnetopause crossing at 65 R_J to at least 30 R_J. Thus corotation seems to occur when the current sheet is present and also within the turbulent internal

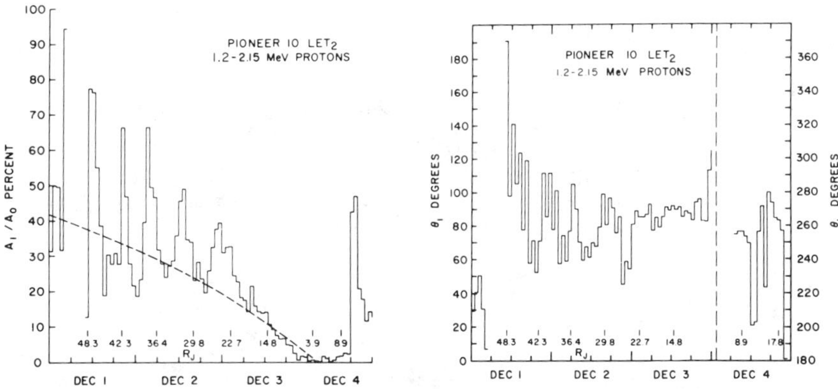

Figure 13 P-10 inbound proton anisotropies. Knowing the proton energy spectrum, one can estimate the directional anisotropy of the proton fluxes, assuming the protons corotate rigidly with Jupiter. The amplitude is plotted in the left panel, with the theoretically expected curve dashed, and the phase at the right. On the inbound pass, a phase of 90° corresponds to that expected from corotation. Various radial disturbances are indicated at the bottom. The protons appear definitely to corotate within 20 R_J inbound, and there is a hint that corotation is preserved out to 40 R_J. The magnetic field on this portion of the pass was not strongly radially extended (Smith et al. 1974a,b) (from Trainor et al. 1974).

boundary layer. A striking confirmation of this fact is shown in Figure 14 (Simpson & McKibben 1976). The data correspond to the brief interval of time when P-11 inbound was within the magnetosphere near 95 R_J; during this brief interval, the angular distribution of 0.5–1.8-MeV protons was consistent with corotation.

On the three passes, P-10 in-, P-11 in-, and P-11 outbound, for which the magnetosphere was evidently nonstationary, the current sheet was encountered on two of the three, and then only before the first or after the last "durable" magnetopause crossing. In all three cases, within the magnetosphere between magnetopause crossings, the field pointed southward and was irregular. At these times, the spacecraft were in the turbulent internal boundary layer. Since this signifies proximity to the magnetopause, neither the temporal duration nor the derived spatial extent of this layer indicates its geometry, since the magnetopause is evidently in motion.

The Signature of the Middle Magnetosphere

While the spatial region between the dipolar radiation belts and the true outer magnetosphere is ill-defined because of time variations, it possesses a characteristic signature in the electron pitch-angle distributions relative to the magnetic field. In the radiation belts, the pitch-angle distributions are peaked perpendicular to the magnetic field, which is consistent with a distribution that is stably trapped in the magnetic mirror field. In the middle magnetosphere (12–25 R_J) the low-latitude P-10 inbound and outbound and the high-latitude P-11 outbound passes revealed electron fluxes peaked along the magnetic field in both directions; the ions were isotropic

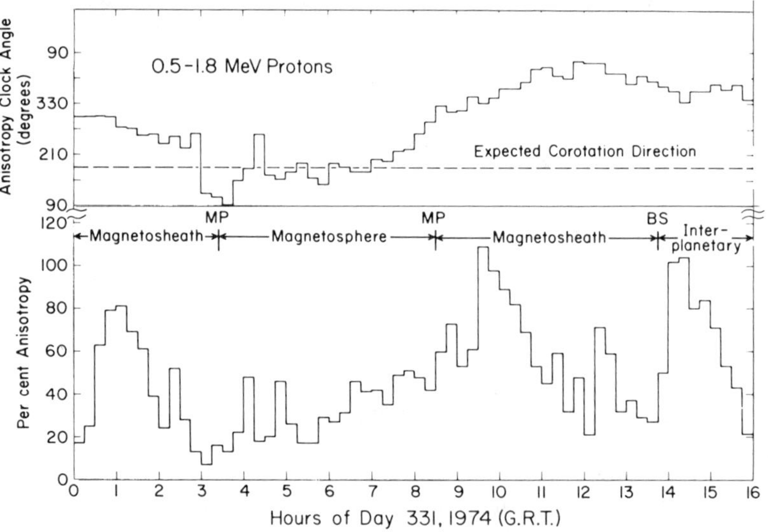

Figure 14 Corotation in the distant magnetosphere. Figures 2 and 3 indicate that P-11 inbound found itself within the magnetosphere for a brief interval of time near 95 R_J. During this interval, the clock angle or phase of the proton angular distribution was consistent with corotation at 95 R_J (from Simpson & McKibben 1976).

(Fillius & McIlwain 1974, Van Allen et al. 1974, Sentman & Van Allen 1976, Trainor et al. 1974). Sentman et al. (1975), using P-11 outbound data, found that both protons and electrons stream along the magnetic field away from Jupiter. When these fluxes reach the magnetic equator they would form bidirectional field-aligned distributions as observed; they invoke (ad hoc) pitch-angle scattering to account for isotropic proton distributions. Thus, either Jupiter, or the inner radiation belts, is a source of energetic particles for the middle magnetosphere. Sentman et al. (1975) suggest, following Nishida (1976), that inner radiation-belt particles that mirror at low altitudes undergo poleward diffusion due to ionosphere electric-field turbulence that violates the second adiabatic invariant. Upon reaching high latitudes, the particles would possess a strongly field-aligned pitch-angle distribution. A difficulty with this interpretation is that diffusion transport should produce a spatially smooth and temporally steady pitch-angle distribution, whereas the observations indicate high temporal and spatial variability.

Jupiter as a Cosmic-Ray Source

Energetic particles generated in the Earth's magnetosphere can escape to the solar wind. They are detectable, for example, upstream of the Earth's bowshock (Fan et al. 1964, Frank & Van Allen 1964, Anderson 1968). It was natural, therefore, to search for energetic particles as the Pioneers approached Jupiter. More than 1 au from Jupiter, both Chenette et al. (1974) and Teegarden et al. (1974) observed sporadic increases of electrons in the few hundred KeV to a few MeV energy range, which became more frequent as Jupiter was approached, and quasi-continuous near Jupiter. Teegarden et al. (1974) then showed that similar electron events observed at Earth's orbit, which had been found not to correlate with solar events, were of Jovian origin, since they have a 13-month periodicity, equal to Jupiter's synodic period. In Figure 15 (Pesses & Goertz 1976) the time of year that the Earth crosses the solar-wind field line that intersects Jupiter's magnetosphere is plotted for the period 1968–1974. The vertical bars indicate when quiet-time electron bursts were present, again suggesting that Jovian electrons were detected at the Earth (Teegarden et al. 1974). McDonald & Trainor (1976) have found simultaneous bursts of electrons on P-11 at 3–4.25 au and IMP-7 at 1 au, allowing for propagation delay. P-11, in the solar wind, apparently detected a flux increase observed when P-10 was outbound in the Jovian magnetosphere. Thus, there is little doubt that Jupiter is a significant source of MeV electrons in the heliosphere.

Given that Jupiter, at times, strongly modulates its energetic electron fluxes at its rotation period, it is natural to ask whether similar modulations are found in the Jovian interplanetary electron bursts. Chenette et al. (1974) reported that within each burst, the electron fluxes are modulated at a 10-hr period and, moreover, the spectral index tracked the 10-hr periodicity even more clearly. Even more remarkably, they claim that the phase of the interplanetary modulations is the same as that found inside the magnetosphere. Figure 16 plots the time-intensity profiles of the 6–30-MeV fluxes (bottom curve) and the ratio of the 3–6-MeV to 6–30-MeV electron fluxes during an interplanetary event measured on P-10. The vertical dashed lines represent times of spectral index maximum for a period of 9 hr, 55 min, 30 sec relative to a spectral index maximum measured when P-10 was subsequently within the magnetosphere. When a new burst commences after an old one dies away, the spectral index maxima have the same phase as in the previous burst. These observations have not been clearly confirmed by the other experiments on board the spacecraft.

The hypothesis that MeV electrons are contained in a flapping sheet in Jupiter's outer magnetosphere, while attractive for P-10 outbound and P-11 inbound, does not completely explain the continuity in phase of interplanetary electron spectral modulations with those in the magnetosphere. Chenette et al. (1974) argue that these facts can best be explained if Jupiter generates energetic electrons periodically in time throughout a significant part of its magnetosphere. No matter how they escape, they would then show phase continuity between the solar wind and magnetosphere.

It appears that a necessary but not sufficient condition for detection of an electron burst upstream of Jupiter is that the interplanetary magnetic field connect the spacecraft and Jupiter's magnetosphere (Smith et al. 1976) just as it must connect

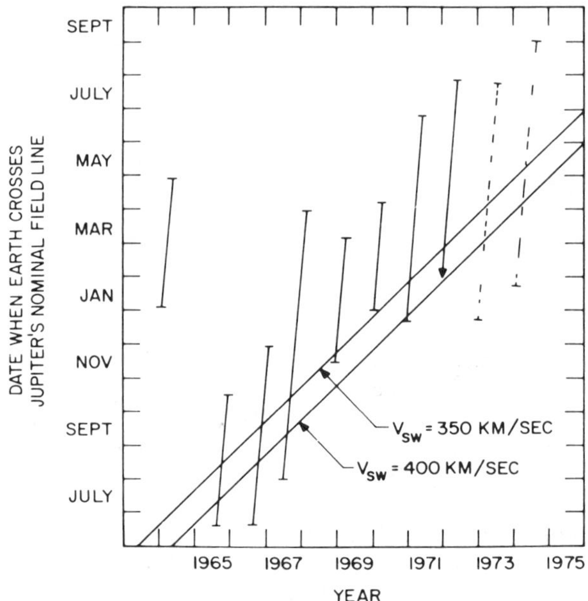

Figure 15 Arrival of Jovian cosmic ray electrons at the Earth. Shown here are the times during the years 1964–1975 when MeV electrons of nonsolar origin attributed to Jupiter arrive at the Earth. The oblique lines indicate when the solar-wind field line threading the Earth also threads Jupiter for solar wind spreads of 350 and 400 km sec^{-1}. The fact that Jovian electrons are seen as much as 3 months after the Earth line threads the nose of Jupiter's magnetosphere suggests that many Jovian electrons come from a tail several au long. The geometry is sketched in Figure 1 (from Pesses & Goertz 1976).

Jupiter and the Earth if a burst is to be detectable at the Earth. The electron intensity maxima are also associated with bursts of hydromagnetic waves in the solar wind (Smith et al. 1976). Vasyliunas (1975) has shown that Jovian cosmic-ray electrons have a minimum intensity and the softest spectrum when the subsolar point on Jupiter is near the longitude of the most frequent non-Io related decametric radio emissions.

The Jovian cosmic-ray electrons immediately pose three questions. When and where are those MeV electrons that escape generated? How and where are they modulated? And how do they escape? In Figure 15, we note that most of the Jovian cosmic-ray events observed at the Earth occur 1–3 months after Jupiter's nominal field line sweeps over Earth. Krimigis et al. (1975), Pesses & Goertz (1976), and Mewaldt et al. (1976) propose that this delay can be explained if one assumes that the electrons come when the Earth interplanetary field line intersects an extended Jovian magnetic tail. Krimigis et al. (1975) estimate a tail length of 4.6 au; Pesses & Goertz (1976) argue that there is an active region between 1 and 2 au. Recently, Barnes et al. (1977) have argued that Jovian cosmic-ray electrons are present through the full 13-month synodic period at lower flux levels than during the 3-month window mentioned above. Thus, electrons may diffuse out of Jupiter's magnetosphere and into the solar wind. A long tail may not be absolutely necessary to

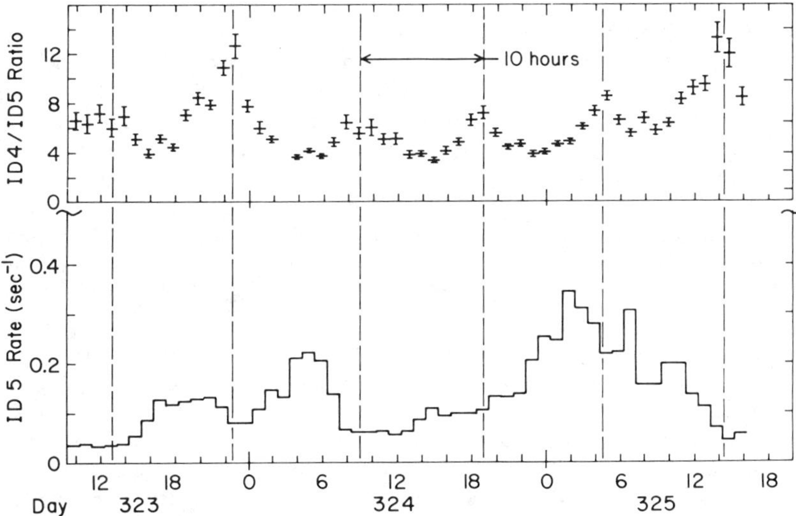

Figure 16 Ten-hr modulations of Jovian electrons in the solar wind. According to Chenette et al. (1974), the spectral index (as indicated here by the ratio of count rates in two energetic electron detectors) is modulated with a 10-hr period—a proof of their Jovian origin. In addition, their phase is identical with the phase encountered within the magnetosphere. The vertical lines indicate predicted phase maxima based on Jupiter's rotation period established when P-10 eventually entered the magnetosphere and extrapolated backwards in time to these solar-wind observations (from Chenette et al. 1974).

explain the observations (Jokipii 1976). However, the P-10 solar-wind plasma detector may have detected the Jovian tail 4 au downstream from Jupiter. At about the expected location, it detected on two occasions a complete dropout in plasma flux (D. S. Intriligator, J. Wolfe, private communications, 1976). However, the P-10 magnetometer was unfortunately not functional, and the energetic particles experiments have not reported any significant perturbations.

Summary

Jupiter's outer magnetosphere is so variable with time that flybys are unable to separate spatial from temporal variations. P-10 and 11 inbound, traversing the same region of space relative to Jupiter, found distinctly different magnetospheres. The longest period during which Jupiter's magnetosphere remained in the same state was 8 days, on P-10 outbound. Spatially separated multiple magnetopause crossings were observed on all four Pioneer passes. The magnetopause was found as much as a factor of two beyond the classical magnetopause radius, though on occasion it approached the classical distance. When the magnetopause was in motion, P-10 and 11 often found themselves in a turbulent boundary layer, characterized by southward turbulent magnetic fields, which has always been found next to the magnetopause. The evidence for periodic crossings of a current sheet in the turbulent layer is less compelling in the internal boundary layer than further within the magnetosphere. There is some evidence that the plasma and field lines tend to corotate with Jupiter in the internal boundary layer.

Time variability is not confined to the region beyond the classical magnetopause. A well-developed current sheet was found between 20 and 80 R_J on P-10 outbound and within 45 R_J on P-11 inbound; within 50 R_J, P-10 inbound and P-11 outbound were more consistent with a weakly distorted dipolar magnetosphere. When P-11 outbound was at sufficiently high latitudes that it should not have encountered a current sheet, it is surprising to us that its field and particle data looked so dipolar.

It will be some time before we understand Jupiter's time-dependent magnetosphere. However, a beginning has been made. Figure 17 is taken from Smith, Fillius & Wolfe (1976). The top two inserts show a compressed display of the UCSD energetic electron fluxes and the magnitude of the magnetic field for the entire P-10 encounter. When P-10 was within the magnetosphere, the magnetic field plot is shaded. The bottom two inserts show the P-11 solar-wind magnetic-field magnitude and solar-wind dynamic pressure during the P-10 encounter, adjusted for solar-wind travel time between P-11 and Jupiter. The solar-wind magnetic-field and dynamic pressure peaks every 7 to 10 days in association with solar-wind sector boundaries. The dynamic pressure at P-11 varies by a factor 10–30. When P-10 first encountered the magnetopause at 98 R_J, the P-11 dynamic pressure was low; the second inbound magnetopause crossing was associated with a pressure peak, as was the first outbound crossing. The distant magnetosphere encounter between 130–150 R_J on P-10 outbound was associated with an anomalously low solar-wind pressure. Thus, the available evidence suggests that when the solar-wind dynamic pressure is low, Jupiter's magnetopause is found well beyond its classical position; when the solar-wind dynamic pressure is high, Jupiter's magnetosphere is compressed and its

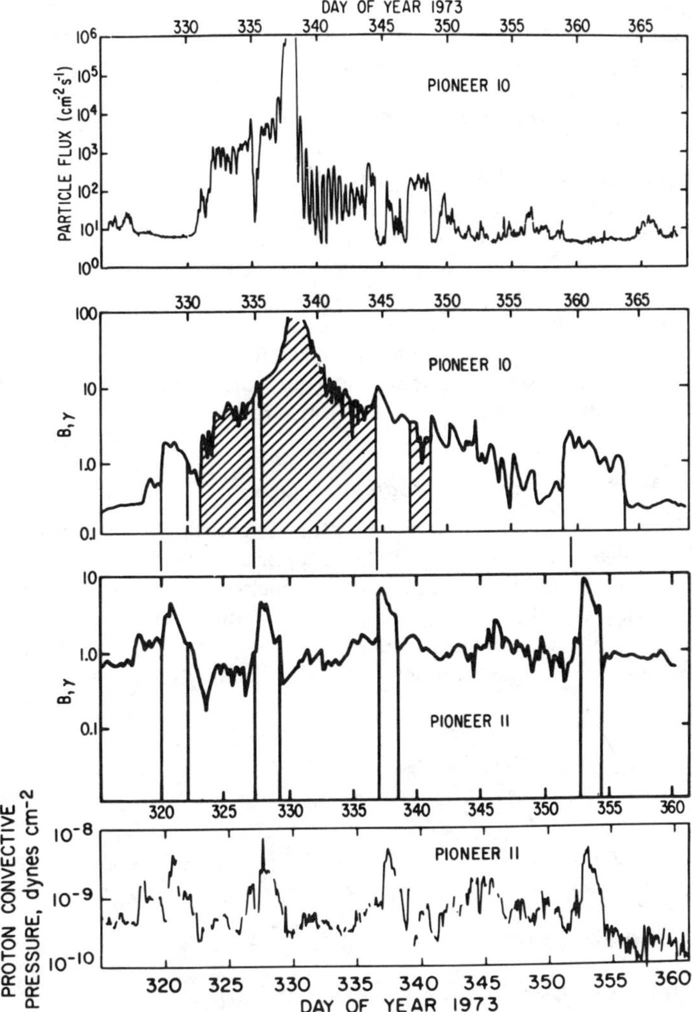

Figure 17 Solar-wind dynamic-pressure modulation of Jupiter's outer magnetosphere. The top two panels show the UCSD energetic electron fluxes and the JPL magnetic field magnitude for P-10 inbound; P-10 was within the magnetosphere in the shaded portions of the diagrams. The bottom two panels show simultaneous P-11 solar-wind data, adjusted for travel time between P-10 and P-11; the magnetic field magnitude and dynamic pressure are shown. Each 7–10 days there is a strong peak in dynamic pressure and magnetic field associated with solar-wind fast streams; the dynamic pressure increases by a factor 10–30 over its "nominal value" of 5×10^{-10} dynes cm^{-2}. When the pressure is nominal, Jupiter's magnetosphere is radially extended; the high-pressure pulses compress the magnetosphere (from Smith, Fillius & Wolfe 1976).

magnetopause may approach its classical position. Solar-wind dynamic pressure variations may therefore induce changes between a disk-like and a more dipolar outer magnetosphere.

5 MODELS OF JUPITER'S OUTER MAGNETOSPHERE

Theoretical models of the Jovian magnetosphere can be divided into three general classes: Earth-like convection models, quasi-static disk models, and radial outflow models. We discuss Earth-like convection models first because they provide a reasonable historical introduction and a context in which the differences between the Jovian and terrestrial magnetospheres are illuminated. We then discuss static disk models in which magnetic tension just balances the corotation centrifugal force and plasma is quasi-permanently trapped. However, Jupiter has at least two internal sources of plasma—the ionosphere and the Galilean satellites. If continuously supplied, this plasma must escape into the solar wind by flowing through the outer magnetosphere.

Earthlike Models of Jupiter's Magnetosphere

The late Neil Brice was the pioneer who first systematically scaled our knowledge of Earth's magnetosphere to Jupiter (Brice 1968, Brice & Ioannidis 1970, Ioannidis & Brice 1971, Brice & McDonough 1973). Kennel (1973) reviewed and extended this work to then hypothetical magnetospheres of Saturn and the still hypothetical magnetospheres of Uranus and Neptune. One starts with the theoretical scaling of steady-state solar-wind parameters from standard theory (Parker 1963):

$$U_r = 400 \text{ km sec}^{-1}, \qquad U_\phi \simeq 0, \tag{1a}$$

$$N = 5\text{–}7/r^2 \text{ cm}^{-3} \simeq 0.2\text{–}0.3 \text{ cm}^{-3}, \tag{1b}$$

$$B_r = 5\,\gamma/r^2 \simeq 0.2\,\gamma, \tag{1c}$$

$$B_\phi = 5\,\gamma/r \simeq 1\,\gamma, \tag{1d}$$

$$B = (B_r^2 + B_\phi^2)^{1/2} \simeq 5\,\gamma/r \simeq 1\,\gamma, \tag{1e}$$

$$NMU_r^2 = 1.25 \times 10^{-8}/r^2 = 5 \times 10^{-10} \text{ dyne cm}^{-2}. \tag{1f}$$

In (1a–f) above, U_r and U_ϕ are the radial and azimuthal components of the solar-wind flow velocity, N is the ion or electron number density, B_r and B_ϕ are the radial and azimuthal magnetic field components, and r is the heliocentric distance in au. The first expressions in (1a–f) above give the scalings normalized to typical parameters at 1 au, and the second expressions give approximate "nominal" values at the orbit of Jupiter, $r \simeq 5$ au.

If the plasma pressure inside the magnetosphere were small or zero, the solar-wind dynamic pressure would compress the dipole field until it is finally balanced by magnetic pressure. Since the current sheet at the interface between the solar wind and dipole region doubles the internal field, pressure balance at the nose stagnation point occurs at a distance D,

$$D = R_J[B_0^2/2\pi\rho u^2]^{1/6}, \tag{2}$$

where B_0 is the equatorial field strength of the surface. For the above "nominal" solar-wind parameters, $D \approx 41\ R_J$, and the magnetic field just inside the magnetopause is 11 γ. For the magnetopause to be at 80–100 R_J, this model requires a factor of 30–50 reduction in ρu^2; at Earth, only 10–20% variations in D are typically observed. Away from the stagnation point the super-Alfvénic, supersonic solar wind exerts a normal pressure, which is well approximated by $\rho u^2 \cos^2 \chi + P$, where χ is the angle of attack and P is the solar-wind thermal pressure. Hydrodynamic calculations indicate that at dawn and dusk the magnetopause is located at 3/2 D (Spreiter & Alksne 1969).

The above model accounts quite well for the static, zeroth order configuration of the Earth's magnetosphere. However, it assumes that the solar wind is unmagnetized and that the magnetopause is free of dissipation. Even though the energy density of the solar-wind magnetic field is only a few percent of the energy density of the flow, reconnection of the magnetic field lines between the solar wind and dipole fields, which requires a dissipative magnetopause, can couple solar-wind energy into the magnetosphere. Directional changes in the solar-wind field vary the reconnection and energy input rates, and thus may be ultimately responsible for many time varying, energetic phenomena within the magnetosphere.

After merging on the dayside magnetopause, dipolar and solar-wind field lines are dragged by the solar wind over and around the magnetosphere and are stretched out into a long magnetic tail. In the ionosphere, the field lines flow from the dayside to the nightside corresponding to a dawn-dusk electric field. In the tail, the merged or open-field lines flow toward the equatorial plane where they reconnect with their partners across a magnetic neutral sheet; in steady state the nose and tail reconnection rates must be equal. The newly closed flux must then flow from the tail reconnection region towards the nose to repeat the cycle. Thus, reconnection has the following general consequences for the Earth's magnetosphere: there is a region of open flux connecting to the polar caps in which cosmic rays have direct access to the surface of the Earth; there is a long magnetic tail containing a magnetic neutral sheet where stored magnetic energy is dissipated and plasma is heated and accelerated towards the Earth; and there is a convective flow of plasma and magnetic flux throughout the interior of the magnetosphere. Figure 18 shows an original sketch (Dungey 1961, Levy et al. 1964) of the flow driven by reconnection in the noon-midnight magnetic meridian plane.

We now construct a convection model based on reconnection for Jupiter. We may estimate the energy dissipation rate due to reconnection as follows. The full solar wind emf Φ across the width $3D$ of Jupiter's classical magnetopause is given by

$$\Phi \simeq 5\ MV \simeq \frac{3\ U_r BD}{C}. \tag{3}$$

A fraction $\beta < 1$ of the solar-wind flux impinging upon the magnetospheric cross section is reconnected; the rest flows around the magnetosphere (Levy et al. 1964). Consequently the emf corresponding to the internal convective flow is $\beta\Phi$.

Phenomenological studies indicate that $\beta \simeq 0.1$–0.2 at Earth. In a steady flow, $e\beta\Phi$ is the largest energy a charged particle can acquire from a betatron acceleration.

The energy dissipation rate $\dot{W}$ is given by computing the total current in the reconnection region of the dayside magnetopause and multiplying by $\beta\Phi$. The current per unit length in the magnetopause is $C\Delta B/4\pi$, where $\Delta B \simeq 11\gamma$ is the jump in magnetic field strength at the magnetopause. The total current I is then approximately $(C\Delta B/4\pi)l$, where l is the effective length of the reconnection region normal to the ecliptic plane, hence

$$W \simeq \frac{C\Delta B}{4\pi} l\beta\Phi \simeq 4 \times 10^{21}\,\beta\,\text{erg sec}^{-1}. \tag{4}$$

At the Earth, $W_E \simeq 10^{19}\,\beta$ erg sec^{-1}. Thus, Jupiter's magnetosphere would be at least two orders of magnitude more energetic than the Earth's in this model.

At the Earth, much of the dissipated solar wind is deposited by energetic particle precipitation and Joule dissipation into the auroral zone atmosphere. The auroral zone is just equatorward of the polar-cap open-field region, and has an area comparable to that of the polar cap. The radius of the polar cap r_{pc} can be estimated by assuming that the open flux in the polar cap approximately equals the dipolar flux that closes beyond the radius D. Then, $r_{pc} \simeq R_J^{3/2}/D^{1/2} \simeq R_J/7 = 10^9$ cm, which corresponds to a colatitude of $\simeq 8°$. At Earth, $r_{pc} \simeq R_E/3$, which corresponds to a colatitude of $\simeq 20°$. Assuming the auroral zone area is πr_{pc}^2, approximately 10^3 $\beta \simeq 10^2$ erg cm^{-2} sec^{-1} of solar-wind energy might be dissipated into the Jovian

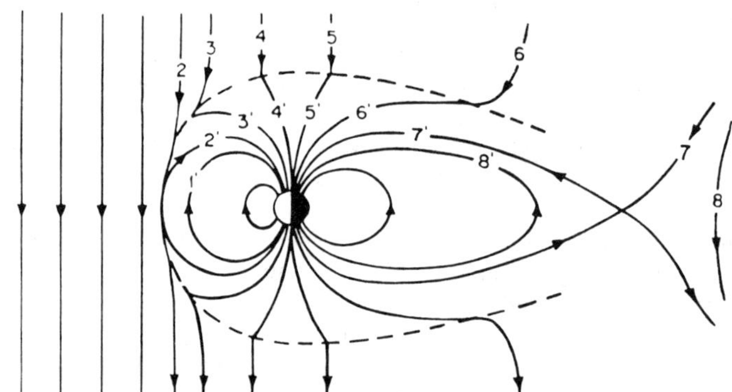

Figure 18 Convection model of a dipolar magnetosphere. This figure, originally drawn for the Earth, shows successive stages in the life cycle of a convecting magnetic flux tube. At 1′ it is closed and flowing towards the nose magnetopause, where it is reconnected with solar-wind field line (2 − 2′) and opened. (Cosmic-ray electrons can leave or enter the magnetosphere on open-field lines.) It is then dragged by the solar wind (3, 4, 5, 6) over the polar caps (3′, 4′, 5′, 6′) to form a long magnetic tail. At 7 it rereconnects with its southern hemisphere partner; at 8 it is closed again and flowing towards the nose. This diagram shows a slice through the noon midnight meridian plane perpendicular to the ecliptic. Since Jupiter's dipole is oppositely directed to Earth's, a *northward* solar wind field is optimum for nose reconnection (from Dungey 1961).

auroral atmosphere. This considerably exceeds the solar flux of UV photon energy deposited in Jupiter's atmosphere and should be the dominant energy source for Jupiter's upper atmosphere and ionosphere in the auroral zone.

Given the polar-cap radius, we may compute, following Dungey (1965), the length of Jupiter's magnetic tail L_T. The convection electric field in Jupiter's polar-cap ionosphere is of the order of $\beta\Phi/2r_{pc}$ in magnitude, directed dusk to dawn corresponding to a transport of flux from the dayside to the nightside. The convection speed is therefore $\beta C\Phi/2r_{pc}B_I$, where B_I is the ionospheric magnetic-field strength. The ionospheric foot of a magnetic field line crosses the polar cap in a time $\tau = 4r_{pc}^2 B_I/C\Phi\beta$. During the time τ, the field line passes from the nose to the tail reconnection region. The entire flow cycle lasts approximately 2τ. The length of the tail is $L_T = U_r\tau$. Inserting our previous estimates, we find that $\tau \simeq 15/\beta$ hours and $L_T = (400/\beta)R_J$. If we assume that the reconnection efficiency $\beta = 0.1$ as at the Earth, then τ is roughly one week. The time τ required to change the convection state of the magnetosphere is comparable with the duration of a given flyby pass and with observed time scales on which Jupiter's magnetosphere does change. With $\tau \simeq 1$ week, $L_T \simeq 4000\ R_J \simeq 2$ au in order-of-magnitude agreement with that estimated for the propagation of Jovian cosmic-ray electrons to Earth. Since the length of Jupiter's magnetic tail and Jupiter's distance from the Sun are comparable, Jupiter's magnetosphere and the solar wind have approximately the same scale. This suggests

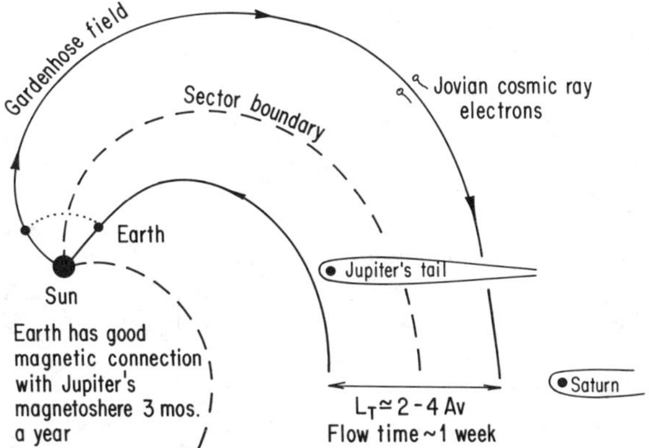

Figure 19 Global geometry of Jupiter's reconnection magnetosphere. According to the theory comparing Jupiter to the Earth, Jupiter's magnetic tail should be several au long, and the convection time the order of a week. Solar-wind field lines connecting the Earth and Jupiter's tail contain bursts of Jovian cosmic ray electrons. The Earth connects to Jupiter's tail for 1/4 of a revolution, about 3 months. The solar wind is typically divided into 4 magnetic sectors, about 1/4 solar rotation period, or 7 days apart; Jupiter's nose can be in one sector, its tail in another, with correspondingly different reconnection boundary conditions. In addition, large dynamic pressures are associated with sector boundary crossings; Figure 21 indicates that these compress at least the nose region. Since all these changes take place on the convection flow time, we expect Jupiter's magnetosphere to be steady only for an exceptionally quiet solar wind.

that Jupiter's magnetosphere varies in time in response to an entirely different set of solar-wind inputs than Earth, whose magnetospheric tail is only 0.04 au long. For example, the solar wind is divided up into a number of sectors, typically four, in which the gardenhose field is directed towards or away from the Sun and reverses as the sector boundary is crossed. Jupiter remains in one sector for one quarter of a solar rotation, or about a week. Thus, the nose of Jupiter's magnetosphere could be in one sector and its distant tail in another, suggesting that Jupiter's magnetosphere rarely achieves a steady connective flow pattern. Magnetic storms, due to a solar wind intensified by a solar flare, last typically only one or two days at the Earth; storms might not last long enough to create a steady flow at Jupiter. The configuration is sketched in Figure 19.

Finally, for later purposes, we need an estimate of the flux of solar-wind ions that circulate in the internal convection flow. The total number of solar-wind ions per second traversing the cross-sectional area of Jupiter's magnetosphere $2.25\,\pi NUD^2 \simeq 7 \times 10^{30}$ sec^{-1}. It is not known precisely how many of the particles circulating in the terrestrial magnetosphere originate in the solar wind and how many in the ionosphere, probably because the ratio varies strongly with solar-wind activity. However, the typical number, 10^{26} sec^{-1}, corresponds to roughly 10^{-3} of the solar-wind flux impinging upon the Earth's magnetospheric cross section. If this scaling applies to Jupiter, 10^{28} particles per second circulate through its magnetosphere.

Corotation and Plasmapause

For a dipole field with aligned magnetic moment and rotation axes, the radially outward electric field due to corotation is

$$E_{CR} = \frac{\Omega R_J B_0}{C}(R_J/r)^2. \tag{5}$$

If we assume that reconnection imposes a uniform dusk-dawn electric field E_C across the magnetosphere, there is a stagnation point ($E_{CR} = E_C$) in the combined corotation-convection flow at local dusk at a radial distance

$$r_p/R_J = [\Omega R_J B_0/CE_C]^{1/2} = [\Omega R_J B_0/\beta U_r B]^{1/2}. \tag{6}$$

By adding corotation and convection electric potentials, the flow streamlines can be determined. The streamline passing through $r = r_p$ forms a closed curve in the spin-magnetic equator, the plasmapause. Inside the plasmapause low-energy particles corotate around Jupiter and cannot escape to the solar wind; if internal plasma sources exist, the plasmasphere should be a high-density region. The minimum plasmapause radius is 0.4 r_p and occurs at local dawn.

If we compare the mean plasmapause radius, 0.7 r_p, with the classical nose radius, we find

$$\frac{0.7\,r_p}{D} = 0.7\left(\frac{\Omega R_J}{\beta B}\right)^{1/2}\left(\frac{2D\,B_0\rho}{U_r}\right)^{1/6} 1/\sqrt{\beta}. \tag{7}$$

Thus Brice & Ioannidis (1970) concluded that the Jovian plasmapause extends beyond the magnetopause; for Earth 0.7 $r_p/D = 0.4$. This is one of the senses in which

Jupiter is a fast, and Earth a slow, rotator. Brice & Ioannidis (1970) argued that near the magnetopause there would be a thin region where all the magnetic flux involved in convection would return to the nose magnetopause. Their sketch of the flow streamlines is shown in Figure 20.

Interaction of the Magnetosphere with the Ionosphere and Atmosphere of Jupiter

Field-aligned currents are a natural feature of the coupling of an external hydromagnetic flow to a condensed conducting central body (Kennel 1975). While field-aligned currents exert no stresses themselves, they communicate stresses between the magnetosphere and atmosphere-ionosphere by forming part of a circuit whose current closes by flowing perpendicular to the magnetic field in the ionosphere, where the conductivity perpendicular to the magnetic field is finite, and in the magnetosphere, where significant plasma pressure gradients support perpendicular currents.

Neglecting convection for the moment, the current system necessary to enforce corotation is shown in Figure 21. The current is in over the poles, out at lower

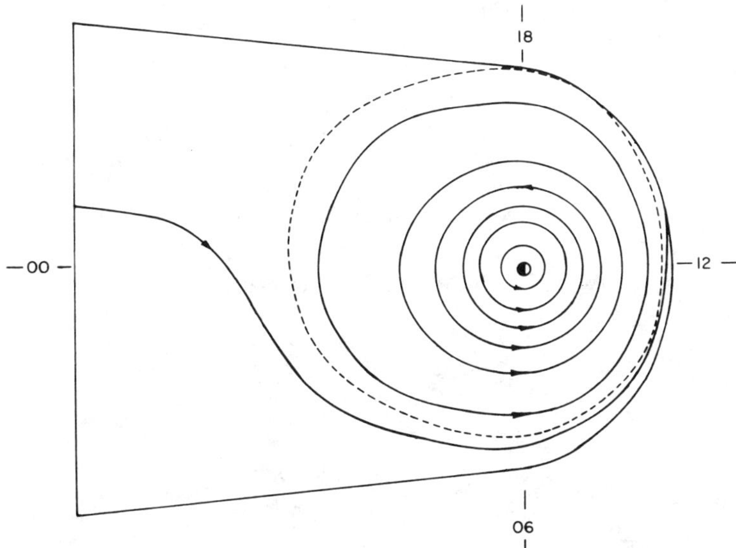

Figure 20 Jupiter's plasmasphere and corotation region. Shown here is a cut through Jupiter's equatorial plane; the outer line represents the classical magnetopause. Adding a uniform convection electric field to Jupiter's corotation electric field permits the flow streamlines to be drawn. They are closed corotating around the planet within the plasmapause-dashed line. All the flux and plasma transported by convection—symbolized by the single flow streamline entering from the magnetic tail—reaches the nose reconnection region in a twin layer between the plasmapause and magnetopause. If there are plasma sources within it, the plasmasphere should be a high-density region; radial transport therein must be diffusive (from Brice & Ioannidis 1970).

latitudes, and closes in the magnetosheath. Neglecting rotation, the topology of the currents threading Jupiter's polar-cap ionosphere, which are demanded by convection, would be that given in Figure 21, which shows a slice through the dawn-evening meridian plane. The dusk-dawn convection electric field drives a dissipative Pedersen current, which must be fed by the field-aligned currents into (out of) the ionosphere at dusk (dawn); the current closes in the magnetosheath. The energy required to sustain these currents against ionospheric dissipation comes from decelerating the solar wind. The correct current topology, which combines convection and corotation systems, has not yet been drawn, but is probably a nonlinear addition of the two.

As the nightside convective flow approaches Jupiter, corotation forces more flux and plasma around in the local morning than in the evening (Figure 20), thus investing angular momentum in a flow that, in the distant tail, possessed none by symmetry. Since the flow encounters the magnetopause before it can symmetrize on the dayside, a net angular momentum is lost from the system to the solar wind. Kennel & Coroniti (1975) estimated the torque T which Jupiter must exert on the flow to be $T \approx \Omega D^2 \dot{M}$, where $\dot{M}$ is the mass flux entrained in the convection flow. Estimating $\dot{M}$ by 10^{28} hydrogen ions per second, we find $T \approx 10^{23}$ dyne cm^{-1}, and the rotational energy dissipation rate $T\Omega \approx 10^{19}$ erg sec^{-1}.

The $\mathbf{J} \times \mathbf{B}$ stress due to ion Pedersen currents is exerted against the neutral atmosphere. In a time $\tau_c = \nu_{in}^{-1} \rho_n/\rho_i$, where ν_{in} is the ion-neutral collision frequency and ρ is the respective mass density, the neutrals are accelerated to the $\mathbf{E} \times \mathbf{B}$ velocity, and, since $\mathbf{E} + 1/c\mathbf{V} \times \mathbf{B} = 0$ in the co-moving plasma-neutral frame, all currents cease (Fedder & Banks 1972). Present solar UV production models (Atreya et al. 1974) indicate that the Jovian ionosphere is similar to the Earth's F region in that ν_{in} is small compared to the ion cyclotron frequency and that the neutral

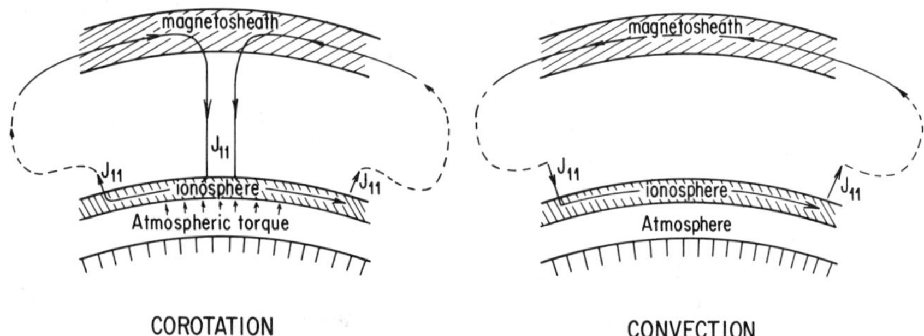

Figure 21 Polar cap current systems. The left panel suggests the topology of the polar cap ionospheric current systems required to despin Jupiter's atmosphere and enforce corotation on the magnetosphere; the right panel shows the current system corresponding to the convection of field lines antisunward over the polar cap (see Figure 22). The combined corotation convection system is some nonlinear addition of the two.

gas at ionospheric altitudes has little inertia. For these models Coroniti (1974) found $\tau_c \approx 40$ min so that convection electric fields might rapidly stimulate ionospheric neutral wind systems. Kennel & Coroniti (1975) pointed out that since $\Omega\tau_c \ll 1$, the ionospheric neutral atmosphere may possess insufficient inertia to enforce corotation on magnetospheric field lines. Assuming an eddy diffusion coefficient of 10^8 cm^{-2} sec^{-1}, the upper limit in ionospheric models, they found that the upward diffusion of angular momentum from deeper layers in the atmosphere could only exert a torque of 10^{21} dyne cm^{-1}, a factor 10^2 less than the estimated torque that a convecting magnetosphere would exert on the polar ionosphere. Hence magnetospherically imposed torques may despin the polar ionosphere.

The observations of corotating protons in the outer magnetosphere suggest that present models of the polar Jovian ionosphere need revision. There seem to be three possibilities for enforcing corotation. First, unstable strong shears in the neutral wind speed induced by convection might lead to a large eddy viscosity and an enhanced upward angular momentum transport. Second, since particle precipitation is expected to deposit 100 erg cm^{-2} sec^{-1} in the polar region, the ionosphere could extend much deeper into the neutral atmosphere than predicted by solar *UV* models, thus increasing the ionospheric inertia. Third, the neutral wind systems postulated by Brice & McDonough (1973) to account for the radial diffusion in the radiation belts could transport angular momentum from low to high latitudes at a rate sufficient to balance magnetospheric torques. In any case the coupling between the magnetosphere, ionosphere, and neutral atmosphere is undoubtedly energetically significant and highly nonlinear and acts as a controlling influence on Jovian plasma dynamics.

Kivelson & Winge (1976) found that the helium-vector magnetometer on P-11 inbound recorded a structured transient perturbation with the characteristic signature of a field-aligned current near the L-shell of Ganymede. They find its interpretation ambiguous; it could have been caused by turbulence in the wake of Ganymede or by coupling of ionospheric and magnetospheric flows. To date, this is the only report of a field-aligned current in the Jovian magnetosphere.

Quasi-static Disk Model of the Jovian Magnetosphere

As presently construed, the convection model has dipolar geometry and corotation flow out to the distant plasmapause, which occurs just inside the magnetopause. However, even a low-density cold plasma in the distant magnetosphere would have a corotation dynamic pressure that exceeded the containing magnetic pressure. Hence the magnetic field, even inside the plasmapause, would be distorted away from dipolar. There have been three attempts to calculate a quasi-static magnetic disk supported by centrifugal forces motivated by the Pioneer observations of thin current sheets. Before turning to the calculations of disk structure, we discuss briefly the parameters deduced from observation of the "hinged current sheet" model of Smith et al. (1974a). If the sheet's hinge is at 20 R_J and if it moves rigidly beyond 20 R_J, the rotation of Jupiter's oblique dipole causes the sheet to move through the spin equatorial plane at a speed of approximately 30 km sec^{-1}, which is larger than the spacecraft speed. Combined with the 1–2 hr duration of a typical sheet crossing, this leads to a sheet thickness of 2 R_J. The observed magnitudes of the magnetic-field reversals then imply a linear current density of 10^{-2} A m^{-1}, which when extended over 60 R_J as for P-10 outbound gives 10^8 A total current, leading to a quasi-uniform northward perturbation field of about 10 γ in the inner magnetosphere. Regardless of the ultimate correctness of this model, it probably gives a reasonably rough estimate of the physical magnitudes involved.

A number of authors (Ellis 1965, Gledhill 1967, Melrose 1967, Mendis & Axford 1974) pointed out the importance of centrifugal forces to Jupiter's outer magnetosphere. Gleeson & Axford (1976) attempted a partially self-consistent model of the disk for an aligned dipole. They considered the disk to be sufficiently thin that the magnetic field outside is due to a thin sheet of azimuthal current. The plasma within the sheet is contained by balancing centrifugal forces with magnetic tension, which requires a component of the magnetic field normal to the sheet. Radial pressure gradients are neglected. They then model, ad hoc, the radial distribution of the azimuthal sheet current and calculate the resulting magnetic field and the cold plasma density profiles it would contain. They sketch a number of sample magnetic field configurations. Since several of their configurations contain field lines that are not closed, they suggest that magnetic connection would occur, rendering their disks unsteady.

Goertz (1976b) has produced a calculation similar to Gleeson and Axford's, but allows the dipole and spin axes to be misaligned and considers a plasma pressure comparable with the corotational energy density. Goertz et al. (1976) had argued that the hinged-disk model of Smith et al. (1974a,b) (see Figure 14) is less consistent with the observations than a current sheet confined to Jupiter's magnetic equatorial plane. Recall that they required field lines to be closed within the disk and open outside to account for the strong 10-hr energetic particle modulations on this pass. Goertz (1976b) writes the magnetic field as the sum of a dipole plus a disk perturbation, parameterizes the pressure by $P = k(r, z)\rho\Omega r^2/z$, and solves the hydromagnetic force-balance equations in the thin-disk approximation. He then fits his model according to P-10 outbound data and finds that even with finite pressure, the disk remains thin (1–2 R_J). The plasma density consistent with this fitting is roughly $6 \times 10^7 (R/R_J)^{-5.4}$, which disagrees with the cold plasma density of $\simeq 80$ cm^3 measured by Frank et al. (1976) by an amount possibly reconcilable with theoretical and observational inaccuracy.

Barish & Smith (1975) pursued another tack in quasi-static current-sheet modeling. They did not solve hydromagnetic equations, but sought an analytic representation of the magnetic field that is consistent with a partial cut through the P-10 data, namely a field in the magnetic equator that falls off as r^{-3} within 20 R_J and as r^{-2} beyond with a confined current "sheet" near the equator. Figure 22 shows a drawing of their derived configuration. It seems to agree with observation in two ways. First, the field is southward at the magnetopause (taken here to be at 100 R_J). Secondly, the above computation, performed before the P-11 encounter, was compatible with the P-11 outbound pass, which showed a weakly distorted dipole at 30–35° latitude.

Several comments are now in order. The models described above require that corotation extends to the magnetopause. At 100 R_J the corotation speed is 1100 km $\sec^{-1}$ corresponding to an ion energy of 7.5 μ KeV, where μ is the mass in units of the hydrogen mass. The corotation speed exceeds the solar-wind speed, is oppositely directed to the solar-wind velocity at local dawn, and adds to it at local evening.

None of the models have an azimuthal field component B_ϕ, which is consistent with portions of the P-10 inbound and P-11 outbound passes. However, as in standard solar-wind theory, an azimuthal field—which requires a radial current—is a signature of outward angular momentum transport. This in turn requires that there be a net outward flow of plasma supplied by Jupiter or its inner magnetosphere. If the radial flow velocity is sub-Alfvénic and subsonic, the flow dynamic pressure is negligible and the quasi-static disk models may be approximately valid.

If the "hinged current sheet" of Smith et al. (1974a) flapped rigidly beyond the hinge, no sheet crossings would have been observed at large distances on P-10 outbound. Smith et al. (1974b) proposed that Alfvén waves propagating radially outward might cause it to ripple sufficiently to permit sheet crossings at large distances. Eviatar & Ershkovich (1976) assumed that the dipole wobble excites Alfvén waves propagating outward along field lines in the sheet. They assume further that the waves are not convected by a super-Alfvénic outflow in the sheet. From the observed magnetic variations and the 10-hr periodicity, they estimate an Alfvén speed and from that a plasma density $n \simeq 4 \text{ cm}^{-3}$, roughly the same as that estimated by Wolfe et al. (1974) from pressure balance with no radial outflow.

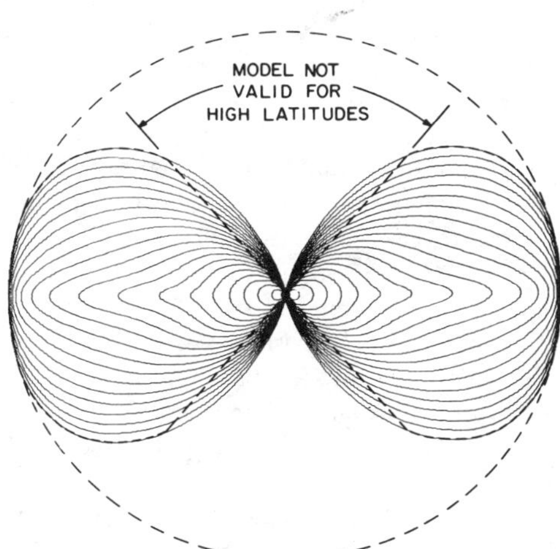

Figure 22 Model of Jupiter's magnetic disk and magnetosphere. Shown here is a static magnetic field model fit analytically by Barish & Smith (1975) to the general behavior of the P-10 outbound data. This model is in rough accord with observation both of a southward field near the magnetosphere and of a dipolar structure at high latitudes, and therefore helps us to visualize the magnetosphere; however, it neglects all dynamics (from Barish & Smith 1975).

Jovian Plasma Sources

What are the sources of the plasma in Jupiter's magnetosphere? The Earth has two possible plasma sources, the ionosphere and the solar wind, and Jupiter three—solar wind, ionosphere, and satellites, particularly Io. An experimental determination of plasma origin throughout the magnetosphere would resolve many theoretical difficulties. Surprisingly enough, even at the Earth, estimates of the strength of the solar-wind source are quite difficult, primarily because diffusion and/or transport across the magnetopause and in the magnetic tail are not well understood. If Earth-like scaling prevailed, 10^{28} solar-wind particles might cycle through Jupiter's magnetosphere each second. It is somewhat easier to estimate the Jovian ionospheric source. This is because at Earth, the proton gravitational escape energy is about 1 eV, and the ion temperature at the top of the ionosphere is 0.1–0.5 eV. Within the Earth's plasmasphere, subsonic motions adjust the plasma pressure in space to the ionospheric boundary conditions. Beyond the plasmapause, and particularly on open-field lines where the plasma pressure is low, ionospheric plasma expands supersonically into space in a miniature version of the solar wind—the polar wind. By contrast, at Jupiter, the proton energy to escape to infinity is about 19 eV, and the Jovian ionosphere is thought to be cold. Straightforward thermal ionospheric expansion seems difficult at Jupiter. However, Ioannidis & Brice (1971) and Michel & Sturrock (1974) observed that photoelectrons from solar-UV ionization of the Jovian upper atmosphere would have energies in the 5-20 eV range. If, as these electrons expand along field lines, they are scattered either by Coulomb collisions or plasma turbulence onto magnetically trapped orbits, they would pull protons along with them by electrostatic ambipolar couplings. The photoelectron energies are certainly sufficient to bring ions to the equatorial plane in the inner magnetosphere; for instance, to escape to $L = 2\ R_J$, about 7 eV is required. If the escaping photoelectrons are in strong diffusion (Kennel 1969), so that they are scattered across their loss cone in less than their transit time between conjugate ionospheres, the ion injection flux approaches its upper limit, which equals the photoelectron flux leaving the ionosphere. At the Earth, this photoelectron flux $F \simeq 7 \times 10^8$ cm^{-2} sec^{-1}. Since the solar UV flux is reduced by ~ 25 at Jupiter, the photoelectron flux there is $\simeq 2.5 \times 10^7$ cm^{-2} sec^{-1} (Ioannidis & Brice 1971, Michel & Sturrock 1974). Recognizing that photoelectrons are produced on Jupiter's dayside only, the maximum number of protons $\dot{N}$ delivered to Jupiter's magnetosphere by photoelectrons is $\dot{N} \simeq 2\pi R_J^2 F \simeq 10^{28}$ sec^{-1}, approximately equal to the much less well-founded estimate of the solar-wind source.

Ioannidis & Brice (1971) and Michel & Sturrock (1974) calculated the equilibrium plasma density in the equatorial plane, taking into account one significant difference between Earth's and Jupiter's magnetospheres. The terrestrial synchronous orbit is at 6.6 R_E, beyond the usual position of the Earth's plasmapause, whereas the Jovian synchronous orbit is at 2.2 R_J. Thus centrifugal forces play no role within the Earth's plasmasphere, and an important one throughout Jupiter's magnetosphere. The component of centrifugal force parallel to the magnetic field is directed towards the equator, which flattens the plasma equilibrium distribution into a disk, concentrating the density near the equator. Ioannidis & Brice's (1971) model of the equatorial plasma density produced densities of <1 cm^{-3} in the inner radiation belts within the orbit of Io ($L = 6$); beyond Io the equatorial density increased rapidly until it reached $\simeq 10^2$ cm^{-3} at $L = 8$, whereupon the corotational energy approaches the equatorial magnetic-energy density. They then invoked the centrifugally driven interchange instability (Hill 1976) to radially diffuse plasma outward, whereupon the density profile follows at L^{-4} law. Thus arose the idea that plasma expanding onto field lines connecting to Jupiter's inner magnetosphere would be a source for Jupiter's outer magnetosphere.

The wisdom of hindsight permits a more modern view of this idea. The observations of the Jovian radiation belts are compatible with a radial diffusion mechanism that derives its energy not from magnetospheric plasma but from atmospheric winds. Thus, ionospheric plasma will be injected onto flux tubes that are always executing flux-tube interchange motions driven by winds.

Using the Jovian photoelectron flux calculated by Swartz et al. (1975), McDonough et al. (1975) updated the diffusive equilibrium calculations of Ioannidis & Brice (1971). Because of differences in atmospheric composition, the Jovian photoelectron flux above 10-eV is an order of magnitude smaller than the scaled terrestrial flux, though the flux below 10-eV is larger. Thus, the above calculation produced smaller equilibrium densitites than that of Ioannidis & Brice (1971). The effect of this revision upon disk and radial outflow models is discussed by Prakash & Brice (1975).

Two comments are now in order. First, precipitation of energetic electrons into Earth's atmosphere produces copious fluxes of backscattered secondaries of several hundred eV. If the outward secondary flux exceeds the inward primary flux, ionospheric plasma could again expand into the magnetosphere. Evaluation of this mechanism requires a detailed calculation of secondary production in a hydrogen atmosphere, which is at present difficult to do because the peak in the electron precipitation energy flux has not been determined experimentally. The convection model of the magnetosphere predicts that ~ 100 erg cm^{-2} sec^{-1} will be deposited into the high-latitude auroral zones; moreover, at mid-latitudes, $L \sim 10$, Coroniti (1975) has estimated that the precipitation energy fluxes from electrons with $E > 1$ MeV could be a few tenths of an erg cm^{-2} sec^{-1}. The energy flux at lower energies is likely to exceed this value. Thus, there is good reason to consider the effects of energetic electron precipitation upon the Jovian ionosphere in general, and upon secondary electron production in particular.

Now for our second comment. Dessler & Hill (1975) have pointed out a possible rotational asymmetry in the Jovian ionospheric plasma source due to the large higher multipoles in Jupiter's surface magnetic field (Smith et al. 1975, Acuña & Ness 1976). Hence flux tubes of the same equatorial area connect to different ionospheric areas. The flux tubes connecting to the largest ionospheric area (smaller B_I) would fill most rapidly. Dessler & Hill (1975) note that the surface magnetic field, averaged over latitude $B_I(\phi)$, varies by a factor of 1.5 with longitude ϕ. Defining the flux-tube content $\eta = \int n\,dl/B$, where dl denotes integration along the flux tube, then $\eta = F/B_I(\phi)$. They assume that plasma so injected radially diffuses until centrifugal instability is achieved, whereupon it escapes. Let this radial diffusion time be τ and let $B_I(\phi) = B_0 f(\phi)$, where B_0 represents an appropriately averaged dipole field. The flux-tube content is $n\tau$; from this, one estimates the equatorial density and the critical L_A where the corotation energy density equals the magnetic

$$L_A = [B_0^2 f(\phi)/8\pi M\Omega^2 R_J \tau F]^{1/4}.$$

Since $\tau = 1$ yr (Coroniti 1975), $L_c \simeq 40\ f^{1/4}(\phi)$. Dessler and Hill estimated $\tau \simeq 10^3$ hours, whereupon $L_A = 80\ f^{1/4}(\phi)$; $f^{1/4}(\phi)$ varies about 10% with longitude. We comment later upon the possible significance of this.

Frank et al. (1976) have derived some information on the plasma density in the inner magnetosphere ($L < 6$) from the solar-wind plasma detector. Their measured ion energies (>100 eV) do not fit the picture of cold ions expanding from the ionosphere. Moreover, the measured densities exceed the Ioannidis-Brice densities by a large factor in the inner magnetosphere. Goertz (1976a) has reexamined the arguments about flux-tube filling, maintaining that instabilities from colliding streams of the equator may be able to produce the required densities. Nonetheless, it seems to us that there is presently no decisive proof that the cold plasma in Jupiter's magnetosphere comes predominantly from the ionosphere.

Siscoe & Chen (1976) propose that Io's neutral hydrogen ring may be a source for plasma within Io's orbit. Judge & Carlson (1974) observed a torus of neutral hydrogen centered radially on Io's orbit and extending a significant portion of the way around it. The measured plasma densities of Frank et al. (1976) are sufficient to account for the neutral hydrogen losses by charge exchange needed to explain the azimuthal extent of the ring. However, charge exchange is not a plasma source since it leaves the proton density constant. Photoionization, though slower than charge exchange, produces about 10^{25} new hydrogen ions per second. If the electron temperature exceeds about 10 eV, collisional ionization could exceed the photo-ionization rate. They point out that the measured flux-tube content η, which is proportional to nL^4, peaks at Io and is consistent with an Io source and radial diffusion losses. Atmospherically driven radial diffusion at the rate demanded by radiation belt observations yields plasma densities within $L = 6$ in reasonable accord with observation. However, the true test of radial diffusion transport in this case is not the spatial dependence of the flux-tube content, but the spatial dependence of the phase space density (Goertz, private communication, 1977). The Siscoe-Chen model predicts that the plasma at $L = 3$ would have 800 eV if it has 100 eV at Io.

In addition to hydrogen, Io emits heavy atoms such as sodium and potassium and perhaps others (Brown et al. 1975, Trafton et al. 1974, Mekler & Eviatar 1974, Trafton 1975, Carlson et al. 1975, Bergstrahl et al. 1975, Kupo et al. 1976, Bergstrahl et al. 1977). Since the main loss mechanism for neutral atoms is probably electron impact ionization (Carlson et al. 1975, Eviatar 1976, private communication), the Jovian magnetosphere must contain a heavy ion plasma. A. Eviatar et al. (private communication, 1976) suggested that the Jovian sodium plasma should have 3 separate energy components—cold, thermal, and energetic plasma, depending upon whether ionization takes place near Io, in the outer magnetosphere, or in the solar wind. Hill & Michel (1976) have suggested that these heavy ions are a significant source of plasma for Jupiter's outer magnetosphere. They point out that ions originating from Io are confined to a disk of about 1 R_J thickness about the Jovian equator and are probably more confined than ionospheric ions. They did not estimate the heavy ion fluxes delivered to the outer magnetosphere.

The first direct observation of a heavy ion in Jupiter's magnetosphere was reported by Kupo et al. (1976). The forbidden doublet of singly ionized sulfur was observed on a large number of plates over several months. The line intensities are consistent with a sulfur ion density of about 10^2 cm^{-3} and require electron densities of 5×10^2 cm^{-3} in the sulfur cloud, which extends out to 8 R_J. Neutral sodium has also been observed (Kupo et al. 1976); its cloud extends from 4–10 R_J near Io, but cuts off sharply at about 7 R_J on the other side of Io's orbit. The total ion and neutral populations are about 7×10^{33} and 10^{30} respectively. If the sulfur ions radially diffuse on the time scale of about a year that is determined by the radiation belt, then about 2×10^{26} sulfur ions sec^{-1} are delivered to the outer magnetosphere, a mass flux roughly equal to the upper limit mass flux of ionospheric hydrogen. For $Te \simeq 10$ eV, the impact ionization time for sulfur is $\sim 5 \times 10^4$ sec, so the neutral-to-ion ratio in the cloud should be less than 10^{-3} in rough accord with observation.

In summary, both Jupiter's ionosphere and Io are possible sources of cold plasmas. Both sources are concentrated within $L < 10$; 90% of Jupiter's surface area lies on field lines connecting to $L < 10$. The ionosphere produces an upper limit of 10^{28} hydrogen ions per second from photoionization, and Io produces at least 10^{25} H ions sec^{-1}, 10^{26} sulfur ions sec^{-1}, and probably other heavy ions. A fraction of this plasma radially diffuses into Jupiter's outer magnetosphere on time scales of months to a year as determined by energetic particle measurements. As far as the outer magnetosphere is concerned, its inner magnetosphere sources, however intrinsically variable in time, will appear constant on scales shorter than the radial diffusion time scale. Therefore, variations in these sources probably cannot account for the gross changes in Jupiter's outer magnetosphere observed to take place in a week.

Radial Outflow Models

Since the centrifugal force confines the cold plasma to near the magnetic equator, thus inhibiting its precipitation loss to the atmosphere, the most likely cold plasma sinks possible are radial diffusion inward to Jupiter and outward into the middle and outer magnetosphere. In diffusing outward, conservation of the first two adiabatic invariants implies that the plasma thermally cools while it gains corotational kinetic energy. In the theory of stellar winds (Mestel 1968), a cold plasma can break out into a radially outflowing wind if the corotational energy density approximately equals the magnetic energy density at the critical point of the flow. This "magnetic sling" stellar wind solution led Michel & Sturrock (1974) to propose that the Jovian internal plasma sources would centrifugally drive a planetary wind similar to stellar and pulsar winds. This model has been elaborated in a series of papers by the Rice group (Hill et al. 1974a,b, Dessler & Hill 1975, Hill & Michel 1976, Carbary et al. 1976).

In this section we present the physical content of this model. We first discuss a cold, centrifugal wind and compute the critical Alfvén point. We tie the critical Alfvén radius to the plasma density at Io which has now been measured. We then compare the dynamic pressure at the Alfvén point with the dynamic pressure in the solar wind to decide whether a sub-Alfvénic or super-Alfvénic flow is expected beyond the critical point. Then we consider the confinement of Jupiter's planetary wind by the solar wind.

Centrifugally Driven Winds

Equating corotational and dipolar equatorial magnetic energy densities leads to the following expression for the Alfvén radius $r_A = L_A R_J$:

$$n_A L_A^8 = B_I^2/4\pi\mu M_H \Omega^2 R_J^2,$$

where n_A is the density at the Alfvén point and μ is the mean molecular weight of the diffusing plasma. If the outward radial diffusion is without loss (free), the flux-tube content is preserved so that nL^4 is constant. Then $n_A L_A^4 = g n_0 L_0^4$, where $n_0 \simeq 10^2$ cm^{-3} is the number density at Io, $L_0 = 6$, and $g < 1$ takes into account the possibility that some ions are lost in diffusing to the Alfvén point. Therefore

$$L_A = [B_I^2/4\pi\mu M_H \Omega^2 R_J^2 n_0 L_0^4]^{1/4} = 41\ \Delta_0^{-1/4},$$

where $\Delta_0 = \mu g n_0/100$. The above probably underestimates L_A because $g \leqq 1$ and the measured magnetic field usually exceeds dipolar. Increasing μ by 16 decreases L_A by a factor 2. The uncertainties therefore compensate.

Beyond the critical point L_A, radially diffusing plasma cannot be contained by the magnetic field; we therefore expect it to break out into a hydromagnetic outflow. The outflow may either be a super-Alfvénic wind or a sub-Alfvénic breeze, depending upon the confining pressure of the solar wind. In this section, we discuss the nature of the flow along the Jupiter-Sun line, and in the next, its nature at all local times.

We have shown that the critical radius r_A and the classical nose radius D may be comparable. D is determined by balancing the stagnation pressure of the shocked solar-wind flow with the dipolar magnetic pressure; r_A, by balancing rotational energy density with dipolar magnetic pressure. If $r_A \simeq D$ the rotational energy density is comparable with the solar-wind stagnation pressure. In a super-Alfvénic wind, the radial and azimuthal velocities are comparable to one another and to the Alfvén speed at the critical Alfvén point (Weber & Davis 1967). Therefore, the rough equality of r_A and D implies that the dynamic pressure of a potential planetary wind and the dynamic pressure of the nominal solar wind are comparable. As Figure 17 indicated, the dynamic pressure of the solar wind is highly variable in the outer solar system. If the dynamic pressure is low, so that $D > r_A$, we expect a super-Alfvénic outflow along the subsolar line. When the solar-wind dynamic pressure is high, so that $D < r_A$, we do not expect a super-Alfvénic wind along the subsolar line. However, since the inner magnetosphere is a source, there must be a sub-Alfvénic wind to transport away the plasma diffusing across r_A. The variations in solar-wind dynamic pressure may induce transitions between a super-Alfvénic Jovian wind and a sub-Alfvénic breeze (Coroniti & Kennel 1977).

The two-dimensional structure of the radial outflow in the local noon meridian plane containing the Sun-Jupiter line is uncertain. However, it seems reasonable to expect that a super-Alfvénic wind would stretch out its magnetic field into a disk-like configuration, whereas the sub-Alfvénic breeze would not, because good communication with the ionosphere is maintained. If so, variations in solar-wind dynamic pressure might also induce changes between disk-like and more dipolar outer magnetospheres.

Figure 17, and the discussion leading up to it, offers some support for this hypothesis, which cannot be definitively tested until the flows in Jupiter's outer magnetosphere are measured. P-10 inbound entered the Jovian magnetosphere about 45° from the Jupiter-Sun line. Just before it entered, it measured a solar-wind dynamic pressure a factor of 6 below nominal, suggesting that a super-Alfvénic wind was possible at that time. If so, the planetary-wind dynamic pressure would have scaled as $(r_A/r)^2$ so that pressure equality with the solar wind at the subsolar point would have been achieved at $r = \sqrt{6}\ R_A \simeq 100\ R_J$, roughly the distance of which P-10 first briefly entered the magnetosphere. Since the azimuthal structure of a planetary wind confined by the solar wind is not known quantitatively, and since the magnetosphere was varying when P-10 first encountered it (P-10 later re-encountered a dipolar magnetosphere within 45 R_J), the above estimates are only suggestive of a possible agreement.

The structure of the magnetopause depends upon whether the planetary outflow is a super-Alfvénic wind or a sub-Alfvénic breeze. If it is the former, an internal fast shock must occur near the pressure-equality radius. The shock decelerates the flow below the Alfvén speed; along the subsolar line, the post-shock sub-Alfvénic wind should be decelerated to a stagnation point. The flows of solar and planetary wind origin would be separated by a tangential discontinuity that is the magnetopause. The structure of the internal shock is not understood; however, the planetary-wind flow between the internal shock and magnetopause should contain both large-scale hydromagnetic and small-scale plasma turbulence. The distance between the magnetopause and internal shock could be about 10% of the radius of curvature of the shock, in perhaps a few R_J, in steady state. The internal shock is not needed for the sub-Alfvénic breeze, and Jupiter's magnetopause might then more nearly resemble the Earth's.

Heliospheric Models of Radial Outflow

For a hypersonic, hyper-Alfvénic solar wind, the pressure normal to the magnetopause is approximately $\rho u^2 \cos^2 \psi + P$, where P is the solar wind static pressure at Jupiter. In the magnetic tail, $\cos^2 \psi \rightarrow 0$ so the pressure within the magnetosphere approaches P asymptotically. Since $P/\rho u^2 \ll 1$, a centrifugal super-Alfvénic flow would be nearly always expected in the tail, and only for weak solar winds at the nose (Hill et al. 1974a,b).

The problem of visualizing the planetary-wind outflow pattern in the two dimensions of the equatorial plane is similar to the problem of the solar wind's heliosphere moving supersonically through the interstellar medium (Axford 1972). The closest analogy is with the case of low solar-wind dynamic pressure, where a super-Alfvénic planetary wind is expected at all local times. The outflow is sketched in the accompanying Figure 23, which shows schematic flow boundaries in the equatorial plane. The bow shock decelerates the solar wind. The magnetopause is the boundary between shocked solar and planetary winds. In the absence of turbulent mixing, there could be a sharp composition gradient across the magnetopause. The internal shock decelerates the planetary wind at the nose and deflects the planetary wind in the tail, so that eventually all the material emitted by Jupiter's inner magnetosphere escapes tailward. The Alfvén surface is the boundary between the corotation-dominated diffusion and wind regimes. Because the mathematical solution for this magnetosphere involves calculating four free boundaries self-consistently, the above sketch is necessarily schematic.

The case of high solar-wind pressure, where the outflow on Jupiter's dayside may be sub-Alfvénic, has no analog in heliosphere physics. A sketch of a possible configuration is shown in Figure 24. In this sketch, the subsonic region extends directly to the magnetopause on the dayside. In this region, the plasma should still dominantly corotate, and so we would expect most of it to exit on the local evening side, bulging the magnetopause somewhat. The internal shock has moved back into the tail and now only deflects the radial outflow into the antisolar direction.

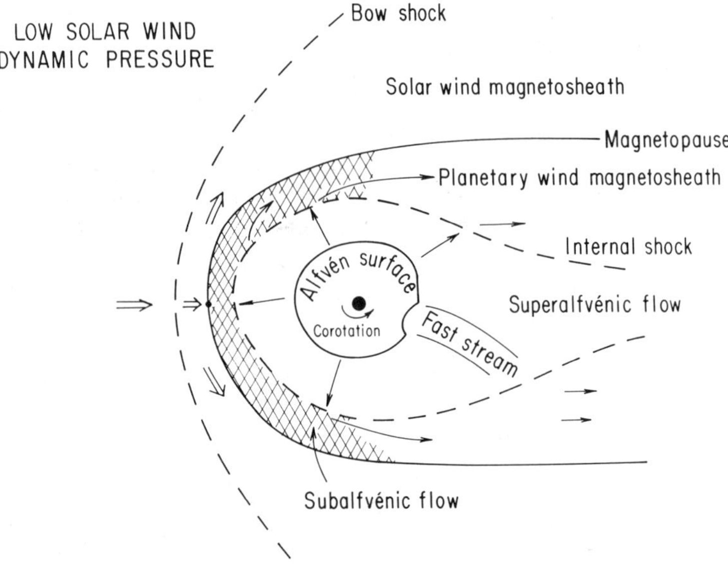

Figure 23 Heliospheric model of super-Alfvénic Jovian wind. Shown here is a cut through the equatorial plane. The Alfvén surface is at approximately 40 R_J. For low solar-wind dynamic pressure, a super-Alfvénic wind is generated at all local times; it is terminated by an internal shock, which decelerates the planetary wind and deflects it into the tail. The whole planetary wind flow is confined by the solar wind; shocked planetary and solar wind are separated in ideal MHD by a tangential discontinuity, the magnetopause. This whole picture neglects magnetic field line reconnection.

Both Figures 23 and 24 have a small "dimple" in the Alfvén surface that represents an important idea of Hill et al. (1974b) and Carbary et al. (1976). They proposed that the longitudes of Jupiter's weakest surface field correspond to the strongest ionospheric source and the nearest Alfvén radius. When the longitude of the strong source rotates into Jupiter's nightside, the planetary wind would be strongest. (Since the assumption that the surface magnetic-field properties determine the wind properties is reminiscent of how solar-wind fast streams are generated, we have called this a Jovian fast stream.) Thus the planetary-wind strength should be modulated with a 10-hr period. When the solar wind pressure is low, the fast stream could penetrate the dayside.

Carbary et al. (1976) suggest that since the fast stream is first generated on closed-field lines, reconnection will occur near the Alfvén point as its source rotates past local dusk, permitting a bubble of plasma to flow downstream with the planetary wind. Since reconnection can generate energetic particles, they suggest that

1. Reconnection will be modulated with a 10-hr period.
2. The particles generated beyond the Alfvén point by reconnection will reach the solar wind. Thus, the particle fluxes escaping to the solar wind will have a 10-hr modulation.
3. The particles generated within the Alfvén point by reconnection will radially diffuse inward. Since the corotation speed exceeds the gradient and curvature drift speeds, the 10-hr modulated pulse of energetic particles diffusing inward will be only slowly dispersed by differential drifts. They suggest these pulses are

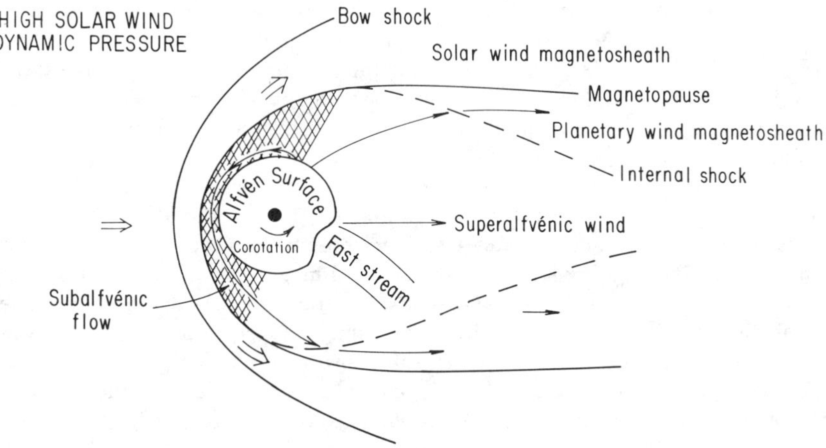

Figure 24 Heliospheric model of sub-Alfvénic Jovian breeze. When the solar-wind dynamic pressure is high, it is possible that only a sub-Alfvénic breeze, which would not radially extend the field lines, develops on the dayside. However, the planetary wind should still be super-Alfvénic in the tail. The large observed dynamic pressure variations could induce wind-breeze changes on Jupiter's dayside.

responsible for the observed modulations of energetic particles in the outer magnetosphere.

4. Since the energetic particles escaping to the solar wind and diffusing inward are generated by the same reconnection event, they should have the same modulation phase. In this way, they explain the results of Chenette et al. (1974).

We have used the liberty provided by a review format to take advantage of hindsight and discuss the Rice model of Jupiter's magnetosphere. We have added several features: first, we have tied the estimate of the Alfvén radius to the measured plasma density at Io; we have discussed the models in a heliospheric context; and finally, we have suggested that variations in solar wind dynamic pressure (magnetic storms) may turn on and off a super-Alfvénic wind on the dayside.

Discussion

We present below our personal summary of the current state of Jupiter magnetospheric modeling. We have discussed three classes of Jupiter models: first, Earth-like reconnection models, including Ioannidis & Brice's (1971) picture of the ensuing internal convection pattern for an inertia and pressure-free plasma; second, static rotating disk models; and third, radial outflow models—either super-Alfvénic, sub-Alfvénic, or mixed. In addition, we discussed the extent to which Jupiter and Io can feed the outer magnetosphere with plasma. All models agree on two points: first, Jupiter's large dipole field can enforce corotation to radial distances the order of 40 R_J, provided the atmosphere can enforce corotation at ionospheric levels. Corotation to such large distances seems to be observed experimentally. Second, all the models predict a transition layer within the magnetopause on the dayside. In Ioannidis and Brice's model, it is the convection layer beyond the corotation region; in radial outflows, it is the region between the critical point and the magnetopause for sub-Alfvénic flows, and is the region between the internal shock and the magnetopause for super-Alfvénic flows. Static disk models also imply a transition near the magnetopause where solar-wind stresses perturb the static force balance.

Since all the models as posed are time independent, whereas Jupiter's magnetosphere is time dependent, each model can apply at best to a portion of the data. As stated, Earth-like models can only apply to those times when Jupiter's magnetosphere was essentially dipolar within the classical magnetopause radius D, such as P-10 inbound within 50 R_J and perhaps P-11 outbound. The reconnection model is consistent with the length of the Jovian magnetic tail inferred from Jovian energetic particle propagation arguments and also with the time scales ($\simeq 1$ week) associated with changes in state of the Jovian magnetopause. The Earthlike models do not account for magnetopause encounters beyond, say, 60 R_J or for the radially extended fields.[1] The disk models do account for radially extended fields and aid

[1] However, Kennel & Coroniti (1975) did suggest that convection driven by strong tail reconnection could drive a super-Alfvénic flow, which would extend the field lines on the dayside. They pointed out that radially extended magnetopauses have been observed at Earth during magnetic storms.

us in visualizing the magnetic configuration within the critical point; however, they require that there are no sources and sinks of plasma, which seems unlikely. If plasma and angular momentum radially diffuse outward, the magnetic field will be swept back out of meridian planes, an effect not considered in static disk models.

It appears that Jupiter's inner magnetosphere and Io supply about 10^{28} hydrogen masses per second, motivating models of radial outflows which commence beyond the critical point. Super-Alfvénic winds, because they extend the magnetic field, are appropriate to such passes as P-10 outbound and P-11 inbound, at least conceptually. However, there exist no accurate solutions for winds carrying an entrained magnetic flux, and so no quantitative statements about vertical disk structure are possible. If our current estimates are correct, strong solar-wind dynamic pressure could suppress, at times, a super-Alfvénic wind on the dayside, converting the flow to sub-Alfvénic. Something like this is needed to account for the difference between P-10 and P-11 inbound, both of which traversed Jupiter's magnetosphere near 1000 LT, yet P-10 inbound found little, and P-11 inbound considerable, evidence for a disk.

Future progress in understanding Jupiter's outer magnetosphere may come in part from a unification of the above models, if attention is paid to encompassing the gross time dependences observed. In our opinion, unified models should take into account the following general principles:

1. Jupiter's ionosphere and satellite Io probably inject dynamically significant quantities of plasma into the outer magnetosphere. If radial diffusion is the transport mechanism, these internal plasma sources are constant over a time scale of a few months to a year, and so variations in the source cannot account for the observed magnetospheric variability. While the winds so produced could be azimuthally asymmetric, producing a 10 hr modulation, it is difficult to account for the 1-week time scales observed.
2. The nature of the wind produced at the critical point depends on exterior solar-wind boundary conditions. Changes in solar-wind dynamic pressure may change the dayside flow from sub- to super-Alfvénic, with concomitant changes in magnetic field from quasi-dipolar to radially extended.
3. Experience with the Earth's magnetosphere suggests that it would be perilous to neglect reconnection of solar wind and Jovian field lines, the resulting formation of the magnetic tail, and the internal convection pattern demanded by solar wind reconnection. Kennel & Coroniti (1977) have attempted to integrate planetary wind and reconvection driven flows in Jupiter's magnetosphere.
4. The scale of Jupiter's convection magnetosphere suggests that rarely, if ever, will it be steady.

In our opinion, the most pressing theoretical need is for qualitative and subsequently quantitative models of Jupiter's outer magnetosphere that account for both its 10-hr modulations and also its gross time variability. On the experimental side, we need to measure the density, pressure, flow velocity and composition of the plasma within Jupiter's magnetosphere, and relate these measurements to variations in the solar wind and of Jupiter's magnetospheric structure. Plasma waves were not measured by Pioneers 10 and 11. These questions will be addressed in part by the forthcoming Mariner-Jupiter-Saturn (MJS) mission. However, MJS will not address the over-riding problem. Since the time scale for gross changes in Jupiter's outer magnetosphere is comparable with a flyby duration, a flyby is so severely time limited that it probably cannot resolve the cause of the time dependences. Hence we must wait for a Jupiter orbiter to sort out the various possibilities. It should be worth it, for Jupiter's magnetosphere—a source of modulated relativistic particles and radio emissions—may by chance sample a wide variety of the physical states available to a rotating magnetized radial outflow confined by external plasma pressure. Our understanding of this class of hydromagnetic objects could be greatly enriched by further intensive study of Jupiter's magnetosphere.

ACKNOWLEDGMENTS

It is a pleasure to acknowledge conversations with A. Eviatar, M. G. Kivelson, and F. L. Scarf. We thank A. J. Dessler and C. K. Goertz for their careful criticism. We are particularly grateful to A. Eviatar, M. Kivelson, D. Intriligator, J. A. Simpson, and E. J. Smith for making information available to us in advance of publication. This work was supported by NASA grant no. NGL05-007-190 S8, and by the Cal Tech President's Fund.

Literature Cited

Acuña, M. H., Ness, N. F. 1976. See Gehrels 1976, p. 830
Anderson, K. A. 1968. *J. Geophys. Res.* 73: 2387–98
Atreya, S. K., Donohue, T. M., McElroy, M. B. 1974. *Science* 184:154
Axford, W. I. 1972. In *Solar Wind*, ed. C. P. Sonett, P. J. Coleman, J. M. Wilcox, p. 609. NASA
Barish, F. D., Smith, R. A. 1975. *Geophys. Res. Lett.* 2:269
Barnes, C. W., Chenette, D. L., Conlon, T. F., Pyle, K. R., Simpson, J. A. 1977. Interplanetary accelerations of nucleons and modulations of Jovian electrons between 1 and 10 AU by co-rotating regions of solar origin. *Ap. J. Lett.* In press
Beck, A. J., ed. 1972. *Proc. Jupiter Radiat. Belt Workshop, NASA Tech. Memo. 33-543*
Berge, G. L., Gulkis, S. 1976. See Gehrels 1976, p. 621
Bergstrahl, J. T., Matson, D. L., Johnson, T. V. 1975. *Ap. J.* 195:L131
Bergstrahl, J. T., Young, J. W., Matson, D. L., Johnson, T. V. 1977. *Ap. J.* 211:L51
Brice, N. M. 1968. *Jupiter's Outer Atmosphere,* Cornell-Sydney Univ. Astron. Center unpubl. preprint CSUAC 124
Brice, N. M., Ioannidis, G. A. 1970. *Icarus* 13:173
Brice, N. M., McDonough, T. R. 1973. *Icarus* 18:206
Brown, R. A., Goody, R. M., Murcray, F. J., Chaffee, F. H. 1975. *Ap. J.* 200:L49
Burke, B. F., Franklin, K. L. 1955. *J. Geophys. Res.* 60:213
Carbary, J. F., Hill, T. W., Dessler, A. J. 1976. *J. Geophys. Res.* 28:5189
Carlson, R. W., Matson, D. L., Johnson, T. V. 1975. *Geophys. Res. Lett.* 2:469
Carr, T. D., Desch, M. D. 1976. See Gehrels 1976, p. 683
Chenette, D. L., Conlon, T. F., Simpson, J. A. 1974. *J. Geophys. Res.* 79:3551
Coroniti, F. V. 1974. *Ap. J. Suppl. Ser.* 27: 261
Coroniti, F. V. 1975. *Space Sci. Rev.* 17:837
Coroniti, F. V., Kennel, C. F. 1977. *Geophys. Res. Lett.* Submitted for publication
Dessler, A. J., Hill, T. W. 1975. *Geophys. Res. Lett.* 2:567
Drake, F. D., Hvatum, H. 1959. *Astron. J.* 64:329
Dungey, J. W. 1961. *Phys. Rev. Lett.* 6:47
Dungey, J. W. 1965. *J. Geophys. Res.* 70: 1753
Ellis, G. R. A. 1965. *Radio Sci.* 65d:1513
Eviatar, A., Ershkovich, A. I. 1976. *J. Geophys. Res.* 81:4027
Fan, C. Y., Gloeckler, G., Simpson, J. A. 1964. *Phys. Rev. Lett.* 13:149

Fedder, J. A., Banks, P. M. 1972. *J. Geophys. Res.* 77:2328
Field, G. B. 1959. *J. Geophys. Res.* 64:1169
Fillius, R. W., McIlwain, C. E. 1974. *J. Geophys. Res.* 79:3589
Formisano, V., ed. 1975. *The Magnetospheres of Earth and Jupiter.* Dordrecht: Reidel
Frank, L. A., Van Allen, J. A. 1964. *J. Geophys. Res.* 69:4923
Frank, L. A., Ackerson, K. L., Wolfe, J. H., Mihalov, J. D. 1976. *J. Geophys. Res.* 81:457
Gehrels, T., ed. 1976. *Jupiter.* Univ. Ariz. Press. 1254 pp.
Gledhill, J. A. 1967. *Nature* 214:155
Gleeson, L. J., Axford, W. I. 1976. *J. Geophys. Res.* 81:3403
Goertz, C. K. 1976a. *J. Geophys. Res.* 81:2007
Goertz, C. K. 1976b. *J. Geophys. Res.* 81:3368
Goertz, C. K., Jones, D. E., Randall, B. A., Smith, E. J., Thomsen, M. F. 1976. *J. Geophys. Res.* 81:3393
Hill, T. W. 1976. *Planet. Space Sci.* 24:1151
Hill, T. W., Dessler, A. J., Michel, F. C. 1974a. *Geophys. Res. Lett.* 1:3
Hill, T. W., Carbary, J. F., Dessler, A. J. 1974b. *Geophys. Res. Lett.* 1:333
Hill, T. W., Michel, F. C. 1976. *J. Geophys. Res.* 81:4561
Ioannidis, G. A., Brice, N. M. 1971. *Icarus* 14:360
Jokipii, J. R. 1976. *Geophys. Res. Lett.* 3:281
Judge, D. L., Carlson, R. W. 1974. *Science* 183:317
Kennel, C. F. 1969. *Rev. Geophys.* 7:379
Kennel, C. F. 1973. *Space Sci. Rev.* 14:511
Kennel, C. F., Coroniti, F. V. 1975. *Space Sci. Rev.* 17:857
Kennel, C. F., Coroniti, F. V. 1977. Submitted to *Geophys. Res. Lett.*
Kennel, C. F. 1975. *Comments Astrophys. Space Phys.* 6:3, 71
Kivelson, M. G. 1976. *Proc. Boulder Sol. Terr. Phys. Symp.*, ed. D. J. Williams, Vol. II, p. 836. Am. Geophys. Union
Kivelson, M. G., Winge, C. R. 1976. *J. Geophys. Res.* 81:5853
Krimigis, S. M., Sarris, E. T., Armstrong, T. P. 1975. *Geophys. Res. Lett.* 2:561
Kupo, I., Mekler, Y., Eviatar, A. 1976. *Astrophys. J.* 205:L51
Levy, R. H., Petschek, H. E., Siscoe, G. L. 1964. *AIAA J.* 2:2065
McClain, E. F., Sloanaker, R. M. 1959. *Proc. IAU Symp. No. 9—URSI Symp. 1*, ed. R. N. Bracewell, p. 61. Stanford Univ. Press
McDonald, F. B., Trainor, J. H. 1976. See Gehrels 1976, p. 961
McDonough, T. R., Prakash, A., Swartz, W. E. 1975. *EOS Trans. AGU* 56:428
McKibben, R. B., Simpson, J. A. 1974. *J. Geophys. Res.* 79:3545
Mayer, C. H., McCullough, T. P., Sloanaker, R. M. 1958a. *Ap. J.* 127:11
Mayer, C. H., McCullough, T. P., Sloanaker, R. M. 1958b. *Proc. IRE* 46:260
Mekler, Y., Eviatar, A. 1974. *Ap. J.* 193:L151
Melrose, D. D. 1967. *Planet. Space Sci.* 15:381
Mendis, D. A., Axford, W. I. 1974. *Ann. Rev. Earth Planet. Sci.* 2:419–74
Mestel, L. 1968. *MNRAS* 138:359
Mewaldt, R. A., Stone, E. C., Vogt, R. E. 1976. *J. Geophys. Res.* 81:2397
Michel, F. C., Sturrock, P. A. 1974. *Planet. Space Sci.* 22:1501
Nishida, A. 1976. *J. Geophys. Res.* 81:1771
Northrup, T., Goertz, C. K., Thomsen, M. F. 1974. *J. Geophys. Res.* 79:3579
Parker, E. 1963. *Interplanetary Dynamical Processes.* New York: Interscience
Pesses, M. E., Goertz, C. K. 1976. *Geophys. Res. Lett.* 3:228
Piddington, J. H. 1967. Univ. Iowa Rep.
Prakash, A., Brice, N. 1975. *The Magnetospheres of Earth and Jupiter*, ed. V. Formisano, pp. 411–23. Dordrecht: Reidel
Radhakrishnan, V., Roberts, J. A. 1960. *Phys. Rev. Lett.* 4:493
Roberts, J. A., Stanley, G. J. 1959. *Publ. Astron. Soc. Pac.* 71:485
Roederer, J. G. 1970. *Dynamics of Geomagnetically Trapped Radiation.* Berlin: Springer
Sakurai, K. 1976. *Planet. Space Sci.* 24:1207
Sentman, D. D., Van Allen, J. A., Goertz, C. K. 1975. *Geophys. Res. Lett.* 2:465
Sentman, D. D., Van Allen, J. A. 1976. *J. Geophys. Res.* 81:1350
Simpson, J. A., Hamilton, D., Lentz, G., McKibben, R. B., Mogro-Campero, A., Perkins, M., Pyle, K. R., Tuzzolino, A. J., O'Gallagher, J. J. 1974a. *Science* 183:306
Simpson, J. A., Hamilton, D. C., McKibben, R. B., Mogro-Campero, A., Pyle, K. R., Tuzzolino, A. J. 1974b. *J. Geophys. Res.* 79:3522
Simpson, J. A., McKibben, R. B. 1976. See Gehrels 1976, p. 738
Siscoe, G. L., Chen, C.-K. 1976. *Io, a source of Jupiter's inner plasmasphere, UCLA Dept. Atmos. Sci. preprint 1976*
Sloanaker, R. M. 1959. *Ap. J.* 133:649
Smith, E. J., Davis, L., Jones, D. E., Colburn, D. S., Coleman, P. J., Dyal, P., Sonett, C. P. 1974a. *Science* 183:305
Smith, E. J., Davis, L., Jones, D. E., Coleman, P. J., Colburn, D. S., Dyal, P., Sonett, C. P., Frandsen, A. M. A. 1974b.

J. Geophys. 79:3501
Smith, E. J., Davis, L., Jones, D. E., Coleman, D. J., Colburn, D. S., Dyal, P., Sonett, C. P. 1975. *Science* 188:451
Smith, E. J., Davis, L. R., Jones, D. E. 1976. See Gehrels 1976, p. 788
Smith, E. J., Fillius, R. W., Wolfe, J. 1976. Private communication, to be published 1977
Smith, R. A. 1976. See Gehrels 1976, p. 1146
Spreiter, J. R., Alksne, A. Y. 1969. *Rev. Geophys.* 7:11
Swartz, W. E., Reed, R. W., McDonough, T. R. 1975. *J. Geophys. Res.* 80:495
Teegarden, B. J., McDonald, F. B., Trainor, J. H., Webber, W. R., Ruelof, E. 1974. *J. Geophys. Res.* 79:3615
Trainor, J. H., McDonald, F. B., Stillwell, D. F., Teegarden, B. J., Webber, W. R. 1975. *Science* 188:462
Trainor, J. H., McDonald, F. B., Teegarden, B. J., Webber, W. R., Roelof, E. C. 1974. *J. Geophys. Res.* 79:3600
Trafton, L., Parkinson, T., Macy, W. 1974. *Ap. J.* 190:L85
Trafton, L. 1975. *Nature* 258:690
Van Allen, J. A. 1959. *J. Geophys. Res.* 64:1683–89
Van Allen, J. A., Baker, D. N., Randall, B. A., Sentman, D. D. 1974. *J. Geophys. Res.* 79:3559
Vasyliunas, V. M. 1975. *Geophys. Res. Lett.* 2:87–88
Weber, E. J., Davis, L. Jr. 1967. *Ap. J.* 148:217
Wolfe, J. H., Mihalov, J. D., Collard, H. R., McKibben, D. P., Frank, L. A., Intrilligator, D. S. 1974. *J. Geophys. Res.* 79:3489

Ann. Rev. Astron. Astrophys. 1977. 15 : 437–78

THEORIES OF SPIRAL STRUCTURE

Alar Toomre
Department of Mathematics, Massachusetts Institute of Technology, Cambridge, Massachusetts 02139

> Much as the discovery of these strange forms may be calculated to excite our curiosity, and to awaken an intense desire to learn something of the laws which give order to these wonderful systems, as yet, I think, we have no fair ground even for plausible conjecture.
>
> Lord Rosse (1850)
>
> A beginning has been made by Jeans and other mathematicians on the dynamical problems involved in the structure of the spirals.
>
> Curtis (1919)
>
> Incidentally, if you are looking for a good problem . . .
>
> Feynman (1963)

1 INTRODUCTION

The old puzzle of the spiral arms of galaxies continues to taunt theorists. The more they manage to unravel it, the more obstinate seems the remaining dynamics. Right now, this sense of frustration seems greatest in just that part of the subject which advanced most impressively during the past decade—the idea of Lindblad and Lin that the grand bisymmetric spiral patterns, as in M51 and M81, are basically compression waves felt most intensely by the gas in the disks of those galaxies. Recent observations leave little doubt that such spiral "density waves" exist and indeed are fairly common, but no one still seems to know why.

To confound matters, not even the N-body experiments conducted on several large computers since the late 1960s have yet yielded any decently *long-lived* regular spirals. By contrast, quite a few such model disks have developed bar-like or oval structures that have endured almost indefinitely in their interiors. This finding is undoubtedly a major step toward understanding various bar phenomena that complicate the observed spirals, but it also reminds us rudely just how little we really comprehend: To avoid the rapid nonaxisymmetric instabilities associated with this bar-making, the experimental "stars" require (or otherwise they acquire) random velocities proportionately much larger than those observed among stars near the

Sun, or than those suggested by edge-on views of other galaxies. It is this dilemma, of course, that has fueled much of the recent speculation about massive but largely unseen halos—though the alternative that some very hot inner disks or spheroids might already cure those instabilities seems not yet to have been excluded firmly. Whatever that answer may be, it is clear that the spiral dynamics gets no easier when even the most rapid but least visible members of a galaxy need to be suspected of helping distort the axial symmetry.

Still, one must begin somewhere. Happily, this remains a subject where it makes sense to start almost at the beginning.

2 SOME OF LINDBLAD'S IDEAS

As Oort (1966) remarked in a graceful obituary, Bertil Lindblad's "long series of papers, beginning already in Uppsala in 1927, and continued until his death, bears witness to the struggle with the spiral problem. It is natural that in this field, on which at that time nothing was ripe for harvesting, he did not immediately find the right path." Oort was referring mostly, of course, to Lindblad's long obsession with *leading* spiral arms (cf his 1948 Darwin lecture). Lindblad believed such arms to consist of material shed recently from unstable circular orbits just outside a spinning central nebula that was shrinking and flattening with time in the manner envisaged by Jeans (1929) for the entire Hubble sequence. In that setting, unlike Chamberlin (1901) or Jeans himself with tidal forces from outside, Lindblad sought an *intrinsic* reason for the frequent two-armed symmetry of spirals. Already in his earliest papers he thought he had found one in the well-known dynamical instability of the Maclaurin spheroids, which generates a growing triaxial shape that rotates in space, in a wavelike fashion, only half as fast as the fluid itself. Lindblad expected the spillage to occur mostly from the two ends of the longer equatorial axis, which he felt would in reality rotate even more slowly.

Such was the origin of Lindblad's lifelong interest in waves in galaxies. During the many years that he continued to believe in the transient leading arms, this interest developed mostly into a fascination with bars. He often wrote of the latter in terms like "vibrations" or "modes" or "density waves"—although the last phrase for him usually meant only motions like the rotation of a lumpy rigid body or of a Jacobi ellipsoid, hardly true waves at all. In retrospect, none of Lindblad's extensive labors to calculate such bar modes seems worth recounting here. Still quite remarkable, however, are his repeated guesses such as "the most promising approach to a solution of the problem of spiral structure appears to be by way of the barred spiral nebulae," or that "in the ordinary spirals . . . a plainly developed barred structure may not appear, but there is likely to be a characteristic density wave $s = 2$, which causes a departure from rotational symmetry and which incites the formation of spiral structure by its disturbing action on the internal motions in the system" (Lindblad 1951).

Much of this early work obviously dates from an era before 21-cm astronomy and even Baade's explicit recognition of the two stellar populations. Yet that alone does not explain why, as Oort put it, the attempts at spiral-making via gravity

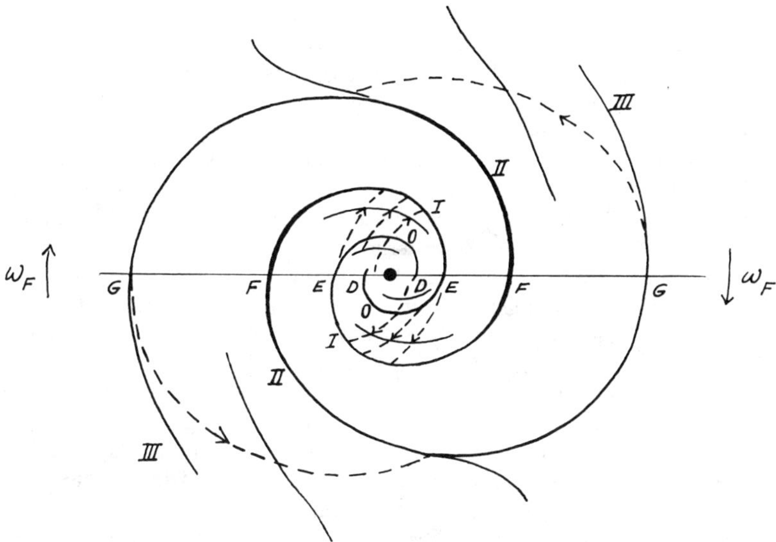

Figure 1 Lindblad's (1964) sketch of his "circulation theory." (Reprinted with permission from *Astrophys. Norv.*)

"remained for 35 years practically a monopoly of the Stockholm Observatory." A subtler reason was that by the 1950s it had become a little too obvious, to Baade (1963, p. 67) and many others, not only "that the primary phenomenon in the spiral structure is the dust and gas," but also "that we could forget about the vain attempts at explaining spiral structure by particle dynamics. It must be understood in terms of gas dynamics and magnetic fields." Amidst such a tide of facts and opinion, Lindblad stood almost alone in not renouncing his hopes for basically gravitational explanations of spiral structure—though he did finally concede, in a review article (Lindblad 1962) still worth reading, that his old leading-arm models are "not reconcilable with modern evidence." Rather more reconcilable, it seems now, are two further notions that occurred to Lindblad during his last decade.

One of these, stemming from the mid-1950s, was the idea of "dispersion orbits," which we return to in the next section. The other notion is pictured in Figure 1, taken from the second of two papers in which Lindblad (1963, 1964) speculated explicitly "on the possibility of a quasi-stationary spiral structure."

To give "the typical spiral structure . . . a certain steadiness in time" despite the differential rotation, Lindblad proposed that the two gaseous arms in Figure 1 are simply a *pattern* that rotates rigidly in space with a clockwise angular speed ω_F. Outward of the two (corotation) points F, where the orbital speed equals ω_F, gas elements move in the retrograde direction, as viewed in the rotating frame; inside F they tend to overtake the pattern. Nevertheless, Lindblad argued, the pattern persists. The self-gravity of the upper arm II, for instance, ensures that any interstellar matter that joins it at, say, the 12 o'clock position will drift slowly outward in radius *along the arm* as its differential rotation carries it a quarter-turn to the leftward point G, where at last it will part company with arm II/III. Then, "thanks to the attraction from [the other arm] II, [such material] will return towards that arc in a shower of orbits, of which one is indicated by the dotted curves." Inside the corotation radius, Lindblad envisaged analogously that the gas would drift inward while sliding clockwise along either of the spiral arcs I. Subsequently it would tend "to be assimilated in an outward motion" toward the next arm, to complete the cycle. Also drawn in Figure 1 are some inclined interarm "branches," which Lindblad suspected to be characteristic of the material in transit.

With this simple sketch Lindblad displayed an intuitive grasp of the gas trajectories involved in spiral shock waves that was not to be surpassed markedly until the work of Fujimoto (1968) and Roberts (1969). Nor was his reasoning here entirely qualitative: If only in a rough and laborious manner, Lindblad reckoned that enough extra force to maintain arms of 18° pitch angle could be provided by an arm mass totalling one-tenth of the entire mass of a galaxy. In short, this "circulation theory" was a big step forward from anything that preceded it, including even Oort's (1956) suggestion that "perhaps one side of an arm collects material while the other side evaporates it, so that the arm holds a constant position."

As it happened, apart from those gas orbits, this hypothesis of quasi-stationary spiral structure was in turn outdistanced soon by the QSSS hypothesis of Lin & Shu (1964, 1966) deliberately bearing the same name. Already these two papers furnished much more detail and clarity in support of their own conjecture that somehow "the matter in the galaxy can maintain a [spiral] density wave through gravitational interaction in the presence of the differential rotation," and that "this density wave provides a spiral gravitational field which *underlies* the observable concentration of young stars and the gas." Lin and Shu took the big extra step of including—and from their 1966 paper onwards, of also calculating far more plausibly—the gravitational forces to be expected from tightly wrapped collective waves among ordinary disk stars. Contrary to what is often supposed, however, these major improvements made or inspired by Lin had remarkably little to do with any "density waves" ever considered by Lindblad. Instead, they relate much more to that other idea which Lindblad termed "dispersion orbits." It is just those orbits, as we will now see following mostly Kalnajs (1973), that can be turned (literally) into a quick introduction to the basic mechanics still envisaged by Lin (e.g. 1971, 1975) and Shu (1973) for the bisymmetric spiral waves.

3 NEARLY KINEMATIC WAVES

The classic paper in which Maxwell (1859) explored the stability of Saturn's rings focused especially on the topic of infinitesimal disturbances to a configuration of N equal small masses rotating in centrifugal equilibrium at the vertices of an N-sided regular polygon, around a heavy mass imagined *fixed* at the center. Maxwell separated this problem into $N/2$ simpler problems, each involving a sinusoidal [$\sin(m\theta)$ or $\cos(m\theta)$] dependence of the radial and tangential displacements of the various ring particles upon their longitudes θ. For each integer wavenumber $0 < m < N/2$, he demonstrated that such a polygon can exhibit four distinct modes of distortion confined to its own plane—and that all four are pure waves travelling around the ring, provided those satellite masses are sufficiently small.

Lindblad (1958, 1961) extended this analysis, with some minor extra approximations, to the $m = 1$ and $m = 2$ perturbations of an essentially continuous ring of particles rotating with angular speed Ω in the fixed, non-Keplerian force field of an idealized galaxy. He likewise obtained four basic modes for each m, all of which are indeed wavelike when the ring mass is almost nil. Two of these modes, the ones he again called "density waves," then reduce to nearly frozen sinusoidal disturbances to the ring density, rotating in space with almost exactly the material speed Ω. As such they need not concern us further here (though the reader may wish to emulate Maxwell by resolving the little paradox of why the obvious tangential attractions from such orbiting mass concentrations are not immediately destabilizing). Much more relevant is the slower of the two remaining modes, or "waves of deformation": As Figure 2 illustrates for wavenumber $m = 2$, the limiting zero-mass version of that mode rotates in space with angular speed $\Omega - \kappa/m$, where κ is the so-called epicyclic frequency, or the angular rate of radial oscillation of each individual particle in the given central force field.

It is essentially this slowly advancing kinematic wave (rather than its fast partner with speed $\Omega + \kappa/m$), composed of many separate but judiciously-phased orbiting test particles, that Lindblad meant by his term "dispersion orbit." That name itself now seems only a quaint relic of his hope from the mid-1950s that the gas layer

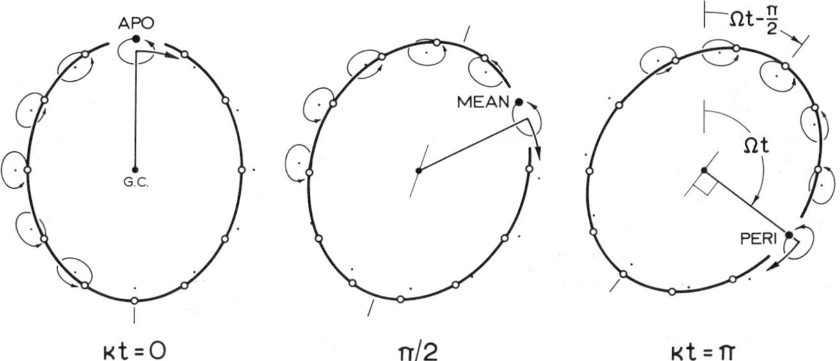

Figure 2 Slow $m = 2$ kinematic wave on a ring of test particles, all revolving clockwise (like the 12 shown) with mean angular speed Ω in strictly similar and nearly circular orbits. The small elliptical "epicycles," traversed counterclockwise in the above sequence of snapshots separated in time by exactly one-quarter of the period $2\pi/\kappa$ of radial travel along each orbit, depict the apparent motions of these particles relative to their mean orbital positions or "guiding centers." Drawn for the case $\kappa = \sqrt{2}\Omega$—or one where the rotation speed $V(r) = r\Omega(r) = \text{const}$ at neighboring radii—the diagram emphasizes that the oval *locus* of such independent orbiters advances in longitude considerably more slowly than the particles themselves. That precession rate equals $\Omega - \kappa/2$, as one can verify at once by comparing the last frame with the first.

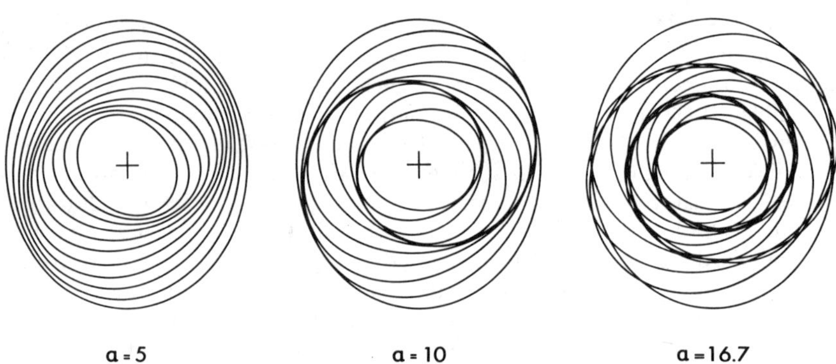

Figure 3 Three kinematic density waves of spiral form (Kalnajs 1973), obtained by merely superposing some kinematic ring waves such as shown in Figure 2.

in a galaxy might spontaneously tend to aggregate or "disperse" into just a few such distorted rings (an idea that might still apply somewhat to the shearing away of star associations born into eccentric orbits). But not at all quaint is the main reason for Lindblad's interest in this dispersal: He had just noticed that, whereas the circular speed $\Omega(r)$ varies markedly with radius r in a typical galaxy, the wave speed $\Omega(r)-\kappa(r)/m$, *in the bisymmetric case* $m = 2$ *only*, remains fairly constant over a considerable range of intermediate radii. This fact greatly intrigued Lindblad—who did not need to be told that strict constancy would banish wrapping-up worries or that the nicest spirals tend to have two arms. Yet astonishingly, that is about as far as he ever got. Per Olof Lindblad agrees that somehow it never occured very explicitly to his father—despite writing in 1958 that "it is likely that the spiral arms are largely osculating to dispersion orbits"—to combine already those "orbits" into any long-lived *spiral* patterns such as he himself later proposed in Figure 1!

This logical gap between Lindblad's orbits and the density waves of Lin and Shu was closed emphatically only by Kalnajs (1973), who presented the three striking geometric examples reproduced in Figure 3. Each of these pictures consists of just 11 ellipses shifted systematically in longitude according to the rule $2\theta = \alpha \ln$ (major axis). None of them pretends to contain any actual dynamics, nor any trace of the helpful extra densities expected near all the major axes already from Figure 2. Kalnajs devised these three diagrams simply to dramatize how easy it is, in retrospect, to concoct trailing or leading spiral loci of modest or even very pronounced crowding from nothing more than *ad hoc* superpositions of individual finite-amplitude ring waves resembling Lindblad's $m = 2$ dispersion orbits. Such spiral patterns are inherently wavelike and thus, quite apart from any deeper questions of origin or the preference for one shape over another, it is clear that these beautiful patterns would last forever *if* all the separate ovals could somehow be persuaded to precess not just with the small and nearly equal speeds $\Omega-\kappa/2$ but (*a*) at exactly the same rate and (*b*) without change of amplitude. The big task is to do all this persuading.

4 LIN-SHU DENSITY WAVES

When Lin and Shu almost succeeded in providing such assurances with their WKBJ estimates a decade ago, they concentrated largely—and understandably—upon condition (*a*), that of avoiding all wave shear. Even more understandably, they focused upon tightly wrapped waves, as a major analytical shortcut. Unlike Lindblad with his few or single rings, Lin and Shu reasoned that the disturbance gravity forces should be predominantly *radial* in any real two-armed spiral that is wound about as tightly as the one shown in the last frame of Figure 3. And these forces, they reckoned, ought to affect the collective oscillations of the disk stars and gas in any given orbiting neighborhood very nearly as much as they would there have affected a strictly axisymmetric density wave of the same (large) radial wavenumber.

4.1 *Axisymmetric Vibrations*

Even today, this reviewer knows of no analytical theory for interactive radial vibrations with amplitudes approaching those in Figure 3. However, the theory for short axisymmetric waves of *infinitesimal* amplitude goes back at least to Safronov (1960). Among other things, Safronov recognized almost explicitly that the self-gravity of any thin, cold disk of surface mass density μ tends to *reduce* the local frequency ω of short axisymmetric waves of wavenumber k from the epicyclic value κ to that implied by

$$\omega^2 = \kappa^2(r) - 2\pi G\mu(r)\,|\,k\,|. \tag{1}$$

In providing this formula, Toomre (1964) in essence only repaired a coefficient error made by Safronov. Thereupon, he very nearly contributed one of his own: To avoid the severe Jeans instabilities implied by Equation (1) for all wavelengths $\lambda = 2\pi/k$ shorter than

$$\lambda_{\text{crit}}(r) = 4\pi^2 G\mu(r)/\kappa^2(r), \tag{2}$$

the root-mean-square radial velocity σ_u of any thin disk consisting only of stars with a Schwarzschild distribution of random velocities must exceed

$$\sigma_{u,\min}(r) = 3.358 G\mu(r)/\kappa(r), \tag{3}$$

as Kalnajs kindly corrected this writer (by about 30%) before publication. It then emerged, ironically, that Kalnajs in 1961 had omitted a factor 2π from an analogous estimate—and this had dampened his own interest in such collective effects!

Kalnajs (1965, p. 50) redeemed himself with a flawless first derivation of the (implicit) mathematical relationship between the local wavenumber k and frequency ω for an axisymmetric wave involving stars with an arbitrary rms speed $\sigma_u = Q\sigma_{u,\min}$, rather than just $\sigma_u = 0$ as in the simple Equation (1). Lin & Shu (1966) soon rederived the same as the crucial $m = 0$ subcase of their own—far more widely known—dispersion relation. In its original integral form, their version looks rather different

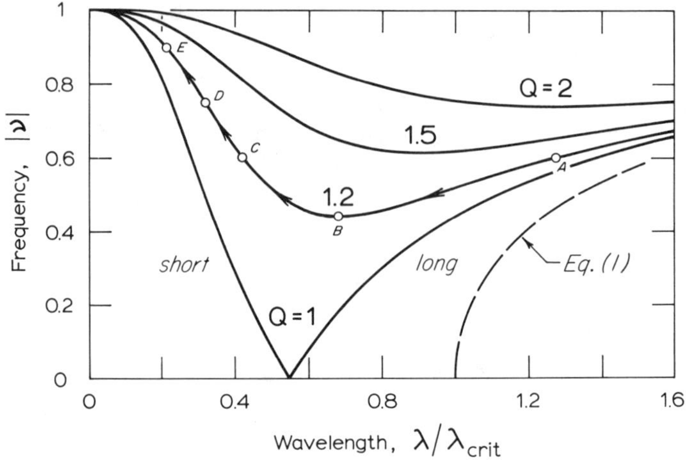

Figure 4 The Lin-Shu-Kalnajs dispersion relation for axisymmetric density waves of local frequency $\omega = \nu\kappa$ and modest radial wavelength λ in a thin, rotating disk of stars endowed with Q times the minimum random motions required by Equation (3).

from the series expression by Kalnajs [cf Toomre 1969, Equation (11)], but both imply the identical curves summarized in Figure 4 for several values of the stability parameter $Q \geqq 1$ at which the Jeans instabilities are no longer possible.

The key dynamical fact in Figure 4 is that the local self-attraction can again only reduce the frequency ω of the small-amplitude *collective* vibrations to some fraction $|\nu|$ of the epicyclic frequency κ, which itself continues to characterize the finite-amplitude radial and tangential excursions of the *individual* stars whose net density changes comprise that wave. This lowering of the wave frequency, as all three authors recognized, is greatest at intermediate wavelengths and tends to zero at both extremes. The important return of ω to κ as $\lambda \to 0$ occurs because, although all stars obviously sense the sinusoidal force field, only those whose maximum radial speeds U are less than about κ/k manage to contribute to the wave density a net amount greater than 60% of what they would have contributed if they had merely orbited in concentric circles before the wave came along. That ratio of cooperativeness indeed reaches zero already at $U = 1.841\ \kappa/k$ for $|\nu| = 1$, and at $2.405\ \kappa/k$ for $\nu = 0$; moreover, from stars with $U = \pi\kappa/k$—or ones whose apocenter and pericenter distances differ by exactly one wavelength—one even expects a bit of cussedness amounting roughly to a *negative* one-tenth contribution to the wave density (Kalnajs 1964, private communication).

4.2 *Matching of Precession Rates*

Together with various kin involving gas and/or thickness corrections, the axisymmetric LSK dispersion relation from Figure 4 has played a fundamental role in nearly all of the "asymptotic" explorations of tightly wrapped (i.e. almost axisymmetric) $m = 2$ spiral waves that Lin, Shu and several collaborators have conducted during the past decade. That role has been remarkably direct: As Lin & Shu (1964) first illustrated using the much less plausible Equation (1), the above reduction of that essential frequency from κ to $|\nu|\kappa$ implies an analogous change, from $\Omega \pm \kappa/2$ to $\Omega \pm |\nu|\kappa/2$, in the speeds of precession of the *net* density contributions expected from the many separate stars belonging to any given mean orbital radius but gyrating around various epicycles. And it was precisely in the seeming nuisance that ν depends upon the local radial wavelength λ (or upon the local pitch angle of the trailing or leading spiral) that Lin and Shu, already in 1964, spotted the grand opportunity of simply *adjusting* those more realistic precession rates to the strict constancy with radius that had eluded Lindblad.

Among much else, it needed to be demonstrated, of course, that this desired "tuning" of $\Omega \pm |\nu|\kappa/2$ could actually be accomplished simultaneously, for any single pattern speed Ω_p, over long enough stretches of radii and with plausible spiral arm spacings. That this is indeed possible without running badly afoul of observations was first shown for the Galaxy by Lin & Shu (1967; see also Lin et al. 1969) and for M33, M51, and M81 by Shu, Stachnik & Yost (1971; see also Tully, 1974, Roberts et al. 1975, Rots 1975, Rogstad et al. 1976).

The essential points of those feasibility studies may be gleaned from Figure 5b. The five curves in that diagram describe, for $N = 2, 3, 4, 5, 6$, the usual signed fractions

$$\nu(r;\Omega_p) = 2[\Omega_p - \Omega(r)]/\kappa(r) \equiv \nu_N(r) \tag{4}$$

of the local epicyclic frequency whereby an intended $m = 2$ wave of hypothetical pattern speed $\Omega_p = \Omega(Nr_0)$ would appear "Doppler shifted" to an observer orbiting with speed $\Omega(r)$ in one rather typical model galaxy. That model is a thin disk of surface mass density $\mu(r) = \mu_0 \exp(-r/r_0)$ analogous to the simple law that Freeman (1970a) among others has stressed to be characteristic of many observed *light* distributions. Its dimensionless rotation curve $\tilde{V}(r)$ in Figure 5a, based on a formula given by Freeman, mainly just corroborates that the Sun belongs somewhere near $r = 3r_0$ in this crude facsimile of our Galaxy.

Given that $|\nu| < 1$, Figure 5b implies first that only pattern speeds from an intermediate range, roughly between $\Omega(3r_0)$ and $\Omega(5r_0)$, make it even plausible to contemplate any single $m = 2$ spiral structure extending from the deep interior well beyond our make-believe Sun. The higher speed limit is here set by the unwillingness of the self-gravitating disk to carry waves that precess even more rapidly than the fast kinematic waves corresponding to $\nu = +1$, whereas the lower limit arises from the similar expectation that the collective effects near each radius can only raise the slow wave speeds above Lindblad's value $\Omega - \kappa/2$. As usual, the radii at which a prescribed Ω_p exactly equals $\Omega \pm \kappa/2$ are here referred to as the outer and inner Lindblad resonances.

A second type of basic restriction is represented by the shaded $|\nu| < 0.44$ strip in Figure 5b. It warns that already when the stability index Q exceeds unity by as little as 20%, there exists a sizable forbidden annulus around each proposed corotation radius. Lin-Shu waves of obviously low apparent frequency might still "tunnel" across any such region (cf Toomre 1969, Figure 3), but they will there not find any *local* hospitality in the WKBJ sense. Clearly that gap only widens for $Q = 1.5$ and 2.0, since Figure 4 then forbids $|\nu| < 0.61$ and 0.74, respectively.

This second restriction has been much underplayed in all of the wave fittings cited above, which have in essence just postulated that $Q = 1$, upon including those gas corrections, thickness effects, etc. Observations of nearby disk stars suggest very roughly that $1.2 \lesssim Q \lesssim 2$ (cf Toomre 1974a); elsewhere Q remains totally unmeasured and well-nigh unmeasurable. On the theoretical side, quite apart from any global instabilities, one interesting difficulty with presuming that $Q \cong 1$ in the gas-rich parts of a galactic disk is that the effective masses of any large chunks of interstellar matter—or, loosely speaking, the various bits and pieces of arms present in even the grandest spirals—should then be strongly enhanced by cooperative density "wakes" from passing stars (cf Figure 13). The result, as Julian (1967) perhaps overcautioned, can be a growth of the stellar random motions (or Q itself) at a rate well in excess of any estimated by Spitzer & Schwarzschild (1953; see also Woolley & Candy 1968) from encounters with "cloud complexes." If only for the last reason, it seems prudent to concentrate mostly on those features of Figure 5b that would survive even if $Q = 1.5$, or at the very least if $Q = 1.2$.

Contrary to what Lindblad (1964) surmised in Figure 1, such features evidently include next to nothing from beyond any of the corotation radii $r = Nr_0$, since the upper strip $0.44 < \nu < 1$ in Figure 5b indicates for each N that one can hope to tamper with the fast kinematic waves only over very limited stretches of radii, of order $r_0/2$, when $Q \geqq 1.2$. This nominal extent seems absurdly small, compared even with the "short" waves of lengths λ of order $2r_0$ implied there by Figures 4 and 5a. What is more, even when $Q = 1$, the ranges of radii corresponding to $0 < \nu < 1$ remain modest also in models such as shown in Figures 2 and 4 of Shu, Stachnik & Yost (1971)—who themselves proceeded to ignore those modified fast waves on the more enigmatic grounds that "if Lin (1970) is correct, only the region where $\Omega - \kappa/2 < \Omega_p < \Omega$ shows an organized spiral pattern." In a sequel, Roberts, Roberts & Shu (1975) likewise took no active interest in that short outer segment, now apparently because "the theoretical behavior of spiral density waves near corotation is more complicated, even in the linear regime, than has been previously assumed." One way or another, this reviewer agrees that the $\nu > 0$ waves do not seem destined for much glory.

This leaves only the bottom half of Figure 5b—or more exactly, only its lowest quarter or so—if we are to continue this exercise in pessimism with $Q = 1.2$ and 1.5. Down there at last we meet one very reassuring fact in the sizable range of radii over which especially the ν_5 curve still manages to remain within even the narrow $Q = 1.5$ corridor $-1 < \nu < -0.61$. Of course, this welcome news leans heavily upon the familiar near-constancy of $\Omega - \kappa/2$. After all, not much help is needed anyway from the self-gravity when (*a*) those slow kinematic speeds are as nearly uniform as Lindblad noted them to be near the "turnover" radii of most rotation curves, and provided also that (*b*) one contemplates pattern speeds like $\Omega(5r_0)$ that exceed those elementary speeds only slightly. In such circumstances, the comparatively modest help available from the LSK dispersion relation when $Q = 1.2$ or 1.5 can still be worth a great deal.

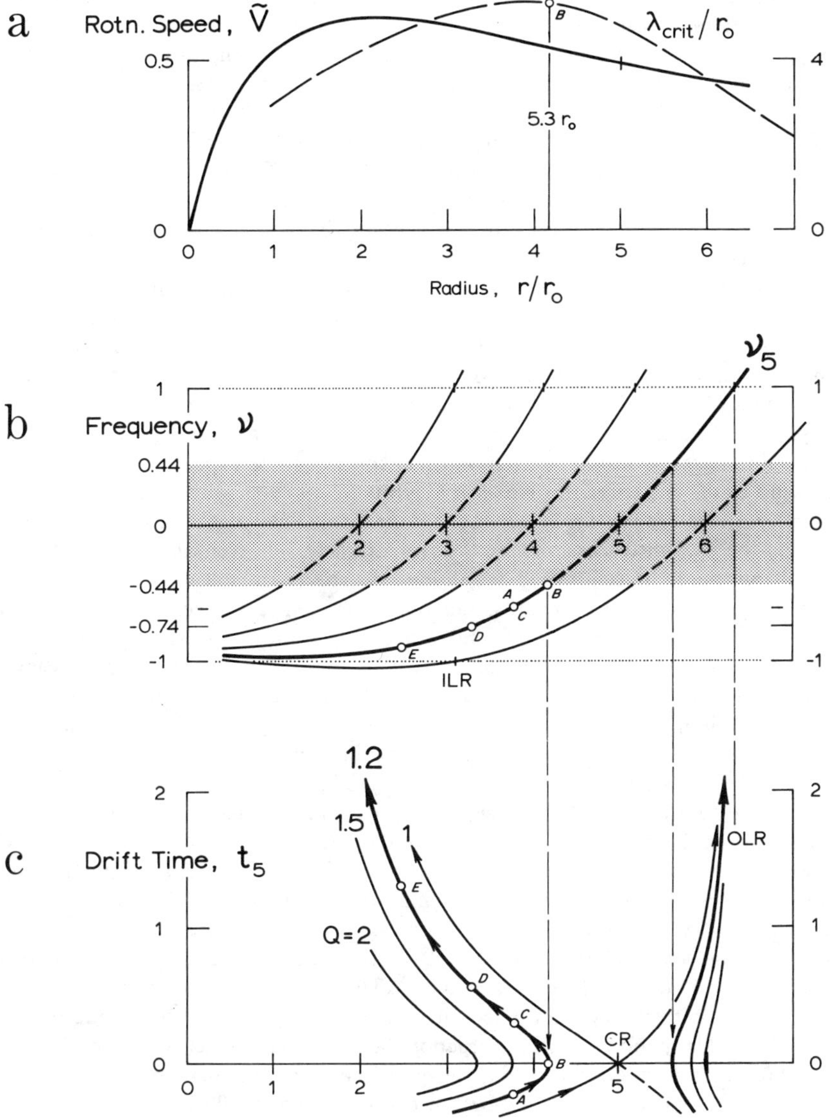

Figure 5 Theoretical data for a thin exponential disk of scale length r_0: (*a*) Rotation speed $\tilde{V}$ and stability length λ_{crit}, (*b*) dimensionless local frequencies $\nu_2, \ldots, \nu_6$, (*c*) group velocity drift times $t_5(r; Q)$, all plotted as functions of r. The shaded frequencies are locally unattainable when $Q = 1.2$. CR means corotation; ILR and OLR are some inner and outer Lindblad resonances. The points marked *A* through *E* appear also in Figure 4.

In short, at least when $Q \geqq 1.2$, the only reasonable overall wave speeds Ω_p for two-armed spiral structures as *extensive* as those seen in M51 and M81—and less obviously, perhaps also in our Galaxy—seem to be those that are slow enough to place corotation on, but not beyond, the outskirts of those systems. [This conclusion, by the way, appears to be completely independent of any "concept of the initiation of density waves by the Jeans instability in the outer regions of normal spiral galaxies (Lin 1970)" that Shu, Stachnik & Yost (1971) felt they were testing.] Almost as a corollary, however, the exact radius of inner Lindblad resonance (ILR), if indeed any, seems much more jittery. It must also be quite sensitive to modifications of the rotation curve itself in the interior.

As for other angular wavenumbers, notice that neither such ILR sensitivity nor any extensive near-resonance with $\Omega - \kappa/m$ could have arisen for $m > 2$, since a mere rescaling of the v axis in Figure 5b reveals that the ILR and OLR radii would lie much closer to each other already when $m = 3$ or 4. This is the simple but compelling reason why Lin (1967) urged that no one should expect *multi*-armed structures to be capable of rotating rigidly as very extensive wave patterns—though he obviously slipped in phrasing it that "only *two*-armed spirals can be expected." In fact, just a halving of all labels on the v axis in Figure 5b explains why Lin, Yuan & Shu (1969) wrote later that "the possibility that one-armed galaxies exist cannot be ruled out by the present discussion" for $Q = 1$. Curiously, however, at least the $Q \gtrsim 1.3$ exclusions ought indeed to rule out all local prospects for slow $m = 1$ waves.

Finally, as regards wavelengths, only one thing seems fairly certain. It is that the large but typical values of the reference length $\lambda_{crit}(r)$ shown in Figure 5a imply that the observed interarm distances can be accommodated in disks of dominant mass only with waves belonging to the *short* side of Figure 4—or to the side dominated by the random epicyclic motions, rather like the $|v| = 0.6$ choice C instead of A in that diagram. This important insight stems uniquely from Lin & Shu (1966, 1967). Indeed the analogous axisymmetric waves were known already to Kalnajs (1965), but it was Lin and Shu who first appreciated and pointed out vividly that just those short collective waves can be remarkably *useful.*

A contrary view was voiced by Marochnik, Mishurov & Suchkov (1972). These authors stressed, in effect, that the original long-wave hopes of Lin & Shu (1964) may yet deserve resurrecting. To illustrate that even such waves can be made tight enough, Marochnik et al. immobilized everything in the Galaxy except a uniform "population I" disk with $\mu = 40\ M_{\odot}\mathrm{pc}^{-2}$. That use of the old trick of reducing λ_{crit} seems a bit extreme (and indeed it was technically weak in admitting neither any shock dissipation nor the refraction of some yet shorter waves back from corotation), but it does provide one good closing reminder: Given our poor knowledge of galactic halos and/or hot disks, no one should be hugely surprised if the *effective* length λ_{crit} from examples such as in Figure 5 should need to be reduced by factors of order 2, at least for tightly wrapped waves and especially toward the interior. This would still not upset the impression that the most prominent real waves are apt to be "short" ones. However, it does make it almost meaningless to debate now whether one model or function $Q(r)$ or pattern speed Ω_p results in a photogenic diagram that fits observations better than another.

4.3 *Group Velocity*

After this long discussion of what might be feasible, it needs to be said plainly that none of these WKBJ studies of what are sometimes still called "self-sustained" density waves has yet come close to demonstrating the long-term survival of any such spiral patterns. Even today, it is not just the *origin* but even the *persistence* (cf Oort 1962) of these waves that remains distressingly unexplained; all we know for sure now is that the two topics are nearly inseparable!

In an early volume of this series, Lin (1967) wrote instead that his work with Shu "essentially amounts to establishing the possible existence of certain density waves of the spiral form, sustained by self gravitation. These are collective modes in the combined gaseous and stellar system." Even at that time, it was clear that the WKBJ approximations underlying the basic Lin-Shu theory become invalid near (and beyond) any radius where $|\nu| = 1$—and that there, as Prendergast (1967) remarked in an excellent short review, "the solution gets very difficult, and one has not yet succeeded in guiding it properly through the Lindblad resonance." Nevertheless, it seemed to Lin that his success with Shu in adjusting the speeds $\Omega \pm |\nu| \kappa/2$ over most of the radii *between* those resonances had in large measure already established a "theory free from the kinematical difficulty of differential rotation."

The fallacy of such reasoning became evident in 1969, when it was noticed that all this tuning of precession speeds had concerned only the *phase* velocity of what might logically be just a traveling wave (in the usual technical sense) rather than a genuine global mode. Lin and Shu had completely overlooked that in repairing one serious defect they had actually created another: An inevitable price for altering those speeds of precession in a wavelength-dependent manner via the (very sensible) radial forces is a *group* velocity, likewise directed radially. And it is this second kind of wave transport, familiar enough from countless other situations, that sooner or later even here plays havoc with the amplitudes—which the kinematic ovals in Figures 2 and 3 at least had the decency to keep invariant. Lin (1970) agreed that this "consideration . . . brings the problem even into sharper focus," and Shu (1970) conceded that the "amplification of trailing waves" announced by Lin & Shu (1966; see also Lin 1967) had resulted from a similar misunderstanding.

The demonstration that the wave amplitudes—or more exactly, the action densities—are indeed conveyed radially is fairly complicated, and seems best left to the papers of Toomre (1969), Shu (1970), Dewar (1972), and Mark (1974a). To dispel any needless mystery, however, let us quickly rederive the parallel effect of the group velocity upon the slow precession rates

$$\Omega_p(r, t) \equiv \Omega(r) - |\nu| \kappa(r)/2 \tag{5}$$

for a trailing, two-armed spiral wave pattern in which Ω_p varies mildly from radius to radius. In that case, the slight wave shear $\partial\Omega_p/\partial r$ near some radius r will slowly loosen or tighten that part of the spiral pattern, at a rate $\partial k/\partial t = -2\,\partial\Omega_p/\partial r$, where again $k(r, t)$ is the radial wavenumber. More to the point, as k changes with time, so must Ω_p itself at that radius, according to Equation (5), at the rate

$$\partial\Omega_p/\partial t = -\tfrac{1}{2}\kappa(r)[\partial|\nu|/\partial k]\,\partial k/\partial t = c_g\,\partial\Omega_p/\partial r, \tag{6}$$

where

$$c_g(r, t) = \kappa(r)[\partial \mid \nu \mid / \partial k]_{r, Q=\text{const}}. \tag{7}$$

The latter is just the group velocity, here reckoned positive when inwards. It deserves that name because Equation (6) implies that a traveler moving with the (gradually varying) speed c_g radially toward the center would find Ω_p changing not at all in his vicinity. Conversely, the Ω_p's expected in the not-too-distant future at any fixed radius are just the current values at nearby exterior radii, all as if prerecorded on a magnetic tape that is about to be played.

The severity of this radial drifting of *all* local properties of a trailing $m = 2$ wave pattern of approximate speed $\Omega_p \cong \Omega(5r_0)$ within the exponential disk is indicated by the four pairs of "characteristic" curves in Figure 5c. For each choice of Q, these curves report the radii which the imagined traveler would have gotten to at various times t_5 reckoned in multiples of the full rotation period $2\pi/\Omega(5r_0)$. Given that Ω_p is conserved along each curve, the dimensionless frequency ν and wavelength $\lambda/\lambda_{\text{crit}}$ at successive radii are implied by Figures 5b and 4 exactly as before.

Since these curves are so similar, let us concentrate now on just the inner member of the $Q = 1.2$ pair. Starting off, for instance, at the point marked C (with $\nu = -0.6$, $\lambda = 0.42\ \lambda_{\text{crit}}$, and in terms of Figure 3, $\alpha = 11$) at $r = 3.8\ r_0$, notice that just one rotation period seems to be required for all this information to drift to point E (with $\nu = -0.9$, $\lambda = 0.21\ \lambda_{\text{crit}}$ and $\alpha = 16$) at $r = 2.5\ r_0$. A halving of λ_{crit} everywhere would obviously double such drift times, but otherwise this relatively tight spiral actually errs on the tranquil side: Had we kept $\lambda_{\text{crit}}(r)$ unchanged but selected instead the generally more open wave whose $\Omega_p \cong \Omega(4r_0)$, similar WKBJ estimates would have yielded a total elapsed time of 1.9 in the present units for the entire drift from the equivalent of point B to the very center. Except versus the old straw man of material arms, such a "persistence" seems not very impressive. Even the kinematic spirals from Figure 3 would last about that long with the speeds $\Omega - \kappa/2$ suggested by Lindblad.

4.4 *Forcing and/or Feedback*

Serious though they were, these group velocity criticisms were directed only at the faulty demonstration of longevity, not at the basic *idea* of nearly permanent spiral waves. As this reviewer has remarked before, no such radial propagation actually excludes really long-lived spiral wave patterns in a galaxy. If those patterns are to persist, the above simply means that fresh waves must somehow be created to take the place of older waves that drift away and presumably disappear.

That these short waves must indeed decay eventually—by Landau damping in an ILR region, at least in *linear* theory, and provided such a resonance is both (*a*) accessible and (*b*) unsaturated—was established firmly by Mark (1971, 1974a) and Lynden-Bell & Kalnajs (1972), who also corrected Toomre's (1969) inept remark that there should be similar damping "already in transit." At larger amplitudes, however, as Kalnajs (1972a; see also Simonson 1970) seems to have been the first to notice, most of the short-wave damping may well occur en route, via the gas-dynamical shocks yet to be discussed in Section 5.

Incidentally, either form of damping argues that these density waves must be trailing. The short leading waves admitted equally well by the original analyses of Lin and Shu would have the opposite (or outward) group velocity, and we would then be stuck with a task analogous to trying to create water waves that travel *away* from a beach!

It is hard enough to tell what causes the incoming waves of a trailing sense. Where could such fresh and relatively open spiral waves conceivably originate in an isolated galaxy? Broadly speaking, the safest answer today remains as obvious (and as painfully vague) as it did eight years ago: They still seem most likely to be a "by-product of some truly large-scale distortion or instability involving an entire galaxy," or the "consequences of some yet more basic density asymmetries—e.g. disturbances such as mildly bar-like waves or oval distortions which may be hard to detect but which cannot even remotely be approximated as tightly wrapped waves" (Toomre 1969; see also Lin 1970, especially p. 389, and Feldman & Lin 1973 for some related opinions).

These unconscious echoes of Lindblad's remarkable insight from 1951 stemmed largely from a single hint conveyed by diagrams such as Figure 5c: At least in a formal sense, the shorter and shorter trailing waves *C, D,* and *E* seem to be the descendants of other, much looser and *superposed* trailing waves such as *A* from the long-wave side of Figure 4. The latter have an outward group velocity, and they seem to *refract* into the short waves only near the "caustic" or turn-around radius *B,* where the group velocity changes sign and the amplitude no doubt tends to be larger than usual. That hint remains very murky, however, owing to the large wavelengths, $\lambda = 0.69$ and $1.28\ \lambda_{\text{crit}}$, expected already for *B* and *A*. Using the full density of the exponential disk, these lengths correspond to would-be $\alpha = 7$ and 3.5 logarithmic spirals from Figure 3. For wave *B,* such WKBJ estimates might still be crudely trustworthy (after the usual extra care to be taken near such a caustic), but no $\alpha = 3.5$ spiral can be deemed tightly wrapped. Indeed, it is not even clear that the latter can be modeled adequately with any modified local dispersion relations (cf Lynden-Bell & Kalnajs 1972). It is no disaster, of course, that these antecedent waves seem to pass so soon beyond the assured competence of WKBJ analyses. But it certainly is a pity.

It seems only fair to add that such skepticism has not been universal. For instance, Shu, Stachnik & Yost (1971) claimed that even the long waves seem self-evident in M51. Lately, the same kinds of waves have figured anew in the three-wave amplification processes (cf the solid $Q = 1$ curve branches in Figure 5c) near corotation that have been studied extensively by Mark (1974b, 1976a,b), and in the unstable spiral modes announced by Lau, Lin & Mark (1976), using just such "amplifiers."

Mark (1976c) and Lau, Lin & Mark (1976) made one very interesting point in noting that *if* the stability index $Q(r)$ can be arranged to rise steeply (and yet gently!) enough toward the center of a disk, then their asymptotic theory predicts a second *refraction* just outside the inhospitable interior, with the incoming short trailing waves reverting there into the outgoing long ones. In effect, those authors required the "forbidden" strip in Figure 5b to widen rapidly enough toward the left to intercept again a given $v_N(r)$ curve, and yet mildly enough not to violate the assumptions of WKBJ theory. This may not be easy, but if it can reasonably be arranged, such a second refraction at last promises a *simple* feedback cycle for which Lin (1970; see also some apt criticisms by Marochnik & Suchkov 1974 on pp. 94–95 of their review) has long been searching. The damping from the gas would continue, but hardly any from the ILR. Unfortunately, the big question is not whether the WKBJ idea gets stretched out of all proportion by the long waves involved in that scheme—an issue that Lau, Lin, and Mark did not resolve by offering an example in which the active density is everywhere less than one-half of the full surface density implicitly required by their adopted rotation curve. The real dilemma is that one just cannot tell, without embarking upon genuinely full-scale analyses such as we have yet to discuss, whether the special conditions (including $Q \simeq 1$ near the CR, and good dissipation near the OLR) that seem to permit these weakly unstable WKBJ modes might not also admit other, more powerful non-axisymmetric instabilities that would completely swamp the former.

Speaking of wave amplifiers, it should not be forgotten that Goldreich & Lynden-Bell (1965; see also Julian & Toomre 1966) showed long ago, via a different sort of local analysis in which azimuthal forces were retained and played a vital role, that the rapid swinging of open leading waves into trailing ones by the differential rotation has one remarkable consequence: Largely for kinematic reasons arising from temporary near-resonance with κ, this swinging manages to tap the abundant energy of differential rotation—and it amplifies those waves easily a hundred times more per full cycle than the typically two-fold growths obtained by Mark (essentially from a slow and almost axisymmetric shrinkage of already tightly wrapped waves through their most nearly Jeans-unstable wavelengths while drifting with the group velocity.) And that is just one reason why other conceivable but not-entirely-WKBJ feedback cycles—including, for instance, even a slight *reflection* of ingoing short trailing waves into short leading ones that propagate away from our internal "beach"—cannot be dismissed lightly.

Lau & Mark (1976) recently announced "a new physically significant enhancement of the growth rate," itself indebted "to the combined effects of differential rotation and of azimuthal forces." It is easily shown, however, that the key amplifying term labeled T_1 in Equations (4) and (5) of that paper exactly matches the only shear-dependent destabilizing term evident in the important Equation (76) of Goldreich and Lynden-Bell.

All things considered, only cumbersome "global" mode analyses and/or numerical experiments seem to offer any real hope of completing the task of providing the wave idea of Lindblad and Lin with the kind of firm *deductive* basis that one likes to associate with problems of dynamics. Before proceeding to such studies and their own joys and frustrations in Section 7, however, it seems well to reflect that another famous structural problem would not have progressed nearly as far as it has during the past decade or two if geologists had insisted that one should first establish theoretically from basic laws that the Earth's mantle must convect in such and such patterns. The same is true of our subject: As reviewed briefly in the next section, the galactic analogues of magnetic stripes in mid-ocean, chains of fresh volcanoes, and zones of intense crushing of relatively superficial material have lately done just about all the persuading one could possibly desire that some grand driving forces indeed are at work.

5 SHOCKS AND OTHER CONSEQUENCES

The first inkling that the interstellar material in certain galaxies may be repeatedly experiencing shock waves on a very large scale predates even the sketch by Lindblad (1964) reproduced in Figure 1. In the early 1960s, Prendergast often expressed the view that the intense, slightly curving dust lanes seen within the *bars* of such SB galaxies as NGC 1300 and 5383 are probably the result of shocks in their contained gas, which he believed to be circulating in very elongated orbits. In such "geostrophic" flows of presumed interstellar clouds with random motions, Prendergast (1962) wrote that "it is not clear what is to be taken for an equation of state," but he knew that "we should expect a shock wave to intervene before the solution becomes multivalued."

5.1 *Spiral Shock Waves*

As regards the normal spirals, Lin & Shu (1964) stressed from the start that since the gas has relatively little pressure, its density contrast "may therefore be expected to be far larger than that in the stellar components" when exposed to a spiral force field such as they had just postulated. That hint remained largely dormant, however, until Fujimoto (1968, printed tardily and in Russian; see also Prendergast 1967) combined these last two lines of thought. He showed that, when a supposedly isothermal gas layer slides past what Prendergast termed a "gravitational washboard," its steady response includes shock waves already at a modest (although finite) amplitude of that periodic forcing. To be sure, Fujimoto made one fairly serious error of analysis (spelled out by Roberts 1969, p. 140) and perhaps another of judgment—the latter in placing corotation in our Galaxy at a mere 5 kpc, rather as Lindblad had imagined. But Fujimoto's goal was commendable: he suspected that not just "the high-density hydrogen gas in the spiral arms" but also "the dark lanes observed on the concave sides of the bright arms of external galaxies may be due to the present shock wave."

This inspired guess was followed up by Roberts (1969). Together with Lin, he improved the analysis and added the further suggestion that the shock wave somehow also "triggers" the formation of the numerous young star associations that accentuate optical photographs of spiral galaxies. Both these themes are summarized neatly in the two often-copied diagrams reproduced anew in Figure 6. Roberts himself (see also Lin, Yuan & Shu 1969, pp. 734, 737) was very vague as to how the sudden increase of the mean density and pressure of the interstellar material could actually provoke the required gravitational collapse of dense preexisting gas clouds. But this hardly matters. The crucial point is that before the shock idea there had been no defensible explanation at all for the striking geometrical fact, first noticed by Baade in the late 1940s, that the main H II regions in large spirals tend to define considerably crisper and narrower arms than the rest of the

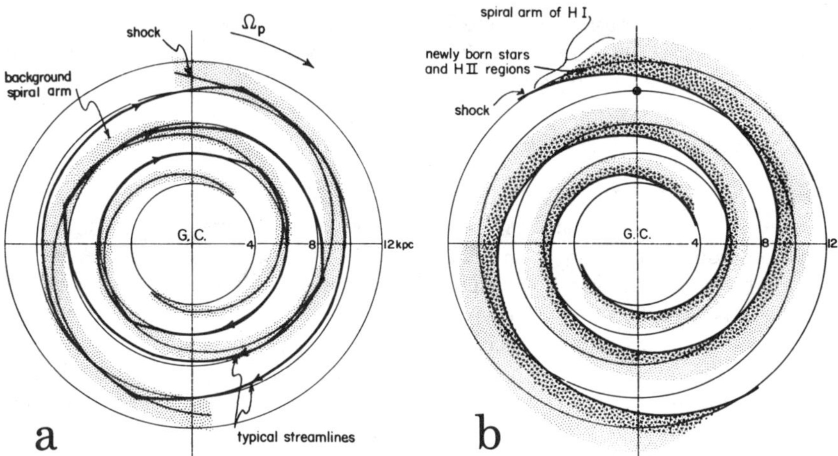

Figure 6 Large-scale spiral shock waves added by Roberts (1969) to a density-wave pattern much like that favored by Lin and Shu for our Galaxy. (*a*) Typical kinked-oval streamlines, viewed from a frame rotating with the presumed pattern speed Ω_p. (*b*) Implied relative locations of the shock fronts, the brightest new stars, and the main gaseous arms.

visible material. Often these emission nebulosities "are strung out like pearls along the arms" (Baade 1963, p. 63; see also Carranza, Crillon & Monnet 1969 and Arp 1976 for two fine modern examples). And like the dust, these highly luminous chains seem biased toward the inner edges of arms, though perhaps not quite as much.

As both Lin (1970) and Roberts (1970) rightly emphasized at the Basel symposium, the occurrence of dust lanes and O and B stars typically (though by no means always) on the *inner* or concave sides of (presumably) *trailing* spiral arms offers strong corroboration not just for the shock waves but indeed also for the theoretical expectation that the basic wave patterns should rotate more *slowly* than most of the disk material.

So much for the "volcanoes" and folded sediments of our subject. Its magnetic stripes are the bright spiral-like ridges in the radio-continuum map of M51 obtained by Mathewson, van der Kruit & Brouw (1972) using the Westerbork telescope. These ridges coincide much better with the major dust lanes of that galaxy than with its optical arms. Here indeed was "first-class observational evidence" for widespread shock-wave compression of a gaseous medium containing both cosmic rays and a weak magnetic field—a circumstance that should lead naturally (cf Pooley 1969) to a spiral pattern of enhanced synchrotron emission. Some difficulties with this picture, resulting mainly from an oversimplified description of the compression by Mathewson et al., have been reviewed already by Kaplan & Pikel'ner (1974) and by van der Kruit & Allen (1976). Hardly more needs to be said about this wave test. It speaks eloquently for itself.

The last of the three principal confirmations to date of the spiral-wave idea came, in this reviewer's opinion, also from Westerbork, this time via 21-cm line studies of M81. As reported by Rots & Shane (1975), the two main optical arms of that galaxy indeed agree well with unmistakable major arms of neutral hydrogen. In the words of perhaps the most vehement critic of the density-wave hypothesis, who himself favored a magnetic explanation, "if so, then it is generally agreed that gravitational forces must be invoked and the density-wave or some similar theory appears inevitable" (Piddington 1973).

These major items of evidence in support of spiral shocks have been accompanied by a host of other observational tests and "applications" of the density-wave hypothesis. Their conclusions have ranged from very encouraging to highly optimistic; none has proved severely contradictory. Much of this (and the above) work has been described at length in several other reviews, including those by Burton (1973), P. O. Lindblad (1974), and Wielen (1974). Rather than repeat here, let us simply close with the reminder that even in such reading one must constantly beware of the old trap of treating often-repeated presumptions as established facts. Perhaps the best warning comes from the confusing situation in our own Galaxy, where the view is far from perfect: it is not at all clear yet whether the famous 3-kpc arm is to be regarded as an ILR terminus of an inward-traveling wave, let alone as the result of violent activity in the nucleus. At least this writer still wonders seriously whether de Vaucouleurs (1964, 1970; see also Lindblad 1951, Kerr 1967) did not come closer to the truth in suggesting that it may instead be "gas streaming along a bar tilted by about 30 to 45° to the sun-center line" in the inner regions of our possibly SAB-type Galaxy.

5.2 *Pendulum Analogy*

In rushing through these important observations, we did not dwell in simple terms upon why one should expect the gas shocks to develop. Roberts (1970, Figure 3) offered an analogy in which a particle gets caught and is dragged some distance radially by the potential well of a given spiral arm. Lin and Shu have often remarked that the density contrast in each subsystem should be roughly proportional to the inverse of its mean-square random velocity. And Kalnajs (1973), somewhat like Prendergast originally, has stressed that the answer lies instead in the attempted crossings of elementary rings or orbits such as shown in the $\alpha = 16.7$ frame of Figure 3. Which, if any, of these helpful hints should one trust?

The first explanation turns out to be essentially the same as what Lindblad wrote about Figure 1—but it does not correspond closely to the sinusoidal gravity forces that Roberts used in his calculations. The second, though fine at large speeds, would clearly be wrong if carried to the zero-pressure extreme. And even the third reason, though most nearly right, does not distinguish plainly enough between the free and forced trajectories of various particles.

A closer analogy to the tightly wrapped waves adopted by Roberts (1969) and Shu, Milione & Roberts (1973) is the (supposedly endless and continuous) row of identical pendula in Figure 7 devised largely by Kalnajs. If we assume that these pendula are long enough to act like harmonic oscillators and that they do not yet bump into each other, then of course their displacements $\xi(r, t)$ from the equilibrium positions r obey the equation

$$\partial^2\xi/\partial t^2+\kappa^2\xi = C \sin [k(r+\xi)-\omega t], \tag{8}$$

when that system is exposed to the traveling sinusoidal force field given on the right. Equation (8) is exactly equivalent to the zero-pressure limit of the locally-approximated Equations (11) of Shu et al. Obviously, it also mimics the radial dynamics of stars and other collisionless objects in a galaxy subjected to strictly *axisymmetric* forcing of a wavelength short enough that variations of the epicyclic frequency κ can be ignored.

Requiring $\partial(r+\xi)/\partial r > 0$ everywhere, the linearized forced solutions

$$\xi(r,t) = C(\kappa^2-\omega^2)^{-1} \sin (kr-\omega t), \tag{9}$$

obtained after omitting ξ from the right-hand side of Equation (8), imply that one should not expect collisions as long as the forcing amplitude $C < (\kappa^2-\omega^2)/k$. This inference is corroborated by the accurately calculated Figure 7a, where C was set to one-half that nominal critical value, and where $|\omega/\kappa| = 3/4$, about as inferred by Lin and Shu for their presumed wave in the solar vicinity.

By definition, at larger amplitudes where conflicts or "shocks" should first arise, Equation (9) is no longer accurate, since $k\xi = \mathrm{O}(1)$ or the response is already nonlinear. [The same criticism, by the way, applies also to any use of the LSK

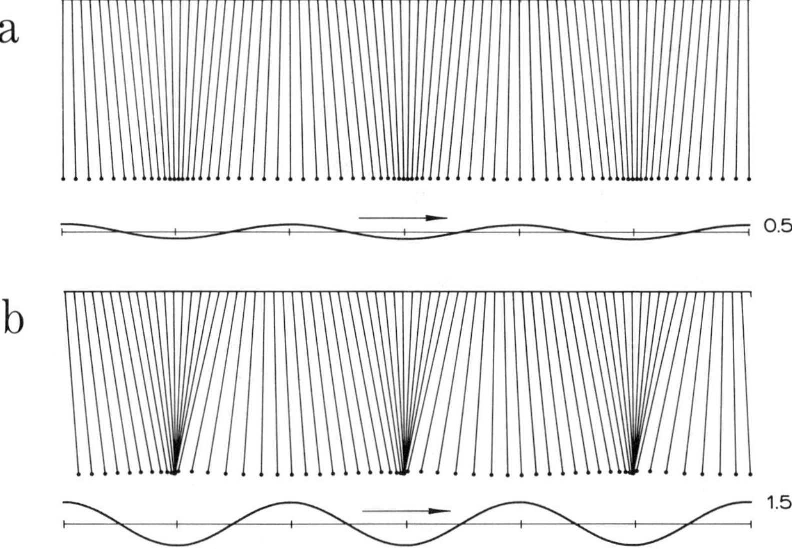

Figure 7 Forced traveling waves in a "curtain" of pendula subject to Equation (8) and inelastic collisions. The sinusoids depict, with 5-fold exaggeration, the respective forcing potentials amounting to (*a*) 0.5 and (*b*) 1.5 times the nominal critical value from the text. The frequency is $\omega = 3\kappa/4$.

dispersion relation whenever *strong* gas shocks are expected. The fact that the perturbation to the overall star density may then still be only of order 5% is misleading. The slow-moving stars expected to contribute the bulk of that extra density would already oscillate in a distinctly nonlinear manner, as illustrated beautifully by Wielen (1975) in his discussion of spiral-forced periodic orbits near the Sun.] Nevertheless, for the above frequency ratio—which lies about as far as possible from either the Lindblad resonance $\omega = \pm\kappa$ or the so-called "ultra-harmonic" resonances $\omega = \pm\kappa/2, \pm\kappa/3, \ldots$ discussed by Shu et al. and also Woodward (1975)—the accurate critical amplitude $C_{\text{crit}} = 0.430\ \kappa^2/k$ happens to agree fortuitously well with the linear estimate 0.438.

Beyond such values, of course, there is no substitute for numerical integration. Figure 7b shows one such result, obtained from forcing steadily with an amplitude 1.5 times the latter estimate. It was supposed in this calculation that the pendula (almost like lead balls, or exactly like gas particles in an isothermal shock of infinite Mach number) undergo perfectly *inelastic* collisons. Contrary to a first impression, however, these individual oscillators do not stick to each other in the intense pile-ups seen in Figure 7b. It is simply the ram pressure of the new arrivals from the right that persuades the early comers to linger for a while (19.6% of the full cycle, to be exact) in these peculiar traffic jams near the minima of the imposed potential, until at last they can slip away quietly to the left.

Compared with this naive rederivation, the inclusion of a uniform "effective speed of sound," $a =$ typically 8 km sec^{-1}, in the analyses of Fujimoto, Roberts, and especially Shu, Milione & Roberts (1973) seems to have had two main effects. One is that, understandably, such a finite intrinsic speed softens the density contrast across a shock and also widens the zone of crowding behind it. In particular, if $\omega/ka < 1$—or whenever the forcing appears subsonic in a mean sense—shocks arise only with difficulty (cf Figure 6 of Shu et al.) and tend to yield "broad regions of relatively low gas compression." As Roberts, Roberts & Shu (1975) stressed, this may well account for the thick and "massive" arms seen in smaller spirals like M33.

The other and perhaps more surprising effect arises when the speed $c = \omega/k$ of the imposed force sinusoid appears moderately supersonic: The finite gas pressure then makes it possible to obtain shocks from considerably weaker forcing than in our $a \to 0$ hypersonic limit. It also causes the shocks from a given forcing to be markedly stronger in terms of impact velocity. For example, as Shu et al. reported for $r = 10$ kpc, $\nu = -0.72$, and Mach number $c/a = 1.63$, the first cusp or zero-strength shock occurs in their isothermal gas when the forcing amplitude C is just under a fraction $F = 1.0\%$ of the central force $\Omega^2 r$. By contrast, the pendula in the same setting would not have bumped until $F = 4.0\%$. Moreover, using $F = 5\%$ as again recommended by those authors for the nearby Lin-Shu wave, their $a = 8$ km sec^{-1} gas enters each shock with a relative speed $u_1 = 26$ km sec^{-1} and exits with speed $u_2 = a^2/u_1$, whereas our $a = 0$ pendula would have managed to crash only with speed $u_1 = 12$. With an intermediate sound speed $a = 4$, by the way, one would already find that $u_1 = 22$—which underscores that the extra shock-making declines only slowly with increasing Mach number.

Technically speaking, there is no real mystery about this perverse supersonic behavior. Those stronger shocks seem to occur essentially because the pressure term in Equation (11) of Shu et al. raises the natural frequencies of our pendula, so to speak, markedly closer to the first ultraharmonic resonance—and no doubt it also contributes something akin to the classical Riemann steepening of free acoustic waves (which is a process discussed briefly by Woodward 1975 amidst his nice time-dependent shock calculations). Nevertheless, this reviewer cannot help wondering whether the shock-making might not have been seriously exaggerated by the isothermal model itself. Can we be sure that this "bouncy" inviscid gas of Fujimoto and the later workers mimics well enough the variously damped and heated multi-phase stew that is the real interstellar medium? The question matters because the increased impact speed u_1 seems a very mixed blessing in the context we are about to describe.

5.3 *Damping*

If the forcing of the very inelastic pendula in Figure 7b were to cease abruptly, we would know instinctively that their present strong collisions (as opposed to many lesser "aftershocks") would endure only perhaps another half-cycle. And, if instead of stopping suddenly, this forcing stemmed from the collisionless but not entirely random swinging to-and-fro of other, much more massive pendula (not shown here) that interact gravitationally with ours, then it would be just as clear that the steady drain of mechanical energy via the visible crashes would in the long run cause those unseen collective oscillations to decay as well.

The first of these thought experiments obviously mimics the demise of free axisymmetric shock waves in a purely gaseous disk. Their rapid damping may seem irrelevant, since it is the second case that at first sight imitates much better any tightly wrapped density/shock waves in a real galaxy, where the gas is well known to comprise only a small fraction of the total mass. Kalnajs (1972a) pointed out, however, that a serious fallacy may lurk even here.

The crucial point, in the context of our Galaxy, is that relatively few disk stars manage to take part significantly in waves of such short, $\lambda \cong 4$ kpc length as Lin, Shu and Roberts adopted for the solar vicinity. Using the conventional value of κ, one calculates easily for $\lambda = 4$ kpc that only about 12% (or 40%) of the stars from a Schwarzschild distribution with $\sigma_u = 40$ km sec^{-1} would have random motions small enough to ensure more than the 60% (or 0%) participation discussed at the end of Section 4.1. Thickness corrections would severely reduce even that. This helps explain why Lin (1970) reported that the nearby wave amplitude $\mu = 11\ M_\odot\ \mathrm{pc}^{-2}$ corresponding to a disturbance gravity $F = 5\%$ should consist almost equally of excess gas and star density. Conversely, however, this argument cautions—ignoring the group transport of fresh waves for the moment—that the Roberts shocks might still be damping the local Lin-Shu wave almost as fiercely as one would have guessed for the gas by itself!

Much paraphrased, that is what worried Kalnajs—who himself, like Roberts & Shu (1972) in a not very convincing rebuttal, couched it all in the more expert but also more opaque terms of net torques and negative densities of wave angular momentum. To reassess this vital matter, it is simpler to pretend again that we are dealing with exactly axisymmetric density waves $\mu'(r, t) = \mu_w \cos k(r-ct)$. Let us also suppose, although perhaps more dubiously, that their amplitudes μ_w are small enough for infinitesimal estimates of the density E_w of wave energy per unit area to remain reasonably accurate.

For such self-consistent waves in an extremely thin isothermal gas layer of unperturbed surface density μ, this energy density would equal $E_w = \frac{1}{2}\mu c^2 \cdot (\mu_w/\mu)^2$, quite independent of the (horizontal) sound speed a. The analogous formula for the LSK waves in a razor-thin stellar sheet is more ponderous [cf Toomre 1969, Equation (33)], but it can be recast as just a multiple $H(\lambda/\lambda_{\rm crit}; Q) > 1$ of that elegant gas result. Some typical values of H are 1.17, 1.55, 2.66, 4.74, 16.86 for the $Q = 1.2$ points A, B, C, D, E in Figure 4; the steep rise of that correction factor toward the smallest wavelengths is helpful, for it means that such short stellar waves tend to contain considerably more kinetic energy (and hidden commotion) than one might have supposed from their phase speed c and net density μ_w alone.

To be compared with the available E_w is the loss ΔE of mechanical energy per unit area and per time interval $T_s = \pi/(\Omega-\Omega_p)$ between successive shocks, here imagined to be steady. As follows strictly from time-averaging over one such cycle, the variable $v(\eta)$ in Equation (11) of Shu, Milione & Roberts (1973), this loss amounts to $\Delta E = \frac{1}{2}\mu_g[u_1^2 - u_2^2 - 2a^2 \ln(u_1/u_2)] \equiv \frac{1}{2}\mu_g u_E^2$, where μ_g is the mean density of the supposedly isothermal gas. At that rate of depletion, the wave energy would last a time

$$T_E = (E_w/\Delta E)T_s = (\mu_w^2/\mu_g\mu)(c/u_E)^2 H T_s. \tag{10}$$

As implied already, $c = \omega/k = 13$ km sec^{-1} and $\mu_w = 11\ M_\odot \text{pc}^{-2}$ for the specific $\lambda = 3.7$ kpc, $\nu = -0.72$, $F = 5\%$ wave at $r = 10$ kpc defined in Table 1 of Roberts and Shu. Assuming $Q = 1$ and ignoring thickness and gas corrections for brevity, those data also imply $H = 5.30$, $\mu = 89\ M_\odot \text{pc}^{-2}$, and $T_s = 2.7 \times 10^8$ yr. Locally, the choice $\mu_g = 5\ M_\odot \text{pc}^{-2}$ seems conservative; together with the above items, and $u_1 = 26$, $a = 8$ km sec^{-1} discussed earlier, it yields a remarkably short $T_E = 2.0 \times 10^8$ yr. This rough appraisal seems, if anything, a little harsher than $T_E = 0.8$ to 1.5×10^8 yr claimed by Kalnajs—who used a perhaps slightly too pessimistic $\mu_g = 10$, and whose low estimate of 0.8 also suffered almost twofold from an erroneous neglect of the $\ln(u_1/u_2)$ term in ΔE mentioned only parenthetically in his paper.

The counterproposal by Roberts and Shu that the relevant decay time works out more like 6 to 9×10^8 yr referred in fact to the entire interior annulus between $r = 5$ and 10 kpc, where at least the unperturbed stars seem more abundant. Moreover, those authors adopted a mean gas density $\mu_g =$ only $3\ M_\odot \text{pc}^{-2}$, a value strangely at odds with the $6\ M_\odot \text{pc}^{-2}$ wave amplitude (cf Yuan 1969, p. 897) in the gas alone imagined to provide half of the local $F = 5\%$ forces in the first place. In retrospect, that mean density also seems low in comparison with the dramatically increased interior amounts of molecular hydrogen inferred somewhat tentatively by Gordon & Burton (1976) from CO data. Using instead $\mu_g = 8$ and $12\ M_\odot \text{pc}^{-2}$, as hinted by Gordon and Burton for $r = 8$ and 6 kpc without even including any He, it also appears that $T_E = 1.8$ and 1.9×10^8 yr for the $F = 5.1$ and 5.7% waves considered by Roberts and Shu at those two interior radii.

Like the earlier criticism involving the group velocity, all this may not be quite as devastating as it sounds. For one thing, the increased stellar participation (and faster c_g) in more open spirals should help considerably. Second, as hinted already, the isothermal model may exaggerate the shock losses. And third, one would suppose in any case that those losses should be largely self-correcting upon some reduction of amplitude—given that the shock-making must be a threshold phenomenon that sets in only beyond some minimum level of forcing. The pendula make these last two points very nicely: If only we could be sure that the nearby gas behaves in that naive fashion under the $F = 5\%$ forces, we would reckon that $T_E = 5.1 \times 10^8$ yr, whereas below $F = 4\%$ there would be no damping at all. More annoyingly, however, the idealized $a = 8$ km sec^{-1} isothermal gas under $F = 4\%$ forcing would still imply $T_E = 2.1$, 2.2, and 3.6×10^8 yr at $r = 10$, 8, and 6 kpc. Alas, it is simply not known yet which story is more to be believed.

What does seem broadly clear, nonetheless, from this insight by Kalnajs is that at least the tighter Lin-Shu waves in any fine "normal" spiral should—during the very process of creating their own luminous froth while getting shorter and thus more intense—be almost as capable of dissipating most of their arriving mechanical energy as any foaming water waves incident upon a gently sloping beach. Better still, just as with the ocean waves that start breaking well offshore, it looks as if this *almost insatiable* damping may not even "scour" very badly the vicinity of an ILR itself. Ironically, though, Kalnajs was wrong in his closing remark that "it takes about 10^9 yr for help to arrive" in the solar region via the group velocity. Another look at Figure 5c near its point D should persuade that his estimate is too large, probably at least by a factor 2. In this respect, Roberts and Shu scored a good point in replying that it may be "no accident that the 'damping time' and the 'propagation time' should be comparable." Indeed not. That is exactly what one finds with those breakers off the beach.

Not to be confused with this remarkably potent wave damping is the related fact that any orbiting gas that persists in overtaking a pattern of trailing spiral shocks must on the average drift inward as well. Such a net transfer of angular momentum from the gas to the stars seems to have been noticed first by Pikel'ner (1970; see also p. 172 of Simonson 1970, Kalnajs 1972a, Shu et al. 1972, and Section 6 of the review by Kaplan & Pikel'ner 1974). It meant not only that Roberts (1969) was slightly mistaken in claiming the gas orbits to be closed in the present Figure 6a. Much more important, this inexorable radial evolution implies that even the neatest spiral structures can at best be only quasi-steady. As Pikel'ner noted also, such inward drifting of the gas in very open spirals should not even require many revolutions, given strong shocks.

5.4 *Spiral Shocks without Spiral Forcing*

It would be regrettable if all this well-deserved emphasis upon, first, the tightly wrapped waves of density and force alike, and then the very nonlinear gas dynamics which the latter can soon provoke, were to leave the impression that it is the only conceivable route to making spiral shocks in a galaxy. Such it definitely is not, as Figure 8 now dramatizes.

To produce this diagram, Sanders & Huntley (1976) began their computer simulation with a uniform disk of almost cold "gas" without any self-gravity. It merely revolved differentially in an inverse 5/4-power central force field. Then, rather gradually, some fairly modest (up to $\pm 13\%$ tangential, $\pm 3.25\%$ radial) extra forces were introduced, as if coming from an *oval* distortion to the density of the unseen background stars. These imposed $m = 2$ forces without any spirality of their own were presumed to rotate at a steady rate, placing corotation and the two Lindblad resonances at the radii indicated. The resulting spiral wave stirred up amidst the shearing gas is pictured in Figure 8, approximately one revolution after the forcing had become steady. That response itself, however, remains only quasi-steady—owing essentially to the radial drifting (and its analogue outside CR) known already to Pikel'ner. The shocks as such are none too clear in this reproduction, but at least they can be located by comparing the concave and convex sides of the arms between the ILR and CR.

Artificial though it is, Figure 8 stresses beautifully that the manner in which Lindblad's "characteristic density wave . . . incites the formation of spiral structure by its disturbing action on the internal motions" may not even be so subtle as to require, as an intermediary, the spiral gravity *forces* from any vaguely Lin-Shu type density waves apt to be excited at the same time. Similar warnings have often been uttered by Kalnajs (1970, 1971, 1973)—e.g. "the density wave [in an unstable mode illustrated in his 1970 paper] is essentially a bar-like distortion of the central region

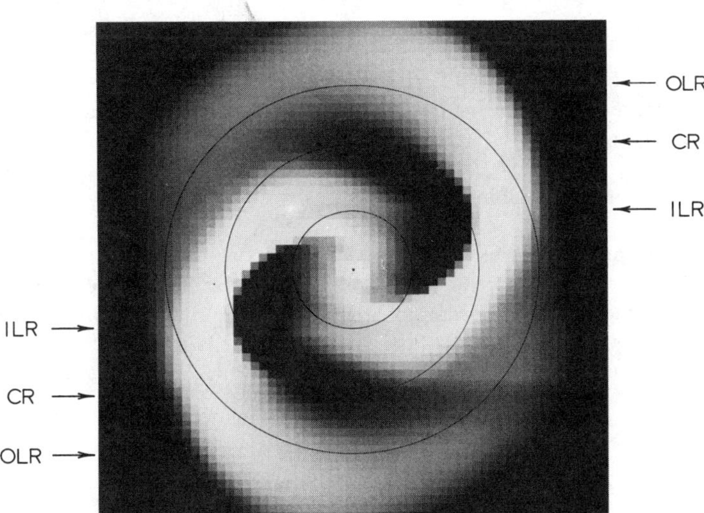

Figure 8 Quasi-steady distribution of density in a shearing gas disk exposed to a mildly oval gravitational potential (Sanders & Huntley 1976). Both that forcing and the gas again rotate clockwise.

of the galaxy which drives the gas." Such alarm signals were also fairly explicit (cf Figure 11) in the N-body experiments of Miller, Prendergast & Quirk (1970) to be discussed below. Lately, Lynden-Bell (1974, 1975; see also 1972, with Kalnajs) has stressed anew that "it is not at all clear to me that the gravity of the 'spiral' is 'important' for the formation of spiral shocks. My advice to observers is to believe those parts of the theory based around propagating shocks and shock-induced star formation, and to be very wary of those parts that rely heavily on the self-gravity of anything but a slight oval distortion or a bar shape."

Faced with such advice, this reviewer remains ambivalent. On the one hand, it seems eminently sensible to presume that the gas forcing might be rather direct in SB/SAB spirals like NGC 1097 (cf Arp 1976 again, though where is corotation?) and perhaps even in such milder SAB types as NGC 4321 = M100. Yet the same warnings seem overdone for the "best" SA spirals, essentially for two reasons. One is that, as Sanders and Huntley explained nicely, their kind of immediate forcing can shift arm longitudes only 90° per resonance radius—and the three familiar resonances are simply too few to meet such hopes in NGC 5364 or M51 or perhaps even M81. The other reason is that Zwicky (1955; see also Sharpless & Franz 1963) really seems to have been onto something with his "composite photography" of M51. As Schweizer (1976) reported from a detailed photometric reexamination of this and several other well-known spirals, the thick "red arms" discovered by Zwicky indeed represent *broad* spiral variations of surface brightness (though not of color!), as if from pronounced waves of similar spirality among the background disk stars as well.

Classic examples like those continue to argue strongly for something much more like the Lin-Shu-Roberts waves than the picture in Figure 8. Yet, as we now know, such waves themselves seem desperately in need of help from larger scales. Hence the main question to be distilled from the last few paragraphs might just be whether the much-desired "crushing" of the gas by forces stemming ultimately from major bars, ovals, long waves, or whatever is best regarded in any given instance as direct, indirect, or some shade in-between. Differences of opinion here among us theorists may merely be akin to those expressed by the blind men who examined the elephant.

It seems right to end this long discussion of galactic shocks almost where it began. A fascinating computer study of (again admittedly impermanent) gas motions in some *strongly* bar-like potentials—with the shocks there falling indeed mostly at the observed locations of the dust lanes that tantalized Prendergast long ago—has recently been published by Sørensen, Matsuda & Fujimoto (1976).

6 A QUICK SYNOPSIS

Lin (1967) has long stressed that he and Shu—with their QSSS hypothesis that a wavelike spiral force field somehow exists and keeps on rotating—jumped deliberately right into the *middle* of a difficult problem, hoping that from this "focal point in the development of a theory" it might be possible to progress faster by working "in both directions." By this technique, much more defensible here than in mountaineering, these and quite a few subsequent explorers landed essentially at point *B* of Figure 9. As discussed in Section 4, the route "downhill" from *B* toward first principles at *A* has since proved surprisingly treacherous; at least, no one has yet managed to descend safely all the way. On the other hand, the going "uphill" from *B* toward the gas arms and spiral shocks at *C* has, as seen in Section 5, turned out to be delightful despite some crevasses. Of course, a nasty segment from *C* to *D* still lies ahead—though even it has lost some of its former terror, chiefly to the helpful notion of a two- or multi-phase interstellar medium favored since the late 1960s (cf Shu et al. 1972, Salpeter 1976). Happily, that task belongs mostly to other specialists. For wave theorists, the big challenge remains much more the logical gap from *A* to *B*. It had better get closed firmly lest those reconnaisance parties far ahead become stranded.

This thumbnail sketch says nothing about the many ragged or patchy or "multi-armed" spiral features that both outnumber and complicate the generally-agreed two-armed "grand designs" among what Hubble called normal spirals. Worse still, our little summary seems even to have forgotten that "barred spirals are not rare oddities of nature. Among 994 bright spirals, about 2/3 show bar structure: 31% SA, 28% SAB, 37% SB, 4% S," to quote Freeman (1970b; see also Sandage 1975) and indirectly de Vaucouleurs.

The first flaw is a real one, to be scarcely remedied in Section 8. As for the bars, however, it is increasingly clear that they, too, belong in Figure 9 as closely related wave phenomena. They are basically to be thought of, it seems now, not as the stiffly revolving Jacobi ellipsoids that one still tends to see quoted, nor as the Dedekind ellipsoids in which the fluid circulates along ellipses but the shape stays put, but more like the in-between Riemann ellipsoids with both rotation and circulation that Chandrasekhar (1969, for a good summary) did much to resurrect and for which Freeman (1966) devised some elegant stellar-dynamical analogues. Of course, the real bars are not isolated entities but tend to be immersed deep within reasonably normal disks, as Freeman (1970b, 1975) stressed in his reviews. An even more vital clue, as he wrote already in 1970, is that unlike the air gap in Hubble's tuning fork, "the transition SA-SAB-SB is so smooth that it seems inconceivable

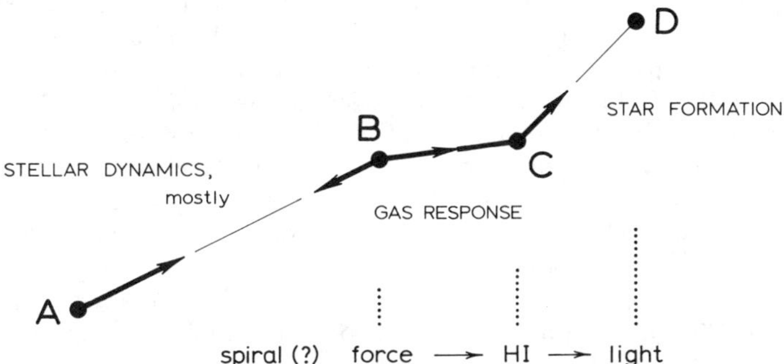

Figure 9 Status of the wave theories. As yet, only the thickened portions of this schematic route seem reasonably secure.

that the structure-producing processes should be essentially different for SA and SB systems." These remarks much deserve to be kept in mind as we now examine briefly in Section 7 the difficult and as yet inconclusive struggle upward from *A* via various large-scale computations.

7 GLOBAL INSTABILITIES

With the sole exception of the completely flattened and uniformly rotating Maclaurin-Freeman disks—for which Bryan (1889), Hunter (1963), and Kalnajs (1972b) together showed how to obtain the entire spectrum of modes, even when "hot," using only pencil, paper, and Legendre polynomials—all existing studies of the very large-scale or "global" behavior of disk-like model galaxies have been heavily numerical. Some of these studies, devoted again to *linear* instabilities and/or modes, have at least made a mild pretense of analysis. Others have amounted frankly to brute-force time-integrations of the *nonlinear* equations of motion of up to about 10^5 gravitationally interacting particles, with their only elegance consisting of various "checkerboard" schemes for rapidly solving the Poisson equation. As might be expected, the mode searches have tended to be much less expensive and, within their limited domain, also more accurate than these large *N*-body experiments. Yet it is indeed the latter that have taught us far more of what we really needed to know.

7.1 *Tendency toward Bar-Making*

Three of those important lessons are summarized well in Figure 10, which is due to Hohl (1971). One lesson was that disks of stars with random motions barely sufficient to avoid the Jeans instabilities still tend to be grossly unstable in an $m = 2$ sense. Secondly, there is nothing long-lived about the *spiral* structures that ensue among such "stars." And a third point was that the much hotter stable configurations into which these experimental disks rather quickly convert themselves tend often—though not always—to include some sort of *oval* shapes that continue to revolve indefinitely in their interiors. Essentially the same conclusions were reached already in the numerically coarser but physically yet more interesting experiments of Miller, Prendergast & Quirk (1970), which typically included also a subsystem of dissipative "gas" whose own strikingly different behavior is shown, as yet a fourth topic, in Figure 11. Let us consider these four topics in 3-4-2-1 order.

The tendency toward bar-making evident in both figures deserves exceptional emphasis because it is the one corner of our subject where the quest for durable *wavelike* distortions of theoretical galaxies can with some justice be said to have succeeded already. This does not mean that such a finite-amplitude lock-in phenomenon is at all well understood yet. However, it has now arisen in enough situations [including the much-softened 1000-particle integrations by Dzyuba & Yakubov (1970) and—in a milder sense—also the pioneering *N*-body experiments of P. O. Lindblad (1960)] that even this skeptic feels convinced that it represents the basic recipe for the survival of the bars and oval shapes seen in the interiors of actual SB and SAB galaxies.

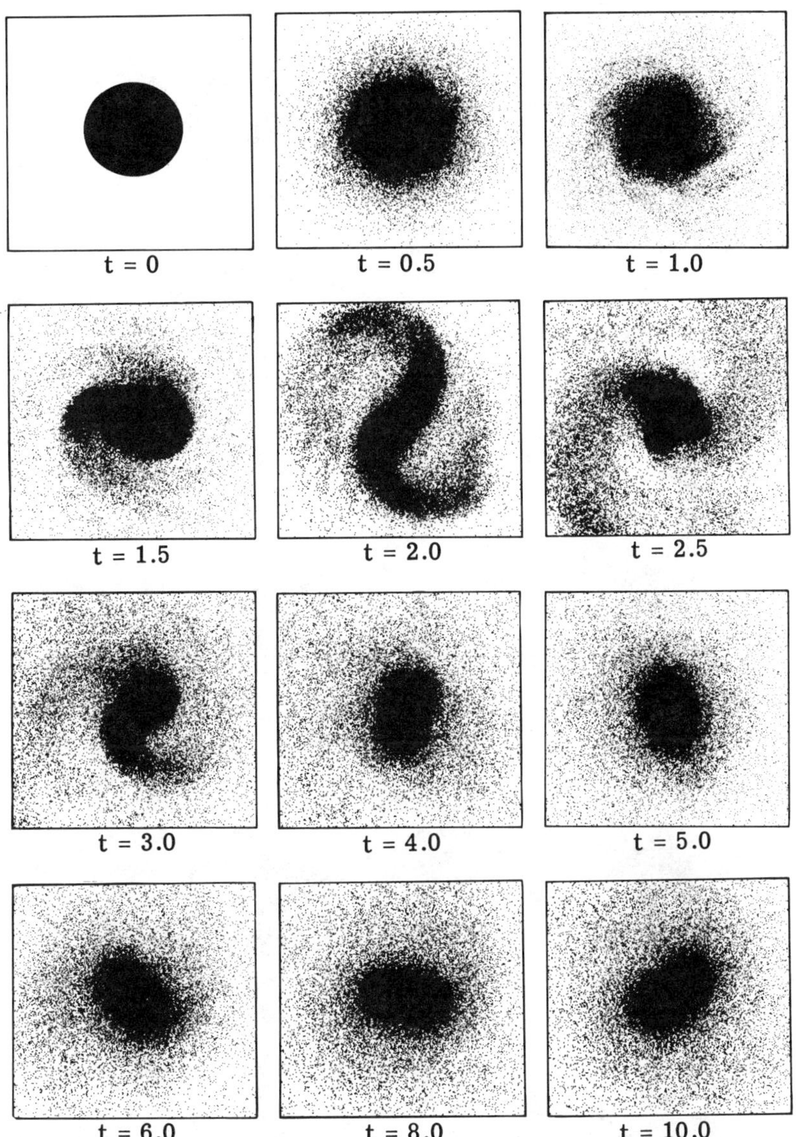

Figure 10 Evolution of a uniformly rotating disk of 10^5 mass points (Hohl 1971). Its surface density began like that of a cold Maclaurin disk; time is reckoned in rotation periods of the latter. Gaussian random velocities were preassigned from Equation (3); hence that model started only roughly in equilibrium, but it was nevertheless essentially stable in an axisymmetric sense, as Hohl also showed.

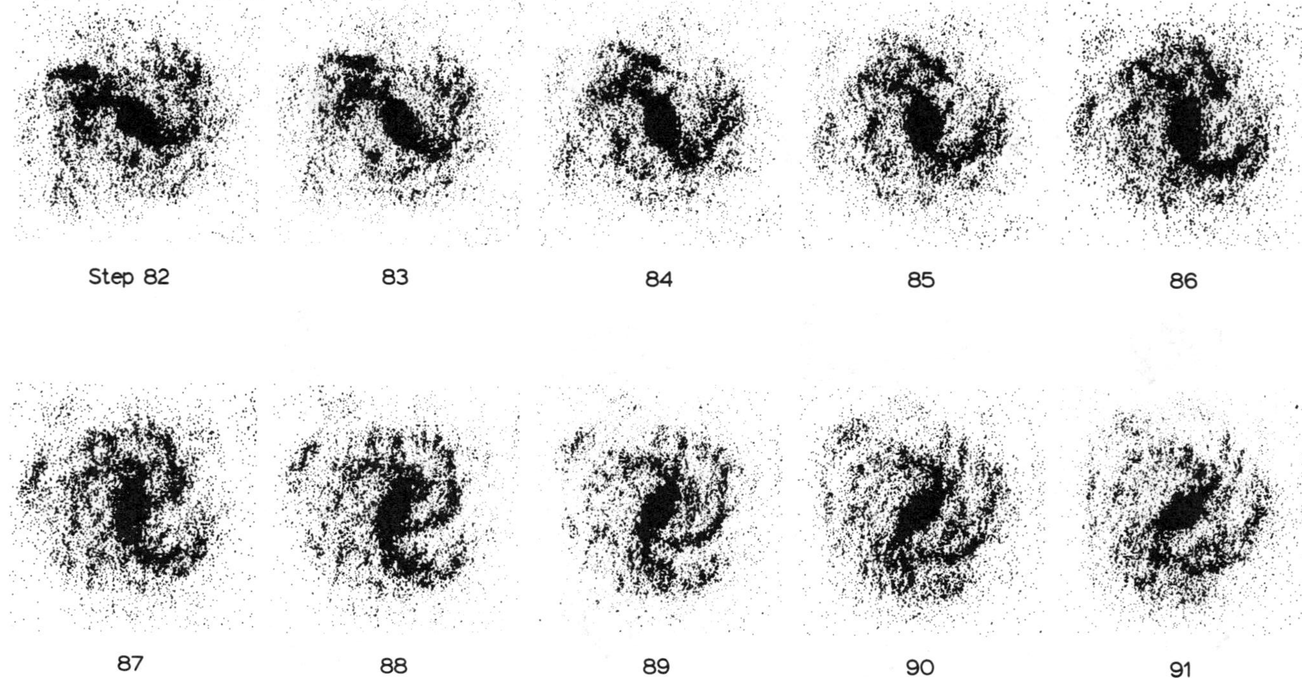

Figure 11 Evolving patterns in the "gas" component of a rotating disk consisting of some 10^5 collisionless "stars." Shown here are the contents of only the central half-squares from Figures 3 and 4 of Miller, Prendergast & Quirk (1970). During this short sequence, the total mass of "gas" decreased from 16.4 to 16.0 per cent of the whole, owing to gradual conversion of such material into additional "stars."

The main thing to notice is that the "bars" in Figures 10 and 11 revolve *in the same direction but more slowly* than the computer particles that undoubtedly circulate (and gyrate) within them. It is, of course, hopeless to try and trace any of the latter in these diagrams, but at least Figure 11 shows clearly enough that some of the material lumps orbiting outside the bar actually manage to overtake it. For Figure 10, Hohl reported and the stroboscopic effect confirms that the rotation period of the oval blob seen in the later frames equals 2.25 times the original spin period—whereas the orbital period of any relevant interior particle probably remains less than 1.5 even then. What we have here, in short, is again something very akin to the slow advance of Lindblad's "dispersion orbit" in Figure 2. To be sure, there are now many players, plus the hot and massive "star" disk that is not even shown in Figure 11. All these interact with each other in an appallingly nonlinear and nonlocal manner amidst much random motion. Yet, remarkably, the outcome falls short of complete axial symmetry. Instead, it seems that the various precessing orbits even here, helped no doubt by the familiar near-constancy of $\Omega - \kappa/2$, still manage well enough to "trap" each other—for reasons that Contopoulos (1970, 1975), Vandervoort (1973), and especially Lynden-Bell & Kalnajs (1972) made rather plausible—to yield the oval shapes that persist near the center. Characteristically, the angular speeds of even these special kinds of density waves, although slow, again exceed slightly the typical values of $\Omega - \kappa/2$ (cf Quirk 1971, Figure 6).

A second but already considerably vaguer sort of encouragement for wave theorists came from the *spiral* structure in the "gas" outside the obvious bar shape in Figure 11. A lot of this structure is chaotic, and even the nice "third" arm seen shearing rapidly past the 3 o'clock position at step 88 can in essence be only a transient material feature. Besides all that "noise," however, these pictures together with the magnified frames 101–103 and the yet later steps 121–130 shown by Miller et al. convey also a fuzzy but persistent image of a reversed *S*, with the bar serving only as its middle segment. This must indeed be a wave. Similar remarks apply to the barred and vaguely two-armed shapes obtained by Hohl (1971, Figure 17) upon "cooling" one previously featureless hot stellar disk in a gentle but admittedly artificial manner.

Miller, Prendergast and Quirk wrote that "the spiral pattern [or did they chiefly mean just a strong *impression* of spirality?] lasted for about three complete revolutions"—that is, approximately from step 60 till 160, by which time the "pattern . . . finally became too close to be seen in our pictures" and even the bar had mostly disappeared. It would be delightful to infer from this half-success that the bar was somehow essential for preserving that spiral. Regrettably, these authors and especially Quirk (1971) concluded after much thoughtful discussion and Fourier analysis that such an interpretation (among others) would be much too simplistic. They agreed that some sort of a stellar "asymmetry seems to 'drive' the pattern." Yet the mild, warped-oval $m = 2$ shape analyzed in detail in Quirk's Figures 3 and 4 plainly involved *comparable* disturbance densities from the gas and stars alike, and it also extended at least 3 times as far in radius as the obvious bar in our Figure 11. Thus, despite some resemblance to the situation later idealized in Figure 8 by Sanders and Huntley, it was hard to tell here just what had caused what.

The total lack of *spiral* shapes of respectable duration not only in Figure 10 but also in every other purely stellar-dynamical experiment conducted with sizable fractions of "mobile" mass (e.g. some by Hockney & Brownrigg 1974, or others by Hohl using initially Gaussian and exponential disks) is one of those results that almost speaks for itself. It is conceivable, of course, that some milder instabilities, which might themselves have led to more enduring spirals, were thwarted in these experiments by a kind of overheating from the fierce initial behavior. This seems unlikely, however, because of Hohl's extra tests with that artificial cooling: Whenever it was switched off again, the spiral features soon disappeared, no doubt speeded by the reason that follows.

7.2 *Ostriker-Peebles Criterion*

As is well known by now, what remained even after the efforts at cooling, by Miller (1971, Figure 6) as well as by Hohl, were some astonishingly hot disks of make-believe stars, with more random motion than actual rotation. Speaking not only personally, the trickle of such reports from about 1968 onwards came indeed as a rude surprise, especially when augmented by the similar analytical findings of Kalnajs (1972b) as to what it takes to stabilize fully those Maclaurin-Freeman disks. It was of little comfort to recall that "a question which the present discussion [of the short and mostly axisymmetric instabilities] leaves completely unanswered is . . . whether or not a given disk might prefer to develop into a bar-like structure" (Toomre 1964). The trouble with such face-saving was that this writer also believed then (and still does now, though much more timidly!) that temptations like the bar-making *ought* to belong mainly to the inner and more uniformly rotating parts of a disk. Out here in this Galaxy—as hinted by a probably exaggerated near-agreement (cf Toomre 1974a) between the observed and required random motions of the nearby K and M dwarfs, and also by the absence of true nonaxisymmetric instabilities of a *local* sort (cf Julian & Toomre 1966)—it had seemed a safe bet that a stability index like $Q = 1.5$ should curb all lesser misbehavior, and yet not exclude the Lin-Shu waves. Instead, the experimenters kept on reporting $Q = 2$, 3, and upwards.

In retrospect, Miller (1971) was clearly too modest in still entertaining the idea that this severe heating arose mostly from procedural flaws in the simulations. Since then, as Miller (1975; see also Hohl 1975) has rediscussed, fears of much-increased relaxation effects in these very thin disks—though quite serious in principle (Rybicki 1971)—were in practice largely laid to rest not only by noting that the near-gravity was much softened in the actual computer codes, but also by checking with further experiments (Hohl 1973, Miller 1974) either known or constrained to remain axisymmetric. In addition, Hohl (1972) confirmed experimentally for the uniformly rotating exact models what Kalnajs already knew from his analyses: they must be hot indeed to avoid the bar-making.

Most convincing of all was the discovery by Ostriker & Peebles (1973; reviewed thoroughly by Bardeen 1975) that the ultimate and/or stable forms of these disks of collisionless particles share the empirical rule or criterion

$$T/W \equiv (\text{organized K.E.})/(\text{negative P.E.}) \lesssim 0.14 \qquad (11)$$

that Ostriker had come to expect for the secular stability of various rotating models of stars. With that, it was suddenly beyond all reasonable doubt that the various investigations were telling essentially the same story—and that it was not a story of numerical errors.

Even today, however, it remains much less certain what this surprising unanimity actually *demands* of seemingly disk-like galaxies like our own. A lot of "heat" is clearly needed, but must it be hidden in faint, massive halos, as Ostriker and Peebles suggested? Or might some very hot inner disks or "spheroidal components" suffice already? Here probably the most important thing to remember is that the Ostriker-Peebles criterion is still not a theorem but just a fine summary of the existing N-body experiments. Reasonable though they seem, those experiments have in fact not been very diverse. One strong bias has been toward disks initially in fairly uniform rotation. Another and possibly even more serious limitation may have been the fact that nearly all constituents of those experimental disks were either prearranged or "cooled" again to rotate in the *same* direction. This concern may seem bizarre until we recall (*a*) that no one would set out to construct a stable spherical galaxy from stars with only one sense of angular momentum, and (*b*) that the average disk galaxy indeed contains a spherical component. Almost none of the existing N-body simulations, it seems, has either included or managed to develop any sizable fraction of stars orbiting backwards near the center. As Kalnajs (1977) has just explained, such mobile retrograde stars ought to interfere considerably with any bar-making tendency.

In short, the *need* for proper halos has still not been established firmly in our subject—though this is neither to imply that they would be unwelcome, nor to dispute the growing evidence from flat rotation curves (cf Roberts 1975, Krumm & Salpeter 1977) in favor of much extra mass of low luminosity. In fact, halos of interest to disk theorists would not even have to be inordinately massive. Roughly speaking, it seems (cf Kalnajs 1972b, Bardeen 1975, Hohl 1976) that already the adoption of a *rigid* halo—or any other frozen background distribution—of total mass comparable to that of the mobile disk within it cures at least the violent bar-making. Of course, immobilizing all but 10 or 20% of the mass does even better. As Hohl (1970) first illustrated and Hockney & Brownrigg (1974) confirmed, experiments starting with cold disks then yield not exactly "a spiral density wave, unchanged for at least 10 rotations" (as the latter authors claimed extravagantly) but something else almost as interesting: What one tends to find are multi-armed structures that survive only in a (gradually evolving) *statistical* sense—though indeed for quite a few revolutions—while their detailed features wind up and yet somehow keep on reappearing!

It is doubtful, however, whether any postulated halos, especially if they rotate to some extent, can be relegated safely to an altogether passive role. For one thing, as several authors including Mark (1976c) have suggested in the logical footsteps of Sweet (1963), some "two-stream" instabilities might well arise between such systems and the cooler disks embedded within them. And more important, even those very hot collections of stars might, like the one shown in Figure 10 or hoped for in Figure 8, still prefer oval distortions of their own.

7.3 *Unstable Spiral Modes*

The main alternative to these fascinating but expensive N-body experiments has been to pretend that a galaxy consists not of a finite number but a *continuum* of stars and/or gas particles, and to try and determine the fates and shapes of various large-scale but *small-amplitude* perturbations to disk-like equilibrium configurations of such material. Serious nonaxisymmetric efforts to explore the overall stability and assorted global modes of shearing disk galaxies in this manner go back at least to Hunter (1965) and Kalnajs (1965). Since then, as summarized in part by Hunter (1972) and Bardeen (1975), a lot of additional cleverness and computing money has been expended—indeed far more than is immediately evident from the literature. Yet the overall progress in this area has been painfully slow.

That situation should improve when several major ongoing studies, particularly the ones by Kalnajs (1971, 1976) and Bardeen (1975), reach print in final form. These promise many interesting comparisons with WKBJ estimates, stability criteria, resonant particle effects, and ideas about angular momentum transfer. Here it seems foolish to try and anticipate such conclusions. Instead, it may be more instructive to look backward for a moment and to ask why all this laudable global-mode calculating has taken so long.

The reasons involve both principles and complexity. For one thing, it has gradually dawned upon most workers that stable self-gravitating disks (especially ones composed of collisionless stars) may not even possess many *discrete* normal modes of the sort that one associates with church bells and Cepheids. The vibrational spectra of our systems seem, by and large, to be *continuous*. The associated modes must look disagreeable indeed near any resonance radii—somewhat like the van Kampen (1955) modes from plasma physics—as Hunter (1969) demonstrated for certain $m \neq 0$ waves near their corotation radii, and as Erickson (1974) showed even for some seemingly innocuous $m = 0$ vibrations. Hunter also cautioned of similar complications near edges of low density.

A second difficulty has been that, even if strictly oscillatory discrete $m = 2$ modes do exist, it remains unproven that they include any of *spiral* form. Indeed, such truly permanent spiral waves of infinitesimal amplitude now seem hardly possible. This personal opinion stems much more from the formidable resonance effects and gravitational torques discussed, for instance, by Lynden-Bell & Kalnajs (1972) than from the weak "anti-spiral theorem" of Lynden-Bell & Ostriker (1967)—which in essence states only that the existence of a steady trailing mode implies a twin of leading shape, and vice versa, because of time-reversibility.

The above frustrations have not referred especially to *unstable* spiral modes. (Quite reasonably, it is the milder versions of just such modes from nondissipative linear theory that are widely thought now to offer the best hope for some sort of conversion into *quasi-steady* spiral patterns of finite amplitude—and this is where threshold effects such as the nonlinear shock damping stressed in Section 5.3 not only assume a special importance but may harbor a second vital reason as to why good spirals seem to need gas.) The problem here has been much more that one has tended to find such unstable modes, even discrete ones, far too fiercely and far too often! Just as in the N-body studies, much of this nuisance has doubtless had good physical causes, but also as with those experiments, some may even have stemmed from ingenious tricks that backfired.

One such example comes from a labor-saving idea that occurred to Miller (1971, 1974) among others. He noticed that the use of the "softened" gravitational potential $\phi = -GM(b^2+d^2)^{-1/2}$ rather than the precise $\phi = -GM/d$ for the interaction of two particles separated by a distance d would yield the dispersion relation

$$\omega^2 = \kappa^2(r) - 2\pi G\mu(r)\,|k|\,\mathrm{e}^{-|k|b} \tag{12}$$

instead of Equation (1) for short axisymmetric vibrations of a cold thin disk. The exponential mimics the "reduction factor" of Lin & Shu (1966), and so this simple relation indeed resembles the one in Figure 4. Toward the short wavelengths in particular it seems much more representative of the $\omega \rightarrow \kappa$ behavior of waves in a stellar disk than would be implied by the analogous formula

$$\omega^2 = \kappa^2(r) - 2\pi G\mu(r)\,|k| + a^2k^2 \tag{13}$$

(cf Safronov 1960, Hunter 1972) for a thin gas layer with "honest" gravity but an imagined sound speed a.

The irony here is that Erickson (1974), using just this presumably superior gravity shortcut (to avoid all the stellar-dynamical bother with the Boltzmann equation in his global-mode calculations for some Gaussian disks) indeed obtained numerous spiral instabilities resembling the $m = 2$ mode in Figure 12. Unfortunately, he also found himself unable to turn off all such growing modes, no matter how large he chose that cutoff length b. By contrast, Bardeen (1975) adopted the seemingly weaker gas idealization and managed to stabilize such disks altogether, at least for energy ratios $T/W \lesssim 0.26$ corresponding to dynamical (rather than secular) stability of Maclaurin spheroids. With these experiences in mind—and also noting that Kalnajs (1970 onwards) has not yet succeeded in obtaining fully stable stellar disks, apart from his *tour de force* with the Maclaurin-Freeman models—it is no wonder that mode calculators remain a little unsure of just where they stand.

To conclude this litany of surprises, some signs of a counterexample to the Ostriker-Peebles criterion have surfaced recently in the work of Zang (1976) on the unstable modes of a class of exact stellar disks with random motions. These disks themselves, which were noticed also by Bisnovatyi-Kogan (1975), epitomize "flat" rotation curves in assuming $V(r) = \text{const} = V_0$ at all radii; this means that the surface density $\mu(r) = V_0^2/2\pi Gr$, or that the total mass grows infinite linearly with the radius r. The latter property, and also the central singularity of μ itself, need not concern us too much, since both can doubtless be remedied with plausible cutoffs. One big advantage of this model is that it is everywhere self-similar. Another is that its velocity distribution function

$$f(u,v,r) = F(E,J) = \text{const}\cdot J^q\,\mathrm{e}^{-E/\sigma_u^2}, \tag{14}$$

with $E = \frac{1}{2}(u^2+v^2)+V_0^2\ln r$, $J = rv$, and $q+1 = V_0^2/\sigma_u^2$, can be written so compactly for $J > 0$ (with $f = 0$ otherwise).

As far as axisymmetric (or $m = 0$) modes are concerned, Zang found this class of models to become globally stable when $\sigma_u \geqq 0.378 V_0$, in good agreement with the "local" estimate given by Equation (3). Already for that critical value of the velocity dispersion, however, Zang was unable to locate numerically, despite much systematic searching, any two- or multi-armed (i.e. $m \geqq 2$) instabilities. This important result needs to be qualified in two ways. One is that Zang literally examined not the full models given by Equation (14), but ones where he had immobilized the central regions—as if the central "bulge" were particularly hot and therefore unresponsive—by multiplying that f by the factor $[1+(J_0/J)^n]^{-1}$; he did this mostly to avoid immense angular speeds as $r \to 0$, but also to provide a length scale J_0/V_0. The other proviso is that Zang managed to interrogate his difficult integral equation, patterned closely after the one by Kalnajs (1971), only for *discrete* unstable modes, which he assumed to grow like e^{st} while rotating with an arbitrary pattern speed Ω_p. An unstable continuum seems unlikely, but Zang could provide no proof.

Thus, to be precise, Zang's inability to locate any $m = 2$ instabilities referred only to discrete modes of modified disks with $n = 1$ or 2—or ones whose centers had been "carved out" fairly gently. As that truncation was made more abrupt, some unstable modes actually emerged; one such mode, for index $n = 4$, is pictured in Figure 12. Modes like that were definitely indebted, however, to the artificial *sharpness* of the inner boundary. Although superficially similar to the modes inferred by Lau, Lin & Mark (1976) from WKBJ theory, the attractive but unrealistic $m = 2$ spiral wave in Figure 12 grew an order of magnitude faster than then expected by those authors.

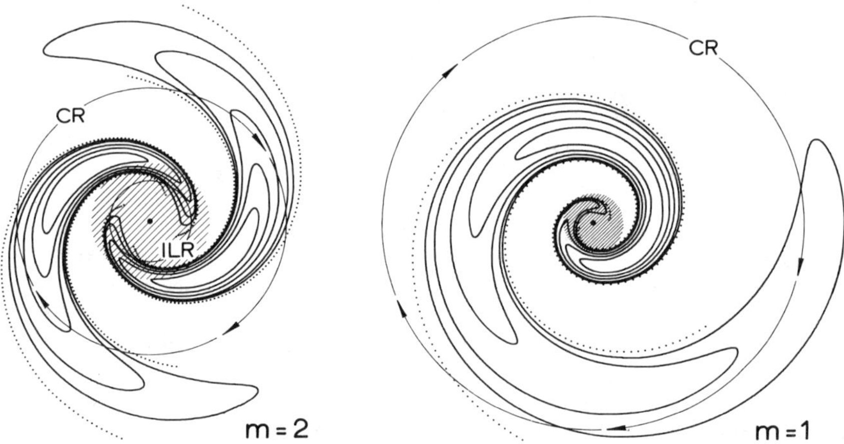

Figure 12 Two most unstable spiral modes obtained by Zang (1976) for centrally cut-out versions of the $V(r) = V_0$ disks. In both cases, only the positive parts of the disturbance density are shown here, using contour lines set at 80, 60, 40, 20, and 10% of peak values; dots mark the nodes. The shaded areas have radius $r = J_0/V_0$. The $m = 2$ mode (referring to the model with velocity dispersion $\sigma_u = 0.378 V_0$ and cut-out index $n = 4$) has pattern speed $\Omega_p = 0.439$ and growth rate $s = 0.127$, both expressed as multiples of V_0^2/J_0. The analogous data for the $m = 1$ mode (obtained for $n = 2$) are $\Omega_p = 0.141$ and $s = 0.066$.

Even this story, however, proved to contain a nasty surprise. Regardless of whether he tried cut-out indices $k = 1, 2, \ldots$, Zang remained plagued with numerous *one*-armed unstable modes! One of them is also shown in Figure 12. Unlike the few two-armed modes, this $m = 1$ instability could not be attributed very plausibly either to the frozen center or even to the fact that such a core—in a deliberate rerun of a rare old error by Maxwell—was here kept mostly fixed. Zang concluded ruefully that with this curious swap of difficulties "we may merely have jumped from the frying pan into the fire."

8 SHEARING BITS AND PIECES

As our last topic, it needs to be said briefly but aloud that there seems nothing fundamentally wrong with the simple idea that much of spiral structure—which, after all, is often very ragged and confusing, as in most of M101—consists more nearly of *things* rather than waves, to use Prendergast's (1967) apt word. The difficulties with this notion are mainly just quantitative, once one excludes the "grand designs." Obviously all shearing "things" in a galaxy will briefly exhibit some spirality, like cream poured into a newly-stirred cup of coffee. Hence the real task remains instead to decide how and whether the material in question can keep on making suitable fresh bits and pieces to replace those older fragments that wrap up and disappear.

The only promising site for such *recurrent* instabilities is, of course, the gas layer of a galaxy. Already von Weizsäcker (1951) thought so, with his vague theory of "turbulence . . . produced by nonuniform rotation." Jeans (1929, p. 379) did almost the same when he wrote of the "chaos of moving clusters" in the outer parts of spiral nebulae, and remarked that "the theory of gravitational instability makes it easy to understand how these large clusters come to exist."

In fact, the regeneration of large new fragments in a gas disk is not quite so easy. Von Weizsäcker in effect forgot that our kind of shear flow is supported much more by gravity forces than by pressure gradients, and Jeans did not realize that rotating disks also have a *maximum* critical length for gravitational instability, such as λ_{crit} from Equation (2), which need not be at all comparable to overall dimensions. Indeed that length can be distressingly short if the "active" surface density μ_g of the subsystem of interest is only a small fraction of the total, and if the stellar disk cooperates hardly at all. For example, even if we adopt $\mu_g = 10\ M_\odot \mathrm{pc}^{-2}$ for the nearby gas in this Galaxy, its λ_{crit} works out as only 1.7 kpc—and then an effective sound speed $a \geqq 4\ \mathrm{km\ sec}^{-1}$ should already curb all Jeans instabilities [cf Equation (13)].

All this was well appreciated by Goldreich & Lynden-Bell (1965) in what remains the only extensive modern exploration of the theme that Sc galaxies consist largely of "a swirling hotch-potch of pieces of spiral arms." To overcome the above difficulties, these authors suggested that "spiral arm formation should not be regarded as an instability in [just] the gas but rather as an instability of the whole star-gas mixture which is triggered by an increase in gas density" in a volumetric sense—or, what amounts to the same thing, by any cooling or damping that lowers the sound speed a in the gas. Even those plans, however, had two serious flaws. One was the fact that the fiercely amplified shearing waves studied by Goldreich and Lynden-Bell were, strictly speaking, no true instabilities. The other flaw was the presumption that strong help from the undamped stars would continue despite all their repeated heating by the postulated clumpings of the gas; this idea rested mostly upon wishful thinking. Nevertheless, in that severe amplification, Goldreich and Lynden-Bell offered one real nugget of a discovery that greatly softens the last criticism. It is shown in a more refined form in Figure 13.

This old diagram reemphasizes that a shearing stellar (or gas) disk that is locally quite stable can still *respond* with remarkable vigor to the nonaxisymmetric gravity forces from any fortuitously bound, orbiting gas lumps. Notice that the excess mass that pauses in the trailing "wakes" illustrated here exceeds the imposed mass by about an order of magnitude, even though $Q = 1.4$ plus a modest softening due to thickness were assumed in this example. Notice especially that this arm-like shape does not undergo any shearing. It is itself a steady density *wave,* albeit a forced one. In short, Figure 13 renews hope that here may be a reason why some of the observed spiral irregularities extend, as in M101, over quite sizable fractions of the radius despite the fact that the gas content tends to be hardly of order 10 per cent. It also cautions that even the "hotch-potch" is apt to be a little more wavelike in character than one might perhaps have supposed.

Any reader who (rightly) finds these remarks unpersuasive should look again at Figure 11 and especially at the experiments of Hohl (1970) and Hockney & Brownrigg (1974), all conducted with active masses totaling only 10 or 20%. He

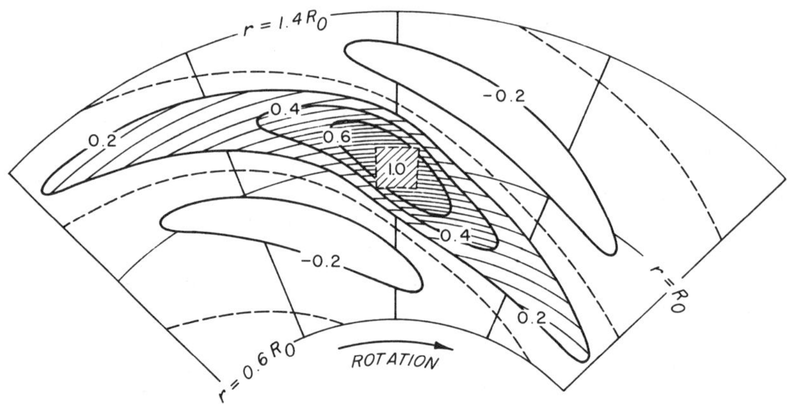

Figure 13 Wavelike ridge of excess density in a $Q = 1.4$, $V(r) = \text{const}$ disk of stars shearing past a small orbiting mass point (Julian & Toomre 1966). By comparison, that mass itself would yield unit density if spread uniformly over the little square. This diagram stems from small-amplitude theory with several "local" approximations; all subsequent overtakings of, and by, stars have been neglected here. The self-gravity of stars was included, however, and the assumption that $\lambda_{crit} = R_0$, made only in plotting, possibly underestimates the true extent of this quasi-steady forced wave.

should ask how he would explain those examples of "material clumping [that] is periodically destroyed by differential rotation and regenerated by gravitational instability" (Lin & Shu 1964), if not by something akin to the amplification process of Goldreich and Lynden-Bell. Of course, the bigger question remains: Why should enough disk material *still* be "cool" enough in its random motion to act even as cooperatively as shown in Figure 13? Are we, for instance, underestimating the amounts of gas present, or else the recent history either of infall of gas from outside or its return from dying stars, including those in any halos?

All in all, this topic of various "secondary" spiral features seems much overdue for renewed attention. Not even von Weizsäcker's idea can yet be dismissed entirely: some instabilities might indeed be largely hydrodynamic. It still needs to be explored, for instance, whether the well-known lumpiness of the major arms themselves on about a 1 or 2 kpc scale might not stem mainly from some sort of Kelvin-Helmholtz instability (rather than Jeans instability, or even the Parker = magnetic Rayleigh-Taylor kind advocated by Shu 1974) of the strong shear flows that are inevitable when gas from a whole range of radii crowds together behind a spiral shock.

Finally, what about spiral structures provoked from outside? A quick answer is that this old tidal idea indeed seems to work fine—in a small handful of obscure cases like Arp 295 (Stockton 1974) and NGC 2535/36 or 3808. In those instances just as in some simple computations (cf Toomre 1974b), the two "material arms" are rather open and the one toward the companion often looks grotesquely straighter and longer than the other. It remains far less certain, however—despite much speculation going back at least to Chamberlin (1901) in the case of M51, and again to Lindblad (1941) with M81/82—that the familiar "grand designs" of M51, M81, and NGC 5364 are indebted in any serious way to the (admittedly worrisome) proximity of major companions.

For reasons of tightness and extent of spiral structure, plus the Westerbork detection of ample H I also *between* the main arms of M81, those famous examples cannot possibly be tidal in the naive kinematic sense implied above. Here the tidal process, if indeed any, must have been a lot subtler. The first step would have had to be some major distortion of the outer disk or halo. This in turn may or may not have induced a relatively transient spiral wave in the proper disk, traveling inward with the group velocity. If only some decently cool but stable models of galaxies existed already, it would be fairly easy to check whether such a scheme makes any sense dynamically. Until then, however, claims such as by Tully (1974) that "a density wave has been generated [in M51] by the innermost material clumping arising from the encounter" with NGC 5195 strike even this sympathetic reviewer as very premature. In fact, judging from the *broad* outer contours of M51 toward the south—especially evident in one very deep photograph and in various H I data that Lynds and Shane have kindly shown—it seems that Toomre & Toomre (1972, Figure 20a) erred in reckoning the present epoch of that encounter to be as late as "$t = 2.4$." A better guess appears to be $t = 1.6$ or 1.8, in the same units. Given that those outer distortions become pronounced only from $t = 1$ onward, this shorter elapsed time gravely aggravates a difficulty noted already by Lin (1975), namely that "the time required for the propagation of such an influence into the central regions is too long." Hence it looks more and more as if even in M51 the main *spiral* structure is in essence just a badly-dented version of what would have been there anyway.

To end on a more positive note, it has emerged recently that density/shock waves of roughly circular shape—now indeed caused by a second galaxy—seem to be present in the rare but very striking oddities known as ring galaxies. As Fosbury & Hawarden (1977) have just illustrated for the fine example of the "Cartwheel" discovered by Zwicky (1941), it seems there again that simple mechanical crushing has evoked intense star formation in the knots of H II regions which mark that ring itself—and also that the *residue* of this indelicate processing includes the usual "feathers" or "interarm branches" or secondary spiral arms!

ACKNOWLEDGMENTS

Among several kind friends, I am most indebted here to Agris Kalnajs—and to editors Burbidge, Goldberg, and Layzer for much patient nudging. I am also very grateful for all the support of the NSF.

Literature Cited

Arp, H. C. 1976. *Ap. J.* 207: L147
Baade, W. 1963. *Evolution of Stars and Galaxies.* Cambridge: Harvard Univ. Press
Bardeen, J. M. 1975. *Proc. IAU Symp.* 69: 297
Bisnovatyi-Kogan, G. S. 1975. *Pis'ma Astron. Zh.* 1(9): 3. Transl. 1976. *Sov. Astron. Lett.* 1: 177 (from Russian)
Bryan, G. H. 1889. *Philos. Trans. R. Soc. London Ser. A* 180: 187
Burton, W. B. 1973. *Publ. Astron. Soc. Pac.* 85: 679
Carranza, G., Crillon, R., Monnet, G. 1969. *Astron. Astrophys.* 1: 479
Chamberlin, T. C. 1901. *Ap. J.* 14: 17
Chandrasekhar, S. 1969. *Ellipsoidal Figures of Equilibrium.* New Haven: Yale Univ. Press
Contopoulos, G. 1970. *Ap. J.* 160: 113
Contopoulos, G. 1975. In *La Dynamique des Galaxies Spirales,* ed. L. Weliachew, p. 21. Paris: CNRS
Curtis, H. D. 1919. *J. Wash. Acad. Sci.* 9: 217
de Vaucouleurs, G. 1964. *Proc. IAU Symp.* 20: 195
de Vaucouleurs, G. 1970. *Proc. IAU Symp.* 38: 18
Dewar, R. L. 1972. *Ap. J.* 174: 301
Dzyuba, B. M., Yakubov, V. B. 1970. *Astron. Zh.* 47: 3. Transl. 1970. *Sov. Astron.* 14: 1 (from Russian)
Erickson, S. A. 1974. *Vibrations and Instabilities of a Disk Galaxy with Modified Gravity.* PhD thesis. Mass. Inst. Technol., Cambridge, Mass. 179 pp.
Feldman, S. I., Lin, C. C. 1973. *Stud. Appl. Math.* 52: 1
Feynman, R. P. 1963. *Lectures on Physics* 1: p. 7-7. Reading, Mass: Addison-Wesley
Fosbury, R. A. E., Hawarden, T. G. 1977. *MNRAS* 178: 473
Freeman, K. C. 1966. *MNRAS* 134: 1; 134: 15
Freeman, K. C. 1970a. *Ap. J.* 160: 811
Freeman, K. C. 1970b. *Proc. IAU Symp.* 38: 351
Freeman, K. C. 1975. In *Galaxies and the Universe,* ed. A. Sandage, M. Sandage, J. Kristian, p. 409. Chicago: Univ. Chicago Press
Fujimoto, M. 1968. *Proc. IAU Symp.* 29: 453
Goldreich, P., Lynden-Bell, D. 1965. *MNRAS* 130: 125
Gordon, M. A., Burton, W. B. 1976. *Ap. J.* 208: 346
Hockney, R. W., Brownrigg, D. R. K. 1974. *MNRAS* 167: 351
Hohl, F. 1970. *Proc. IAU Symp.* 38: 368
Hohl, F. 1971. *Ap. J.* 168: 343
Hohl, F. 1972. *J. Comp. Phys.* 9: 10
Hohl, F. 1973. *Ap. J.* 184: 353
Hohl, F. 1975. *Proc. IAU Symp.* 69: 349
Hohl, F. 1976. *Astron. J.* 81: 30
Hunter, C. 1963. *MNRAS* 126: 299
Hunter, C. 1965. *MNRAS* 129: 321
Hunter, C. 1969. *Stud. Appl. Math.* 48: 55
Hunter, C. 1972. *Ann. Rev. Fluid Mech.* 4: 219

Jeans, J. H. 1929. *Astronomy and Cosmogony.* Cambridge: Univ. Press; 1961. New York: Dover
Julian, W. H. 1967. *Ap. J.* 148:175
Julian, W. H., Toomre, A. 1966. *Ap. J.* 146:810
Kalnajs, A. J. 1965. *The Stability of Highly Flattened Galaxies.* PhD thesis. Harvard Univ., Cambridge, Mass. 129 pp.
Kalnajs, A. J. 1970. *Proc. IAU Symp.* 38:318
Kalnajs, A. J. 1971. *Ap. J.* 166:275
Kalnajs, A. J. 1972a. *Astrophys. Lett.* 11:41
Kalnajs, A. J. 1972b. *Ap. J.* 175:63
Kalnajs, A. J. 1973. *Proc. Astron. Soc. Aust.* 2:174
Kalnajs, A. J. 1976. *Ap. J.* 205:745; 205:751
Kalnajs, A. J. 1977. *Ap. J.* 212:637
Kaplan, S. A., Pikel'ner, S. B. 1974. *Ann. Rev. Astron. Astrophys.* 12:113
Kerr, F. J. 1967. *Proc. IAU Symp.* 31:239
Krumm, N., Salpeter, E. E. 1977. *Astron. Astrophys.* 56:465
Lau, Y. Y., Lin, C. C., Mark, J. W.-K. 1976. *Proc. Natl. Acad. Sci. USA* 73:1379
Lau, Y. Y., Mark, J. W.-K. 1976. *Proc. Natl. Acad. Sci. USA* 73:3785
Lin, C. C. 1967. *Ann. Rev. Astron. Astrophys.* 5:453
Lin, C. C. 1970. *Proc. IAU Symp.* 38:377
Lin, C. C. 1971. In *Highlights of Astronomy,* ed. C. de Jager, 2:88. Dordrecht: Reidel
Lin, C. C. 1975. In *Structure and Evolution of Galaxies,* ed. G. Setti, p. 119. Dordrecht: Reidel
Lin, C. C., Shu, F. H. 1964. *Ap. J.* 140:646
Lin, C. C., Shu, F. H. 1966. *Proc. Natl. Acad. Sci. USA* 55:229
Lin, C. C., Shu, F. H. 1967. *Proc. IAU Symp.* 31:313
Lin, C. C., Yuan, C., Shu, F. H. 1969. *Ap. J.* 155:721
Lindblad, B. 1941. *Stockholm Obs. Ann.* 13: No. 10
Lindblad, B. 1948. *MNRAS* 108:214
Lindblad, B. 1951. *Publ. Obs. Univ. Mich.* 10:59
Lindblad, B. 1958. *Stockholm Obs. Ann.* 20: No. 6
Lindblad, B. 1961. *Stockholm Obs. Ann.* 21: No. 8
Lindblad, B. 1962. *Proc. IAU Symp.* 15:146
Lindblad, B. 1963. *Stockholm Obs. Ann.* 22: No. 5
Lindblad, B. 1964. *Astrophys. Norv.* 9:103
Lindblad, P. O. 1960. *Stockholm Obs. Ann.* 21: No. 4
Lindblad, P. O. 1974. *Proc. IAU Symp.* 58:399
Lynden-Bell, D. 1974. In *Proc. Eur. Astron. Meeting, 1st, Athens,* 3:114. Berlin: Springer
Lynden-Bell, D. 1975. See Contopoulos 1975, p. 91
Lynden-Bell, D., Kalnajs, A. J. 1972. *MNRAS* 157:1
Lynden-Bell, D., Ostriker, J. P. 1967. *MNRAS* 136:293
Mark, J. W.-K. 1971. *Proc. Natl. Acad. Sci. USA* 68:2095
Mark, J. W.-K. 1974a. *Ap. J.* 193:539
Mark, J. W.-K. 1974b. *Proc. IAU Symp.* 58:417
Mark, J. W.-K. 1976a. *Ap. J.* 203:81
Mark, J. W.-K. 1976b. *Ap. J.* 205:363
Mark, J. W.-K. 1976c. *Ap. J.* 206:418
Marochnik, L. S., Mishurov, Yu. N., Suchkov, A. A. 1972. *Astrophys. Space Sci.* 19:285
Marochnik, L. S., Suchkov, A. A. 1974. *Usp. Fiz. Nauk* 112:275. Transl. 1974. *Sov. Phys. Usp.* 17:85 (from Russian)
Mathewson, D. S., van der Kruit, P. C., Brouw, W. N. 1972. *Astron. Astrophys.* 17:468
Maxwell, J. C. 1859. In 1890. *Scientific Papers,* 1:288. Cambridge: Univ. Press
Miller, R. H. 1971. *Astrophys. Space Sci.* 14:73
Miller, R. H. 1974. *Ap. J.* 190:539
Miller, R. H. 1975. See Contopoulos 1975, p. 45
Miller, R. H., Prendergast, K. H., Quirk, W. J. 1970. *Ap. J.* 161:903
Oort, J. H. 1956. *Sci. Am.* 195(3): 100
Oort, J. H. 1962. In *Interstellar Matter in Galaxies,* ed. L. Woltjer, p. 234. New York: Benjamin
Oort, J. H. 1966. *Q. J. R. Astron. Soc.* 7:329
Ostriker, J. P., Peebles, P. J. E. 1973. *Ap. J.* 186:467
Piddington, J. H. 1973. *Ap. J.* 179:755
Pikel'ner, S. B. 1970. *Astron. Zh.* 47:752. Transl. 1971. *Sov. Astron.* 14:602 (from Russian)
Pooley, G. G. 1969. *MNRAS* 144:101
Prendergast, K. H. 1962. See Oort 1962, p. 217
Prendergast, K. H. 1967. *Proc. IAU Symp.* 31:303
Quirk, W. J. 1971. *Ap. J.* 167:7
Roberts, M. S. 1975. *Proc. IAU Symp.* 69:331
Roberts, W. W. 1969. *Ap. J.* 158:123
Roberts, W. W. 1970. *Proc. IAU Symp.* 38:415
Roberts, W. W., Roberts, M. S., Shu, F. H. 1975. *Ap. J.* 196:381
Roberts, W. W., Shu, F. H. 1972. *Astrophys. Lett.* 12:49
Rogstad, D. H., Wright, M. C. H., Lockhart, I. A. 1976. *Ap. J.* 204:703
Rosse, Earl of. 1850. *Philos. Trans. R. Soc. London,* p. 503

Rots, A. H. 1975. *Astron. Astrophys.* 45:43
Rots, A. H., Shane, W. W. 1975. *Astron. Astrophys.* 45:25
Rybicki, G. B. 1971. *Astrophys. Space Sci.* 14:15
Safronov, V. S. 1960. *Dokl. Akad. Nauk* 130:53. Transl. 1960. *Sov. Phys. Dokl.* 5: 13 (from Russian)
Salpeter, E. E. 1976. *Ap. J.* 206:673
Sandage, A. 1975. See Freeman 1975, p. 1
Sanders, R. H., Huntley, J. M. 1976. *Ap. J.* 209:53
Schweizer, F. 1976. *Ap. J. Suppl.* 31:313
Sharpless, S., Franz, O. G. 1963. *Publ. Astron. Soc. Pac.* 75:219
Shu, F. H. 1970. *Ap. J.* 160:99
Shu, F. H. 1973. *Am. Sci.* 61:524
Shu, F. H. 1974. *Astron. Astrophys.* 33:55
Shu, F. H., Milione, V., Gebel, W., Yuan, C., Goldsmith, D. W., Roberts, W. W. 1972. *Ap. J.* 173:557
Shu, F. H., Milione, V., Roberts, W. W. 1973. *Ap. J.* 183:819
Shu, F. H., Stachnik, R. V., Yost, J. C. 1971. *Ap. J.* 166:465
Simonson, S. C. 1970. *Astron. Astrophys.* 9: 163
Sørensen, S.-A., Matsuda, T., Fujimoto, M. 1976. *Astrophys. Space Sci.* 43:491
Spitzer, L., Schwarzschild, M. 1953. *Ap. J.* 118:106
Stockton, A. N. 1974. *Ap. J.* 190:L47
Sweet, P. A. 1963. *MNRAS* 125:285
Toomre, A. 1964. *Ap. J.* 139:1217
Toomre, A. 1969. *Ap. J.* 158:899
Toomre, A. 1974a. In *Highlights of Astronomy*, ed. G. Contopoulos, p. 457. Dordrecht: Reidel
Toomre, A. 1974b. *Proc. IAU Symp.* 58:347
Toomre, A., Toomre, J. 1972. *Ap. J.* 178:623
Tully, R. B. 1974. *Ap. J. Suppl.* 27:449
van der Kruit, P. C., Allen, R. J. 1976. *Ann. Rev. Astron. Astrophys.* 14:417
Vandervoort, P. O. 1973. *Ap. J.* 180:739
van Kampen, N. G. 1955. *Physica* 21:949
von Weizsäcker, C. F. 1951. *Ap. J.* 114:165
Wielen, R. 1974. *Publ. Astron. Soc. Pac.* 86: 341
Wielen, R. 1975. See Contopoulos 1975, p. 357
Woodward, P. R. 1975. *Ap. J.* 195:61
Woolley, R., Candy, M. P. 1968. *MNRAS* 139:231; 141:277
Yuan, C. 1969. *Ap. J.* 158:889
Zang, T. A. 1976. *The Stability of a Model Galaxy.* PhD thesis. Mass. Inst. Technol., Cambridge, Mass. 204 pp.
Zwicky, F. 1941. *Applied Mechanics* von Kármán Anniv.: 137
Zwicky, F. 1955. *Publ. Astron. Soc. Pac.* 67: 232

Ann. Rev. Astron. Astrophys. 1977. 15: 479–504

INTERSTELLAR SCATTERING AND SCINTILLATION OF RADIO WAVES

Barney J. Rickett
Department of Applied Physics and Information Science, University of California, San Diego, La Jolla, California 92093

INTRODUCTION

In the development of radio astronomy there has been an interesting interplay between the studies of radiosources (including radio galaxies, quasars, and pulsars) and the way in which their signals are distorted by scintillation and scattering along the line of sight.

Hewish (1975) has reviewed how his early studies of radio galaxies led to the identification of ionospheric scintillation, then to angular scattering in the outer corona, and then to interplanetary scintillation (IPS). Recognizing that only very small diameter sources could show IPS, he turned this around to probe source structure in the range 0.1″ to 1.0″ at meter wavelengths. The small angular sizes of many quasars have been determined by this technique (Little & Hewish 1968, Cohen et al. 1967). In pursuing this technique Hewish and his colleagues built a very large antenna array for an IPS survey at 81 MHz (Hewish & Burnell 1970). As is now well known, it was with this antenna that Jocelyn Bell and he opened a whole new field with the discovery of pulsars (Hewish et al. 1968). The extreme regularity of the pulse timing contrasts with the erratic variations in their intensities. The slow component in these intensity variations was identified as yet another propagation process, namely interstellar scintillation (ISS) (Scheuer 1968, Rickett 1969). Once again this scintillation can be turned around to search for radiosources smaller than about 0.001″. At present such tests have been negative on all sources except pulsars (Condon & Backer 1975). Another consequence of interstellar scattering is that small-diameter radiosources appear broadened. This amounts to about 0.1″ at meter wavelengths. Readhead & Hewish (1972) used an IPS technique to explore how this angular scattering is distributed in our Galaxy. Yet another scattering effect is that an emitted impulse (from a pulsar) is broadened in time, since the scattered waves have a longer propagation path. This was first observed for the Crab pulsar by Rankin et al. (1970).

This paper reviews the observations and theory of interstellar scattering and scintillations (ISS). The reader is referred to Coles, Rickett & Rumsey (1974) and Jokipii (1973) for reviews of IPS. ISS is clearly of relevance to studies of the interstellar medium; however, it is also important to radio astronomers who observe radio signals that have travelled through the medium.

Early theoretical discussions of the scattering and scintillation (Scheuer 1968, Salpeter 1969) pinpointed many of the basic physical processes, but were based on simple approximations. Since that time the theory has developed considerably in sophistication and generality. Different developments have been motivated by optical and radio applications. Ratcliffe (1956) gave a lucid review of the theory applicable to ionospheric scintillation at radio frequencies, while Tatarski (1961) pioneered the theory of optical scintillation in the earth's atmosphere. The key differences were that the radio waves were assumed to be scattered in a thin layer of plasma inhomogeneities that were described by a Gaussian spatial wavenumber spectrum; the optical waves were assumed to be scattered all along the propagation path by atmospheric inhomogeneities that were described by a power law spatial wavenumber spectrum. Of course, the Kolmogorov spectrum for the inertial range of neutral turbulence was predicted for the atmosphere. One of the major goals of this paper is to review the ISS observations in the light of modern theory for extended spatially homogeneous, power law inhomogeneities. We present formulas, with attention to the various numerical factors, for the experimentally observable quantities and compare them with the data. A general discussion of the model is postponed until the final section. Many of the previous interpretations have been based on the Gaussian spectrum (Rickett 1970, Lang 1971b, Sutton 1971, Backer 1975). Our analysis differs from the power law analyses of Little & Matheson (1973) and of Lee & Jokipii (1976) in the inclusion of a much wider set of data. Other theoretical points concern the geometry (screen or extended medium), whether the source is a plane wave, an extended source, or a point source within the medium, and whether the medium is effectively frozen or undergoes temporal changes.

The next section introduces without any derivations the theory of the various ISS phenomena mentioned above.

THEORY: THE CORRELATION FUNCTIONS

The theory of ISS is concerned with a scattering medium extended along the propagation direction. In what follows a plane wave is assumed incident on such a medium; as it propagates the wave is scattered and is then observed at a point within the medium. This correctly models the observations of small-diameter extragalactic radiosources. However, pulsars are better described as point sources within the medium; the theory for spherical waves is more difficult and has not been solved in general. Before the theoretical results are applied to the observations, estimates of the correction from plane to spherical waves is presented.

Consider a plane monochromatic wave travelling in the $+z$ direction incident on a weakly irregular medium that fills the half-space $z > 0$. The medium is an irregular plasma with electron density $N_T(\mathbf{r})$ at position $\mathbf{r} = (x, y, z)$ such that the observing frequency f is everywhere much greater than the plasma frequency, and so the total refractive index $n_T(\mathbf{r}) = 1 - r_e\lambda^2 N_T(\mathbf{r})/2\pi$, where λ is the free space wavelength and r_e is the classical electron radius. We write the fluctuation in refractive index as

$$n(\mathbf{r}) = -r_e\lambda^2 N(\mathbf{r})/2\pi, \tag{1}$$

where $n = n_T - \langle n_T \rangle$, $N = N_T - \langle N_T \rangle$, and $\langle\ \rangle$ denotes an assembly average, which is equivalent to an (x, y) spatial average since the fluctuations are assumed to be statistically homogeneous in the (x, y) plane. In particular,

$$R_N(\mathbf{r}, z_1) = \langle N(\mathbf{r}_1)N(\mathbf{r}_1 + \mathbf{r})\rangle \tag{2}$$

depends only on $\mathbf{r}$, with (possibly) a slow separable dependence on z_1.

The following assumptions are also made about the medium and are well justified for the interstellar (and interplanetary) media.

(*a*) The deviations in refractive index $n(\mathbf{r})$ are everywhere much less than one.

(*b*) The smallest scale represented by R_N is much larger than λ. Hence the angular spectrum is very narrow (less than about 10^{-5} radians for interplanetary and interstellar scintillations).

(*c*) The total distance through the medium is much greater than the largest scale in R_N.

We now write the electric field (with a plane monochromatic incident wave) as the real part of $u(\omega, \mathbf{r}) \exp(i\omega t - ikz)$, where u is the complex Fourier component at angular frequency ω $(= 2\pi f = kc)$ at a position $\mathbf{r}$, referenced to the unscattered plane wave. Under the assumptions *a*–*c* above, the wave equation reduces to the "parabolic equation" (e.g. Barabanenkov et al. 1971),

$$-2ik\frac{\partial u}{\partial z} + \nabla^2 u = -2nk^2 u = 4\pi r_e N u, \tag{3}$$

where $k = 2\pi/\lambda$, $\nabla^2 = \partial^2/\partial x^2 + \partial^2/\partial y^2$, and (1) has been used.

Questions concerning the observable properties of the field in a plane z can nearly all be reduced to questions about the various correlation functions of u. We thus define the following, where $\mathbf{s} = (x, y)$ is a vector in the plane:

$$\begin{aligned} \gamma_0 &= \langle u(\omega, \mathbf{r})\rangle, \\ \gamma_1(\mathbf{s}) &= \langle u(\omega, \mathbf{r})u^*(\omega, \mathbf{r}+\mathbf{s})\rangle = \gamma_2(0, \mathbf{s}), \\ \gamma_2(\Delta\omega, \mathbf{s}) &= \langle u(\omega, \mathbf{r})u^*(\omega+\Delta\omega, \mathbf{r}+\mathbf{s})\rangle, \\ \gamma_4(\Delta\omega, \mathbf{s}) &= \langle |u(\omega, \mathbf{r})|^2\, |u(\omega+\Delta\omega, \mathbf{r}+\mathbf{s})|^2\rangle. \end{aligned} \tag{4}$$

Note that because the medium is homogeneous in (x, y) these correlations are only dependent on z and ω in addition to their explicit dependencies.

Differential equations for the propagation of such correlations have been given by various authors (e.g. Tatarski 1971, Uscinski 1968, 1974, Rumsey 1974). The results of Lee (1974) are of particular interest and are followed here, since they include non-zero $\Delta\omega$; his equation (37) is a generalized set of differential equations for such correlations of arbitrary order. His results are valid under the Markov approximation, which neglects the correlation of n in the z direction. The Markov approximation is valid if the largest z scale in $n(\mathbf{r})$ is much smaller than the z scale for a substantial change in $u(\mathbf{r})$. Rumsey (1974) and Lee (1974) have demonstrated that a very narrow angular spectrum, (*b*) above, is a sufficient condition for its validity. It is not even necessary for the irregularities in n to have a normal distribution (Lee 1974).

The equations for γ_0 and for $\gamma_1(\mathbf{s})$ can be solved directly:

$$\begin{aligned}\gamma_0 &= \exp\left[-\lambda^2 r_e^2 \int_0^z A(0)\,\mathrm{d}z'/2\right],\\ \gamma_1(\mathbf{s}) &= \exp\left\{-\lambda^2 r_e^2 \int_0^z [A(0)-A(\mathbf{s})]\,\mathrm{d}z'\right\},\end{aligned} \tag{5}$$

where

$$A(\mathbf{s}) = \int_{-\infty}^{\infty} R_N(\mathbf{s}, \zeta, z')\,\mathrm{d}\zeta.$$

In what follows we consider two models for the inhomogeneities as specified by their spatial power spectrum, which is the three-dimensional Fourier transform of $R_N(\mathbf{s}, \zeta, z')$,

$$M_{3N}(\mathbf{q}, \kappa_z, z') = (2\pi)^{-3} \iint \mathrm{d}\mathbf{s}\,\mathrm{d}\zeta R_N(\mathbf{s}, \zeta, z') \exp(-i\mathbf{q}\cdot\mathbf{s} - i\kappa_z\zeta), \tag{6}$$

where the three-dimensional wavenumber is (q_x, q_y, κ_z). The formulas for a spherically symmetrical power law model are

$$M_{3N} = C_N^2(z')(\kappa^2+\kappa_1^2)^{-\alpha/2};\ \kappa^2 = q^2+\kappa_z^2, \tag{7}$$

$$A(s) = 4\pi^2 C_N^2(z')(s/2\kappa_1)^{(\alpha-2)/2} \mathrm{K}_{(\alpha/2)-1}(\kappa_1 s)\Gamma(\alpha/2)^{-1}, \tag{8}$$

$$\gamma_1(s) = \exp[-\lambda^2 B_P(z) s^{\alpha-2}] \quad \text{for} \quad \kappa_1 s \ll 1, 2<\alpha<4, \tag{9}$$

where

$$B_P(z) = 4\pi^2 r_e^2 \int_0^z C_N^2(z')\,\mathrm{d}z' \Gamma(2-\alpha/2)(\alpha-2)^{-1}\Gamma(\alpha/2)^{-1} 2^{2-\alpha}.$$

The corresponding formulas for a Gaussian model are

$$M_{3N} = G_N^2(z')\exp(-\kappa^2/\kappa_0^2), \tag{10}$$

$$A(s) = G_N^2(z') 2\pi^2\kappa_0^2 \exp(-s^2\kappa_0^2/4), \tag{11}$$

$$\gamma_1(s) = \exp[-\lambda^2 B_G(z) s^2] \quad \text{for} \quad \kappa_0 s \ll 1, \tag{12}$$

where

$$B_G(z) = \pi^2 r_e^2 \kappa_0^4 \int_0^z G_N^2(z')\,\mathrm{d}z'/2.$$

For a power law model with $\alpha \geqq 4$, (9) is not valid and $\gamma_1(s)$ becomes a Gaussian when $\kappa_1 s \ll 1$. For many observations $\gamma_1(s)$ is all that is important, in which case no distinction can be made between power law models with $\alpha \geqq 4$ and the Gaussian. The term *square law structure function* is used to refer to these cases together, since $\int [A(0)-A(s)]\,\mathrm{d}z$ is the structure function for $\int N(x,y,z)\,\mathrm{d}z$.

The above formulas apply for a plane wave incident on the medium at $z = 0$. Ishimaru (1976) gives $\gamma_1(s)$ when the source is a point at $\mathbf{r} = 0$ and the baseline is centered about $(\mathbf{s} = 0, z)$:

$$\gamma_{1\,pt}(s) = \exp\left\{-\lambda^2 r_e^2 \int_0^z [A(0)-A(z's/z)]\,\mathrm{d}z'\right\}. \tag{13}$$

This leads to modified expressions for the constants B_P (9) and B_G (12) as follows:

$$B_{P\,pt} = B_P/(\alpha-1); \qquad B_{G\,pt} = B_G/2, \tag{14}$$

where the level of turbulence C_N^2 or G_N^2 is assumed independent of distance.

ANGULAR SCATTERING

Interferometer Observations

A radio interferometer with a projected baseline $\mathbf{s}$ measures the time-averaged product of predetection voltages from two spaced antennas. If both antennas are much smaller than the transverse scale in u and the time or bandpass averaging can be regarded as equivalent to an assembly average, then the resultant fringe amplitude is an estimate of $\gamma_1(\mathbf{s})$, which is often called the mutual coherence function. This second assumption, however, must be examined carefully.

Baselines of relevance to ISS (and IPS) are 100–5000 km, typical of VLBI experiments. In such observations the signals over bandwidth B are recorded (in one-bit format). The two data tapes are subsequently played back into a correlator in which the products are formed. A variable delay and variable local oscillator frequency are introduced into one signal and adjusted to take out the timing and Doppler differences due to the earth's motion. A small residual frequency offset generates "fringes," whose amplitude and phase can be estimated over time interval T. These define the complex visibility function:

$$\Gamma(t_1,\mathbf{s}) = \int_{t_1}^{t_1+T} v(\mathbf{r},t)v^*(\mathbf{r}+\mathbf{s},t)\,\mathrm{d}t, \tag{15}$$

where

$$v(\mathbf{r},t) = \int_{-\infty}^{\infty} H(\omega-\omega_0)N(\omega,\mathbf{r})u(\omega,\mathbf{r})\exp(i\omega t)\,\mathrm{d}\omega.$$

$H(\omega-\omega_0)$ is the bandpass transfer function of width $2\pi B$ centered at ω_0; $N(\omega,\mathbf{r})$ is the Fourier component at ω of the source white noise fluctuations observed at $\mathbf{r}$; $u(\omega,\mathbf{r})$ is the electric field response for each incident plane monochromatic wave. In practice $u(\omega,\mathbf{r})$ is also a function of time as the medium drifts and/or changes. If the integration time T is much longer than the time for such changes in u, then the time average in (15) is equivalent to an assembly average. If the reverse is true then it follows that

$$\Gamma(t_1, \mathbf{s}) \approx \int d\omega \, |H(\omega-\omega_0)|^2 N(\omega, \mathbf{r}) N^*(\omega, \mathbf{r}+\mathbf{s}) u(\omega, \mathbf{r}) u^*(\omega, \mathbf{r}+\mathbf{s}), \tag{16}$$

where u is typical of the time t_1. The $d\omega$ integral will include a very large number of stochastic fluctuations of $N(\omega, \mathbf{r})N^*(\omega, \mathbf{r}+\mathbf{s})$, which will approximate $\langle N(\omega, \mathbf{r})N^*(\omega, \mathbf{r}+\mathbf{s})\rangle$. Similarly the product of u's will approximate the required assembly average if there are many independent flucutations of $u(\omega, \mathbf{r})$ across the bandwidth $2\pi B$. The frequency dependence of u is characterized by the width $\Delta\omega_c$ of $\gamma_2(\Delta\omega, 0)$; so the interferometer output will approximate $\gamma_1(\mathbf{s})$ if $2\pi B \gg \Delta\omega_c$. Estimates of $\Delta\omega_c$ are presented in a subsequent section and these conditions are discussed by Cohen & Cronyn (1974). If the above conditions do apply, then $\Gamma_1(t_1, \mathbf{s}) \approx V(\mathbf{s})\gamma_1(\mathbf{s})$, where $V(\mathbf{s})$ is the source visibility function. The derived brightness distribution would then be the convolution of the source brightness [Fourier transform of $V(\mathbf{s})$] and the apparent distribution due to scattering,

$$P(\boldsymbol{\theta}) = \int d\mathbf{s} \, \gamma_1(\mathbf{s}) \exp(-ik\boldsymbol{\theta}\cdot\mathbf{s})/4\pi^2. \tag{17}$$

Observations

There have been two sets of experiments with sufficient frequency averaging and from which $P(\boldsymbol{\theta})$ has been studied. The first is the series of VLBI observations of the Crab nebula pulsar PSR 0531-21. Mutel et al. (1974) summarize these observations as a graph of apparent angular width θ_d of the assumed circular Gaussian form of $P(\theta)$ against wavelength. They concluded that $\theta_d \propto \lambda^{2.05 \pm 0.25}$. The theoretical form of $P(\theta)$ is Gaussian, scaling as λ^2 only for the square law structure function models discussed in the previous section. In particular the Gaussian spectrum leads to

$$P_G(\theta) \propto \exp(-\theta^2/\theta_{SG}^2); \quad \theta_{SG} = \lambda^2 (B_{Gpt})^{1/2} \pi^{-1} \propto \lambda^2 z^{1/2}. \tag{18}$$

For a power law exponent $2 < \alpha < 4$ the shape of $P(\theta)$ is not exactly Gaussian so we need an algorithm to relate the effective Gaussian angular width with Equation (9) for $\gamma_1(\mathbf{s})$. In an idealized observation $\gamma_1(\mathbf{s})$ is estimated at several values of the baseline $\mathbf{s}$ and the angular width inferred by fitting an assumed Gaussian visibility $\exp(-k^2 s^2 \theta_{SP}^2/4)$ to the data. If the data were actually given by the power law form (9) and the fitted Gaussian equalled it at, say, the e^{-1} level, the inferred scattering radius θ_{SP} would be

$$\theta_{SP} = \pi^{-1}\lambda(\lambda^2 B_{Ppt})^{1/(\alpha-2)} \propto \lambda^{\alpha/(\alpha-2)} z^{1/(\alpha-2)}. \tag{19}$$

Mutel et al. (1974) used such an argument to conclude that the Crab observations were consistent with a density spectrum model of a Gaussian or power law with $\alpha \geqq 3.5$. However, in many actual observations (e.g. Vandenberg 1974) only one value of the baseline was available ($s = b$). Thus equating (9) to the Gaussian visibility yields $\theta_{SP} = \lambda^2 (B b^{\alpha-4})^{1/2} \pi^{-1}$. Thus for fixed b a range of wavelengths should give $\theta_{SP} \propto \lambda^2$ precisely, for all values of $\alpha > 2$, so that no conclusions about

α could be drawn. W. A. Coles (private communication, 1976) pointed to this difficulty and has analyzed the data from the Crab in a more systematic fashion. He plots the observed estimates of log $[1/\gamma_1(s)]$ against the actual baselines (s) used in each experiment and, comparing the results with (9), he concludes that the data allow any $\alpha \gtrsim 3$.

The second investigation of angular scattering comes from the extensive survey of angular diameters at 81.5 MHz by Readhead & Hewish (1972, 1974) and at 151 MHz by Duffett-Smith & Readhead (1976) using IPS measurements. Their data are shown in Figure 1 from which they deduce the apparent diameter θ_e (at e^{-1}) for an extragalactic point source viewed through the interstellar medium in a direction normal to the Galactic plane. They found $\theta_e = 0.15 \pm 0.05''$ at 81.5 MHz. If the interstellar medium is assumed to be homogeneous, this result can be used to predict the angular diameter for the Crab pulsar and compared with the observed value. As in subsequent discussions pulsars are assumed to be point sources within a homogeneous Galactic disc distribution of irregularities; the half-thickness of the disc is taken as 400 pc (corresponding to the 13 pc cm^{-3} and average electron density 0.03 cm^{-3} assumed by Duffett-Smith & Readhead normal to the disc).

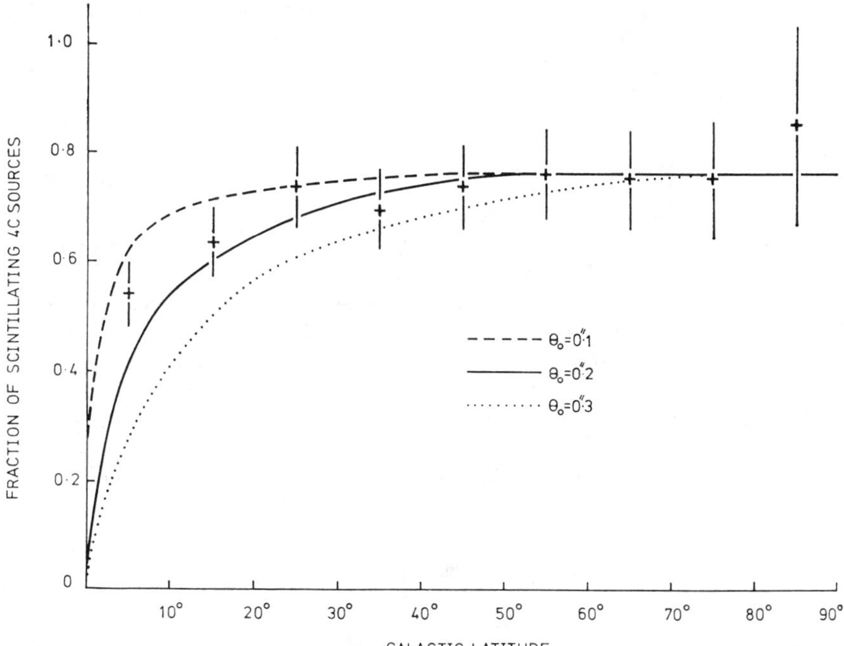

Figure 1 Fraction of strongly scintillating 4C sources at 81.5 MHz in each 10° latitude interval (Duffett-Smith & Readhead 1976). The curves assume a uniform disc distribution of irregularities and are indexed by the apparent diameter of a source observed at right angles to the Galactic plane. A somewhat modified interpretation is given in the Discussion.

Assuming a 2 kpc distance for the Crab, (19) and (18) lead to predicted full widths at half-maximum of $0.19 \pm 0.05''$ and $0.25 \pm 0.05''$ for the Kolmogorov and square law structure function cases respectively. These are in fair agreement with the values $0.2 \pm 0.1''$ measured by Bell & Hewish (1967) at 81.5 MHz and $0.15 \pm 0.01''$ measured by Armstrong et al. (1973) at 73.8 MHz.

Visibility Scintillations

Another aspect of interferometer observations occurs when the time and frequency averaging discussed above are not equivalent to the assembly average of scattering media. In such cases the fringe visibility $\Gamma(t_1, \mathbf{s})$ equals that for the source, but multiplied by a slowly varying amplitude and phase due to variations in u. The theory has been presented by Cronyn (1972) for scattering in a thin screen. In the weak scattering limit he shows that the phase flucutation spectrum is proportional to the density spectrum modified by a filter function that suppresses spatial wavenumbers $q < 2/s$ for a baseline s in the direction in which the medium drifts. Thus for a power law density spectrum the phase fluctuations are dominated by wavenumbers $\sim 1/s$ and, if the drift speed is about v, typical time scales for phase fluctuations are about s/v. This leads to a time scale of seconds for IPS and minutes for ISS. Mutel (1975) reports some observations for the former case and extends the analysis to an extended weak scattering medium. It is to be hoped that VLBI observations of pulsars will be analyzed for ISS visibility scintillations, though additional theoretical developments for strong scattering are still necessary.

PULSE BROADENING

It was recognized early in the investigation of pulars that propagation through the irregular interstellar medium would lengthen the observed pulse widths (Salpeter 1969). In the notation of the previous section, the predetection voltage from a single antenna has a complex envelope function $v(\mathbf{r}, t)$ as given in (15), where $H(\omega - \omega_0)$ is the bandpass response (only non-zero for a bandwidth $2\pi B$ centered on ω_0); $N(\omega)$ is the Fourier component at ω of the emitted signal, which for a perfect impulse is $N(\omega) = 1$. The detected signal is $J(\mathbf{r}, t) = |v(\mathbf{r}, t)|^2$ which if averaged over many pulses gives

$$\langle J(t) \rangle = \iint \mathrm{d}\omega_1 \, \mathrm{d}\omega_2 \, H(\omega_1 - \omega_0) H^*(\omega_2 - \omega_0) \gamma_2(\omega_1 - \omega_2, 0) \exp\left[i(\omega_1 - \omega_2)t\right]. \tag{20}$$

If $\gamma_2(\omega_1 - \omega_2, 0)$ depends only on the difference $\omega_1 - \omega_2$ within the bandwidth, then integration over ω_1 yields the convolution (*):

$$\langle J(t) \rangle = G_1(t) * |h(t)|^2, \tag{21}$$

where

$$G_1(t) = \int_{-\infty}^{\infty} \gamma_2(\Delta\omega, 0) \exp(i\Delta\omega t) \, \mathrm{d}(\Delta\omega),$$

$$h(t) = \int_{-\infty}^{\infty} H(\omega - \omega_0) \exp(i\omega t)\, d\omega.$$

In practice the scattered pulse shape $G_1(t)$ is also broadened by dispersion across the bandwidth caused by the total electron content along the line of sight (dispersion measure) and by the emitted pulse shape and by the post-detector smoothing (see Hankins & Rickett 1975 for a discussion of the instrumental effects).

Lee & Jokipii (1975) developed the theory of pulse broadening (including dispersion, which is represented by the dependence of average refractive index on ω). Their paper presents an elegant theory derived from solving for $\gamma_2(\Delta\omega, \mathbf{s})$. Several other authors (e.g. Lovelace 1970, Cronyn 1970, Uscinski 1974) have contributed to the theory, but have relied on a variety of more restrictive assumptions than Lee and Jokipii. We thus follow the latter authors whose equation (22a) gives a differential equation for the propagation of $\gamma_2(\Delta\omega, \mathbf{s})$ when the fractional bandwidth $\Delta\omega/\omega_0$ is small; in our notation it becomes

$$\frac{\partial \gamma_2}{\partial z} + \frac{ic\Delta\omega}{2\omega_0^2} \nabla^2 \gamma_2 + \lambda^2 r_e^2 \left[A(0)\left(1 + \frac{\Delta\omega^2}{2\omega_0^2}\right) - A(\mathbf{s}) \right] \gamma_2 = 0. \tag{22}$$

The Thin Screen Case

The ∇^2 term represents the effect of diffraction, while the term in $\Delta\omega^2$ represents the dispersive nature of the refractive index fluctuations. Lee & Jokipii solve (22) for the thin screen case by first ignoring the diffraction term within the screen (of thickness L); then with $A = 0$ in free space beyond the screen, (22) is solved with the boundary conditions found for the screen. Their result at z beyond the screen is

$$G_1(t) = \exp(-t^2/t_R^2) * G_2(t), \tag{23}$$

where

$$G_2(t) = P\left[\theta = \left(\frac{2ct}{z}\right)^{1/2}\right] \quad \text{for} \quad t \geqq 0,$$

$$= 0 \quad \text{for} \quad t < 0$$

and

$$t_R = [A(0)L/2]^{1/2} \lambda^2 r_e / 4\pi c.$$

The first Gaussian term represents the effect of slow wandering in the arrival time averaged over the assembly of scattering media and is discussed below. $G_2(t)$ is identical to that obtained by ray optics (Cronyn 1970) where the extra delay for a scattered ray at angle θ is $z\theta^2/2c$.

Also obtainable from (40) of Lee & Jokipii is an expression for the pulse shape as detected in the visibility function of an interferometer of baseline $\mathbf{s}$. This puts mathematically the idea that scattered rays arrive at angles that increase with time after the unscattered component. Thus an interferometer would see a shorter pulse than a single antenna, since it resolves out the delayed signals. $G_2(\mathbf{s}, t)$ is the scattered pulse shape for the visibility function over baseline $\mathbf{s}$.

$$G_2(\mathbf{s},t) = P(\theta)J_0(ks\theta)\,|_{\theta=(2ct/z)^{1/2}}. \tag{24}$$

The Fourier transform of (24) with respect to t gives an expression for $\gamma_2(\Delta\omega, \mathbf{s})$ in the thin screen case.

For the particular spectral models (17) and (9) or (12) can be substituted into (23) to give the scattered pulse shape $G_2(t)$. For the square law structure function models (12) this gives the truncated exponential, much discussed in the literature,

$$\begin{aligned} G_2(t) &= \exp(-t/t_D) \quad \text{for} \quad t \geqq 0, \\ &= 0 \quad \text{for} \quad t < 0, \end{aligned} \tag{25}$$

where $t_D = zB_G\lambda^4/2\pi^2 c$, z is the distance to the screen, and B_G is given by (12) with the screen thickness L in place of z. This equation has been the basis of much first-order discussion of pulsar broadening. Before comparison with the data the screen model must be related to the real extended medium. This is done by assuming that the pulsar is at distance z and the extended medium of thickness $z(=L)$ is collapsed to a thin screen located at $z/2$. A further correction is necessary to account for the fact that the pulsar is a point source, whereas the above theory is for a plane wave. This is done simply for ray optics as for example in Figure A1 of Duffett-Smith & Readhead (1976); we assume that this path length ratio (1/2) between point source and plane wave applies to (25). Finally, converting (18) to the apparent half-power diameter θ_d for a point source, the thin screen theory gives $t_D = 0.18z\theta_d^2/c$. Similar formulas can be written down for the Kolmogorov or other power law spectra ($2 < \alpha < 4$), but the pulse shapes cannot be expressed in simple functions. Of interest are the asymptotic expansions for the pulse shape $G_2(t)$, which is proportional to $t^{-\alpha/2}$ at large t; $\gamma_2(\Delta\omega, 0)$ is proportional to $\Delta\omega^{-1}$ at large $\Delta\omega$ independent of α (as for the square law structure function).

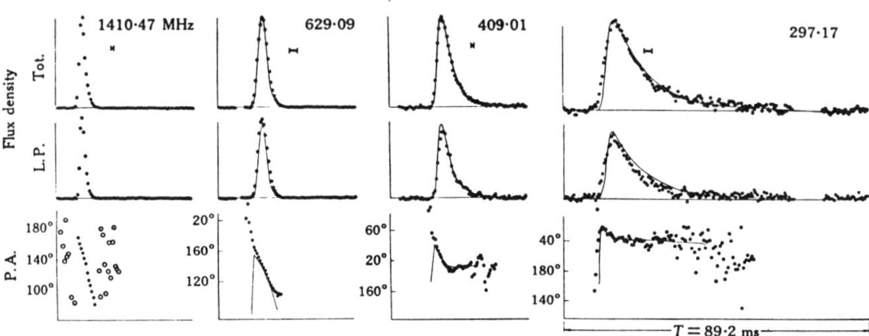

Figure 2 Average values of total intensity, linearly polarized intensity (L.P.), and polarization angle (P.A.) of PSR 0833-45 for the frequencies (in MHz). The filled circles are observed values; the curves represent values obtained by convolution of the 1410 MHz data with a truncated exponential of width—$(300/f_{\text{MDz}})^4 \times 9.4$ msec—and also with a resolution function of width indicated by the horizontal bar. The time scale is the same for each plot and is shown by the 89.2 msec pulse period (from Komesaroff, Hamilton & Ables 1972).

Figure 2 (Komesaroff, Hamilton & Ables 1972) demonstrates the success of the thin screen scattering model in describing the pulses from PSR 0833-45. The intensity and linear polarization profiles are shown at four frequencies. The lines are the thin screen theory, $t_D \propto \lambda^4$; however, good agreement would also be obtained by assuming $t_D \propto \lambda^{4.4}$ appropriate to the Kolmogorov model.

Delay Fluctuations

Before proceeding with a more precise theory for an extended medium for comparison with the observations, we return to discuss the delay fluctuation term $\exp(-t^2/t_R^2)$ in (23). In practice pulse shapes are computed from a time average over T from pulses emitted repetitively at the pulsar period, and not from $\langle\ \rangle$ an assembly average of scattering media. We may estimate the effect of a finite time average as follows.

The term $\exp(-t^2/t_R^2)$ is the solution to (22) with the diffraction term ignored and $s = 0$, and so represents the smearing caused by fluctuations in the dispersion measure. To examine the instantaneous delay we look at (3) and, ignoring diffraction, obtain a simple solution for the phase deviation of u at position (x, y, z), $\varphi(x, y) = \int_0^z r_e \lambda N(x, y, z')\,dz'$. The associated group delay deviation is $\tau(x, y) = \partial\varphi/\partial\omega = \varphi(x, y)/\omega$. In order to study the time variations we regard the medium as moving rigidly at speed v in the x direction; thus the delay at time t will be $\tau(t) = -\varphi vt, y)/\omega$. In computing an average pulse shape over interval t_1 to $t_1 + T$, the average arrival time $\tau_a(t_1, T) = T^{-1}\int_0^T \tau(t_1+t)\,dt$ will be the time reference. Fluctuations $\tau(t) - \tau_a(t_1, T)$ will cause the pulse to be smeared. An estimate of effective smearing time is t_{RT} defined by

$$t_{RT}^2 = T^{-1}\int_0^T \langle[\tau(t+t_1) - \tau_a(t_1, T)]^2\rangle\,dt, \tag{26}$$

and after some reduction and using the definition of $A(s)$ in (5),

$$t_{RT}^2 = r_e^2\lambda^2 z\omega^{-2}T^{-2}\int_0^T [A(0)T - 2A(vT)(T-t)]\,dt, \tag{27}$$

and hence for a power law from (8) and (9):

$$t_{RT} \approx \lambda^2[2B_P/\alpha(\alpha-1)]^{1/2}(vT)^{(\alpha-2)/2}/2\pi c, \quad \text{if} \quad vT \ll \kappa_0^{-1} \tag{28}$$

and

$$t_{RT} \approx \lambda^2[A(0)z/2]^{1/2} r_e/4\pi c, \quad \text{if} \quad vT \gg \kappa_0^{-1}.$$

The importance of delay fluctuations is estimated by comparing t_{RT} and t_D, leading to

$$t_{RT}/t_D \approx 7{\cdot}4[\alpha(\alpha-1)/2]^{-1/2}[vT/(\lambda z)^{1/2}]^{(\alpha-2)/2}[s_c/(\lambda z)^{1/2}]^{(6-\alpha)/2}, \tag{29}$$

where s_c is the correlation scale for the field defined by $\gamma_1(s_c) = 1/e$. In all published observations from which a scattered pulse shape has been estimated (i.e. $T \sim 10^3$ sec, $v \lesssim 10^5$ m sec^{-1}) this ratio is always much less than unity. Consequently we can ignore the term $\exp(-t^2/t_R^2)$ in expressions for $G_1(t)$.

The Extended Medium

Whereas the thin screen case led to some analytical results, the more interesting case of an extended medium must be solved numerically. Nevertheless both diffraction [∇^2 term in (22)] and dispersion in the irregularities [$\Delta\omega^2$ term in (22)] are occurring in the extended medium as in the screen. However, these two effects cannot be separated and (22) must be solved with both terms present. As for the screen, the assembly average pulse shape is the convolution $G_1(t) = G_2(t) * \exp(-t^2/t_R^2)$. However, a discussion similar to the one above can be applied to show that the delay fluctuations can be ignored in practice.

Lee & Jokipii (1975) present numerical solutions for $G_2(t)$ for the square law structure function and for the Kolmogorov ($\alpha = 3.67$) models. These are essentially indistinguishable except in the tail where the Kolmogorov falls slightly more slowly. Williamson (1972) presented a ray theory of pulse broadening, in which the probability distribution of angular scattering was assumed to be Gaussian (corresponding to the square law structure function models) and the medium was either concentrated in a screen or extended. His theoretical curve is compared with that of Lee & Jokipii in Figure 3. Surprisingly they are significantly different in the rising portion; the ray theory pulse rises much more slowly than that of the wave theory. In Williamson's notation the wave theory rise time is about one-half that of the ray theory for the same decay time.

Williamson's calculation is for a point source in the medium, whereas Lee & Jokipii's is for a plane wave incident on the medium; their different pulse shapes could be due to this difference. If this is not the reason, the discrepancy could throw

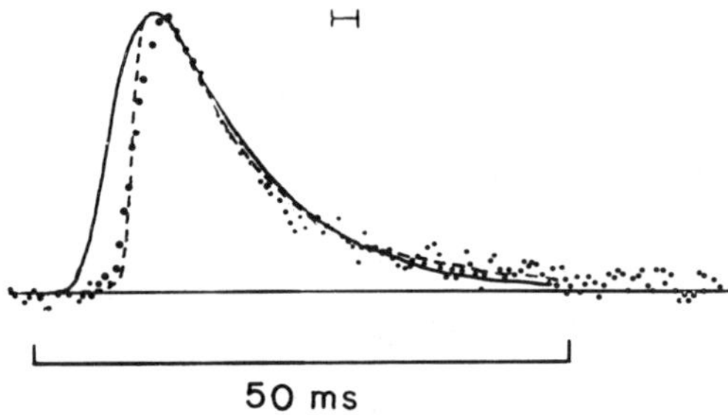

Figure 3 Pulse shape of PSR 0833-45 at 327 MHz from Komesaroff, Hamilton & Ables (1972), plotted as filled circles. The solid line is the theoretical curve from the ray calculation of Williamson (1974) for uniform Gaussian inhomogeneities (this curve includes the effect of instrumental smoothing). The dashed line is the theoretical curve from the wave theory calculation of Lee & Jokipii (1975) for uniform Kolmogorov inhomogeneities (this curve does not include the instrumental smoothing represented by the short horizontal bar). The theoretical curves were adjusted to fit the decaying portion of the pulse. The difference between the Gaussian and Kolmogorov curves in the wave theory is insignificant compared to the discrepancy between the wave and ray theories.

doubt on Williamson's (1974) conclusion. He compared observations of the Vela pulsar (PSR 0833-45) by Komesaroff, Hamilton & Ables (1972) with his theory and found that the observed rise time was much shorter than that predicted for an extended medium. Only by confining the medium to a thin screen could he match the data. Figure 3 shows the observations as well as the theoretical curves of Williamson and Lee and Jokipii for a uniform extended medium. The latter fits the data well within the experimental errors, implying that the scattering for this pulsar could well be extended over much of the line of sight. Similarly the data for PSR 1858+03, 1933+16, and 1946+35 from Lang (1971a) are consistent with Lee and Jokipii's extended medium model. The Crab pulsar is discussed separately below.

One of the goals of this paper is to compare the observations with a quantitative model for the uniform extended medium. We therefore give formulas for the width of the scattered pulse from the numerical solutions of Lee & Jokipii (1975). The time t_D between pulse maximum and $1/e$ on the decay is taken from their Figures 2 and 4 and expressed in our notation:

$$t_D = 0.42z\lambda^4 B_G/4\pi^2 c, \quad \text{Gaussian}; \tag{30}$$

$$t_D = 0.32z\lambda^2(\lambda^2 B_P)^{1.2}/4\pi^2 c, \quad \text{Kolmogorov}. \tag{31}$$

Here z is the pulsar distance; the plane wave values for B_G and B_P from (12) and (9) should be used because the numerical factors have already been reduced by one half to convert from plane wave to point source ray paths.

The Crab Pulsar

Pulse broadening from the Crab pulsar has been extensively studied by Rankin and Counselman in a series of papers on the scattered pulse shape versus frequency and time since 1969. Their chief conclusions were that there are two components to the scattering; one varies over the course of years and the other is steady. In Rankin & Couselman (1973) they suggested that the varying component is caused by scattering in the Crab nebula, which changes as the nebula expands and re-arranges; the steady component is identified with general interstellar scattering extended along the line of sight.

In a recent paper Isaacman & Rankin (1977) have reanalyzed all the Crab data from 1969–1974, basing their analysis on a Kolmogorov spectrum instead of the Gaussian assumption of the previous papers. Their analysis of the frequency dependence of scattered pulse profile widths yielded $f^{-4.4\pm0.3}$ (in fact they analyzed the phases of the Fourier components in the profiles at frequencies from 73.8 to 430 MHz). This frequency-scaling is that of the Kolmogorov model (31) and yet is still consistent with the Gaussian (30). More evidence for the Kolmogorov model came from the 73.8 MHz data, which in previous (Gaussian) analyses had been consistently discrepant. Isaacman & Rankin modeled the net scattering by convolving two Kolmogorov screen components with $1/e$ pulse widths $S_1(f/10^8)^{-4.4}$ and $S_2(f/10^8)^{-4.4}$. Their Figure 6 shows the variations in parameters S_1, S_2, and dispersion measure from 1969 to 1974. Whereas they point out that their data is just consistent with a variable thick slab model in which the effective S_1/S_2 would be constant, they favor the earlier model of a nebular origin for a time variable S_2 and an interstellar origin for a constant S_1 ($=2.5 \pm 1.5$ msec). In their analysis they also found that the observed pulse rise times were significantly faster than the extended model of Williamson (1972). However, as discussed in the previous section the wave theory of Lee & Jokipii (1975) gives rise times about one half of the extended medium ray theory of Williamson. Thus the steady component (S_1) is still compatible with extended interstellar scattering.

We may quantitatively compare the angular diameter and pulse broadening by combining equations (31) and (18), which is first converted into θ_d, the apparent half-power diameter of a point source at distance z:

$$t_D = 0.1z\theta_d^2/c. \qquad (32)$$

This equation, which applies to the Kolmogorov model, is only accurate to about 20%, since the method for converting from plane wave to point source is approximate. Scaling the 111 MHz observations of Vandenberg (1974) to 100 MHz (as $f^{-2.2}$), and with $z = 2$ kpc, we predict $t_D = 3.7$ msec in good agreement with the observed $S_1 = 2.5$ msec; this further supports the extended medium model for the steady component.

The angular broadening from the variable nebular scattering should be very much less because the nebula is so close to the source. This agrees with the conclusions of Vandenberg, who also reported no variations in visibility comparable to the variations of S_2, and no decrease of visibility across the scattered profile to be expected (24) for a single scattering location. The same picture emerges from the constancy of the pulsar angular diameter observed by IPS at 73.8 MHz (Kaufman 1976) between 1972 and 1974, while the total pulse broadening varied by up to one hundred times in 1974 (Lyne & Thorne 1975). The apparent diameter increase of three times claimed by Brown (1976) is in conflict both with this picture and with the observations of Kaufman (1976).

Discussion

To summarize the long and complicated story for the Crab pulsar, the pulse broadening, which is well described by the Kolmogorov model, is dominated by variable scattering presumed to be in the Crab nebula. On the other hand, the angular broadening is dominated by a steady interstellar scattering that is compatible with either the extended medium Kolmogorov or square law structure function models. The success of the Kolmogorov model for the pulse broadening thus does not reflect on the general interstellar medium.

In order to test the extended Kolmogorov model further we make use of the various pulse broadening observations to deduce an effective C_N^2 for each pulsar. It is assumed that the medium is homogeneous between the pulsar (at a distance taken from Table III of Taylor & Manchester 1975) and the earth, and (31) and (9) are used to derive C_N^2. The values are given in Table I and plotted in Figure 6. However, we postpone discussion until the other estimates of C_N^2 are introduced in the next section.

SCINTILLATIONS OF INTENSITY VERSUS TIME, SPACE, AND FREQUENCY

Theory

It was not until the discovery of pulsars that radio sources were found of sufficiently small angular diameter to show intensity scintillations in the interstellar medium. In practice, the intrinsic fluctuations in pulsar signals confuse the observations but are sufficiently smoothed by averaging over about one minute to reveal the scintillation variations, with typical time scales of about ten minutes. Expressing the smoothed intensity in a receiver bandwidth B mathematically yields $I_B(\mathbf{r}) = |v(\mathbf{r}, t)|^2 * p(t)$, where $p(t)$ represents the time averaging and v is given by (15) in which $N(\omega)$ comes from the pulsar. The smoothing is equivalent to source assembly averaging, $\langle |N(\omega)|^2 \rangle$ is assumed flat over the bandpass, and hence we find that intensities add:

$$I_B(\mathbf{r}) = \int_{-\infty}^{\infty} |H(\omega_1 - \omega_0)|^2 \, |u(\omega_1, \mathbf{r})|^2 \, \mathrm{d}\omega_1. \tag{33}$$

We are concerned with the variations in I_B, when the movement of the scattering medium convects the intensity pattern past a fixed observer. We describe the scintillations by the correlation function,

$$R_{IB}(\mathbf{s}) = [\langle I_B(\mathbf{r}) I_B(\mathbf{r}+\mathbf{s}) \rangle - \langle I_B \rangle^2] \langle I_B \rangle^{-2}, \tag{34}$$

and, in particular, by the scintillation index $m = [R_{IB}(0)]^{1/2}$. From (33) and (4) we obtain

$$R_{IB}(\mathbf{s}) = \iint \mathrm{d}\omega_1 \, \mathrm{d}\omega_2 \, |H(\omega_1 - \omega_0)|^2 \, |H(\omega_2 - \omega_0)|^2 \gamma_{\Delta I}(\omega_1, \omega_2, \mathbf{s}), \tag{35}$$

where

$$\gamma_{\Delta I}(\omega_1, \omega_2, \mathbf{s}) = \gamma_4(\omega_1 - \omega_2, \mathbf{s}) - \gamma_1(0)^2.$$

In weak scintillation ($m \ll 1$) several authors (e.g. Tatarski 1961) have shown that we can add the intensity pattern of successive thin layers in the medium. In such a case the cross-spectrum, defined by

$$\tilde{\gamma}_{\Delta I}(\omega_1, \omega_2, \mathbf{q}) = \int_{-\infty}^{\infty} \mathrm{d}\mathbf{s} \, \gamma_{\Delta I}(\omega_1, \omega_2, \mathbf{s}) \exp(i\mathbf{q} \cdot \mathbf{s})(2\pi)^{-2}, \tag{36}$$

is given by

$$\tilde{\gamma}_{\Delta I}(\omega_1, \omega_2, \mathbf{q}) = \frac{32\pi^3 r_e^2 c^2}{\omega_1 \omega_2} \int_0^z \mathrm{d}z' \sin\left[\frac{q^2(z-z')c}{2\omega_1}\right] \sin\left[\frac{q^2(z-z')c}{2\omega_2}\right] M_{3N}(\mathbf{q}, 0, z'), \tag{37}$$

which for $\omega_1 = \omega_2$ yields the standard Born result in which the $\sin^2$ term (Fresnel filter) cuts off wavenumbers below about $[2\omega/c(z-z')]^{1/2}$. This result is valid for a plane incident wave if m^2 calculated from it is small compared to one. Following Rumsey (1976) we call this quantity U; it is equal to m^2 when the Born approximation is valid. For a point source at distance z, the term $(z-z')$ in (37) are replaced by $(z-z')z'/z$. If we do this and apply the extended homogeneous power law model for M_{3N} we find

$$U_P = 8\pi^2 r_e^2 \lambda^2 \overline{C}_N^2 z^{\alpha/2} (\lambda/2\pi)^{(\alpha-2)/2} \sin(\pi\alpha/4)\Gamma(2-\alpha/2)\Gamma(\alpha/2)^2/(\alpha-2)\Gamma(\alpha). \tag{38}$$

For a plane wave the equivalent U_P is 2.5 times greater (when $\alpha = 3.67$).

In practice most ISS observations have been made below 1 GHz, yielding scintillation indices near unity. Only the data of Downs & Reichley (1971) at 2.3 GHz and of Backer (1975) at 2.7 GHz have revealed weak scintillation. From Backer's Figure 7 of m versus dispersion measure (DM), we infer that $m \approx 0.5$ at $DM = 3.5$ pc cm^{-3}. Hence using (38) with $U_P = m^2 = 0.25$, a frequency of 2.5 GHz and a distance $z = 3.5 \times 10^{18}$ m ($\langle N_T \rangle = 0.031$ cm^{-3}), another estimate of the local C_N^2 can be obtained and is added to Table 1.

Strong Scintillations

As U increases, the index m approaches unity, and for $U \gg 1$ all spectral models predict $m = 1$, providing that the bandwidth integration represented in (35) does not limit the scintillations. In order to model the bandwidth suppression we need an expression for $\gamma_{\Delta I}$ at large U. In the limit $U \gg 1$, the statistics of the fluctuating field $u(\omega, \mathbf{r})$ tend toward normal (as a consequence of the central limit theorem), the magnitude $|u|$ takes a Rayleigh distribution, and hence the intensity $|u|^2$ takes an exponential distribution. Under this "Rayleigh limit" all the higher-order correlations of u can be expressed in terms of the second-order correlations. In particular for $U \gg 1$ we find

$$\gamma_{\Delta I}(\omega_1, \omega_2, \mathbf{s}) \approx |\gamma_2(\omega_1 - \omega_2, \mathbf{s})|^2, \tag{39}$$

which, if applied to (35), will yield an expression for the scintillation index $m_B = [R_{IB}(0)]^{1/2}$ in a bandwidth B. Evidently if B includes many independent variations in $|u|^2$, specified by the width of $\gamma_2(\Delta\omega, 0)$, the index will be suppressed. Observations have been made of the bandwidth B_h such that m_B is reduced to 0.5. This revealed a steep dependence of B_h on λ and on DM (Rickett 1970, Backer 1974); thus we need theoretical expressions for the shape and width of $\gamma_2(\Delta\omega, \mathbf{s})$.

The scintillations of stronger pulsars have been observed more directly with multichannel spectrometers (e.g. Rickett 1970, Ewing et al. 1970). If each spectrum is autocorrelated versus frequency and then averaged, estimates of $\gamma_{\Delta I}(\Delta\omega, 0)$ are obtained (e.g. Rickett 1970, Lang 1971b). As discussed in the previous section, analytical results for $\gamma_2(\Delta\omega, \mathbf{s})$ can be obtained for the thin screen (e.g. equation 24), but numerical solutions must be used for the extended medium. The calculations of Lee & Jokipii (1975) for the extended Kolmogorov model can now be compared with observed $\gamma_{\Delta I}$ curves (J. W. Armstrong and B. J. Rickett, in preparation).

We can characterize the spectral width of the scintillations by the frequency separation for which $(\gamma_{\Delta I}\Delta\omega_I, 0) = 0.5$. In view of (39) and (21) there is evidently an inverse relationship between $\Delta\omega_I$ and the scattered pulse width t_D, as first pointed out by Salpeter (1969). The precise relation will depend on the shapes of $\gamma_{\Delta I}(\Delta\omega, 0)$ and $G_2(t)$. Sutton (1971) showed that for the thin Gaussian screen $\Delta\omega_I t_D = 1$. We can use the Lee & Jokipii (1975) computation to obtain this product for the extended Kolmogorov model,

$$\Delta\omega_I t_D = 0.6. \tag{40}$$

This strictly applies for a plane incident wave, but we assume that the correction to a point source would not change the product. Thus we use (40) and (31) to obtain $\Delta\omega_I$ for a point source in an extended Kolmogorov medium,

$$\Delta\omega_I = 7.5\pi^2 c\lambda^{-2} z^{-1} (\lambda^2 B_P)^{-1.2}. \tag{41}$$

In addition we can relate $\Delta\omega_I$ to B_h from the model of $\gamma_{\Delta I}(\Delta\omega, 0)$. Lovelace (1970) showed that for the thin Gaussian screen $2\pi B_h = 4\pi\Delta\omega_I$. Little (1968) and Rickett (1970) came to similar relationships from less rigorous theory. From Lee (1976) we take a value of $2\pi B_h = 10\Delta\omega_I$ in order to combine B_h observations with those of $\Delta\omega_I$ in a comparison with the extended homogeneous Kolmogorov model.

Comparison of Observations and Theory

Figure 4 shows observations of $\Delta\omega_I$ and t_D^{-1} plotted against dispersion measure (DM) for various pulsars. For $DM \lesssim 20$ $\Delta\omega_I$ decreases roughly as DM^{-2}, but for $20 \lesssim DM \lesssim 400$ the relationship steepens considerably (approaching DM^{-4}). This conclusion was first noted by Sutton (1971). If we take DM as proportional to z

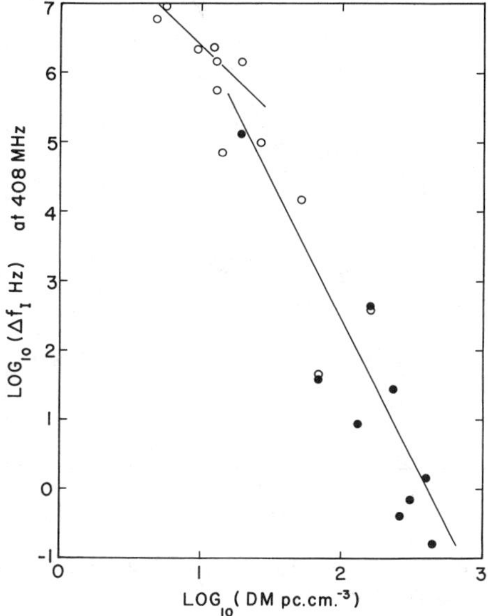

Figure 4 A logarithmic plot of the decorrelation frequency width Δf_I against dispersion measure. The values have been scaled to 408 MHz according to $f^{4.4}$; the open circles are from measurements of Δf_I or B_h, the filled circles are from t_D and (40). The two lines have slopes of -2 and -4 and are only fitted by eye. See Table 1 for the data sources.

(the pulsar distance) the square law structure function models predict $\Delta\omega_I \propto DM^{-2}\lambda^{-4}$, while the power law models predict $\Delta\omega_I \propto DM^{-\alpha/\alpha-2}\lambda^{-2\alpha/\alpha-2}$ [from (41) and (9)]. Thus the slope of -4 at high DM values is at odds with both the Kolmogorov and Gaussian models. Only a power law ($\alpha \sim 2.7$) could match this slope. However, such a value of α would predict $\Delta\omega_I \propto \lambda^{-8}$, which does not agree with the variation $\propto \lambda^{-4.0\pm0.5}$ shown in Figure 5 (from Backer 1974, PSR 0833-45). This λ dependence allows $\alpha \gtrsim 3.5$. Our conclusion is different from that of Little & Matheson (1973), who took the nominal λ^{-4} scaling to require $\alpha \geqq 4$.

The steep decrease of $\Delta\omega_I$ at large DM in fact comes from observations of t_D and the reciprocal relationship (40). If this were wrong, the discrepancy could be explained. However, the theory leaves very little flexibility (the factor 0.6 would

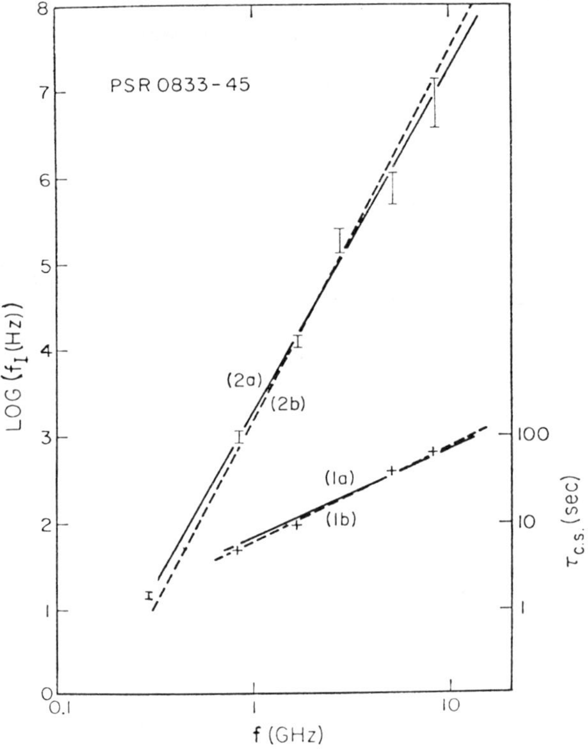

Figure 5 Observed values of the bandwidth for decorrelation of interstellar scintillations of PSR 0833-45. The quantity (denoted by B_h in the text) is plotted as a vertical bar, whose length indicates the range of values observed. The + symbol shows the decorrelation time for the scintillations. The data are from Backer (1974) except for the point at 300 MHz, which is derived from pulse broadening observed by Komesaroff, Hamilton & Ables (1972). The lines have slopes of 4.0 (2a), 4.4 (2b), 1.0 (1a), and 1.2 (1b), corresponding to (*a*) Gaussian and (*b*) Kolmogorov spectra (Lee & Jokipii 1976).

change little with other models). The one experimental determination confirms the relation. For PSR 0833-45 Backer (1974) measured $\Delta\omega_I = 6.9 \times 10^3$ rad sec^{-1} at 837 MHz (derived from a B_h measurement). Scaling the value found by Komesaroff et al. (1972) for t_D from 300 to 837 MHz gives $\Delta\omega_I t_D = 1.07$. We conclude that the steep $\Delta\omega_I/DM$ relation is real and so have to find what other departure from the theory might be responsible.

The most obvious candidate is the homogeneity of the medium; we will examine this by assuming the homogeneous Kolmogorov model and comparing C_N^2 with DM. In these calculations we use distances tabulated by Taylor & Manchester (1975) that are derived from DM but include the estimated effect of H II regions in the line of sight. Observations of $\Delta\omega_I$ by Lang (1971b) are included as well as those derived from B_h by Rickett (1970) and Backer (1974). We have not used observations in which $\Delta\omega_I$ was estimated by eye rather than from a correlation analysis of the spectra. The results are added to Table 1 and discussed in the last section after we have added the data from the pattern structure.

It is worth commenting at this stage on the suggestion of Bonazzola & Celnikier (1975) that the steepening in $\Delta\omega_I$ vs DM might be caused by the transition from weak to strong scattering. Their theory is completely ad hoc and cannot be correctly formalized to predict the observed $\Delta\omega_I \propto DM^{-4}\lambda^{-4}$. As they point out, however, the evidence for strong scattering ($m \simeq 1$) at low DM (~ 10 pc cm^{-3}) and $\lambda \sim 0.50$ m is not very convincing because of possible contamination of ISS with intrinsic fluctuations. It is interesting here that the Kolmogorov spectrum cannot give a narrow $\Delta\omega_I$ ($\ll \omega_0$) in weak scintillation. Combining (41) modified for any α ($2 < \alpha < 4$) and (38) and (9), we find for $U \gg 1$

$$U_P \approx 1.8(\omega_0/\Delta\omega_I)^{(\alpha-2)/2}, \tag{42}$$

where the numerical factor depends weakly on α and is given for $\alpha = 3$. For $U \ll 1$, (37) shows $\Delta\omega_I \sim \omega_0$. Hence if the observed values of $\Delta\omega_I$ ($\ll \omega_0$) at low DM are in weak scattering ($U \ll 1$) then the spectral model cannot be a power law $2 < \alpha < 4$, but must have a sharp cutoff as in the Gaussian model. This may well be a critical test between the Gaussian and power law models. If careful measurements reveal situations in which $m < 1$ and $\Delta\omega_I \ll \omega_0$ the power law models are definitely ruled out. However, a theory still has to be found to explain the high DM behavior $\Delta\omega_I \propto DM^{-4}\lambda^{-4}$.

Spaced Receiver Observations

Soon after the discovery of pulsars, searches were made for differences in intensity variations between two spaced antennas, which would be a hallmark of scintillation. However, Slee, Komesaroff & McCulloch (1968) and Conklin et al. (1968) chose the first discovered pulsar PSR 1919+21 for study. This pulsar has slow intrinsic variations at radio frequencies below about 200 MHz, which masked the ISS in their experiments. Subsequently, other experimenters have observed scintillations with time shifts of 10–100 sec over baselines of several thousand kilometers (Lang & Rickett 1970, Galt & Lyne 1972, Rickett & Lang 1973, Slee et al. 1974).

These experiments successfully detected temporal and spatial differences in the

Table 1 Estimates of C_N^2 for $\alpha = 3.67$

Pulsar (source)	Quantity observed	Value	Freq. (MHz)	Ref.[a]	Distance (pc)	Eq. no.	C_N^2 ($m^{-6.67}$)	DM (pc cm^{-3})	$\langle N_T \rangle$ (cm^{-3})
IPS Survey	θ_d	0.12″	81.5	DS	400	19	2.4×10^{-4}	13	0.03
0531+21	θ_d	0.2″	81.5	BH	2,000	19	2.8×10^{-4}	57	0.028
Cyg X−3	θ_d	2″	408	A	4,000	19	2.4	—	—
Sag A	θ_d	1.5″	960	D	10,000	19	14	—	—
0823+26	t_D	20 ms	40	C	790	31	1.3×10^{-4}	19	0.024
1933+16	t_D	67 ms	111	LA	6,000	31	3.6×10^{-4}	158	0.026
1946+45	t_D	32 ms	318	LA	5,400	31	1.1×10^{-2}	129	0.024
1859+03	t_D	187 ms	318	LA	15,000	31	7.5×10^{-3}	402	0.027
2002+31	t_D	10 ms	318	LA	7,800	31	2.1×10^{-3}	233	0.03
1323−62	t_D	30 ms	750	KA	10,300	31	2.9×10^{-2}	310	0.03
1557−50	t_D	16 ms	750	KA	9,000	31	5.6×10^{-2}	270	0.03
1641−45	t_D	40 ms	750	KA	15,000	31	4.8×10^{-2}	450	0.03
0833−45	t_D	9.4 ms	300	K	500	31	2.5×10^{-1}	69	0.14
0833−45	$B_h/10$	1.1 KHz	837	BA	500	41	2.2×10^{-1}	69	0.14
0823+26	Δf_I	0.5 MHz	318	LB	790	41	1.7×10^{-5}	19	0.024
0834+06	Δf_I	0.5 MHz	318	LB	480	41	4.3×10^{-5}	13	0.027
1133+16	Δf_I	2 MHz	318	LB	180	41	8.2×10^{-5}	4.8	0.027
1237+25	Δf_I	0.7 MHz	318	LB	370	41	5.4×10^{-5}	9.3	0.025
1919+21	Δf_I	0.8 MHz	318	LB	420	41	3.7×10^{-5}	12	0.029
1933+16	Δf_I	0.1 MHz	1420	G	6,000	41	3.8×10^{-4}	158	0.026
0329+54	Δf_I	0.1 MHz	408	R	2,000	41	3.0×10^{-5}	27	0.014
0809+74	$B_h/10$	0.12 MHz	151	R	200	41	4.6×10^{-5}	5.8	0.029
0834+06	$B_h/10$	0.55 MHz	408	R	480	41	9.9×10^{-5}	13	0.027
1749−28	$B_h/10$	15 KHz	408	R	1,700	41	1.9×10^{-4}	51	0.030
2016+28	$B_h/10$	70 KHz	408	R	480	41	5.4×10^{-4}	14	0.029
—	m	0.5	2500	BB	110	38	4.5×10^{-4}	3.5	0.032
1642−03	s_I	8×10^6 m	327	S	160	44	2.6×10^{-3}	36	0.22
1749−28	s_I	8×10^6 m	327	S	1,700	44	2.5×10^{-4}	51	0.031

[a] References for Table 1:
A, Anderson et al. (1972); BA, Backer (1974); BB, Backer (1975); BH, Bell & Hewish (1967); C, Craft (1970); D, Davies et al. (1976); DS, Duffett-Smith & Readhead (1976); G, Guelin, Encrenaz & Bonazzola (1974); K, Komesaroff, Hamilton & Ables (1972); KA, Komesaroff et al. (1973); LA, Lang (1971a); LB, Lang (1971b); R, Rickett (1970); S, Slee et al. (1974).

ISS pattern but have raised more questions than they answered. A basic difficulty is that the pattern scale is large ($\gtrsim 10^4$ km) and the drift speeds (~ 100 km sec^{-1}) are such that in a typical 2–3 hr observing period only 5–100 independent intensity variations are included; thus correlation scales and time shifts are statistically very uncertain. The most effective experiment was that of Galt & Lyne (1972) who observed PSR 0329+54 for 24 hr simultaneously at two antennas. The earth's rotation gave all baseline orientations and so a changing time shift. They thus deduced both a pattern speed (350 km sec^{-1}) and its direction. This very high speed was the first real evidence that some pulsars have a large proper motion. Subsequently, interferometer observations (Anderson, Lyne & Peckham 1975) have confirmed that 0329 is moving very rapidly (170 km sec^{-1}). The discrepancy is probably caused by the influence of temporal changes in the pattern and intrinsic fluctuations, both of which bias ISS velocities upwards. This point is raised by Slee et al. (1974), who discuss and apply a technique for estimating time offsets that gives the velocity component along the baseline even in the presence of temporal and intrinsic variations. Armstrong & Coles (1972) discussed the same technique for interplanetary scintillations.

In spite of this improved analysis our knowledge of the spatial and temporal structure of the ISS pattern remains very sparse. Of the seven pulsars observed by Slee et al. only three gave measurable velocities. More confusing still, one of them (PSR 1642-03) gave apparent velocities that changed in magnitude and direction over a few hours. Similar changes had earlier been detected by Rickett & Lang (1973) for PSR 1133+16. The latter pulsar is now known to have a large proper motion (~ 380 km sec^{-1}) according to Manchester, Taylor & Van (1974). Uscinski (1975) has presented a possible explanation in terms of velocity shear in the medium.

Evidently, the theoretical problem is now very severe: a rapidly moving point source is seen through an irregular medium that is extended and drifting at a range of velocities with respect to the observer. This is an area for future theoretical analysis, in which the effects of refraction from the larger scales as well as diffraction will also have to be included. As pointed out by Rickett & Lang (1973), such situations may also explain the drifting spectral features observed in the frequency-time plots of the ISS from some pulsars (e.g. Ewing et al. 1970). Again this latter phenomenon has no adequate theoretical explanation.

Temporal Variations

The most ambitious description of the temporal ISS behavior has been attempted by Backer (1975), who has computed the 400 MHz temporal ISS spectra. He deduced a spectral width Δf_s under the assumptions that $m = 1$ and that the intensity spectra are Gaussian. Whereas the latter assumption is probably not correct and some of his width estimates are confused by intrinsic variations, we assume that his Δf_s data do statistically represent the ISS spectra. He found that Δf_s increases linearly with DM (varying by a factor of 3 either way).

A frozen pattern gives a direct relation between spatial and temporal spectra. However, the spaced receiver experiments demonstrate that the pattern is often not frozen. Nevertheless, we can again argue that the temporal spectrum widths will statistically reflect the product of typical drift speeds and spatial spectrum widths in the ISS pattern: $\Delta f_s \propto \Delta q_s v$.

The theoretical spatial spectrum is $\tilde{\gamma}_{\Delta I}(\omega, \omega, \mathbf{q})$ defined by (36). In the Rayleigh limit we can apply (39), giving $\gamma_{\Delta I}(\omega, \omega, \mathbf{s}) = [\gamma_1(\mathbf{s})]^2$, which for the Kolmogorov spectrum (9) yields

$$\gamma_{\Delta I}(\mathbf{s}) = \exp(-2B\lambda^2 s^{\alpha-2}). \tag{43}$$

If the spatial scale s_I is defined by the condition $\gamma_{\Delta I}(s_I) = 1/e$,

$$s_I = (2B\lambda^2)^{-1/(\alpha-2)} \propto z^{-1/(\alpha-2)}\lambda^{-2/(\alpha-2)}. \tag{44}$$

Hence $\Delta q_s \propto s_I^{-1} \propto \lambda^{1/(\alpha-2)} z^{2/(\alpha-2)}$ for the power law models and $\Delta q_s \propto \lambda z^{1/2}$ for the square law structure function models via equivalent formulas. The roughly linear *DM* dependence of Δf_s is thus at odds with both the Kolmogorov and Gaussian models, and requires $\alpha \sim 3$. However, the predicted λ dependence of $\Delta f_s \propto \lambda^2$ for $\alpha = 3$ then disagrees with the observations of Rickett (1970) for PSR 0329+54 and of Backer (1974) for PSR 0833-45. The latter author (see Figure 5) finds effectively $\Delta f_s \propto \lambda^{1.0 \pm 0.2}$, which is consistent with both the Kolmogorov and Gaussian models. These interpretational difficulties echo those for $\Delta\omega_I$, where the λ dependence fits both models, but the *DM* dependence is too steep. Backer (1975) reached a similar conclusion though did not explore any of the power law models.

In general the Δf_s values are rather uncertain and the velocity is not known, so quantitative estimates of C_N^2 cannot be extracted. As a final estimate of C_N^2 we use (44) based on the conclusions of Slee et al. (1974) that the pattern scale s_I is about 8000 km for PSR 1642-03 and PSR 1749-28 at 326 MHz. These values are also entered in Table 1.

DISCUSSION

The homogeneous extended Kolmogorov model is very ambitious in the sense that it has only one parameter, C_N^2, from which theory predicts the magnitude of all the scattering phenomena (angular widths, pulse widths, weak scintillation index, intensity scintillation scales versus distance and frequency). Figure 6 tests the success of the model by plotting the values of C_N^2 calculated from each observation (Table 1) against the distance through the medium (z). The homogeneous model should give a single value of C_N^2 independent of z; by contrast the data show a large scatter and a definite increase in C_N^2 at distances over about 1 kpc.

Particularly notable in this figure is the very high value of C_N^2 for the Vela pulsar PSR 0833-45. Similarly, the average electron density (0.14 cm^{-3}) derived from the dispersion measure and distance is higher than the value (0.03 cm^{-3}) typical of other directions. The Gum nebula is presumed to be responsible.

With this in mind we can test the alternative hypothesis that $C_N/\langle N_T \rangle$ is constant, corresponding to an rms density deviation that is a fixed fraction of the average. However, Table 1 reveals a narrow range in $\langle N_T \rangle$ in contrast to the 40 to 1 range in C_N. It could be argued that the C_N values are wrong because the distance estimates are wrong, since distances are simply derived from *DM* and an assumed constant $\langle N_T \rangle$ (after first correcting for large H II regions). If we do the opposite, constrain $C_N/\langle N_T \rangle$ to be constant and deduce z, then $\langle N_T \rangle$ could be estimated for each direction. However, this yields $\langle N_T \rangle$ increasing with z, again in contradiction with the simple homogeneous model. Excluding Vela, the few pulsars for which z is known independently support the interpretation in Table 1, in which $\langle N_T \rangle$ is approximately constant and so C_N^2 increases with distance through interstellar medium. It is certain at least that the scattering medium is not uniform on a scale of 100–1000 pc.

Of course, the spatially homogeneous model assumes that the extent of the medium is much larger than the "outer scale" (κ_1^{-1}). If this were similar to the 100–1000 pc propagation paths, then the homogeneity assumption would fail. Such a model is indeed reasonable since κ_1^{-1} would then be determined by the scale of spiral arms or the disc thickness. If also the sun were in a local minimum of scattering, the effective C_N^2 would depend on the actual path and tend to increase with path length. In a related discussion, Lee & Jokipii (1976) assumed $\alpha = 11/3$ and $\kappa_1^{-1} \sim 30$ pc and derived an rms electron density equal to or a little larger than the mean; however, the allowable range of α makes such estimates uncertain by at least an order of magnitude.

Future observations should be of a uniformly selected set of pulsars to guard against selection effects, which may have biased the results of Figure 6. In particular, the large scattering (C_N^2) for distant pulsars is inferred from pulse length measurements; this technique may have selected the highly scattered pulses because they are easier to distinguish from intrinsic pulse widths.

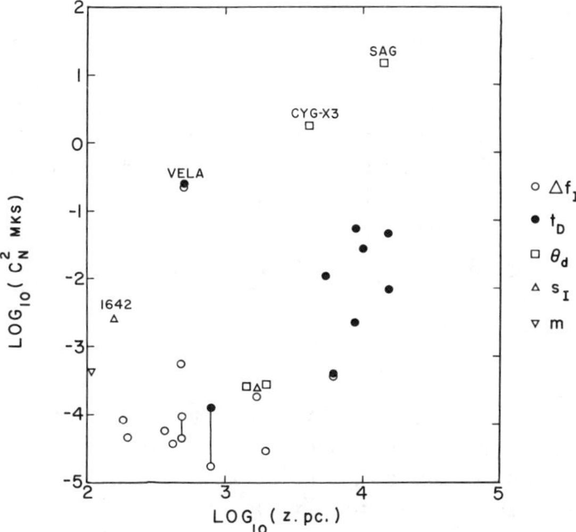

Figure 6 A logarithmic plot of C_N^2, derived by assuming a uniform distribution of Kolmogorov inhomogeneities versus distance through the scattering medium. Different symbols identify the various types of observation as listed in Table 1. The angular broadening point at 1500 pc corresponds to the estimated path length through the Galactic disk at the 15° latitude for measurements shown in Figure 1. The unusually large scattering for Vela and PSR 1642-03 are probably caused respectively by the Gum Nebula and the H II region surrounding the O-star ζ Oph (Prentice & ter Haar 1969).

Sources Other than Pulsars

In this paper we are largely concerned with a model for the scattering irregularities. Also of interest are the effects of ISS on radio signals from extragalactic sources. There are two such effects.

First, angular scattering partially decorrelates the signals at the two ends of an interferometer. The reduction in fringe visibility can be directly estimated from (9); a value of C_N^2 appropriate to the actual path length z in the scattering medium should be used here since the medium is not uniform. Indeed the angular broadening deduced by Duffett-Smith & Readhead (1976) only applies at Galactic latitudes $\lesssim 15°$. Normal to the plane $C_N^2 \sim 5 \times 10^{-5}$ m$^{-6.67}$, from which we predict a scattered diameter of only 0.04″ at 81.5 MHz through 400 pc of medium. There are two further observations in which interstellar broadening has been invoked. Anderson et al. (1972) found an apparent diameter of 2″ at 408 MHz for the variable radio source associated with Cygnus X-3; at its assumed 4 kpc distance this corresponds to a linear diameter about ten times greater than that allowed by its 5-day time scale of variability. The authors thus suggest that the diameter is broadened to 2″ by interstellar scattering. Assuming this to be the correct interpretation, we have added to Table 1 the deduced value of C_N^2; this is indeed large but about equal to that of PSR 0833-45. Similarly, the apparent diameter of the small component discovered in the Galactic center by Balick & Brown (1974) may be caused by scattering. Davies et al. (1976) estimate its diameter to be $1.5 \pm 0.3''$ at 960 MHz, which corresponds to the large value $C_N^2 = 14$ m$^{-6.67}$ for the 10 kpc line of sight toward the Galactic center. Much of this extra scattering maybe occurring in the dense nuclear region. The latter authors also report a possible transient source appearing for a time less than 40 days with a similar diameter but displaced by 53″. Again the variability requires a scattering explanation for the angular size, although it should be noted that if most of the scattering is occurring at a small distance L from the source, a minimum value of L is required to give a time variation Δt and apparent diameter θ. Thus we find $(c\Delta t/z\theta) \gtrsim z\theta/2L$, which gives $L \gtrsim 0.3$ pc in this case. For easy reference we give the relation between C_N^2, the distance z_{pc}, the wavelength λ_m (meters), and apparent full diameter at half-maximum θ_d (seconds of arc) for the uniform Kolmogorov medium:

$$C_N^2 = 1.0 \times 10^2 (\theta_d'')^{5/3} (z_{pc})^{-1} (\lambda_m)^{-11/3} m^{-6.67}. \tag{45}$$

The second effect is the slow intensity fluctuations detectable from sufficiently small sources. Rumsey (1976) and Budden & Uscinski (1970) have developed the appropriate theory for extended scattering media. Different limits apply in weak and strong scattering, such that for source diameters $10^{-5''}$ to $10^{-3''}$, ISS is more likely to be detectable in weak scintillation at high frequencies (3–10 GHz). Condon & Backer (1975) observed 16 small-diameter sources and found no ISS above a few percent at 2.7 and 8.1 GHz. Making the Gaussian assumption they derived lower limits near $10^{-3''}$ for the effective source diameters. A power law model would not yield significantly different diameters. The low-frequency variability of some radio sources, such as 3C454.3 at 408 MHz (Hunstead 1972) cannot be explained as ISS, since their measured diameters are too large and fluctuation times are too long.

CONCLUSION

The initial investigation of ISS, soon after the discovery of pulsars, established the basic phenomena. However, most interpretations at that time were based on the Gaussian spectrum model. Since then the theory has been developed considerably, following the lead of Tatarski (1971) in dealing with extended irregular media with power law spectra. The major observations can now be predicted accurately for both spectral models. This review has relied heavily on the work of Lee & Jokipii (1975), but almost all the tractable results are for weak scintillation or the Rayleigh limit of strong scintillation. To this extent the work of Budden & Uscinski (e.g. 1970, 1972) is completely accurate since it involves an equivalent assumption of normally distributed electric fields. Future theoretical developments are needed in the area between weak and strong scintillation, in exact solutions for a point source and in time variations of intensity when there are velocity gradients distributed along the propagation path.

In analyzing the published observations we find that they are compatible with both the Gaussian model and the more realistic power law model with any $\alpha \gtrsim 3.5$ (including the Kolmogorov value $\alpha = 11/3$). The limited accuracy and sophistication of the data are now the limiting factors in distinguishing between the models. The only major conclusion is that the interstellar irregularities are not homogeneously distributed on scales of 100–1000 pc on either model. This was already the conclusion of Backer (1975) for the Gaussian model. With improved theory we now need better data. I hope this review may stimulate at least the publication of existing data if not some new observations.

ACKNOWLEDGMENTS

The author thanks W. A. Coles, V. H. Rumsey, and J. W. Armstrong for many scintillating discussions and the NSF for support under grants AST 76-03745 and ATM-13451 A02.

Literature Cited

Anderson, B., Conway, R. G., Davis, R. J., Peckham, R. J., Richards, R. J., Spencer, R. E., Wilkinson, P. N. 1972. *Nature Phys. Sci.* 239:117–18

Anderson, B., Lyne, A. G., Peckham, R. J. 1975. *Nature* 258:215–16

Armstrong, J. W., Coles, W. A. 1972. *J. Geophys. Res.* 77:4602

Armstrong, J. W., Coles, W. A., Kaufman, J. J., Rickett, B. J. 1973. *Ap. J.* 186:L135–38

Backer, D. C. 1974. *Ap. J.* 190:667

Backer, D. C. 1975. *Astron. Ap.* 43:395

Balick, B., Brown, R. L. 1974. *Ap. J.* 194: 265–70

Barabanenkov, Ju. N., Kravtsov, Yu. A., Rytov, S. M., Tatarski, V. I. 1971. *Soviet Phys.* 13:551–75

Bell, S. J., Hewish, A. 1967. *Nature* 213:1214

Bonazzola, S., Celnikier, L. M. 1975. *Astron. Ap.* 45:185–92

Brown, A. G. F. 1976. *MNRAS* 176:53P–55P

Budden, K. G., Uscinski, B. J. 1970. *Proc. R. Soc. London Ser. A* 316:315–39

Budden, K. G., Uscinski, B. 1972. *Proc. R. Soc. London Ser. A* 330:65–77

Cohen, M. H., Cronyn, W. M. 1974. *Ap. J.* 192:193–97

Cohen, M. H., Gundermann, E. J., Hardebeck, J. E., Sharp, L. E. 1967. *Ap. J.* 147: 449

Coles, W. A., Rickett, B. J., Rumsey, V. H. 1974. In *Solar Wind Three,* ed. C. T. Russell, pp. 417–31. Los Angeles: Univ. Calif.

Condon, J. J., Backer, D. C. 1975. *Ap. J.* 197: 31–38
Conklin, E. K., Howard, H. T., Craft, H. D. Jr., Comella, J. M. 1968. *Nature* 219: 1238
Craft, H. D. Jr. 1970. PhD thesis. Cornell Univ.
Cronyn, W. M. 1970. *Science* 168: 1453–55
Cronyn, W. M. 1972. *Ap. J.* 174: 181–200
Davies, R. D., Walsh, D., Browne, I. W. A., Edwards, M. R., Noble, R. G. 1976. *Nature* 261: 476–78
Downs, G. S., Reichley, P. E. 1971. *Ap. J.* 163: L11
Duffett-Smith, P. J., Readhead, A. C. S. 1976. *MNRAS* 174: 7–17
Ewing, M. S., Batchelor, R. A., Friefeld, R. D., Price, R. M., Staelin, D. H. 1970. *Ap. J.* 162: L169–72
Galt, J. A., Lyne, A. G. 1972. *MNRAS* 158: 281–90
Guelin, M., Encrenaz, P., Bonazzola, S. 1974. *Astron. Ap.* 14: 387
Hankins, T. H., Rickett, B. J. 1975. *Methods Comp. Phys.* 14: 55–129
Hewish, A. 1975. *Science* 188: 1079–83
Hewish, A., Bell, S. J., Pilkington, J. D. H., Scott, P. F., Collins, R. A. 1968. *Nature* 217: 709
Hewish, A., Burnell, J. 1970. *MNRAS* 150: 141
Hunstead, R. W. 1972. *Ap. Lett.* 12: 193–200
Isaacman, R., Rankin, J. M. 1977. *Ap. J.* Preprint
Ishimaru, A. 1976. *Topics in Applied Physics*, ed. J. W. Strobehn. New York: Springer
Jokipii, J. R. 1973. *Ann. Rev. Astron. Astrophys.* 11: 1
Kaufman, J. J. 1976. PhD thesis. Univ. Calif., San Diego
Komesaroff, M. M., Ables, J. G., Cooke, D. J., Hamilton, P. A., McCulloch, P. M. 1973. *Ap. Lett.* 15: 169–74
Komesaroff, M. M., Hamilton, P. A., Ables, J. G. 1972. *Aust. J. Phys.* 25: 759–77
Lang, K. R. 1971a. *Ap. Lett.* 7: 175–78
Lang, K. R. 1971b. *Ap. J.* 164: 249–64
Lang, K. R., Rickett, B. J. 1970. *Nature* 225: 528–30
Lee, L. C. 1974. *J. Math. Phys.* 15: 1431–35
Lee, L. C. 1976. *Ap. J.* 206: 744–52
Lee, L. C., Jokipii, J. R. 1975. *Ap. J.* 201: 532–43
Lee, L. C., Jokipii, J. R. 1976. *Ap. J.* 206: 735–43
Little, L. T. 1968. *Planet. Space Sci.* 16: 749
Little, L. T., Hewish, A. 1968. *MNRAS* 138: 393
Little, L. T., Matheson, D. N. 1973. *MNRAS* 162: 329–38
Lovelace, R. V. E. 1970. PhD thesis. Cornell Univ., Ithaca, NY
Lyne, A. G., Thorne, D. J. 1975. *MNRAS* 172: 97
Manchester, R. N., Taylor, J. H., Van, Y. Y. 1974. *Ap. J.* 189: L119–22
Mutel, R. L. 1975. PhD thesis. Univ. Color., Boulder
Mutel, R. L., Broderick, J. J., Carr, T. D., Lynch, M., Desch, M., Warnock, W. W., Klemperer, W. K. 1974. *Ap. J.* 193: 279–82
Prentice, A. J. R., ter Haar, D. 1969. *MNRAS* 146: 423
Rankin, J. M., Comella, J. M., Craft, H. D. Jr., Richards, D. W., Campbell, D. B., Counselman, C. C. III. 1970. *Ap. J.* 162: 707–25
Rankin, J. M., Counselman, C. C. III. 1973. *Ap. J.* 181: 875–89
Ratcliffe, J. A. 1956. *Rep. Prog. Phys.* 19: 188
Readhead, A. C. S., Hewish, A. 1972. *Nature* 236: 440
Readhead, A. C. S., Hewish, A. 1974. *Mem. R. Astron. Soc.* 78: 1
Rickett, B. J. 1969. *Nature* 221: 158–59
Rickett, B. J. 1970. *MNRAS* 150: 67–91
Rickett, B. J., Lang, K. R. 1973. *Ap. J.* 185: 945–50
Rumsey, V. H. 1974. *IEEE Conf. Publ.* 114: 1–6
Rumsey, V. H. 1976. *Radio Sci.* 11: 545–49
Salpeter, E. E. 1969. *Nature* 221: 31–33
Scheuer, P. A. G. 1968. *Nature* 218: 920
Slee, O. B., Ables, J. G., Batchelor, R. A., Krishna-Monhan, S., Venugopal, V. R., Swarup, G. 1974. *MNRAS* 167: 31–47
Slee, O. B., Komesaroff, M. M., McCulloch, P. M. 1968. *Nature* 219: 342
Sutton, J. M. 1971. *MNRAS* 155: 51
Tatarski, V. I. 1961. *Wave Propagation in a Turbulent Medium.* Transl. R. A. Silverman, p. 195. New York: McGraw-Hill
Tatarski, V. I. 1971. *Effects of the Turbulent Atmosphere on Wave Propagation,* p. 232. Transl. Isr. Prog. Sci. Transl., US Dept. Commerce. Washington DC: Natl. Sci. Found.
Taylor, J. H., Manchester, R. N. 1975. *Ap. J.* 80: 794–806
Uscinski, B. J. 1968. *Proc. R. Soc. London Ser. A* 307: 471
Uscinski, B. J. 1974. *Proc. R. Soc. London Ser. A* 336: 379–92
Uscinski, B. J. 1975. *MNRAS*
Vandenberg, N. R. 1974. PhD thesis. Univ. Maryland, College Park
Williamson, I. P. 1972. *MNRAS* 157: 55–71
Williamson, I. P. 1974. *MNRAS* 166: 499–512

Ann. Rev. Astrophys. 1977. 15 : 505–40

CLUSTERS OF GALAXIES

Neta A. Bahcall[1]
Princeton University Observatory, Princeton, New Jersey 08540

1 INTRODUCTION

The distribution of galaxies on the celestial sphere, as seen on wide-field sky photographs, shows a high degree of clumping. Most galaxies are believed to belong to groupings ranging in richness from numerous poor associations to rich clusters of galaxies like the Coma and Virgo clusters. The rich clusters can be easily recognized, even at relatively great distances; their study is thus important both for an intrinsic understanding of clusters of galaxies as fundamental systems and for a determination of the large-scale structure of the universe. Numerous studies of clusters of galaxies over the past decade have added many important facts to our knowledge of their optical, X-ray, and radio properties. In the present article I review the observational properties of clusters of galaxies and mention some of the theoretical interpretations. The subjects reviewed are: catalogs (Section 2), static properties (Section 3), dynamics (Section 4), X-ray emission (Section 5), and radio emission (Section 6). Previous review papers on clusters of galaxies include those by Abell (1965, 1976) and van den Bergh (1976b).

I assume (except where specified otherwise) a Hubble constant of 50 km sec^{-1} Mpc^{-1}, with $h_{50} \equiv H_0$ (km sec^{-1} Mpc^{-1})/50 used to indicate the conversion to other values of H_0.

2 CATALOGS

The classical way of selecting a cluster for inclusion in a catalog is by observing an enhancement in the surface number density, σ, over the uniform background number density σ_{bg}, i.e.

$$\langle \sigma/\sigma_{\text{bg}} \rangle \geqq N, \tag{1}$$

where N can be specified as desired. If N is chosen to be too low ($N \sim 1$) then all galaxies would belong to very few large groups, while if it is too high then most galaxies would belong to the field and only a few clusters would be identified. In compiling a catalog, a density enhancement can be defined for either a fixed linear extent of the cluster and a minimum mean density within it (as in the Abell catalog;

[1] Research supported by NSF grant AST 76-15386 and NASA grant NSG-7256.

see below) or for a variable extent determined directly by the observed surface density at its borders (as in the Zwicky catalog; see below). Cataloged sizes and populations of clusters depend, of course, on the definitions used in the identification process; therefore, such parameters should be taken only as operational prescriptions intended to describe the range and extent specified by the catalog. If a cluster is defined as a system containing all galaxies with velocities within a specific velocity range in the general direction of the cluster, then its extent and population can be determined only by studying the velocities of all galaxies in this direction (see Sections 3.1 and 3.6).

Two main catalogs of rich clusters of galaxies are available: the catalog of rich clusters by Abell (1958) and the Catalog of Galaxies and Clusters of Galaxies by Zwicky et al. (1961–68). Both authors identified clusters on the Palomar Sky Survey plates but used different criteria for the inclusion of clusters in their catalogs.

The Abell criteria are summarized as follows: 1. a cluster must contain at least 50 members in the magnitude range m_3 to m_3+2, where m_3 is the magnitude of the third brightest galaxy; 2. the $\gtrsim 50$ members should be contained within a circle of radius $3h_{50}^{-1}$ Mpc (see Equation 2 below) around the center of the cluster; 3. the cluster redshift should be in the range $0.02 \lesssim z \lesssim 0.20$; 4. the cluster should lie north of declination $-27°$ and within the region of complete identification given in Table 1 of Abell (1958). The catalog lists 2712 clusters (1682 of which comprise a complete sample) found over $\sim 3 \times 10^4$ square degrees of projected area of the sky. For each cluster the catalog lists: estimated center position in 1855 coordinates (Sastry & Rood 1971 precessed all positions to the 1950 epoch); the photo-red magnitude of the tenth brightest galaxy; the distance group (i.e. the distance estimated from the apparent magnitude of the tenth brightest galaxy, m_{10}); and the richness classification (see Section 3.1).

The criteria for inclusion in the Zwicky catalog are less strict, and the catalog therefore contains more entries than the Abell catalog. The selection criteria are 1. the cluster must contain at least 50 galaxies in the magnitude range m_1 to m_1+3, where m_1 is the magnitude of the brightest galaxy; 2. these galaxies must lie within the cluster's contour, defined as the isopleth where the projected density of galaxies is about twice that of the neighboring field; 3. no limit on the cluster redshift is specified, but aggregates such as the Virgo cluster, which cover very large areas of the sky, are not included in the contour maps; 4. the clusters must lie north of declination $-3°$ and within the areas given in the introduction (p. IV) to Volume VI of the catalog.

Listed for each cluster in the six volumes of the catalog are: the estimated central position in 1950 coordinates; the cluster classification (see Section 3.2 below); population (see Section 3.1); the cluster diameter; and the distance estimate (from $z \leqq 0.05$ to $z > 0.2$). Maps of the isopleths of all clusters as observed on the Palomar Sky Survey fields are presented in the catalog. Also given are the estimated photographic magnitudes for all the individual galaxies brighter than $m_p = 15.7^m$.

The Zwicky catalog contains systems that are less rich than those of Abell (compare the first criteria for each above), and hence a much larger total number of clusters. There are also detailed differences (especially in the extent of clusters) between the two catalogs, again mostly resulting from the different definitions used in the identification process.

Small lists of clusters of galaxies (mostly either in the southern hemisphere or at $z > 0.2$, ranges not covered by the Abell or Zwicky catalogs) include, among others, those of Klemola (1969), Snow (1970), Sersic (1974), Rose (1976), and Lugger (1976) for clusters in the southern hemisphere; Humason, Mayall & Sandage (1956), Sandage, Kristian & Westphal (1976), Gunn & Oke (1975), and Spinrad (1977). Charts and catalogs of the surface density of galaxies over the surface of the sky have been compiled by Shane and collaborators (see Shane & Wirtanen 1954, 1967, and references therein) from galaxy counts on the Lick survey photographs. Catalogs of small groups of galaxies are given by Holmberg (1937), Sandage & Tammann (1975), de Vaucouleurs (1976), and, perhaps most systematically, by Turner & Gott (1976a).

3 STATIC PROPERTIES

In the following subsections I discuss several observed static properties of rich clusters: 1. richness; 2. classification schemes; 3. galactic content; 4. cD galaxies; 5. density profiles; 6. sizes; and 7. the optical luminosity function.

3.1 *Richness*

Richness is a measure of the number of member galaxies in a cluster within a certain distance from the cluster center; it is thus also a measure of a mean number density in the cluster. The richness of clusters varies over a very wide range, from the rich and dense clusters that contain thousands of members, down to low-density groups like our Local Group, and even to double galaxies.

The total number of member galaxies in a cluster depends on the definition of a "cluster" (e.g. all galaxies with velocities within the observed velocity dispersion range or only those with crossing times shorter than a Hubble time). The estimated total number depends strongly on the assumed extent of the cluster, the background corrections applied, and the extrapolation to faint members—none of which can be determined with great accuracy. Therefore, in order to define a richness parameter for clusters, one adopts operational definitions for the size, magnitude range, and background.

Zwicky and his collaborators define and list populations of clusters in their catalog. They define population as the number of galaxies visible on the red Palomar Sky Survey plate, corrected for the mean field count, that are located within the isopleth of twice the field density. Because of the boundary at twice the field density, these populations depend systematically on the cluster redshift (Abell 1962, Scott 1962). Another distance dependence of the Zwicky populations results from the larger magnitude range included for the nearby clusters as compared with the distant ones. For a meaningful comparison of the populations of different clusters, the effects of distance have to be removed.

Abell (1958) introduced a different and largely distance-independent definition for the richness of a cluster: the number of galaxies within a fixed magnitude range (brighter than m_3+2, where m_3 is the photo-red magnitude of the third brightest cluster member), and a fixed linear boundary, a circle of radius

$$R_A = 1.7/z \text{ arc min} = 3h_{50}^{-1} \text{ Mpc}, \tag{2}$$

where z is the redshift of the cluster. The populations are corrected by a background count in a nearby field. The redshifts of the clusters, all in the range of 0.02 to 0.2, were estimated by Abell from the magnitude of the tenth brightest member in each cluster. The Abell populations, defined as above, range from 50 to over 300 members; the less populated clusters are much more numerous than the more richly populated ones (see Tables 4 and 7 of Abell 1958). A slight correlation between richness and distance was found (Just 1959) in the Abell catalog. However, this effect is most likely due to an incompleteness of about 10% in the most distant clusters in the Abell catalog, rather than to evolution.

Sandage & Hardy (1973) also determine populations for a number of clusters; they include galaxies within a circle of radius 2 Mpc ($H_0 = 54$ km sec^{-1} Mpc^{-1}, $q_0 = 1$) about each cluster center and within 2.5 mag from the third brightest galaxy.

Small groups probably contain a substantial fraction of all galaxies (van den Bergh 1962, de Vaucouleurs 1976, Turner & Gott 1976a,b). Classically, only a few percent ($\sim 10\%$) of the galaxies in the universe are believed to be members of rich clusters such as the Virgo and Coma clusters (van den Bergh 1961a, Minkowski 1963). Holmberg (1962) claims from a statistical study of nearby galaxies that a little over half of all galaxies (53%) are clustered in groups from double systems and up. The multiplicity function, i.e. the luminosity function of groups of galaxies with various N values (Equation 1), has been studied recently by Gott & Turner (1977b). The relative number of galaxies in small groups versus rich clusters, as well as the possibility that almost all galaxies have at least one companion, are still open questions. A detailed spectroscopic study aimed at determining redshifts of a complete sample of galaxies over an extended area of the sky is needed in order to determine these ratios. Recently Chincarini & Rood (1975, 1976) and Tifft & Gregory (1976) began such studies by investigating a region around the Coma cluster. The pronounced clustering of redshifts found by these authors may be an indication, if it occurs in other areas as well, for a much lower percentage of field galaxies than generally has been assumed; this would imply that nearly all galaxies are members of superclusters or isolated groups. (See also Section 3.6 on cluster sizes and the correlation-function studies of Peebles and collaborators).

3.2 *Classification Schemes*

Clusters of galaxies can be organized into a one-parameter sequence analogous to stellar spectral type or galaxy Hubble type. The sequence ranges from early- to late-type clusters, or equivalently, from regular to irregular clusters. The early (regular) clusters are believed to be systems that are dynamically more evolved than the late (irregular) clusters. Many cluster properties (shape, concentration, dominance of bright galaxies, galactic content, density profile, mass-segregation, and radio and X-ray emission) are correlated with position in the cluster sequence. A summary of the sequence and its related properties, as well as the correlations among the various classification schemes based on these properties, is presented in Table 1. The classification schemes are based on the following properties: the morphological appearance of the cluster (Zwicky, Rood-Sastry), the dominance of bright galaxies (Bautz-Morgan, Rood-Sastry), and the galactic content of clusters (Morgan, Oemler). I discuss below the definitions and details of these classification schemes.

One of the early systematic classifications was described by Zwicky in his Catalogue of Galaxies and Clusters of Galaxies. Zwicky and his collaborators divide clusters into three groups: "compact," "medium-compact," and "open." A "compact" cluster contains a single outstanding concentration among its bright members, with at least ten galaxies appearing in contact. A "medium-compact" cluster has either a single concentration in which the ten brightest galaxies are separated by several of their own diameters or several distinct concentrations, some of which may be compact. An "open" cluster contains no very obvious condensations.

The main features of the Zwicky classification, as well as several of the newer systems discussed below, can be represented in terms of the "regular" and "irregular" clusters (Abell 1965, 1976), defined according to their central concentration and degree of circular symmetry. "Regular" clusters have populations of at least 10^3 in the brightest six magnitude range, high central concentration, and circular symmetry (examples are the Coma and the Corona Borealis clusters). "Irregular" clusters show little or no circular symmetry and no strong central concentration. The galactic content of these groups is also different: the regular clusters contain almost entirely E and SO galaxies, whereas the irregular clusters usually contain galaxies of all types, including appreciable numbers of spirals and irregulars (see below). A summary of the main properties of regular and irregular clusters (labelled also as early and late types, respectively), is given in Table 1 together with an "intermediate" group introduced here in order to make the transition between the regular and irregular types smoother and more descriptive of the observations. There is some overlap among the three groups; the relations given in the table are for "typical" cases.

A refinement of the regular/irregular classification scheme for clusters of galaxies was developed by Rood & Sastry (1971), who based their detailed scheme on the distribution of the ten brightest members. The Rood-Sastry (RS) system can be represented by a "tuning-fork" diagram:

$$\begin{array}{lll} & & \mathrm{L-F} \\ & \nearrow & \\ \mathrm{cD-B} & & \\ & \searrow & \\ & & \mathrm{C-I} \end{array} \tag{3}$$

where the cluster types fall into an ordered sequence with criteria varying systematically from one end to the other. The general classification criteria are

Table 1 Classification schemes of clusters and related characteristics

Property/class	Regular (Early)	(Intermediate)	Irregular (Late)	Reference[a]
Zwicky type	Compact	Medium-Compact	Open	1
Bautz–Morgan type	I, I–II, II	(II), II–III	(II–III), III	2
Rood–Sastry type	cD, B, (L, C)	(L), (F), (C)	(F), I	3
Content	Elliptical-rich	Spiral-poor	Spiral-rich	4,6
~E:SO:S Ratio	3:4:2	1:4:2	1:2:3	6
Symmetry	Spherical	(Intermediate)	Irregular shape	1,5
Central concentration	High	Moderate	Very little	1,6
Central profile	Steep gradient	(Intermediate)	Flat gradient	6
Mass segregation	Marginal evidence for $m-m(1) \lesssim 2$ mag	Marginal evidence for $m-m(1) \lesssim 2$ mag	No segregation	6
Radio emission	~50% detection rate[b]	~50% detection rate[b]	~25% detection rate[b]	7
L_R	High	Low	Low	
X-ray emission	~33% detection rate[c]	~8% detection rate[c]	~8% detection rate[c]	8
L_x	High	Intermediate	Low	
Examples	A2199, Coma	A194, A539	Virgo, A1228	

[a] References: 1. Zwicky et al. 1961–1968; 2. Bautz & Morgan 1970, Bautz 1972; 3. Rood & Sastry 1971; 4. Morgan 1962; 5. Abell 1962; 6. Oemler 1974; 7. Owen 1975; 8. Bahcall 1974b.

[b] To the limit of 0.2 f.u. at 1400 MHz (ref. 7) and for Abell distance groups $\leqq 3 (z \lesssim 0.07)$.

[c] To the limit of the 3U catalog and for Abell distance groups $\leqq 3 (z \lesssim 0.07)$.

as follows:

cD-type (= supergiant):	the cluster is dominated by a cD galaxy (e.g. A401, A2199);
B-type (= binary):	the cluster is dominated by a bright "binary" system (e.g. Coma);
L-type (= line):	three or more of the ten brightest members are arranged in a line (e.g. Perseus);
C-type (= core):	at least four of the ten brightest members are located with comparable separations in the cluster core (e.g. A2065);
F-type (= flat):	several of the brightest ten galaxies are distributed in a flattened configuration (e.g. A397);
I-type (= irregular):	an irregular distribution of the galaxies with no well-defined center (e.g. A1228).

Rood & Sastry (1971) have classified all 110 Abell clusters of distance groups $\leqq 3$. They find the following frequency distribution according to type:

cD: 21% B: 9% L: 9% C: 14% F: 18% I: 29%.

Each arm of the tuning-fork diagram contains about one-third of all clusters.

Bautz & Morgan (1970) developed a classification system that depends on the relative contrast of the brightest galaxy to the other galaxies in each cluster. The Bautz-Morgan (BM) system has the following five-part form:

Type I:	the cluster is dominated by a single, centrally located, cD galaxy (e.g. A2199);
Type II:	the brightest members are intermediate in appearance between cD galaxies (which have extended envelopes) and normal giant ellipticals (e.g. Coma);
Type III:	the cluster contains no dominant galaxies (e.g. Virgo, Hercules); and two intermediate types, Type I-II and Type II-III.

Classifications of various clusters on the BM system have been given by Bautz & Morgan (1970), Bautz (1972), Bautz & Abell (1973), Sandage & Hardy (1973), Corwin (1974), McHardy (1974), and Leir & van den Bergh (1977). Leir & van den Bergh (1977) pointed out that the BM classification system (as well as the RS system) may be affected systematically by distance because of the K-dimming reduction of the surface brightness of the large envelopes of distant cD galaxies relative to the other galaxies in the cluster. Also, a misidentification of a bright galaxy in the projected central area of the cluster can change the classification from one extreme type to another. In any classification system based on the brightest members (e.g. BM and RS systems) a careful investigation for possible contamination by field galaxies should always be carried out.

Bautz & Abell (1973) and Sandage & Hardy (1973) find that the absolute magnitude of the first brightest galaxy changes with the BM type. From magnitude residuals from the Hubble diagram Sandage and Hardy show that the first-ranked galaxy is absolutely brighter in Type I clusters than in Type III by $\langle \Delta M_v \rangle = 0.6$ mag. They also find that the second- and third-ranked galaxies are fainter by 0.5 mag in Type I compared to Type III clusters. This inverse effect is a result of the fact that BM Type I clusters are *defined* as having the highest luminosity contrast between the first-ranked and fainter galaxies; the extent of the contrast was measured by Sandage and Hardy. It would be useful to investigate the possible correlation of the brightest cluster galaxies with other classification schemes shown in Table 1, e.g. the Rood-Sastry scheme.

Both the RS and the BM classes seem to be independent of cluster richness but are correlated with the central density in the cluster (the latter is higher in BM Type I than in BM Type III clusters). (See Table 1 and appropriate sections for the dependencies of various properties on these classification systems).

A different type of classification scheme is based on the galactic content of clusters of galaxies (cf. Morgan 1962, Oemler 1974). This scheme is found to be tightly correlated with the structural appearance of the clusters and thus with the previously discussed schemes. Morgan (1962) introduced the classification of clusters according to the morphological type of their bright members. More recently Oemler (1974) classified a sample of 15 clusters into three types according to their galactic content. Although the clusters studied display a wide range of characteristics, most of their features can be summarized by dividing them into the following groups. 1. *Spiral-rich* clusters have a composition similar to that of the field, with a high proportion of spiral galaxies. They are also irregular in appearance, with low mean densities, no tendency to central concentration, flat central density gradients, and no segregation by morphological type or evidence of relaxation by two-body encounters. 2. *cD-clusters* are dominated by central supergiant galaxies and have no spirals in their cores. They contain a much higher proportion of ellipticals in the central regions than other cluster types; and they are dense, centrally concentrated, spherical in appearance, have steep central density gradients, and show evidence of two-body relaxation. 3. *Spiral-poor* clusters are intermediate between 1 and 2, with a composition dominated by SO galaxies. They show segregation by mass and morphological type, but they are not as regular, compact, or centrally condensed as the cD clusters.

Typical percentages of ellipticals, SO's, and spirals in the three types of rich clusters, as well as in the field, are summarized in Table 2. The ratio of spiral to elliptical galaxies in a cluster may depend on magnitude (cf Shapley 1950) and on distance from the cluster center (Section 3.3). The brightest few galaxies of any cluster, however, are almost always of the E or SO type.

Table 2 Galactic content of clusters

	E	SO	S	(E+SO)/S	
cD Clusters:	35%	45%	20%	4.0	e.g. Coma, A2199
Spiral-poor:	15%	55%	30%	2.3	e.g. A194, A400, A539
Spiral-rich:	15%	35%	50%	1.0	e.g. Hercules, A1228, A1367, A2197
Field:	15%	25%	60%	0.7	e.g. de Vaucouleurs 1959, van den Bergh 1962, Faber & Gallagher 1976

3.3 *Galactic Content*

Rich dense clusters are dominated by elliptical galaxies; in the low-density general field spirals occur most frequently (Hubble 1936, Zwicky 1938). Galaxies of type SO are generally distributed in the same way as ellipticals. Some data on the relative frequency with which various types of galaxies occur in clusters are given by, among others, van den Bergh (1962, 1975) and Oemler (1974). According to van den Bergh, the percentages of E+SO galaxies in rich clusters, poor clusters, and the field, are, respectively, 56%, 20%, and 24% [see his definitions of rich and poor clusters (van den Bergh 1962)]; the appropriate percentages for S+Ir-type galaxies are given as 38%, 14%, and 48%. According to van den Bergh, 93% of cD galaxies are in rich clusters and 6% in poor ones (compare Morgan, Kayser & White 1975, Albert, White & Morgan 1977). According to the above study (van den Bergh 1975), Seyfert galaxies are rare in either rich or poor clusters; they occur mostly in the field.

Some evidence suggesting that cluster galaxies (E, SO, and S) may have, on the average, less gas than similar galaxies in the field is summarized by van den Bergh (1976a, and references therein; see also Davies & Lewis 1973). Van den Bergh also introduces a new sequence of "anemic spirals," which occur most frequently in rich clusters and have characteristics intermediate between the gas-rich normal spirals and gas-poor systems of type SO. The differences between normal spirals, anemic spirals, and SO's were tentatively interpreted in terms of the influence of environment on the evolution of flattened galaxies.

A study of the density profiles of E, SO, and S galaxies in clusters (Oemler 1974) indicates that in cD and spiral-poor clusters the projected density of spirals decreases toward the center (see Figure 9 of Oemler) and is consistent with a zero space density of spirals in the core. Other studies of Coma (Rood et al. 1972, Rood 1974a, Gregory 1975, and Chincarini & Rood 1975) also show a tendency for the relative number of spirals to decrease toward the cluster center. A recent investigation of six clusters by Melnick & Sargent (1977) indicates that the ratio of the surface densities of spirals to SO's increases with distance from the cluster center.

Spitzer & Baade (1951) suggested that the absence of spiral galaxies in the dense regions of clusters may be a result of collisions between the cluster galaxies and the consequent stripping of interstellar matter from them; the SO galaxies are interpreted as spirals from which the gas and dust have been removed. Such collisions are most frequent in the dense cores of regular clusters where one would expect to find many SO's but no spirals, as is indeed indicated by the data. However, with the revisions in the distance scale (and assuming the galaxy size remains ~5–10 kpc), the Spitzer-Baade mechanism may not predict a sufficiently rapid conversion of spirals into SO's. Gunn & Gott (1972) suggested that SO galaxies lost their interstellar gas as a result of ram pressure produced when the parent galaxies travel with high velocities through an intracluster gas (see also Tarter 1975). The correlation found (Melnick & Sargent 1977) between velocity dispersion and the proportion of SO galaxies in some clusters is consistent with this picture. An alternative suggestion (Faber & Gallagher 1976) is that the large number of SO's in clusters is due to initial conditions rather than some external stripping mechanism.

3.4 *cD Galaxies*

The term cD galaxy was introduced by Matthews, Morgan & Schmidt (1964) to describe a galaxy with the nucleus of a giant elliptical surrounded by an extended, slowly decreasing envelope. The cD galaxies are of large extent (they are the largest known galaxies) and high luminosity, and frequently contain multiple nuclei. Several lists of Abell clusters that contain cD galaxies have been published (e.g. Matthews, Morgan & Schmidt 1964, Morgan & Lesh 1965; see also Bautz & Morgan 1970, Bautz 1972, Rood & Sastry 1971, Leir & van den Bergh 1977). Approximately 20% of rich clusters of galaxies contain a dominant central cD galaxy. Recently Morgan, Kayser & White (1975) and Albert, White & Morgan (1977) investigated the possible existence of cD galaxies in poor clusters and listed suspected cD galaxies in some small groups (cf also van den Bergh 1975). Since all previous examples of cD's were located in rich clusters, the newly reported objects are important if they are really cD's and if they are valid members of the

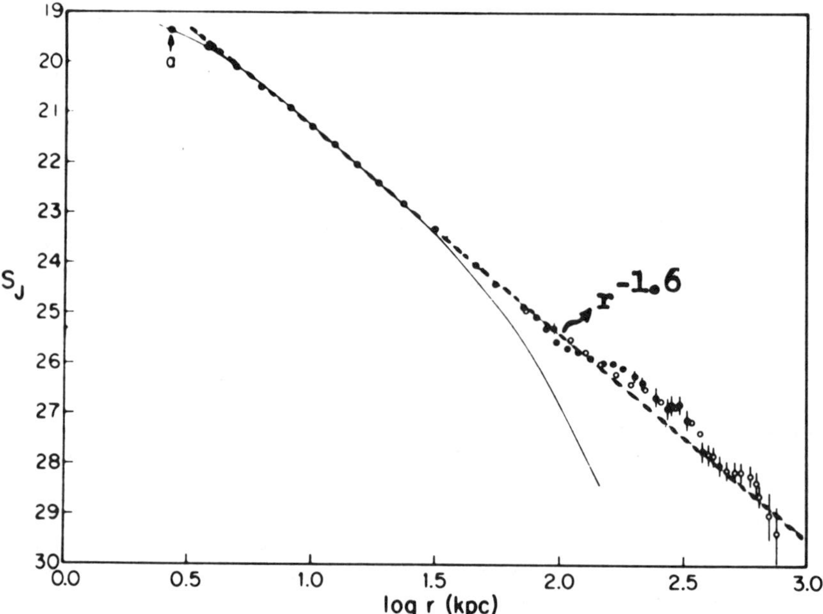

Figure 1 Surface brightness profile of the cD galaxy in A2670 (measured by Oemler 1973a). S_J is in mag (arc sec)$^{-2}$; open circles are green (J) magnitudes and filled circles are red magnitudes shifted by +1.1 mag. The solid line is the profile of a normal elliptical galaxy with a length scale a. The dashed line represents the relation $\sigma(r) \propto r^{-1.6}$ suggested by Equation 4 (Section 3.4).

groups. Oemler (1976) studied the profile of one of the suggested cD galaxies, NGC4073, located in the poor cluster MKW4; its profile is similar to some of the smaller cD galaxies, with a radius of about 10^2 kpc.

Detailed galaxy photometry has been presented by Oemler (1973a) for the cD galaxy in the compact cluster A2670. The observed surface-brightness profile out to 1 h_{50}^{-1} Mpc, as well as a comparison profile of an elliptical galaxy, is shown in Figure 1. No measurable color gradient appears to exist in the regions where two colors are available ($\sim$60 to 400 kpc), suggesting that the stellar content remains roughly the same. The inferred profile of the outer envelope and its extent are affected by the outer envelopes of other bright galaxies and by uncertainties in the sky level. It is possible that a still larger envelope, with a small gradient, extends beyond 1 Mpc and fills the entire cluster. For such large continuum envelopes, it is difficult to distinguish observationally between light belonging to the cD envelope and light belonging to the cluster as a whole. The envelope profile of the cD galaxy (see Equation 4) seems to match the cluster profile quite well. If the cD profile is extrapolated beyond the reliable measurements, then the envelope contains about 35% of the total cluster light. The total absolute visual magnitude obtained by Oemler for the measured part of the cD galaxy in A2670 (i.e. $\leqq$0.8 Mpc) is -25.2 mag for $h_{50} = 1$.

From a study of the profiles and magnitudes of another set of cD galaxies Oemler (1976) finds that their measurable extent ranges from about 100 kpc for NGC4073 to over 2 Mpc for A1413 (see also Morgan & Lesh 1965, who noted that the cD galaxy in A1413 is the largest of all known cD galaxies). I find that the general behavior of the surface-brightness profile can be roughly described [excluding the innermost core ($\lesssim$5 kpc) and the outermost cutoff] as a single power law of the form

$$\sigma(r) \propto r^{-1.6} \quad \text{(cD galaxy),} \tag{4}$$

where $\sigma(r)$ is the surface-brightness distribution. Oemler finds that the total luminosity of the cD galaxy, $L_{\rm gal}$, and the luminosity of its extended envelope, $L_{\rm env}$, are correlated with the cluster luminosity, $L_{\rm cl}$. This correlation is roughly $L_{\rm gal} \propto L_{\rm cl}^{1.25}$ and $L_{\rm env} \propto L_{\rm cl}^{2.2}$.

The profiles can be decomposed, at least for purposes of discussion, into a bright giant elliptical plus an extended low-surface-brightness halo. The possibility that the extended envelopes are produced by some processes in the cluster as a whole has been discussed recently by several authors. Gallagher & Ostriker (1972), Oemler (1973a), and Richstone (1974) have suggested that the envelopes are simply intergalactic material moving in the potential well of the cluster; the intergalactic material is supposed to have been torn from the outer envelopes of cluster members by tidal interactions. However, no strong evidence for tidal stripping of galaxies is observed (cf Oemler 1976). Different authors (e.g. Lecar 1975, White 1976a, Ostriker & Tremaine 1975, and others) have explored the effects of dynamical friction. Galaxies that fall to the cluster center may be torn apart by their mutual tidal forces; their remnants can produce a cloud of intergalactic stars. Both models for cD galaxies depend on general cluster processes; according to these explanations, the cD nucleus, and the diffuse halo, are naturally found together at the bottom of the cluster potential well. Sandage & Hardy (1973) noted that the presence of the cD galaxy does not depend on cluster richness; it does, however, depend, as suggested by the models, on the cluster central density which is highest in BM Type I clusters. Another consequence of the debris hypothesis is that the galaxies in the cluster decrease in luminosity as the cD grows at their expense; such an effect is present in BM I clusters and was measured by Sandage and Hardy (see Section 3.2).

The galaxy NGC4839 in the Coma cluster is an odd case of a possible cD galaxy that does not sit at the bottom of the potential well of the cluster. The galaxy has an extended halo of the usual cD shape (Oemler 1976), but it is not the brightest member of the cluster and it is located at a projected distance ~ 1.7 Mpc from the cluster center. Theories of the origin of cD galaxies may also need to account for this object.

The mean absolute metric magnitude of cD galaxies within a radius of 43 kpc, corrected for aperture effect, K-dimming, and galactic reddening, is (Sandage 1976)

$$\langle (M_v)_c \rangle_{\mathrm{cD}} = -23.7 \pm 0.1 + 5 \log h_{50}. \tag{5}$$

The mean mass of such cD galaxies is $10^{13}\ M_\odot$ for $M/L = 50$ visual solar units. A similar mass for these galaxies is estimated by other investigators (e.g. Minkowski 1961, Jenner 1974). Wolf & Bahcall (1972) derived formulae on the basis of computer simulations and analytic calculations that permit the estimate, with probable errors, of masses of individual dominant galaxies and of binary galaxies in groups or clusters. They calculated dynamical masses for ten cD galaxies in rich clusters, obtaining masses $\sim 10^{13}\ M_\odot$ but with rather large uncertainties. The cD galaxies, as expected, are at the high end of the mass spectrum of galaxies.

Galaxies of the cD type frequently contain double or multiple nuclei (see, e.g. lists by Matthews, Morgan & Schmidt 1964, Rood & Sastry 1971). The bright central cD galaxy in A2199, NGC6166, contains four components in a large diffuse envelope. Minkowski (1961) measured the radial velocities of the three brightest components and found a spread of roughly 2×10^3 km sec^{-1}, consistent with the interpretation of a multiple system of galaxies within a common envelope. The center of gravity velocity of NGC6166 is close to the average velocity of the cluster as a whole. Jenner (1974) studied a sample of ten supergiant galaxies with double nuclei; he determined the redshifts and angular separations of the components of each galaxy. The relative radial velocities of the components in the sample range from 1 ± 72 to 1871 ± 29 km sec^{-1}, with an average of 560 km sec^{-1}.

The use of cD galaxies as standard candles in the classical cosmological tests is discussed in Section 3.7. The correlation between the presence of cD galaxies and radio and X-ray emission is discussed in Sections 5 and 6.

3.5 *Density Profiles*

The density profile of the galaxy distribution in a cluster can be described by a three-parameter function:

$$\rho(r) = \rho_0 f_1(r, R_c, R_h), \tag{6a}$$

$$\sigma(r) = \sigma_0 f_2(r, R_c, R_h), \tag{6b}$$

where $\rho(r)$ and $\sigma(r)$ are, respectively, the space and projected profiles (of mass, number, or brightness) and f_1 and f_2 are functions that best fit the observed profiles. The three parameters are: 1. the central density (ρ_0 or σ_0); 2. a central scale length [the core radius R_c is defined by $\sigma(R_c) = \sigma_0/2$]; and 3. a measure for the limit of the cluster, R_h (a halo or cutoff radius important at large distances from the cluster center). The reason why several different models can be made to fit the data is that the density profiles of clusters of galaxies can be determined reliably only down to about $10^{-2}\sigma_0$, and three parameters are adequate to describe the observed distributions in this range. It is difficult to extend accurately the profiles of rich clusters to still lower values of density because of the large uncertainties in the clumpy background distributions. The observed density profile exhibits a relatively smooth falloff from a high central density to a low-density tail at the outskirts of the cluster. Secondary maxima or subclustering are sometimes superimposed on a generally smooth and monotonic profile.

Data on the projected density profiles in a number of clusters are given by several authors (e.g. Shane & Wirtanen 1954, Zwicky 1957, Omer, Page & Wilson 1965, Clark 1968, Noonan 1971, 1974a, Bahcall 1971, 1972a,b, 1973a,b, 1974a, 1975a, Rudnicki 1963, Rudnicki & Baranowska 1966a, b, Rood et al. 1972, Rood & Sastry 1972, Oemler 1974, Austin & Peach 1974a,b, Rood 1975, Chincarini & Rood 1975, 1976, Gregory & Tifft 1976, and references quoted therein). Several three-parameter functions have been successfully fitted to the data; the profiles most commonly used are the bounded isothermal function (Zwicky, Bahcall), the King function, and the de Vaucouleurs function (all discussed below).

Zwicky (1957) and I (Bahcall 1972a,b, 1973a,b, 1974a, 1975a) obtained good fits to the observed distribution of galaxies with a projected bounded Emden isothermal gas sphere distribution. I (Bahcall 1972a, 1975a) fit the surface distribution $\sigma_{obs}(r)$ of cluster members (after correcting for background) with the Emden isothermal function, $\sigma_{iso}(r)$, via:

$$\sigma_{obs}(r) = \alpha[\sigma_{iso}(r/\beta) - C], \tag{7}$$

where α is a normalization factor related to the central density, β is the scale-length parameter, and C is a constant cutoff. The core radius is $R_c = 3\beta$, and the cutoff $C = \sigma_{iso}(R_h/\beta)$. The core radii defined in this way all have a similar linear value (Bahcall 1975a) of 0.25 h_{50}^{-1} Mpc for all rich regular clusters I have studied so far. I also find that the cutoff parameter C has a small spread in values in my sample; C is approximately 1.5% of the isothermal central density. The observed galaxy distributions in all fifteen clusters I have studied are compared with a projected bounded Emden isothermal gas sphere model in Figure 2.

Avni & Bahcall (1976) have studied recently, by Monte-Carlo techniques, the statistics of the agreement between observed and model profiles. By comparing simulated and observed rich clusters of galaxies they find that the observed distributions fit the model profiles somewhat better than one might expect on the most naive considerations—largely because of the process of analysis and the limited statistics.

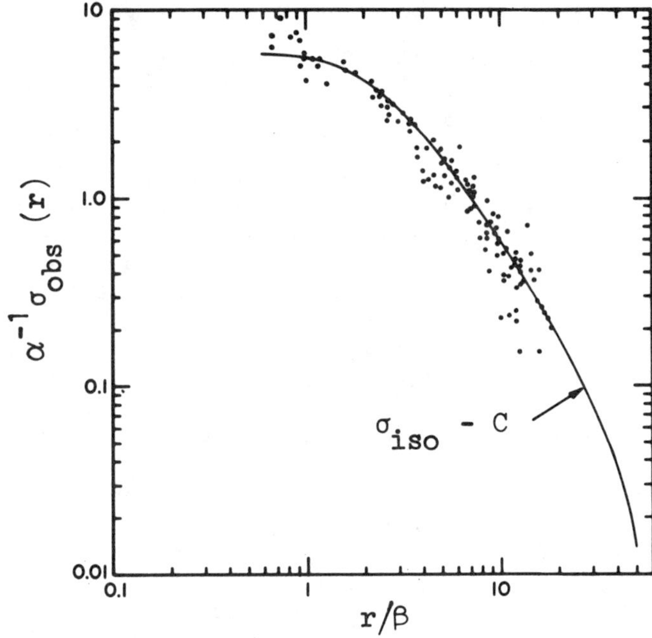

Figure 2 Projected galaxy distributions in 15 regular clusters (studies by Bahcall 1975a) plotted as a function of distance from the cluster centers. All distributions are normalized with their best fit parameters α (density normalization) and β (distance scale). The solid line is the projected bounded Emden isothermal gas sphere model with a cutoff $C = 0.1$.

The best fits to the observed distributions are obtained with a *bounded* isothermal model (i.e. $C > 0$). Hence, the slope of the projected distribution is *not* -1.0 as for an unbounded isothermal sphere (however, cf Yahil 1974), but changes with distance from the cluster center (see Figure 2) as follows: $\sigma(r \lesssim \beta) \simeq$ constant, $\sigma(4\beta \gtrsim r \gtrsim 25\beta) \propto r^{-1.6}$, and $\sigma(r \gtrsim 30\beta) \propto r^{-3}$ or r^{-4} (depending on the exact value of C). Large uncertainties due to field galaxies make it very difficult to determine the cluster profile in the outermost regions. Recent measurements of the density profile in the Coma cluster (Rood 1975, Gregory & Tifft 1976) out to a distance of a few degrees from the center suggest roughly the same slopes as expected from the bounded isothermal model (~ -1.8 in the inner parts and ~ -3.0 in the outer regions of the cluster). The projected distribution in the outer extent of the Coma supercluster (Chincarini & Rood 1975, 1976) indicates a less steep slope (~ -1.3). The outer envelopes of the 15 clusters studied by Oemler (1974) have steep slopes with a projected gradient of ~ -3.0, in agreement with the bounded isothermal model.

Another analytic representation of the galaxy distribution profile in clusters is given by King (1966, 1972). These profiles were devised to describe the surface-brightness distributions in globular clusters and elliptical galaxies, but they can also be applied to clusters of galaxies. The distribution in the *central* regions of the King models can be approximated by

$$\rho(r) = \rho_0(1+r^2/R_c^2)^{-3/2}, \tag{8a}$$

$$\sigma(r) = \sigma_0(1+r^2/R_c^2)^{-1}, \tag{8b}$$

where the parameters are the same as in Equation (6). The projected and space central densities are related through $\sigma_0 = 2R_c\rho_0$. The profile in the outer region of a cluster involves the halo radius, which physically corresponds to a distance beyond which high-velocity galaxies escape from the cluster; see King (1972) for details of the profile in these regions.

De Vaucouleurs (1948, 1960) finds that the surface density of galaxies in the Coma cluster follows closely a two-parameter law of the form $\log \sigma(r)/\sigma(r_e) = -3.33 [(r/r_e)^{1/4} - 1]$, where r_e is the effective radius of the cluster.

The shape of the surface-brightness profile of clusters is correlated with the galactic content (cf Oemler 1974). Spiral-poor and cD clusters have higher central concentrations and steeper density gradients than irregular clusters, although the central gradients are never as steep as the slope -3 observed in the envelope. Oemler also notices that midway in the profiles there is a plateau or local minimum at about 0.4 R_G (where R_G is the gravitational radius of the cluster; Section 3.6) which grows in importance as one moves from spiral-rich to cD clusters. It is most prominent in some of the individual cD clusters (A1904, A2199, and A2670). This feature is more significant in the three-dimensional mass profiles (see Press 1976). Similar secondary maxima (or local minima) were reported previously by other investigators in several clusters (Sharov 1959, Omer, Page & Wilson 1965, Clark 1968, Bahcall 1971, Austin & Peach 1974a). It is observed in both the luminosity distribution and the number density distribution. I have shown (Bahcall 1971) that the secondary maximum (in Ursa Major II) appeared in all four quadrants of the appropriate ring, thus ruling out the possibility of localized subclustering. This interesting feature in the projected distribution is statistically not larger than about a 2σ effect in any one case, although it is probably a real feature because it occurs in a number of different clusters.

By fitting observed densities to the isothermal model I have estimated central densities in rich, regular clusters of galaxies (Bahcall 1973b, 1974a, 1975a). In my study of 15 regular-type clusters, I find (Bahcall 1975a) that the central number densities, N_0, are all similar (although the scatter is large) with values $N_0(\Delta m = 3) \simeq 200 \pm 100\ h_{50}^3$ galaxies Mpc^{-3} for the three brightest magnitudes. Extrapolating this density to fainter magnitudes (down to ~ 8 mag below the brightest member) using the Abell luminosity function (Section 3.7), I obtain the following central number density of galaxies in rich regular-type clusters:

$$N_0(\Delta m = 8) \simeq 3 \times 10^3\ h_{50}^3 \text{ galaxies Mpc}^{-3}. \tag{9}$$

The *total mass density* in the central region of a cluster is given by (22) of Section 4.4; this density includes all matter at the center of the cluster (the virial density).

If this density is all in galaxies with the above number density, then the average mass of the galaxies is $\sim 10^{12}\ M_{\odot}$. If, on the other hand, only about 10% of the central density ρ_0 is due to galaxies, then their average mass would be $\sim 10^{11}\ M_{\odot}$.

3.6 *Sizes*

The total "size" of a galaxy cluster is a matter of definition. Since the outer envelope of a cluster does not exhibit an obvious sharp edge, the "total cluster size" is not a uniquely defined property; one extreme view is that each cluster merges into the low-density envelopes of other clusters. Some of the more commonly used size definitions include the following: a gravitational radius, R_G; the innermost characteristic size, the core radius R_c; a "mean" or "effective" size determined from the galaxy distribution; a halo size determined from the tail of the observed density profile as it approaches the background; and a dynamical size determined by including all galaxies within the cluster's velocity range. I discuss below these size parameters.

The gravitational radius of a cluster may be defined as

$$R_G \equiv \frac{2GM}{3v_r^2} \simeq 3h_{50}^{-1}\ \text{Mpc} \times (M/10^{15}\ M_{\odot})/(v_r/10^3\ \text{km sec}^{-1})^2, \tag{10}$$

where M is the cluster mass and v_r is the observed radial velocity dispersion. R_G, defined as in (10), is the radius at which the gravitational energy approximately equals the kinetic energy of a galaxy moving in the cluster.

Several characteristic sizes that involve the inner part of a cluster (rather than its envelope) have been suggested in the literature. Zwicky (1957) and I (Bahcall 1975a) studied the core radius [defined as $\sigma(R_c) \equiv \sigma_0/2$; see Section 3.5] from fits of the observed density profile to a bounded isothermal model. In a systematic study of 15 rich regular clusters I find (Bahcall 1975a) that the clusters studied have very similar core radii with an average value of

$$R_c = (0.25 \pm 0.04) \times h_{50}^{-1}\ \text{Mpc}. \tag{11}$$

I also find that the core radius can be used statistically as a distance indicator for regular-type clusters and that it can constrain a different combination of cosmological models and evolutionary corrections than does the more classical redshift-magnitude test. The dependence of R_c on the cluster redshift as I have determined it is shown in Figure 3. The core radius involves mainly the central parts of a cluster and hence is only slightly affected by the uncertain field density and other factors that make it difficult to measure halo sizes. The dependence of the inferred core radius on the actual position of the cluster center is also rather minor because the study is done for the regular (cD, B) clusters, which have strong central concentrations and well-defined centers. The accuracy with which core radii can be inferred for clusters of galaxies with realistic backgrounds was determined by Avni & Bahcall (1976) using computer simulations.

Additional scale lengths resulting from various mathematical representations of the observed density distribution are available. The effective radius, r_e, can be determined from the de Vaucouleurs law (e.g. Schechter 1977); Noonan (1972b)

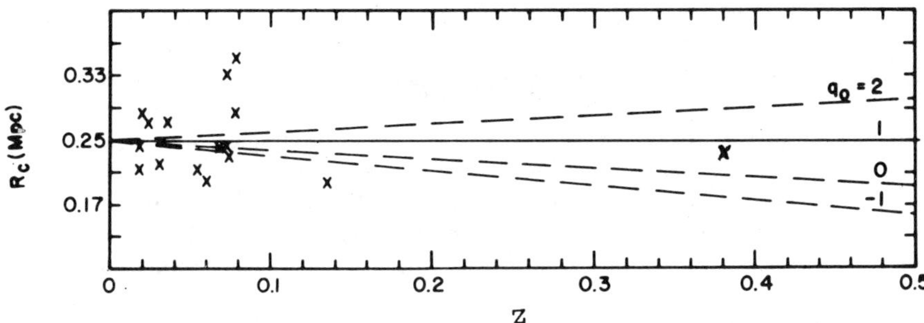

Figure 3 Linear core radii of 16 clusters (determined by Bahcall 1972b, 1975a) plotted as a function of redshift. The lines are the theoretically expected dependence of a standard linear size on redshift for different values of q_0.

raised some objections to the use of r_e and suggested a different parameter, the "mean radius" of a cluster, which he defined as the mean distance of cluster members from the line-of-sight through the cluster center. Noonan (1974b) determined the mean radii of 23 clusters of galaxies from galaxy counts. The mean radii of clusters, as well as the effective radii, are much more sensitive to the uncertain background-density contribution than are the more centrally located core radii. Austin & Peach (1974a) define another measure of angular radius by reducing the observed value of the mean radius to that which would be observed for a cluster of standard radius on the assumption that the core radius and cluster population uniquely determine the intrinsic cluster radius. Recently Hickson (1976) studied the harmonic mean separation of the brightest galaxies in a cluster and found that this scale exhibits relatively little dispersion from cluster to cluster.

It is difficult to determine accurately a halo size from the observed tailing-off of the density profile to the background. Operational definitions of the background are easy to state (e.g. a leveling-off of the density; radial velocities outside the cluster's range) but hard to apply. Clumpiness, and the fact that faint members may extend further than bright galaxies, complicate the determination of the background. A crude halo size can be estimated from fits of the observed distribution with an appropriate analytic function. Fits with a bounded isothermal model with a typical cutoff of $C \simeq 0.015\ \sigma_{\text{iso}}(0)$ (Bahcall 1975a) yield a cutoff (or halo) radius $[\sigma_{\text{iso}}(R_h/\beta) \equiv C]$

$$R_h \simeq 20\ R_c \simeq 5\ h_{50}^{-1}\ \text{Mpc}. \tag{12}$$

One may also determine a size from the velocity profile by including either all galaxies in an area of the sky within the assumed cluster velocity range or only those with crossing times not larger than a Hubble time.

A good example of some of the difficulties in determining sizes from the tails of the density and velocity profiles is provided by the most extensively studied cluster, Coma ($z = 0.023$). Different investigators have reported discrepant results for the size and the relative distribution of Coma's bright and faint galaxies (see Zwicky

1957, Noonan 1961, Omer, Page & Wilson 1965, Abell 1963, Rood et al. 1972, Rood 1975, and Yahil 1974). The discrepancies are caused mostly by the different assumptions about the density of field galaxies around the Coma cluster. Chincarini & Rood (1975, 1976) found from a velocity profile of Coma that the cluster extends out to 14° west of its center, corresponding to a cluster halo (or a supercluster) with a radius of at least 35 h_{50}^{-1} Mpc. Any such cluster extension beyond 3° is apparently not symmetric about the center of Coma (Tifft & Gregory 1976). Galaxies in these distant regions have crossing times longer than a Hubble time and presumably mostly reflect the initial conditions of the cluster formation.

The following general picture emerges. The main body of the Coma cluster is contained within a radius of about 100′ ($= 4\ h_{50}^{-1}$ Mpc) while the low-density outskirts of the cluster extend to larger radii (possibly $\gtrsim 10$ Mpc). A low-density, highly irregular supercluster is detected to a radial distance of the order of 35 h_{50}^{-1} Mpc from the center of Coma. A similar hierarchy of main concentration, low-density outskirts, and a less-regular surrounding supercluster might be typical for many of the rich regular-type clusters. Superclusters of radius ~ 35 Mpc would occupy $\lesssim 10\%$ of space for a density of Abell clusters of $\sim 6 \times 10^{-7}\ h_{50}^{3}$ clusters Mpc^{-3} (see Section 4.6).

From a power-spectrum analysis of the distribution of clusters in the Abell catalog, Hauser & Peebles (1973, Peebles 1974) find evidence for superclustering with small scale (of the order of 100 Mpc). This observed structure corresponds to an average of 2 to 3 clusters per supercluster. Inhomogeneities in the large-scale distribution of clusters in space are also observed (see also references quoted therein).

Peebles and collaborators (Peebles 1973, 1974, Peebles & Hauser 1974, Peebles & Groth 1975) performed statistical analyses of various catalogs of galaxies (the Zwicky, Shane and Wirtanen, and Jagellonian catalogs) and estimated the two- and three-point correlation functions for the galaxy distributions. They find that the correlation functions vary as simple powers of the angular scale. The index of the power law for the three-point correlation function is approximately twice the index for the two-point function, in agreement with the assumption that the clustering satisfies a simple scaling over the studied range of length. These authors conclude that their results are evidence that, in a useful approximation, galaxy clustering has no preferred scales in the range of 0.06 to 40 h_{50}^{-1} Mpc that has been studied so far. Although there can (and do) exist clusters that appear as distinct entities, Peebles and collaborators suggest that such systems are selected from a continuous distribution. They argue that only a small fraction of the galaxies can be involved in any preferred scale; that is consistent with clusters having characteristic core radii that contain only a fraction of their total populations.

Zwicky and collaborators (see Zwicky & Karpowicz 1966 and references therein) discussed the sizes of the largest clusters in the Zwicky catalog which correspond to a projected density of about twice that of the field. They report that, although there is considerable scatter in the data, the largest clusters of all types have about the same size with a diameter of the order of 10 h_{50}^{-1} Mpc. Since this size corresponds to Zwicky's isopleth, the actual extent of the clusters may be larger than 10 Mpc. Abell (1958) defined, for use in his catalog, a fixed linear cluster radius of 3 h_{50}^{-1} Mpc

[Equation (2)] that is considerably larger than the main concentration of the cluster but may not include all of its low density envelope.

3.7 *The Luminosity Function*

The luminosity function of galaxies in a cluster, defined in its integrated form as the number of galaxies $N(\leqq m)$ brighter than magnitude m, has been studied observationally by various investigators (e.g. Abell 1962, 1976, Rood 1969, Bautz & Abell 1973, Oemler 1974, Krupp 1974, and others). The range of magnitudes covered by present observations includes about the seven brightest magnitudes of a cluster. Several analytic representations for the cluster luminosity function have been proposed; I discuss below the functional forms due to Zwicky, Abell, and Schechter. These functions are compared with each other and with some observations in Figure 4. All of the observational studies I discuss yield luminosity functions of the same general shape.

Zwicky (1957) suggested that $N(\leqq m) = K(10^{0.2m} - 1)$ where K is a constant.

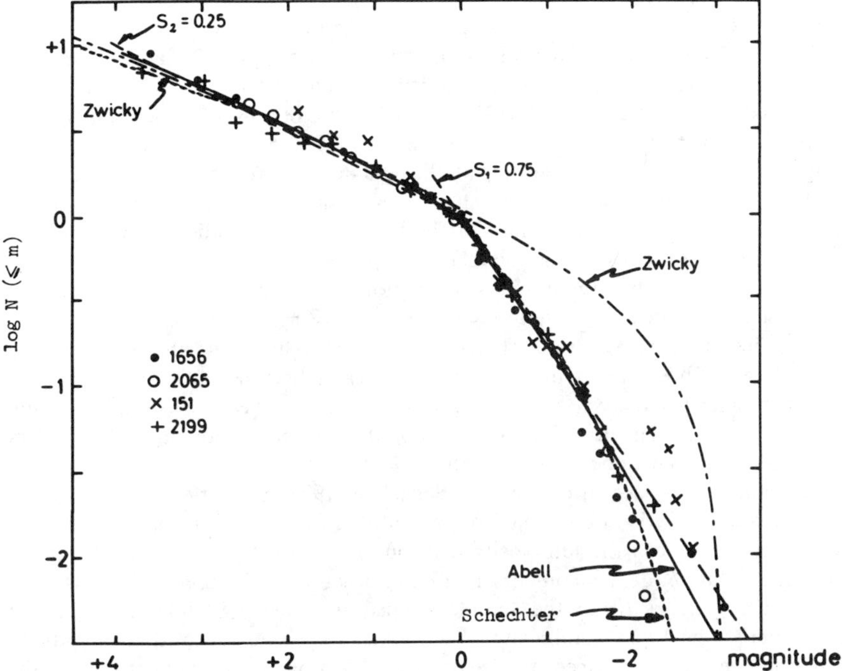

Figure 4 Logarithmic intergrated luminosity functions of four rich clusters (Abell 1976). The zero points for magnitude and number are arbitrary. The labelled lines are the functions due to Zwicky, Abell (Equation 13; also an average line marked "Abell"), and Schechter (from Equation 14). The Abell and Schechter functions are practically identical except in the brightest (i.e. cD) region.

Although the Zwicky function represents the faint end of the observed luminosity function rather well, it is not steep enough at the bright end (Figure 4). We note that the slope of the faint end of the luminosity function may depend on the photometric system used since fainter galaxies tend to be bluer than brighter galaxies.

Abell has applied extrafocal photographic photometry (Abell & Mihalas 1966) to determine "total" visual magnitudes of galaxies and luminosity functions of several clusters (Abell 1962, 1976, Bautz & Abell 1973). He represents the logarithm of the integrated luminosity function by two intersecting straight lines; the magnitude m^* at which the two lines intersect characterizes the point at which a change in slope occurs in the luminosity function. If all clusters have the same shape for their luminosity function, then m^* may be a useful "standard candle" for determining relative distances of clusters (Abell 1962, 1976, Bautz & Abell 1973, Austin et al. 1975, see also Schechter 1976, Schechter & Press 1976). The Abell luminosity function can be represented by (cf Figure 4)

$$\begin{aligned} \log N(\leqq m) &= K_1 + S_1 m \qquad (m \leqq m^*), \\ \log N(\leqq m) &= K_2 + S_2 m \qquad (m > m^*), \end{aligned} \tag{13}$$

where the constant slopes have the preliminary values (Abell 1976) of $S_1 = 0.75$ and $S_2 = 0.25$. (Zwicky's function has a similar slope, $S_2 = 0.20$, at faint magnitudes). From data of nine clusters studied by Bautz & Abell (1973) and three additional clusters, Austin et al. (1975) obtain the main absolute visual magnitude at the discontinuity to be $\langle M_V{}^* \rangle = -21.16 \pm 0.15 + 5 \log h_{50}$. This absolute magnitude is obtained by integrating all the light down to apparent visual intensities of 25 mag (arc sec)$^{-2}$ and is corrected for K-dimming and galactic absorption. The dependence of m^* on redshift for the twelve clusters studied shows a rather small scatter around a straight line relation expected for $q_0 = 1$.

Krupp (1974) studied the luminosity functions of E and SO galaxies in twenty-seven rich clusters over a visual magnitude range of nearly six magnitudes below the brightest member. He finds that these luminosity functions all have the same general form, shown in Figure 4. Oemler (1974) studied the luminosity functions of fifteen rich clusters over the five brightest magnitudes of each cluster. Composite differential luminosity functions for the spiral-rich, spiral-poor, and cD clusters show small differences between these three classes.

Schechter (1976, see also Press & Schechter 1974, Schechter & Press 1976) suggested an analytic approximation for the luminosity function that shows good agreement with both the luminosity distribution of bright nearby galaxies (from de Vaucouleurs & de Vaucouleurs 1964) and a composite luminosity distribution for cluster galaxies (from the sample studied by Oemler 1974). The shape of Schechter's integrated luminosity function is very similar to that of Abell's function (see Figure 4), except that Schechter's function is somewhat steeper at the bright end. The analytic expression for the number of cluster galaxies in the luminosity interval L to $L+dL$, $N(L)dL$, is given by Schechter as

$$N(L)dL = N^*(L/L^*)^{-5/4} \exp(-L/L^*) d(L/L^*), \tag{14}$$

where L^* is a characteristic luminosity (with an equivalent magnitude M^*) at which the luminosity function exhibits a rapid change in slope in the log N-log L

plane. The parameter N^* is proportional to the cluster luminosity and is a measure of its richness. From fits to the data for thirteen of the clusters studied by Oemler (1974), Schechter determined the characteristic absolute magnitude: $M^*_{J\,(24.1)} = -21.4 \pm 0.1 + 5 \log h_{50}$ [where $J \simeq V + 0.35\,(B-V)$; see Oemler 1974]. The above expression gives a good approximation to the cluster luminosity function only when the *cD galaxies are excluded.* The cD galaxies have higher luminosities than given by the bright end of Schechter's function; their luminosities range up to $10L^*$ for the values of the parameters given above. According to Schechter, the data for most of the clusters in Oemler's sample suggests that M^* and the slope at the faint end of the distribution (i.e. $-5/4$) are constant from cluster to cluster; however several clusters (e.g. A2670, A665) may have differently shaped luminosity functions, and there may be some dependence of M^* on cluster type (cf Richstone 1974).

Schechter's analytic expression (14) predicts a correlation of absolute magnitude of first-ranked cluster galaxies with cluster richness that is not observed in the data (Sandage 1976 and references therein). Schechter & Peebles (1976) suggest that the apparent absence of a correlation with richness may be understood by a combination of statistics and selection effects. Although no appreciable dependence on richness is observed, the magnitude of the first-ranked galaxy appears to be correlated with the cluster central density (Section 3.4).

The bright end of the luminosity function is of great importance for use as a standard candle in the classical cosmological tests (Sandage & Hardy 1973, Sandage 1976, Sandage et al. 1976, Gunn & Oke 1975, Peebles 1968, Schechter & Peebles 1976, and references quoted therein). As stressed by Sandage, the brightest galaxies in rich clusters exhibit a remarkably small luminosity dispersion of only about ± 0.35 mag about $\langle M_V \rangle (1) = -23.2 + 5 \log h_{50}$. Small corrections to $M_V(1)$ that depend on the cluster BM type and richness are given in Sandage (1976) and references quoted therein. The shallowness of the correlation of the absolute magnitude of the brightest cluster galaxy with cluster richness requires, according to Sandage (1976), that *if* $M(1)$ is governed by the general luminosity function for fainter members, then the slope of the function at $M(1)$ must be very steep (~ 5.5). No observed luminosity functions of either clusters or field galaxies show such a steep slope. Sandage (1976, see also Peach 1969, Tremaine & Richstone 1977) thus suggest that the luminosity of brightest cluster galaxies is determined by some special physical processes. Peebles (1968, Schechter & Peebles 1976, and references therein) has argued however that the small luminosity dispersion of $M(1)$ has no particular physical significance and it can arise statistically when the dominant member is drawn at random from a sample of galaxies that obeys a universal luminosity function.

The magnitude interval in the cluster luminosity function between $M(1)$, at the bright end, and M^*, at the discontinuity point, is typically $M^*_v - M_v(1) \simeq 2$ to 3 mag, depending on the cluster BM type and richness. It would be of interest to determine both M^* and $M(1)$ for the *same* data sample and to compare their dependences on various cluster parameters (e.g. richness and structural type).

The luminosity function of galaxies in small groups was studied by Turner & Gott (1976b), who found the data to be well fit by a function similar to (14).

4 DYNAMICS

4.1 *Characteristic Times*

I discuss in this subsection several characteristic time scales of clusters of galaxies: the crossing, two-body relaxation, dynamical friction, collision, and bremsstrahlung cooling times. The numerical values are given for a Hubble constant of 50 km sec^{-1} Mpc^{-1} and are proportional to h_{50}^{-1}.

A galaxy travelling through a cluster with a velocity v will cross a radius R in a cluster in a crossing time given by

$$T_{CR} = R/v \simeq 6 \times 10^8 \text{ yr} \times [(R/\text{Mpc})/(v_r/10^3 \text{ km sec}^{-1})], \tag{15}$$

where v_r is the observed radial velocity (spherical symmetry, $v^2 = 3v_r^2$, is assumed throughout). The crossing time for a typical distance of 10 Mpc is $\sim 6 \times 10^9$ yr, only slightly shorter than the Hubble time (2×10^{10} yr). Galaxies at the outer regions of large superclusters ($R \gtrsim 35$ Mpc) have crossing times $\gtrsim 2 \times 10^{10}$ yr and therefore have not yet had time to travel through the center of the cluster.

The two-body relaxation time for galaxies in clusters, which measures the time in which collisions can produce a large alteration in the original velocity distribution, is given by (see Chandrasekhar 1942, Spitzer & Hart 1971)

$$\begin{aligned} T_R &= v^3/4\pi G^2 M_g^2 N \ln\Lambda \\ &\sim 2 \times 10^{10} \text{ yr} \times [(v_r/10^3 \text{ km sec}^{-1})^3/(M_g/10^{12} M_\odot)^2(N/10^3 \text{ gal Mpc}^{-3})\ln\Lambda], \end{aligned} \tag{16}$$

where v and M_g are the galaxy velocity and mass, N is the number density in the cluster, and $\ln\Lambda$ is the ratio of maximum to minimum impact parameters. If relaxation is due to the dynamical friction of a galaxy moving through a homogeneous and isotropic background distribution of lighter bodies, the term NM_g in Equation (16) should be replaced by ρ_{bg}, the background mass density in the cluster; thus $T_{DF} = v^3/(4\pi G^2 M_g \rho_{bg} \ln\Lambda)$. The dynamical friction time is related to the two-body relaxation time through the ratio of mass density in bright galaxies to the density in the background of lighter bodies. Galaxies relax faster the larger their mass and the higher the galaxy density is in their vicinity. In the central parts of regular clusters, where $N \sim 3 \times 10^3$ gal Mpc^{-3} [see (9)], the two-body relaxation time may be $\sim 10^9$ yr for the massive galaxies ($M_g \sim 10^{12}$ $M_\odot$); some relaxation could have occurred in these central regions. No appreciable two-body relaxation is expected for much lighter galaxies.

The observable effects of relaxation are spatial and velocity segregation of galaxies according to their masses. The data (Section 4.2) suggest that a partial evolution toward energy equipartition has actually taken place for the most massive galaxies at the centers of highly concentrated clusters. No significant segregation is observed for the less bright (i.e. less massive) galaxies. Outside the cores of regular clusters, and throughout most of the irregular clusters, the relaxation time is long ($> 2 \times 10^{10}$ yr); no appreciable relaxation is expected to have occurred in these regions, which is consistent with the available data (Section 4.2).

The average time between successive collisions of a galaxy with other cluster members is given by

$$T_{\text{coll}} = [2^{1/2} v N \pi R_g^2]^{-1} \sim 10^9 \text{ yr} \times [(v_r/10^3 \text{ km sec}^{-1})(N/10^3 \text{ gal Mpc}^{-3})(R_g/10 \text{ kpc})^2]^{-1}, \tag{17}$$

where R_g is the galaxy radius. The effect of gravitational focusing is negligible since $v^2 \gg GM_g/R_g$. In the dense central regions of a regular cluster, the collision time is typically 10^8 to 10^9 yr for a galaxy with a radius of ~ 10 kpc. In the lower density parts of a regular cluster (or in irregular clusters), where $N \lesssim 100$ gal Mpc^{-3}, the time between collisions is much longer ($> 10^{10}$ yr). For a galaxy radius larger than 10 kpc, the collision time is reduced considerably.

Richstone (1974) studied the collisions of galaxies in dense clusters and found that if the galaxies start out with extended halos 500 kpc in extent that encompass all the mass in the cluster, about 90% of the mass and 25% of the luminosity is liberated in a Hubble time; this intergalactic material may be identified with the halo of the cD galaxy. The classification schemes discussed in Section 3.2 and Table 1 are ordered in this scenario according to the increasing importance of collisional effects on the cluster galaxies (compare also Section 4.3 below). Collisions in small dense groups were recently studied by Hickson, Richstone & Turner (1977).

The cooling time of the intracluster gas by bremsstrahlung emission is given by

$$T_{BR} = 9 \times 10^7 \text{ yr} \times T_8^{1/2} n_e^{-1}, \tag{18}$$

where T_8 is the gas temperature in 10^8 K, and n_e is the electron density in particles cm^{-3}. For a typical gas temperature of 10^8 K, as expected from the observed velocities in the cluster and as suggested by the observed X-ray emission, and a typical density of $\sim 10^{-3}$ cm^{-3} (see Section 5), the cooling time is rather long ($\sim 10^{10}$–10^{11} yr). Therefore, gas of this density that is heated to 10^8 K will stay hot for roughly the cluster lifetime.

4.2 *Mass Segregation*

If a cluster has relaxed through two-body interactions and has approached a state of equipartition of energy, the fainter (presumably less massive) galaxies should occupy a larger volume of space than the brighter ones. Unfortunately, segregation in the distribution of galaxies of different brightnesses is difficult to measure accurately. In the outer regions of clusters, the distribution of the faint galaxies is strongly dependent on the assumed field density, which is clumpy and usually uncertain (see Sections 3.5 and 3.6); a slightly different choice of field density can completely change the inferred distribution of faint cluster members in the outer regions. Near the cluster center, on the other hand, the accuracy with which the distribution can be determined is limited by the statistics of small numbers. Despite these observational difficulties it seems that a partial energy equipartition of the brightest galaxies ($\Delta m \lesssim 2$ mag) of regular-type clusters is indicated by the data.

Various observations suggest that in some clusters the fainter galaxies may have a wider projected distribution than the brighter ones (see, e.g. Zwicky 1942, 1957: Virgo, Coma; Rudnicki 1963: Perseus; Abell 1963, Rood 1969, Rood & Abell 1973, Bahcall 1973a, White 1977: Coma; Zwicky & Humason 1964: A194; Hodge, Pyper, & Webb 1965: Fornax; Rudnicki & Baranowska 1966b: A1185 and A1213; Kwast 1966: A1060; Noonan 1972a: A1413, but compare Austin & Peach 1974b; Bahcall 1972a: A31; Noonan 1974: A234 and C1.0024+1654B; and others). However, many of these results are marginal because of one or more of the following complications: 1. very uncertain (or incorrectly assumed) field density of galaxies; 2. fluctuations of small-sample statistics; 3. subclustering; 4. different symmetry in the distributions of bright and faint galaxies; and 5. color or type segregation. Other investigations of different clusters have shown no statistically significant mass segregation between the bright and faint galaxies.

In the case of the best studied cluster, Coma, only the brightest galaxies ($\Delta m \sim 2$) indicate a possible stronger concentration at the center of the cluster ($R \lesssim 12'$); (see Rood 1969, Rood & Abell 1973, Bahcall 1973a, Oemler 1974). Velocity measurements of galaxies in Coma (Rood et al. 1972), as well as in some other clusters (e.g. A194: Zwicky & Humason 1964; A1367: Tifft & Tarenghi 1975; see however Perseus: Chincarini & Rood 1971) indicate that the velocity dispersion of the brightest few galaxies in the core is lower than that of the fainter galaxies—a finding consistent with energy equipartition of the brightest galaxies; the statistical significance of each case is rather poor, however. Further studies of velocity profiles as a function of magnitude are needed.

Another indication of incomplete energy equipartition was derived by Oemler (1974) from a study of 15 clusters of all three classification types (cD, spiral-poor, and spiral-rich clusters). The mean distance of galaxies from the cluster center as a function of magnitude was determined by Oemler for the brightest 3 magnitudes in each cluster. The results (see Figure 10 of Oemler) suggest that some mass segregation is present in the brightest $\sim$1–2 magnitudes of the cD and spiral-poor clusters. No evidence for segregation at magnitudes fainter than 2 mag below the brightest member was observed in this sample. The irregular, spiral-rich clusters show no evidence of mass segregation at any of the studied magnitude intervals. This is another indication that little two-body dynamical evolution has occurred in these clusters.

4.3 *Dynamical Evolution*

The hypothesis that clusters of galaxies develop from small density perturbations in the early Universe has been investigated by many workers (see Gunn & Gott 1972 and references therein). The collapse time scale of the perturbation is given by

$$T_{\mathrm{collap}} \simeq \pi(R^3/2GM)^{1/2} \sim 10^9\ \mathrm{yr} \times [(R/\mathrm{Mpc})^3/(M/10^{15}\ M_\odot)]^{1/2}, \tag{19}$$

where M and R are the cluster mass and radius. For typical values similar to those of the main body of the Coma cluster ($R = 4\ h_{50}^{-1}$ Mpc, $M = 4 \times 10^{15}\ h_{50}^{-1}\ M_\odot$), the collapse time is $\sim 5 \times 10^9\ h_{50}^{-1}$ yr. Since this is shorter than the Hubble time, it is believed that rich regular-type clusters such as Coma are collapsed systems that have undergone violent quasi-relaxation (Lynden-Bell 1967). Numerical models of the collapse (Peebles 1970, see also Aarseth 1969, White 1976b and references therein) indicate that in the period $T_{\rm collap} < t < 1.5\ T_{\rm collap}$, dynamical evolution is violent and a density distribution differing only slightly from the present distribution in the cluster is established; after $1.5\ T_{\rm collap}$ the cluster is essentially quiescent.

The distinction between loose, irregular clusters and compact, regular ones could be explained (cf Gunn & Gott 1972) simply if the latter are older, and the former younger, than their collapse time. The observed differences between the regular and irregular clusters (i.e. in mean density, symmetry of distribution, mass segregation, and composition) all support this view. The intermediate clusters may have ages, measured in units of their collapse time, between those of regular and irregular clusters. Since the intermediate clusters seem to have a higher SO/E ratio than do the regular clusters, a mechanism for efficiently transforming SO galaxies into ellipticals is needed *if* the regular clusters have evolved from the intermediate ones.

4.4 *Masses and Mass-to-Light Ratios*

The total mass of a cluster can be expressed in terms of the observed properties R_e and v^2 as (Zwicky 1933, 1957, Smith 1936, Schwarzschild 1954)

$$M = v^2 R_e/G \simeq 0.7 \times 10^{15}\ M_\odot \times [(v_r/10^3\ \text{km sec}^{-1})^2\,(R_e/\text{Mpc})], \tag{20}$$

where R_e is the effective mean radius (see below). In writing equation (20) one assumes that the observed velocity dispersion applies to all the mass in the cluster and that all the mass is distributed in the same way as the galaxies from whose distribution one determines the potential energy. The effective mean radius R_e is given by Schwarzschild (1954) as

$$R_e \equiv GM^2/E_G = 2\left[\int_0^R S(x)\,dx\right]^2 \Big/ \int_0^R S^2(x)\,dx, \tag{21}$$

where E_G is the gravitational energy of the cluster, $S(x)$ is the number of galaxies in a strip of unit width at a perpendicular distance x from the cluster center, and R is the limiting radius of the cluster. The virial theorem does not provide an accurate determination of the total mass if the mass is largely in the poorly observed outer regions. [The total mass computed from (20) and (21) contains a major contribution from the outer envelope if the projected density decreases like r^{-2} or slower, corresponding to r^{-1} or slower in the strip density.]

An alternative approach depends only on the densities and velocities in the well-observed region. The central density distribution can be combined with the radial velocity dispersion at the cluster center, $v_{r,c}$ to give the average total mass density at the center,

$$\begin{aligned}\rho_0 &= 9v_{r,c}^2/4\pi G R_c^2 \qquad (\propto h^2)\\ &\simeq 3 \times 10^{15}\ M_\odot\ \text{Mpc}^{-3} \times [(v_{r,c}/10^3\ \text{km sec}^{-1})/(R_c/0.25\ \text{Mpc})]^2\end{aligned} \tag{22}$$

regardless of the form in which the mass exists. The total mass of the cluster included within a distance R is thus

$$M(\leqq R) = 4\pi\rho_0 \int_0^R [\rho(r)/\rho_0] r^2 \, dr, \tag{23}$$

where $\rho(r)$ is the spatial distribution of the galaxies given by either an isothermal sphere model or by King's (1972) formula. The mass of any part of the cluster for which the density distribution is observationally well determined can be estimated from (23).

Another method is to determine directly the mass-to-light ratio of the cluster core. The central mass-to-light ratio in the cluster is (see Rood et al. 1972)

$$(M/L)_c = 2\rho_0 R_c/\sigma_{0,L} = 9v_{r,c}^2/2\pi G\sigma_{0,L}R_c \simeq 133\, h_{50}\, M_\odot/L_\odot$$
$$\times\, [(v_{r,c}/10^3 \text{ km sec}^{-1})^2/(\sigma_{0,L}/10L_c \text{ pc}^{-2})(R_c/0.25 \text{ Mpc})], \tag{24}$$

where $\sigma_{0,L}$ is the central surface luminosity density (the space luminosity density is $\sigma_{0,L}/2R_c$). The product of the projected central density and core radius, $\sigma_{0,L}R_c$, is insensitive to variations in the cluster profile outside the core (Bahcall 1975a). The central mass-to-light ratio of a cluster can be determined from (24) by use of the observed radial velocity dispersion of galaxies in the core and the product $\sigma_{0,L}R_c$ obtained from the central surface distribution; both of these parameters can be reliably measured.

Estimates of the mass of the Coma cluster were made by the above three methods. For use in Equation (20), Rood et al. (1972) obtain $R_e = 3.68° = 9.15\, h_{50}^{-1}$ Mpc for Coma from an average of four different strip counts analyses. For the observed velocity dispersion of $v_r = 861$ km sec^{-1}, the virial mass of Coma is calculated to be

$$M(\text{coma}) = 4.7 \times 10^{15}\, h_{50}^{-1}\, M_\odot. \tag{25}$$

The total luminosity of the Coma cluster is (Rood et al. 1972) $L(\text{coma}) = 3 \times 10^{13}\, h_{50}^{-2}\, L_\odot$. The Coma cluster mass-to-light ratio is thus

$$M/L(\text{coma}) \simeq 150\, h_{50} M_\odot/L_\odot. \tag{26}$$

Oemler (1973b) obtains $M/L = 195 \pm 38\, h_{50}\, M_\odot/L_\odot$ and $L = 0.94 \times 10^{13}\, h_{50}^{-2}\, L_\odot$ for the Coma cluster.

The second method [Equations (22) and (23)] can be used by applying the measured density profile (Rood et al. 1972) for $r \leqq 200'$; one finds $M(\text{Coma}, \leqq 200') \simeq 5 \times 10^{15}\, h_{50}^{-1}\, M_\odot$, in rough agreement with the virial method mass.

Equation (24), assuming $v_{r,c} = 1060$ km sec^{-1}, $R_c = 0.25$ Mpc, and $\sigma_{0,L} = 9.2\, L_\odot\, p_c^{-2}$ (cf Rood et al. 1972) yields $(M/L)_c$ (coma) $= 160\, h_{50}\, M_\odot/L_\odot$. This result is similar to that obtained by the virial theorem (26), but is independent of the densities and velocities in the envelope.

Masses and mass-to-light ratios of various rich clusters have been calculated by different investigators (see Oemler 1973b, Rood 1974b, Bahcall 1974a, and references therein). The masses, luminosities, and M/L ratios of rich clusters of galaxies typically fall within the following ranges:

$$M \sim 10^{15\pm1}\,h_{50}^{-1}\,M_\odot,\ L \sim 10^{12}-10^{13}\,h_{50}^{-2}\,L_\odot,\ M/L \sim 50-500\,h_{50}\,M_\odot/L_\odot, \quad (27)$$

with an average of

$$\langle M/L \rangle \sim 200\,h_{50}\,M_\odot/L_\odot. \quad (28)$$

A dependence of the virial mass-to-light ratio of rich clusters on the velocity dispersion in the cluster was reported by Rood (1974b, see his Figure 1); according to Rood, clusters with higher velocity dispersions tend to have a higher M/L ratio.

The masses and mass-to-light ratios of small groups of galaxies are more difficult to study since the exact membership and dynamical state of individual groups are hard to determine. For various studies of the masses of small groups of galaxies see Gott & Turner (1977a) and references therein.

4.5 *The Missing-Mass*

Zwicky (1933) and Smith (1936) were the first to point out that galaxies in clusters exhibit unexpectedly large velocity dispersions. Cluster masses derived from velocity dispersions with the aid of the virial theorem, assuming all matter in the cluster is distributed like the galaxies, are an order of magnitude greater than the conventionally estimated sum of the masses of the individual galaxies. This so-called "missing mass" problem has produced, over the past four decades, numerous observational and theoretical studies, including many contradictory results and conclusions.

The range of suggested M/L ratios is large: from $\sim 1\ M_\odot/L_\odot$ to $\sim 500\ M_\odot/L_\odot$ ($h_{50} = 1$). Studies of individual galaxies using optical and 21-cm rotation curves and velocity dispersions within the visible parts of galaxies yield mass-to-light ratios that are typically $\sim 8\ M_\odot/L_\odot$ for spirals and irregulars and $\sim 30\ M_\odot/L_\odot$ for ellipticals and SO's (cf Roberts 1969, King & Minkowski 1972, Chiano & Reinhardt 1973, and references therein; see also Faber & Jackson 1976, who give a ratio of $\sim 7\ M_\odot/L_\odot$ for the cores of elliptical and SO galaxies). Binary and group studies typically yield ratios in the range ~ 20 to $100\ M_\odot/L_\odot$ (Page 1962, de Vaucouleurs 1976, Turner 1976, and references therein). Rich clusters of galaxies usually show the largest $\langle M/L \rangle$ ratios with an average value of $\sim 200\ M_\odot/L_\odot$ (Section 4.4).

Turner (1976) studied the dynamics and mass-to-light ratio of a sample of binary galaxies, paying special attention to the removal of selection effects. He finds a rather large total M/L ratio of late-type galaxies, estimating an average value $\sim 65\ M_\odot/L_\odot$ when measured to projected separations ~ 100 kpc. The less certain value for ellipticals and SO's is $\sim 130\ M_\odot/L_\odot$. This result extends the classical "missing-mass" problem for groups and clusters to binary systems.

The mass discrepancy can be understood if spiral galaxies possess dark but very massive extended halos of ~ 100 kpc in radius and containing ~ 10 times the disc mass. It is known from studies on Schmidt plates that many galaxies possess large low-luminosity halos (see Arp & Bertola 1971, Kormendy & Bahcall 1974); it is not definitely known whether these halos have the very high mass-to-light ratios required if they contain most of the galaxy mass. The massive-halo hypothesis was suggested and discussed by Ostriker & Peebles 1973, Ostriker et al. 1974, and Einasto et al. 1974 (cf however Burbidge 1975). Various direct (Ostriker et al. 1974, Rogstad & Shostak

1972, Roberts & Rots 1973, Turner 1976) and indirect (Ostriker & Peebles 1973, Kaljnas 1972) indications that galaxies possess such massive halos have been reported. The rotation-curve data may reflect only the small-radius mass, while group or cluster data reflect the *total* mass. If the galaxies themselves contain all the hidden mass then two-body relaxation is fast enough (Section 4.1) to produce substantial equipartition of energy between light and massive galaxies. The lack of observed mass segregation in the Coma cluster (except for the brightest 2 mag galaxies; Section 4.2) implies (White 1977) that only a small fraction of the cluster mass can be attached to the individual galaxies. This is consistent with estimates by Richstone (1974) that cluster galaxies are stripped of their halos by collisions.

Ambartsumian (1961) suggested that the mass inferred from the virial theorem might be incorrect if clusters have positive energy and are expanding. While this hypothesis could be valid for special systems, it probably does not apply to rich regular clusters in general. The short crossing times in the cluster's cores ($\sim 10^8$ yr), coupled with their observed high central concentrations, as well as the high percentage of elliptical galaxies in clusters that have not been distributed into the general field, are arguments against expansion.

If a cluster is bound, and the "missing-mass" exists as an invisible, nonhalo matter, then the distribution of this missing mass should be similar to the galactic distribution (Rood et al. 1972). If the mass were much less centrally concentrated than the galaxies it could not bind the cluster core; if it were much more centrally concentrated than the galaxies it could not hold in the distant envelope of the cluster that exhibits nearly the same velocity dispersion as the rest of the cluster.

The missing mass is probably not in dwarf galaxies since an extrapolation of the luminosity function (Section 3.7) suggests that faint galaxies contribute little to the cluster luminosity and mass.

Evidence for the existence of ionized intracluster gas is provided by cluster X-ray and radio emission (Sections 5 and 6; see also Field 1973). However, the inferred amount of the intracluster gas is much too small to bind the clusters. The amount of neutral hydrogen gas in clusters (De Young & Roberts 1974) seems to be negligible ($< 10^{11}\ M_\odot$) (cf Smart 1973). Davidsen, Bowyer & Welch (1973) and others showed that the Coma cluster cannot be bound by relatively cool ionized intracluster gas. Lea et al. (1973) calculated the density of hot ($\sim 7 \times 10^7$ K) gas that could produce the observed X-ray emission in Coma by bremsstrahlung and found a total gas mass of $(5 \pm 3) \times 10^{14}\, h_{50}^{-3/2}\ M_\odot$. This is comparable to the conventional mass of the galaxies in Coma and contributes only about 10% of the virial mass for $h_{50} = 1$. Similar values have been obtained for other X-ray clusters.

4.6 *Mean Galaxy and Cluster Mass Density in the Universe*

The contribution to the mean mass density in the universe of matter associated with galaxies can be estimated from an average mass-to-light ratio, $\langle M/L \rangle$, and the total luminosity density due to galaxies.

In the simplest Friedman cosmologies with $\Lambda = 0$, the parameter $\Omega \equiv \rho_0/\rho_{\rm crit}$ determines whether the universe is closed and will eventually recollapse ($\Omega > 1$) or open and will expand forever ($\Omega < 1$). ρ_0 is the present mean mass density in the universe, and $\rho_{\rm crit}(= 3H_0^2/8\pi G = 5 \times 10^{-30}\, h_{50}^2$ g cm^{-3}) is the critical mass density needed to close the universe. Note $\Omega = 2q_0$.

If Ω_G is the contribution due to matter associated with galaxies (which also includes any intergalactic matter falling into clusters and groups), then $\Omega_G \lesssim \Omega$. The parameter Ω_G can be determined from

$$\Omega_G = \rho_G/\rho_{\rm crit} \simeq 0.14 \times [(\rho_L/10^8\, L_\odot\, {\rm Mpc}^{-3})(\langle M/L\rangle/100)], \tag{29}$$

where $\rho_G = \rho_L\langle M/L\rangle$ is the density associated with galaxies and ρ_L is the mean luminosity density of galaxies. Ω_G is independent of H_0. Conventional estimates of $\rho_L \simeq 1.6 \times 10^8$ $L_\odot$ Mpc^{-3} (Oort 1958, Shapiro 1971) and $\langle M/L\rangle \sim 10$ $M_\odot/L_\odot$ yield $\Omega_G \simeq 0.02$. Gott & Turner (1976a), in a recent reevaluation, obtain $\rho_L = 4.7 \times 10^7\, h_{50}$ $L_\odot$ Mpc^{-3} and $\langle M/L\rangle = 120\, h_{50}$ $M_\odot/L_\odot$, and hence

$$\rho_G = 4 \times 10^{-31}\, h_{50}^2\ {\rm g\ cm}^{-3}, \quad \Omega_G = 0.08. \tag{30}$$

The mass density in the universe due to rich (Abell) clusters, $\rho_{\rm Abell}$, can easily be estimated. The total number of Abell clusters expected over the whole sky (to within 1200 h_{50}^{-1} Mpc) is $\sim 4 \times 10^3$; thus the number density is

$$N_{\rm Abell} \sim 0.6 \times 10^{-6}\, h_{50}^3\ {\rm clusters\ Mpc}^{-3}. \tag{31}$$

The mean mass density in the form of Abell clusters (for an average cluster mass M) is

$$\begin{aligned} \rho_{\rm Abell} &\simeq 4 \times 10^{-32}(M/10^{15}\, M_\odot)\, h_{50}^2\ {\rm g\ cm}^{-3}, \\ \Omega_{\rm Abell} &\simeq 0.01(M/10^{15}\, M_\odot). \end{aligned} \tag{32}$$

5 X-RAY EMISSION

The majority of all presently identified extragalactic X-ray sources are associated with clusters of galaxies (Kellogg et al. 1973, Bahcall & Bahcall 1975, Rowan-Robinson & Fabian 1975, Gursky & Schwartz 1977). A list of all X-ray sources that are currently believed to be associated with clusters of galaxies is presented in this volume by Gursky and Schwartz. These authors review in detail the X-ray aspects of the subject. I give below a brief summary of the main characteristics of clusters of galaxies that contain X-ray emission, with an emphasis on the relation of X-ray emission to optical properties.

The suggested associations between clusters and X-ray sources are based primarily on positional coincidences between the error boxes of the X-ray sources and the observed clusters. By analyzing with Monte Carlo studies the positional correlations between 3U X-ray sources (Giacconi et al. 1974) and the Abell clusters, Bahcall & Bahcall (1975) found that practically all the suggested identifications with nearby clusters ($z \lesssim 0.07$) are valid associations, while correlations with distant Abell clusters ($z \gtrsim 0.14$) are, with the Uhuru-size error boxes, mostly accidental superpositions. X-ray detectors with better spatial resolution ($\lesssim$ a few arc min) for relatively faint sources ($\lesssim$ several Uhuru flux units) are required to test the possible identification of faint X-ray sources with distant clusters.

The cluster X-ray sources, when strong enough to be studied in detail, are found to be extended; a typical core radius (of the gas producing the X-rays on the thermal bremsstrahlung model) is $\sim 0.5\ h_{50}^{-1}$ Mpc (Lea et al 1973, Kellogg & Murray 1974). The X-ray emission thus includes the cluster core and could extend throughout the visible cluster. In contrast, other extragalactic sources (isolated galaxies, QSO's) do not exhibit a finite size with present techniques. No single galaxies, with the possible exception of cD galaxies (Section 3.4), are known to have optical sizes comparable to those of the extended cluster X-ray emission. In some cases, such as the Perseus cluster (Wolff et al 1974), a point-like X-ray source coincident with the central bright and active galaxy is superimposed on the extended source.

The X-ray luminosities of the cluster sources are in the range 10^{43}–$10^{45}\ h_{50}^{-2}$ ergs sec^{-1}; for comparison, the X-ray luminosities of single galaxies (e.g. NGC 4151, NGC 5128) are $\lesssim 10^{42}$ ergs sec^{-1} and that of the QSO 3C273 is $\sim 10^{46}$ ergs sec^{-1}. The cluster X-ray luminosity is typically a few percent of its optical luminosity.

I have studied the correlation between optical and X-ray properties of clusters of galaxies (Bahcall 1974b, 1975b), and I arrived at the following main conclusions. 1. A bright and active radio galaxy or a cD galaxy is present near the centers of most clusters for which the identification seems likely (i.e. small error boxes). Most small error boxes contain the central parts of the optical clusters; a few do not contain the cluster center but do contain an active galaxy. The percentage of Abell clusters that contain X-ray sources is a strong function of the classification type of the cluster. Comparison of the frequency distribution of Rood-Sastry classes within all 110 Abell clusters of distance groups $\leqq 3$ indicates that a higher fraction of the concentrated cD-B clusters, 33%, are observed to be X-ray sources compared with only 8% for the irregular clusters. 2. A strong correlation is found (see Figure 1 of Bahcall 1974b) between the X-ray luminosity and the classification type of the clusters. cD-B clusters have higher X-ray luminosities ($L_x \gtrsim 2 \times 10^{44}$ ergs sec^{-1}) than the intermediate and irregular clusters ($L_x \lesssim 2 \times 10^{44}$ ergs sec^{-1}). 3. When the active radio galaxy is located near the center of the cluster, the X-ray luminosity is higher than when the active galaxy is located away from the center. 4. The galaxy distribution in the clusters has a typical core radius of $\sim 0.25\ h_{50}^{-1}$ Mpc, which is about a factor of two smaller than the core radii of the gas obtained by Kellogg & Murray (1974), assuming a thermal bremsstrahlung model. The shapes of the galaxy and the gas distributions are similar within the present uncertainties. 5. There is a higher probability for the richer clusters (richness class 2 as compared with richness 0 and 1) to be X-ray sources, but the various richness classes exhibit the same range of X-ray luminosity.

Several production mechanisms have been suggested that could account for the observed X-rays from clusters of galaxies:

1. Inverse Compton radiation from the scattering of relativistic electrons off the microwave background radiation (Felten & Morrison 1966, Brecher & Burbidge 1972, Bridle & Feldman 1972, Perola & Reinhardt 1972).
2. Thermal bremsstrahlung from a hot intracluster gas (Felten et al. 1966, Ruderman & Spiegel 1971, Gunn & Gott 1972, Solinger & Tucker 1972, Lea et al. 1973, Silk 1973, 1976). The hot intracluster gas could be either primordial or could be heated by: (*a*) infalling clouds (Gunn & Gott 1972); (*b*) random motions of the

galaxies; and (*c*) active galaxies. Arguments that support a thermal emission model have been reviewed by Silk (1973).

3. Emission from a population of compact sources, like those known in our own Galaxy (Katz 1976, see also Fabian et al 1976).

The X-ray emission from clusters of galaxies probably includes contributions from all the mechanisms listed above. The thermal bremsstrahlung hypothesis is favored as the major contribution by many authors (see e.g. reviews by Gursky 1973 and Silk 1973), and by some of the data including the recent detection of an iron emission feature in the Perseus cluster (Mitchell et al. 1976, Serlemitsos et al. 1977).

Predicted correlations of the X-ray luminosity with the velocity dispersion of galaxies in the cluster have been suggested by several authors (Solinger & Tucker 1972, Silk 1976, Yahil & Ostriker 1973, Katz 1976). Such correlations however cannot be used at present to select a particular model for the X-ray emission due to the large uncertainties in the available data.

Models for the temperature and density distribution of intracluster gas can be divided into three categories: hydrostatic distributions (Lea et al. 1973, Gull & Northover 1975, Cavaliere & Fusco-Femiano 1976), infall of primordial material (Gunn & Gott 1972), and galactic winds driven from the cluster (Yahil & Ostriker 1973). Silk (1976) and Cowie & Binney (1977) discuss the possibility that cooling of the gas in the central regions regulates the infall of material and hence the central density of the gas. Typical values for the temperature and central density of the intracluster gas for all of these models are

$$T \sim 10^7\text{–}10^8 \text{ K}, \quad \eta_0 \sim 10^{-3} \text{ particles cm}^{-3}. \tag{33}$$

Additional evidence for the existence of intracluster gas is provided by radio observations (see Section 6).

6 RADIO EMISSION

The association of clusters of galaxies with radio sources was discovered by Mills (1960). Van den Bergh (1961b) confirmed and strengthened this conclusion by studying sources from the 3C catalog. Since that time radio emission from clusters of galaxies has been studied extensively. Radio surveys in the direction of selected samples of clusters of galaxies have been published by several authors (e.g. Fomalont & Rogstad 1966, Owen 1974a, 1975, Jaffe & Perola 1975, Riley 1975, Colla et al 1975, and related references therein). I give below a brief summary of the main properties of the cluster radio emission and indicate the relation of this emission to other cluster properties that are discussed in this article.

1. Clusters of galaxies contain radio emission associated with individual radio galaxies as well as (at least occasionally) extended low-frequency radio emission from the central parts of the clusters (see Willson 1970, Jaffe, Perola & Valentijn 1976, and references therein).
2. A large fraction ($\sim$20%) of all strong radio galaxies is located in rich clusters of galaxies (e.g. van den Bergh 1961b, Matthews et al 1964, Fomalont & Rogstad 1966, and others). This fraction may be explained by the fact that bright elliptical

galaxies, which are the major class of strong radio galaxies, are more concentrated in rich clusters than in the field. On this picture, any bright elliptical galaxy, whether in a cluster or in the field, has roughly the same chance of being a radio emitter (Rogstad & Ekers 1969, Jaffe & Perola 1976).

3. Powerful radio galaxies occur much more frequently in BM Type I clusters than in any other class (Wills 1966, Tovmassian & Shirbakyan 1974, Guthrie 1974, McHardy 1974, Owen 1975). This is most likely due to the presence of the optically bright cD galaxy, which is usually the powerful radio source in that type of cluster (e.g. Riley 1975).
4. In general, when a cluster contains an intrinsically strong radio source, this source is associated with one of the most luminous galaxies in the cluster (Riley 1975).
5. The probability that a rich cluster is a radio source is at most weakly correlated with its Abell richness (e.g. van den Bergh 1961b, Owen 1975).
6. Radio sources in clusters of galaxies tend to have steeper radio spectra ($\alpha > 1.2$ where flux density $\propto |\text{frequency}|^{-\alpha}$) than galaxies outside clusters ($\propto < 1.2$) (van den Bergh 1965, Baldwin & Scott 1973, Slingo 1974, Guthrie 1976).
7. Radio galaxies in BM Type I or II clusters tend to have steeper radio spectra than those in Type III clusters (McHardy 1974, Roland et al. 1976).
8. The average separations of double radio sources are smaller by a factor of more than two for sources found in clusters than for double sources outside clusters (De Young 1972; see however Hooley 1974).
9. Some radio galaxies in clusters exhibit a trailing radio structure, known as head-tail (or tailed) radio sources (Ryle & Windram 1968, Miley et al. 1972, Jaffe & Perola 1973, 1974, Rudnick & Owen 1976). Such head-tail structure has been observed also in a small group (Schilizzi & Ekers 1975, Colla et al 1975), but has not yet been observed with certainty for galaxies outside clusters (although this could conceivably be due to the fact that the appropriate radio observations have so far been concentrated in rich clusters).
10. Clusters of galaxies that are strong X-ray emitters are also usually radio sources; a correlation between the radio and X-ray properties of the clusters has been suggested (Owen 1974b, cf Rowan-Robinson & Fabian 1975). Costain et al. (1972) and Baldwin & Scott (1973) suggest that X-ray clusters often contain a radio source with a steep spectrum; this X-ray radio-spectra dependence can be understood from point 7 above plus the correlation (Bahcall 1974b) of X-ray clusters and BM Type I classification.

The observations summarized in points 6–9 (and possibly 10) are believed to be the result of the influence of an intracluster gas on the radio galaxies and their relativistic electrons. Since this gas is believed to be denser in the central parts of BM Type I and II clusters than in Type III clusters, it could account for the observed dependence of the radio spectrum on the cluster BM type as described in 7 above.

ACKNOWLEDGMENTS

I am grateful to a number of colleagues for helpful discussions and valuable suggestions, especially J. N. Bahcall, J. R. Gott, W. J. Jaffe, J. I. Katz, A. Oemler, J. P.

Ostriker, P. J. E. Peebles, H. J. Rood, P. Schechter, J. Silk, and E. L. Turner. I also thank the many people who sent preprints and reprints.

Literature Cited

Aarseth, S. J. 1969. *MNRAS* 144:537

Abell, G. O. 1958. *Ap. J. Suppl.* 3:211

Abell, G. O. 1962. In *Problems of Extra-Galactic Research,* ed G. C. McVittie, p. 213. New York: Macmillan.

Abell, G. O. 1963. *Astron. J.* 68:271

Abell, G. O. 1965. *Ann. Rev. Astron. Astrophys.* 3:1

Abell, G. O. 1976. In *Galaxies and the Universe,* eds. A. Sandage, M. Sandage, J. Kristian, p. 601. Chicago: Univ. Chicago Press. 818 pp.

Abell, G. O., Mihalas, D. M. 1966. *Astron. J.* 71:635

Albert, C. E., White, R. A., Morgan, W. W. 1977. *Ap. J.* 211:309

Ambartsumian, V. A. 1961. *Astron. J.* 66:536

Arp, H., Bertola, F. 1971. *Ap. J.* 163:195

Austin, T. B., Peach, J. V. 1974a. *MNRAS* 167:437

Austin, T. B., Peach, J. V. 1974b. *MNRAS* 167:591

Austin, T. B., Godwin, J. G., Peach, J. V. 1975. *MNRAS* 171:135

Avni, Y., Bahcall, N.A. 1976. *Ap. J.* 209:16

Bahcall, J. N., Bahcall, N. A. 1975. *Ap. J. Lett.* 199:L89

Bahcall, N. A. 1971. *Astron. J.* 76:995

Bahcall, N. A. 1972a. *Astron. J.* 77:550

Bahcall, N. A. 1972b. *Ap. J.* 180:696

Bahcall, N. A. 1973a. *Ap. J.* 183:783

Bahcall, N. A. 1973b. *Ap. J.* 186:1179

Bahcall, N. A. 1974a. *Ap. J.* 187:439

Bahcall, N. A. 1974b. *Ap. J.* 193:529

Bahcall, N. A. 1975a. *Ap. J.* 198:249

Bahcall, N. A. 1975b. *Ann. NY Acad. Sci.* 262:361

Baldwin, J. E., Scott, P. F. 1973. *MNRAS* 165:259

Bautz, L. P. 1972. *Astron. J.* 77:1

Bautz, L. P., Morgan, W. W. 1970. *Ap. J. Lett.* 162:L149

Bautz, L. P., Abell, G. O. 1973. *Ap. J.* 184:709

Brecher, K., Burbidge, G. R. 1972. *Ap. J.* 174:253

Bridle, A. H., Feldman, P. A. 1972. *Nature Phys. Sci.* 235:168

Burbidge, G. 1975. *Ap. J. Lett.* 196:L7

Cavaliere, A., Fusco-Femiano, R. 1976. *Astron. Astrophys.* 49:137

Chandrasekhar, S. 1942. *Principles of Stellar Dynamics.* New York: Dover. 313 pp.

Chiao, R. Y., Reinhardt, M. 1973. *Astron. Astrophys.* 22:257

Chincarini, G., Rood, H. J. 1971. *Ap. J.* 168:321

Chincarini, G., Rood, H. J. 1975. *Nature* 257:294

Chincarini, G., Rood, H. J. 1976. *Ap. J.* 206:30

Clark, E. E. 1968. *Astron. J.* 73:1011

Colla, G., Fanti, C., Fanti, R., Gioia, I., Lari, C., Lequex, J., Lucas, R., Ulrich, M. H. 1975. *Astron. Astrophys. Suppl.* 20:1

Corwin, H. G. 1974. *Astron. J.* 79:1356

Costain, C. H., Bridle, A. H., Feldman, P. A. 1972. *Ap. J. Lett.* 175:L15

Cowie, L. L., Binney, J. 1977. *Ap. J.* 215:723

Davidsen, A., Bowyer, S. C., Welch, W. 1973. *Ap. J. Lett.* 186:L119

Davies, R. D., Lewis, B. M. 1973. *MNRAS* 165:231

de Vaucouleurs, G. 1948. *Ann. Astrophys.* 11:247

de Vaucouleurs, G. 1959. *Handb. Phys* 53:275

de Vaucouleurs, G. 1960. *Ap. J.* 131:585

de Vaucouleurs, G. 1976. See Abell 1976, p. 557

de Vaucouleurs, G., de Vaucouleurs, A. 1964. *Reference Catalog of Bright Galaxies.* Austin: Univ. Texas Press. 268 pp.

De Young, D. S. 1972. *Ap. J. Lett.* 173:L7

De Young, D. S., Roberts, M. S. 1974. *Ap. J.* 189:1

Einasto, J., Kaasik, A., Saar, E. 1974. *Nature* 250:309

Faber, S. M., Gallagher, J. S. 1976. *Ap. J.* 204:365

Faber, S. M., Jackson, R. E. 1976. *Ap. J.* 204:666

Fabian, A. C., Pringle, J. E., Rees, M. J. 1976. *Nature* 263:301

Felten, J. E., Morrison, P. 1966. *Ap. J.* 146:686

Felten, J. E., Gould, R. J., Stein, W. A., Woolf, N. 1966. *Ap. J.* 146:955

Field, G. B. 1973. *IAU Symp. No. 63,* Krakow, Poland.

Fomalont, E. B., Rogstad, D. H. 1966. *Ap. J.* 146:528

Gallagher, J. S., Ostriker, J. P. 1972. *Ap. J.* 77:288

Giacconi, R., Murray, S., Gursky, H., Kellogg, E., Schreier, E., Matilsky, T., Koch, D., Tananbaum, H. 1974. *Ap. J. Suppl.* 27:37

Gott, J. R., Turner, E. L. 1976a. *Ap. J.* 209:1

Gott, J. R., Turner, E. L. 1977a. *Ap. J.* 213:309

Gott, J. R., Turner, E. L. 1977b. *Ap. J.* 216: In press

Gregory, S. A. 1975. *Ap. J.* 199:1
Gregory, S. A., Tifft, W. G. 1976. *Ap. J.* 206:934
Gull, S. F., Northover, K. J. E. 1975. *MNRAS* 173:585
Gunn, J. E., Gott, J. R. 1972. *Ap. J.* 176:1
Gunn, J. E., Oke, J. B. 1975. *Ap. J.* 195:255
Gursky, H. 1973. *Pub. Astron. Soc. Pac.* 85:493
Gursky, H., Schwartz, D. 1977. *Ann. Rev. Astron. Ap.* 15:541
Guthrie, B. N. C. 1974. *Astrophys. Space Sci.* 27:489
Guthrie, B. N. C. 1976. Preprint
Hauser, M. G., Peebles, P. J. E. 1973. *Ap. J.* 185:757
Hickson, P. 1976. The structure of clusters of galaxies and the angular size redshift test. PhD thesis. Calif. Inst. of Technol. 82 pp.
Hickson, P., Richstone, D. O., Turner, E. L. 1977. *Ap. J.* 213:323
Hodge, P. W., Pyper, D. M., Webb, C. J. 1965. *Astron. J.* 70:559
Holmberg, E. 1937. *Ann. Obs. Lund. No. 6*
Holmberg, E. 1962. See Abell 1962, p. 187
Hooley, J. 1974. *MNRAS* 166:259
Hubble, E. P. 1936. *The Realm of the Nebulae.* New Haven: Yale Univ. Press. 210 pp.
Humason, M. L., Mayall, N. U., Sandage, A. R. 1956. *Astron. J.* 61:97
Jaffe, W. J., Perola, G. C. 1973. *Astron. Astrophys.* 26:423
Jaffe, W. J., Perola, G. C. 1974. *Astron. Astrophys.* 31:223
Jaffe, W. J., Perola, G. C. 1975. *Astron. Astrophys. Suppl.* 21:137
Jaffe, W. J., Perola, G. C. 1976. *Astron. Astrophys.* 46:275
Jaffe, W. J., Perola, G. C., Valentijn, E. A. 1976. *Astron. Astrophys.* 49:179
Jenner, D. C. 1974. *Ap. J.* 191:55
Just, K. 1959. *Ap. J.* 129:268
Kaljnas, A. J. 1972. *Ap. J.* 175:63
Katz, J. I. 1976. *Ap. J.* 207:25
Kellogg, E., Murray, S., Giacconi, R., Tananbaum, H., Gursky, H. 1973. *Ap. J. Lett.* 185:L3
Kellogg, E., Murray, S. 1974. *Ap. J. Lett.* 193:L57
King, I. R. 1966. *Astron. J.* 71:64
King, I. R. 1972. *Ap. J. Lett.* 174:L123
King, I. R., Minkowski, R. 1972. In *External Galaxies and Quasi-Stellar Objects,* ed. D. S. Evans, p. 87. Dordrecht: Reidel. 549 pp.
Klemola, A. R. 1969. *Astron. J.* 74:804
Kormendy, J., Bahcall, J. N. 1974. *Astron. J.* 79:671
Krupp, E. C. 1974. *Publ. Astron. Soc. Pac.* 86:385
Kwast, T. 1966. *Acta Astron.* 16:45
Lea, S. M., Silk, J., Kellogg, E., Murray, S. 1973. *Ap. J. Lett.* 184:L105
Lecar, M. 1975. *IAU Symp. No. 69*, Besancon, France.
Leir, A. A., van den Bergh, S. 1977. *Ap. J Suppl.* 34:381
Lugger, P. M. 1976. Senior thesis. Harvard Univ.
Lynden-Bell, D. 1967. *MNRAS* 136:101
Matthews, T. A., Morgan, W. W., Schmidt, M. 1964. *Ap. J.* 140:35
McHardy, I. M. 1974. *MNRAS* 169:527
Melnick, J., Sargent, W. L. W. 1977. *Ap. J.* 215:401
Miley, G. K., Perola, G. C., van der Kruit, P. C., van der Laan, H. 1972. *Nature* 237:269
Mills, B. Y. 1960. *Aust. J. Phys.* 13:550
Minkowski, R. 1961. *Astron. J.* 66:558
Minkowski, R. 1963. *Proc. Natl. Acad. Sci. USA* 49:779
Mitchell, R. J., Culhane, J. L., Davison, P. N. J., Ives, J. C. 1976. *MNRAS* 175:29P
Morgan, W. W. 1962. *Ap. J.* 135:1
Morgan, W. W., Lesh, J. R. 1965. *Ap. J. Lett.* 142:1364
Morgan, W. W., Kayser, S., White, R. A. 1975. *Ap. J.* 199:545
Noonan, T. 1961. *Publ. Astron. Soc. Pac.* 73:212
Noonan, T. 1971. *Astron. J.* 76:182
Noonan, T. 1972a. *Astron. J.* 77:9
Noonan, T. 1972b. *Astron. J.* 77:134
Noonan, T. 1974a. *Astron. J.* 79:358
Noonan, T. 1974b. *Astron. J.* 79:775
Oemler, A. 1973a. *Ap. J.* 180:11
Oemler, A. 1973b. PhD thesis. Calif. Inst. Technol.
Oemler, A. 1974. *Ap. J.* 194:1
Oemler, A. 1976. *Ap. J.* 209:693
Omer, G. C., Page, T. L., Wilson, A. G. 1965. *Astron. J.* 70:440
Oort, J. H. 1958. *Distribution of Galaxies and the Density in the Universe.* Brussels: Inst. Int. Phys., Solvay
Ostriker, J. P., Peebles, P. J. E. 1973. *Ap. J.* 186:467
Ostriker, J. P., Peebles, P. J. E., Yahil, A. 1974. *Ap. J. Lett.* 193:L1
Ostriker, J. P., Tremaine, S. D. 1975. *Ap. J. Lett.* 202:L113
Owen, F. N. 1974a. *Astron. J.* 79:427
Owen, F. N. 1974b. *Ap. J. Lett.* 189:L55
Owen, F. N. 1975. *Ap. J.* 195:593
Page, T. 1962. *Ap. J.* 136:685
Peach, J. V. 1969. *Nature* 223:1140
Peebles, P. J. E. 1968. *Ap. J.* 153:13
Peebles, P. J. E. 1970. *Astron. J.* 75:13
Peebles, P. J. E. 1973. *Ap. J.* 185:413
Peebles, P. J. E. 1974. *Astron. Astrophys.* 32:197

Peebles, P. J. E., Hauser, M. G. 1974. *Ap. J. Suppl.* 28:19
Peebles, P. J. E., Groth, E. J. 1975. *Ap. J.* 196:1
Perola, G. C., Reinhardt, M. 1972. *Astron. Astrophys.* 17:432
Press, W. H. 1976. *Ap. J.* 203:14
Press, W. H., Schechter, P. 1974. *Ap. J.* 187:425
Richstone, D. O. 1974. Collisions of galaxies in dense clusters: morphological effects. PhD thesis. Princeton Univ. 59 pp.
Riley, J. M. 1975. *MNRAS* 170:53
Roberts, M. S. 1969. *Astron. J.* 74:859
Roberts, M. S., Rots, A. H. 1973. *Astron. Astrophys.* 26:486
Rogstad, D. H., Ekers, R. D. 1969. *Ap. J.* 157:481
Rogstad, D. H., Shostak, G. S. 1972. *Ap. J.* 176:315
Roland, J., Vernon, P., Pauliny-Toth, I. I. K., Preuss, E., Witzel, A. 1976. *Astron. Astrophys.* 50:165
Rood, H. J. 1969. *Ap. J.* 158:657
Rood, H. J. 1974a. *Publ. Astron. Soc. Pac.* 86:99
Rood, H. J. 1974b. *Ap. J.* 194:27
Rood, H. J. 1975. *Ap. J.* 201:551
Rood, H. J., Sastry, G. N. 1971. *Publ. Astron. Soc. Pac.* 83:313
Rood, H. J., Sastry, G. N. 1972. *Astron. J.* 77:451
Rood, H. J., Page, T. L., Kintner, E. C., King, I. R. 1972. *Ap. J.* 175:627
Rood, H. J., Abell, G. O. 1973. *Astrophys. Lett.* 13:69
Rose, J. A. 1976. *Astron. Astrophys. Suppl.* 23:109
Rowan-Robinson, M., Fabian, A. C. 1975. *MNRAS* 170:199
Ruderman, M. A., Spiegel, E. A. 1971. *Ap. J.* 165:1
Rudnick, L., Owen, F. N. 1976. *Ap. J. Lett.* 203:L107
Rudnicki, K. 1963. *Acta Astron.* 13:230
Rudnicki, K., Baranowska, M. 1966a. *Acta Astron.* 16:55
Rudnicki, K., Baranowska, M. 1966b. *Acta Astron.* 16:65
Ryle, M., Windram, M. D. 1968. *MNRAS* 138:1
Sandage, A. 1976. *Ap. J.* 205:6
Sandage, A., Hardy, E. 1973. *Ap. J.* 183:743
Sandage, A., Tammann, G. A. 1975. *Ap. J.* 197:265
Sandage, A., Kristian, J., Westphal, J. A. 1976. *Ap. J.* 205:688
Sastry, G. N., Rood, H. J. 1971. *Ap. J. Suppl.* 23:371
Schechter, P. 1976. *Ap. J.* 203:297
Schechter, P. 1977. *Astron. J.* Submitted for publication
Schechter, P., Press, W. H. 1976. *Ap. J.* 203:557
Schechter, P., Peebles, P. J. E. 1976. *Ap. J.* 209:670
Schilizzi, R. T., Ekers, R. D. 1975. *Astron. Astrophys.* 40:221
Schwarzschild, M. 1954. *Astron. J.* 59:273
Scott, E. L. 1962. See Abell 1962, p. 269
Serlemitsos, P. J., Smith, B. W., Boldt, E. A., Holt, S. S., Swank, J. H. 1977. *Ap. J. Lett.* 211:L63
Sersic, J. L. 1974. *Astrophys. Space Sci.* 28:365
Shane, C. D., Wirtanen, C. A. 1954. *Ap. J.* 119:91
Shane, C. D., Wirtanen, C. A. 1967. *Publ. Lick Obs.* 22:1
Shapley, H. 1950. *Publ. Univ. Michigan Obs.* 10:79
Shapiro, S. 1971. *Astron. J.* 76:291
Sharov, A. S. 1959. *Astron. Zh.* 36:807
Silk, J. 1973. *Ann. Rev. Astron. Astrophys.* 11:269
Silk, J. 1976. *Ap. J.* 208:646
Slingo, A. 1974. *MNRAS* 168:307
Smart, N. C. 1973. *Astron. Astrophys.* 24:171
Smith, S. 1936. *Ap. J.* 83:23
Snow, T. P. 1970. *Astron. J.* 75:237
Solinger, A., Tucker, W. 1972. *Ap. J. Lett.* 175:L107
Spinrad, H. 1977. Preprint
Spitzer, L., Baade, W. 1951. *Ap. J.* 113:413
Spitzer, L., Hart, M. 1971. *Ap. J.* 171:399
Tarter, J. 1975. PhD thesis. Univ. Calif., Berkeley
Tifft, W. G., Tarenghi, M. 1975. *Ap. J. Lett.* 198:L7
Tifft, W. G., Gregory, S. A. 1976. *Ap. J.* 205:696
Tovmassian, H. M., Shirbakyan, M. S. 1974. *Astrofizika* 10:29
Tremaine, S. D., Richstone, D. O. 1977. *Ap. J.* 212:311
Turner, E. L. 1976. *Ap. J.* 208:304
Turner, E. L., Gott, J. R. 1976a. *Ap. J. Suppl.* 32:409
Turner, E. L., Gott, J. R. 1976b. *Ap. J.* 209:6
van den Bergh, S. 1961a. *Z. Astrophys.* 53:219
van den Bergh, S. 1961b. *Ap. J.* 134:970
van den Bergh, S. 1962. *Z. Astrophys.* 55:21
van den Bergh, S. 1963. *Publ. Astron. Soc. Pac.* 75:498
van den Bergh, S. 1965. *Ap. J.* 141:1579
van den Bergh, S. 1975. *Ap. J. Lett.* 198:L1
van den Bergh, S. 1976a. *Ap. J.* 206:883
van den Bergh, S. 1976b. *Vistas Astron.* Submitted for publication
White, S. D. M. 1976a. *MNRAS* 174:19

White, S. D. M. 1976b. *MNRAS* 177:717
White, S. D. M. 1977. *MNRAS* 179:33
Wills, D. 1966. *Observatory* 86:140
Willson, M. A. G. 1970. *MNRAS* 151:1
Wolf, R. A., Bahcall, J. N. 1972. *Ap. J.* 176:559
Wolff, R. S., Helava, H., Kifune, T., Weisskopf, M. C. 1974. *Ap. J. Lett.* 193:L53
Yahil, A. 1974. *Ap. J.* 191:623
Yahil, A., Ostriker, J. P. 1973. *Ap. J.* 185:787
Zwicky, F. 1933. *Helv. Phys. Acta* 6:10
Zwicky, F. 1938. *Publ. Astron. Soc. Pac.* 50:218
Zwicky, F. 1942. *Ap. J.* 95:555
Zwicky, F. 1957. *Morphological Astronomy.* Berlin: Springer–Verlag
Zwicky, F., Herzog, E., Wild, P., Karpowicz, M., Kowal, C. T. 1961–1968. *Catalogue of Galaxies and Clusters of Galaxies.* 6 volumes. Pasadena: Calif. Inst. Technol.
Zwicky, F., Humason, M. L. 1964. *Ap. J.* 139:269
Zwicky, F., Karpowicz, M. 1966. *Ap. J.* 146:43

Ann. Rev. Astron. Astrophys. 1977. 15: 541–68

EXTRAGALACTIC X-RAY SOURCES

Herbert Gursky and Daniel A. Schwartz
Center for Astrophysics, Harvard College Observatory/Smithsonian Astrophysical Observatory, Cambridge, Massachusetts 02138

INTRODUCTION

Discovery

The earliest sounding rocket observations of cosmic X rays in the 1–10 keV energy range yielded a clear indication that significant fluxes of X rays were originating far beyond the Galaxy. In addition to the detection of discrete sources apparently belonging to our Galaxy, there was found an excess background of X rays that was more or less uniform around the sky (Giacconi et al. 1962, Bowyer et al. 1964). This background is now known to be isotropic to a high degree of precision. It must consist to some extent of the superposition of discrete, extragalactic sources out to z of 2 at least, and may also consist of a true diffuse component originating in the intergalactic medium.

Within a few years of these earliest observations, weak, discrete sources of X rays were localized which could be credibly identified as originating in powerful, active galaxies—objects that a priori might be expected to be intrinsically strong X-ray sources because of their known luminosity of nonthermal origin. The first of these sources to be found was M87 (Byram, Chubb & Friedman 1966, Bradt et al. 1967), followed by 3C273 and NGC 5128 (Bowyer et al. 1970). Although these detections were only marginal at the time, they were quickly confirmed and extended by observations from the *Uhuru* satellite launched late in 1970 (Giacconi et al. 1971a, Kellogg et al. 1971). In addition, sources were discovered originating from the Coma and Perseus clusters of galaxies and from the Seyfert Galaxy, NGC 4151 (Gursky et al. 1971a, Forman et al. 1972). The emission from Perseus was found at about the same time during a sounding rocket flight (Fritz et al. 1971).

Subsequent observations from *Uhuru* have revealed clusters of galaxies as a distinct class of X-ray sources and a large number of weak sources that cannot be identified with prominent optical or radio objects. These could be extragalactic because of their more or less uniform distribution in galactic latitude, and thus may be called "X-ray galaxies," (Giacconi et al. 1971b). Remarkably, few prominent active galaxies have been reliably identified as X-ray sources.

There is by now a wide variety of data from several satellites, complemented by sounding rockets and balloon observations on the extragalactic X-ray sources. In contrast to emission from galactic X-ray sources where only a small number of stars are seen to be X-ray sources at present levels of sensitivity, we see X-ray emission from at least one object in most of the major categories of galaxies.

For purposes of this review, we only discuss those extragalactic objects that are likely to be intrinsic X-ray sources. We ignore objects such as M31, where the observed emission is likely to comprise the summed X-ray luminosity of bright X-ray sources, such as seen in our own Galaxy, or the discrete sources observed in Magellanic Clouds, which are very likely single, discrete galactic sources; indeed, one, SMC X-1, is a pulsing, binary X-ray star.

Observational Constraints

Although the experimental aspects of X-ray astronomy are outside the scope of this article, it is useful to discuss certain elements of the discipline that bear on the discussions here. For more extensive hardware reviews see Peterson (1975) and Gursky & Schwartz (1974).

The data reported here are, by and large, obtained with sealed proportional counters operating in the 1–10 keV energy range and with angular collimation of order 1°. Typically, the available detector area is $\sim 10^3$ cm^2. Since these devices always record single counts, corresponding to energy deposition in the proportional counters (whether X rays or not), fluxes are reported in counts/sec. Analysis of the amplitude of the counter signals can yield broad-band spectral data with a resolution of 20–30% in this energy range. The conversion to absolute units is always dependent on experiment and source; in the case of the *Uhuru* satellite, the conversion is approximately 1 count/sec = 2×10^{-11} ergs cm^{-2} sec^{-1} (2–10 keV). The comparison between experiments is never very precise; typically, no better than 20%.

The extragalactic X-ray sources are usually found at the limiting sensitivity of the individual experiments; thus, count statistics rather than systematic errors tend to dominate the uncertainties. The faintest detected sources have been in the range of ~ 2 *Uhuru* count/sec. It will be difficult to improve this figure by any great margin with mechanically collimated proportional counters. The reason is simply that this number, in the case of *Uhuru*, is $\sim 1/4$ of the intrinsic background in the detector; thus, improvement can only come as the square root of increased aperture or integration time. A more severe limit will be source confusion that will dominate at about 1/10 of the present limits of sensitivity on a scale of 1 deg^2.

A consequence of the low statistical significance is poor positional accuracy, ranging typically from ~ 0.1 deg^2 at best to many deg^2 at worst. A few sources have been observed with much higher angular precision by the modulation collimator technique.

Typically, we know only that X-ray sources exist. Data are available on spectra, angular structure, and intensity variation for a few sources, but only a fraction of the sources are optically identified—more on the basis of statistical arguments than on precise angular correlation. However, X-ray instruments that utilize focusing optics are becoming more widespread. As is noted later, these instruments provide much greater sensitivity and higher quality data than do mechanically collimated instruments.

ACTIVE GALAXIES

The active galaxies that are found to be measurable X-ray sources are listed in Table 1. The remarkable feature of this list is that it contains only one example of each of the major kinds of identifiable active galaxies. This is presumably a reflection of an observational bias: the sources are seen at close to the lower limit of present X-ray sensitivities. However, it also indicates the pervasiveness of X-ray production as a complement to nonthermal emission from galaxies.

NGC 5128

Significant X-ray emission from the vicinity of NGC 5128 was first reported by Bowyer et al. (1970). From *Uhuru* data, Kellogg et al. (1971) were able to determine that the emission originated from the galaxy itself, and not from the major radio lobes that extend for several degrees away from the galaxy. By now, results from the SAS-3 satellite demonstrate that the bulk of the emission originates from a point source within ~15 arc sec of the center of the galaxy (Delvaille et al. 1977, in preparation). There is no report of extended emission from elsewhere in the galaxy or from the radio lobes.

Table 1 X-ray emission from active galaxies

Object	Class of Galaxy	X-ray source[a]	2–10 keV flux (10^{-11} ergs cm^{-2} sec^{-1})	D (Mpc)	L(2–10 keV) (ergs sec^{1})	Observed X-ray range (keV)
NGC 4151	Seyfert	3U1207+39	8.5 ± 1.4	20	3.8×10^{42}	2–100
Cen A	giant radio source	3U1322−42	14 to 90	5	0.4 to 2.5×10^{42}	1–1000
3C 273	quasar	3U1224+02	7.1 ± 0.8	950	7×10^{45}	2–8
M87	radio source in cluster	3U1228+12[b]	~9[b]	22	5×10^{42}	0.25–2
NGC 1275	peculiar Seyfert in cluster	3U0316+41[b]	~16[b]	110	2×10^{44}	0.25–2
3C 120[c]	Radio/optical variable, Seyfert	U0432+05	9.5 ± 1.5	190	4×10^{44}	2–8
3C 390.3[d]	N Galaxy, compact radio source	3U1825+81[d]	7 to 20	336	9 to 25×10^{44}	2–30

[a] 3U designates third *Uhuru* catalog (Giacconi et al. 1974), U designates next *Uhuru* catalog (Forman et al. 1977, in preparation).
[b] Discrete source within extended cluster.
[c] Identified by SAS-3 (Schnopper et al. 1977, in preparation).
[d] Identified by Charles et al. (1975), outside of *Uhuru* error box.

The localized character of the emission is also demonstrated by the rapidity of the observed variations in intensity, illustrated in Figure 1: the intensity has been found to vary by about a factor of 5 over a period of 4 years. There is one report of variability by almost a factor of two within a week (Winkler & White 1975), and of a decrease subsequent to the 1973 outburst (Culhane, Davison & Stark 1976).

The spectrum of the X rays is relatively flat. In the low-energy range the spectrum is consistent with a power law with an index of -1.4 ± 0.4 between 1 and 20 keV (Grindlay et al. 1975b). Between 20 and 50 keV, Mushotzky et al. (1976) report an index of -1.2 ± 0.2, based on data from OSO-7. They also report an increase of the 10–100 keV X rays of a factor $\geqq 2$ from 1972 to 1973, which is consistent with the lower energy X rays. Hall et al. (1976) report significant emission extending to several MeV, but with a steeper spectrum: an index of -1.99 ± 0.04.

Finally, Grindlay et al. (1975a) report the detection of a significant flux of photons with energy above 10^{12} eV using the Cerenkov air shower technique. The spectral index in this energy range is steeper yet, about -2.7.

Thus, NGC 5128 is an impressive source of high-energy photons. The total power emission is $\sim 10\%$ of the total nonthermal emission originating from the center of the galaxy.

No other extragalactic X-ray source has yielded so much information. It is generally presumed that the core region in the galaxy, seen in the red as a starlike image (Kunkel & Bradt 1971), is the source of the high-energy photons. This interpretation is supported by the position of the X-ray source and the observed intensity variations. It is possible to develop self-consistent models of the high-energy photons from this core region by either a synchrotron or Compton process or a combination of the two (Tucker et al. 1973, Perola & Tarenghi 1973, Grindlay 1975, Mushotzky et al. 1977). The idea is simply that there must be high-energy electrons present to account for the observed radio emission. These electrons may yield X rays by synchrotron radiation depending on the magnetic field strength or by Compton scattering from the observed infrared photons depending on the size of the infrared source. Grindlay (1975) has argued that both processes occur and that the X-ray

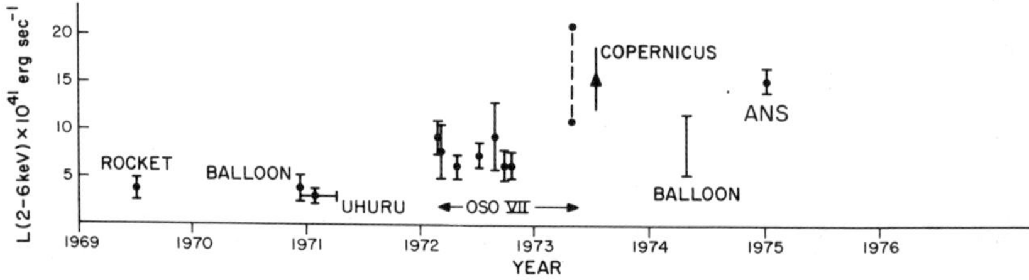

Figure 1 Light curve of Cen A (from Grindlay et al. 1975b) for the 2–6 keV region, extrapolated as necessary. Data sources are as follows: Rocket (Bowyer et al. 1970), "Balloon 1971" (Lampton et al. 1972), *Uhuru* (Tucker et al. 1973), OSO-VII (Winkler & White 1975), Copernicus (Davison et al. 1975), "Balloon 1974" (Haymes 1975), and ANS (Grindlay et al. 1975b). The dashed OSO-VII data for 1973.3 is an observation of a 50% increase in a five-day period.

photons are produced by the synchrotron process and the 10^{12} eV γ rays are produced by Compton scattering of electrons from the X-ray photons.

As is true of other active galaxies, the ultimate source of energy that powers this complex array of high- and low-energy photons is still a mystery. Fabian et al. (1976) have described the nucleus of NGC 5128 as a massive black hole accreting matter. The X rays would be the result of thermal bremsstrahlung, but there is no simple way to generate high-energy electrons.

If analogies have any merit, the nucleus of NGC 5128 looks more like a giant Crab Nebula rather than a giant Cygnus X-1, which favors the synchrotron-Compton models, rather than the bremsstrahlung models of this system.

NGC 4151

NGC 4151, a classical Seyfert Galaxy, was first detected as an X-ray source in the 1–10 keV energy range by *Uhuru* (Gursky et al. 1971). It has since been seen by other satellites in this energy range (Culhane 1976) and from a sounding rocket experiment (Margon et al. 1975). Higher energy photons have been detected from this object up to 100 keV (Baity et al. 1975). These show an unusually flat spectrum so that most of the energy appears to be in the 10–100 keV range. There is not yet evidence for intensity variability; however, if the emission were constant, fewer than 1% of all Seyferts could be similar sources to avoid conflict with the X-ray background at 100 keV.

The X-ray emission from this galaxy may be associated with the Seyfert phenomenon. The X-ray power is comparable to the nonthermal emission seen from the core region at other wavelengths; thus, the emission can be interpreted as an extension of the synchrotron spectrum believed to give rise to the optical continuum, or as thermal bremsstrahlung from a hot component ($10^7 - 10^9$ K) of the gas in the nucleus.

Ulmer & Murray (1976) searched, without success, for X-ray emission from Seyfert Galaxies as a class of objects. It follows that the X-ray luminosity of NGC 4151 is higher by at least a factor of 3 than that of Seyferts such as NGC 3227 and NGC 1566, and by a factor of 10 than NGC 4051.

3C 273

X-ray emission was first detected from 3C 273 in the 1–10 keV energy range by the Naval Research Laboratory (Bowyer et al. 1970); the result was confirmed by *Uhuru* (Kellogg et al. 1971). The object has also been observed from other satellites. There is no evidence for variability in the emission, nor are there reports of hard X-ray emission from this object. Results from SAS-3 locate the emission at 3C 273 within a 30″ uncertainty (Schnopper, presented at meeting on "X-Ray Astronomy" at M.I.T., 29 January 1976).

3C 273 remains the only quasar associated with an X-ray source with a high degree of certainty. A number of other associations have been suggested, as is discussed below. Thus, the situation remains as it is with NGC 4151; we cannot state that X-ray emission is a normal by-product of the quasar phenomena or that 3C 273 is simply a singular object.

Cygnus A

This object is near the center of the 0.23-deg^2 *Uhuru* error box 3U1957+40 (Giacconi et al. 1974), and also near the center of a cluster of galaxies. A recent observation by the Astronomical Netherlands Satellite (ANS) shows that any point-source flux at the Cyg A position is less than about 25% of the total source flux (Brinkman et al. 1977). An extended source of > 12 arc min diameter is allowed, and therefore it seems more likely that 3U1957+40 is another example of cluster X-ray sources. This is consistent with the discussion of Longair & Willmore (1974), which supports a thermal bremsstrahlung emission mechanism and is probably consistent with the size of < 10 arc min deduced by Longair and Willmore for a specific assumed spectrum and surface-brightness profile. We note that a flux from Cyg A of only 1% of the cluster flux would still make its intrinsic X-ray power 4 to 10 times greater than Cen A.

Other Active Galaxies

The galaxies M87 and NGC 1275 are discussed below in conjunction with X-ray emission from the Virgo and Perseus clusters. There is not yet specific evidence that the X-ray emission is associated with a compact source in the galaxy rather than an enhancement of the cluster X-ray mechanism.

The SAS-3 satellite has recently identified 3C 120 as an X-ray source (Schnopper et al. 1977), which is associated with a revised *Uhuru* source position (Forman et al. 1977, in preparation). This object is a well-known radio and optically variable Seyfert; however, its X-ray luminosity is at least 100 times greater than can be associated with Seyferts as a class. Walmsley (1969) discussed the possibility of X-ray emission from this object due to Compton scattering of electrons off the synchrotron photons. A sufficient density of radio photons to produce observable X-ray emission would exist only for a short time period following the peak of a radio outburst.

3C 390.3 has been observed by Charles et al. (1975) and up to 30 keV by Mushotzky et al. (1977). Those authors suggest an identification with the source 3U 1825+81 (also suggested by Giacconi et al. 1974), but it is not included in the 90% *Uhuru* error box of 0.69 deg^2. In any case, the observations from the *Copernicus* Satellite (Charles et al. 1975) indicate an increase of at least a factor of 3 in about 2 years compared with *Uhuru*.

3C 120 and 3C 390.3 are interesting in having radio luminosities intermediate between Cen A or NGC 4151 and the quasar 3C 273. Mushotzky et al. (1977) suggest that the underlying feature of all these sources is compact, synchrotron self-absorbed gigahertz emission, with the X rays arising from Compton scattering. The luminosities are in fact comparable to those for rich clusters of galaxies, as discussed below.

THE UNIDENTIFIED HIGH-LATITUDE SOURCES

The *Uhuru* survey, besides revealing a large number of X-ray sources along the Milky Way that are obviously galactic in nature, and a few individual sources that could be credibly identified with prominent single galaxies and clusters of galaxies, demonstrated the existence of 42 weak sources at galactic latitudes $>20°$ whose origin could not be easily surmised. The apparent space density of these objects is about 1.6×10^{-3} deg^{-2} to a limiting sensitivity of 3 *Uhuru* counts/sec ($\sim 5 \times 10^{-11}$ ergs cm^{-2} sec^{-1} in the range 2–6 keV). Matilsky et al. (1973) argued that they were extragalactic since their distribution was more or less isotropic. A local galactic population would have resulted in a measurable band of diffuse radiation centered on the galactic equator. Also, the superimposed effect of unseen sources of this kind could account for the diffuse X-ray background. Holt et al. (1974) pointed out that the current data did not rule out all possible galactic distributions; however, Schwartz, Murray & Gursky (1976) argued that the constraints imposed by background fluctuations make any attempt to explain all the high-latitude sources as galactic very ad hoc.

The high-latitude sources ($|b| \gtrsim 20°$) from the Third *Uhuru* Catalog (Giacconi et al. 1974) that are not identifiable with galactic objects or clusters of galaxies (see below) are listed in Table 2. The purpose of this table is to present the objects that are known not to be of the types discussed elsewhere in this article. Some objects not

Table 2 Unidentified high galactic latitude sources

3U Name	l	b	Error box size (deg^2)	Flux (1.7×10^{-11} ergs cm^{-2} sec^{-1})	Possible identification
0012−05	100.0	−66.2	0.23	4.9	Galaxy, $m_p \sim 14$ outside error box[a]
0042+32	121.5	−29.8	0.14	7.0	Red galaxy, $m_p \sim 14.5$[a]
0055−79	302.6	−37.7	0.18	2.2	
0138−01	149.4	−61.4	0.2	6.2	QSO NAB 0137−01[a]
0431−10	205.9	−35.0	4.4	3.0	
0510−44	250.0	−35.9	18.0	2.0	PIC A
0527−05	208.8	−20.7	1.4	4.2	M42 = Orion Nebula
0530−37	241.6	−31.0	1.6	2.5	Group of galaxies[b]
0705−55	265.7	−19.9	1.0	3.2	
0917+63	151.0	40.7	0.19	4.0	Blank[a]
1231+07	260.7	69.3	1.2	6.7	IC 3576[c]
1237−07	298.1	55.3	2.8	1.3	NGC 4428, 4433, 4487
1410−03	339.2	53.7	0.14	3.5	NGC 5506 (emission lines[a]), 5507
1443+43	74.7	62.2	0.15	3.0	Poor group of galaxies[a]
1645+21	40.6	36.4	1.1	6.1	
1825+81	113.2	27.9	0.69	2.7	3C 390.3
1843+67	97.9	25.7	2.1	3.5	
1904+67	97.8	23.6	1.1	5.0	
2041+75	109.4	−19.9	1.2	3.4	
2128+81	116.1	21.8	1.1	1.5	

[a] Bahcall et al. (1975) [b] Lugger (1976) [c] Margon et al. (1972)

in this table may have incorrect identifications and might properly belong here. Also, we have limited ourselves to the 3U catalog as the most complete, single survey published to date. The possible identifications listed in the table result from searching catalogs of prominent extragalactic objects—normal and peculiar galaxies, Seyfert Galaxies, quasars, and radio sources. The positional uncertainty of these sources ranges from ~ 0.1 deg^2 to several square degrees. Thus the total area of sky comprising the most likely position of these sources amounts to tens of square degrees, and it is not surprising that candidate objects are too numerous to exclude chance juxtaposition.

There have been several attempts to make identifications of extragalactic objects with these weak, high-latitude sources. Bahcall et al. (1975) examined the fields associated with 10 of these sources with a total positional uncertainty of 2.8 deg^2. Three color plates were taken of this region and examined for QSO's down to a limiting magnitude of $+18$. The results of this search are shown in Table 2. The possible candidates include a 17th mag quasar (which falls within a reduced error box found by Schreier et al. 1975) and the emission-line galaxy NGC 5506. The conclusion from this study was that the unidentified X-ray sources, as a class, did not coincide with known, or relatively nearby extragalactic objects; specifically, there was not a strong correlation with radio-quiet quasars as suggested by Setti & Woltjer (1973). This was also a conclusion reached by Bahcall & Bahcall (1975) who examined the positional correlation between 64 high-latitude sources with all Abell Clusters and with galaxies ($m_p \leqq 15.7m$) in the Zwicky catalog.

The results can be stated in terms of the ratio of X-ray to optical luminosity. Normal galaxies, clusters of galaxies, Cen A, 3C 273, and NGC 4151 all have $L_x/L_{op} \leqq 1$ (cf Kellogg 1973). The null results of Bahcall et al. (1975) can be expressed as $L_x/L_{op} > 5$ if galaxies and > 25 if quasars prove to be the identifications for the objects in Table 2. This requirement of greater X-ray than optical luminosity was actually realized upon the first attempts to identify *Uhuru* sources and led Giacconi et al. (1971b) to introduce the term "X-ray Galaxy."

Some of the sources may have inaccurate positions or intensities because of source-confusion noise. Schwartz et al. (1976) estimate that this effect introduces no more than a 25% reduction in numbers of sources at greater than 3 counts/sec.

Another aspect of this subject is the appearance of *high-latitude* transient sources that may be extragalactic. These are distinct from the *galactic* transient sources of which about a dozen examples are now known, and which exhibit a small dispersion in latitude and a nova-like X-ray intensity falloff on a time scale of days to months. Four high-latitude transient sources have been reported. Their galactic latitude ranges from 26° to 65° and the time scale for the X-ray emission ranged from less than 1 hour to many days. The first of these were reported from the Ariel 5 satellite (Ricketts, Cooke & Pounds 1976) and from the SAS-3 satellite (Rappaport et al. 1976). The Ariel source, designated A1103+38, is at a latitude of 65° and positional uncertainty of 0.3 deg^2. The source was present for several months at a very low level (2–3 *Uhuru* counts/sec) with the exception of a ten-day interval. During one of those ten days, it "flared" significantly to ~ 44 *Uhuru* counts/sec. The SAS-3 source, designated as MX2346$-$65, lies at a latitude of $-51°$ and is positioned with an accuracy of 1.5 deg^2.

The source persisted for less than 2200 sec at a level of ~100 *Uhuru* counts/sec. The Ariel 5 satellite has revealed at least two other high-latitude transients (Cooke 1976) designated A0000+28 and A0353–40. The latter source may have flared on two occasions, 200 days apart.

In view of the known intensity variability of emission from the cores of external galaxies and quasars, it is possible that these high-latitude transients are the emission from low optical-radio luminosity extragalactic sources. However, the positional uncertainties of these sources are so great that it is difficult to make a case for an extragalactic origin based on candidate identifications. MK421, a possible BL Lac object, is in the error box of A1103+38; however, the bright reflection nebulae NGC 5367 is in the error box of A0353–40, leading to the idea that a flare star may be responsible for the observed X-ray emission.

X-RAY EMITTING CLUSTERS OF GALAXIES

Identification

With the discovery that clusters of galaxies are extended, intrinsically luminous sources of X rays (Gursky et al. 1971b, 1972, Forman et al. 1972) with a wide range of luminosities and sizes (Kellogg et al. 1973, Kellogg & Murray 1974), we have found a process that is characteristic of this fundamental unit as a whole. We may well expect the X-ray clusters to provide a fruitful channel of cosmological data regarding the formation and evolution of clusters and of galaxies, and regarding the large-scale structure of the Universe. Furthermore, new radio and optical data from clusters of galaxies have revealed new information on their structure.

Much of the basic X-ray data appears in two reviews by Kellogg (1973, 1974). Other identifications of X-ray sources with Abell clusters are given by Heinz et al. (1972), Bahcall (1974a), Rowan-Robinson & Fabian (1975), Elvis et al. (1975), Cooke & Maccagni (1976), and are reviewed by Bahcall & Bahcall (1975). Suggested identifications of X-ray sources with Southern sky clusters are given by Disney (1974), Vidal (1975a, b), Melnick & Quintana (1975), and Lugger (1976, 1977). Searches for X-ray sources in Abell clusters were carried out by Kellogg et al. (1973) and Cushman (1976).

Table 3 gives the clusters identified as X-ray sources, and other, more distant, clusters that fall inside *Uhuru* error boxes but for which the possibility of chance coincidence cannot be neglected. The Table contains three categories of identification: *A*. Confident identifications with nearby Abell clusters, *B*. Confident identifications with other clusters, *C*. Allowed identifications with clusters. We distinguish the Abell clusters because they are a well-defined astronomical object for which a statistical catalog (Abell 1958) exists. Category *C* contains distant clusters, which are sufficiently numerous that chance positional coincidences would be expected (cf Bahcall & Bahcall 1975); X-ray sources detected at less than 2.5 *Uhuru* counts/sec ($=4.2 \times 10^{-11}$ ergs sec^{-1} cm^{-2} from 2 to 7 keV), for which source-confusion noise may be making a dominant contribution (Fabian 1975, Schwartz, Murray & Gursky 1976); and unpublished identifications suggested by Cushman (1976; Peterson et al. 1977) and based on scanning 98 of the 111 clusters with distance class $\leqq 3$ in the *Uhuru* data. Effects of confusion noise and coincidence with

Table 3 X-ray emitting clusters of galaxies

Cluster[a]	X-ray source[b]	2–7 keV Flux (1.7×10^{-11} ergs cm^{-2} sec^{-1})	Distance[c] (Mpc)	X-ray luminosity (10^{44} ergs sec^{-1}, 2–7 keV)	Dispersion velocity (3-dimensional km sec^{-1})	Morphology[l]	Error box size[f] (deg^2)
A. Identifications with Abell Clusters							
A347	3U0227+43	4.2 ± 0.8	116	1.1		$R = 0$	13.0
A401	3U0254+13	3.4 ± 0.6	452	13		$R = 2$, cD	0.22
A426	3U0316+41	47.4 ± 0.8	110	11	2420 ± 490	$R = 2$, L	0.0123
A496		2.5 ± 0.9	216	2.2		$R = 1$, cD	
A576		2.6 ± 0.9	242	2.9		$R = 1$, I	
A754	3U0901−09	4.4 ± 0.9	322	8.8		$R = 2$, cD	0.11[g]
A1367	3U1144+19	3.6 ± 0.6	123	1.0	1730 ± 270	$R = 2$, F	0.13
A1656	3U1257+28	14.8 ± 0.6	138	5.4	1560 ± 220	$R = 2$, F	0.0110
A2052	MX1514+06	6.7 ± 0.7	211	5.7		$R = 0$, cD	0.5[h]
A2199	3U1639+40	4.0 ± 0.8	185	2.6	1460 ± 390	$R = 2$, cD	0.55[h]
A2241	3U1706+32	4.1 ± 0.8	380[d]	11		$R = 0$	9.8
A2256	3U1706+78	3.2 ± 0.6	324	6.5		$R = 2$, B	0.0536
A2319	3U1921+43	6.3 ± 0.6	329	13		$R = 1$, cD	0.015[h]
A2666	3U2346+26	7.0 ± 1.3	164	3.6		$R = 0$, cDp	7.0
(A2634)			184	4.5		$R = 1$	
B. Identifications with Other Clusters							
SC0002−308	3U0001−31	3.2 ± 0.6	366[e]	6.8		$R = 0$, L	5.1
SC0141−340	U0143−33	6.2	384[e]	15		$R = 1$, I	0.29[h]
SC0417−558	3U0400−59	3.8 ± 0.8	18	0.024	370[m] ± 160	$R = 0$, I	9.0
SC0430−616	3U0426−63	2.6 ± 0.6	361	5.3		$R = 3$, C	0.87
3C129	3U0446+44	5.5 ± 1.0	128	1.7			0.0534
SG0550−316	U0550−31	3.2 ± 0.6	198	2.4		$R < 0$, L	0.24[h]

SC0627 – 544	3U0624 – 55	3.4 ± 0.6	301	4.9		$R = 2$, C	0.25
Virgo	3U1228 + 12	21.7 ± 0.6	22	0.20	1070 ± 110		0.0211
Cen	3U1247 – 41	6.2 ± 0.6	68	0.60	1160 + 460		0.0506
SC1251 – 288	3U1252 – 28	4.5 ± 0.6	384[l]	13		$R = 0$, cD	0.13
SC1329 – 314	MX1329 – 31	5.9 ± 0.8	438[e]	18		$R = 0$, Bb	1.0[i]
SC1345 – 301	MX1347 – 32	5.7 ± 0.8	86	0.67	1040[m] ± 201	$R = 2$, C	1.0[i]
SC1834 – 770	3U1849 – 77	3.0 ± 0.7	116	0.78		$R = 1$, I	0.15
Cyg A	3U1957 + 40	5.6 ± 1.6	337	12			0.23
CA2013 – 710	3U1959 – 69	2.8 ± 0.6	76	0.32	460[m] ± 230	$R = 0$, C	0.87
SC2146 – 594	MX2140 – 60	3.0 ± 0.8	300[e]	4.3		$R = 2$, F	1.7[i]
		C. Positional Coincidences with Clusters					
A71		2.1 ± 0.8	375[d]	5.7		$R = 0$	
A85	3U0026 – 09	4.3 ± 1.2	394[d]	13		$R = 1$	1.2
A104	3U0032 + 24	6.8 ± 1.4	434[d]	25		$R = 1$	18.0
A127	3U0057 – 23	2.1 ± 0.6	1036[d]	43		$R = 1$	1.2
A262	3U0151 + 36	2.4 ± 0.6	101	0.47		$R = 0$, I	0.94
SC0316 – 458	3U0302 – 47	3.3 ± 0.9	474[e]	14		$R = 0$, L	2.0
CA0340 – 538	3U0328 – 52	1.7 ± 0.6	300[e]	3.0		$R = 2$, cD	18.0
(CA0329 – 527)			342	3.8		$R = 4$, F	
(CA0325 – 539)			360[e]	4.3		$R = 1$, F	
A478	3U0405 + 10	3.4 ± 0.6	896[d]	52		$R = 2$	0.1[j]
A504	3U0440 + 06	5.6 ± 1.0	896[d]	86		$R = 1$	0.5
SG0528 – 356	U0528 – 35	4.0	1060[e]	54		$R < 0$, I	0.44[h]
A873	3U0943 + 71	4.0 ± 0.7	896[d]	62		$R = 3$	5.3
(A875)			896[d]	62		$R = 0$	
(A864)			896[d]	62		$R = 1$	
(A1037)			1040[d]	83		$R = 0$	
SG0959 – 359	U0957 – 37	4.0	480[e]	15		$R < 0$, cD	0.87[h]
A1060	3U1044 – 30	2.2 ± 0.9	65	0.18	1340 ± 500	$R = 1$, C	0.8[h]
A1192	3U1109 + 59	2.4 ± 0.6	987[d]	45		$R = 0$	6.3
A1314		2.6 ± 0.8	201	2.0		$R = 0$	

Table 3 (continued)

Clusters[a]	X-ray source[b]	2–7 keV flux (1.7×10^{-11} ergs cm^{-2} sec^{-1})	Distance[c] (Mpc)	X-ray luminosity (10^{44} ergs sec^{-1}, 2–7 keV)	Dispersion velocity (3-dimensional km sec^{-1})	Morphology[l]	Error box size[f] (deg^2)
A1797	3U1349+24	3.8±1.0	738[d]	40		$R = 1$	12.0
(A1819)			813[d]	48		$R = 1$	
(A1841)			940[d]	65		$R = 1$	
A2142	3U1555+27	5.1±0.9	455[d]	20		$R = 2$, B	0.05[k]
A2147	U1601+16	2.1±0.7	266	2.1	1870±30	$R = 1$	0.23[h]
A2204	3U1623+05	2.6±0.6	775[d]	30		$R = 3$	12.0
(A2210)			775[d]			$R = 1$	
A2255		1.8±0.6	473	7.7	2140±570	$R = 2$	0.21[h]

[a] A = Abell (1958); CA = Melnick & Quintana (1975); SC or SG = Lugger (1977). Indented clusters also fall in error box, and we cannot say which cluster emits X-rays.

[b] 3U = Giacconi et al. (1974); a blank indicates X-ray emission from the Abell cluster but not crossed lines of positions (Kellogg et al. 1973, Cushman 1976); MX = Markert et al. (1976); U = Forman et al. 1977, in preparation.

[c] Based on a Hubble constant 50 km sec^{-1} Mpc and measured redshifts (Noonan 1973, Vidal 1975a, b, Sandage 1975, Disney 1974, Zwicky 1971, Whiteoak 1972, Spinrad 1975, Rudnick & Owen 1976, Baade & Minkowski 1954, Jenner 1974), except as noted.

[d] Estimated distance based on 10th brightest galaxy according to $\log z = -4.4818 + 0.21\, m_{10}$ (Rowan–Robinson 1972).

[e] Estimated (Lugger 1976) from the brightest galaxy according to Sandage (1973).

[f] From 3U Catalog (Giacconi et al. 1974) except as noted.

[g] Pye et al. (1976).

[h] Jones & Forman 1977, in preparation.

[i] Markert et al. (1976).

[j] Elvis et al. (1975).

[k] Cooke & Maccagni (1976).

[l] For A clusters, from Abell (1958) and Rood & Sastry (1971). For SG, SC, CA, from Lugger (1976, in preparation 1977).

[m] Calculated by Lugger (1976, in preparation 1977), from data of Sandage (1975).

other X-ray source lines of position have not yet been analyzed for this latter survey.

The first two columns give the cluster and X-ray source name. The intrinsic luminosity between 2 and 7 keV is computed from the equivalent *Uhuru* flux from 2 to 7 keV and the distance, without redshift or K-correction. The dispersion velocity has been multiplied by $\sqrt{3}$ to give a space velocity. The relative error in dispersion velocity is given as $[2/(n-1)]^{1/2}$, when n galaxies have measured redshifts. The morphology includes richness class R as defined by Abell (1958) and type defined by Rood & Sastry (1971). The last column gives the area of the X-ray source error box.

Notes on individual clusters are as follows: A376 has previously been listed based on a single sighting at 3.2 σ. This has recently been revised to 2.8 σ (C. Jones 1976, private communication), and we therefore do not list it here. A2199 falls within a reduced error box, which now excludes A2197 (Forman et al. 1977, in preparation). A2147 falls within a reduced error box, which now excludes A2151 and A2152 (Forman et al. 1977, in preparation). Based on the small error box size, we assume that A262, A478, A1060, A2142, A2147, and A2255 are confident identifications.

Table 4 contains clusters for which some additional X-ray information is available. The angular size a is the observed effective core radius if an isothermal Emden sphere model of the gas is fit to the angular distribution of emission. In this model (cf Lea et al. 1973) the particle density at a distance r from the center is given by

$$n(r) = n_0[1+r^2/(Da)^2]^{-3/2}, \tag{1}$$

with D the distant to the source. The next column gives the average surface brightness within this core radius in units of the surface brightness of the diffuse cosmic X rays $I_B = 3 \times 10^{-8}$ ergs cm^{-2} sec^{-1} ster from 2 to 6 keV. The temperature results from fitting an optically thin, isothermal bremsstrahlung model to the X-ray spectral data. The density is n_0 in the above fit. Additional data on clusters appears in the preceding article by Bahcall (1977).

There are at least 40 confident identifications in Table 3, and we may begin to make qualitative statements about the class even though selection effects are certainly significant. There is a wide range of intrinsic properties: luminosities vary from $< 10^{43}$ to $> 10^{45}$ ergs sec^{-1}, intrinsic core radii from ~ 50 kpc to 1.5 Mpc, and surface brightness from 3 to 500 times the cosmic background. The temperature ranges of 10^7–10^8 K are a selection effect, following from the observed bandwidth of 2–10 keV.

For the Abell clusters (Table 3*A*) it was originally noticed that absolute luminosity increased with richness class (Gursky et al. 1972) and correlated with the presence of a cD or B morphology (Bahcall 1974b). Addition of the Southern Sky clusters (Table 3*B*) tends to wash out the correlation (Lugger 1976, 1977) in the sense that any type of cluster might have been an emissivity in the range 3×10^{44} to 10^{45} ergs sec^{-1}. The absence of $R \geqq 2$ or cD, B clusters with X-ray emission below 10^{44} ergs sec^{-1} might reflect the fact that such cluster types are less common and merely too distant to have been detected with current X-ray sensitivities. Clusters of luminosity $\leqq 3 \times 10^{44}$ ergs sec^{-1} could be expected to be detectable above 2.5 counts/sec only if they are within Abell distance classes $\leqq 2$. Of the five such $R = 2$ clusters in Abell's catalog, A1367 and A2199 are detected with $L < 3 \times 10^{44}$ ergs sec^{-1}, and a limit 2.2×10^{44} ergs sec^{-1} can be placed on A2151.

Table 4 Properties of X-ray clusters

Cluster	X-ray source	X-ray size[a] (arc min radius)	X-ray surface brightness ($I_B = 3 \times 10^{-8}$ cm^{-2} sec^{-1} ster^{-1})	Effective temperature (keV)	Central density[e] (#/cm^3)	Fe line emission[b] (ratio to cosmic abundance)
A262	3U0151+36	45^{+40}_{-20}	2.8			
A401	3U0254+13	<24	>14			
A426	3U0316+41	15 ± 2	480	7.0 ± 0.4[b]	4.5×10^{-3}	0.45 ± 0.08
3C129	3U0446+44	<21	>20			
A1060	3U1044−30	<15	>14	2.6 ± 0.8[c]		
A1367	3U1144+19	<22	>10			
Virgo	3U1228+12	25 ± 4	48	2.8 ± 0.2[b]	5×10^{-3}	0.40 ± 0.21
Cen	3U1247−41	$16 \pm ^{7}_{5}$	34	7.8[d]		
A1656	3U1257+28	16 ± 3	82	9.3 ± 1.3[b]	3×10^{-3}	0.38 ± 0.23
A2199	3U1639+40	<14	>28			
A2256	3U1706+78	16 ± 9	18			
A2319	3U1921+43	<14	>46			

[a] Kellogg & Murray (1974).
[b] Serlemitsos et al. (1977).
[c] Ives & Sanford (1976).
[d] Mitchell et al. (1975).
[e] Lea et al. (1973).

The existence of radio emission from the X-ray clusters was immediately noted (Forman et al. 1972, Bridle & Feldman 1972, Costain, Bridle & Feldman 1972). In fact, the Perseus cluster was originally identified with the radio source 3C84 (Fritz et al. 1971) and the Virgo cluster with M87 (Byram, Chubb & Friedman 1966). Owen (1974) found radio emission at 1400 MHz from 9 of the sources in Table 3*A* and further suggested that the radio emission was correlated with the cD, B morphological types. We still cannot infer any causal connection between the X-ray emission and the general existence of the radio source. They might both be independent features of rich clusters of galaxies.

Miley et al. (1972) interpreted the "head-tail" radio sources in the Perseus (Ryle & Windham 1968), Coma (Willson 1970), and 3C129 (MacDonald et al. 1968) clusters as trails due to motion of the galaxies in the intracluster medium. This implied particle densities of the order of 10^{-3} to 10^{-4} cm^{-3}, which are comparable to the densities deduced if thermal bremsstrahlung is assumed to provide the X-ray emission. Radio "head-tail" morphology has also been observed in the X-ray–emitting Abell clusters 401, 1314, 1367, 2241, 2255 (Valleé & Wilson 1976), 2142 (Harris 1976), and 2256 (Rudnick & Owen 1976). In this case, it is very suggestive that both the X-ray source and the radio trails depend on a dense (10^{-3} to 10^{-5} cm^{-3}) and hot intracluster medium.

The sources of Table 3*A* may be used to estimate roughly the X-ray luminosity function for all *Abell* clusters (Schwartz 1977). Table 5 summarizes the results. The luminosity function is expressed as a volume density averaged over one magnitude range in absolute luminosity. Formal errors are about a factor of 2.5 for the 2.4 to 15×10^{44} ergs sec^{-1} range, considering errors in the fluxes, distances, and limited numbers of sources. In the range 15 to 95×10^{44} ergs sec^{-1} we have used the three clusters in Table 3*C* for which the X-ray error box is $\leqq 2$ deg^2, and flux $\geqq 2.5$ counts/sec, to set the lower end of the range, and all the Abell clusters in 3*C* to set the upper end. We can numerically represent Table 5 in the form

$$\phi(L) = 1.3 \times 10^{-6}(L_{44})^{-2.9} \text{ clusters Mpc}^{-3}\,(10^{44} \text{ ergs sec}^{-1})^{-1}, \tag{2}$$

where L_{44} is the luminosity in units of 10^{44} ergs sec^{-1}. In this case, we must truncate the distribution below $L_A = 0.8 \times 10^{44}$ ergs sec^{-1} to avoid a space density greater than that of all Abell clusters, $\sim 10^{-6}$ Mpc^{-3}. Physically, of course, there must be a gradual flattening out of $\phi(L)$ below a few times 10^{44} ergs sec^{-1}. We can set another

Table 5 Luminosity function of Abell clusters

$L(10^{44}$ ergs)	No. of clusters	$\phi(L)\,\#/\text{Mpc}^3$	Volume emissivity (ergs sec^{-1} Mpc^{-3})
2.4 – 6	5	1.2×10^{-7}	4.2×10^{37}
6 – 15	6	3.2×10^{-8}	2.5×10^{37}
15 – 38	1 to 3	1.8 to 4×10^{-9}	0.7×10^{37}
38 – 95	2 to 4	0.6 to 1.7×10^{-9}	0.6×10^{37}

firm lower limit at $L_B = 5 \times 10^{42}$ ergs sec^{-1}, below which extrapolation of Equation 2 would result in a greater diffuse background than is observed.

Emission Mechanisms

Four classes of emission mechanisms may be invoked to interpret the observations of X rays from clusters of galaxies. The two most probable are thermal bremsstrahlung from intracluster gas or Compton scattering of high-energy electrons on the microwave background photons. Katz (1976) has suggested that a population of stellar-type sources spread throughout the cluster might provide the emission. We also now know that the discrete sources NGC 1275 and M 87 each contribute to the cluster emission.

The various thermal bremsstrahlung theories have been most widely discussed, and have proven satisfactory. The observations in Perseus (Mitchell et al. 1976), Virgo, and Coma of a spectral feature at ~6 keV that is interpreted as iron-line emission (Serlemitsos et al. 1977) gives very strong evidence for this mechanism. The existence

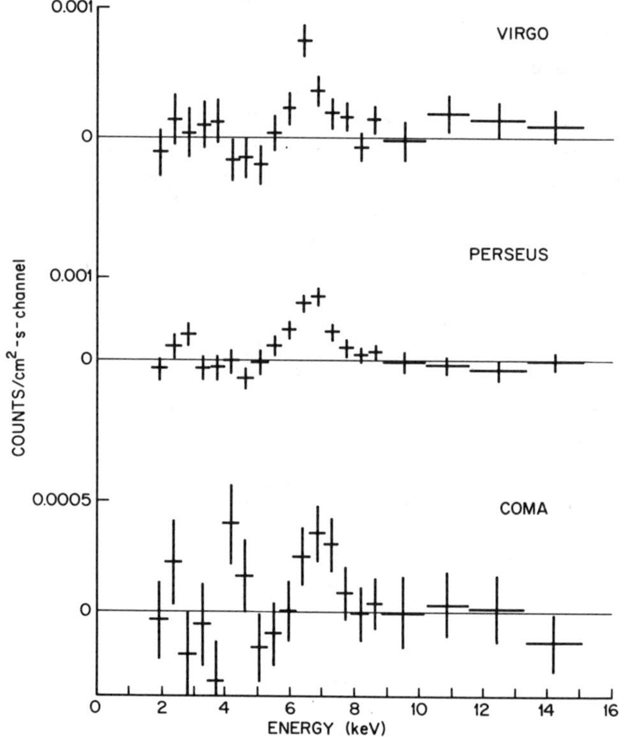

Figure 2 (From Serlemitsos et al. 1977). Spectral data from Virgo, Perseus, and Coma, showing an Fe-line feature at 7 keV. The predicted counting rate due to the best fitting thermal bremsstrahlung spectrum has been subtracted in each case.

of iron shows that the gas has been through a cycle of stellar evolution. Figure 2 shows the excess emission at ~7 keV, above the best fit thermal bremsstrahlung continuum. In Perseus and Virgo we might speculate that all the Fe is produced in the massive, active galaxies NGC 1275 and M 87, respectively. However, in Coma there does not appear to be such an active galaxy, and since the inferred iron abundance is of the order observed in our own solar system we infer that the intracluster gas participated in the initial condensation of the cluster, and has gone through an epoch of star formation similar to the gas that eventually formed the galaxies.

Thermal bremsstrahlung from intracluster gas of temperature $\sim 10^8$ K and density $\sim 10^{-3}$ particles cm^{-3} can explain all the X-ray emission. More detailed models predict temperature and density profiles across the cluster, which will be observable by X-ray telescopes. The parameters of Table 4 are derived on the simplest possible model of an isothermal gas (Lea et al. 1973). A significant result follows immediately: for the Coma and Perseus clusters about 10% of the virial mass is in the form of this hot gas. Therefore, an intergalactic hot plasma cannot bind the cluster, but is significant in that its mass is approximately the same as that observed to be in galaxies. An adiabatic model of the gas distribution (Gull & Northover 1975), where the gas temperature at any point in the cluster is proportional to the gravitational potential at that point, can accommodate a somewhat larger mass since the gas at the outer part of the cluster is cooler and produces few X-rays. In such a model we still do not expect the entire virial mass to be in the form of hot gas—the model as worked out so far is in fact inconsistent in this case since it is assumed that the galaxies establish the potential well and that the hot gas does not significantly perturb it.

The primary questions concern the origin of the gas and the source of the heating. The gas may simply arise as part of the dynamical process of forming the cluster (cf Lea 1976), the cluster may continuously accrete from an intercluster medium (Gunn & Gott 1972), or gas may be expelled from active galaxies (Yahil & Ostriker 1973).

Solinger & Tucker (1972) showed that since thermal bremsstrahlung emission is proportional to density squared, the X-ray luminosity $L_x \propto \sigma^4$, where σ is the rms dispersion of the space velocities of the galaxies, relative to the cluster center of mass. From the virial theorem, $\sigma^2 \propto M$, the total mass of cluster. They showed this to be consistent with existing data and also predicted X-ray emission from Abell 2199 and 2147 prior to its discovery.

Figure 3 plots all the clusters in Table 3 for which σ is measured. The extent along the σ axis gives a 90% confidence region for the measurement of σ^2, based on the number of galaxies with measured redshifts. Certainly $L_x = k\sigma^4$ cannot be said to fit as an exact prediction; however, a correlation with σ to the (4 ± 1) power has continued to hold.

This correlation is not surprising, because many different schemes of heating the gas lead to similar predictions. The fourth power relation corresponds to temperature T independent of σ. This could result if the gas is heated by a process intrinsic to an active galaxy, such as a wind (Yahil & Ostriker 1973), or by cosmic rays (Ipavich 1975). Alternatively, the heating could be related to the dynamic processes in the cluster, for example galactic wakes (Ruderman & Spiegel 1971) or accretion (Gott & Gunn 1971).

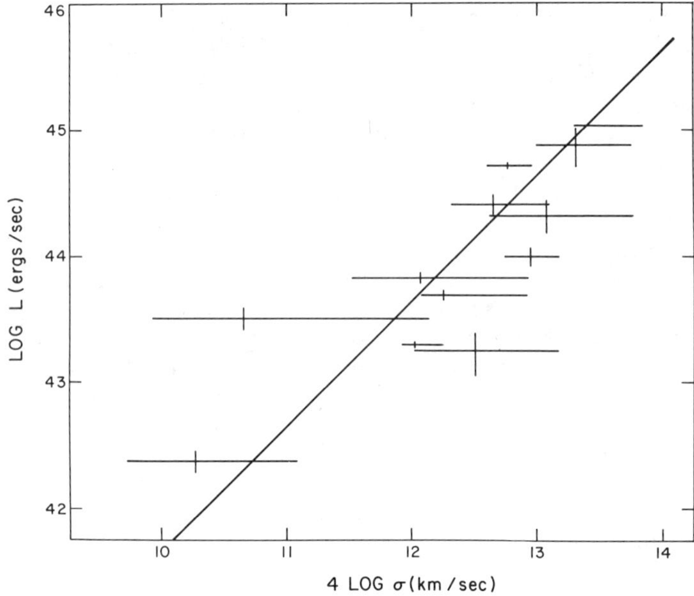

Figure 3 Log of the X-ray luminosity plotted against log of the velocity dispersion to the fourth power. The straight line is the best fit of $L = K\sigma^4$.

In these cases the energy of the gas particles is the same as the kinetic energy of the galaxies per particle, or $T \propto \sigma^2$ and $L \propto \sigma^5$. A third power dependence on σ would result from $T \propto \sigma^2$ and $M \propto \sigma^2$, but with the density M/V roughly constant from cluster to cluster.

The correlation of radio emission and X-ray emission (Owen 1974) forces us to consider seriously inverse Compton scattering of the high-energy electrons against the microwave background photons (Bridle & Feldman 1972, Costain, Bridle & Feldman 1972, Harris & Romanishin 1974). There is much motivation for this discussion despite the fact that the Fe line emission observed in Coma, Virgo, and Perseus (Serlemitsos et al. 1977) establishes thermal bremsstrahlung as the radiation mechanism. We can constrain the average magnetic fields even if we take the X-ray emission as an upper limit to the Compton scattered flux. The diffuse X-ray background has a harder spectrum than the clusters. Even if inverse Compton X-rays in clusters are negligible between 2 and 10 keV, they might be related to the origin of the background at higher energies because they have a flatter spectrum. Of course, some of the clusters or other radio sources may emit X rays by this mechanism.

An electron of energy E (GeV) will scatter an average photon in a black-body distribution at $T = 2.7$ K up to an energy E_x(keV) (cf Felten & Morrison 1966) where

$$E = 0.558\sqrt{E_x}. \tag{3}$$

The same electron in a magnetic field of $B(\mu G)$ will produce synchrotron radio emission at an average frequency

$$\nu(\mathrm{MHz}) = 16.1\ B\ E^2. \tag{4}$$

Thus, a 4-keV X ray arises from a 1.1-GeV electron, which will also produce 20-MHz emission in a typical equipartition field of 1 μG. The actual radiation by an electron of energy E covers a wide range of E_x or ν (cf Blumenthal & Gould 1970, Blumenthal & Tucker 1974a); however, the above approximations are adequate when applied to power-law spectra of electrons $dN/dE = K\,E^{-m}$ (electrons cm^{-3} GeV). In this case, the emitted spectrum goes as $E_x^{-\alpha}$ or $\nu^{-\alpha}$, where $\alpha = (m-1)/2$. The ratio of power emitted as inverse Compton, P_c, or as synchrotron radiation, P_s, is given approximately by the ratio of energy densities in the respective target fields:

$$R = \frac{P_c}{P_s} = \frac{aT^4}{H^2/8\pi}.$$

For the $T = 2.7$ K microwave photons, $R = 1$ when $H = 3 \times 10^{-6}$G.

Radio flux densities for the Abell clusters 347, 401, 1367, 1656, and 2256 extrapolated down to 2 MHz (cf Harris & Romanishin 1974) are about 10^{-3} of the X-ray flux. We can immediately infer that the average magnetic field throughout the extended radio source must be $H \geqq 10^{-7}$ G, or else an X-ray flux larger than is observed would result. On the other hand, if $H \geqq 3 \times 10^{-7}$, less than 10% of the X rays would arise from Compton scattering of the radio electrons. Therefore, we have constrained the average magnetic field to fairly tight limits if we assume the Compton mechanism. These values compare with $\sim 10^{-6}$ G for the typical fields if equipartition of energy between the magnetic field and cosmic rays applies.

The radio and X-ray fluxes can be compared more carefully using the ratio of the spectral flux densities, at a fixed frequency ν, defined as the amplitude ratio

$$\frac{k_c}{k_s} = \frac{dF_c(\nu)}{d\nu} \bigg/ \frac{dF_s(\nu)}{d\nu} = (2.47 \times 10^{-19}) \frac{(4.99 \times 10^{-3})\alpha T^{3+\alpha}}{H^{\alpha+1}} \frac{F(m)}{a(m)}, \tag{5}$$

where the functions $F(m)$ and $a(m)$ are defined and tabulated by Blumenthal & Gould (1970). Then, if we observe Compton X rays at energy E_x arising from the $T = 2.7$ K background, and synchrotron radio waves at frequency ν,

$$\frac{h\,dF_c(E_x)}{dE_x} \bigg/ \frac{dF_s(\nu)}{d\nu} = \frac{k_c}{k_s}\left(\frac{h\nu}{E_x}\right)^{\alpha}. \tag{6}$$

Substituting k_c/k_s from Equation 5 results in H as the only unknown parameter. Tucker et al. (1973) have used this argument to show that $H \gtrsim 10^{-6}$ G for the Cen A radio lobes.

In summary, it seems fair to say that there is no specific evidence that we observe inverse Compton X rays from clusters of galaxies. The most compelling argument for Compton emission is the need for a surprisingly strong magnetic field, $H \sim 0.3\ \mu G$, extended throughout a large volume in order to explain the observed radio emission and not give X rays (Harris & Romanishin 1974). However, we do not regard the lower limits of $H > 0.1\ \mu G$ needed to prevent greater X-ray luminosities than are observed as a substantial alleviation to this problem.

X-Ray Structure

The three strongest clusters, Perseus, Virgo, and Coma, were soon measured to be extended rather than pointlike. This made them natural targets for rocket observations with focusing X-ray telescopes. As the telescopes have developed from one-dimensional to two-dimensional focusing, increasing detail has been revealed. Both X-ray and optical asymmetry increases from Coma to Virgo to Perseus.

Without two-dimensional imaging, the general approach has been to fit a model of the spatial distribution, convoluted with the detector angular response, to the observed counting rates. In one of the earliest such studies (Lea et al. 1973), the X-ray emission was assumed to be axially symmetric and a one-dimensional profile was fit as a function of distance from the cluster center. The model used was isothermal bremsstrahlung (Equation 1 above) with central density n_0, core radius a, and the background counting rate R_B as free parameters. Although the model fit very well, Lea et al. (1973) pointed out that R_B was in excess of the general sky background and therefore might indicate additional emission. Based on unpublished *Uhuru* results, we can establish an excess emission of $\sim 4 \pm 1\%$ at 2° from the center of the Perseus cluster X-ray emission.

VIRGO/M87 The Virgo Cluster extended emission is centered on the giant radio galaxy M87. This case is unique in that the X-rays are displaced $\sim \frac{1}{2}°$ from the cluster center as determined by the elliptical galaxy counts. A point source with 20% to 50% of the Virgo flux was found at the location of M87 (Catura et al. 1972, Malina et al. 1976). The "point source" has recently been shown to be a discrete but extended source identified with M87 (Gorenstein et al. 1977a, in preparation). The intrinsic luminosity is $\sim 5 \times 10^{42}$ ergs sec^{-1}, comparable to Cen A or NGC 4151.

PERSEUS/NGC 1275 NGC 1275 is another case of coming full circle from an original identification as a discrete X-ray source (Fritz et al. 1971, Gursky et al. 1971a), to identification as an extended source associated with the Perseus cluster (Forman et al. 1972), and to a final identification as a discrete component within the cluster (Fabian et al. 1974). The discrete source has between 10 and 25% of the cluster emission below several keV, or a luminosity of 1–4 $\times 10^{44}$ ergs sec^{-1}.

The extended emission has prompted models with at least three components (Wolff et al 1974, 1976, Malina et al. 1976, Cash et al. 1976): a large halo of $\sim$50 arc min diameter, as reported by Lea et al. (1973); a "disk" component extended east-west along the line of galaxies; and a small diffuse source of $\sim$5 arc min around NGC 1275. However, we must remember that this description is first of all a numerical model. The physical model of thermal bremsstrahlung is "one" component determined by the detailed gravitational potential of the cluster and the dynamics of gas flow.

Figure 4 (Gorenstein & Harnden 1976; Gorenstein et al. 1977b in preparation), the first X-ray "photograph" of a celestial source, directly measures the cluster structure. The photograph is preliminary and has not been corrected for the telescope

Figure 4 X-ray photograph of the Perseus cluster (from Gorenstein & Harnden 1976, Gorenstein et al. 1977b, in preparation). Scale is about 1.5 arc min per pixel. A single bright source at the center coincides with NGC 1275.

resolution of about 4 arc min. The gray scale display here does not show the single bright element centered on NGC 1275 that is present in the raw data. Although the figure presents raw data, superposed according to the aspect solution but with no other corrections, everything shown is real in the sense that the non–X-ray and diffuse cosmic backgrounds are both negligible, within about the central 40 arc min of the picture. We see that the emission is not spherically symmetric, but it is much smoother than is implied by the three-component model mentioned above.

THE DIFFUSE X-RAY BACKGROUND

The origin of the diffuse X-ray background still remains unexplained. Schwartz & Gursky (1973; 1974) have given the most recent complete review of this topic. Measurements of the spectral distribution have been reviewed by Pal (1973) and Horstman et al. (1975), and theories of the origin by Silk (1970, 1973), Felten (1973), and Rees (1973). It is not obvious that this topic even falls within the mandate of a review of extragalactic sources. However, the known sources must contribute at least 10% of the background, and even if there is a truly continuous intergalactic component, its energy must arise from compact objects, since it surely is not a relict from the equilibrium of matter and radiation in the early universe.

Spectral Intensity

We may numerically represent the spectrum (Schwartz & Gursky 1974) by either an exponential shape

$$\frac{dN}{dE} = \frac{4.1}{E} e^{-E/35},$$

or a power-law shape

$$\frac{dN}{dE} = \begin{cases} 8.5E^{-1.40} & 1 \leqq E \leqq 21\,\text{keV} \\ 167E^{-2.38} & E \geqq 21\,\text{keV}. \end{cases}$$

The units are photons (keV cm^{-2} sec^{-1} ster^{-1}).

Below a few keV and above a few hundred keV the intensity clearly turns up above either spectral representation. The soft X-ray background is generally recognized to be mostly galactic in origin (cf Gorenstein & Tucker 1976). In the 0.5–50 MeV gamma-ray region, it is extremely difficult even to establish the cosmic intensity, and measurement of the isotropy over the sky is still needed before establishing the origin. If the exponential shape is to be considered as due to thermal bremsstrahlung, then consideration of the Gaunt factor implies a plasma temperature of $\sim 4.4 \times 10^8$ K (Field & Perrenod 1977).

Volume Emissivity

The isotropy (Schwartz 1970) of the X-ray background shows that it is extragalactic. Because of the transparency of intergalactic space (Rees 1969), we know the X-rays can travel freely from a redshift of at least $z \sim 7$, and because the spectrum is relatively flat up to 20 keV the redshift losses are not severe up to $z \gtrsim 1$. If we postulate a uniform volume emissivity $B(E)$ ergs Mpc^{-3} sec^{-1} keV^{-1}, then the observed intensity is approximately $I(E) = \frac{1}{8\pi} R_H B(E)$, where $R_H = 6000$ Mpc. Integrated over 2–10 keV, we have the volume emissivity $B = 2.2 \times 10^{39}$ ergs sec^{-1} Mpc^{-3} (Schwartz & Gursky 1974).

By comparing this emissivity to that for the classes of extragalactic sources discussed above, Schwartz & Gursky (1974) estimate that 22%, within a factor of about 2, of the diffuse background can be accounted for in terms of known discrete sources. We would now revise this to 17%, a negligible correction considering the uncertainties, based on upper limits to emission from Seyfert galaxies discussed above. Possible evolution of the sources (Silk 1968, Rowan–Robinson & Fabian 1975) can be invoked to provide the X-ray background. For the present, however, most X-ray observers are taking a conservative viewpoint (i.e. no evolution) and are searching for new classes of objects to make up the remaining 1 to 2×10^{39} ergs sec^{-1} Mpc^{-3}. Alternately, there may exist a truly metagalactic component to the background. This could arise either from Compton scattering of cosmic ray electrons on the microwave background photons (cf Felten & Morrison 1966), or from thermal bremsstrahlung from a hot intergalactic plasma (cf Field & Henry 1964, Field 1972). A cosmological theory of scattering on optical photons from supernova explosions predicts the required spectrum and intensity (Hogan & Layzer 1977).

Fluctuations

Measurements of the isotropy of the diffuse X-rays in the 7–40 keV region (Schwartz 1970) over half the sky placed a limit of $\sim 1\%$ on the amplitude of large scale ($\gg 20°$) variations. This showed that the background was not associated with our Galaxy or local supercluster and set a limit of 800 km sec^{-1} on the velocity of the Sun relative to this background based on the unobservability of the Compton–Getting effect. The *Uhuru* data (unpublished) in the 2–10 keV region give about the same percentage limits, down to scales much larger than 5°.

The *Uhuru* observations have revealed fluctuations on a scale of one beam width (5° × 5° collimator), which are in excess of Poisson counting statistics and must be intrinsic to the sky (Schwartz et al. 1976). Analysis of further data now gives $\delta I/I = 3.6 \pm 0.5\%$, for an effective solid angle $\Omega \cong 0.004$ ster. Scheuer (1957, 1974) has shown how the fluctuations can be related to the $\ln N$ vs $\ln S$ curve. If $dN/dS = kS^{-5/2}$, we find that $k \leqq 30$ $ster^{-1}$ $(c/sec)^{3/2}$. Because the fluctuations are intrinsic to the sky, they will provide the "noise" against which future measurements of the isotropy will be made.

We can now establish the properties of the unidentified high galactic latitude sources if we assume they comprise the remaining $\sim 80\%$ of the X-ray background. If they have an average intrinsic luminosity j_0 and number density n_0, then their volume emissivity is $B = n_0 j_0$ and the constant above is $k = \frac{1}{2} n_0 (j_0/4\pi)^{3/2}$. This gives $j_0 = 6 \times 10^{42}$ ergs sec^{-1} and $n_0 = 3 \times 10^{-4}$ Mpc^{-3}. These are actually upper limits on j_0 and lower limits on n_0, if we consider that the given class of sources only makes up part of the $\ln N$ vs $\ln S$ source population.

PROSPECTS FOR FUTURE OBSERVATIONS

The use of focusing techniques should provide a qualitative improvement in the observational data base. As an example, Harnden et al. (1977), detected Algol as a point source of X rays during 25 sec of observing time using a focusing telescope, compared to the discovery observation by Schnopper et al. (1976) on this object, which required 3×10^4 sec of observing with a modulation collimator.

In Figure 5 we illustrate the observational capabilities for the near future ($\leqq 10$ years). On the left-hand ordinate we indicate the luminosity distance D defined as $F(E) = P[E(1+z)]/4\pi D^2$. F is the energy flux received at energy E, and P is the power emitted at $E(1+z)$. The dependence of D on the redshift z depends significantly on the cosmological model for $z \gtrsim 1$. To avoid this model dependence, the right-hand ordinate gives the "effective Euclidean volume" V_E in a cone of 1 deg^2 solid angle. By definition, the number of sources observable per deg^2 is $N = n_0 V_E$, where n_0 is the local volume density of sources, and for no source evolution. The quantity V_E is very weakly dependent on the cosmological model, i.e. on the deceleration parameter q_0. For small redshift $\frac{1}{3}(\pi/180)^2 D^3 = V_E$. Of course, the maximum distances are only approximate, depending on the exact telescope quality, the detector backgrounds, and an assumed 10^5 sec observation.

The HEAO-B observatory is now in the final stages of preparation for launch in June 1978. The telescope has an effective area of ~ 200 cm^2, and two imaging

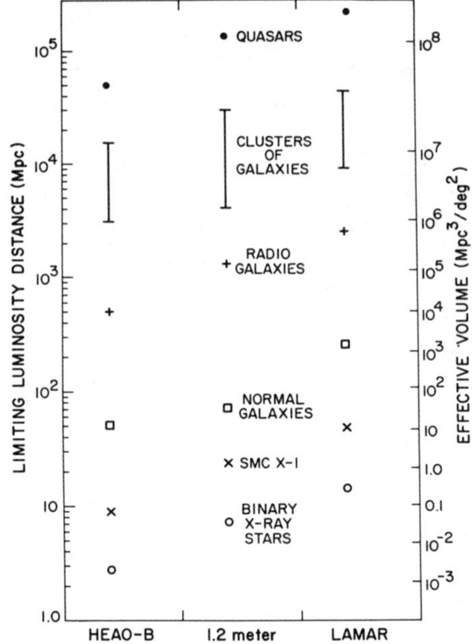

Figure 5 Estimates of sensitivities of X-ray telescope observatories. The maximum distances to which objects can be seen, in 10^5 sec of pointing, are plotted for HEAO-B (under construction for a June 1978 launch) and for the 1.2-meter high-resolution telescope and LAMAR observatories (as recommended by the National Academy of Sciences, 1975). The LAMAR resolution must be assumed to be 20 to 30 arc sec to avoid source confusion. The effective volume (right hand scale) in a cone of 1 deg^2 is only weakly dependent on the cosmological model.

detectors. The highest sensitivity is obtained with an imaging proportional counter with ~ 1.5 arc min resolution and 70% quantum efficiency. A high-resolution imager gives 1 arc sec resolution with $\sim 10\%$ quantum efficiency. The other two observatories are within the state of existing technology and are being considered as future space missions (National Academy of Sciences 1975). The 1.2-meter telescope would have $\sim 10^3$ cm^2 and a negative electron affinity detector (Van Speybroeck et al. 1974) with $\gtrsim 50\%$ quantum efficiency. The Large Area Modular Array of Reflections (LAMAR) observatory would be modular in concept, starting with 10^4 cm^2 and ~ 3 arc min resolution and capable of extension to larger areas and 20 arc sec resolution. The LAMAR observatory might ultimately be limited by source confusion.

Cosmological Studies

By "cosmological" we mean only modest redshifts from a few tenths ($D \sim 10^3$ Mpc), where nonEuclidean effects become measurable, to a few units ($D \sim 2 \times 10^4$ Mpc), where we might be observing the formation of galaxies or clusters. From Figure 5 we can obviously gather much data from this region. The information might be classified as the nature and evolution of objects, or as geometric studies of the fundamental structure of space.

The connection between the "activity" manifested by quasars, Seyfert galaxies, and strong radio sources and an evolutionary sequency through which all galaxies pass has long been tantalizing (cf Burbidge & Burbidge 1967). The contribution of X-ray astronomy may be in the selection of statistically unbiased samples to great distances, since the problem of selection against normal galaxies and stars is immensely alleviated. We may also hope that the X-ray observations, in conjunction with the radio and optical data, may help in elucidating the energy source and mechanism of the explosive or other supraluminous phenomena. We recognize in general that X-ray astronomy is totally dependent on optical observations to establish a redshift. However, in the case of the cluster sources, measurement of the received energy of Fe XXV/XXVI emission may provide the redshift (Boldt 1976). We might seek to measure the abundance of Fe as a function of cosmic time. Somewhat less ambitious would be the measurements of average intracluster gas temperatures and densities as a function of time, assuming the confirmation of thermal bremsstrahlung as the emission mechanism. The condensation of the intracluster gas into an X-ray emitting plasma may be fundamentally connected to the condensation into the galaxies of the cluster. These events may be directly observable if they take place at an epoch corresponding to $z \lesssim 5$.

The maximum luminosity of binary star X-ray sources seems to be about 10^{38} ergs sec^{-1} (Margon & Ostriker 1973). These authors suggest it is governed by the Eddington limit at which radiation pressure counteracts the gravitational accretion energy source. If this limit does apply, then measurements of the apparent fluxes of binary X-ray sources may give independent estimates of the distance scale out to ~ 30 Mpc. This is sufficiently far to measure the Hubble constant. Measurements of the isotropy of the X-ray background (Schwartz 1970) will be extended at least to the point of measuring the velocity of the Sun relative to an effective rest frame at a redshift $z \sim 1$. Perhaps the most powerful contribution of X-ray astronomy will be in finding the clusters of galaxies at redshift $\geqq 1$. Schwartz (1976) has shown that counts of the numbers of X-ray clusters as a function of redshift provides a test for the curvature of the universe (i.e. open vs closed cosmology), which is more powerful than the classical tests of apparent magnitude or apparent size vs redshift. Because the Abell cluster luminosity function must flatten below 10^{44} ergs sec^{-1}, we need only estimate possible evolutionary effects to an accuracy of about a factor of two in order to establish counts within $\sim 50\%$, which in turn serves to distinguish either a $q_0 = 0$ (open) or $q_0 \geqq 1$ (closed) universe from the borderline $q_0 = \frac{1}{2}$ case.

One of the early reviews of X-ray astronomy predicted that it would serve "... either ... as one of the most fruitful channels of data about the largest scales of the universe, or ... the detailed history of certain unusual stars or binaries" (Morrison 1967). We hope that this review, together with that of Blumenthal & Tucker (1974b) on compact, galactic X-ray sources, clearly shows that X-ray astronomy will do both. We may predict that the next reviews of extragalactic X-ray sources will each cover one of the topics dealt with here, and will chiefly address questions of origin and evolution which we are yet too timid to ask.

ACKNOWLEDGMENTS

We thank many colleagues who have sent preprints, and have discussed their work with us, especially H. Schnopper for communicating the identification of 3C120 in advance of publication, and P. Serlemitsos for results on Fe line emission in clusters. We also especially thank C. Jones for giving us many positions, identifications, and error boxes from the fourth *Uhuru* catalog in advance of publication. We thank P. Gorenstein for results on the Virgo and Perseus clusters, including Figure 4, in advance of publication. Capabilities of future X-ray observatories have been derived from discussions with P. Gorenstein, H. Tananbaum, and L. Van Speybroeck.

Literature Cited

Abell, G. 1958. *Ap. J. Suppl.* 3, No. 31:211
Baade, W., Minkowski, R. 1954. *Ap. J.* 119: 206
Bahcall, N. A. 1974a. *Nature* 252:661
Bahcall, N. A. 1974b. *Ap. J.* 193:529
Bahcall, N. A. 1977. *Ann. Rev. Astron. Astrophys.* 15:505
Bahcall, J. N., Bahcall, N. A. 1975. *Ap. J. (Lett)* 199:L89
Bahcall, J. N., Bahcall, N. A., Murray, S. S., Schmidt, M. 1975. *Ap. J. (Lett.)* 199:L9
Baity, W. A., Jones, T. W., Wheaton, W. A., Peterson, L. E. 1975. *Ap. J. (Lett.)* 199:L5
Blumenthal, G. R., Gould, R. J. 1970. *Rev. Mod. Phys.* 42:237
Blumenthal, G. R., Tucker, W. H. 1974a. In *X-Ray Astronomy,* ed. R. Giacconi, H. Gursky, Chap. 3, p. 99. Dordrecht: Reidel
Blumenthal, G. R., Tucker, W. H. 1974b. *Ann. Rev. Astron. Astrophys.* 12:23
Boldt, E. 1976. *Ap. J. (Lett.)* 208:L15
Bowyer, S., Byram, E. T., Chubb, T. A., Friedman, H. 1964. *Nature* 201:1307
Bowyer, S. C., Lampton, M., Mack, J., de-Mendonca, F. 1970. *Ap. J. (Lett.)* 161:L1
Bradt, H., Mayer, W., Naranan, S., Rappaport, S., Spada, G. 1967. *Ap. J. (Lett.)* 150:L199
Bridle, A. H., Feldman, P. A. 1972. *Nature Phys. Sci.* 235:168
Brinkman, A. C., Heise, J., denBoggende, A. J. F., Grindlay, J., Gursky, H., Parsignault, D. 1977. *Ap. J.* 214: In press
Burbidge, G. R., Burbidge, E. M. 1967. *Quasi-Stellar Objects,* pp. 180–187. San Francisco: Freeman
Byram, E. T., Chubb, T. A., Friedman, H. 1966. *Science* 152:66
Cash, W., Malina, R., Wolff, R. S. 1976. *Ap. J. (Lett.)* 209:L111
Catura, R. C., Fisher, P. C., Johnson, H. M., Meyerott, A. J. 1972. *Ap. J. (Lett.)* 177:L1
Charles, P. A., Longair, M. S., Sanford, P. W. 1975. *MNRAS* 170:17p
Cooke, B. A. 1976. *Nature* 261:564
Cooke, B. A., Maccagni, D. 1976. *MNRAS* 175:65
Costain, C. H., Bridle, A. H., Feldman, P. A. 1972. *Ap. J. (Lett.)* 175:L15
Culhane, J. L. 1976. *Bull. Am. Astron. Soc.* 8:445
Culhane, J. L., Davison, P. J. N., Stark, J. P. 1976. *MNRAS* 174:35p
Cushman, P. B. 1976. Unpubl. senior honors thesis. Dept Phys. Harvard Univ.
Davison, P. J. N., Culhane, J. L., Mitchell, R. J., Fabian, A. C. 1975. *Ap. J. (Lett.)* 196: L23
Delvaille, J. P., Epstein, A., Schnopper, H. 1977. In preparation
Disney, M. J. 1974. *Ap. J. (Lett.)* 193:L103
Elvis, M., Cooke, B. A., Pounds, K. A., Turner, M. J. L. 1975. *Nature* 257:33
Fabian, A. C. 1975. *MNRAS* 172:149
Fabian, A. C., Maccagni, D., Rees, M. J., Stoeger, W. R. 1976. *Nature.* 260:683
Fabian, A. C., Zarnecki, J. C., Culhane, J. L., Hawkins, F. J., Peacock, A., Pounds, K. A., Parkinson, J. H. 1974. *Ap. J. (Lett.)* 189: L59
Felten, J. E. 1973. In *I.A.U. Symp. No. 55, Non-Stellar X-Ray and Gamma-Ray Astronomy,* eds. H. Bradt, R. Giacconi, pp. 258–75. Dordrecht: Reidel
Felten, J. E., Morrison, P. 1966. *Ap. J.* 146: 686
Field, G. B. 1972. *Ann. Rev. Astron. Astrophys.* 10:227–60
Field, G. B., Henry, R. C. 1964. *Ap. J.* 140: 1002
Field, G. B., Perrenod, S. C. 1977. *Ap. J.* In press
Forman, W., Jones, C., Cominsky, L., Julien, P., Murray, S., Peters, G., Tananbaum, H., Giacconi, R. 1977. To be submitted to *Ap. J. Suppl.*
Forman, W., Kellogg, E., Gursky, H., Tananbaum, H., Giacconi, R. 1972. *Ap. J.*

178:309
Fritz, G., Davidsen, A., Meekins, J. F., Friedman, H. 1971. *Ap. J. (Lett.)* 164:L81
Giacconi, R., Gursky, H., Paolini, F., Rossi, B. 1962. *Phys. Rev. Lett.* 9:439
Giacconi, R., Kellogg, E., Gorenstein, P., Gursky, H., Tananbaum, H. 1971a. *Ap. J. (Lett.)* 165:L27
Giacconi, R., Murray, S., Gursky, H., Kellogg, E., Schreier, E., Matilsky, T., Koch, D., Tananbaum, H. 1974. *Ap. J. Suppl.* 27, No. 237:37
Giacconi, R., Murray, S., Tananbaum, H., Kellogg, E., Gursky, H. 1971b. *Am. Astron. Soc. Bull.* 3:477
Gorenstein, P., Fabricant, D., Topka, K., Tucker, W., Harnden, F. R. Jr. 1977a. In preparation
Gorenstein, P., Fabricant, D., Topka, K., Tucker, W., Harnden, F. R. Jr. 1977b. In preparation
Gorenstein, P., Harnden, F. R. Jr. 1976. *Bull. Am. Astron. Soc.* 8:445
Gorenstein, P., Tucker, W. 1976. *Ann. Rev. Astron. Astrophys.* 14:373
Gott, J. R., Gunn, J. E. 1971. *Ap. J. (Lett.)* 169:L13
Grindlay, J. E. 1975. *Ap. J.* 199:49
Grindlay, J. E., Helmken, H. F., Hanbury Brown, R., Davis, J., Allen, L. R. 1975a. *Ap. J. (Lett.)* 197:L9
Grindlay, J. E., Schnopper, H., Schreier, E. J., Gursky, H., Parsignault, D. R. 1975b. *Ap. J. (Lett.)* 201:L133
Gull, S. F., Northover, K. J. E. 1975. *MNRAS* 173:585
Gunn, J. E., Gott, J. R. 1972. *Ap. J.* 176:1
Gursky, H., Kellogg, E. M., Leong, C., Tananbaum, H., Giacconi, R. 1971a. *Ap. J. (Lett.)* 165:L43
Gursky, H., Kellogg, E., Murray, S., Leong, C., Tananbaum, H., Giacconi, R. 1971b. *Ap. J. (Lett.)* 167:L81
Gursky, H., Schwartz, D. A. 1974. See Blumenthal & Tucker 1974a, pp. 25–48
Gursky, H., Solinger, A., Kellogg, E., Murray, S., Tananbaum, H., Giacconi, R., Cavaliere, A. 1972. *Ap. J. (Lett.)* 173:L99
Hall, R. D., Meegan, C. A., Walraven, G. D., Djuth, F. T., Haymes, R. C. 1976. *Ap. J.* 210:631
Harnden, F. R. Jr., Fabricant, D., Topka, K., Flannery, B. P., Tucker, W. H., Gorenstein, P. 1977. *Ap. J.* 214:418
Harris, D. E. 1976. *IAU Highlights Astron.*
Harris, D. E., Romanishin, W. 1974. *Ap. J.* 188:209
Haymes, R. C. 1975. *IAU Circ.* No. 2780
Heinz, C., Markert, T., Clark, G. W., Lewin, W. H. G., Schnopper, H. W., Sprott, G. 1972. *IAU Circ.* No. 2466
Hogan, C., Layzer, D. 1977. *Ap. J.* 212:360
Holt, S. S., Boldt, E. A., Serlemitsos, P. J., Murray, S. S., Giacconi, R., Kellogg, E. M., Matilsky, T. A. 1974. *Ap. J. (Lett.)* 188:L97
Horstman, H. M., Cavallo, G., Moretti-Horstman, E. 1975. *Riv. Nuovo Cimento,* 5:255
Ipavich, F. 1975. *Ap. J.* 196:107
Ives, J. C., Sanford, P. W. 1976. *MNRAS* 176:13p
Jenner, D. C. 1974. *Ap. J.* 191:55
Jones, C., Forman, W. 1977. In preparation
Katz, J. I. 1976. *Ap. J.* 207:25
Kellogg, E. M. 1973. See Felten 1973, pp. 171–83
Kellogg, E. M. 1974. See Blumenthal & Tucker 1974a, pp. 321–57
Kellogg, E., Gursky, H., Leong, C., Schreier, E., Tananbaum, H., Giacconi, R. 1971. *Ap. J. (Lett.)* 165:L49
Kellogg, E. M., Murray, S. S. 1974. *Ap. J. (Lett.)* 193:L57
Kellogg, E., Murray, S., Giacconi, R., Tananbaum, H., Gursky, H. 1973. *Ap. J. (Lett.)* 185:L13
Kunkel, W. E., Bradt, H. V. 1971. *Ap. J. (Lett.)* 170:L7
Lampton, M., Margon, B., Bowyer, S., Mahoney, W., Anderson, K. 1972. *Ap. J. (Lett.)* 171:L45
Lea, S. M. 1976. *Ap. J.* 203:569
Lea, S. M., Silk, J., Kellogg, E. M., Murray, S. S. 1973. *Ap. J. (Lett.)* 184:L105
Longair, M. S., Willmore, A. P. 1974. *MNRAS* 168:479
Lugger, P. M. 1976. Unpublished senior honors thesis. Dept. Astron. Harvard Univ.
Lugger, P. M. 1977. Submitted to *Ap. J.*
MacDonald, R. H., Kenderdine, S., Neville, A. C. 1968. *MNRAS* 138:259
Malina, R., Lampton, M., Bowyer, S. 1976. *Ap. J.* 209:678
Margon, B., Lampton, M., Bowyer, S., Cruddace, R. 1975. *Ap. J.* 197:25
Margon, B., Ostriker, J. P. 1973. *Ap. J.* 186:91
Margon, B., Spinrad, H., Heiles, C., Tovmassian, H., Harlan, E., Bowyer, S., Lampton, M. 1972. *Ap. J. (Lett.)* 178:L77
Markert, T. H., Canizares, C. R., Clark, G. W., Li, F. K., Northridge, P. L., Sprott, G. F., Wargo, G. F. 1976. *Ap. J.* 206:265
Matilsky, T., Gursky, H., Kellogg, E., Tananbaum, H., Murray, S., Giacconi, R. 1973. *Ap. J.* 181:753
Melnick, J., Quintana, H. 1975. *Ap. J. (Lett.)* 198:L97
Miley, G. K., Perola, G. C., Vander Kruit, P. C., Van Der Laan, H. 1972. *Nature* 237:269

Mitchell, R. J., Charles, P. A., Culhane, J. L., Davison, P. J. N. 1975. *Ap. J.* (*Lett.*) 200: L5
Mitchell, R. J., Culhane, J. L., Davison, P. J. N., Ives, J. C. 1976, *MNRAS* 175: 29p
Morrison, P. 1967. *Ann. Rev. Astron. Astrophys.* 5: 325
Mushotzky, R. F., Baity, W. A., Peterson, L. E. 1977. *Ap. J.* 212: 22
Mushotzky, R. F., Baity, W. A., Wheaton, W. A., Peterson, L. E. 1976. *Ap. J.* (*Lett.*) 206: L45
National Academy of Sciences. 1975. *Opportunities and Choices in Space Science 1974,* pp. 43 ff., 60–64. Washington, D.C.: Space Sci. Board
Noonan, T. W. 1973. *Astron. J.* 78: 26
Owen, F. N. 1974. *Ap. J.* (*Lett.*) 189: L55
Pal, Y. 1973. See Felten 1973, pp. 279–302
Perola, G. C., Tarenghi, M. 1973. *Astron. Astrophys.* 25: 461
Peterson, L. E. 1975. *Ann. Rev. Astron. Astrophys.* 13: 423
Peterson, P., Bahcall, J., Murray, S. 1977. In preparation
Pye, J. P., Cooke, B. A., Elvis, M. 1976. *Nature* 262: 195
Rappaport, S., Buff, J., Clark, G., Lewin, W. H. G., Matilsky, T., McClintock, J. 1976. *Ap. J.* (*Lett.*) 206: L139
Rees, M. J. 1969. *Ap. Lett.* 4: 113
Rees, M. J. 1973. See Felten 1973, pp. 250–57
Ricketts, M. J., Cooke, B. A., Pounds, K. A. 1976. *Nature* 259: 546
Rood, H. J., Sastry, G. N. 1971. *Publ. Astron. Soc. Proc.* 83: 313
Rowan–Robinson, M. 1972. *Astron. J.* 77: 543
Rowan–Robinson, M., Fabian, A. C. 1975. *MNRAS* 170: 199
Ruderman, M. A., Spiegel, E. A. 1971. *Ap. J.* 165: 1
Rudnick, L., Owen, F. N. 1976. *Ap. J.* (*Lett*) 203: L107
Ryle, M., Windram, M. D. 1968. *MNRAS* 138: 1
Sandage, A. 1973. *Ap. J.* 183: 731
Sandage, A. 1975. *Ap. J.* 202: 563
Scheuer, P. A. G. 1957. *Proc. Cambridge Philos. Soc.* 53: 764
Scheuer, P. A. G. 1974. *MNRAS* 166: 329
Schnopper, H. W. 1976. High Energy Astrophys. Div. Meet. on X-Ray Astron., M.I.T., 29 Jan. 76.
Schnopper, H. W., Delvaille, J. P., Epstein, A., Helmken, H., Murray, S. S., Clark, G., Jernigan, G., Doxsey, R. 1976. *Ap. J.* (*Lett.*) 210: L75
Schnopper, H. W., Epstein, A., Delvaille, J. P., Tucker, W., Doxsey, R., Jernigan, G. 1977. *Ap. J.* (*Lett.*) 215: In press
Schreier, E., Schnopper, H., Gursky, H., Parsignault, D. R. 1975. *Ap. J.* (*Lett.*) 201: L33
Schwartz, D. A. 1970. *Ap. J.* 162: 439
Schwartz, D. A. 1976. *Ap. J.* (*Lett.*) 206: L95
Schwartz, D. A. 1977. *Ap. J.* Submitted for publication
Schwartz, D. A., Gursky, H. 1973. In *International Symposium and Workshop on Gamma-Ray Astrophysics,* eds. F. W. Stecker & J. I. Trombka, p. 15. NASA SP-339
Schwartz, D. A., Gursky, H. 1974. See Blumenthal & Tucker 1974, pp. 359–88
Schwartz, D. A., Murray, S. S., Gursky, H. 1976. *Ap. J.* 204: 315
Serlemitsos, P. J., Smith, B. W., Boldt, E. A., Holt, S. S., Swank, J. H. 1977. *Ap. J.* (*Lett.*) 211: L63
Setti, G., Woltjer, L. 1973. See Felten 1973, pp. 209–10
Silk, J. 1968. *Ap. J.* (*Lett.*) 151: L19
Silk, J. 1970. *Space Sci. Rev.* 11: 671
Silk, J. 1973. *Ann. Rev. Astron. Astrophys.* 11: 269
Solinger, A., Tucker, W. 1972. *Ap. J.* (*Lett.*) 175: L107
Spinrad, H. 1975. *Ap. J.* (*Lett.*) 199: L1
Tucker, W., Kellogg, E., Gursky, H., Giacconi, R., Tananbaum, H. 1973. 180: 715
Ulmer, M. P., Murray, S. S. 1976. *Ap. J.* 207: 364
Valleé, J. P., Wilson, A. S. 1976. *Nature* 259: 451
VanSpeybroeck, L., Kellogg, E., Murray, S., Duckett, S. 1974. *IEEE Trans. Nucl. Sci., NS-21,* 408
Vidal, N. V. 1975a. *Publ. Astron. Soc. Pac.* 87: 625
Vidal, N. V. 1975b. *Astron. Astrophys.* 42: 145
Walmsley, C. M. 1969. Unpublished PhD thesis, University of California, San Diego, pp. 90–101
Whiteoak, J. B. 1972. *Aust. J. Phys.* 25: 233
Willson, M. A. G. 1970. *MNRAS* 151: 1
Winkler, P. F., Jr., White, A. E. 1975. *Ap. J.* (*Lett.*) 199: L139
Wolff, R. S., Helava, H., Kifune, T., Weisskopf, M. C. 1974. *Ap. J.* (*Lett.*) 193: L53
Wolff, R. S., Mitchell, R. J., Charles, P. A., Culhane, J. L. 1976. *Ap. J.* 208: 1
Yahil, A., Ostriker, J. P. 1973. *Ap. J.* 185: 787
Zwicky, F. 1971, *Catalog of Selected Compact Galaxies and of Post-Eruptive Galaxies,* p. 369. Zurich: L. Spreich

AUTHOR INDEX

A

Aannestad, P. A., 267, 278, 279, 282, 288
Aaronson, M., 280
Aarseth, S. J., 257, 529
Abell, G. O., 239, 505-11, 522-24, 528, 549, 552, 553
Ables, J. G., 21, 23, 24, 38, 488-91, 496-500
Abraham, F. F., 275, 277
Abt, H. A., 142
Ackerson, K. L., 422, 426
Acton, L. W., 180, 364
Acuña, M. H., 425
Adams, J. B., 114, 115, 117
Adams, T. F., 69, 71, 73-75, 79-83, 85, 87
Adlington, R. H., 217
Agrawal, P. C., 43
Aguirre, C., 78
Aitken, D. K., 280
Albert, C. E., 513, 514
Albritton, D. L., 204
Alexander, M. E., 146
Alfvén, H., 63, 286
Alksne, A. Y., 415
Allen, C. W., 201, 207
Allen, D. A., 72-75, 81, 85, 87, 88
Allen, L. R., 544
Allen, M., 289
Allen, R. J., 72-74, 454
Aller, L. H., 199, 201, 206, 269, 270
Allison, A. C., 223
Alloin, D., 74
Alme, M. L., 146, 147
Altenhoff, W. J., 344
Altschuler, M. D., 159, 161, 169, 373-75
Ambartsumian, V. A., 532
Ambrož, P., 159
Amnuel, P. R., 147
Anders, E., 272, 278
Anderson, B., 40, 498, 502
Anderson, D. L., 119
Anderson, J. D., 98
Anderson, K., 544
Anderson, K. A., 409
Anderson, K. S., 72-75, 91
Anderson, L., 142
Anderson, R. A., 220, 222, 224
Andrillat, Y., 72, 74, 75
Angel, J. R. P., 72, 86, 87, 193
Anketell, J., 224
Antipova, L. I., 285
Arakelian, M. A., 72-75, 78, 87, 88
Ardavan, H., 178
Armstrong, B. H., 207, 209, 211
Armstrong, J. W., 486, 499
Armstrong, T. P., 411
Arnett, W. D., 144, 177, 270
Arnold, J. O., 199, 219-21, 226
Arons, J., 189
Arp, H. C., 72, 73, 75, 454, 462, 531
Arrhenius, G., 278, 286
Asbridge, J. R., 103
Ash, M. E., 33, 98
Assousa, G. E., 182
Athay, R. G., 367-70, 372, 376, 380, 381, 383
Atreya, S. K., 420
Audouze, J., 270, 286
Auman, J. R., 72, 74
Aumann, H. H., 358
Auriere, M., 114, 115, 117
Austin, T. B., 517, 519, 521, 524, 528
Avni, Y., 517, 520
Avrett, E. H., 364, 367, 368, 370
Axford, W. I., 422, 429
Ayres, T. R., 367, 368

B

Baade, W., 175, 439, 454, 513, 526, 552
Babcock, H. D., 153, 156
Babcock, H. W., 153, 156
Backer, D. C., 479, 480, 494, 496-503
Bahcall, J. N., 516, 531, 533, 547-49
BAHCALL, N. A., 505-40; 510, 517-21, 528, 530, 533, 534, 536, 547-49, 553
Baity, W. A., 88, 544, 546
Baker, D. N., 400, 409
Baker, J. C., 188, 251
Balasubramanian, V., 21
Balasubrahmanyan, V. K., 64
Baldwin, J. A., 89
Baldwin, J. E., 536
Baldwin, R. B., 122
Balick, B., 182, 288, 351, 502
Baliunas, S., 364
Bame, S. J., 103
Bania, T. M., 309, 311, 324, 326-28, 330, 335, 336
Banks, P. M., 224, 420
Bar, V., 63
Barabanenkov, J. N., 481
Baranowska, M., 517, 528
Barbaro, G., 129
Barbieri, C., 72, 74
Bardeen, J. M., 468-71
Barish, F. D., 422, 423
Barlow, M. J., 278, 285, 287, 289
Barnes, C. W., 411
Barrett, J. W., 72, 75
Barrow, R. F., 204
Barshay, S. S., 272
Bartoe, J. D. F., 369, 380
Bass, A. M., 222, 224
Bassett, W. A., 120
Batchelor, R. A., 332, 494, 497-500
Batson, R. M., 116
Batten, A. H., 139
Baum, W. A., 238
Bautz, L. P., 510, 511, 514, 523, 524
Baym, G., 35
Beaver, E. A., 72, 86, 87
Beck, A. J., 390
Becker, K. H., 224
Becker, R. H., 25, 180
Beckers, J. M., 49, 50, 60, 61, 369, 370, 380, 381
Becklin, E. E., 72-75, 80, 82, 86, 87, 90, 283, 297-99, 351, 355-58
Beebe, R. F., 272
Behannon, K. W., 102-4
Behring, W. E., 380
Belchman, C. A., 281

Bell, A. R., 190
Bell, B., 374
Bell, R. A., 212
Bell, S. J., 19, 479, 486, 498
Belton, M. J. S., 101, 105, 113, 114, 117
Bender, C. F., 220
Benfield, A. E., 119
Bennett, R. G., 217, 219, 220, 222, 224
Benson, R. S., 135
Berg, R. A., 147
Berge, G. L., 391
Bergeman, T. H., 222
Bergstrahl, J. T., 426
Berkhuijsen, E. M., 186
Bernstein, I. B., 179
Bershader, D., 216
Bersohn, C. K., 220
Berthel, R. O., 226
Bertola, F., 258, 531
Beuermann, K. P., 176
Bieging, J. H., 336, 337
Biermann, L., 61, 252
Biermann, P., 42, 138, 139, 141, 142
Billingsley, F. P., 214
Binney, J., 258-60, 535
Biskamp, D., 63
Bisnovatyi-Kogan, G. S., 471
Black, G., 222
Blackwell, D. E., 198
Blake, G. M., 179
Blandford, R., 147, 189
Blandford, R. D., 33
Blumenthal, G. R., 84, 91, 559, 565
Bodenheimer, P., 142, 144
Bohannan, B., 141
Bohlin, J. D., 47, 62, 190, 369, 370, 375
Bohlin, R. C., 280
Bohuski, T. J., 286
Boksenberg, A., 74, 81, 85, 88
Boland, B. C., 380
Boldt, E. A., 25, 180, 535, 547, 554, 556, 558, 565
Bolton, J. G., 75, 331, 332
Bonazzola, S., 497, 498
Bond, H. E., 74, 90
Bonilha, J. R. M., 285
Booth, R. S., 349, 352
Borgman, J., 351, 355
Boriakoff, V., 22, 26, 30
Borken, R., 180, 187
Born, M., 208
Borst, W. L., 222
Boss, A. P., 101
Bourene, M., 222
Bovier, P., 249
Bowyer, S., 532, 541, 543-45, 547, 560
Boynton, P. E., 35, 37
Bracessi, A., 74
Bradt, H., 25, 541
Bradt, H. V., 544
Branch, D., 212
Brandt, J. C., 176
Brannen, E., 217
Braun, W., 222
Bray, R. J., 367-70
Brecher, K., 534
Brett, R. A., 90
Brice, N. M., 391, 414, 418, 419, 421, 424, 425, 432
Bridge, H. S., 103
Bridle, A. H., 38, 534, 536, 555, 558
Briggs, F. H., 116
Brinkman, A. C., 364, 546
Brinkman, W., 220
Broadfoot, A. L., 101, 102
Brocklehurst, B., 204, 220
Broderick, J. J., 70, 73, 90, 484
Broida, H. P., 200, 216, 218, 224, 225
Brooks, J. W., 332
Brooks, N. H., 219, 220
Brophy, J. H., 224
Brouw, W. N., 454
Brouwer, D., 99
Brown, A. G. F., 492
Brown, R. A., 426
Brown, R. L., 351, 502
Browne, I. W. A., 498, 502
Browne, J. C., 223
Brownrigg, D. R. K., 468, 469, 474
Bruck, Y. M., 21
Brueckner, G. E., 47, 170, 369, 381
Brunn, D. L., 98
Bryan, G. H., 464
Brzozowski, J., 224, 225
Budden, K. G., 502, 503
Buff, J., 181, 548
Bumba, V., 155, 156, 159, 164, 172
Bunner, A. N., 187
Burbidge, E. M., 75, 79, 89, 565
Burbidge, G. R., 69, 72-75, 80, 89-91, 137, 531, 534, 565
Burger, H. L., 135
Burke, B. F., 389
Burke, J. R., 183, 278, 286
Burnell, J., 479
Burnett, K., 221
BURNS, J. A., 97-126; 100, 120
Burton, W. B., 267, 268, 280, 289, 301, 306, 310, 324-28, 336, 339, 342, 455, 459
Bussoletti, E., 279
Butler, C. P., 155
Bychkov, K. V., 184, 185
Byram, E. T., 541, 555

C

Cameron, A. G. W., 118, 142, 267, 269, 278, 286, 289
Camichel, H., 117
Campbell, D. B., 21, 24, 27-29, 32, 479
Campbell, M. F., 341, 359, 360
Candy, M. P., 446
Canfield, R. C., 380, 381
Canizares, C. R., 552
Cann, M. W. P., 204
Cannon, R. D., 90
Capelle, G., 216, 218
Capps, R. W., 72, 280-82, 288
Carance, M., 221
Carbary, J. F., 427, 429, 431
Carlquist, P., 63
Carlson, D. J., 224
Carlson, R. W., 222, 227, 426
Caroff, L. J., 282
Carr, T. D., 391, 484
Carranza, G., 454
Carrasco, L., 289
Carrington, T., 224
Carruthers, G. R., 193
Carswell, R. F., 74, 88
Carusi, A., 278
Caruso, R., 218
Carver, G. H., 223
Cash, W., 180, 560
CASSEN, P., 97-126; 120, 121
Caswell, J. L., 38, 40, 41, 191
Catura, R. C., 364, 560
Cavaliere, A., 535, 549, 553
Cavallo, G., 561
Celniker, L. M., 497

Cerny, D., 219, 225
Chaffee, F. H., 426
Chamberlain, J. W., 224
Chamberlin, T. C., 438, 475
Chan, Y.-W. T., 89
Chandrasekhar, S., 51, 248, 259, 463, 526
Chao, J. K., 104
Chapman, C. R., 122, 123
Chapman, G. A., 49, 50, 377, 379
Charles, P. A., 543, 546, 554, 560
Chase, R. C., 170
Chase, S. C., 114, 115
Chau, W. Y., 146
Chen, C.-K., 426
Chenette, D. L., 409-11, 432
Cheng, C. C., 379
Chervenak, J. G., 222
Chevalier, R., 223
CHEVALIER, R. A., 175-96; 176, 177, 179-83, 185, 186, 188-90, 192, 272, 286, 287
Chiao, R. Y., 531
Chicada, Y., 336, 339
Chincarini, G., 508, 513, 517, 518, 522, 528
Chiosi, C., 129, 146
Chiu, B. C., 176
Christopher, L., 103
Chubb, T. A., 541, 555
Chupp, E. L., 63
Churchwell, E. B., 346-48, 351, 352
Chuvaev, K. K., 72, 73, 75
Clark, B. G., 70, 72, 73, 90, 352
Clark, D. H., 38, 40, 41, 191
Clark, E. E., 517, 519
Clark, G., 548, 563
Clark, G. W., 549, 552
Clark, T. A., 73, 90
Clayton, D. D., 285
Clerc, M., 220
Cleve, M., 222
Cocke, W. J., 26
Code, A. D., 285
Cogdell, J. R., 116
Cohen, C. J., 99
Cohen, J. G., 282
Cohen, M., 282, 285
Cohen, M. H., 70, 73, 90, 352, 479, 484
Cohen, R. J., 302, 309, 311-15, 332-35, 337, 339
Colburn, D. S., 104, 120, 168, 398, 399, 402-4, 406, 407, 421-23, 425
Coleman, P. J., 398, 399, 402-4, 406, 407, 421-23, 425
Coleman, P. L., 25
Coles, W. A., 480, 486, 499
Colgate, S. A., 176
Colla, G., 535, 536
Collard, H. R., 395, 423
Collin-Souffrin, S., 72, 74
Collins, R. A., 19, 479
Colombo, G., 99-101
Comella, J. M., 21, 479, 497
Cominsky, L., 543
Condon, J. J., 479, 502
Conklin, E. K., 497
Conlon, T. F., 409-11, 432
Conti, P. S., 141
Contopoulos, G., 467
Conway, R. G., 498, 502
Cook, T. J., 221
Cooke, B. A., 88, 548, 549, 552
Cooke, D. J., 498
Cooper, D. M., 199, 219
Coppi, B., 63
Coradini, A., 278
Cordes, J. M., 29-32
Cornett, R. H., 182
CORONITI, F. V., 389-436; 390, 391, 420, 421, 425, 426, 428, 432, 433
Corwin, H. G., 511
Costain, C. H., 536, 555, 558
Costero, E., 269
Cotton, W. D., 90
Counselman, C. C. III, 21, 90, 145, 479, 491
Cowie, L. L., 179, 181, 184, 535
Cowley, A. P., 141
Cowling, T. G., 61
Cox, D. P., 372
Coxon, J. A., 212, 219
Craft, H. D. Jr., 21, 27, 479, 497, 498
Creek, D. M., 204
Crillon, R., 454
Cronyn, W. M., 484, 486, 487
Crosley, D. R., 224
Cross, M. A., 63
Crosswhite, H. M., 224
Cruddace, R., 545
Crutcher, R. M., 289, 290
Cudaback, D. D., 281
Cugnon, P., 296, 313
Culhane, J. L., 180, 535, 544, 545, 556, 560
Cuny, Y., 367, 368
Curtis, H. D., 437
Curtis, S. A., 102
Cushman, G. W., 381
Cushman, P. B., 549, 552
Cuzzi, J. N., 116
Czernichowski, A., 226
Czyzak, S. J., 269, 270, 278

D

Dalby, F. W., 217, 219, 220, 222, 224
Dale, T. M., 141
Dalgarno, A., 223
Danielson, G. E., 105, 107, 113-15, 117
Danielson, R. E., 319, 320
Danylewych, L. L., 219-21
Darchy, B. F., 72-74
Das, G., 214
Davenport, J. W., 222
Davidsen, A., 532, 541, 555, 560
Davidson, K., 89, 286
Davies, J. G., 34, 37, 39
Davies, M. E., 105, 113, 114, 116, 117
Davies, R. D., 74, 309, 311, 313-15, 349, 352, 498, 502, 513
Davis, B. D., 224
Davis, D. R., 123
Davis, J., 544
Davis, J. H., 116
Davis, J. M., 370, 379
Davis, L., 398, 399, 402-4, 406, 407, 421-23, 425
Davis, L. Jr., 428
Davis, L. R., 394, 395, 400, 401, 411
Davis, M., 240, 261
Davis, R. J., 498, 502
Davis, S., 204
Davison, P. J. N., 180, 535, 544. 554. 556
Day, K. L., 281
De, B. R., 278
deBoer, K. S., 271
de Bruyn, A. G., 72-75, 87, 317
De Cuyper, J. P., 146, 147
deFieter, L. D., 383
Degen, V., 204
DeGreve, J. P., 131, 134, 143-47
Deinzer, W., 47

De Jager, C., 64
Delache, P., 372
DeLoore, C., 131, 134, 143-47, 363
Delvaille, J. P., 543, 563
Demarque, P., 129
deMendonca, F., 541, 543-45
Dempsey, M. J., 282
Demtroder, W., 217
den Boggende, A. J. F., 364, 546
Denisyuk, E. K., 74
Desch, M., 484
Desch, M. D., 391
Despain, K. H., 274
Dessler, A. J., 425, 427, 429, 431
Deubner, F. L., 62, 380, 381
Deutsch, A. J., 201
de Vaucouleurs, A., 70, 72, 74, 75, 79, 237, 524
de Vaucouleurs, G., 70, 72-75, 79, 98, 114, 117, 237, 238, 240, 250, 455, 507, 508, 512, 519, 524, 531
Devine, N., 97, 116
de Vries, M., 351, 355
Dewar, R. L., 449
De Young, D. S., 72, 532, 536
De Zafra, R. L., 224
Dibay, E. A., 73-75, 80, 87
Dicke, R. H., 47, 245, 262
Dickey, J., 270, 288
Dickinson, D. F., 72, 74
Dickinson, R., 98
Dieke, G. H., 224
D'Incan, J., 219, 221, 225
Dinger, A. S., 274
Disney, M. J., 74, 237, 238, 549, 552
Diuth, F. T., 544
Dixon, R. N., 220
Dmitriev, A. I., 223, 225, 226
Dodd, W. W., 179, 180, 189
Dodson, H. W., 155
Dollfus, A., 100, 114, 115, 117
Dongre, M. B., 223
Donn, B., 272, 275, 277, 278, 281, 283
Donohue, T. M., 420
Dorfeld, W. G., 276, 278, 283
Dorling, E. B., 63
Doroshenko, V. T., 73
Doroshkevich, A. G., 243, 245
Doschek, G. A., 380, 381
Doughty, N. A., 142
Downes, D., 297, 340, 344, 345, 347, 348, 350, 351
Downs, G. S., 21, 35, 494
Doxsey, R., 25, 180, 563
Draine, B. T., 279, 283, 287, 288
Drake, F. D., 27, 116, 390
Drake, J., 213
Dressler, K., 217
Drobyshevski, E. M., 137, 141
Dube, P. S., 226
Duckett, S., 564
Duffet-Smith, P. J., 38, 485, 488, 498, 502
Dufour, R. J., 182
Duin, R. M., 190, 191
Dulk, G. A., 62
Duncombe, R. L., 98
Dungey, J. W., 63, 415-17
Dunphy, P. P., 63
Dupree, A. K., 364
Dupuis, R. A., 224
Durney, B., 47, 49
Duthie, J. G., 147
Duvall, T. L., 162, 168
Dyal, P., 398, 399, 402-4, 406, 407, 421-23, 425
Dyce, R. B., 99, 100
Dyck, H. M., 72, 281, 282
Dyer, E. P., 380
Dyne, P. J., 224
Dzurisin, D., 104, 105, 111-13
Dzyuba, B. M., 464

E

Early, D., 298, 299
Eddington, A. S., 6
Eddy, J. A., 46, 169
Edelson, R. E., 98
Edwards, M. R., 498, 502
Edwards, P. L., 90
Effantin, C., 221, 225
Egan, W. G., 282
Eggen, O. J., 142, 249, 254
Eggleton, P. P., 138, 142, 143
Ehlers, J., 138
Eilek, J. A., 72, 74
Einasto, J., 260, 531
Ekers, J., 75
Ekers, R. D., 297, 336, 339, 350, 351, 536
Elander, N., 219, 224, 226
Elias, J. H., 325, 341
Ellerman, F., 153
Elliot, J. L., 90
Ellis, D., 84
Ellis, G. R. A., 391, 422
Elmergreen, B. G., 187, 224, 258
Elvis, M., 88
Emery, R. J., 358
Encrenaz, P., 498
Engelman, R., 221, 224
Engstrom, S. F. T., 380
Epstein, A., 543, 563
Epstein, R. I., 244
Erickson, S. A., 470, 471
Erickson, W. C., 193
Ergma, E. V., 144
Erkes, J. W., 182
Erman, P., 224
Ershkovich, A. I., 423
Esposito, P. B., 98, 101
Evans, R. G., 364
Eviatar, A., 423, 426, 427
Ewart, G. M., 35
Ewing, M. S., 494, 499
Ezer, D., 142

F

Fabbiano, G., 145
Faber, S. M., 238, 251, 255, 258, 262, 512, 514, 531
Fabian, A. C., 533, 536, 544, 545, 549, 561, 562
Fabian, W., 223
Fabricant, D., 560, 561, 563
Fahlman, G. G., 72, 74
Fairall, A. P., 72, 75
Fairbairn, A. R., 222, 226
Falgarone, E., 38, 288
Falk, S. W., 177, 286
Fall, S. M., 260
Falle, S. A. E. G., 182, 183
Fan, C. Y., 409
Fanti, C., 535, 536
Fanti, R., 535, 536
Fanucci, J., 179
Farmer, A. J. D., 223
Faulkner, J., 138
Fedder, J. A., 420
Feder, J., 275
Federico, C., 278
Feldman, P. A., 534, 536, 555, 558
Feldman, S. I., 451
Feldman, U., 380, 381
Feldman, W. C., 169, 374
Felenbok, P., 222
Felten, J. E., 534, 558, 561, 562

Ferch, R., 286
Ferguson, D. C., 26
Ferguson, H. I. S., 226
Few, R. W., 332-34, 337
Feynman, R. P., 437
Fichtel, C. E., 25
Field, G. B., 187, 188, 235, 240, 241, 267, 270, 271, 290, 390, 532, 562
Filius, R. W., 409, 410, 412, 413
Filseth, S., 224
Fink, E. H., 217, 219, 222
Firth, J. C., 380
Fish, R. A., 237
Fisher, J. R., 21
Fisher, P. C., 560
Fix, J. D., 275
Fjeldbo, G., 98, 101
Flannery, B. P., 135, 137, 142, 563
Flechas, J., 214
Flower, D. R., 372
Fomalont, E. B., 336, 535
Ford, W. K. Jr., 73, 75
Forman, W., 541, 543, 549, 552, 555, 560
Formisano, V., 391
Forrest, D. J., 63 G
Forrest, W. J., 280, 283, 288
Fosbury, R. A. E., 74, 81, 85, 88, 476
Foukal, P. V., 369, 370, 375, 379-81, 383
Fowler, L. A., 33, 34
Fowler, R. G., 222
Frandsen, A. M. A., 398, 399, 403, 406, 407, 422, 423
Frank, J., 343
Frank, L. A., 395, 409, 422, 423, 426
Franklin, F. A., 99
Franklin, K. L., 389
Franz, O. G., 462
Fraser, P. A., 218, 221, 225, 226
Frazier, E. N., 49, 368, 379
Frederick, C. L., 358
Freeman, K. C., 235, 238, 240, 249, 254, 445, 463
Frey, N., 222
Fricker, P. E., 120
Friedland, A. B., 63
Friedman, H., 541, 555, 560
Friedman, M., 63
Friefeld, R. D., 494, 499
Fritz, G., 541, 555, 560
Frogel, J. A., 280
Fujimoto, M., 440, 453, 462
Fujita, Y., 274
Fukao, S., 63
Fukui, Y., 336, 339
Fulchignoni, M., 278
Fung, P. C. W., 373
Furth, H. P., 63
Fusco-Femiano, R., 535

G

Gabriel, A. H., 64, 371-73, 377-80
Gallagher, J. S., 285, 512, 514, 515
Galt, J. A., 497, 499
Garber, J. A., 221
Garcia-Munoz, M., 64, 192
Gardner, F. F., 331, 332, 336, 350
Gardner, J., 188
Garmire, G., 43
Garrison, R. F., 201
Garvin, D., 225
Gatley, I., 358
GAULT, D. E., 97-126; 104, 105, 107, 109, 111-14, 117, 122-24
Gaustad, J. E., 281, 282
Gaydon, A. G., 204, 215
Geballe, T. R., 308, 353
Gebel, W., 460, 463
Gehrels, T., 26, 391
Gehrz, R. D., 279
Geisel, S. L., 285
Geldzahler, B. J., 352
Gelernt, B., 224
Gerassimenko, M., 379
German, K. R., 224
Gerola, H., 73, 369
Gezari, D. Y., 325, 341
Giacconi, R., 88, 370, 533, 541, 543-49, 552, 553, 555, 559, 560
Giannone, P., 131, 132
Giannuzzi, M. A., 131, 132
Gierasch, P. J., 47
Gillespie, B., 156
Gillett, F. C., 72, 74, 280, 281, 283, 288
Gilliam, L. B., 159
Gilman, P. A., 46, 47
Gilman, R. C., 278, 282
Gilra, D. P., 281
Gingerich, O., 367, 368
Gioia, I., 535, 536
Gisler, G. R., 262
Gissler, R. C., 222
Glasco, V. B., 223, 225, 226
Glaspey, J. W., 72, 74
Glass, I. S., 73, 74
Gledhill, J. A., 391, 422
Gleeson, L. J., 422
Glencross, W. M., 63, 384
Gloeckler, G., 409
Goad, L., 344
Godwin, J. G., 524
Goertz, C. K., 398, 399, 401-3, 409-11, 422, 426
Gohel, V. B., 226
Gold, R. E., 374
Gold, T., 100, 261
Goldberg, J. N., 138
Goldgerg, L., 200, 221
Goldreich, P., 34, 100, 101, 139, 283, 452, 473
Goldsmith, D. W., 270, 460, 463
Goldstein, R. M., 73, 90, 99, 116
Goldstein, S., 52
Goldstein, S. J., 288
Golub, L., 59, 370
Gómez-Gonzáles, J., 38
Goody, R. M., 224, 426
Gopal-Krishna, 351
Gordon, M. A., 268, 280, 289, 310, 324, 325-27, 459
Gorenstein, P., 25, 175, 180, 541, 560-63
Gosling, J. T., 62
Goss, W. M., 35, 297, 350, 351
GOTT, J. R. III, 235-66; 235, 239, 240, 242-50, 252-58, 260, 262, 507, 508, 513, 525, 528, 529, 531, 533, 557
Gould, R. J., 534, 559
Graham, D. A., 22, 23, 30
Grasdalen, G. L., 279, 285, 286, 288, 289
Green, S., 218, 220
Greenberg, J. M., 279, 281, 282
Greenstein, G., 35, 36
Greenstein, J. L., 141
Gregory, S. A., 508, 513, 517, 518, 522
Grevesse, N., 219, 220, 221, 225
Grewing, M., 271
Grindlay, J. E., 544, 546
Grinfeld, R., 204
Gronenschild, E. H. B. M., 364

Gross, P. G., 131, 132
Grossman, L., 272
Groth, E. J., 33-37, 246, 261, 522
Gubbins, D., 47, 103
Guélin, M., 38, 498
Guest, J. E., 104, 105, 107-14, 119, 122-24
Gulkis, S., 391
Gull, S. F., 178, 179, 189, 190, 535, 557
Gullahorn, G. E., 33, 34, 36
Gundermann, E. J., 479
Gunn, J. E., 34, 40, 237, 239, 240, 247, 255-58, 262, 263, 507, 513, 525, 528, 529, 557
Gurman, J. B., 379
GURSKY, H., 541-68; 88, 144, 533, 535, 541-49, 552, 553, 555, 559-63
Gurzadyan, G. A., 179
Guseinov, O. H., 147
Guthrie, B. N. C., 536
Guyer, R. A., 35

H

Haaks, D., 224
Habing, H. J., 270
Hackney, K. R., 90
Hackney, R. L., 90
Hackwell, J. A., 279
Haddad, G. N., 223
Hainebach, K. L., 191
Hale, G. E., 153
Hall, D. S., 138, 139, 141
Hall, R. D., 25, 544
Halmann, M., 221
Hamberger, S. M., 63
Hameen-Antilla, K., 117
Hamilton, D. C., 400, 402
Hamilton, P. A., 21, 23, 24, 35, 488-91, 496-98
Hanbury Brown, R., 544
Hankins, T. H., 29, 30, 487
Hanks, T. C., 119
Hansen, R. T., 168
Hansen, S. F., 168
Hapke, B., 105, 107, 113-115, 117
Hardebeck, J. E., 479
Hardee, P. E., 182
Hardy, E., 508, 511, 516, 525
Harlan, E., 547
Härm, R., 129, 285
Harnden, F. R., 25
Harnden, F. R. Jr., 180, 560, 561, 563
Harper, D. A., 72, 86, 87, 285, 358
Harrington, J. A., 204, 224, 226
Harrington, R. S., 99
Harris, D., 114, 117
Harris, D. E., 555, 558, 559
Harrop, W. J., 204
Hart, M., 526
Hartle, R. E., 102, 103
Hartmann, W. K., 122
Harvey, J., 156
Harvey, J. W., 46, 49, 155, 169, 373-75
Harvey, K. L., 59, 155
Harvey, P. M., 341, 359, 360
Hartwick, F. D. A., 258
Harwit, M., 87, 286
Hasson, V., 222
Hauser, M. G., 325, 341, 522
Havas, P., 138
Hawarden, T. G., 476
Hawkins, F. J., 560
Hayashi, C., 278
Hayatsu, R., 278
Haycock, S. C., 212
Haymes, R. C., 25, 544
Hazlehurst, J., 142
Heaps, H. S., 225
Hearn, A. G., 363
Hébert, G. R., 204, 217, 218, 220, 225, 226
Hedelund, J., 226
Hedeman, E. R., 155
Hefferlin, R., 214
Heiles, C., 182, 280, 287, 288, 547
Heinz, C., 549
Heise, J., 134, 364, 546
Helava, H., 180, 534, 560
Helfand, D. J., 24
Helmken, H., 563
Helmken, H. F., 544
Henneker, W. H., 218
Henriksen, R. N., 146
Henry, R. C., 364, 562
Herbert, F., 104
Herbig, G. H., 201, 281, 286
Herman, R. C., 221, 225
Heron, S., 217
Herring, J. R., 63
Herzberg, G., 201, 204, 208, 219, 225
Herzog, E., 506, 510
Hesse, K. H., 24, 26
Hesser, J. E., 217, 219, 220, 222
Hewish, A., 19, 479, 485, 486, 495
Hewner, F., 64
Heyvaerts, J., 63, 64
Hickson, P., 263, 521, 527
Higbie, P. R., 63
Hilgeman, T., 282
Hill, G., 141
Hill, H. A., 47
Hill, J. K., 289
Hill, T. W., 425-27, 429, 431
Hills, J. G., 84, 141
Hinteregger, H. F., 73, 90
Hirth, J. P., 275, 277
Hobbs, L. M., 272
Hockney, R. W., 468, 469, 474
Hodder, R. V., 225
Hodge, P. W., 79, 84, 528
Hodges, C. A., 109
Hodges, R. E. Jr., 102
Hoffmann, W. F., 341, 359, 360
Hofmeister, E., 129
Hogan, C., 562
Hoggan, T., 224
Hohl, F., 250, 259, 464, 465, 467-69, 474
Holland, R. F., 222
Hollenbach, D. J., 289
Holmberg, E., 507, 508
Holt, S. S., 25, 180, 535, 547, 554, 556, 558
Holweger, D. G., 367
Holzer, T. E., 376
Hong, S. S., 279, 281, 289
Hooker, W. S., 226
Hooley, J., 536
Hornstein, S., 220
Horstman, H. M., 561
Hoshi, M. N., 351
Hougen, J., 211
Howard, H. T., 98, 101, 497
Howard, K. A., 105
HOWARD, R., 153-73; 46, 49, 153, 155, 156, 159, 161-64, 166, 168-70, 172
Hoyle, F., 90, 267, 283, 285
Hubbard, E. C., 99
Hubbard, R., 84
Hubble, E. P., 235, 237, 238, 513
Huber, K., 204
Huber, M. C. E., 216, 369, 370, 375, 380

Hubisz, J., 212, 225
Huchra, J., 72, 75, 78, 79
Huchra, J. P., 79, 82
Hudson, J. B., 276, 278, 283
Hudson, J. P., 272, 281
Hudson, R. D., 224, 227
Huffman, D. R., 281
Huguenin, G. R., 21, 24, 26, 27, 29-32
Hulse, R. A., 33, 34, 37, 38
Humason, M. L., 63, 79, 507, 528
Hundhausen, A. J., 375
Huneke, J. C., 113
Hunstead, R. W., 502
Hunt, F. R., 217
Hunter, C., 464, 470, 471
Hunter, C. E., 281
Huntley, J. M., 321, 461
Huron, B., 217
Hutchings, J. B., 141, 287
Hutchinson, D. P., 35, 37
Hutton, L. K., 90
Hvatum, H., 390
Hyland, A. R., 285

I

Iben, I. Jr., 129, 144
Iguchi, T., 336, 339
Ilovaisky, S. A., 41
Imhof, R. E., 222, 223
Innanen, S. E. H., 204
Innes, K. K., 217
Intrilligator, D. S., 395, 423
Ioannidis, G. A., 391, 414, 418, 419, 424, 425, 432
Ipavich, F., 557
Irvine, W., 114
Ishimaru, A., 483
Isaacman, R., 491
Isaacson, L., 220, 222
Ivanisevich, G., 71
Ives, J. C., 88, 535, 554

J

Jackson, B. V., 375, 377
Jackson, R. E., 237, 255, 258, 531
Jackson, W. M., 220
Jaffe, D. T., 370
Jaffe, W. J., 535, 536
Jaggi, R. K., 63
Jain, D. C., 222
James, T. C., 222
Jameson, R. F., 72
Jancarik, J., 63
Janin, G., 249
Jaques, S., 62
Jarmain, W. R., 218, 221, 225, 226
Jauncey, D. L., 70, 73, 90
Jeans, J. H., 438, 473
Jefferts, K. B., 323, 327, 336, 337
Jenkins, E. B., 187, 193, 267, 270-72, 279, 287
Jenner, D. C., 516, 552
Jensen, E. B., 238, 251-53, 259, 260, 262
Jernigan, G., 543, 563
Jeunehomme, M., 216, 220
Joeveer, M., 260
Johnson, C. E., 222, 223
Johnson, H. M., 320, 364, 560
Johnson, H. R., 272
Johnson, S. E., 216, 219
Johnson, T. V., 426
Jokipii, J. R., 412, 480, 487, 490-92, 494-96, 501, 503
Jones, A. V., 224
Jones, B., 280
Jones, B. B., 30, 380
Jones, B. J. T., 235, 246
Jones, C., 543, 552
Jones, D. E., 394, 395, 398-404, 406, 407, 411, 421-23, 25
Jones, E. M., 187, 188
Jones, T. W., 72, 86-89, 281, 283, 545
Jordan, C., 364, 369, 372, 379, 380, 383
Jordan, S. D., 369
Joss, P., 270
Joy, A. H., 63
Joyce, R. R., 74
Judge, D. L., 222, 227, 426
Julian, W. H., 34, 446, 452, 468, 474
Julien, P., 543
Jura, M., 272
Jurkevich, I., 73
Just, K., 508

K

Kaasik, A., 260, 531
Kaburaki, O., 382, 383
Kafatos, M., 176, 180, 183
Kahn, F. D., 176, 178
Kaifu, J., 339
Kalkofen, W., 367, 368
Kalnajs, A. J., 238, 440, 442, 443, 448, 450, 451, 458, 460-62, 464, 467-72
Kane, S. R., 63
Kaplan, S. A., 454, 460
Karifu, N., 336, 339
Karpowicz, M., 506, 510, 522
Kato, T., 339
Katz, A., 85
Katz, J. L., 277, 535, 556
Kaufman, J. J., 486, 492
Kaufman, M., 240
Kaufmann, P., 73
Kaula, W. M., 99
Kayser, S., 513, 514
Keenan, P. C., 201
Kellermann, K. I., 70, 72, 73, 87, 90, 352
Kellogg, E., 88, 532-35, 541, 543-49, 552-55, 557, 560, 564
Kenderdine, S., 190, 555
KENNEL, C. F., 389-436; 391, 414, 419-21, 424, 428, 432, 433
Kerr, F. J., 182, 280, 315, 316, 455
Kerridge, J. F., 278
Kerridge, S. J., 288
Kestenbaum, H. L., 25
Khabazin, Y. G., 144
Khachikian, E. Y., 69, 71-75, 80, 81, 88, 89
Kifune, T., 534, 560
Kihara, T., 245, 246
Killeen, J., 63
Kineyko, W. R., 218
King, I. R., 237, 258, 513, 517-19, 522, 528, 530-32
Kinman, T. D., 301
Kinsey, J. L., 224
Kintner, E. C., 513, 517, 522, 528, 530, 532
Kippenhahn, R., 47, 129, 132, 134-36
Kirshner, R. P., 185, 193, 379
Kivelson, M. G., 404, 421
Klaasen, K., 107, 114, 115, 117
Klaasen, K. P., 100
Klein, K. M., 116
Kleinmann, D. E., 74, 285
Klemola, A. R., 507
Klemperer, W., 222
Klemperer, W. K., 484
Klemsdal, H., 214
Klepczynski, W. J., 98
Kliore, A. J., 98, 101

Knacke, R. F., 72, 74, 278, 281
Knapp, G. R., 182, 280
Kniffen, D. A., 25
Knight, C. A., 73, 90
Koch, D., 533, 543, 546, 547, 552
Koch, R. H., 142
Kockarts, G., 224
Kohl, K., 134
Kojoian, G., 72-74
Komesaroff, M. M., 21, 23, 488-91, 496-98
Kondo, A., 286
Kondo, Y., 141
Koornneef, J., 351, 355
Kopal, Z., 128, 137, 139
Kopilov, I. M., 72, 75
Kopp, R. A., 370, 372, 376, 381
Kormendy, J., 238, 531
Koski, A. T., 72-75, 81-86
Kostkowski, H. J., 225
Kotov, V., 168, 169
Kovács, I., 211, 217
Kowal, C. T., 506, 510
Kozlovskaya, S. V., 117
Kraft, R. P., 128, 138, 142, 143
Krassner, J., 288
Krause, F., 47
Kraushaar, W. L., 187
Kravtsov, Y. A., 481
Krieger, A. S., 59, 155, 169, 170, 370, 373, 374, 379
Krimigis, S. M., 411
Krishna-Monhan, S., 497-500
Kristian, J., 25, 71, 507, 525
Krumm, N., 469
Krupenie, P. H., 204, 221, 222
Krupp, E. C., 523, 524
Kruskal, M., 179
Ku, W., 25
Kudo, A., 279
Kuhi, L. V., 72
Kuiper, G. P., 97, 128
Kulsrud, R. M., 179, 192
Kumar, S., 101, 102
Kunkel, W. E., 544
Kuperus, M., 363, 372
Kupo, I., 426, 427
Kurfess, J. D., 25
Kurucz, R. L., 367, 368
Kusserow, H. U. V., 47
Kuz'menko, N. E., 214, 225
Kuznetsova, L. A., 214
Kuzyakov, Y. Y., 214, 225
Kwan, J., 280
Kwast, T., 528

L

Lacy, J. H., 308, 353
Ladenberg, R., 216
Laegrid, N., 279
Lagerqvist, A. 219
Laksham, S. V. J., 223
Lambert, D. L., 225, 226
Lamers, H. J., 271
Lamers, H. J. G. L. M., 144, 145, 147, 364
Lampton, M., 541, 543-45, 547, 560
Landini, M., 363, 379
Lang, K. R., 480, 491, 494, 497-99
Langhoff, S. R., 219, 226
LaPaglia, S. R., 223
Lapp, M., 223
Large, M. I., 37
Lari, C., 535, 536
Larimer, J. W., 272
Laros, J. G., 25
Larson, R. B., 141, 188, 237, 240, 250-53, 286
Lasher, G., 176
Lassettre, E. N., 222
Lau, Y. Y., 451, 452, 472
Laulicht, I., 221
Lauque, R., 72-74
Lauterborn, D., 129, 131, 133
Lawrence, G. M., 217, 222
Layzer, D., 562
Lazareff, B., 50
Lazarus, A. J., 103, 373, 374
Lea, S. M., 532, 534, 535, 553, 554, 557, 560
Leacock, R. J., 90
Learner, R. C. M., 224, 225
Lebovitz, N. R., 142
Le Calve, J., 222
Lecar, M., 144, 145, 515
Lee, L. C., 222, 227, 480-82, 487, 490-92, 494-96, 501, 503
Lee, T., 119
Lefèvre, J., 289
Leibacher, J., 363, 382, 383
Leighton, R. B., 16, 47, 153, 155
Leir, A. A., 511, 514
Lengel, R. K., 224
Lentz, G., 400, 402
Leong, C., 88, 541, 543, 545, 549, 560
Lepping, R. P., 102-4
Lequex, J., 38, 41, 288, 535, 536
Lerche, I., 181
Leroy, R. J., 212
Lesh, J. R., 514, 515
Leung, C. M., 324
Levato, H., 138
Levine, A., 25
Levine, R. H., 169, 370, 373-75, 379, 381, 383, 384
Levy, D. H., 221
Levy, G. S., 98
Levy, R. H., 415
Levy, S. G., 142
Lewin, W. H. G., 548, 549
Lewis, B. M., 72-74, 513
Lewis, B. R., 223
Lewis, J. S., 117, 119, 272
Li, F. K., 552
Libby, D. R., 224
Light, E. S., 319, 320
Lillie, C. F., 279
Lin, C. C., 440, 443, 445, 446, 448, 449, 451-54, 458, 462, 471, 472, 475
Lin, D. N. C., 137
Lindblad, B., 259, 320, 438, 439, 441, 446, 453, 466, 464, 475
Lingenfelter, R. E., 192
Linhart, J. G., 63
Linke, R. A., 324
Linsky, J. L., 364, 367-70, 379
Linton, C., 218, 226
Lipovetsky, V. A., 72-75, 78
Liszt, H. S., 218-20, 224-27, 324, 327, 328, 336, 339, 342
Little, A. G., 346
Little, L. T., 479, 480, 495, 496
Liu, B., 218
Liu, H. -S., 100
Liu, L., 120
Livingston, W., 49, 156
Lo, K. Y., 325, 341, 352
Lockhart, I. A., 336, 337, 351, 445
Lockwood, G. W., 282
Loeser, R., 367, 368
Loewenstein, R. F., 72, 86, 87
Lohsen, E., 35, 36
Lokan, K., 223

Longair, M. S., 543, 546
Longmore, A. J., 72
Lord, H. C., 272
Lothe, J., 275
Loughhead, R. E., 367-70
Lovas, F. J., 204
Lovelace, R. V. E., 487, 495
Lovell, B., 63
Low, B. C., 63
Low, F. J., 72-75, 79, 86-89, 285, 355, 358
Lowinger, T., 300, 301
Lubow, S. H., 137, 142
Lucas, R., 535, 536
Lucy, L. B., 137, 142
Lugger, P. M., 507, 547, 549, 552, 553
Luk, C. K., 220
Lüst, R., 53
Lutz, B. L., 219, 220
Luyten, W. J., 63
Lynch, M., 484
Lynden-Bell, D., 88, 248, 249, 251, 254, 343, 351, 450, 451, 462, 467, 470, 473, 529
Lynds, B. T., 267, 289
Lynds, C. R., 74
Lynds, R., 261
Lyne, A. G., 22-24, 30, 34, 37-40, 492, 497, 499
Lyot, B., 114
Lyutyi, V. M., 72, 75, 80, 90

M

MacAlpine, G. M., 75, 85
Maccacaro, T., 88
Maccagni, D., 545, 549, 552
MacDonald, R. H., 555
MacFarlane, M., 25
Mack, J., 541, 543-45
MacPherson, G. J., 75
MacQueen, R. M., 62
Macy, W., 426
Mader, G. L., 311
Magni, G., 278
Magun, A., 62
Maheshwari, R. C., 223, 225
Mahoney, W., 544
Maier, B., 222
Main, R. P., 214, 224, 226
Malin, M. C., 104, 105, 111-13
Malina, R., 560
Mallama, A. D., 141
Mallia, E. A., 223, 225
Malmberg, C., 219
MANCHESTER, R. N., 19-44; 20-27, 30-41, 492, 497, 499
Mandelman, M., 219
Mandeville, J. C., 278
Mansfield, V. N., 178, 180, 182, 186
Maran, S. P., 176, 186
Maranan, J., 223
Marandino, G. E., 73, 90
Marchetti, M. A., 223
Margon, B., 544, 545, 547, 565
Mariska, J. T., 373, 375
Mark, J. W. -K., 449-52, 469, 472
Markarian, B. E., 72-75, 78, 81, 87
Markert, T. H., 540, 552
Marlow, W. C., 216
Marochnik, L. S., 448, 452
Marram, E. P., 222
Marscher, A. P., 176, 183
Marsell, A. L., 226
Marshall, A., 224
Martin, A. H. M., 340, 351
Martin, J. W., 59
Martin, P. G., 72, 86, 87
Martin, S. F., 59, 155
Martin, W. L., 73-75, 78, 98
Martres, M. J., 64, 155
Martynov, D. Y., 128
Mashburn, J., 214
Mason, B., 120
Mason, B. J., 278
Mason, G. M., 64, 192
Massevitch, A. G., 131, 134, 144
Massey, H., 64
Mast, J., 288
Matheson, D. N., 480, 496
Mathews, W. G., 84, 188, 251
Mathewson, D. S., 317, 321, 454
Mathis, J. S., 84
Matilsky, T., 533, 543, 546-48, 552
Matson, D. L., 426
Matsuda, T., 462
Matsui, T., 119
Matteson, J. L., 25
Matthews, K., 86, 283, 357, 358
Matthews, T. A., 514, 516, 535
Mauder, H., 142
Maunder, E. W., 46
Max, C., 189
Maxwell, A., 344, 345
Maxwell, J. C., 440
Mayall, N. U., 79, 507
Mayer, C. H., 389
Mayer, W., 25, 541
Maza, J., 177, 286, 287
Mazzucato, E., 63
McCallum, J. C., 218-23, 225-27
McCammon, D., 187
McCauley, J. F., 105
McClain, E. F., 390
McClintock, J., 548
McClintock, W., 364
McCluskey, G. E. Jr., 141
McCord, T. B., 114, 115, 117
McCray, R., 183
McCulloch, P. M., 21, 23, 24, 497, 498
McCullough, T. P., 389
McDonald, F. B., 400, 404, 406, 407, 409
McDonald, J. K., 217
McDonough, T. R., 414, 421, 425
McEachran, R. P., 218
McElroy, M. B., 101, 420
McEwen, D. J., 221
McGee, J. D., 80
McGee, R. X., 331, 332
McGimsey, B. Q., 90
McGregor, A. T., 223, 226
McHardy, I. M., 511, 536
McIlwain, C. E., 409
McIntosh, P. S., 374
McKee, C. F., 144, 145, 180, 181, 184, 187
McKellar, A. R. W., 223
McKibben, D. P., 395, 423
McKibben, R. G., 392, 393, 396, 400-2, 407, 408
McLean, A. D., 218
McLean, A. I. O., 24
McLinn, J. A., 72
McMath, R. R., 221
McWhirter, R. W. P., 217, 372, 373, 376, 380
Mebold, U., 304-6, 308
Meegan, C. A., 25, 544
Meekins, J. F., 541, 555, 560
Meier, D. L., 240
Mekler, Y., 426, 427
Melnick, J., 513, 549, 552
Meloy, D. A., 187, 272
Melrose, D. D., 391, 422
Melzer, J. E., 204
Mendis, D. A., 286, 422

Mengel, J. G., 129, 131, 132
Menzel, D., 114
Merrill, K. M., 72, 74, 280-84, 288
Merrill, P. W., 201
Mestel, L., 427
Meszaros, P., 241
Metcalf, H., 224
Mewaldt, R. A., 411
Mewe, R., 364
Meyer, C., 278
Meyer, F., 50
Meyer, V. D., 222
Meyer-Hofmeister, E., 134-37, 142
Meyrick, G., 277
Mezger, P. G., 346-49, 351, 352
Michel, F. C., 424, 426, 427, 429
Michels, H. H., 218
Mihalas, D., 258
Mihalas, D. M., 524
Mihalov, J. D., 395, 422, 423, 426
Miley, G. K., 72, 536, 555
Milione, V., 455, 457, 459, 460, 463
Milkey, R. W., 369, 382
Miller, J. S., 83, 182
Miller, R. H., 236-38, 259, 462, 464, 466, 468, 471
Mills, B. Y., 175, 182, 346, 535
Milman, A. S., 324
Miner, E. D., 114, 115
Minkowski, R., 258, 309, 508, 516, 531, 552
Mirabel, I. F., 296, 313
Mishurov, Y. N., 448
Mitchell, R. J., 180, 535, 544, 554, 556, 560
Miyaji, S., 135
Miyaji, T., 336, 339
Miyazawa, K., 336, 339
Miyoshi, K., 246
Mizutani, H., 119
Mochnacki, S. W., 142
Modica, A. P., 224
Moe, O. K., 379, 380
Moffett, T. J., 63
Mogro-Campero, A., 400, 402
Mohanty, B. S., 225
Mohanty, D. K., 21
Mohler, O. C., 221
Moller, J., 238, 251-53, 259, 260, 262
Mon, J. P., 219
Monk, P., 380
Monnet, G., 454
Monsignori-Fossi, B. C., 363, 379
Moore, C. E., 200
Moore, G., 180
Moore, H. J., 109
Moore, J. H., 220
Moore, R. L., 373
Moore, W. E., 43
Moos, H. W., 364
Moretti-Horstman, E., 561
Morgan, D. J., 267
Morgan, W. W., 70, 510-16, 535
Morimoto, M., 336, 339
Morreal, J., 220
Morris, G. A., 21
Morrison, D., 114-16
Morrison, P., 176, 534, 558, 562, 565
Morton, D. C., 72, 134, 141, 258, 270, 300
Moskalenko, N. I., 221
Moss, D. L., 142
Mullan, D. J., 7
Müller, E. A., 367, 368
Muncaster, G. W., 26
Munch, G., 114, 115
Munro, R. H., 369, 375, 377
Murcray, F. J., 426
Murray, B. C., 105, 113, 114, 117, 122-24
Murray, J. B., 100, 104, 105, 111-13
Murray, S., 88, 532-35, 541, 543, 545-49, 552-54, 557, 560, 563, 564
Murthy, B. N., 223, 226
Murthy, N. S., 219, 223
Mushotzky, R. F., 544, 546
Mutel, R. L., 484, 486

N

Nagane, K., 336, 339
Nagaraj, S., 219
Nakagawa, Y., 63, 64, 278, 379, 383
Nandy, K., 267
Nanos, G. P., 35, 37
Naranan, S., 541
Narayanan, P. S., 225
Nather, R. E., 25
Natta, A., 288
Neff, S. G., 141
Nelson, B., 141
Neo, S., 135
Ness, N. F., 102-4, 169, 179, 425
Netzer, H., 74, 83
Neugebauer, G., 72-75, 80, 82, 86, 87, 90, 114, 115, 283, 285, 297-99, 325, 341, 351, 355, 356-58
Neupert, W. M., 64, 379, 383
Neville, A. C., 555
Newkirk, G. Jr., 159, 161, 169
Newton, L. M., 35
Ney, E. P., 282
Ng, A. T. Y., 98
NICHOLLS, R. W., 197-234; 199, 204, 207-9, 211-13, 217-27
Nicolas, K. R., 380
Niell, A. E., 70, 72, 73, 90
Nishida, A., 409
Nishida, S., 285
Nobili, L., 129
Noble, R. G., 498, 502
Noci, G., 374
Nolte, J. T., 373, 374
Nomoto, K., 135
Noonan, T. W., 517, 520-22, 528, 552
Norman, E. B., 191
Norris, J., 131, 132
Northover, K. J. E., 535, 557
Northridge, P. L., 552
Northrup, T., 403
Notni, P., 88
Novick, R., 25
NOYES, R. W., 363-87; 49, 153, 367-70, 372, 375, 379, 380
Nussbaumer, H., 74

O

Obayashi, T., 64
O'Dell, S. L., 89, 90
Oemler, A., 237, 239, 254, 257, 262, 510, 512-19, 523-25, 528, 530
Oesterwinter, C., 99
O'Gallagher, J. J., 400, 402
Ogawa, M., 227
Ögelman, H. B., 25, 186
Ogilvie, K. W., 102, 103
Oke, J. B., 72, 73, 75, 76, 82, 84-87, 507, 525
O'Keefe, J. A., 100
Oldenberg, O., 223
O'Leary, B., 105, 113, 114, 117

Olsen, E. T., 72, 75
Olthof, H., 269
Omer, G. C., 517, 519, 522
OORT, J. H., 295-362; 288, 297, 303, 306, 307, 311, 317, 321, 327, 338, 438, 440, 449, 533
ÖPIK, E. J., 1-17; 6-11, 13, 15, 16, 186
Oppenheimer, J. R., 208
Ornstein, L. S., 220
Orrall, F. Q., 375, 376
Ortenberg, F. S., 223, 225, 226
Osmer, P. S., 73, 74, 79, 89
Oster, L., 28
Osterbrock, D. E., 69, 72-75, 81-86, 91, 182, 269, 382, 383
Ostriker, J. P., 34, 40, 42, 142, 144, 187, 238, 239, 250, 252, 254, 257-60, 263, 468, 470, 515, 531, 532, 535, 557, 565
Owen, F. N., 510, 535, 536, 552, 555, 558

P

Pacini, F., 87
Paczyński, B., 128, 132-35, 137-39, 143, 144
Page, C. G., 28
Page, T., 193, 531
Page, T. L., 513, 517, 519, 522, 528, 530, 532
Pal, Y., 561
Palmer, H., 277
Panagia, N., 288
Paolini, F., 541
Papanastassiou, D. A., 113, 119, 122
PARKER, E. N., 45-68; 45-52, 55, 57-63, 65, 384, 414
Parkinson, J. H., 560
Parkinson, T., 426
Parkinson, W. H., 200, 221
Parsignault, D. R., 544, 546, 548
Partridge, R. B., 35, 37, 240
Pastoriza, M., 73
Paterson, J. A., 211, 217
Pathak, A. N., 227
Patterson, N. P., 380
Pauliny-Toth, I. I. K., 352, 536
Pauls, T. A., 346-48, 351, 352
Payne, R. R., 22, 24, 34, 36
Peach, J. V., 517, 519, 521, 524, 525, 528
Peacher, J., 220, 224
Peacock, A., 560
Peale, S. J., 100, 101
Pearse, R. W. B., 204, 215
Peckham, R. J., 40, 498, 502
Peebles, P. J. E., 238-42, 245, 246, 249, 254, 257, 261, 262, 468, 522, 525, 529, 531, 532
Peery, B. F., 201
Peimbert, M., 179, 269
Pelling, R. M., 25
Penman, J. M., 281
Penner, S. S., 221, 226
Penston, M. J., 72-75, 80, 86, 90
Penston, M. V., 72-75, 80, 81, 85, 86, 88, 90
Penzias, A. A., 311, 314, 323-25, 331, 336, 337, 340, 341
Perkins, M., 400, 402
Perley, R. A., 193
Perola, G. C., 72, 534, 536, 544, 555
Perrenod, S. C., 562
Persson, S. E., 280
Pery-Thorne, A., 224
Pesses, M. E., 409-11
Peters, G., 543
Peters, W. L., 33, 34, 321
Peterson, L. E., 88, 542, 544, 546
Peterson, P., 549
Peterson, S. D., 78
Pethick, C., 35
Petrasso, R., 170
Petrosian, V., 244, 282, 288
Petschek, H. E., 63, 415
Pettengill, G. H., 99, 100
Petterson, J. A., 144, 145
Phillips, J. G., 204
Phillips, M. M., 72-75, 83-86
Phillis, G. L., 61
Piddington, J. H., 382, 383, 391, 455
Pierce, A. K., 221
Pike, C. D., 80
Pikel'ner, S. B., 182, 184, 185, 454, 460
Pikkarainen, T., 117
Pikoos, C., 114
Pilkington, J. D. H., 19, 479
Pilling, M. J., 222
Pineau des Forets, G., 372
Pines, D., 35
Pipher, J. L., 288
Pizzo, V., 373
Plastinin, Y. A., 214
Plaut, L., 297
Plavec, M., 128, 130-33, 137, 139, 141
Pneuman, G. W., 373, 376, 381
Podosek, F. A., 113
Polidan, R. S., 130-33, 137, 141
Pooley, G. G., 454
Popkie, H. E., 218, 223
Popper, D. M., 141
Porco, C. C., 278, 281
Postal, R. B., 98
Pottash, S. R., 269, 271, 272
Pound, G. M., 275
Pounds, K. A., 548, 549, 552, 560
Prakash, A., 425
Prata, S. W., 63
Pratt, J. P., 138
Pravdo, S. H., 25, 180
Prendergast, K. H., 137, 236-38, 319, 449, 453, 462, 464, 466, 473
Prentice, A. J. R., 501
Press, W. H., 246, 519, 524
Preuss, E., 352, 536
Price, M. L., 226
Price, R. M., 494, 499
Priest, E. R., 63, 64
Pringle, J. E., 137, 251, 535
Pronik, I. I., 72-75, 90
Ptak, R., 84, 85
Punsky, J. J., 73, 90
Purcell, E. M., 279, 282
Purcell, J. D., 47, 62, 170, 369, 370
Purton, C. R., 72-74
Pye, J. P., 552
Pyle, K. R., 400, 402, 411
Pyper, D. M., 528

Q

Quintana, H., 73, 549, 552
Quirk, W. J., 462, 464, 466, 467

R

Radhakrishnan, V., 35, 390

Radler, K. H., 47, 49
Rai, D. K., 218, 226
Randall, B. A., 398, 399, 400, 401, 409, 422
Rank, D. M., 308, 353
Rankin, J. M., 21, 24, 27-29, 32-34, 36, 479, 491
Rao, T. V. R., 223
Rappaport, S., 25, 180, 181, 541, 548
Ratcliffe, J. A., 480
Raymond, J. C., 182, 372
Rayrole, J., 64
Read, F. H., 222, 223
Read, P. L., 193
Readhead, A. C. S., 38, 479, 485, 488, 498, 502
Reasenberg, R. D., 98
Redman, R. O., 141
Reed, R. W., 425
Rees, M. J., 128, 235, 240-46, 252, 257, 263, 343, 535, 545, 561, 562
Reese, E. J., 99
Reeves, E. M., 200, 221, 368-70, 373, 375, 380
Refsdal, S., 132, 138, 141
Reichley, P. E., 21, 35, 494
Reid, M. J., 100
Reimers, D., 144
Reinecke, R., 22, 23
Reinhardt, M., 531, 534, 536
Rense, W. A., 381
Reynolds, T. R., 117, 120
Reynolds, R. T., 120, 121
Reznikov, B. I., 137
Rhoderick, E. H., 217
Rich, J. C., 222
Richards, D. W., 21, 34, 36, 479
Richards, R. J., 498, 502
Richstone, D. O., 42, 72, 247, 263, 515, 525, 527, 532
Richter, G. M., 88
Richtmyer, R. D., 185
Ricke, F. F., 223
RICKETT, B. J., 479-504; 29, 30, 479, 480, 486, 487, 494, 495, 497-500
Ricketts, M. J., 548
Rieke, G. H., 72-75, 79, 86-89, 355
Riley, J. M., 535, 536
Rinehart, R., 73, 90, 344
Ritchings, R. T., 26, 29, 34, 35, 41
Ritter, H., 137, 143
Roark, T. P., 277
Roberts, B., 61, 63
Roberts, D. H., 21, 37, 39, 41
Roberts, J. A., 390
Roberts, M. S., 72, 239, 257, 445, 446, 457, 469, 531, 532
Roberts, W. W., 440, 446, 453-60, 457-60, 463
Robertson, D. S., 73, 90
Robertson, J. A., 142
Robertson, J. W., 189, 190
Robinson, B. J., 331, 332
Robinson, D., 218, 221, 222
Robinson, D. W., 220
Robinson, E. L., 143, 147
Robinson, G. W., 289
Robinson, L. B., 73
Robinson, R. D., 62
Roederer, J. G., 398
Roelof, E. C., 169, 373, 374, 400, 406, 407, 409
Rogers, A. E. E., 73, 90
Rogers, J., 222
Rogerson, J. B. Jr., 144, 147, 364
Rogstad, D. H., 297, 336, 337, 350, 351, 445, 532, 535, 536
Roland, J., 536
Romanishin, W., 558, 559
Romano, G., 72
Romney, J. D., 73, 90
Ronnang, B. O., 90
Rood, H. J., 506, 508-11, 513, 514, 516-18, 522, 523, 528, 530-32, 552, 553
Rose, J. A., 507
Rosen, B., 204, 217
Rosenberg, I., 182, 190
Rosenblum, A., 138
Rosenbluth, M. N., 63
Rosino, L., 72
Rosner, R., 373, 376
Ross, H. N., 352
Ross, J. E., 199, 269, 270
Rosse, (Earl of), 437
Rossi, B., 541
Roth, M. L., 138, 141
Rothery, R. W., 221, 225
Rothschild, R. E., 25, 180
Rots, A. H., 445, 455, 532
Rougoor, G. W., 301-4, 306, 307, 310
Rouse, P. E., 221, 224
Routly, P. M., 272, 287
Routly, P. McR., 258
Roux, F., 219, 221, 225
Rowan-Robinson, M., 533, 536, 549, 552, 562
Roxburgh, I. W., 142
Rozhdestvenskii, D. S., 216
Rubin, R. J., 221, 225
Rubin, V. C., 73, 75
Rucinski, S. M., 142
Ruderman, M., 35
Ruderman, M. A., 534, 557
Rudnick, L., 536, 552, 555
Rudnicki, K., 517, 528
Ruelof, E., 409
Ruiz, M. T., 301
Rumsey, V. H., 480-82, 493, 502
Ruskin, A. D., 226
Russell, H. N., 267
Russell, K. C., 275
Russell, R. W., 281, 282, 288
Rust, D. M., 60, 63, 64, 379, 383
Rybicki, G. B., 468
Rydbeck, O. E. H., 90
Ryle, M., 536, 555
Rytov, S. M., 481

S

Saar, E., 531
Sackmann, I. J., 274
Safronov, V. S., 119, 443, 471
Sakurai, K., 401
SALPETER, E. E., 267-93; 178, 180, 182, 186, 188, 270, 275, 276, 278, 279, 282, 283, 286, 288, 289, 343, 463, 469, 480, 486, 494
Sancisi, R., 186
Sandage, A. R., 73-75, 79, 81, 85, 88, 90, 235, 240, 249, 254, 279, 299, 318, 463, 507, 508, 511, 516, 525, 552
Sanders, R. H., 300, 301, 304-8, 311, 314, 319, 321, 322, 327, 328, 336, 339, 342, 351, 461
Sanders, W. T., 187
Sandquist, A., 332, 336, 337, 339, 351
Sanford, P. W., 88, 543, 546, 554
Sankaranayanan, S., 225
Sanner, F., 283
Santiago, J. J., 278
Saraber, M. J. M., 311
Sargent, W. L. W., 69, 72-76, 78, 79, 81, 82, 85-88, 90, 258, 513

Sarma, N. V. G., 351
Sarris, E. T., 411
Sartori, L., 176
Saslaw, W. C., 69, 72, 91
Sastry, G. N., 506, 509-11, 514, 516, 517, 552, 553
Sauval, A. J., 199, 219-21, 225
Savage, B. D., 279
Savedoff, M. P., 147, 288
Sawyer, C., 49
Scalo, J. M., 286, 289
Schadee, A., 219, 223-25
Schatten, K. H., 102, 103, 155, 169
Schechter, P., 246, 254, 520, 524, 525
Scherrer, P. H., 168, 169
Scherrer, V. E., 47, 170, 369
Scheuer, P. A. G., 182, 479, 563
Schild, R. E., 72
Schilizzi, R. T., 73, 90, 352, 536
Schindler, K., 63
Schlüter, A., 53, 61
Schmahl, E. J., 369, 370, 375, 380
Schmahl, G., 377
Schmeltekopf, A. L., 204
Schmidt, H. U., 50
Schmidt, M., 220, 222, 254, 514, 516, 535, 547-49
Schnopper, H. W., 543-45, 548, 549, 563
Schönhardt, R. E., 26
Schoonveld, L., 221
Schramm, D. N., 120, 191, 240
Schreier, E., 533, 541, 543-48, 552
Schreier, E. J., 145
Schrijver, J., 364
Schroeder, M., 114, 115
Schröter, E. H., 49, 50, 60
Schubert, G., 121
Schultz, G. V., 279
Schultz, P. H., 109
Schumacher, R. J., 369
SCHWARTZ, D. A., 541-68; 533, 542, 547-49, 555, 561-63, 565
Schwartz, K., 104, 120
Schwarz, U. J., 297, 336, 339, 350, 351
Schwarzschild, M., 70, 129, 283, 285, 319, 320, 446, 529
Schweizer, F., 462
Schwenker, R. P., 217
Scott, D. H., 105, 109
Scott, E. L., 507
Scott, J. S., 189, 190
Scott, P. F., 19, 479, 536
Scott, R. L., 90
Scoville, N., 280, 283
Scoville, N. Z., 268, 289, 323, 325, 327, 331, 332, 334, 336, 337, 340, 341
Scudder, J. D., 103
Seares, F. H., 153
Searle, L., 72, 73, 75, 82, 86, 87
Sedov, L., 180
Seel, R. M., 204, 220
Segalovitz, A., 327
Seidel, B., 98, 101
Seidelmann, P. K., 98
Seielstad, G. A., 76
Seiradakis, J. H., 34, 37, 39
Seitel, S. C., 222
Selmes, R. A., 72-75, 80, 86, 90
Sentman, D. D., 400, 407, 409
Serlemitsos, A. T., 64
Serlemitsos, P. J., 25, 180, 535, 547, 554, 556, 558
Sersic, J. L., 73, 507
Sesplaukis, T. T., 98
Setti, G., 548
Severny, A. B., 156, 168
Seyfert, C. K., 90
Sgro, A., 184, 185
Shaffer, D. B., 70, 72, 73, 90, 352
Shah, M. P., 226
Shane, C. D., 507, 517
Shane, W. W., 311, 317, 455
Shapiro, I. I., 33, 73, 90, 98-100
Shapiro, M. M., 269
Shapiro, P., 278
Shapiro, P. R., 187, 188
Shapiro, S. L., 90, 240, 262, 533
Shapley, H., 512
Sharma, A., 218
Sharov, A. S., 519
Sharp, L. E., 479
Sharpe, H. N., 120
Sharpless, S., 288, 462
Shaver, P. A., 336, 339
Shectman, S. A., 75
Sheeley, N. R., 47, 49, 62, 374
Sheeley, N. R. Jr., 156, 158, 169, 170, 369, 370, 375
Shelef, M., 277
Shelton, D. H., 25
Shemansky, D. E., 101, 102
Shen, B. S. P., 72, 73, 75
Sheridan, K. V., 62
Shields, G. A., 72, 73, 75, 76, 83-87, 269
Shin, J. B., 225
Shine, R. A., 369, 379
Shirbakyan, M. S., 536
Shirley, D. L., 98
Shklovsky, I., 180
Shortridge, K., 74, 81, 85
Shostak, G. S., 532
Shu, F. H., 137, 142, 440, 443, 445, 446, 448, 449, 451, 453, 455, 457-60, 463, 471, 475
Shukla, M. M., 223
Shuler, K. E., 221, 225
Shull, J. M., 271
Shurmann, S., 288
Sieber, W., 21-24, 26, 28
Siegfried, R. W. II., 117, 119, 120
Sienkiewicz, R., 133
Silberberg, R., 269
Silk, J., 176, 183, 187, 188, 193, 235, 243, 263, 271, 272, 278, 286, 287, 289, 532, 534, 535, 553, 554, 557, 560-62
Silk, J. K., 59, 170, 370
Siluk, R. S., 188, 272
Silver, J. A., 224
Silvers, S. J., 222
Simkin, S. M., 80
Simmons, J. D., 204
Simon, G. W., 49, 153, 159
Simon, M., 74
Simon, T., 72, 114
Simons, S., 286, 289
Simonson, S. C., 450, 460
Simonson, S. C. III., 311
Simpson, E. A., 198
Simpson, J. A., 64, 192, 392, 393, 396, 400-2, 407-11, 432
Sinclair, M. W., 315, 332
Singh, I. D., 223, 225
Singh, N. L., 226
Singh, O. N., 225
Singh, P. D., 227
Singh, R. B., 218
Siscoe, G. L., 103, 415, 426
Skerbele, A., 222
Skumanich, A., 368, 379

Slanger, T. G., 222
Slee, O. B., 497-500
Slingo, A., 536
Sloanaker, R. M., 389, 390
Smarr, L. L., 147
Smart, N. C., 532
Smerd, S. F., 62
Smirnov, A. D., 225
Smith, A. G., 90
Smith, A. J., 222
Smith, B., 100
Smith, B. A., 99
Smith, B. W., 186, 372, 535, 554, 556, 558
Smith, D. F., 63
Smith, E. J., 394, 395, 398-404, 406, 407, 410-13, 421-23, 425
Smith, F. G., 26, 29, 34, 35, 41
Smith, H. E., 238, 251
Smith, L. F., 134
Smith, M. G., 72-74, 78, 79, 89, 286
Smith, R. A., 391, 422, 423
Smith, R. L., 274
Smith, S., 529, 531
Smith, W. B., 33, 98
Smith, W. H., 217-20, 222-27
Smythe, C., 368, 379
Sneden, C., 272
Snellen, G. H., 25
Snow, T. P., 270, 290, 507
Sofia, S., 147
Soifer, B. T., 280-82, 288
Solinger, A., 181, 534, 535, 549, 553, 557
Solomon, P. M., 268, 289, 323, 325, 327, 331, 332, 334, 336, 337, 340, 341
Solomon, S. C., 100, 111, 117, 119, 120
Somon, J. P., 178
Sonett, C. P., 104, 120, 398, 399, 402-4, 406, 407, 421-23, 425
Sonnerup, B. U. O., 63
Sorensen, S. -A., 462
Soru-Escaut, I., 64
Soru-Iscovici, I., 155
Soter, S., 139
Sotirovski, P., 201
Souffrin, S., 72, 74, 75
Spada, G., 25, 541
Sparks, W. M., 285
Spencer, R. E., 498, 502
Spiegel, E. A., 534, 557
Spindler, R. J., 212, 221, 223, 225
Spinrad, H., 72, 73, 238, 251, 507, 547, 552
Spitzer, L., 248, 258-60, 267, 270, 272, 278, 279, 287, 288, 372, 446, 513, 526
Spitzer, L. Jr., 175, 188, 194
Spitzmesser, D. J., 73, 90
Spreiter, J. R., 415
Sprott, G. F., 549, 552
Spruit, H. C., 61
Sramek, R. A., 40, 72-74, 87
Stachnik, R. V., 445, 446, 448, 451
Staelin, D. H., 494, 499
Stanley, G. J., 390
Stark, A. A., 259
Stark, J. P., 544
Starrfield, S., 285
Stebbins, R. T., 47
Stecher, T. P., 176, 272, 279, 281, 283
Steenbeck, M., 47
Stein, R. F., 183, 363, 382, 383
Stein, W. A., 89, 90, 280, 281, 283, 284, 534
Stella, G., 224
Stellmacher, G., 377
Stelzried, C. T., 98
Stenflo, J. O., 49, 50, 153, 159, 164, 367-69, 377, 379, 382
Stephenson, A., 104, 121
Stevens, A. E., 226
Stevenson, D. J., 103
Stewart, R. T., 62
Steyer, T. R., 281
Stillwell, D. F., 404
Stix, M., 47, 49
Stockman, H. S., 72, 86, 87, 193
Stockton, A. N., 475
Stoeckly, R., 142
Stokes, N. R., 235, 249
Stone, E. C., 411
Stoner, R. E., 84, 85
Stothers, R., 138, 139
Straka, W. C., 182
Strangway, D. W., 120
Strittmatter, P. A., 89, 90, 138
Strom, K. M., 238, 251-53, 259, 260, 262, 279, 286, 289
STROM, R. G., 97-126; 104, 105, 107-11, 113, 114, 117, 119, 122-24, 190
Strom, S. E., 238, 251-53, 259, 260, 262, 286, 279, 289
Stuart, R. V., 279
Studier, M. H., 278
Stuhlinger, T., 141
Sturrock, P., 63
Sturrock, P. A., 424, 427
Suarez, C. B., 204, 223
Suchard, S. N., 204, 217, 221
Suchkov, A. A., 448, 452
Sugimoto, D., 135
Sulentic, J. W., 72-75
Sullivan, J. D., 373, 374
Sulzmann, K., 218, 226
Summa, C., 129, 146
Summers, A. L., 117, 120
Sunyaev, R. A., 243, 245, 247
Suomi, V., 105, 113, 114, 114, 117
Suri, A. N., 63
Sutantyo, W., 145
Sutherland, R. A., 222, 224
Sutton, J. M., 480, 494, 495
Svalgaard, L., 162, 168, 169
Svestka, Z., 64, 170
Swank, J. H., 25, 180, 535, 554, 556, 558
Swartz, W. E., 425
Swarup, G., 351, 497-500
Sweet, I. A., 62, 64, 469
Sweetnam, D. N., 98, 101
Swenson, G. W. Jr., 73, 90
Swings, P., 201
Sykorá, J., 159
Syrovatski, S. I., 64

T

Tabak, R. G., 277
Takeushi, H., 119
Talbot, R. J., 270
Tammann, G. A., 42, 177, 240, 507
Tanaka, K., 63
Tananbaum, H., 88, 533, 541, 543-49, 552, 553, 555, 559, 560
Tandberg-Hanssen, E., 381
Tang, F., 370
Tapia, S., 142
Tapscott, J. W., 70
Tarafdar, S. P., 284
Tarenghi, M., 528, 544, 555
Tarter, C. B., 286
Tarter, J., 513
Tatarczyk, B., 224

Tatarski, V. I., 480, 481, 493, 503
Tatum, J. B., 210, 211
Taylor, D. J., 238, 251
Taylor, G. I., 178
TAYLOR, J. H., 19-44; 20, 21, 24, 26, 27, 30-34, 36-41, 492, 497, 498
Taylor, K., 185, 193
Teegarden, B. J., 400, 404, 406, 407, 409
Telesco, C. M., 72, 86, 87, 285
Temesvary, S., 61
Tempra, W., 222
Tera, F., 122
Terebizh, V. U., 73
ter Haar, D., 501
Terzian, Y., 176, 183, 270, 285, 288
Teukolsky, S. A., 33
Thaddeus, P., 220, 336
Theys, J. C., 183
Thomas, G. E., 102
THOMAS, H.-C., 127-51
Thompson, D. J., 25
Thompson, L. A., 238, 251-53, 259, 260, 262
Thomsen, M. F., 398, 399, 401, 403, 422
Thonemann, P. C., 372, 373, 380
Thonnard, N., 75
Thorne, D. J., 24, 492
Thuan, T. X., 42, 235, 238, 246, 248, 250-54, 259, 260, 262, 300
Tifft, W. G., 72-75, 508, 517, 518, 522, 528
Tilford, S. G., 204
Timothy, A. F., 59, 155, 169, 170, 370, 373, 374
Timothy, J. G., 369, 370, 375, 380
Tinsley, B. M., 42, 177, 270, 286, 240
Tohline, J. E., 75, 91
TOOMRE, A., 437-78; 238, 250, 261, 443-46, 449-52, 459, 468, 474, 475
Toomre, J., 261, 475
Topka, K., 560, 561, 563
Totsuji, M., 245
Tousey, R., 47, 62, 170, 369, 370, 375, 380, 381
Tovmassian, H. M., 72-74, 87, 536, 547
Townes, C. H., 308, 353
Trafton, L., 426
Trainor, J. H., 400, 404, 406, 407, 409
Trask, N. J., 104, 105, 107-11, 113, 114, 117, 119, 122-24
Tremaine, S. D., 258-60, 515, 525
Trimble, V., 40, 269
Tritton, K. P., 72, 75
Troland, T. H., 27
Trotter, D. E., 46, 159, 161, 169, 375
Truran, J. W., 285, 286
Tsai, C., 63
Tscharnuter, W., 141
Tsuda, T., 63
Tsuji, T., 199, 272
Tucker, W. H., 91, 175, 180, 372, 383, 534, 535, 543, 544, 557, 559-63, 565
Tully, R. B., 445, 475
Turner, E. L., 239, 247, 257, 260, 262, 263, 507, 508, 525, 527, 531-33
Turner, G., 113
Turner, K. C., 296, 313
Turner, M. J. L., 549, 552
Turner, R. F., 380
Tutukov, A. V., 131, 134
Tuzzolino, A. J., 400, 402
Tyler, G. L., 98
Tyte, D. C., 204, 217, 218, 227

U

Uchida, Y., 382, 383
Udal'tsov, V. A., 26
Ulich, B. L., 116
Ulmer, M. P., 88, 545
Ulmschneider, P., 363, 368, 372, 382
Ulrich, M. H., 69, 73-75, 535, 536
Ulrich, R. K., 16, 130-33, 135, 137, 141
Ulrych, T. J., 72
Upson, W. L., 212
Urey, H. C., 119, 120
Uscinski, B. J., 481, 487, 499, 502, 503
Usher, P. D., 73, 75
Ustimenko, B. Y., 21

V

Vaiana, G. S., 59, 155, 170, 370, 373-76, 379, 383
Valentijn, E. A., 535
Vallee, J. P., 555
Van, Y. Y., 40, 499
van Albada, G. D., 317
Van Allen, J. A., 380, 400, 407, 409
van de Hulst, H. C., 267, 279, 281, 288, 289
Vandenberg, N. R., 484, 492
van den Bergh, S., 69, 79, 176, 177, 179, 180, 183, 189, 191, 286, 287, 505, 508, 511-14, 535, 536
van den Heuvel, E. P. J., 128, 134, 139, 144, 145, 147
van der Bergh, S., 300
van der Kruit, P. C., 73-75, 87, 302, 311, 315, 317, 318, 321, 454, 536, 555
van der Laan, H., 191, 555, 536, 555
Vandervoort, P. O., 467
Van Dyck, R. S., 222
van Flandern, T. C., 99
Van Hoosier, M. E., 62, 369, 370, 380
Van Horn, H. M., 147
Van Hoven, G., 63
van Kampen, N. G., 470
Van Maanen, A., 153
Van Pee, M., 218
VanSpeybroeck, L., 564
Van't Veer, F., 142
van Woerkom, A. J. J., 99
Vasyliunas, V. M., 63, 181, 411
Vauclair, G., 141
Vaughan, A. E., 37
Vedder, J. F., 278
Vennik, J., 260
Venugopal, V. R., 38, 497-500
Verges, J., 221
Vernazza, J. E., 367-70, 375, 380
Vernon, P., 536
Verschuur, G. L., 272
Vidal, N. V., 549, 552
Vilas, F., 114, 115, 117
Vilhu, O., 142
Visvanathan, N., 25, 86
Vogel, S. N., 62, 370
Vogt, R. E., 411
von Weizsäcker, C. F., 473
Vorontsov-Velyaminov, B. A., 71
Vorpahl, J. A., 63
Vrba, F. J., 279, 289
Vrscay, E. R., 212

Wagner, W. J., 159
Wagoner, R. V., 282
Wahl, A. C., 214
Walborn, N. R., 70
Waldmeier, M., 168
Walker, G. A. H., 72
Walker, G. H., 277
Walker, M. F., 70, 80, 84
Wallace, L. V., 204
Wallerstein, G., 187, 193, 271, 287
Walmsley, C. M., 271, 546
Walraven, G. D., 25, 544
Walsh, D., 349, 352, 498, 502
Wampler, E. J., 72, 73, 75, 84
Wannier, P. G., 324
Ward, M. J., 88
Ward, W. R., 99, 100
Wargo, G. F., 552
Warner, B., 25, 84
Warnock, W. W., 484
Wasserburg, G. J., 113, 119, 122
Watson, W. D., 267, 279, 289
Watson, W. S., 222
Webb, C. J., 528
Webber, W. R., 400, 404, 406, 407, 409
Webbink, R. F., 131, 133, 135, 142, 143, 147
Weber, D., 221
Weber, E. J., 428
WEEDMAN, D. W., 69-95, 71-75, 78-83, 85-89
Weeks, D. M., 198
Wehner, G. K., 279
Weidemann, V., 282
Weidenschilling, S. J., 119
Weigert, A., 129, 131-34, 138, 141
Weiler, K. W., 189
Weinberg, S., 240
Weiss, N. O., 50, 159, 172
Weisskopf, M. C., 180, 534, 560
Welch, W., 532
Welge, K. H., 217, 219, 222
Weliachew, L. N., 336
Wells, W. C., 222
Wentink, T., 220, 222, 225
Wentzel, D. G., 192, 383
Werner, M. W., 279, 282, 289, 325, 341, 358
Westbrook, R., 286
Westbrook, W. E., 325, 341
Westphal, J. A., 25, 507, 525
Wetherill, G. W., 122, 123
Weymann, R. J., 82, 84
Whang, Y. C., 102-4
Wheaton, W. A., 88, 544, 545
Wheeler, J. C., 144, 145, 147
Whelan, J. A. J., 141, 142, 144
Whipple, F. L., 268
Whitaker, E. A., 122, 123
White, A. E., 544
White, J. U., 220
White, R. A., 513, 514
White, S. D. M., 515, 528, 529, 532
Whitehurst, R. N., 239, 257
Whiteoak, J. B., 336, 337, 350, 351, 552
Whiting, E. E., 208, 211, 217, 219, 221
Whitney, A. R., 73, 90
Whitworth, A. P., 289
Wicke, V. G., 222
Wickramasinghe, N. C., 272, 281-86
Wiehr, E., 377
Wielebinski, R., 22-24, 26, 30
Wielen, R., 455, 457
Wiemer, W., 279
Wilcox, D. M., 220, 224
Wilcox, J. M., 155, 162, 168, 169
Wild, P., 506, 510
Wilhelms, D. E., 105, 107, 109, 110
Wilkinson, D. T., 35, 37, 240
Wilkinson, P. N., 498, 502
Will, C. M., 99
Williams, I. P., 286, 289
Williams, R. E., 82
Williamson, F. O., 187
Williamson, I. P., 490-92
Willis, A. G., 72-75
Willmore, A. P., 546
Willner, S. P., 324, 355, 357, 358
Wills, D., 536
Willson, M. A. G., 535, 555
Wilson, A. G., 517, 519, 522
Wilson, A. S., 72-75, 87, 189, 555
Wilson, C. P., 235, 237
Wilson, J. R., 146, 147
Wilson, L., 107, 114, 115, 117
Wilson, P. R., 50, 59
Wilson, R., 364, 372, 373, 380
Wilson, R. E., 141
Wilson, R. W., 323, 324, 336, 337
Windram, M. D., 536, 555
Winge, C. R., 404. 405
Winkler, P. F. Jr., 544
Wirtanen, C. A., 507, 517
Wiskerchen, M., 104
WITHBROE, G. L., 363-87; 369, 370, 372, 373, 375, 379-81
Witt, A. N., 279
Wittels, J. J., 90
Witzel, A., 352, 536
Wojslaw, R. S., 201
Wolf, R. A., 516
Wolfe, J., 410, 412, 413
Wolfe, J. H., 395, 422, 423, 426
Wolff, R. S., 25, 180, 534, 560
Wollman, E. R., 308, 353, 354
Wolnik, S. J., 226
Woltjer, L., 84, 175, 185, 189, 191, 548
Wood, G. E., 98
Woodgate, B. E., 193
Woodward, P. R., 185, 457, 458
Woolf, N. J., 72, 86, 87, 267, 280, 282, 534
Woolley, R., 446
Wordsworth, R. W., 72, 75
Wright, E. L., 74
Wright, M. C. H., 445
Wrixon, G. T., 304-8, 311, 314, 322, 336
Wu, C. S., 102
Wulf-Mathies, C., 271
Wyndham, J. D., 73-75
Wynn-Williams, C. G., 86, 358

Y

Yahil, A., 518, 522, 531, 535, 557
Yakubov, V. B., 464
Yamamoto, T., 285
Yamasaki, A., 141
Yankulova, I. M., 72-75
Yeates, C. M., 103
Yen, J. L., 73, 90
Yesipov, V. F., 73-75, 87
York, D. J., 271
Yoshimine, M., 218, 220

Yoshimura, H., 47, 49, 159, 164-66
Yost, J. C., 445, 446, 448, 451
Young, A., 114, 141
Young, E., 74
Young, J. W., 426
Young, R. E., 121
Yu, G., 65
Yu, J. T., 241
Yuan, C., 445, 448, 453, 459, 460, 463
Yungelson, L. R., 131, 133-35, 137, 144

Z

Zahn, J.-P., 138
Zaikowski, A., 278, 281
Zambetta, A. M., 279
Zang, T. A., 471, 472
Zappala, R. R., 283
Zare, R. N., 204
Zarnecki, J. C., 560
Zasov, A. V., 72-75, 80
Zeissig, G., 34
Zeldovich, Y. B., 241-43, 245
Zemelman, M., 78
Ziólkowski, J., 131, 132, 137, 141, 144
Zipf, E. C., 222
Zirin, H., 50, 63, 370
Zirker, J., 375-77
Zlobin, V. N., 26
Zohar, S., 116
Zombeck, M. V., 370, 373, 374
Zweibel, E., 192
Zwicky, F., 72, 175, 462, 476, 506, 513, 517, 520-23, 528, 529, 531, 552
Zygielbaum, A. I., 98
Zyrnicki, W., 226

SUBJECT INDEX

A

Aluminum hydride
  transition probability data for, 217
Aluminum oxide
  transition probability data for, 217, 218
Andromeda nebula
  core structure of, 319
  radiation around, 299, 300
Asteroids
  "cataclysmic events" from, 122, 123

B

Big-bang cosmology
  galaxy formation and, 240
Boron hydride
  transition probability data for, 218
Boron oxide
  transition probability data for, 218

C

C_2
  transition probability data for, 219
Calcium
  solar abundance of, 269, 270, 272
Calcium oxide
  transition probability data for, 219
Carbon
  cosmic abundance of
    dust formation and, 274
  solar abundance of, 269
Carbon hydride
  transition probability data for, 219, 220
Carbon monoxide
  cosmic dust formation and, 272, 274
  distribution in Galaxy, 310, 311, 323
  role in galaxy inner regions, 323-31, 337, 338, 341-43
  transition probability data for, 221, 222
Carbon sulfide
  transition probability data for, 222, 223
Cas A
  age of, 189
  energy in relativistic protons, 192
  Fermi-type acceleration
    in turbulent medium and, 190
  knots in, 180
  map of, 190
  optical emission from, 185
  recent increase in flux, 193
  second class of young supernova remnants and, 189
cD galaxies
  properties of, 514-28, 534, 536
CH^+
  transition probability data for, 220
Chlorine oxide
  transition probability data for, 219
Chromium oxide
  transition probability data for, 219
Chromosphere, solar
  see Sun, chromosphere of
CN
  transition probability data for, 220, 221
CO^+
  transition probability data for, 222
Comets
  molecular spectra of, 200
Corona, solar
  see Sun, corona of
Corundum

cosmic dust formation and, 277, 278
Cosmic energy
dissipation of, 192
Cosmic rays
acceleration sites for
supernovae remnants and, 191
Jupiter as source of, 409-12
Cosmic rays, galactic
element abundance and, 269
Cosmic X rays
see X-ray sources, extragalactic
Crab pulsar
characteristic age of, 39
irregularities in periods of, 35-37
Crab pulsar PSR 0531 + 21
interpulses from, 21
optical emission from, 25
pulse broadening from
source of, 491, 492
type of radiation emitted by, 24
VLBI observations of, 484-86
Crater formation
history of, 8-10
Cygnus A
X-ray emission from, 546
Cygnus loop
dense shell model for, 181
line intensities of, 182
radio supernova remnant and, 191
regional spectral differences in, 182
X-ray and optical observations of
cloud interaction model and, 185

D

Dust grains, formation and destruction of, 267-90
chemical physics of, 272-82
condensation curves for solids, 272, 273
condensation temperatures, 273
condensing solids and, 272-74
cooling track to condensation, 276
"dirty grains" formation and, 278
grain destruction by radiation, 279
grain-grain collisions, 278, 279
grain-grain sputtering and, 279
grain reprocessing, 278
homogeneous nucleation theory, 275-77
photodesorption by UV photons, 279
surface nucleation and, 277, 278
thermochemistry and, 272-75
circumstellar dust, 282-86
dust flow paths and, 283
dust nucleation and grain growth, 283
grain reprocessing and, 286
mass loss from stars and, 282, 283
novae and planetary nebulae, 284-86
protostellar nebulae and star formation, 286
red giants, supergiants, and infrared stars, 282-84
theoretical models for dust envelopes, 283
element abundances in, 268-72
absorption lines of interstellar medium, 270-72
abundance and gas velocity, 272
solar and "cosmic" abundances, 268-70
grain optical properties, 279-82
grain models, 281, 282
interstellar extinction and, 279, 280, 282
interstellar observations of, 279-81
types of grains, 281
interstellar medium and, 286-90
cloud-cloud collisions and, 287, 288
grain destruction by shock waves, 288
molecular clouds and H II regions, 288, 289
shock-wave effects on dust in, 287
supernova remnants and, 286, 287
turbulent velocities and, 289
turnover rate of elements in, 289
overview of, 267, 268
"cosmic" abundance of elements, 268
molecules in dust, 290
reviews of, 267
turnover rates for formation and destruction, 268
types of dusts, 268
turnover rates for formation and, 268

E

Earth
atmosphere of
energetic electrons entering, 425
auroral zone of, 416
magnetic storms on, 418
magnetosphere of, 390
convection model of, 416
energetic particles generated in, 409
magnetic tail of, 418
magnetopause and, 394
model for, 415
plasma sources in, 424
plasmasphere of, 424
radiation belt of, 390, 391
solar wind at, 416, 418
Electrons, relativistic
study by emitted radiations, 189

F

Formaldehyde
galactic inner region and, 323, 331-34, 338, 341
Fosterite
dust condensation and, 273, 274

G

Galactic center, 295-360
contour map of central 1", 298
emission at radio frequencies, 346-55
high-velocity ionized gas around Sagittarius A, 352
nonthermal emission, 348, 349

radio sources near galactic center, 346, 349
Sagittarius A, 350-55
thermal emission, 346-48
expanding H I features, 309-16
causes of gas expulsion from center, 311
cloud ejection in opposite directions, 311
Cugnon's object, 314, 315
diagrams of, 312, 314
differences in expansion rates of arms, 311
gaseous masses and, 310, 319
inclination of layer of expanding H I complexes, 315, 316
inner Lindblad resonance and, 311
integrated brightness temperature for fast atomic hydrogen, 316
Jodrell Bank latitude-velocity maps, 311, 312
latitude-velocity contours, 312
massive clouds ejected from nucleus, 315
noncircular orbits near center, 312
possible situation and motion of features, 310
Rougoor's "expanding arm," 310
smaller features of nucleus, 313, 314
van der Kruit's feature XII, 313, 314
gravitational field and nuclear disk, 297-309
atomic hydrogen density in galactic plane, 307
circular velocity in central region, 306
compact center mass upper limit, 318
dense thin ring in, 306, 307
disk rotation and light distribution, 306
disk thickness and, 307
gravitation field determination possibility, 308, 309
hydrogen atoms with high velocities and longitude, 314
infrared intensities and, 300
interstellar gas and, 301
mass distribution from 2.2μ radiation, 300, 301
mass-to-light ratio, 300, 301
near-infrared radiation and, 297-300
radial motions and, 302, 303
radiation origin and, 299
rotating nuclear disk, 301-8
surface brightness distribution of, 298-300
velocity-longitude contours in, 305, 306
velocity-longitude contours of hydrogen in galactic planes, 302, 303
infrared core, 355-60
emission from, 355-60
extinction in, 358
maps of, 355-57, 360
"ridge" in, 358
maps of galactic center, 356, 357, 360
molecular clouds and total gas density, 323-45
abundance of carbon and oxygen and, 325
carbon monoxide and, 323
center in nonthermal radiation, 344
central black hole and, 343
cloud circulation and, 343, 345
distribution and motions of molecules in nuclear disk, 327-34
distribution of thermal radiation, 345
expanding molecular ring, 334, 335
extinction of galactic disk, 324
hydroxyl and formaldehyde, 323, 331-34
inner region structure and, 323
+40 Km sec^{-1} feature, 335-40, 342
longitude-velocity contour map of CO emission, 331
longitude-velocity map of hydroxyl, 333
masses and dynamics of, 341-45
molecular density in Galaxy, 325, 327
Sagittarius B2 and, 340, 341
surface brightness and, 325
swept-up gas and, 342
total gas density from carbon monoxide, 323-31
nucleus position and distance, 297
RR Lyrae variables and, 297
overview of, 295-97
expulsive phenomena from, 296
gravitational field and mass distribution, 295
H I features of, 296
infrared core of, 297
potential, circular velocity, time revolution, and density of, 295
radial motions and deviation from galactic plane origins, 316-22
central bar in equatorial plane, 320
disk's magnetic field, 322
ejection or gravitational field and, 321, 322
expulsion hypothesis problems and, 321
gravitational field hypothesis, 319-22
hypotheses about, 316, 317
isotropic explosion of gas, 319
observations concerning, 316, 317
orbits of expelled gas clouds, 318
some expulsion models, 318, 319
spiral structures and, 317, 318
stages of nuclear activity in, 317
synchrotron emission and, 317
tilting of central plane of symmetry, 300
violent activities in galaxies, 317
Galactic coronae
hot gas from supernovae and, 188
Galactic winds

properties of, 188
Galaxies
evolution of
collapse processes and, 248
supernovae winds and, 188
spiral arms of
see Spiral structure theories
Galaxies, clusters of, 505-36
catalogs of, 505-7
criteria for inclusion, 505, 506
dynamics of, 526-33
average time between collisions, 527
bremsstrahlung emission, 527
characteristic times, 526, 527
collapse in evolution, 529
cooling time of intracluster gas, 527
dynamical evolution, 528, 529
dynamical friction time, 526
incomplete energy equipartition, 528
intracluster gas, 532
masses and mass-to-light ratios, 529-31
massive-halo hypothesis, 531, 532
mass segregation, 527, 528
mean galaxy and cluster mass density in universe, 532, 533
missing-mass, 531, 532
two-body relaxation time, 526
velocity dispersions, 531
radio emission and, 535, 536
sources of, 535, 536
X-ray emission and, 536
static properties of, 507-25
"anemic spirals," 513
average mass of galaxies, 520
Bautz-Morgan classification, 510-12
bounded isothermal function, 517-19
cD galaxies, 514-16
classification schemes of, 508-12
core radius of clusters, 520-22
definition of "cluster," 507, 508
density profiles, 516-20
De Vaucouleurs function, 519
envelopes of galaxies, 515, 516
galactic content of, 512-14
galaxy distributions, 322
gravitational radius of clusters, 520
halo size in clusters, 521, 522
King function, 518, 519
location of spirals in clusters, 513
luminosity function, 523-25
"mean radius" of clusters, 521
Morgan classification, 512
multiplicity function, 508
Oemler classification, 512
populations for clusters, 508
power-spectrum analysis of distribution, 527
redshifts of galaxies, 508
regular-irregular classification, 509
richness, 507, 508
Rood-Sastry classification, 509-12
sizes of clusters, 520-23
superclusters, 508, 522
surface brightness profile, 519
Zwicky classification, 509, 510, 524
X-ray emission of, 523-35
classification type of clusters and, 534
intracluster gas and, 535
optical properties and, 534
production mechanisms for, 534, 535
velocity dispersion of galaxies and, 535
Galaxies, elliptical
Seyfert galaxies and, 79, 89
Galaxies, radio
Seyfert galaxies and, 79
Galaxies, ring
density shock waves from, 476
Galaxies, Seyfert, 69-91
Balmer decrement and line profiles, 81-85
decrement problem in, 83
Doppler broadening and, 84
electron scattering and, 84
gas acceleration mechanism and, 84, 85
hydrogen energy levels and, 83, 85
line profile broadening and, 84
models for line profiles, 84, 85
origin of, 83, 85
reddening problem and, 83, 84, 88
rotation models and, 84
conclusions concerning, 91
energy sources in, 91
continuum radiation and luminosities, 85-88
dust effects on, 86-88
infrared sources of, 86-88
luminous ultraviolet sources in, 87
polarization in optical continuum, 86
radiation mechanisms and, 85, 86
radio emission by, 87
Seyfert power-law spectrum and, 86
spectra turnover points, 87
X-ray emission from, 88
emission-line spectra of, 80-82
classification of galaxies by, 81, 82
surveys of, 81, 82
morphology of, 76-80, 83
annular structures and, 77, 80
redshift controversy and, 80, 82
spiral arms and, 80
QSO relations of, 88-91
"Christmas tree" models of, 90
emission-line variations and, 90, 91

luminosity variation in, 89
redshifts in, 88, 89
relativistic expansions in, 90
source luminosity and, 90
synchrotron radiation and, 90
reviews on, 69, 70
Seyfert galaxies currently known, 70, 71
nuclei of, 70, 83
photographic data concerning, 71
spectroscopic definition of, 70, 71, 76, 77
surveys of, 71-80
elliptical galaxies and, 79
Markarian galaxies and, 79
radio galaxies and, 79
variability of, 90
Galaxies, spiral
Seyfert galaxies among, 79, 89
see also Spiral structure theories
Galaxy
main optical arms of
neutral hydrogen arms and, 455
structure of
galaxy collapse of, 249
Galaxy formation, recent theories in, 235-64
collapse picture of, 247-50
changing gravitational potential in, 249
collapse of a sphere of stars, 249
collapse time and, 248
dissipationless collapse, 248
final states of, 249, 250
free-face trajectories of stars and gas, 250
gravitational attraction and, 247
spheroidal vs. disk components, 250
star formation and, 248
violent relaxation in, 249
conclusion concerning, 263, 264
dissipationless collapse models for elliptical galaxies, 253-60
collapse of, 253, 260
cosmological infall and, 255
drag terms in, 260
dynamical friction, 259
elliptical isophotes and, 253
ellipticals, origin of, 254, 259
ellipticals, triaxial, 259
envelope formation in, 255
heavy-hole theory, 257, 258
gravitational potential and, 257
metal abundance of ellipticals, 259
metallicity gradients in, 260
spirals, origin of, 254
star formation and, 254-56
surface brightness distribution and, 255, 258
tidal stealing in, 255
tidal torques and, 254
two-body relaxation and, 260
velocity dispersion of halo stars, 258
violent relaxation and, 259
dissipation models for ellipticals, 250-53
bremsstrahlung cooling time and, 252
disk colors, 252, 253
elliptical isophotes and, 252
gas escape from, 251
gas viscosity role in, 251, 252
spiral galaxy formation and, 252
star formation and, 250
stellar metal abundance and, 251, 252
galaxies from small inhomogeneities in early universe, 240-47
adiabatic perturbations and, 243-45
angular momentum and, 246
baryon fluctuations and, 243, 244
big-bang cosmology, 240
covariance function role in, 246
density fluctuations in, 243
Einstein-de Sitter model, 240
formation of bound protoclusters of galaxies, 245
Friedmannian model of, 242
galaxy stripping and, 247
gravitational instability growing mode and, 241, 242, 246
helium and deuterium abundances, 242, 244
isothermal fluctuations in, 243, 245
luminosity function and, 247
microwave background and, 245
multiplicity functions and, 246
mutual tidal interactions of galaxies, 245, 246
primordial fluctuations and, 241
recombination in cosmology, 240, 241
standard open model for, 240
universe expansion, 240-42
observations of galaxies, 235-40
classical model of Galaxy, 238, 239
color gradients in elliptical galaxies, 238
combined halo-disk models and, 239
de Vaucouleurs' law and, 237
disk components of, 238
dynamic models of elliptical galaxies, 236, 237
elliptical galaxies, 235, 237, 238
GO galaxies, 238
Hubble's (1930) law and, 237
irregular galaxies, 238
neutral hydrogen in elliptical galaxies, 238
primeval galaxies, 240
spheroidal components of, 238
spiral galaxies, 238
star formation rate in young galaxies, 240
surface brightness of galaxies, 236-39

types of galaxies, 235
other models for, 261
collision mechanisms of, 262
color vs. absolute magnitude relations, 262
covariance function of elliptical galaxies and, 261, 262
covariance functions of spiral galaxies and, 262
group crossing times for dense groups, 263
histories of large galaxies, 263
luminosity functions and, 262
metal-abundance gradients and, 262
origin of ring galaxies, 261
SO galaxies, 262
tidally interacting galaxies, 261
predictions from astrophysics, 242
reviews on, 235
Galaxy M 31
properties of, 239
Ganymede
L-shell of, 421
Graphite
cosmic dust formation and, 275, 277, 278, 281

H

HD
transition probability data for, 223
Helium
atmosphere of Mercury and, 101, 102
H II regions
dust in, 288
infrared emission from, 288
Hydrocarbons
mid-infrared emission from giants and, 284
Hydrogen
atmosphere of Mercury and, 101, 102
cosmic abundance of, 270
galactic center and, 307, 309, 310, 316, 321, 323, 340
inner galactic regions and, 324, 325, 335
interstellar in galactic center, 301, 304
Hydrogen cyanide
emission of
+40 KM sec^{-1} feature and, 339
Hydroxyl
galactic inner region and, 323, 331-34, 337
transition probability data for, 224, 225

I

Interstellar clouds
molecular spectra of, 200
velocities of, 287
Interstellar dust
see Dust grains, formation and destruction
Interstellar gas
radial velocity distribution of, 270
Interstellar medium
absorption lines in, 270
cloud evaporation in model for, 187
clouds in
origin of kinetic energy in, 188
velocities of, 188
density gradient out of plane of Galaxy, 187
effects of supernovae on, 186-89
hot phase of, 187, 188
hydrogen in, 280
inhomogeneous nature of supernova remnant evolution and, 183, 184
interaction with supernovae
see Supernovae, interstellar medium interactions
radio wave scattering and scintillation in, 479-503
see also Radio waves scattering and scintillation, interstellar
spiral density wave propagation in, 187
supernovae contributions to, 192-94
turbulence in
supernova ejecta causing, 190
Io
atoms emitted by, 426, 427
influence on Jupiter's magnetosphere of, 389
neutral hydrogen ring of
as source for plasma, 426, 427, 432
Iron
abundance of
time measurement by, 565
cosmic abundance of, 268, 269
Mercury interior content of, 103, 117-22
production of
X-ray clusters and, 447

J

Jupiter
cosmic-ray source, 409-12
electron bursts from, 409, 410
ionosphere of, 420, 421
meteorite origin and, 11, 12
plasmapause of, 418
plasmasphere of, 419
radiation belt of, 390, 391
reviews on, 391
radiation from, 389-91
satellites of
synchrotron region sweep-up by, 39
synchrotron hypothesis concerning, 390
Jupiter's magnetosphere, 389, 434
models of outer magnetosphere, 414-34
Alfvén surface "dimple," 431
centrifugal forces and, 422-29
centrifugally driven winds, 428, 429
classes of, 414
convection model, 416, 417, 421, 425, 432
corotation, 419-27, 430-32
corotation and plasmapause, 418, 419
discussion of, 432-34
disk models, 422, 432, 433
earthlike models, 414-18, 432
electron precipitation and, 425
equatorial plasma density model, 425
fast planetary wind and, 431
field-aligned currents in, 419-22
flux tubes and, 425, 426, 428

heavy ions and, 427
heliospheric models of
radial outflow, 429-32
"hinged current sheet"
model, 421, 423
hinged disk model, 422
internal shocks and, 429-
32
Io as plasma source, 426,
427
ionosphere-magnetosphere
current topology, 426
Jovian plasma source, 424-
27
Jovian sodium plasma and,
426
"magnetic sling" stellar
wind, 427
magnetic storms in, 418
magnetic tail and, 415-20,
433
magnetopause in, 415, 416
magnetosphere-ionosphere
interaction, 419-21
photoelectrons and, 424,
425
planetary-wind outflow and,
429
polar-cap ionosphere and,
417, 420
quasi-static model, 421-
23
radial outflow models,
427, 428
reconnection and, 415-17
Rice model, 432
solar wind in, 414-33
static magnetic field model,
422, 423
stellar winds, 427
sub-Alvénic Jovian breeze
and, 430, 431
sulfur clouds and, 427
super-Alvénic Jovian
winds, 430, 433
time variability and, 433,
434
unified models, 433
Pioneer observations of
middle and outer magneto-
spheres, 398-414
cosmic-ray electrons and,
411
current sheet properties
and, 401-3, 412
flapping-sheet model and,
402, 403, 410
garden hose angle and,
398, 399
ionic streams and, 408,
409
Jupiter as cosmic-ray
source, 409-12
magnetic longitudes and,
398, 399
magnetosheath in, 394-
98, 404
magnetosphere lobes and,
403
magnetosphere model
from, 403
magnetosphere regions,
397-403
middle magnetosphere
signature, 408, 409
Pioneer 10 outbound, 398-
403
Pioneers 10 and 11 in-
bound, 406-8
Pioneer 11 outbound pass,
403-8
proton-flux energy spectra
and, 406
proton fluxes and, 400,
401
reconnection theory and,
401
summary of, 412-14
turbulent areas and, 404-
8
Pioneers 10 and 11 findings
on, 389-98
electron fluxes and mag-
netic fields, 396, 397,
400, 401
magnetic field magnitude
in, 395-98
magnetopause in, 394-400,
404, 406-8, 412
on board experiments,
392
overview of data, 394-98
solar-wind dynamic pres-
sure and, 397
time variability of mag-
netic fields, 396, 397,
412, 433
trajectories and, 392,
393
reviews on, 391
satellite Io effects on,
389

L

Lanthanum oxide
transition probability data
for, 223
Local Group
dynamics of, 239, 246

M

Magellanic Cloud, Large
irregular galaxy identity,
238
Magnesium
cosmic abundance of, 268,
269
Magnesium fluoride
transition probability data
for, 223
Magnesium hydride
transition probability data
for, 223, 224
Magnesium oxide
transition probability data
for, 224
Magnesium silicate
condensation in cosmic dust,
273
Magnetic fields
theoretical studies of, 64
Magnetospheres
Jupiter's, 389-434
Mars
ancient "rivers" of
history of, 10, 11
Mercury, 97-124
atmosphere of, 101, 102
helium and hydrogen in,
101, 102
hydrogen from water
photolysis in, 102
loss of heavier ions, 102
nightside temperatures
and, 102
solar wind and, 102
geological history of, 121-
24
"cataclysmic event" and,
122, 123
Caloris basin formation,
123
condensation of solar
nebula in, 122
final epoch in, 124
heavily cratered terrain
in, 122, 123
stages of, 121, 122
images from Mariner flybys,
97
interior of, 117-21
accretion process models
of, 119
aluminum isotope and heat
source, 119, 120
composition and structure,
117, 118
connecting core and, 120,
121

density and pressure models of, 117, 118
heat sources for, 119, 120
iron in, 117-22
models of, 117
molten core models, 120
radioactives absent from, 120
residual dipole model of, 121
silicates in, 117, 120-22
thermal history of, 118-21
magnetic field and magnetosphere, 102-4
fluid dynamo and, 103, 104, 121
induction mechanisms for, 104
iron in interior and, 103
magnetism of iron-bearing rocks and, 103, 104
orientation of field, 103
solar wind and, 103
sources of, 103, 104
substorms in, 103
orbit of, 98, 99
Einstein's general theory of relativity and, 99
prograde precession of perihelion, 99
rotation of, 99-101
Cassini's laws and, 101
extensional surface strains and, 100
lobate scarps and, 100
obliquity and, 100
rotation axis and, 100
solar capture probability and, 101
solar tides and, 100, 101
spin period of, 100, 101
size and mass of, 98
oblations and, 98
radius and, 98
surface features of, 104-17
albedos of, 114, 115
Caloris basin in, 105, 107-16, 122, 123
cartography of, 116, 117
color variations in, 115
composition of materials of, 115
compressive stresses acting on, 111
craters in, 104-6, 109, 111-14, 122
degradational processes in, 112
ejecta systems in, 113, 116
generalized terrain map, 107
heavily cratered terrain in, 105, 106, 114
intercrater plains in, 104, 105, 108
lobate scarps in, 104, 108, 111, 123
major physiographic provinces, 104-9
Mariner flyby pictures of, 104, 106
microwave radiometry of, 116
naming of features of, 116, 117
optical and thermal properties of, 114-16
polarization measurements of, 114, 115
pre-Imbrium pitted plains in, 105
regolith of iron and titanium glasses on, 115
seismic disturbances in, 105
smooth plains in, 106-9, 113
tectonic framework of, 111
tensional stresses acting in, 111
thermal conductivity of, 115, 116
thermal inertia of, 116
volcanic action on, 109, 110, 123, 124
Meteorites
origin of, 11, 12
Mira Ceti
A10 emissions of, 201
Molecules, transition probability data for, 197-227
action
assessment of data on spectra, 204
astrophysically important molecules, 199-204
astronomical locations of molecular spectra, 200
chemical kinetic data, 204
diatomic and small polyatomic molecules, 201
dissociation energy data compiled, 204
emission bands of, 201
metal oxides and hydrides and, 201
molecular spectra identification, 204
molecules and astronomical locations, 202, 203
planetary atmospheres and, 201
reference sources for diatomic molecule data, 204
shock-tube spectroscopy and, 199
spectroscopic data availability, 201-4
thermochemical methods for, 199
conclusion, 227
diatomic molecular transition probability data, 217-27
specific compounds, 217-27
parameters of molecular spectroscopy, 205-17
absorption measurements, 215, 216
band oscillator strength, 210
band strength electronic and vibrational components, 209
Born-Oppenheimer approximation, 208
computer programs for, 212, 213
diatomic molecular transition probability concepts, 207-10
differential extinction in absorption spectra, 206
Einstein A and B coefficients, 205
electronic oscillator strength, 210
electronic transition moments, 208, 213, 214
emission measurements, 214, 215
experimental methods for, 214
Franck-Condon factors, 211-19, 225
general concepts of, 205
Hönl-London factor, 208, 211, 217
"hook method" and, 216
lifetime measurements, 216, 217
Mach-Zehnder interferometry and, 216

optically thin emission
spectroscopy and, 206
optimized valence configuration, 214
oscillator strength and, 206, 207
phenomenological concepts, 205-7
photoelectric spectroscopy, 215
quantum concepts and, 207
r-centroid approximation, 208
r-centroids, 211-13
rotational energy and, 212
shock-tube radiometry, 215
specific molecular potentials, 211, 212
theory and experiment on, 210, 211
transition-strength of molecular line, 208
vibrational wave functions and, 211
spectroscopic studies of, 197, 198
atomic and molecular spectra, 198, 199
electronic, vibrational and rotational changes in, 198, 199
oscillator strength data and, 198
Swings effect and, 199
theoretical research on, 199
Moon
history of
"cataclysmic event" in, 122
crystallization ages of samples of, 122
M31 nebula
stratoscope II observations of, 319, 320

N

Nebulae, planetary
element abundance in, 269
infrared emission from, 285
origin of, 285
surveys of, 309
Nebula, protostellar
dust grain destruction and coalescence in, 278
element outflow from, 272
matter expelled from
dust grain formation and, 286
NGC 4151
X-ray emission by, 545
NGC 5128
X-ray emission by, 543-45
NGC 5128 galaxy
nuclear activity in, 317
Nitrogen
cosmic abundance of, 268, 269
Nitrogen hydride
transition probability data for, 224
Novae
outbursts of
dust grain production and, 284, 285
element abundances and, 285

O

OH^+
transition probability data for, 225
Orion Nebula
dense dark clouds in, 279
element abundance in, 269
Oxygen
cosmic abundance of, 268, 269

P

Perseus/NGC 1275
X-ray emission from, 560
X-ray photograph of cluster, 561
Planets
chemical composition of model for, 117, 118
Pulsar PSR 0031-07
drifting subpulses from, 28
Pulsar PSR 0329 + 54
mode changing in, 24
polarization in pulses of, 30, 31
Pulsar PSR 0809 + 74
drifting pulses from, 28
Pulsar PSR 0833 - 45
characteristic age of, 39, 40
circular polarization in emission of, 24
polarization in emission from, 23
pulses from, 489 - 502
Pulsar PSR 0834 + 06
spectral analysis of pulses, 22, 23
Pulsar PSR 1055 - 52
two-component main pulse of, 21
Pulsar PSR 1133 + 16
orthogonal polarization in pulses, 31
Pulsar PSR 1237 + 25
five-component pulse from, 27
mode changing in, 24
Pulsar PSR 1541 + 09
pulse characteristics of, 21
Pulsar PSR 1857 - 36
polarization of emission of, 23-25
Pulsar PSR 1919 + 21
slow intrinsic variations and, 497
Pulsar PSR 2016 + 28
drifting subpulses from, 27, 29
Pulsars
broadening of signal from, 488
data concerning, 498
intensity scintillations from, 493, 494
proper motion of, 499
radio wave study of, 479
supply of energy for nebula from, 189
Pulsars, recent observations of, 19-66
celestial coordinates of, 32
evolution of, 32-42
ages and origins of pulsars, 39-42
braking index and, 34, 35, 39
characteristic ages of, 39, 40
differences among pulsars, 35
distances and Z-distribution of, 37, 38
galactic distribution of, 37-42
irregular variations in pulsar periods, 35-37
kinematic ages of, 40, 41
luminosity function and galactic population, 39
magnetic and rotational axes and, 35
magnetic delay and, 41
pulsar birthrate, 41, 42

pulsar periods distribution, 33
pulsar timing observations and, 32-37
pulse arrival times, 33
R-distribution of pulsars, 38, 39
secular variations of pulsar periods, 34, 35
supernovae and, 41, 42
individual pulses from, 26-32
drifting subpulses, 27-29
microstructure of pulses, 29, 30
polarization of, 30-32
pulse intensity histograms, 26
pulse-to-pulse fluctuations in, 26, 27
secondary periodicities in, 26
subpulses, 26, 27
observed pulse characteristics, 20-32
classifications of pulsars, 21
components of, 21
high-frequency observations, 24-26
integrated pulse profiles, 20-26
interpulses, 21
polarization and, 23, 24
profile stability of, 24
pulse shapes, 20, 21
spectral properties of, 21-23
types of radiation emitted by, 24, 25
surveys of, 37, 38

Q

Quasars
radio wave study of, 479
Quasar 3C 273
X-ray emission by, 545

R

Radio waves scattering and scintillation, interstellar, 479-503
angular scattering, 483-86
complex visibility function, 483
interferometer observations, 483, 484
observations, 484-86
source visibility function, 484
visibility scintillations, 486
conclusion, 503
discussion of, 500-2
homogeneous extended Kologorov model, 500, 501
nonuniformity of scattering medium, 501
sources other than pulsars, 502
pulse broadening, 486-500
Crab pulsar, 491, 492
delay fluctuations, 489
discussion of, 492
extended medium, 490, 491
pulsar broadening and, 488
pulse shape deviation, 487, 488, 490
theory of, 487, 490
thin screen case and, 487-89
width of scattered pulse, 491
reviews of, 480
scattering of
sources broadening and, 479
scintillation of intensity, 493-500
observation-theory comparisons, 495-97
"Rayleigh limit" and, 494
spaced receiver observations, 497-99
strong scintillations, 494, 495
temporal variations in, 499, 500
theoretical spatial spectrum, 500
theory of, 493, 494
small angular size of pulsars and, 479
quasars and, 479
theory: the correlation functions, 480-83
assumptions about medium, 481
refractive index fluctuations and, 481
square law structure function and, 483
R Coronae Borealis
CN bands in spectra of, 201
Rho Ophinchi cloud complex
large dust grains in, 289
RR Lyrae variables
galactic center and, 297

S

Sagittarius A
compact nucleus of, 351, 352
components of, 351
emission rates from, 350-53
halo surrounding, 351
high-velocity ionized gas around, 352
"infrared core" in, 353
maps of inner region of, 350, 351
mass of, 353
profiles of Neon II line and, 354
Sagittarius B2
gas-dust relations in, 341
nuclear region and, 340, 341
Scintillation, interplanetary
radio waves and, 479
Scintillation, interstellar
radio astronomy and, 479
Scintillation, ionospheric
radio waves and, 479
Seyfert galaxies, 69-91
see also Galaxies, Seyfert
SiH^+
transition probability data for, 225, 226
Silicates
dust composition and, 281
dust emission from stars and, 282, 283
Mercury content and role of, 117, 120-22
Silicates, hydrated
condensation in cosmic dust, 274, 281-84
Silicon
cosmic abundance of, 268, 269
Silicon carbide
spectra of cool stars and, 267
Silicon fluoride
transition probability data for, 225
Silicon hydride
transition probability data for, 225
Silicon oxide

transition probability data
for, 226
Solar activity origin, 45-66
convective zone of, 48, 49
field concentration into flux
tubes, 60-62
Alfvén waves and, 61
cooling cause, 61
dynamic activity of tubes,
62
magnetic field pressure
and, 60, 61
spicules and, 62
temperature effects and,
60, 61
implications of solar activity, 45-47
absence of in 17th century,
46
convection and circulation
in, 46, 47
magnetic flux tubes and,
45, 46
terrestial temperature
and, 47
magnetic buoyancy, 51,
52
upper surface of field,
51, 52
magnetic fields dynamical
dissipation, 62-66
fluid motions in, 62, 63
gigantic flares and, 63, 64
hydromagnetic exchange
instabilities, 63
isotopic abundances and, 64
lines of force direction
and, 65, 66
magnetic field equilibrium,
64, 65
microturbulence in, 63
"neutral point annihilation,"
63
nonequilibrium conditions,
66
magnetic fields, origin of,
47-49
angular velocity gradient
in Sun, 48, 49
cyclonic rotation and,
48, 49
hydrodynamic models of,
49
kinematic dynamo models
of, 47, 49
location of, 48
magnetic flux loss from,
47, 48
solar magnetic fields behavior, 49-51
flux tube role and, 49-52
neutral point annihilation
and, 49
sunspots and, 50, 51
wasp-waisted tubes, 50,
59
sunspots, 50-52, 56, 59
twisted flux tubes, 52-55
buckling of tubes, 55
magnetohydrostatic equi-
librium, 52, 53
tension of tubes, 54
twisting variation along
flux tubes, 55-60
annulus mapping of, 56,
57
boundary confining, 56
buckling of tubes, 59
helical coils of field
migration, 59
Solar chromosphere and
corona
heating mechanisms, 382-
84
chromosphere and, 382
mass and energy flow in,
363-85
energy input mechanisms,
366
energy loss mechanisms,
364, 366
magnetic fields role, 363,
364, 366
major classes of structure
and, 364
overview of, 363-66
physical models of, 363,
364
X-ray photography of,
365
mass and energy flow in
regions with strong
coronal magnetic fields,
377-79
chromosphere and, 377,
379
structures in, 377, 378
transition region and
corona, 379
mass and energy flow in
regions with weak coro-
nal magnetic fields,
366-77
chromosphere and, 366-
70
coronal holes, 373-77
transition layer and corona
in quiet regions, 370-
73
see also Sun, chromo-
sphere of
observational evidence for
wave heating and mass
flows, 380-82
spectral line profiles,
380-82
XUV brightness and, 380,
381
see also Sun, chomosphere
of and Sun, corona of
Solar flares
evolution of, 63, 64
reviews on, 64
Solar system
tidal origin of, 12, 13
Solar wind
acceleration mechanism
for, 377
coronal holes and, 373-76
dynamic pressure of, 428,
429
Earth's radiation belt and,
390
heliosphere of, 429
Jupiter electrons in, 411
Jupiter magnetosphere and,
391
sectors of, 418
Spiral structure theories,
437-76
global instabilities, 464-73
evolution of rotating disk,
465
evolving patterns in gas
component of rotating
disk, 466, 467
global mode calculations,
470-72
halos and, 469
mobile retrograde stars
and, 469
Ostriker-Peebles criterion,
468, 469, 471
rotating disk experiments,
465-69
"softened" gravitational
potential, 471
tendency toward bar-making,
464-68
transient spiral structures,
468
unstable spiral modes,
470-73
Lindblad's ideas, 438-40
circulation theory in, 439,
440
density waves in, 440
dispersion orbits in, 439
leading spiral arms in,
438

quasi-stationary spiral
structure, 440
waves in galaxies in, 438
Lin-Shu density waves, 443-53
angular wave numbers, 448
axisymmetric vibrations, 443, 444
"characteristic" curves, 447, 450
dispersion relation for axisymmetric density waves, 444
forcing and/or feedback, 450-53
"global" mode analyses, 452
group velocity and, 449, 450
matching of precession rates, 444-48
origin of incoming waves, 451
outer and inner Lindblad resonances, 445
short-wave damping and, 450-52
theory for short asymmetric waves, 443
thin exponential disk: theoretical data, 447
wave "amplifiers," 451, 452
wavelengths and, 448, 452
wave refraction, 451, 452
nearly kinematic waves, 440-42
kinematic density waves of spiral form, 442
perturbation of continuous ring, 440, 441
quick synopsis of, 462-64
bar structure, 463
"multi-armed" spiral features, 463
status of wave theories, 463
shearing bits and pieces, 473-76
recurrent instabilities, 473, 474
ring galaxies, 476
"secondary" spiral features, 475
tidal forces and, 475
wavelike density ridges in disk of stars, 474
shocks and other consequences, 453-62
bright areas on concave side of arms, 454
bright ridges and, 454
damping and, 458-60
density distribution in forced gas disk, 461
forced travelling waves in "curtain," 456, 457
origin of shocks, 456-58
pendulum analogy, 455-58
spiral shocks without spiral forcing, 460-62
spiral shock waves, 453-55
Stars
chromospheres and coronae of, 364
death rate of, 42
element abundance in, 270, 271
evolution of
angular momentum and, 249
formation in galaxies, 250, 255, 256
metal-rich vs. metal-poor, 249-51, 257-61
formation of
interstellar matter used for, 286
shells from supernova explosion, 185, 186
spheroidal systems of
origin of, 261
Stars, carbon-rich
cosmic dust formation and, 274, 275
Stars, cool carbon
spectra of, 281
Stars, late-type
molecular spectra of, 200, 201
Stars, reddened
element abundance in, 270, 271
Star structure
convection model of, 6-8
history of, 6
mixing-length concept in, 6-8
Sulfur
clouds of
Jupiter atmosphere and, 427
Sulfur oxide
transition probability data for, 225
Sun
abundance of elements in, 269
chromosphere of
brightness of, 370
dynamic pressure in, 372
empirical models of, 377
energy losses from, 376
heating mechanisms in, 382
homogeneous models of, 367
hydromagnetic waves in, 382
layers of coronal holes in, 369, 370
macrospicules in, 370, 373
mass and energy flow in, 366-70
network in, 368-72
oscillatory motions in, 381
physical conditions in, 367
spicules of, 369, 370, 373
structure of, 366, 367
temperature vs. density in, 371
temperature vs. optical depth in, 368
corona of
current dissipation in, 383, 384
empirical models for, 379
empirical models of holes in, 375, 376
energy balance models of, 373
energy losses from, 376, 381
field lines of, 372
field reconnection and, 384
forbidden spectral lines of, 380
heating mechanism of, 380
holes in, 364, 367, 370, 373-77
loops of, 379, 381
magnetic configuration in holes, 373, 374
magnetic fields in, 370
magnetic heating of, 383, 384
physical properties of holes in, 375
solar wind and holes in, 373-76
spectral line profiles of, 380, 381
temperature density models of, 372
theoretical modes for holes in, 376, 377
wave heating of, 382, 383

X-ray photograph of, 365
molecular spectra of, 200
CO Fourth Positive band system of, 200
quiet regions of
heating requirements and, 377
magnetic configurations in, 370, 372
transition layers of
empirical models for, 379
energy losses from, 376
gas pressure in, 372
magnetic fields in, 370
models for, 371-73
theoretical models for, 376
Supergiants
black-body emission from, 284
Supernova blast waves
element abundance affected by, 272
Supernovae
explosions of
matter returned to interstellar medium by, 286
remnants of
effects on interstellar medium, 286, 287
Supernovae-interstellar medium interactions, 175-94
evolution phases of, 178
explosion of, 176, 177
energy released in, 176, 177
hydrodynamic models of, 176
radiation emitted from, 176
fragments of
Cas A, 190
surface brightness of, 191
future prospects, 192-94
interstellar medium effects of, 192-94
Origin hypothesis for, 175
properties of, 177
ejected mass of, 177
energy radiated from, 177
mean velocity of, 177
stellar population of, 177
Type I vs. Type II, 177
relativistic particles and, 189-92
blast wave effects, 191
cosmic rays and, 191, 192
knots formed by ejecta, 189, 190
old supernova remnants, 190, 191
radio flux from, 191
surface brightness of SNRS, 191
turbulent acceleration for ejecta, 189
young supernova remnants and, 189, 190
remnant evolution, 177-86
adiabatic instability in shocked clouds, 185
adiabatic vs. isothermal models, 181
Ar, S, and O abundance in, 179
blast-wave adiabaticity, 180, 181
blast-wave interaction with cloud, 184, 185
breakup of, 179
changing shape of, 178
cloud evaporation and, 181
cooling region behind shock wave, 183
deceleration of, 179
dense shell model and, 181
electron and ion temperatures and, 180
element abundance in shock waves, 182
emitted radiation from, 182
energy transfer to ISM, 180
expanding hydrogen shells and, 183
fronts classification, 181
heat conduction effects, 180, 181
high electron thermal velocity, 181
infrared emission of shock wave energy, 183
instability, Rayleigh-Taylor, 179
interior temperature profile and, 181
interstellar medium inhomogeneity and, 183, 184
irregular structure cause, 183, 185
jump conditions at conduction front, 181
knot formation, 179, 180
large compressive forces in, 185
magnetic effects of shell expansion, 179
magnetic stresses and, 183
momentum conservation and, 183
observed shock spectra of, 182
old remnant irregular structure, 183-86
radiative shock phase, 181-83
shell formation stage, 182
shock-wave formation, 178-80
star formation and, 185, 186
supernova ejecta, 179, 180
supersonic expanding shell formation, 182
transfer of energy to ISM, 177-79
X-ray emission from, 180
X-rays from blast waves, 185
remnants of, 175
form of, 177
supernovae effects on ISM, 186-89
ionization energy, 186
kinetic energy and, 188, 189
numerical simulations of SNR evolution, 187
thermal energy, 186-88

T

Titanium oxide
transition probability data for, 226
Tycho's remnant
energy balance in, 189
relativistic proton content of, 192
search for fossil H II region in, 176
second class of young supernovae remnants and, 189

U

Universe
curvature of, 565
expansion of, 240-42
future of, 432, 433

V

Vela pulsar PSR 0833 - 45
emissions outside radio band from, 25
irregularities in periods of, 35
Virgo/M 87
X-ray emission by, 560
Volcanoes, lunar and Martian
history of, 10

W

Water-ice
particles in cosmic dust, 282

X

X-ray sources, extragalactic, 544-65
active galaxies and, 543-46
Cygnus A, 546
intensity relations and, 544
NGC 4151, 545
NGC 5128, 543-45
other active galaxies, 546
photon flux from, 544, 545
source of X-rays, 545
X-ray spectrum from, 544, 545
3C 273, 545
clusters of galaxies emitting X-rays, 549-61
angular sizes of, 553, 554
central densities of, 553, 554
compilations concerning, 549-54
Compton scattering of radio electrons, 559
effective temperatures of, 553, 554
electron energy, 558, 559
emission mechanisms, 556-60
gas heating mechanisms, 557, 558
identification of, 549-56
intergalactic hot plasma and, 557
iron production and, 557 558
morphologies of, 550-52
Perseus/NGC 1275, 560
radio and X-ray fluxes, 559
radio emission from X-ray clusters, 555, 558
space velocities of, 550-52
surface brightnesses of, 553-55
thermal bremsstrahlung mechanisms, 556-60
Virgo/M87, 560
X-ray photograph of cluster, 560, 561
X-ray structure, 560, 561
diffuse X-ray background, 561-65
cosmological theory of scattering, 562
fluctuations in, 583
origin of, 561
part attributable to known objects, 562
reviews on, 561
spectral intensity of, 562
volume emissivity of, 562
discovery of, 541
observational constraints, 542, 543
instrumentation for, 542
positional accuracy, 542
prospects for future observations, 563-65
clusters of galaxies at redshift, 565
cosmological studies and, 564, 565
curvature of universe, 565
formation of galaxies and clusters, 564, 565
HEAO-B observatory and, 563, 564
improved instrumentation and, 563, 564
iron abundance and, 565
LAMAR observatory and, 564
unidentified high-latitude sources, 547-49
distribution of, 546-48
transient sources, 548, 549
X-ray galaxies, 548
X-ray-optical luminosity ratio, 548
X-ray galaxies, 541

Y

Yttrium oxide
transition probability data for, 226

Z

Zirconium oxide
transition probability data for, 227

CUMULATIVE INDEXES

CONTRIBUTING AUTHORS VOLUMES 11-15

A

Aannestad, P. A., 11:309-62
Aarseth, S. J., 13:1-21
Acton, L. W., 12:359-81
Allen, R. J., 14:417-45
Arnett, W. D., 11:73-94
Athay, R. G., 11:187-218
Audouze, J., 14:43-79

B

Bahcall, N. A., 15:505-40
Barkat, Z., 13:45-68
Barshay, S. S., 14:81-94
Blumenthal, G. R., 12:23-46
Burns, J. A., 15:97-126
Burrus, C. A., 11:51-72
Burton, W. B., 14:275-306

C

Canuto, V., 12:167-214; 13:335-80
Carson, T. R., 14:95-117
Cassen, P., 15:97-126
Caughlan, G. R., 13:69-112
Chaffee, F. H. Jr., 14:23-42
Chevalier, R. A., 15:175-96
Code, A. D., 11:239-68
Coroniti, F. V., 15:389-436
Counselman, C. C. III, 14:197-214
Cox, J. P., 14:247-73
Culhane, J. L., 12:359-81

D

De Young, D. S., 14:447-74
Duncombe, R. L., 11:135-54

E

Eardley, D. M., 13:381-422
El-Baz, F., 12:135-65

F

Fowler, W. A., 13:69-112

G

Gault, D. E., 15:97-126
Gilman, P. A., 12:47-70
Ginzburg, V. L., 13:511-35
Gorenstein, P., 14:373-416
Gott, J. R. III, 15:235-66
Grasdalen, G. L., 13:187-216
Gursky, H., 15:541-68

H

Harrison, E. R., 11:155-86
Heiles, C., 14:1-22
Howard, R., 15:153-73
Huebner, W. F., 14:143-72

I

Iben, I. Jr., 12:215-56

J

Jauncey, D. L., 13:23-44
Jenkins, E. B., 13:133-64
Jokipii, J. R., 11:1-28

K

Kaplan, S. A., 12:113-33
Keenan, P. C., 11:29-50
Kennel, C. F., 15:389-436
Klepczynski, W. J., 11:135-54

L

Larson, R. B., 11:219-38
Lecar, M., 13:1-21
Leibacher, J., 12:407-35
Lewis, J. S., 14:81-94
Litvak, M. M., 12:97-112
Livingston, W. C., 11:95-114

M

Manchester, R. N., 15:19-44
Marsden, B. G., 12:1-21
Mihalas, D., 11:187-218
Miller, J. S., 12:331-58
Morgan, W. W., 11:29-50

N

Nicholls, R. W., 15:197-234
Novikov, I. D., 11:387-412
Noyes, R. W., 15:363-87

O

O'Dell, S. L., 14:173-95
Oke, J. B., 12:315-29
Oort, J. H., 15:295-362
Öpik, E. J., 15:1-17

P

Palmer, P., 12:279-313
Parker, E. N., 15:45-68
Peale, S. J., 14:215-46
Peimbert, M., 13:113-31
Penzias, A. A., 11:51-72
Peterson, L. E., 13:423-509
Pikel'ner, S. B., 12:113-33
Press, W. H., 13:381-422
Preston, G. W., 12:257-77
Purcell, E. M., 11:309-62

R

Reeves, H., 12:437-69
Rickett, B. J., 15:479-504
Robinson, E. L., 14:119-42
Robinson, L. B., 13:165-85

S

Salpeter, E. E., 15:267-93
Schramm, D. N., 12:383-406
Schroeder, D. J., 14:23-42
Schwartz, D. A., 15:541-68

Searle, L., 12:315-29
Seidelmann, P. K., 11:135-54
Silk, J., 11:269-308
Spitzer, L. Jr., 13:133-64
Steigman, G., 14:339-72
Stein, R. F., 12:407-35
Stein, W. A., 14:173-95
Strittmatter, P. A., 14:173-95, 307-38
Strom, K. M., 13:187-216
Strom, R. G., 15:97-126
Strom, S. E., 13:187-216

T

Taylor, J. H., 15:19-44
Thomas, H.-C., 15:127-51
Tinsley, B. M., 14:43-79
Toomre, A., 15:437-78
Tsytovich, V. N., 11:363-86
Tucker, W. H., 12:23-46; 14:373-416

V

van de Kamp, P., 13:295-332
van den Bergh, S., 13:217-55
van der Kruit, P. C., 14:417-45
Verschuur, G. L., 13:257-93

W

Wallerstein, G., 11:115-34
Weedman, D. W., 15:69-95
Wentzel, D. G., 12:71-96
Whipple, F. L., 14:143-72
Williams, R. E., 14:307-38
Withbroe, G. L., 15:363-87

Z

Zel'dovich, Ya. B., 11:387-412
Zhelezynakov, V. V., 13:511-35
Zimmerman, B. A., 13:69-112
Zuckerman, B., 12:279-313

CHAPTER TITLES VOLUMES 11-15

PREFATORY CHAPTER		
About Dogma in Science and Other Recollections of an Astronomer	E. J. Öpik	15:1-17
SOLAR SYSTEM ASTROPHYSICS		
Comets	B. G. Marsden	12:1-21
Surface Geology of the Moon	F. El-Baz	12:135-65
Chemistry of Primitive Solar Material	S. S. Barshay, J. S. Lewis	14:81-94
Physical Processes in Comets	F. L. Whipple, W. F. Huebner	14:143-72
Mercury	D. E. Gault, J. A. Burns, P. Cassen, R. G. Strom	15:97-126
Jupiter's Magnetosphere	C. F. Kennel, F. V. Coroniti	15:389-436
SOLAR PHYSICS		
Turbulence and Scintillations in the Interplanetary Plasma	J. R. Jokipii	11:1-28
Solar Rotation	P. A. Gilman	12:47-70
The Solar X-Ray Spectrum	J. L. Culhane, L. W. Acton	12:359-81
Waves in the Solar Atmosphere	R. F. Stein, J. Leibacher	12:407-35
The Origin of Solar Activity	E. N. Parker	15:45-68
Large-Scale Solar Magnetic Fields	R. Howard	15:153-73
Mass and Energy Flow in the Solar Chromosphere and Corona	G. L. Withbroe, R. W. Noyes	15:363-87
STELLAR PHYSICS		

Title	Authors	Volume:Pages
Spectral Classification	W. W. Morgan, P. C. Keenan	11:29-50
Explosive Nucleosynthesis in Stars	W. D. Arnett	11:73-94
The Physical Properties of Carbon Stars	G. Wallerstein	11:115-34
The Effects of Departures from LTE in Stellar Spectra	D. Mihalas, R. G. Athay	11:187-218
Post Main Sequence Evolution of Single Stars	I. Iben Jr.	12:215-56
The Chemically Peculiar Stars of the Upper Main Sequence	G. W. Preston	12:257-77
The Spectra of Supernovae	J. B. Oke, L. Searle	12:315-29
Planetary Nebulae	J. S. Miller	12:331-58
The Structure of Cataclysmic Variables	E. L. Robinson	14:119-42
Nonradial Oscillations of Stars: Theories and Observations	J. P. Cox	14:247-73
Recent Observations of Pulsars	J. H. Taylor, R. N. Manchester	15:19-44
DYNAMICAL ASTRONOMY		
Dynamical Astronomy of the Solar System	R. L. Duncombe, P. K. Seidelmann, W. J. Klepczynski	11:135-54
Computer Simulations of Stellar Systems	S. J. Aarseth, M. Lecar	13:1-21
Unseen Astrometric Companions of Stars	P. van de Kamp	13:295-333
Radio Astrometry	C. C. Counselman III	14:197-214
Orbital Resonances in the Solar System	S. J. Peale	14:215-46
Theories of Spiral Structure	A. Toomre	15:437-78
INTERSTELLAR MEDIUM		
Processes in Collapsing Interstellar Clouds	R. B. Larson	11:219-38
Interstellar Grains	P. A. Aannestad, E. M. Purcell	11:309-62
Cosmic-Ray Propagation in the Galaxy: Collective Effects	D. G. Wentzel	12:71-96
Coherent Molecular Radiation	M. M. Litvak	12:97-112
Large-Scale Dynamics of the Interstellar Medium	S. A. Kaplan, S. B. Pikel'ner	12:113-33
Radio Radiation from Interstellar Molecules	B. Zuckerman, P. Palmer	12:279-313
Ultraviolet Studies of the Interstellar Gas	L. Spitzer Jr., E. B. Jenkins	13:133-64
Young Stellar Objects and Dark Interstellar Clouds	S. E. Strom, K. M. Strom, G. L. Grasdalen	13:187-216
The Interstellar Magnetic Field	C. Heiles	14:1-22
The Interaction of Supernovae with the Interstellar Medium	R. A. Chevalier	15:175-96
Formation and Destruction of Dust Grains	E. E. Salpeter	15:267-93
Interstellar Scattering and Scintillation of Radio Waves	B. J. Rickett	15:479-504
SMALL STELLAR SYSTEMS		
Consequences of Mass Transfer in Close Binary Systems	H.-C. Thomas	15:127-51
THE GALAXY		
Nucleo-Cosmochronology	D. N. Schramm	12:383-406
On the Origin of the Light Elements	H. Reeves	12:437-69
High-Velocity Neutral Hydrogen	G. L. Verschuur	13:257-93
The Morphology of Hydrogen and of Other Tracers in the Galaxy	W. B. Burton	14:275-306
The Galactic Center	J. H. Oort	15:295-362
EXTRAGALACTIC ASTRONOMY		
Chemical Composition of Extragalactic Gaseous Nebulae	M. Peimbert	13:113-31
Stellar Populations in Galaxies	S. van den Bergh	13:217-55

Chemical Evolution of Galaxies	J. Audouze, B. M. Tinsley	14:43-79
The Radio Continuum Morphology of Spiral Galaxies	P. C. van der Kruit, R. J. Allen	14:417-45
Extended Extragalactic Radio Sources	D. S. De Young	14:447-74
Seyfert Galaxies	D. W. Weedman	15:69-95
Clusters of Galaxies	N. A. Bahcall	15:505-40
OBSERVATIONAL PHENOMENA		
Diffuse X and Gamma Radiation	J. Silk	11:269-308
Compact X-ray Sources	G. R. Blumenthal, W. H. Tucker	12:23-46
The BL Lacertae Objects	W. A. Stein, S. L. O'Dell, P. A. Strittmatter	14:173-95
The Line Spectra of Quasi-Stellar Objects	P. A. Strittmatter, R. E. Williams	14:307-38
Soft X-Ray Sources	P. Gorenstein, W. H. Tucker	14:373-416
Extragalactic X-Ray Sources	H. Gursky, D. A. Schwartz	15:541-68
GENERAL RELATIVITY AND COSMOLOGY		
Standard Model of the Early Universe	E. R. Harrison	11:155-86
Physical Processes Near Cosmological Singularities	I. D. Novikov, Ya. B. Zel'dovich	11:387-412
Radio Surveys and Source Counts	D. L. Jauncey	13:23-44
Astrophysical Processes Near Black Holes	D. M. Eardley, W. H. Press	13:381-422
Observational Tests of Antimatter Cosmologies	G. Steigman	14:339-72
Recent Theories of Galaxy Formation	J. R. Gott III	15:235-66
INSTRUMENTATION AND TECHNIQUES		
Millimeter-Wavelength Radio-Astronomy Techniques	A. A. Penzias, C. A. Burrus	11:51-72
Image-Tube Systems	W. C. Livingston	11:95-144
New Generation Optical Telescope Systems	A. D. Code	11:239-68
On-Line Computers for Telescope Control and Data Handling	L. B. Robinson	13:165-85
Instrumental Technique in X-Ray Astronomy	L. E. Peterson	13:423-509
Astronomical Applications of Echelle Spectroscopy	F. H. Chaffee Jr., D. J. Schroeder	14:23-42
PHYSICAL PROCESSES		
Interaction of Fast Particles with Waves in Cosmic Magnetoactive Plasma	V. N. Tsytovich	11:363-86
Equation of State at Ultrahigh Densities, I	V. Canuto	12:167-214
Neutrino Processes in Stellar Interiors	Z. Barkat	13:45-68
Thermonuclear Reaction Rates, II	W. A. Fowler, G. R. Caughlan, B. A. Zimmerman	13:69-112
Equation of State at Ultrahigh Densities, II	V. Canuto	13:335-80
On the Pulsar Emission Mechanisms	V. L. Ginzburg, V. V. Zheleznyakov	13:511-35
Stellar Opacity	T. R. Carson	14:95-117
Transition Probability Data for Molecules of Astrophysical Interest	R. W. Nicholls	15:197-234